Telecommunication Networks

Devices, Circuits, and Systems

Series Editor
Krzysztof Iniewski
CMOS Emerging Technologies Inc., Vancouver, British Columbia, Canada

Telecommunication Networks
Eugenio Iannone

Atomic Nanoscale Technology in the Nuclear Industry
Taeho Woo

Nano-Semiconductors: Devices and Technology
Krzysztof Iniewski

Electrical Solitons: Theory, Design, and Applications
David Ricketts and Donhee Ham

Radiation Effects in Semiconductors
Krzysztof Iniewski

Electronics for Radiation Detection
Krzysztof Iniewski

Semiconductor Radiation Detection Systems
Krzysztof Iniewski

Internet Networks: Wired, Wireless, and Optical Technologies
Krzysztof Iniewski

Integrated Microsystems: Electronics, Photonics, and Biotechnology
Krzysztof Iniewski

FORTHCOMING

Optical, Acoustic, Magnetic, and Mechanical Sensor Technologies
Krzysztof Iniewski

Biological and Medical Sensor Technologies
Krzysztof Iniewski

Telecommunication Networks

Eugenio Iannone

CRC Press
Taylor & Francis Group
Boca Raton London New York

CRC Press is an imprint of the
Taylor & Francis Group, an **informa** business

CRC Press
Taylor & Francis Group
6000 Broken Sound Parkway NW, Suite 300
Boca Raton, FL 33487-2742

First issued in paperback 2017

ISBN 13: 978-1-138-07777-5 (pbk)
ISBN 13: 978-1-4398-4636-0 (hbk)

Library of Congress Cataloging-in-Publication Data

Iannone, Eugenio.
   Telecommunication networks / author, Eugenio Iannone.
      p. cm. -- (Devices, circuits, and systems)
   Includes bibliographical references and index.
   ISBN 978-1-4398-4636-0 (hardback)
   1. Data transmission systems. 2. Packet switching (Data transmission) 3. Telecommunication--Switching systems. I. Title.

TK5105.I247 2011
621.39'81--dc23   2011042298

Visit the Taylor & Francis Web site at
http://www.taylorandfrancis.com

and the CRC Press Web site at
http://www.crcpress.com

Contents

Preface

The telecommunication infrastructure is perhaps the most impressive network developed by humankind. Almost all the technical knowledge forming the basic human know-how is exploited in the telecommunication network, from quantum field theory needed to study optical amplifier noise to software architectures adopted to design the control software of the network, from abstract algebra used in error correcting codes and in network design algorithms to thermal and mechanical modeling adopted in the design of telecommunication equipment platforms.

The network is present almost everywhere in the world, allowing seamless communication of sounds and images through a chain of different types of equipment produced by several equipment vendors. Not only is communication carried out smoothly in normal conditions, but its quality is also monitored continuously, allowing it to survive failure of individual components and even to maintain a certain degree of functionality in case of catastrophic events like earthquakes.

The aim of this book is to present the telecommunication network as a whole, adopting a practical approach that examines evidence of not only recent developments and research directions, but also engineering subjects and key market needs that are no less important in guaranteeing the network operation standard.

The great attention to standardization, both in the description of standards and in the bibliography, is functional to part of this strategy.

The only area not covered by this book, out of a generic discussion, is constituted by radio systems. Considering that cellular systems are part of the access area and that radio bridges are used only in emergency situations out of the access area, this mainly impacts the way in which access to the network is performed. It will be the task of a future work to carry out an analysis of the trend toward the integration of mobile and fixed access and its effects.

This book starts with a market analysis of the telecommunication environment needed to individuate trends in services and equipment development. Technical advancements, thus, do not appear as mere improvements, but as answers to precise market needs.

Moreover, attention is devoted throughout the book to issues such as power consumption and real estate, which in some practical cases are more important than engineering key performances like capacity or reach in determining the success of a product.

Quantitative data are provided for all the analyzed systems, inspired by real products and product-level prototypes, to make clear potentialities and limitations of every technology.

Whenever possible, measured data are sustained by mathematical modeling in order to help the reader apply the concepts to his or her own case.

The result of this effort is a book that will help both professionals and advanced students have a global picture of the telecommunication network in order to be able to make the right choices and to always remain updated.

Author

Eugenio Iannone received his university degree (old Italian Laurea) in electronic engineering from Facoltà di Ingegneria, Università La Sapienza, Rome, Italy. He is an executive consultant working mainly for medium and small companies to drive key innovation processes or to transfer technologies born in research institutes and universities to the industrial environment.

Iannone has 15 years of experience in the telecommunication industry and has held several managerial positions. He started his career at Fondazione Ugo Bordoni in Rome as a researcher in the areas of optical transmission systems and optical networks.

He then joined Pirelli Optical Systems in 1997 as a member of the network optical design team in the R&D department. In 2000, Iannone moved to Cisco's DWDM business when Pirelli Optical Systems was acquired by Cisco. He worked as a manager of the network dimensioning group within the marketing and product management department.

Since 2002, Iannone has been a senior vice president of application engineering at Pirelli Labs OI, the Pirelli research and design center for telecommunications, and strategy and marketing director at PGT Photonics, the Pirelli company devoted to telecommunication components and subsystems business.

During the course of his career, Iannone has authored more than 100 papers and several international patents on optical transmission, optical switching, and the architecture of optical networks. In addition, he has published two books for John Wiley & Sons, Inc.: *Coherent Optical Communication Systems* in 1995 and *Nonlinear Optical Networks* in 1998.

1

Introduction

From the very first step in the field of telecommunications, every relevant development has brought about a change in people's lifestyles. The introduction of fiber optics and of the digital mobile phone (the GSM) are examples in this regard.

Within a brief period of about 10 years, fiber optics have increased the capacity of commercial transmission systems from 0.5 to 1280 Gbit/s, creating an abundance of capacity on which the Internet is based.

The diffusion of GSM has transformed the telephone from a location-based communication to a mobile form of personal communication, completely changing the way in which people communicate.

Internet and GSM, among the other prominent developments in the field of telecommunications, have shaped our lifestyles and how we interact, to the extent that sometimes our society is called "the communication society."

The complex and pervasive services that today are offered by telecommunication carriers are based on the telecommunication infrastructure: the network. This network is perhaps the most complex infrastructure deployed by man. It is pervasive, reaching almost all parts of the world, and uses a wide set of different technologies.

As a natural consequence, managing a telecommunication network is a very complex task and to develop and produce equipment or software for the telecommunication network requires a high level of technological and of industrial expertise.

The aim of this book is to identify the unitary design within this complexity, which allows the network to function, to evolve, and, finally, to satisfy the ever-changing needs of customers.

1.1 Book Content and Organization

Even though the scope of the book is to give a unitary view of telecommunications under a technical point of view, it is simply impossible to include in a single book all the various elements that combined form the telecommunication network. Thus, some choices have to be made.

The first clear choice is to exclude all forms of free space communication and related networking techniques. Radio bridges are no longer used as transmission trunks in the network, but only for particular situations, generally in the access area. Almost all the wireless segment of the network is constituted by the cellular infrastructure that is a relevant part of the access network.

Even if a strong push by the market exists toward a services and techniques convergence of wireless and wired access networks, the cellular network has up to now evolved with its own standards and specific equipment.

Even radio technology used in cellular equipment is not utilized in other telecommunication equipment, having, on the contrary, common elements with other applications of the radio technology like radars, free space communications between moving platforms (like airplanes or ships), and similar systems.

Therefore, in Chapter 2, wireless access is considered only to look at wireless services. Readers interested in the wireless technology are encouraged to access the rich technical bibliography.

Also, software technology is considered only marginally in this book, even if its evolution is instrumental in the present transformation of telecommunication engineering.

This will result evident from the great attention devoted to the network control procedures, whose evolution would be impossible without modern software development techniques.

Considering the different areas in which the wireline telecommunication network is divided, the greater attention is reserved to the so-called core network: that part of the network covering geographical areas that are extended over a continent or a big country. Here optical transmission systems convey thousands of Gbit/s over distances of thousands of kilometers, and IP routers of the capacity of several Tbit/s carry out the basic operations of the Internet. Service sources, like video on demand, IP TV servers, soft switches for the IP phone, and so on, are generally located in the core network nodes.

The core network is divided into several subareas, different for capacity, extension and role: backbone, metro core, and regional. The characteristics of these different subnetworks are analyzed in detail.

The end user has the possibility to access the resources of the core network through the access network. Architectures and technologies of the access network, which are key for the carriers' success, are also analyzed, focusing particularly on the evolution from the copper-based access to the fiber-based access.

In any case, the analysis has been presented keeping engineering practice in mind, even when complex concepts of modern physics or of control structures have been considered.

Instead of general and abstract results, the book tries, through a great number of examples, to evidence the practical elements of the design of key communication equipment.

Some of these examples are synthetic exercises of design that are conceived to put the reader in front of a realistic coexistence of all the elements in a single equipment, instead of insisting on the separate analysis of different design points.

In every case, for simple components and for the more complex equipment, realistic values of the system parameters have been reported by collecting information from the products available on the market. In this way, the reader is provided with an extended database of what present-day technology can do, which is necessary to understand, at first glance, if a set of requirements is within the reach of the technology.

In this context, Chapter 2 is devoted to an analysis of the recent history and of the present status of the telecommunication market, so as to be able to derive requirements for the network evolution from market trends.

Chapter 3 gives a general overview of the current networking techniques, while Chapters 4 through 7 review the core network in detail from the technology, network element, and architecture points of view.

Not only are consolidated techniques illustrated, but the main evolution directions are also considered. In any case, a large volume of numerical examples gives solid structure to the different subjects.

While in the first seven chapters we have mainly considered engineering techniques, we enter the research labs in Chapters 8 through 10 to present the latest techniques in the telecommunication field that will shape the long-term network evolution.

The book concludes with a chapter on access optical technology.

1.2 Using This Book

Requisites for a full exploitation of this book's potential are the technical culture typical of an engineer (not necessarily one working in the field) or of an advanced PhD student in engineering or physics as well as a very basic knowledge of telecommunication services.

Following are examples of potential users of this book:

The industry manager, who can have a clear and complete idea of the helping him to consider correctly also key elements that are at the edge of his or her core area of competence. Due to its strong focus on interdependence, the book offers critical help in the decision process and could well become a working instrument for technical and marketing managers.

The designer, who can have a global idea of new technologies so as to channelize his or her work in the most suitable direction and to make design choices having a complete view of the possible alternatives.

The university professor, who can study the connection between market trends and technology evolution among the different technical areas (e.g., technology and network architecture) so as to transmit this to the students. Due to its structure, the book should be useful in carrying out research for an up-to-date university course in tune with the latest realistic trends. Finally, a university professor can also use this book to update himself or herself on aspects of telecommunications that fall within the core area of his or her competencies.

The senior student, who can use this book to decide where to direct his or her activity and to acquire an organized view of the wider telecommunication environment.

Finally, a comment on the organization of the bibliography. We have chosen not to mention who has discovered a certain technique or which research group has first introduced a certain type of analysis. Frequently, the first work in which a concept appears is neither the most complete nor the most useful from a didactical point of view. We have decided to mention only those papers that have helped us in gaining more insight into a topic or in explaining it. Where possible, we have cited complete works, or even books if available.

Acknowledgments

This book would not have been possible without the continuous support and encouragement of my wife Flavia and of my three children Emanuele, Maria Grazia, and Gabriele.

I enjoyed the rare advantage of finding in my wife dedication, understanding, support, and technical help, and I took full advantage from them.

My friends and colleagues have provided a great number of suggestions and discussions that have been invaluable. Among them, I would like to thank, in particular, Stefano Santoni, Massimo Gentili, Marco Romagnoli, and Roberto Sabella for their help in reviewing a few key points of the book.

I would also like to remember Filippo Bentivoglio. At the very beginning of optical networking, Filippo and I worked together in the exciting mission of defining a complete new network layer. A few still valid concepts regarding optical protection (and a few schemes relating to that) derive from our common work.

2

Drivers for Telecommunication Network Evolution

2.1 Market of Telecom Carriers

The telecommunication market has been, for a long time, characterized by a set of very specific features that distinguished it from all other markets.

Up to the early 1990s, there was almost no competition due to the legal monopoly existing in the main western European countries and the practical monopoly existing in the United States.

As a matter of fact, in all the western European countries, telecommunication services were managed by a single carrier, either directly owned by the administration itself or operating under an exclusive contract with the administration [1].

The situation was quite different in the United States, where after the split of old AT&T in 1984, the long-haul network was managed by a group of competing carriers (the new AT&T, MCI, ...) while monopoly remained in the local networks. In particular, any local area was served by different regional bell operating companies (RBOCs), all coming from the original local structure of the old AT&T [2].

The U.S. situation was richer in terms of players, and competition was created in the long-distance segment of the network. However, these competing long-haul carriers were not in control of the services provided to the final users, but simply delivered transport capability to the local carriers. Services managing carriers, RBOCs, were on the contrary, in a practical monopoly situation due to the geographical separation of their areas of business.

Both in the Far East and in eastern Europe, the development of telecommunication services was a step behind the United States and western Europe and, at least in communist countries, was often more influenced by political reasons than by a real need coming from the users.

Also, under the product offering point of view, the telecommunication market was quite a particular one.

The customer base of telecommunication carriers was divided into two big categories—business and consumer customers—where revenues coming from consumers were much higher in the typical case than revenues coming from business customers.

The product offered to the consumer market was essentially built around telephone service. Adding a new feature to the plain telephone (e.g., added value services, a network-located memory to trace lost calls and so on) implied a complex engineering development since it had to be implemented in the software of the telephone digital switches, very complex machines constituting the core of the network.

Due to the lack of real competition and the amount of effort needed to design and upgrade the switches, all telephone carriers offered the customers essentially the same service profile, with a noticeable difference existing only in the United States and western Europe.

As far as business customers are concerned, the product offering dedicated to them was quite richer. Besides plain telephone and dedicated lines to support the multiplexed data coming from corporate switches (Private Automatic Branch eXchange [PABX]), data traffic were also required.

In order to convey data generated from business customers, the technology put at the designer's disposal different techniques from the integrated services digital network (ISDN) and frame relay protocols [3] that were able to embed data in a time division multiplexing (TDM) frame, to the complex and performing asynchronous transmission mode (ATM) [4,5], using both statistical multiplexing and statistical switching.

Generally, frame relay and ATM networks were superimposed on the telephone network, constituting a second, independent telecommunication infrastructure dedicated to business customers.

At the beginning of the 1990s, a big evolution happened both in the market and in the available technologies. The Internet revolution brought about a sudden evolution in transmission and switching techniques and a complete redesign of the telecommunication network. The end result of this huge effort is the network that we have today.

The process to produce this result was not smooth, but involved a rapid and huge growth of the value of the telecommunication market, the so-called telecommunication bubble, and the subsequent market crisis, that in a certain measure still affects the telecommunication market today, more than 10 years from the bubble end.

The main driver of the evolution, started in the early 1990s, was the diffusion of the Internet among small businesses and in the consumer market via the dial-up connection. The impact of this phenomenon is shown in Figure 2.1.

In this figure, the average traffic per day and per active user line in the United States is represented in terms of bit/s/line. Reported on a day (24 h), the traffic of 11 bit/s/line represents an average time of about 2.5 min/day at a transmission speed of 64 kbit/s.

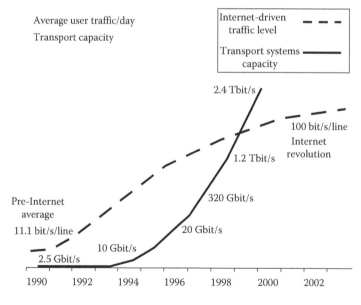

FIGURE 2.1
Average traffic per day and per active user line in the United States is in bit/s. As an example, reported on a day (24 h), the traffic of 11 bit/s represents an average time of about 2.5 min at a transmission speed of 64 kbit/s. (After Iannone, E., Personal Elaboration of Ovum "Capex Conundrum" report, Ovum, s.l., 2003.)

From now on, for simplicity, we will refer to 64 kbit/s as the telephone standard bit rate. All the discussions are easily adapted to the cases in which the standard bit rate was different, for example, 56 kbit/s.

In conclusion, the average traffic in the pre-Internet era was constituted by 2.5 min of phone calls per day per active line.

At the end of the analyzed period (2001–2002), the traffic increased at about 25 min of phone calls per day per active line. Of course, this increase was not due to the fact that the average person increased the duration of the phone calls by a factor 10, but the fact that the phone line was used via the dial-up connection also to browse the Internet.

From Figure 2.1, it is also clear that the traffic curve almost saturates, due to the fact that the average person tends to limit his or her use of the Internet if only the dial-up connection is available. Thus, in order to completely unfold the potentiality of the Internet, it was needed to provide the user with a greater capacity access line.

Figure 2.1 also reports the evolution of the capacity of the most advanced commercial long-haul transport systems in the years of the first Internet revolution. The capacity increased from 2.5 Gbit/s (typical capacity of a high-performance SONET/SDH transport system in 1990) to 2.4 Tbit/s (the capacity of a 10 Gbit/s, 128 channels wavelength division multiplexing [WDM] long-haul system, the most advanced commercial long-haul system in 2002). Perhaps, there is no other example in the history of the engineering of a key technical performance of a widely used product that increased three orders of magnitude in 10 years.

This evolution was possible due to a huge research and development effort that brought to the market, in a very short time, an impressive number of new technologies, from single-mode semiconductor lasers to optical amplifiers and from optical modulators to low-loss WDM passive elements.

All this effort required a great injection of money into the telecommunication market, as shown in Figure 2.2, where the evolution of the capital expenditure of the U.S. carriers is reported in the considered period [6].

Two important points of discontinuity can be observed in the figure, corresponding to specific events in the market.

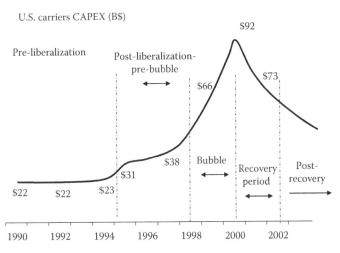

FIGURE 2.2
Evolution of the capital expenditure of U.S. carriers in billion dollars. (After Iannone, E., Personal Elaboration of Ovum "Capex Conundrum" report, Ovum, s.l., 2003.)

The first is related to the U.S. market, in particular to the telecommunication act of 1996, which opened the access market to competition, introducing the concept of access unbundling and abolishing the monopoly of the RBOCs [7,8].

The second discontinuity, located around 1997, is related to the effort to develop the network under the pressure of the first Internet revolution.

Looking at Figures 2.1 and 2.2, a comment immediately emerges. The traffic increased from 1991 to 2002 by a factor of about 10, while the carrier revenues increase was much less, due to the decrease in prices. In the technical field, from 1990 to 2002, the long-haul capacity increased by a factor of 1000, while the access capacity remained almost constant (at least for the great majority of the users still having a 64 kbit/s standard phone line).

This huge increase in the long-haul capacity was paid by an increase of about a factor 4 in the carrier capital expenditure (CAPEX), and by an even higher increase in the network operational expenditures (OPEX). All this capital was not injected in the market by end users, whose individual expenditure for telecommunication services was not increased (and in many cases was even decreased due to the price behavior), but by financial players that were hoping to exploit the financial gain driven by a fast-growing market.

In the absence of a relevant capital injection into the market from end customers, this situation was destined to be only temporary. In a few years, the debt of the telecommunication carriers was so big to prevent them from continuing their investment policy. The carrier CAPEX had a sudden collapse (see Figure 2.2), and a great crisis struck the market.

This was a structural crisis, deeply anchored to the way the market developed. The only possible solution was to induce the end customers to inject more money into the market to sustain its growth.

The key of this strategy was individuated in the introduction of new telecommunication services so as to drive an average revenue per customer (ARPU) increase. In order to do that, it was needed to increase the capacity of the customer access line.

For this reason, after the end of the telecommunication bubble, all the major European and U.S. carriers started a wide renewal of the access network in order to provide more bandwidth to the end user as a support to a new generation of telecommunication services.

Renewing the access network is perhaps the most CAPEX-intensive operation a carrier can undertake, due to the extreme geographical spread of this part of the network, which arrives with an access termination at every user location.

This extremely CAPEX-intensive operation was undertaken by carriers with a difficult financial situation due to the debts generated in the bubble period.

The carrier strategy was also influenced by another element that brought about a deep market change—market liberalization and globalization.

In the United States, the telecommunication act of 1996 [7] extended the competition to the access market allowing the growth of a great number of competitive local exchange carriers (CLEXs) that started operations almost all over the United States. Of course, these new companies were not able to deploy a completely new access network, and competition was allowed by enforcing the incumbent RBOCs to operate network unbundling.

Network unbundling is the procedure allowing a CLEX to deploy its own access equipment in the incumbent access central office and to rent the access line from the incumbent at a price determined by a public authority.

A similar evolution took place in western Europe, where in these years almost all the public incumbent carriers were transformed in private companies and the unbundling was introduced to allow competition.

Many of the new carriers born during the bubble expansion were not able to survive the following crisis, but a few of them were sufficiently stable to survive. The telecommunication market coming out from the bubble in 2002 was definitely a competitive market in all its segments, both in Europe and the United States.

Besides liberalization, after the collapse of the communist governments in eastern Europe and with the evolution of China and other Far East countries, the telecommunication market started to become a globalized one.

China and India, in particular, beginning from the period of the bubble end started to become very important players, both as a potential market for telecommunication equipment and, in the case of China, as the home country of a few key telecommunication equipment vendors.

The growth of the telecommunication market in the Far East was characterized by great speed during the period in which the crisis was hitting harder in Europe and the United States. As a matter of fact, the development of Far East society had to be sustained with the fast development of a modern telecommunication infrastructure, and local administrations were in many cases more than willing to at least partially finance this operation.

To give an example, in 2006, South Korea was the first country in the world to install a widely diffused optical fiber, WDM-based access network [9], using a very new technology, even if standards were not yet consolidated and the selected technology was supplied by a single, small U.S. company. Though at that time probably the business case of this operation was not positive for the Korean carrier, the deployment of the new access network was seen as a strategic step in the modernization of the country.

The combination of the bad financial position of the main carriers, of the presence of competitive carriers based on unbundling, and of the growth of new markets in eastern Europe and the Far East drove the evolution of the access network toward solutions that did not require a change of the copper cable infrastructure already deployed for the telephone service.

This was done by exploiting a technology development that was made available around the same time—with the development of cheap asymmetric digital subscriber line (ADSL) chips, it was possible to convey on the copper access line much more capacity than the 64 kbit/s of the standard telephone line.

To use ADSL on the copper access lines, access equipment had to be changed, passing from access telephone switches to broadband digital subscriber line access multiplexers (DSLAMs), but no new access cable had to be deployed, completely exploiting the available infrastructure. These solutions were thus ideal both to limit carrier CAPEX and to allow unbundling to be implemented.

The period from the telecommunication bubble end in 2001 to the present time can be called the second Internet revolution, dominated by the penetration of ADSL broadband access, by the launch of a certain number of new broadband services, and by the global nature of the telecommunication market.

The progressive penetration of ADSL broadband access has increased the number of users that are able to use advanced Internet services as broadcast and on-demand video services. This situation is depicted in Figure 2.3 [10], where the broadband access penetration in various countries in 2005 is shown via a ball graph, referred to as the case of maximum penetration (the United States with about 49 million broadband lines). Since this is a 2005 data collection, the considered broadband lines are almost all ADSL lines.

From the graph it can be derived, as an example of globalization, that the broadband lines in China were almost 80% with respect to the United States (i.e., about 40 million),

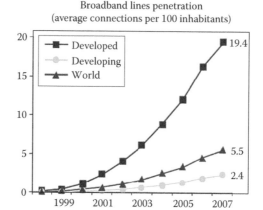

Broadband lines penetration
(average connections per 100 inhabitants)

FIGURE 2.3

Increase of the average broadband lines penetration versus time for developed countries (e.g. USA, Germany, Japan) and for countries under development (e.g. India). (After Public ITU-T report *Measuring the Information Society. The ICT Development Index 2009*, ITU-T publishing, 2009.)

while in Italy, there were about 6.5 million broadband lines (13% with respect to the United States) out of a total of about 20 million access lines.

In spite of this very good penetration of broadband lines, there is not yet the desired increase in ARPU. As a matter of fact, prices of telecommunication services have experienced a great decrease in the last 10 years, and the introduction of new broadband services was not always sufficient to counterbalance it. As a result, the revenues of telecommunication carriers did not experience the desired increase [11].

Much more effective in pushing the carrier revenues was the introduction of mobile services. After the introduction of GSM in 1991, the success of digital mobile telephony in Europe was so big to justify the huge CAPEX needed to deploy a continental network covering almost all the populated areas and with easy roaming among all the relevant carriers [12].

However, up to the present time, mobile and fixed-line services are considered by end customers as separate products, and the success of mobile services does not have a positive influence on fixed-line revenue. Moreover, after a long period of growth, the mobile market started to slow down after 2005.

This effect is shown in Figure 2.4, where the global revenues and CAPEX for telecom carriers are shown versus time from 2006 to 2010 (where the 2008–2010 figures are forecasts) [11].

The average yearly growth rate is about 2%, quite low taking into account that it is positively affected both by the weakness of the U.S. dollar with respect to the euro and the yen and by the mobile revenues that, in the considered period, are still growing faster than those related to fixed services.

As far as CAPEX is concerned, it is almost constant, a further proof of the state of stagnation of the fixed-line telecom market.

Due to these trends, the market of fixed-line services started to experience a structural state of crisis, mainly due to the imbalance between the investment and running costs needed to set up and maintain an up-to-date telecommunication network, and the amount of revenues related to service delivery. This condition is worsened by the fact that many carriers have a high debt that makes it difficult to carry out CAPEX-intensive operations.

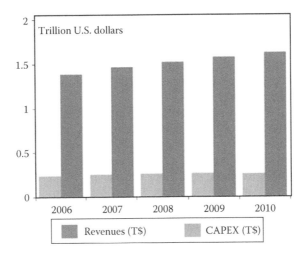

FIGURE 2.4
Global revenues and CAPEX for telecom carriers in trillion dollars versus time. (After Saadat, A., Service Providers Capex, Opex, ARPU and Subscribers Worldwide, Infonetics, s.l., 2007.)

Exceptions are carriers in emerging nations like China, where the market is more dynamic due to the needs of completely new infrastructures and the contribution of local administrations.

In order to move the market out of this state of stagnation, a further evolution is needed in the service bundle offered to customers. Several analysts individuate in a bundle mainly based on centralized and pair-to-pair video services the key to overcome the crisis calling for a further increase in access network capacity.

It is a consolidated opinion that this further evolution of carrier service offering pushes, at least in the most populated areas, a recabling of the access network by some type of fiber optics architecture.

Besides bandwidth increase, the penetration of ultra-broadband connections stresses the capacity of existing copper cables to support a high number of very fast ADSLs. In many cases, copper cables deployed before the diffusion of broadband access presents a non-negligible pair-to-pair interference when ADSL is used. This phenomenon often sets a limit to the number of connections possible in a given area, depending on the quality and length of the cables [13].

Last, but not least, the need of the main carriers to reduce operational costs also drives toward a fiber-optic-based access network. As a matter of fact, optical transmission allows a completely passive access infrastructure to be implemented. Moreover, much longer spans can be realized via passive optical networks (PON), opening the possibility of a delayering of the peripheral part of the network. In delayered architectures, local exchanges are almost completely eliminated in populated areas (e.g., cities), initiating the passive access network directly into the core nodes [14]. Eliminating a certain number of local exchanges further reduces OPEX, at the cost of a higher initial capital investment (CAPEX).

The first important market to move in the direction of a new, fiber-based access network was Japan, where NTT (the local incumbent operator) started to deploy fiber-based access in 2003.

Immediately after Japan, the U.S. market experienced an evolution pushing toward fiber access. In the United States, cable television providers are diffused almost in all the populated areas. When the need of an ultra-broadband access emerged in 2004 and 2005, cable television operators proposed themselves as competitors to traditional telecommunication carriers using cable modems to convey traditional telephony and web browsing on CATV cable infrastructures.

Moreover, in 2005, the U.S. administration introduced an important exception to the liberalization act stating that unbundling cannot be forced on completely new infrastructures, like fiber access networks [15].

This decision was based on the observation that unbundling relied on the fact that copper access was built with the help of the administration when incumbents were monopolists. Thus, they have a key business advantage that has to be compensated to preserve competition.

If a carrier completely rebuilds his network with no administrative help and in a competition regime, it has no advantage and there is no need to force unbundling on the new infrastructure.

The combination of these two facts moved the main RBOX (in particular Verizon, SBC, and Bellsouth) to start large projects for a new fiber-based access. These projects are still open today, even if revised in timing and dimension in pace with market evolution and with the change in the U.S. carriers scenario due to the merger between AT&T, Bellsouth, and SBC on one side and MCI and Verizon on the other side.

Different from the United States, no derogation from the unbundling obligation has been approved in Europe, thereby causing a further delay in the evolution of the access network toward fiber-optic transmission.

In Europe, however, traditional carriers have to face a new competitor—utility companies. These companies are owners of some different, but pervasive infrastructure (e.g., power distribution or water ducts) and in many cases have a rich fiber network deployed using their own network. A certain number of such utilities, attracted by the potential ultrabroadband market, either entered the access market directly or rented to competitive local carrier their infrastructure.

Thus, while in the United States local carriers are disappearing by merging with long-haul companies, in Europe the inverse process is ongoing, with a set of competitive carriers emerging in the access area.

A paradigmatic case is that of Fastweb in Italy, the first carrier in Europe that was born in 1999, deploying a fiber-based access network, that today is present in several areas of north and central Italy [16].

To face this competition, European incumbents are also moving toward fiber-based access, even if no clear decision has been taken, with carriers involved in field tests and preliminary planning.

In this situation, it is clear that the telecommunication market today is menaced by stagnation and by financial weakness of the main carriers. Nevertheless, the push toward new technologies, in particular in the access, is strong both to gain advantage in a complex competitive scenario and to decrease the network operation cost.

2.1.1 Customer Base Impact on the Economics of Carriers

The first step to analyze today's telecommunication market is to look at the final customer.

A tier one carrier has a customer base divided into three categories:

1. Private customers (the so-called retail market)
2. Business customers
3. Wholesale customers (i.e., other carriers renting network resources, including CLEXs using unbundling)

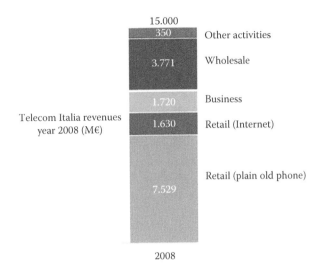

FIGURE 2.5
Revenue of Telecom Italia in 2008. (From the public balance sheet.)

Private customers and business customers buy a service bundle from the carrier, which can vary from customer to customer and from carrier to carrier.

Considering an incumbent carrier, the revenues are mainly from the retail market and are still more from telephone rather than from Internet services.

As an example, in Figure 2.5 [17], the detailed analysis of the 2008 fixed-lines revenue stream of Telecom Italia is reported. From the figure, it is clear that as much as 43% of the global revenue in the retail market is due to telephone services. Revenue in the retail market from Internet services is almost as big as the revenue from business customers, and about half the revenue is due to wholesale.

This clearly states that the old-style products (telephone and wholesale, that quite often is simply capacity leasing) produce by far the bigger revenue.

Moreover, comparing fixed-line revenues with mobile ones (again considering the Telecom Italia case), the impact on the overall carrier revenue stream is 60% and 38%, respectively (with a 2% revenue from other sources) [17].

Since this situation is quite typical, an incumbent European carrier fights its main battle in the retail arena, the wholesale prices being mainly stated by the administrative authority.

Unfortunately, the retail market is also the most difficult; the cost of different telephone services is steadily decreasing, striking carriers hard, and the trend does not seem to stop.

The fast decrease of the cost of telephone services is largely due to the intense price competition that has emerged after the end of the bubble. However, the penetration of broadband lines has a role in this scenario, allowing broadband users to adopt alternative telephonic systems like IP-phone or Skype [18]. These services are free of charge or almost free of charge, and this cooperates to push down the cost of standard phone calls.

However, it is clear that Internet revenues are far from compensating for the loss of telephone incomes. Thus, in this situation, the more the imbalance in the carrier service offering toward Internet, the more the revenue decrease.

The situation is of course completely different for competitive carriers. In this case, the value proposition at the base of their proposal to leave the incumbent is generally a

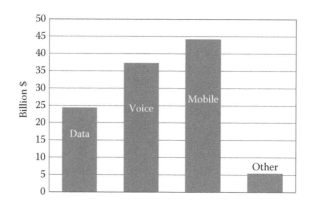

FIGURE 2.6
AT&T revenue in 2008. (From the public balance sheet.)

faster and cheaper access to Internet and the possibility to exploit it to sensibly reduce the cost of telecommunications. Thus, the success of the competitive carrier is linked to its Internet offering.

In the United States, the situation is similar, with greater revenues generated from Internet services. As an example, the domestic revenue of AT&T is analyzed in Figure 2.6, which shows that wireline telephone gives 34% of the revenue while data services give 24% [19].

In the Far East, the situation is quite different. The development of a telephone network is often completely parallel to the development of the Internet.

The case of China Telecom is characteristic: The income of the company due to flat telephone lines is still almost four times the income from broadband lines [20]; however, the use of broadband is growing much faster, and in a few years, it will probably give more revenue when compared to plain telephone.

2.1.2 Broadband Services

After a long period in which the center of telecommunication services was the telephone, with the Internet and with the diffusion of broadband access, a plethora of new telecommunication services have been introduced.

Although every tier one carrier is able to offer the same services, what makes the difference is the bundle that is offered to the end user. As a matter of fact, a good bundle gives the user the impression that he is paying for only what he wants and allows the carrier to deploy efficiently the service-related equipment.

Starting from single services, services offered to the retail market can be divided into three categories:

1. Traditional telephone-based services
2. Data services
3. Multimedia services

Traditional telephone-based services are all the pre-Internet features of the telecommunication network: telephone calls first, then a set of other related features as network-based answering service, calling number identification, running telephone cost control, and so on. These services, which in telephone switched network were performed by digital telephone switches, are carried out by simulation using packet telephone machines in packet-based networks, so that the end user experiences no difference in service level.

All traditional telephone services are designed to be delivered to the user via the 64 kbit/s analog telephone line; however, they are generally also available for broadband users. In this case, either the broadband access line has a traditional telephone line in parallel (as in the ADSL case) or a packet-based simulation is devised when pure packet access is available (as in the case of all-fiber access).

Data services are services based on low-activity data transfer. The main example is web browsing. During web browsing, the activity of the browser is concentrated in short periods of time, when the user downloads web pages. During these periods, the data transfer has to be very fast, since web pages can contain any kind of multimedia elements.

During the largest portion of time, however, the user reads the web pages and the transmission line is not used. The resulting average activity is in general very low, critically depending on the browsing style of the user. However, activities as low as 1% are normal. In this condition, statistical multiplexing on the access network works at best and access resources can be shared among many users.

Other services of this category are instant messaging, e-mail, pair-to-pair file sharing, access to network databases (like online file repository), and a plethora of other services that do not require real-time data transmission.

Multimedia services are, on the contrary, based on real-time data exchange and are dominated by video streaming (either broadcast or on demand) and IP phone (e.g., either audio or video as the service provided by Skype).

It is clear that even if IP-based audio telephone imposes on the network severe requirements in terms of quality of service (QoS), the required bandwidth is not very high, being in general lower than 64 kbit/s, which is needed for traditional telephones. Thus, the class of services requiring a higher bandwidth is constituted by video services.

The bandwidth occupied on the access line by video streaming depends both on the video resolution and on the line code used for video coding.

Under the video resolution point of view, there are a great number of possibilities related to the different possible resolutions of the screen of a PC, but two cases are particularly important. These are the resolutions compatible with the television standards: standard television (SDTV) [21] and high-definition television (HDTV) [21]. As a matter of fact, a television is present in almost all houses in rich countries, and it spreads rapidly in all emerging countries. Moreover, many video services (e.g., streaming film or television online) are conceived to be used by more than one person at the same time, and this is difficult using a PC, while the television is naturally designed to be watched by more than one person.

For these reasons, many carriers also offer PC-less video services to the retail market, directed to people who do not want to use a PC but are of course skilled in using television.

In this case, the service is delivered by connecting to the access line and to the television an equipment called "set-top box" that adapts the format of the signal arriving from the network to the television format and allows the user to select via a simple menu system the video he wants to watch.

Many set-top boxes also incorporate other features, like a hard disk for video storage and web browsing. However, it is a key point to maintain the set-top box simple to use, due to the fact that it has to be suitable for those who, for different reasons, find the PC too complex.

For all these reasons, we will concentrate on the television standards.

TABLE 2.1

Bandwidth Occupied by Digital Video
with Different Compression Codes

Bit Rates per Channel (Mbit/s)	Range	Typical
SDTV	2–5	3
SDTV	1.5–2	1.5
HDTV	15–20	16
HDTV	5–10	8

On the coding side, there are several codes available, and the continuous improvement in code compression of video signals is one the keys to the success of broadband video services [22].

The bandwidth occupied on the access line by different video codes is reported in Table 2.1.

Let us now consider the most important bundle of services that is offered by almost all carriers to the retail user—the so-called triple play.

The triple play is built by putting together one or more phone lines, a package of data services mainly centered on web browsing (even if instant messaging and pair-to-pair are very popular), and one or more contemporary television channels. The need for more than one television channel is related to the fact that frequently more than one television is present in a house and different persons can wish to watch different programs at the same time.

The first important observation is that triple play is intrinsically asymmetric. As a matter of fact, video services are delivered from the network to the user at high speed, while the data stream from the user to the network is constituted only by the commands needed to set the options of the service and is low speed.

This is the main reason for the success of asymmetric access systems like ADSL and GPON in the retail market that efficiently shapes the bandwidth usage on the characteristic of the most bandwidth-consuming services.

The downstream bit rate needed for different triple play configurations is reported in Table 2.2. From the table, it is quite evident that the most bandwidth-demanding configuration is also accommodated with an access bandwidth of 50 Mbit/s, thereby fixing the order of magnitude of the target bandwidth for the next generation access network. In line with this evaluation, many carriers target a future access delivering 100 Mbit/s to every user to maintain scalability in front of the introduction of new services.

Up to now, we have considered real services, that is, products that carriers offer to the retail customer. It is not possible to close the discussion about services for the retail market without also considering pair-to-pair applications.

In a pair-to-pair application, the content of the service is not delivered by the carrier or by another service provider via the carrier network, but it is generated by a customer. For example, in a typical peer-to-peer service, a community of customers store in a network repository the digital version of the films they have bought on DVD. Every customer, under the only condition of a forum subscription, can access the repository and download the content he needs.

It is not possible here to enter into the details of the complex legal issues implied by this content management. As a matter of fact, films, music, books, and so on are protected by

TABLE 2.2

Downstream Bit Rates for All-Digital Triple-Play

	Mbit/s		Mbit/s
Video Services with MPEG-2 Compression			
Internet (average user)	5	Internet (high user)	10
Telephony	0.1	Telephony	0.1
2 SDTV channels	6	2 SDTV channels	6
1 HDTV channel	16	2 HDTV channels	32
Total	27.1	Total	48.1
Video Services with MPEG-4 Compression			
Internet (average user)	5	Internet (high user)	10
Telephony	0.1	Telephony	0.1
2 SDTV channels	3	2 SDTV channels	3
1 HDTV channel	8	2 HDTV channel	16
Total	16.1	Total	29.1

copyright and the boundary of a legal exchange of content between friends can be easily overcome by configuring a copyright violation.

However, the pair-to-pair phenomenon is so huge that it seems difficult to cancel it by simply calling it "informatics piracy." Perhaps, the phenomenon will be regulated both by controlling the exchange sites and by adapting the copyright rules, and it will be established as a telecommunication service in its own right.

Business-oriented services are clearly divided among services for small businesses and services for big firms. In the first case, the service bundle does not differ essentially from that offered to the retail market. Typical differences are more than one telephone number associated with different internal phones, network support for phone conferences with several external numbers, and the substitution of TV channels with a network-based videoconferencing service.

The main difference between retail and small business is, however, the granted service level agreement (SLA) and the security from intrusion and detection of private data.

In the case of small business, an SLA is generally contained in the service fruition contract, both in terms of service availability and service quality.

On the contrary, no specific SLA is granted in retail contracts, considering data services operated on a best-effort ground. In a few European countries, there is an SLA on the plain telephone, surviving from the monopoly period, fixing the maximum out-of-service time.

This SLA was imposed by the administration on the grounds of the necessity to assure telephone service in case of extraordinary need. However, there is almost everywhere a push from the carriers for the cancellation of this regulation. As a matter of fact, with the diffusion of cellular phones, its rationale almost disappears.

In the case of big companies, the service offering is quite different.

Before the Internet revolution, the typical service for big companies was the rent of a dedicated TDM capacity. This capacity was used to connect the firm PABX in a secure manner and to create a private company telephone network.

This kind of offering still exists for companies having its own switches and desiring a complete isolation of its internal network. However, the leasing of a whole physical line is a costly solution and many companies prefer to lease a given capacity, multiplexed on the physical lines via statistical multiplexing with other signals. In this way, the exploitation efficiency of the physical line is much bigger and rent prices are smaller.

Renting a virtual capacity opens two problems: assuring both an agreed SLA via a QoS monitoring and data privacy even if they are statistically multiplexed with other data.

Both these problems are solved using the concept of virtual private network (VPN) [23]. A VPN is implemented enveloping the packets belonging to this virtual network in a public envelop before being injected into the public network. This public envelop masks completely all the characteristics of the packet (such as the specific destination, the nature of the service, and so on), and simply reports the address of the destination within the public network (i.e., the last public node of the packet path) and a set of other data related to the SLA and the QoS.

The concept of VPN has been a really fruitful idea, since it has been used for several applications also different from its original rationale, for example, to manage QoS in carrier class Ethernet networks.

Last, but not least, a carrier managing its own network delivers wholesale services to other carriers.

Wholesale services consist in renting transmission capacity to other carriers in terms either of physical trunk (like copper pairs in the access network) or logical capacity within a multiplexed signal.

Wholesale services can be classified into regulated and unregulated services.

Regulated services are provided in Europe and in the United States by the incumbent carriers to the competitive carriers in order to allow them to carry out their business. The main regulated wholesale service is local loop unbundling. In this case, the incumbent carrier rents to a competitive one both central office space and services to operate the access machines (generally DSLAMs) and a certain number of copper pairs in order to reach end customers.

A similar situation can also happen in transport trunks, where the competitive carrier has to complete his transport network renting capacity from the incumbent.

These services are called regulated since the SLA between the incumbent and the competitive customer (comprising the cost of the service) is regulated by the national communication authority in order to prevent the incumbent from exploiting its dominant position to put the competitor out of the market.

On the contrary, in the case of unregulated services, the capacity rent is negotiated on the ground of a market price by the provider and the customer.

For example, when a carrier rents a set of fibers from a utility company (e.g., the railways company or the energy distribution company), there is no dominant position, and the SLA is simply negotiated by the two players.

2.1.3 Seamless Fixed and Mobile Service Convergence

Up to now, we have talked about services delivered on a fixed access network. The telecommunication customer generally considers fixed and mobile services as two separate entities even if they are done by the same carrier.

This approach implies the difference between the fixed-line phone that is associated with a place (the house of the Boosh family, the office of Mr. Smith, …) and the mobile phone that is associated with a person. With the diffusion of cellular phones, fixed-line phones are less and less useful since a phone call is directed to a specific person, and usually the possibility that another person could answer only because she is near the phone is considered a loss of time in the best case, if not a real problem.

This is the main reason why fixed-line carriers have seen year after year a slow decrease in the number of fixed-line phone subscriptions in favor of mobile subscriptions.

Big carriers like AT&T and all the European incumbents have both the fixed and the mobile access network, and it is natural to conceive a new telephone solution where the difference disappears.

Several such proposals have been done, where the customer has only one, personal phone that is similar to a cellular terminal.

When the terminal is in the range of the customer house network, that can be either a dedicated radio link with the telephone base or a broadband Wi-Fi network, the telephone behaves like a fixed-line phone; when the telephone is far from the house, it behaves like a cellular phone [24,25].

This simple service profile can be implemented in several different ways, either leveraging on network intelligence or leaving all the management to the terminal software.

In any case, the seamless transition from one working mode to another is a key performance, since during a phone call the working mode change has to have a minimal impact on the call quality.

Beyond the simple telephone call, there are other services that are similar on fixed and mobile terminals, for example, a video call or the delivery in real time of images of a sport event.

Starting from these similitudes, several studies have foreseen that in the future network not only telephones but all video services will pass seamlessly from fixed to mobile terminals [26].

For example, in such a network, it would be possible to start a video call using the call-enabled set-top box and the TV screen. After a certain time, it should be possible to pass the call on a mobile terminal without interrupting it. As far as the mobile terminal remains within the reach of the home Wi-Fi, it works like a fixed terminal (with the corresponding lower cost), and when the caller goes out of the Wi-Fi's reach, the terminal connects to the mobile network and changes tariff accordingly. All these changes must happen without a significant drop of the call quality (but for the passage from the TV screen to a handheld terminal, where the quality is impacted from the screen dimension).

Similar transitions can be done for the online transmission of a football match or of a news program, all of which can be watched on television or by using a handheld mobile terminal.

2.1.4 Prices Fall and Overall Market Scenario

At the conclusion of this very rapid analysis of the services delivered by a telecommunication carrier, it is evident that the Internet revolution followed by the penetration of broadband access has generated a big service differentiation and multiplication.

At first, it could seem that telecom carriers must have had a steady increase in revenue, in pace with the increase of the average bandwidth used by the end customer.

The reality is completely different.

As a matter fact, when the Internet revolution started, the "network," as almost all Internet users call it, was seen as a real social revolution. In contrast with the traditional news and entertainment media, the Internet was not broadcasting contents controlled by a small number of entities, but allowed every single user to become a content generator, to publish its photos and films to entertain friends, to comment on political events, or to exchange recipes.

Many people around the world consider the Internet the most powerful instrument to affirm intellectual freedom, to diffuse news that is not in line with the thinking of the local government, and, in general, to react against authority control of information.

However, in order to unleash all this power, two conditions are needed: The Internet would have to reach almost all users in the world with a sufficiently wide bandwidth, and the cost of the Internet would have to be almost zero.

At the end, if the cost is related to the fruition of some content, when the content is simply exchanged among users, it seems reasonable that it is done almost for free, exactly like the exchange of DVDs among friends during a Sunday meeting.

However, this wonderful idea crashed with a telecommunication system regulated by market rules and with the fact that developing Internet involves huge costs both for equipment and infrastructure.

As we have seen, the great effort that was made during the so-called bubble period by telecommunication carriers to upgrade the network in pace with Internet penetration left the carriers in a difficult financial position.

Starting from that period, there was a continuous push from users to develop broadband access, to deliver a wider access bandwidth, and to decrease the cost of connection.

This push was operated in a market where there were too many players, and also the bigger carrier was hungry to increase its share to face its big debt. Thus, a sort of commercial war started, that has lasted up to present times, where in front of a continuous increase of broadband penetration and offered bandwidth, the bandwidth price decreased at a fast pace.

The increase of the access bandwidth is so uniform and continuous that Jacob Nielsen in a paper published in 1998 [27] introduced the Nielsen law of Internet bandwidth that essentially states that access bandwidth increases about 50% a year. This is similar in its formulation to the Moore law of computer power, and it is surprising how much it fits the reality up to now.

An example of the Nielsen law is given in Figure 2.7, where the peak bandwidth offered by different carriers to a residential customer in California is plotted versus time [27].

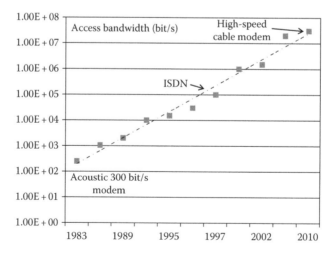

FIGURE 2.7
Plot of peak bandwidth offered to the residential users for Internet surfing in California versus time. (After Nielsen, J., Nielsen's Law of Internet Bandwidth [Online], *Alertbox*, 1998, last update 2010, http://www.useit.com/alertbox/980405.html (accessed: March 5, 2010), ISSN 1548-5552.)

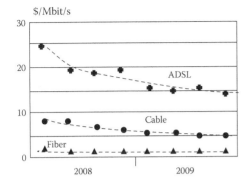

FIGURE 2.8
The U.S. average retail price of 1 Mbit/s versus time. (After Field, D., *Fire the Phone Company: A Handy Guide to Voice Over IP*, Peachpit Press, Berbeley, CA, s.l., 2005, ISBN-13: 978-0321384867.)

In front of this technical effort, the average price for unit bandwidth decreases continuously, so that the increase in offered bandwidth generates almost no revenue increase.

The price reduction is shown in Figure 2.8, where the U.S. average price of 1 Mbit/s in the access line is shown versus the time in a 2 year period. Even in so short a time, the decrease in the price is evident, and the trend has lasted since the early 1990s.

The fast price decrease generated by the Internet revolution also hit the standard telephone.

Even in the Internet era, telephone is by far the telecommunication service that attracts more customers. Thus, the commercial war among carriers also brought about a relevant decrease in telephone prices.

This decrease was also caused by the emergence of IP and mobile phone services. On one side, free IP phone, although not so diffused, constitutes a dangerous competitor for standard telephone, especially for international calls, where the tariff of the main carriers is higher [28,29].

As far as the mobile phone is concerned, as underlined in the previous section, it has the advantage of being a personal communication device. Since the start of the massive diffusion of GSM phones, several people have replaced their fixed-line phones with mobile phones [30].

Under the combined pressure of internal competition among carriers and external competition from service providers using other technologies, telephone prices have been consistently going down since the early 1990s.

As an example, in Figure 2.9, the average price of a phone call in different European countries is shown versus time in the period 1996–2004. All the prices are averaged on real traffic and reported as percentage of the average price in 1996. The price fall is quite evident, and even if its importance is higher in the bubble years and becomes less pronounced after 2002, it lasts even after the period shown in the plot up to the present time.

The behavior of prices is the last element needed to depict a global scenario of the telecommunications market, as it is experienced by European and American carriers.

The main elements influencing such a market are summarized as follows:

- There is a strong market requirement from the end users to increase the access bandwidth and the broadband access penetration since this is perceived as a necessary infrastructure for social growth.

- There is a strong requirement to lower the prices of both telephone and Internet, corresponding both to the need of a larger diffusion of Internet and to a general negative trend in the customers' willingness to spend money on telecommunication services.

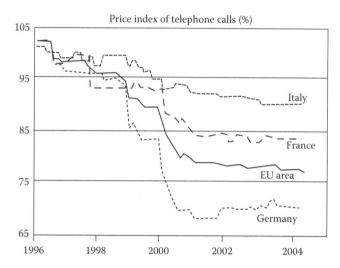

FIGURE 2.9
Average price of a phone call in different European countries versus time in the period 1996–2004. All prices are relative to the average price in the Euro area at the beginning of 1996 (assumed to be 100%). (After Church, S. and Pizzi, S., *Audio Over IP: Building Pro AoIP Systems with Livewire*, Focal Press, Burlington, MA, s.l., 2009, ISBN-13: 978-0240812441.)

- Carriers are almost all in a difficult financial position due to debts, but their net income at present is positive.
- The carrier position is worsened by the price-decreasing trend that does not seem to stop and forces them to continuously introduce new and larger bandwidth services to compensate price fall.
- Carrier income comes mostly from voice services; their expenses are mainly due to the need to sustain and expand broadband services.
- ADSL technologies, allowing broadband services to be implemented without the need of deploying new cables, seem to be at their maximum; in order to attempt the next jump in capacity, new cabling with optical fibers seems necessary, with the need of a huge CAPEX.
- Competition in the carrier market is regulated by the administration of countries via telecommunication authorities.
- Unbundling of copper access networks is present both in Europe and in the United States, but while in Europe it seems that it will remain unchanged for new networks, in the United States, carriers deploying completely new networks will not be obliged to unbundle them.

2.1.5 Telecommunication Market Value Chain

Up to now, we have considered carriers as the players of the telecommunication market since they are in contact with the final customer and manage the telecommunication network. Nevertheless, the market relies on an articulated value chain to provide to carriers the complex technology equipment needed to build the network.

The value chain of the telecommunication market is depicted in Figure 2.11.

The first element of the chain is constituted by component suppliers, the companies' buildings, for example, optical fibers, lasers, integrated electronic circuits (IC), optical amplifiers, and all the components that are needed in telecommunication systems.

Three types of companies are present in this segment:

1. Components divisions of companies that are vertically integrated and build in-house the key components for their systems (e.g., Infinera and partially also Alcatel-Lucent in the measure in which Bell labs are used for design and production)

2. Companies devoted only to a niche class of components for use in telecommunications (e.g., Santur that was in 2010 the market leader for tunable lasers designed for telecommunication system)

3. Big companies producing components for a plethora of applications (e.g., NEC, which produces a huge number of other product ICs for telecommunications)

The situation of this market segment is not easy. In front of the huge CAPEX investments needed to develop the technology, all the price pressure generated at the end customer level goes down along the chain and at the end arrives here.

To present an example, tunable lasers for optical fiber transmission had an average cost of about $2000 in 2004. This cost went down to $800 in 2008 and was around $600 in 2010.

Components are often assembled in subsystems called modules before being inserted in telecommunication systems.

This step is due to the fact that system vendors generally do not want to develop the analog electronics needed to drive individual components, either to avoid the need of a dedicated development group or to avoid risks related to this very specific development.

Thus, modules collect all the components needed for a specified functionality (e.g., in/out interface of an equipment) and close them in a module comprising all the hardware and firmware drivers. Such a module can be driven by the control board of the equipment via a digital interface and is much easier to substitute if needed than bare components integrated in a card.

Modules are often manufactured by component companies, but there are also niche companies acquiring components and building modules with them.

Modules are sold to system companies: The companies manufacturing, for example, WDM transport systems, Ethernet carrier class switches, and IP core routers. The companies in this market segment are generally big companies due to the fact that a telecommunication system involves very complex equipment, comprising hardware and software, and requires the capability of carrying out a complex design and production process. Some of the most well-known companies in the world belong to this market segment: Alcatel–Lucent, Eriksson, Nokia–Siemens, Huawei, NEC, and Fujitsu.

In general, telecommunication carriers do not buy bare equipment, but buy a service consisting in the setup of a network segment. In order to provide this service, the required equipment must be installed in the carrier central offices, connected to the network, set up under a software point of view and finally tested. When all is complete, the new elements can be virtually switched on by the management systems and are ready for real traffic.

These operations are the business of solution providers. A few system vendors like Ericsson are also solution providers, but there are many companies that do only this business. These companies work, in general, in a specific geographical area and with a few

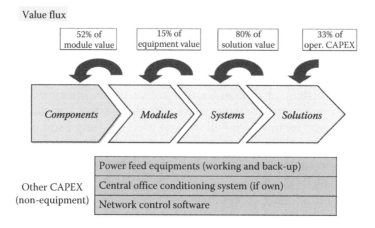

FIGURE 2.10
Value chain of the telecommunication market from physical components (e.g., integrated circuits) to deployed networks. The CAPEX flux is indicated from higher to lower segments and examples are done of carriers' CAPEX spent out of this value chain.

specific customers, due to the great impact both of logistic issues and of specific customer needs on their operations.

We have seen in Figure 2.4 that telecom carriers have a huge CAPEX investment every year, of the order of several trillion U.S. dollars. It is quite interesting to analyze how this great cash flow distributes along the value chain.

In Figure 2.11, the average destination of the U.S. and European carrier CAPEX is shown, using 2008 data, while the distribution of CAPEX along the value chain is represented in Figure 2.10.

The first interesting feature is that only 33% of the carrier CAPEX is injected along the equipment value chain. The other CAPEX is invested in other elements of the network, as reported in the table included in Figure 2.10.

Moreover, while the main part of this value transits up to system manufacturers, only a small part goes down along the hardware value chain toward module manufacturers. This is due to the great design and testing effort needed to design and manufacture telecommunication equipment that imposes on system vendors great investments in terms of development structures and testing grounds.

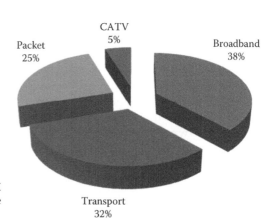

FIGURE 2.11
Average destination of those carriers' CAPEX invested in network infrastructures in Europe and the United States.

The analysis we have carried out in the last two sections is mainly centered on American and European carriers. As far as Far East carriers are concerned, three main differences can be identified.

1. The network has to be developed from green field in a few areas due to the fact that no previous network exists.
2. There is no great attention to boost competition; on the contrary, in a few important cases, the administration supports the "de facto" monopolist with the target of accelerating the creation of an infrastructure that is strategic for the country's growth.
3. Mainly in Japan and in China, local players are practically favored due to both the difficulty entering the market and the very specific requirements.

However, excluding Japan where the market is already consolidated and has strong entry barriers, over a period of time, almost all developing countries will grow from a social and economic point of view, and the market will become more and more global. This will probably cause the market dynamics to be more and more similar to that depicted for the United States and Europe.

2.2 Requirements for Next Generation Networks

2.2.1 Network Operational Costs

In the previous section, we analyzed from different points of view, the CAPEX of telecommunication carriers. However, besides being CAPEX intensive, operating a telecommunication network also requires big OPEX.

A very effective classification of the different sources of OPEX is reported in [31]. In particular, the following OPEX sources are identified:

- Operational costs to maintain the network in a failure-free situation. The cost category comprises rent of space in central offices, power consumption of the equipment, cost of personnel that in normal conditions operate in the central office, and so on.
- Operational costs to keep track of alarms and prevent failures. Control of the network performances via the control plane is a typical cost belonging to this class.
- Costs deriving from failures: on one side, repairing costs like spare management and field interventions and, on the other side, penalties due to SLA violation when this event happens.
- Costs for authentication, authorization, and accounting (AAA).
- Costs for planning, optimization, and all the running activities for the continuous network upgrade both from a hardware and a software point of view.
- Costs related to commercial activities, including new commercial offerings and the launch of new services.

From the list given earlier, it becomes evident that it is difficult to carry out an abstract evaluation of OPEX unless a completely defined specific situation is referred to.

OPEX depends heavily on factors like the kind of equipment the carrier deploys in its network, what equipment is of the most recent generation, what is legacy equipment, and if the control is integrated over all the network or if different parts of the network (e.g., areas where different technologies are deployed) have different control software.

The organizational structure of the carrier also impacts the OPEX, and even the local laws regulating work contracts have their own influence.

Nevertheless, analyzing the carrier balance sheets and working out carrier OPEX evaluation models [32], it becomes clear that OPEX is the main expenditure of a carrier that has an established network.

For example, a good approximation of the OPEX expenditure related to network operations for U.S. carriers is about 70% of the overall company expenditure and a similar situation also exists in Europe. As far as Far East carriers are concerned, two elements contribute to creating a different situation: on one side, the CAPEX is higher (relative to the overall carrier business) due to the need of developing the network in areas in which either it uses old technologies or it is practically not present; on the other side, the impact of the labor cost is smaller. However, analyzing the results of China Telecom, for example, OPEX reduction remains a key for success [20].

2.2.2 Requirements for Next Generation Equipment

The profound transformation in the telecommunication business caused by the Internet revolution, market liberalization, and the post-bubble crisis has a huge impact on technical aspects.

In the following sections, we will summarize the main impacts in terms of changes in telecommunication network requirements.

In the telephone-centric period, the telecommunication network was constituted by superimposed and separated networks.

The most important of these superimposed networks, generating the main part of carrier revenues, was the telephone network. The telephone network was built on a few key elements:

- Deterministic TDM transmission
- Call-based circuit switching
- Very fast protection in the core area (less than 50 ms of protection time end-to-end)
- Very high availability (99.999% that means less than 5 min out of work in a year)

The control plane of such a network, in its most advanced version, was provided by a centralized software control that was able to monitor both equipment and network (e.g., TDM streams exploitation efficiency) and incorporated algorithms for network reconfiguration and upgrade.

Besides the telephone network, there were also other logically distinct networks that exploit the same set of transmission systems intended to provide data services mainly to business customers.

Generally, such networks used ATM or frame-relay protocols and were characterized by a very high availability (required by business customers) and high security.

The control plane of such networks were completely separated by those of the telephone network, being the monitoring and configuration of the common transmission systems operated by a dedicated control plane.

A similar structure is suitable if data traffic is quite a small portion of the overall traffic and data users are well separated by standard telephone users, but it becomes very inefficient if the main part of the traffic is constituted by data and both high-speed data and telephone have to be delivered almost to every user.

Evolving the network in pace with traffic evolution without changing the architecture would be a so CAPEX-intensive operation that carriers have been driven toward a complete network redesign.

At the base, there is the most natural consequence of data traffic domination: a telephone network carrying also data has to be transformed into a data network carrying also telephone.

From an equipment point of view, this means eliminating the separation between data and telephone networks and deploying a unique network whose nodes are based on packet switching. From a transmission point of view, TDM systems are substituted by powerful WDM systems.

This is a technically complex and very CAPEX-intensive operation, lasting several years and still undergoing for many carriers. As far as the equipment is concerned, the final target is to deploy a network architecture that fulfills two main requirements:

1. Scaling with the increase of traffic with minimum CAPEX
2. Relevant OPEX reduction mainly due to the following expectations
   - The power consumption of the packet machines is smaller for a given network traffic load with respect to the equivalent TDM machines.
   - Packet machines are more compact, due to the concentration in a single machine of all the logical functionalities that where performed by a plethora of machines in the old network.

However, the traditional telephone has to be delivered maintaining all the related services in such a way that the end user does not perceive the difference due to the network paradigm change.

This causes the introduction of a new class of equipment whose role is to manage the telephone service exactly like other services (e.g., IP TV, network storage, etc.), through dedicated network equipment.

This equipment relies essentially on a specific software to emulate the hardware performances of a traditional telephone switch and is called soft-switch [33].

Up to now, the requirements of new network equipment have been analyzed. In practical network planning, however, there is another key issue to take into account. The network evolution from one technology to another is a process lasting several years, during which the network is in an hybrid situation where new and old equipment have to coexist without a disruption of services.

Moreover, even when the transition to the new technology is completely carried out, old technology machines that are still working properly are not eliminated from the network but are still used up to the end of their lifetime, in the worst case, moving them away from the most critical part of the network to be deployed where the situation is less critical.

This fact has a consequence: the survival of isles where the TDM technology is still used for a time much longer than the time strictly needed to deploy the packet technology, with the need of implementing in the new machines (either the packet switches/routers or the WDM transmission systems) interfaces that are suitable to integrate them with the TDM equipment.

2.2.3 Requirements for Next Generation Network Control Plane

In the old network, imagining that a new path was needed to connect new ATM interfaces installed in the data network, a long procedure was needed:

- The data network had to be preconfigured locking the resources needed for the new capacity.
- The control plane of the transport network had to be manually programmed in order to provision new paths providing the required new transport capacity.
- When the new paths were provisioned and tested, the transport resources were assigned to the newly created paths that can start working.
- The presence of the required paths is communicated to the control plane of the packet network that is manually induced to set up the wanted capacity.

Generally, since the departments managing the transport network and the packet one were different, the control SWs were located in different buildings and communications between them were carried out by phone by the persons working in the network control centers.

On the other hand, this inefficient provisioning procedure was acceptable due to the rarity of provision events and the possibility that it would take days to carry out a provision.

With the wide diffusion of broadband and the need of delivering wide band services to private customers, packet services began to be the main part of the traffic.

At this point, the structure of a set of superimposed networks is no longer effective due to all the inefficiency related to it.

In particular, in the Internet era, the traffic pattern is much more dynamic and difficult to foresee, both because commercial offerings of the carriers often vary and because customers frequently change both their way of using telecommunication services and the carrier delivering them services (as typical in a competitive market).

With the substitution of ATM and frame relay with IP, the core packet machines (IP routers) [34] have gained greater capacity, so to be able to absorb telephone traffic and to provide a unified switching system to the whole network.

Naturally, this operation has to be performed without affecting the service delivered to customers, thus using specific equipment and software to simulate all the features of the telephone service via a packet network [35].

However, a problem remains. In order to assure very fast recovery from failures, local redundancy has to be provided so that almost half the transport capacity of the network is devoted to protection from failures. This is true not only for the transmission lines (that are the lower availability elements of the network due to the probability of cable cut for civil works or other accidents) but also for the equipment elements like optical interfaces, switching matrixes, and so on.

This strategy, while technically quite effective, is contrary to the main requirement of the new network—it has to be cheap in terms of CAPEX, but, more importantly, it has to sensibly reduce the OPEX.

Unifying the switching technology with the use of the IP protocol stack is an important step in the direction of OPEX reduction.

As a matter of fact, replacing the set of two or more insulated control planes with a single control plan allows operations like provisioning of new packet capacity to be carried out automatically, thereby gaining both a greater provisioning speed and a lower cost in terms of qualified personnel working in the control centers.

This effect is also enhanced by the fact that the IP control plane, whatever technical solution is used, is a distributed intelligence system, working on the set of router control units. This system in standard conditions works without the need of human manual programming and the central control room is reduced to a simple reporting center where the status of the network is depicted to individuate possible critical issues or points of potential traffic congestion.

All these advantages, however, risks to be counterbalanced by the huge waste of capacity due to the need for local redundancy increasing greatly the CAPEX needed to evolve the network.

It is not difficult to imagine that using the distributed IP control plane, it is possible to devise complex algorithms that are able to exploit a small redundant capacity (much less than 50%) in order to reconfigure the paths affected by a failure and recover the correct network connectivity [36,37].

These algorithms, whatever be the used rerouting criterion, are slower than a simple switch between a working and a stand-by redundant capacity and in general do not satisfy the 50 ms requirement for protection fulfilled by the TDM network.

The problem related to the failure recovery time is only one of the issues to be solved to transmit on a packet switched network a service requiring real-time data transmission, and it is to consider that the telephone is not the only service of this type. Video streaming services are an important part of every broadband service offering, and a key point for its success is the perceived quality of the video, especially in the high-definition case.

In order to assure real-time transmission, not only does the recovery time have to be controlled, but the network traversing time also has to be controlled in ordinary condition so as to avoid a too long latency and a disordered arrival of the packets at their destination.

Two steps are needed in order to satisfy technical requirements imposed by the different services while fulfilling the fundamental condition to maintain CAPEX at a minimum and achieve a relevant OPEX reduction.

First of all, traditional requirements of the TDM telephone network have to be analyzed, again taking into account the new situation and the different potentialities of the packet network.

Once a set of minimum network requirements that have to be satisfied are identified, a new network control plan has to be defined that is able to perform the traditional operations of the network management software (called FCAPS from the initials of their description) that will have to be defined along the following lines:

- Fault management
- Configuration management
- Accounting management
- Performance management
- Security management

All these operations have to be performed in pace with the target OPEX reduction, with the characteristics of packet switching and with individual service requirements.

There is a wide agreement in the engineering community that the aforementioned high-level requirements can be specified as follows.

The new packet switched network has to be managed by a control plane having the following characteristics:

- It should be able to collect the different services delivered through the network in QoS classes, that is, collecting in the same class the service that has the same requirements as far as the transmission through the network is concerned.
- It should perform distributed routing of packet in the network via a distributed algorithm running in the control cards of the network equipment; this algorithm has to distinguish different classes of QoS, assuring that each service is delivered fulfilling its specific requirements in terms of delay and packet-ordered arrival at destination.
- It should perform fault-driven rerouting via the exploitation of the free network capacity; also this algorithm has to distinguish different classes of QoS assuring that each service is protected against failures fulfilling its specific requirements in terms of packet loss and recovery time.
- It should allow centralized alarm report and failure diagnostic so as to drive failure repair interventions; this means, among other things, the capability of filtering the alarm chain derived from a single failure so as to identify the source of failure.
- It should allow centralized traffic optimization via traffic engineering algorithms in order to optimize the network from the central control room, when needed.
- It should allow AAA to be performed.

Up to now we have listed the requirements of the new control plane. But naturally, since the equipment must be interoperable with legacy TDM technology, the new control plane has to allow this equipment feature.

2.2.4 Summary of Requirements for Next Generation Networks

From the previous sections, it is clear that the network evolution is driven by the difficult moment that the telecommunication market is traversing, with a stagnation risk even if the main world carriers have a positive EBIT and the users ask for more and more bandwidth.

Since the main structure cost of a telecom carrier is constituted by network OPEX, the main target of the network evolution is to provide enough bandwidth to the end user so as to enable new services that will boost carrier revenues in pace with a substantial reduction of the network OPEX.

Moreover, even if this operation is unavoidably CAPEX intensive, carriers push equipment vendors strongly for a price reduction, compatible with their limited CAPEX spending.

The requirements discussed in the previous sections have been the guidelines for the network evolution of almost all the carriers.

In particular, almost all the carriers have injected, following different technology visions, a huge amount of packet functionalities in the transport network.

At present, it can be asserted that major carriers have completed the evolution of the long distance network and have almost completed that of the metropolitan and regional network. Unified IP processing and WDM transmission are diffused in long distance while different approaches have been adopted to inject packet-switching capabilities in the metro and regional networks, always relying on WDM optical systems for data transmission.

As far as the access network is concerned, while in Asia and the United States there is a mixed situation, where relevant fiber-based access networks coexist with more traditional ADSL-based areas, in Europe fiber-based access is almost absent.

At this point, it should be clear if the great technology evolution that has transformed the old telephone-based network into today's data-based network has achieved or not its goal.

The answer is that the target to diffuse broadband and to lower sensibly the prices of wideband services while taking under control carriers OPEX and CAPEX has been reached, but the market has not started to grow again.

On the contrary, the risk of stagnation is always present, a strong push to consolidation and selection in all the market segments is still at work, and the customers go on asking for much more bandwidth at a much lower cost.

Since no new service is going to be introduced that is able to relevantly increase the ARPU, the only possible solution is to envision a further deep change in the network that will drive lower OPEX with a reasonable amount of CAPEX. This new technology should bring about the so-called next generation network (NGN).

Considering the technologies involved in the vision of NGN, there are a lot of different ideas and there is no consensus around a single solution or even around the fact that one of the proposed solutions could achieve the cost reduction targets while providing the needed bandwidth and service quality.

In the following chapters, we will analyze in detail the present network and the main proposals or its evolution from a technical point of view, keeping always in mind the market situation and the consequent requirements.

References

1. Noam, E., *Telecommunications in Europe (Communication and Society)*, Oxford University Press Inc., s.l., New York, 1992, ISBN-13: 978-0195070521.
2. Sterling, C.H., *Shaping American Telecommunications: A History of Technology, Policy, and Economics*, Lawrence Erlbaum Associates, Mahwah, NJ, s.l., 2005, ISBN-13: 978-0805822373.
3. Stallings, W., *ISDN and Broadband ISDN with Frame Relay and ATM*, 4th edn., Prentice Hall, Upper Saddle River, NJ, s.l., 1998, ISBN-13: 978-0139737442.
4. Kasera, S., *ATM Networks: Concepts and Protocols*, McGraw-Hill, New Delhi, India, s.l., 2006, ISBN-13: 978-0071477321.
5. Pattavina, A., *Switching Theory, Architectures and Performance in Broadband ATM Networks*, Wiley, s.l., Chichester, U.K., 1998, ISBN-13: 978-0471963387.
6. Iannone, E., Personal Elaboration of Ovum "Capex Conundrum" report, Ovum, s.l., 2003.
7. U.S. G.P.O., Public Law No: 104-104, *Telecommunications Act*, 1996.
8. Library of Congress, *S.652—All Congressional Actions w/Amendments*, March 23, 1995 through February 8, 1996.
9. Library of Congress, Korea Telecom deploys Novera Optics DWDM transport solutions, *Fiber Optics Weekly Update*, July 28, 2006.
10. Public ITU-T report *Measuring the Information Society–The ICT Development Index 2009*, ITU-T publishing, 2009.
11. Saadat, A., Service Providers Capex, Opex, ARPU and Subscribers Worldwide, Infonetics, s.l., 2007.
12. Hillebrand, F. (Ed.), *GSM & UMTS: The Creation of Global Mobile Communications*, Wiley, New York, s.l., 2001, ISBN-13: 978-0470843222.

13. Leshem, A., The capacity of next limited multichannel DSL, In *Proceedings of Sensor Array and Multichannel Signal Processing Workshop*, Sitges (Spain) Barcelona, Spain, July 18–21, 2004, Vols. 1–2, pp. 696–700, 2004.

14. Iannone, E., Next generation wired access: Architectures and technologies. In K. Iniewski (Ed.), *Internet Networks: Wired, Wireless, and Optical Technologies (Devices, Circuits, and Systems)*, Taylor & Francis, Boca Raton, FL, s.l., 2009.

15. Library of Congress, Senate report 109-355 29 document, description of the Communications Act of 2006, 2006.

16. Fastweb web site [Online], www.fastweb.it (accessed: March 1, 2010).

17. Telecom Italia Annual report 2008 [Online], www.telecomitalia.it, http://telecom-italia-annual-report-2008.production.investis.com/it-IT/report-on-operations/business-units-of-the-group/domestic/operating-and-financial-data/revenues.aspx (accessed: March 4, 2010).

18. Miller, F.P., Vandome, A.F., McBrewster, J. (Eds.), *IP Phone: IP Phone, Voice over Internet Protocol, Public Switched Telephone Network, Session Initiation Protocol, Skinny Call Control Protocol, Proprietary… Skype, Telephone, Cordless Telephone*, Alphascript Publishing, Mauritius s.l., 2009, ISBN-13: 978-6130274504.

19. AT&T balance sheet 2008, *AT&T Home Site* [Online], 2008, www.att.com (accessed: March 3, 2010).

20. Oliver, C., China Telecom profit falls 34% in first three quarters (China Telecom press release), *Market Watch—The Wall Street Journal Digital Network*, October 2009.

21. Whitaker, J., Benson, B., *Standard Handbook of Video and Television Engineering*, McGraw-Hill Professional, New York, s.l., 2003, ISBN-13: 978-0071411806.

22. Koc, U.-V., Chen, J., *Design of Digital Video Coding Systems (Signal Processing and Communications)*, CRC Press, Boca Raton, FL, s.l., 2001, ISBN-13: 978-0824706562.

23. Mairs, J., *VPNs: A Beginners Guide*, McGraw-Hill/Osborne Media, New York, s.l., 2001, ISBN-13: 978-0072191813.

24. Watson, R., *Fixed/Mobile Convergence and Beyond: Unbounded Mobile Communications*, Newnes, Bostan, MA, s.l., 2008, ISBN-13: 978-0750687591.

25. Ahson, S.A., Ilyas, M. (Eds.), *Fixed Mobile Convergence Handbook*, CRC Press, Boca Raton, FL, s.l., 2010, ISBN-13: 978-1420091700.

26. Sutherland, E., Fixed mobile convergence, *ITU-T Global Symposium for Regulators (GSR)*, February 5–7, Dubai, United Arab Emirates, s.n., 2007.

27. Nielsen, J., Nielsen's Law of Internet Bandwidth [Online], *Alertbox*, 1998, last update 2010, http://www.useit.com/alertbox/980405.html (accessed: March 5, 2010), ISSN: 1548-5552.

28. Field, D., *Fire the Phone Company: A Handy Guide to Voice Over IP*, Peachpit Press, Berbeley, CA, s.l., 2005, ISBN-13: 978-0321384867.

29. Church, S., Pizzi, S., *Audio Over IP: Building Pro AoIP Systems with Livewire*, Focal Press, Burlington, MA, s.l., 2009, ISBN-13: 978-0240812441.

30. Ward, M.R., Woroch, G.A., *Usage Substitution between Mobile Telephone and Fixed Line in the U.S.*, Texas University Report, Department of Economics, Arlington, TX, 2004, http://elsa.berkeley.edu/~woroch/usage%20substitution.pdf (accessed: March 10, 2010).

31. Di Giglio, A., Ferreiro, A., Schiano, M., The emerging core and metropolitan network. In Stavdas (Ed.), *Core and Metro Networks*, Wiley, Chichester, U.K., s.l., 2010, ISBN: 9780470512746.

32. Verbrugge, S., Pasqualini, S., Fritz-Joachim, W., Jager, M., Iselt, A., Kirstadter, A., Chahine, R., Colle, D., Pickavet, M. Demeester. P. Modeling operational expenditures for telecom operators, In *Proceedings of the Conference on Optical network Design and Modelling*, Milano, Italy, s.n., February 7–9, 2005, ISBN: 0-7803-8957-3.

33. Ohrtman, F., *Softswitch: Architecture for VoIP*, McGraw-Hill, New York, s.l., 2002, ISBN-13: 978-0071409773.

34. Doyle, J., Carroll, J., *Routing TCP/IP*, Vol. 1, 2nd edn., Cisco Press, Indianapolis, IN, s.l., 2005, ISBN-13: 978-1587052026.

35. Ahson, S.A., Ilyas, M., *VoIP Handbook: Applications, Technologies, Reliability, and Security*, CRC Press, Boca Raton, FL, s.l., 2008, ISBN-13: 978-1420070200.
36. Somani, A.K., *Survivability and Grooming in WDM Optical Networks*, Cambridge University Press, Cambridge, U.K., s.l., 2004, ISBN-13: 978-0521853880.
37. Goralski, W., *The Illustrated Network: How TCP/IP Works in a Modern Network*, Morgan Kaufmann Series in Networking, Burlington, MA, s.l., 2008, ISBN-13: 978-0123745415.

3

Networks Fundamentals and Present Architectures

3.1 Network Infrastructure Architecture

The telecommunication network is perhaps the most complex infrastructure deployed by humankind. It is pervasive, reaching almost all parts of the world, and exploits a wide set of different technologies.

In this chapter, we will depict the current situation of the most advanced networks, considering only wire-line access. Wireless access is another important part of the telecommunication network, but it is out of the scope of this book.

Not all the carriers reach the same level of technology. Some carriers are not able to invest the required capital expenditure (CAPEX) to be technology updated in all parts of the network; others simply do not experiment a push from their own market toward the use of the most advanced technology.

In describing the present day's situation, we will also review some fundamentals of the telecommunication networks theory to give the reader all the needed elements to become acquainted with the object of this chapter.

3.1.1 Network Topology

A telecommunication network spans a wide geographical area and is composed of different types of equipment.

The basic infrastructure is composed of fiber cables or, in the access area, of copper pair cables. It is no mean task to arrange thousands of kilometers of optical fibers in an underground or submarine cable so that it resists both environmental erosion and shocks due to accidents like civil works near the cable location. A detailed discussion of cable technology is reported in the following bibliographies [1–3] and some information on fiber optics cables is contained in Chapter 4.

All the other equipment composing the network has some degree of onboard intelligence, generally provided by an equipment control card. This intelligence is needed to perform remote monitoring and configuration of the equipment via the network control software.

Every equipment that is controlled by the network control software is called network element.

As far as the geographical configuration is concerned, the network of a tier one carrier can be divided into three areas: access, metro and regional, and core.

A high-level scheme of the telecommunication network is reported in Figure 3.1, while a topological scheme of the core network of a U.S. carrier is reported in Figure 3.2 to provide an example of continental core network, while the topology of a pan-European network can be found in Ref. [4].

The access network is the part of the network connecting the access central offices to the users' physical locations. This is the most branched part of the network, penetrating

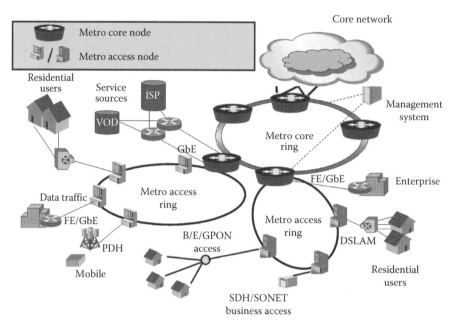

FIGURE 3.1

Topological representation of metro and access networks. Metro network is composed of a two-level ring topology, while different technologies are used in the access area each with its own topology (PON, ADSL, point-to-point Ethernet, and even direct SDH access are depicted in the figure). Finally, a metro node hosting content servers is shown: where VoD represents Video on Demand and ISP some other Internet Service Provider content server.

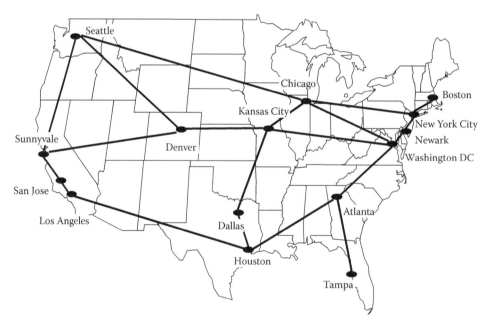

FIGURE 3.2

Simplified scheme of the topology of a long haul North American carrier network.

in metropolitan roads and reaching far into the countryside. The scope of the access network is one directional—to collect the signal generated from the end users and convey it to the peripheral network elements, located in the access central offices and in the other direction to deliver the signal from the central office network elements to the destination customer.

From a cable infrastructure point of view, in copper-based access, every user has a dedicated copper pair, so that no multiplexing between signals from different users happens. In the case of fiber-based access, in order to exploit at best the huge capacity of the optical fiber, signals from different users are multiplexed over part of the fiber infrastructure. This multiplexing is always operated via passive multiplexers/demultiplexers, where passive in this case means that they do not need power feeding and no monitoring by the management software is carried out.

Where a mixed copper–fiber infrastructure is adopted, in the location where the optical fiber leaves the place to copper cables, a network element that adapts the signal modulation to the different transmission media is needed.

As for signal modulation, in the traditional telephone network, an analog signal was transmitted from the access telephone switch located in the central office to the terminal in the user house via a user-dedicated copper pair. In the broadband network, this is not possible since the signal to be delivered is digital.

In order to deliver a broadband digital signal, carriers deploy at every user location (in offices for business users and in homes for residential users) a network element, generally called home gateway (or office gateway).

Generally, the home gateway is a part of the carrier network; thus, even if it is deployed in a private house, it is controlled and configured by the carrier via a remote management software.

The role of the home gateway is both to multiplex all the signals coming from the users appliances (PCs, televisions with set-top-boxes, game consoles, and so on) to transmit it to the network and to mark the separation between the home networking area that is within the customer responsibility, and the public network area that is within the carrier responsibility.

The home gateway has to implement all the functionalities needed to guarantee the quality of the delivered services, so it has to be capable of implementing full quality of service (QoS) functionalities [5].

When all the customers' signals arrive at the central office equipment, they are groomed to efficiently exploit the output transmission line and injected into the metro or regional network.

Thus, in a pure copper or fiber network, the access network is completely passive, and signal processing is performed only at its ends: at user locations and in the central offices.

In case of a mixed infrastructure, another point of processing exists where the two transmission media are joined.

The metropolitan or regional network is the part of the network spanning big cities or regional areas. General diameter of such a network is few hundred kilometers.

The role of the metro network critically depends on the location of the content sources in the network.

As discussed in Chapter 2, besides point-to-point services such as the telephone, immediate messaging, and e-mail, there are services that imply the access to content repositories, such as Internet protocol (IP) TV or video on demand (VOD).

In order to deliver such services, there are network elements that are devoted to store and distribute the content.

In Europe, incumbent carriers have always managed the network spanning a whole country: the same carrier thus manages access, metro, and core areas. In this situation, it is quite natural to place the content sources in the core network nodes: as a matter of fact here there is more space; a huge amount of traffic terminates here being delivered to different metro networks. In addition, monitoring and maintenance are easier.

With the same rationale, in Europe, the equipment that has to accept broadband users in the network and to authenticate their permissions is located in core nodes.

In this case generally, peer-to-peer services also transit through a core node before being delivered to destination, even if both the source and destination are in the same metro area.

In this architecture, the metro network has the role of effectively transporting the signals consolidated by the access equipment to the core.

In the United States on the contrary, traditionally, the local carriers were separated from the core carriers. This means that router boxes (RBOCs), being in the position not to own the core network, were forced to deploy a great part of the service sources and authentication machines into the metro area. Even if the carrier scenario in the United States has changed as a consequence of a few huge mergers in the last few years, the network architecture has not changed.

When the content sources and the authentication machines are located in metro nodes, the metro network task changes slightly, since besides grooming and transport toward core nodes, the metro network also performs direct connections between couples of its nodes.

After access and metro areas, the network is completed by the core, the part of the network whose role is to transport aggregated signals over very long distances and, in many cases, to host in the core central offices the machines for content delivery and users authentication.

The most important network elements in terms of technology are deployed in the core network that in general is characterized by a small number of nodes and very long span transmission systems.

3.1.2 Access Network Architecture

The copper network is a completely passive structure connecting every end user with the local exchange using a dedicated twisted pair [6].

In different parts of the network, different cables are used, composed of a different number of copper pairs. When a type of cable is terminated and the copper pairs are divided into smaller cables, the needed patch panel is contained in a street cabinet, called partition cabinet.

Generally, there are two levels of partition cabinets in an access network, as shown in Figure 3.3. The last patch panel is generally located inside buildings, at least in city arrangements, where every pair is distributed to its own user.

A critical characteristic of a copper-based access network is the average length of the user line. As a matter of fact, the maximum capacity that can be conveyed on a copper pair via digital subscriber lines (DSL) techniques critically depends on the copper pair length (see Figure 3.4, where as an example the bit-rate distance relation is shown for the very-high-speed digital subscriber line2 [VDSL2] format in typical conditions [7]).

Average length of copper lines in different European countries and in the United States is reported in Figure 3.5 [8].

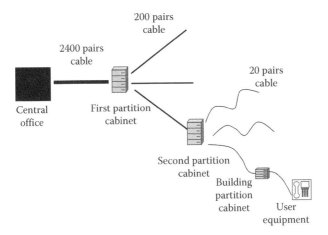

FIGURE 3.3
Block scheme of the copper pair–based access network.

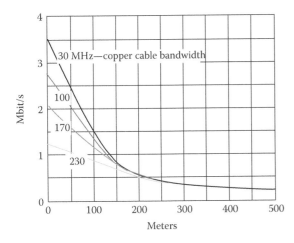

FIGURE 3.4
Example of the dependence of the VDSL2 capacity from the copper pair length in typical conditions.

From the figure, it is possible to see that relevant differences exist among European countries, while the copper length is longer in the United States with respect to all European networks.

This is another reason, besides regulatory differences and harder competition, that explains why U.S. carriers are much more advanced in deploying fiber optics in the access loop.

The exact replica of the copper-based structure using fiber optics would require direct connection of each end user with the local exchange using a dedicated fiber optics cable.

A similar architecture would have several advantages since it would assure maximum transmission capacity, physical separation of signals belonging to different users, and, last but not least, it would allow unbundling to be replicated in the new infrastructure with the same strategy used in the copper network.

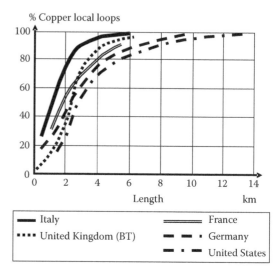

FIGURE 3.5
Statistical distribution of copper pair lengths in different countries.

In practice however this solution is considered unfeasible in many cases due to the fact that it would require a massive deployment of new fiber optic cables in almost every area, comprising city centers.

Among all the activities a carrier can undertake to enhance its network, a new cabling of the cities is by order of magnitude more CAPEX intensive and also the most advanced cabling techniques, quite effective for a limited deployment of new cables, does not change this issue [2,3].

For this reason, the fiber access architecture is devised so as to allow the greatest possible part of the fiber infrastructure to be shared among the maximum number of users.

Even if a smart fiber topology is adopted, capable of exploiting at best the available infrastructure, in many situations, connecting each end user with the local exchange directly via the fiber infrastructure requires still too much cabling.

In order to have a gradual CAPEX, a gradual fiber deployment strategy is often required. This means deploying optical fiber up to some intermediate point between the local exchange and the end user. In this intermediate point, active equipment converts the optical signal in an electrical one and the end user is reached with a traditional copper pair using a suitable xDSL format (generally VDSL2).

Mixed optical–copper architectures are called in different names depending on the distance between the intermediate point and the end user and on the number of users connected to the single intermediate point equipment. In particular, the following nomenclature is generally used (see Figure 3.6) that is tailored mainly for urban areas, where the deployment of next generation access will come first [6].

Fiber to the home (FTTH) is an architecture where there is no copper and the fiber arrives at the end user.

Fiber to the building (FTTB) is an architecture where the intermediate point is in the building basement or just outside a building and the number of end users linked to the same intermediate point is on the order of 20–30. In this case, the intermediate node is generally within 50 m from the user.

Fiber to the curb (FTTC) is an architecture where the intermediate point collects a few buildings and the number of end users linked to the same intermediate point is on the order of 100. In this case, the intermediate node is generally within 300 m from the user.

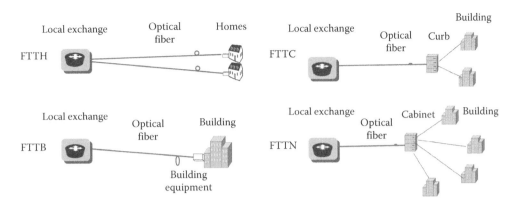

FIGURE 3.6
Topology of various hybrid fiber-copper access architectures.

Fiber to the node (FTTN) or fiber to the cabinet (FTTCab) is an architecture where the inter-mediate point collects many buildings and the number of end users linked to the same intermediate point is on the order of 400. In this case, the intermediate node distance from the user is generally beyond 300 m.

From a capacity point of view, in principle, a single fiber is largely over-dimensioned with respect to the needs of any group of nearby users: hundreds of multigigabit channels can be transmitted on a single fiber using dense wavelength division multiplexing (DWDM). However, DWDM techniques used in the backbone are far too expensive for the access network. In order to reduce costs, the single-fiber capacity is also greatly underused. Thus, adopting a network architecture that optimizes fiber use can be an important issue.

Moreover, in several practical cases either the carrier owns a limited number of fibers or it leases fibers from a third party (e.g., a utility company). In this situation, it is a key point to optimize the number of fibers used in the infrastructure.

The reduction of the number of fibers arriving at the central office also allows smaller optical patch panels to be deployed. Smaller patch panels means lower optical connectors failure rate and lower routing error probability. Moreover, it is much easier to realize automatic patch panels if they are small, further reducing the local exchange maintenance cost.

Last, but not least, reducing the number of fiber arriving at the central office for a given capacity allows the switching machine located in the central office to interface with the network with a small number of high-bit-rate interfaces. Using a small number of high-bit-rate interfaces optimizes the working of the local exchange switch machine, while minimizing its cost [9].

The earlier discussion shows that there are several reasons for pushing toward the reduction of the amount of fiber in the access network. The solution to this problem is constituted by the so-called passive optical network (PON).

In a PON [6], the fiber deployed in the field constitutes a logical tree, with the root in the local exchange and leafs at the end users' locations (see Figure 3.7). At the branching points, an optical passive element is present, either a splitter in time-division multiplexing (TDM) PON or a wavelength-division multiplexing (WDM) mux/demux in WDM-PON.

In the PON tree, every branch can be constituted either by a single fiber or by a fiber couple, depending on the adopted transmission technique. In particular, a single fiber is

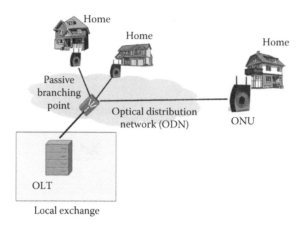

FIGURE 3.7
PON topology.

enough where bidirectional transmission is carried out, otherwise a fiber couple is used, one fiber for the upstream and the second fiber for the downstream.

In order to effectively use a PON infrastructure, the optical signal traveling through the tree root has to be shared among the different end users. This sharing can be realized either via WDM or via TDM multiplexing.

3.1.3 Metro Network and Core Network Architectures

The metro network collects signals from the access central offices and delivers it generally to the core nodes. This is the situation in Europe and in most of Far Eastern countries.

The traffic pattern in which all the traffic is to be transported to the core (there is no traffic directly routed from a metro node to another metro node) is called hub traffic; the core nodes interfacing the metro network are called hubs, and the access nodes are called sources (even if the traffic is bidirectional so that it also routes from the hubs to the sources) [8].

Since metro networks provide connectivity in all cities, there is a higher probability of cable break with respect to long haul systems deployed along great infrastructures like roads and railroads mainly in the countryside.

Metro network connections transmit high-capacity signals where a great number of customer signals are aggregated; thus, effective survivability has to be provided.

Ring topology is suitable for both hub traffic and fast survivability mechanisms [10] and is practically the standard topology in the metro or regional area.

Since the evolution to broadband network started from a telephone-centric network generally maintaining the same central offices, a two-layer architecture is frequently adopted, as shown in Figure 3.8.

A first ring set connects the access nodes with metro intermediate nodes, constituting the so-called metro access network. All the signals processed by the intermediate nodes are collected by a single ring, called metro core ring [11], delivering it to the hubs.

It is quite common, especially in big cities, to have two hubs physically separated in different buildings. If the hub equipment is correctly dimensioned, this assures network survivability in case of a catastrophic event completely destroying a hub.

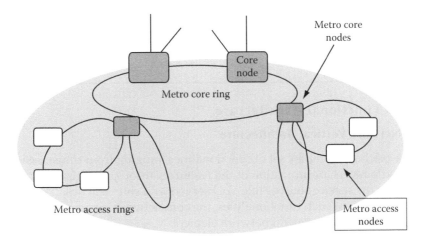

FIGURE 3.8
Two levels multi-ring metro network.

The presence of intermediate nodes also provides a further level of traffic consolidation increasing the exploitation efficiency of the high-capacity metro core and core transmission systems.

As already discussed, in the United States, frequently the traffic managed by metro networks, especially in big cities, is not pure hub traffic, but has a hub component plus a component (called mesh component) constituted by direct connections between metro nodes [11].

Moreover, when the planning of the broadband evolution was done, the main RBOCs were not in charge of the core network. Long haul transmission was carried out by leasing capacity by the long haul carriers.

In order to reduce the leasing cost, the main regional bell operating companies (RBOCs) (first of all Verizon) designed a network that was able to avoid transit through long haul nodes as far as possible. Starting from the two-level ring architecture, the solution is to directly connect different metro core rings to implement all metro-to-metro connections without network blocking and without transit through core machines.

This interconnected ring architecture is also characterized by high survivability, if a suitable mechanism is implemented to protect the ring interconnection [12].

The core network that is in charge of covering very long distances is characterized by a few specific elements:

- All the network nodes are similar (there is no hierarchy as in the metro network).
- Network nodes perform signal processing in terms of
  - Consolidation (grooming and multiplexing)
  - Routing
  - Rerouting to assure survivability
- Every network node can add-drop capacity to-from the core network.
- A set of logical connections between couples of network nodes exists (pure mesh traffic).
- Distances between nodes are big (sometimes several thousand kilometers) requiring long haul transmission.

The core topology, as a consequence, is a mesh of connections between couples of nodes (see as an example Figure 3.2).

3.2 Network Functional Architecture

3.2.1 Core Network Vertical Architecture

The core network has a complex set of functionalities, ranging from transmission through thousands of kilometers, optimization of the resources through multiplexing and grooming, and hosting the service sources like VoD servers and soft switches.

In order to implement such functionalities, the core network is composed of several different equipment, called in general network elements.

The way in which different functionalities are divided among the network elements and the way in which they cooperate to carry out collective tasks constitutes the so-called vertical or functional network architecture.

The remaining of this chapter is devoted to analyzing this architecture, the different alternatives and the standards that have been introduced to allow compatibility among products of different vendors and to realize a quality standard on which carriers can build their business.

In this section, we introduce a synthetic but complete description of the whole architecture to give the reader a reference frame where more detailed information is collocated to make the reader familiar with the main network functionalities.

The network entities are organized in three superimposed networks: the transport network, the packet network, and the service network. This vertical architecture is represented in Figure 3.9.

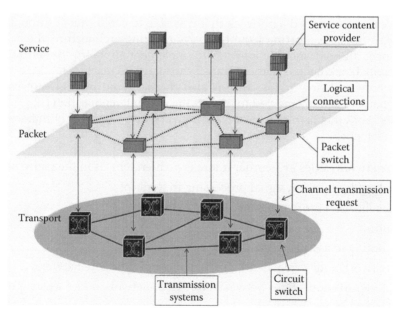

FIGURE 3.9
Vertical architecture of the core network evidencing the transport, the packet, and the service networks.

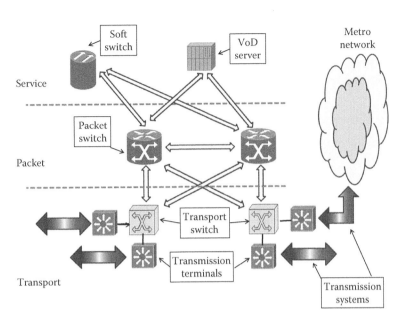

FIGURE 3.10
Scheme of the node of a core network where there are present service, packets, and transport equipment.

At each node of the core network are potentially present three groups of node equipment. One or more network entities belonging to the service network generate service signals. These feed one or more network elements of the packet network that performs statistical multiplexing, besides a great number of other functionalities that we will detail in the following chapters; the packet switches process the signal in small segments, called packets. At their output the packets are aggregated in continuous data fluxes (called channels) that are passed for transmission to the network entities of the transport layer.

The core node architecture formed by these three equipment groups is represented in Figure 3.10.

The architecture of Figure 3.9 is not yet a formal functional layering as it will be introduced in the next section, but it is frequently considered as the basic structure of the core network.

The functional layering can be derived by structuring the three networks with suitable layer models and formally defining opportune interfaces among them.

3.2.1.1 Service Network

The service network is constituted by the machines that customers interface to request the various telecommunication network services.

For example, when requesting a VoD, a customer has to communicate with a video server where all the content is stored to ask the transmission of the desired video.

Service equipment is naturally strongly dependent on the service under consideration, but in any case they interface the packet network by packet interfaces. This means that the service equipment fragments the information that has to transmit through the network into data packets that are delivered to the packet network as individual entities.

Each data packet has a data field, in which the information to be transmitted is encoded, and a header field containing information about the packet itself, such as origin and destination, length and type of service, number within the communication between the user and the service source, and so on.

A typical packet format used by service machines to interface the packet machines is the Ethernet frame, which we will introduce later in this chapter.

3.2.1.2 Packet Network

The main functions of the packet network can be summarized as follows:

- Packet delivery to the correct destination in the order in which they are transmitted
- Individuation and, if possible, correction of errors due to packet processing or transmission
- Services requirements fulfillment in terms of, for example, delay and reliability
- Packets recovery in case of failure in the network
- Packet network performance monitoring and control

In order to deliver packets to the destination, the packet network uses a set of virtual connections among its nodes created by asking permanent TDM connections to the transport network.

In particular, a packet network machine, creates, at its interface with a transport network, a TDM channel and asks the transport network to transport the channel from its origin to its destination.

A TDM channel is a continuous stream of information constituted by a periodical repetition of the TDM frame. An example of TDM frame is reported in Figure 3.11: in the simpler case, it is constituted by a frame header and an information field.

The frame header contains all the information that are needed for the correct transmission of the channel from the origin to the destination in the transport network, for example, the destination address, the frame length, a frame synchronization word allowing receivers to individuate the frame start, an error correcting code (ECC), the position inside the information field of the packets transported by the frame, and so on. We will detail in

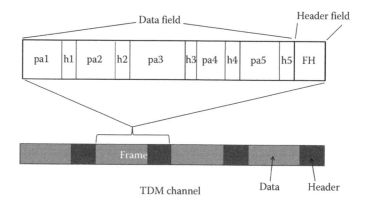

FIGURE 3.11

Scheme of the TDM frame of a channel. In the lower part of the figure the TDM channel is represented as a periodical sequence of identical frames. In the upper part, one frame is analyzed evidencing the data field, where the packets are located, and the header field. In this part of the figure, pa indicated the data part of a transported packet and h the packet header, while the frame header is represented by FH.

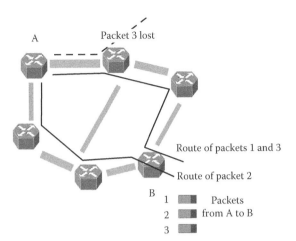

FIGURE 3.12
Routing in a datagram packet network.

Appendix A the structure of one of the most used TDM frame formats, the synchronous digital hierarchy (SDH)/synchronous optical network (SONET) frame.

The information field contains the packet to be transported by the frame.

Once the transport network assures the correct transmission of the channels, the packet network sees them as the transmission resources at its disposal to deliver packets to their destination.

Using channels, the packets can be delivered by the packet network following two main strategies: datagram and virtual circuit.

In a datagram network each packet is managed individually, with no relation with the packets constituting the same communication between the service equipment and the customer.

Thus, consecutive packets of the same communications can be routed through the network via different routes, depending on the network state at that moment (see Figure 3.12).

This technique allows the channels to be exploited at best, since a packet is multiplexed in a specific channel among those useful to transmit it to destination with the only task of saturating the channel so as to optimize its use.

On the other hand, due to the independence of packets in a datagram network, packets from the same connection between user and server can arrive in the wrong order experiencing different routes in the network. Moreover, if a packet gets lost, it is evident only at the end of the communication and duplicated packets (that can exist as a consequence of routing errors) are identified only at destination.

This lack of control on transmitted packet by the datagram causes the service network (or the receiving node in the packet network) to carry out a supplement of work in reordering the arriving packets and asking for retransmission of the lacking one while disregarding the duplicated one.

All these operations can be closed only when all the messages have been delivered, making it difficult to have a short delivery time for messages. Moreover, the delivery time also manifests a strong random variability, due to the fact that duplication, lost, or wrong order are random phenomena depending on instantaneous network status or on errors that cannot be forecasted.

This lack of control on the packets delivery time renders difficult to support real-time services like TV or telephone, rendering datagram difficult to implement in public networks.

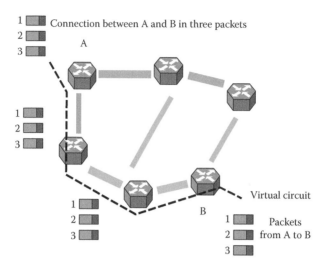

FIGURE 3.13
Routing in a virtual circuit network.

If all the packets belonging to the same communication are routed through the same route in the network, the network is said to implement virtual circuit routing (see Figure 3.13). The path the packets have to follow has to be set up when the communication starts by advertising all the traversed machines that packets with a certain communication ID have to be routed through a certain path. At the end of communication the network has to tear down the path by advertising the network entities that the communication has ended.

This means that a more complex signaling is needed with respect to a datagram network.

Moreover, since the packet routes are stated at the beginning of a communication, it is impossible to change routing to follow the network status changes and channels are exploited less efficiently.

On the other hand, the packets delivery time can be controlled much more effectively and monitoring of a communication in terms of delivered packets, errors done, and so on can be performed much more effectively.

Real-time services require a virtual circuit network and core carrier networks implement virtual circuits, even if IP is datagram oriented, using for example multiprotocol label switching (MPLS).

Returning to the packet network functionalities, error management is carried out either through error correction or through error detection codes whose redundancy is carried out by the packet header.

Fast recovery of data disrupted by a failure is carried out with different methods.

In datagram networks, the packets involved in the failure are simply lost. They will lack at the destination causing a retransmission request. For the datagram routing properties, retransmission will naturally use available resources, avoiding failed equipment.

In a virtual circuit network, virtual circuits will be disrupted by a failure and they have to be recovered. Several methods are available that can be divided into two classes: protection and restoration.

In the protection case, every time a virtual path is set up, another path is also set up for the same connection. This second path has the constraint of not having network resources in common with the first (request that has to be passed to the transport network since the packet one does not have transport equipment visibility).

The second path is used as backup of the first in case of failure. Both hot and cold backup techniques can be used, depending on whether the packet is always sent contemporary on the working and protection path or if the protection path is used only in case of failure.

In case of restoration, spare capacity is planned in the network without connecting its use to a particular failure. When there is a failure, the affected virtual circuits are rerouted using the spare capacity. Also in this case, different types or restoration exist, as detailed in Chapter 8, depending, for example, if the spare capacity is reserved for restoration or, in case of failure, circuits to be rerouted compete with the requests on new circuits.

Restoration is largely the most used recovery method of packet circuits, since protection is left to the transport network.

3.2.1.3 Transport Network

The functionalities of the transport network can be summarized as follows:

- Transmission of channels received from the packet network
- Optimization of transmission systems exploitation via TDM hierarchical multiplexing and grooming
- Verification of the absence of errors in the transmitted TDM frame
- Fast recovery of lost channels due to a failure
- Monitoring and control of channels transmission

Transmission itself is carried out by transmission systems that physically connects the near to the far ends via fiber optics. These systems will be analyzed in detail in Chapters 6 and 9.

Generally, transmission systems transmit a set of very high-speed signals (10 Gbit/s or more) over different optical wavelengths (WDM). The channels coming from the packet networks are not so fast; thus, TDM multiplexing is performed in different steps to create a TDM hierarchy.

The scheme of a simple TDM hierarchy is represented in Figure 3.14: the channels received from the packet network are multiplexed in a faster TDM frame where every data

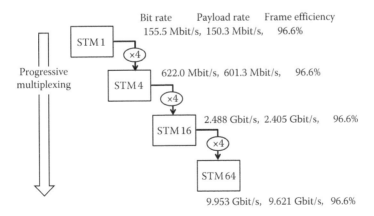

FIGURE 3.14
Scheme of a simple TDM multiplexing hierarchy. A subset of the SDH hierarchy is shown.

field is filled with the frames of the channels to be multiplexed, which are called multiplexing tributaries.

This operation is repeated several times, passing through a set of intermediate-speed TDM channels up to achieving the speed compatible with transmission systems.

The operation of multiplexing is not performed in a single step for different reasons.

Besides the fact that it would be technically difficult to multiplex together a huge number of slow channels in a single very fast channel, in real networks, the speed of the channels created by the packet network is not unique; on the contrary the packet network can create channels at different speeds following the needs.

The creation of a TDM hierarchy allows channels of different bit rate to be received from the packet network and to be accommodated at different levels of the hierarchy.

The most important standards defining TDM multiplexing hierarchies are SDH/SONET and optical transport network (OTN) that we will analyze in some detail later in this chapter.

Even if an effort is done to optimize the use of the transmitted high-speed channels, it is frequent in the transport network that not all the tributaries are really present in the data field (or payload) of the transmitted frame. In this case, the carried information is smaller than its maximum possible value, depending on the bit rate and the frame structure.

This lack of efficiency can be at least partially corrected by means of grooming, that is, the operation consisting in the demultiplexing of a channel in the middle of its route to distribute all or part of the tributaries in other channels.

This operation can be performed all in the transport layer or the cooperation of the packet layer can be required.

An example of grooming in the transport layer is provided in Figure 3.15. In part (a) of the figure, the situation of a transport network receiving channels at 622 Mbit/s (synchronous transport module 4 [STM4] in the SDH language) and transmitting wavelength channels at 2.5 Gbit/s (STM16 in the SDH language) is shown so that up to four STM4 can be multiplexed in a single STM16.

The STM4 channels routed at a certain instant is given by the traffic matrix incorporated into Figure 3.15a: three bidirectional STM4 from A to B and two bidirectional STM4 from B to D. The traffic from A to B is routed through the direct link using a single STM16 and the traffic from B to D is also routed through the direct link using another STM16. The return channel is always routed on the same route.

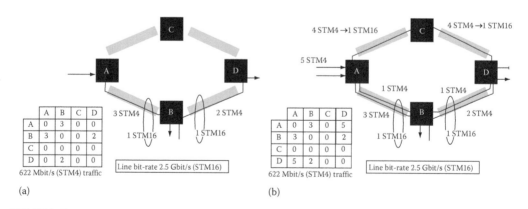

FIGURE 3.15
Example of grooming in the transport network: (a) situation before the new request; (b) situation after the new request of five channels between A and D.

Both the high-speed channels used for transmission are not fully exploited, since their potentiality would be to carry up to four lower speed channels.

In a certain instant, a new connection request arrives from the packet network, consisting in five STM4 connections from A to D, as shown in the traffic matrix present in Figure 3.15b.

If all the capacity would be routed through the route A-C-D, two STM16 would be needed, one of them greatly underutilized since it transports only one tributary.

On the contrary, four STM4 can be routed through the route A-C-D as shown in the figure, the fifth is joined to the three channels directed in B from A, saturating the corresponding STM16 whose destination is node B.

In order to optimize the use of the transmission resources, the STM16 is demultiplexed in B, the channels directed to B are passed to the packet network, and the channels directed to D are joined to the STM16, multiplexing the channels from B to D, thus improving its exploitation efficiency.

In this way, by using a redistribution of the tributaries among the transmitted channels in an intermediate node, the transmission capacity is fully optimized.

A similar operation can be performed by returning the channels at the packet network asking to redistribute them among already existing channels.

The grooming at the packet level is clearly the most efficient, but it requires cooperation between packet and transport network.

Several methods are used to guarantee transmission quality at the transport level. Errors are corrected using Forward Error Correcting Codes (FEC) while channels lost after a failure can be recovered either with protection or with restoration procedures.

Due to the structure of the transport network and to the possible presence of restoration in the packet network, protection is by far the most used survivability technique.

There is a fundamental difference between protection in the transport network and in the packet network. In the packet network, even if virtual circuit routing is used, the channels are used simultaneously by different virtual circuits and no capacity is reserved to a specific virtual circuit. Thus, different circuits compete for the use of the capacity within the rules stated by the QoS management.

This means that even if a virtual circuit is reserved for protection it is used also by other virtual circuits. This exploitation in any case is carried out so that, when protection is activated, there is always an available bandwidth for the recovery of the circuits under failure due to the resources reservation process carried out when the virtual circuits were set up.

In the case of the transport network, when a capacity is reserved for protection, no other TDM channel can occupy it since TDM transmission is continuous in time and different channels cannot coexist in the same bandwidth.

This is why, besides standard 1+1 protection, shared protection is also defined in the transport network to optimize the use of transmission resources.

If share protection is used, the wavelength channels are divided into groups so that all the channels traversing at least one common element are in the same group. The meaning of these groups is that they collect wavelength channels that could be hit by the same failure.

When a wavelength channel is set up, a protection channel is also set up, but this operation is done considering the available protection capacity already reserved for wavelength channels in different groups.

This method is based on the hypothesis that a double failure is quite improbable and on the observation that, since wavelength channels in different groups do not traverse a common element, only a double failure can disable both of them.

Another way to try to use the protection capacity is to use it to route it the so-called low-priority traffic, that is, channels that can be lost in case of failure.

If a failure occurs, the low-priority traffic is stopped and the protection capacity is used to recover standard traffic disabled by the failure.

Even from this simple and incomplete description it is clear that the vertical structure of the core, and often also of the metro, network is quite complex.

The network equipment communicates with each other and with the central network management to carry out complex operations.

In order to go deeper into the description of the network architecture, a more formal method needs to be introduced: the so-called layering. In the next section we will describe in more detail the network control system. After that, using the layering method, we will go in deep into the functional structure of the netowrk starting from the service/packet/transport levels model.

3.2.1.4 Control Plane and Management Plane

In order to coordinate their working, network elements exchange control information both among elements of the same network (e.g., transport elements) and among elements of different networks. These exchange of information and commands follow set of rules that are called protocols.

Such control information have to travel, much like data do, from one network element to the other; thus, a transmission capacity is needed to implement coordination among network elements.

Such transmission capacity can be either embedded into the same network transmitting data (e.g., in the headers of the TDM frames) or provided by a completely separated control network. In both cases, it is possible to define two logical networks having the same nodes locations and the same network elements, but using a different transmission capacity and transmitting different data. The logical network transmitting data is called "data plane" and the logical network conveying protocol-related information is called "control plane."

The control plane is necessary to assure network operation, but generally it is not sufficient. As a matter of fact, the control plane comprises procedures that coordinate elements of network operation, but do not comprise functions for reporting the network status to a central control, where network management operators supervise network working.

Such reporting is needed for various reasons. The first reason is that distributed protocols are fast and easy to program, but as their action is local, they are not generally able to perform a real optimization of the network, as we will see when describing the Transmission Control Protocol/Internet Protocol (TCP*/IP) control plane.

In order not to allow a progressive sliding of the network state from the optimum, for example, in terms of routing or backup capacity assignment, centralized traffic engineering algorithms are needed that allows routing and capacity assignment to be enforced on the network.

Even more important is the need for hardware maintenance and field intervention to substitute malfunctioning hardware parts. As a matter of fact, survivability protocols are able to reroute a signal that is not transmitted in the correct way and they can even identify

---

* It is to be noted that the acronym TCP in this context is used as "Transmission Control Protocol." is also used with the meaning of "Terminal Connection Point" in the general layering theory. The two meanings indicate very different concepts so that no difficulty should arise to identify the correct meaning from the context.

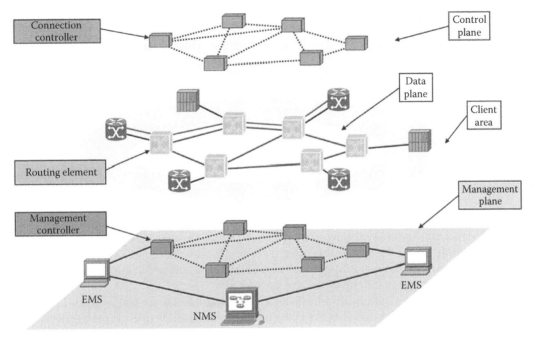

FIGURE 3.16
Network decomposition in data plane, control plane, and management plane. EMS, element management system; NMS, network management system.

the problem underlying the performance degradation, but, in case of out-of-order network elements, for sure, protocols cannot change the failed hardware with a working one.

Another important reason justifying the presence of a centralized network management is the need of a carrier to certify the network performances in front of the service level agreement (SLA) that guarantee to customers a certain performance level.

Finally, when a new service is enabled in the network or a new software release is downloaded over some network elements, all the setup procedure has to be managed by the central network management.

The network management has also the need of exchanging data to and from all the network elements; thus, it needs a transmission capacity. In general, this transmission capacity is different from that assigned to the control plane and, either embedded in the data network or provided with a separated network, it is called management plane.

Summarizing, the architecture of a public telecommunication network spanning a geographical area is composed of three superimposed plans, each plan representing a different network component: the data plane, the control plane, and the management plane.

As shown in Figure 3.16, the three plans can be represented by graphs that have the same nodes, the locations of the central offices, but potentially a different set of links, representing the transmission capability available between a couple of nodes on the considered plan.

Each network generally has its own control and management planes. Thus, in many present networks there are transport management and control planes to manage and control the transport network and packet management and control planes to control and manage the packet network.

As we have seen in the previous chapter, this separation is a source of inefficiency pushing carriers to work to integrate the management plane and the control plane with the target of having only one management and one control over all the network structure.

3.2.2 Network Layering

Making a design of the network starting from the base components is practically impossible, due to the conceptual distance between the functionalities of a single building block (like a transmitter or a filter) and the required network functions (like delivering high-definition television [HDTV] to selected customers or allowing World Wide Web browsing).

In order to design the network, a design method capable of overcoming this separation is needed: this is the layering model. Not only the layer model is a powerful design tool, but it also provides the means of describing in detail the functional architecture of the network using a formal instrument.

This technique was introduced in the field of telecommunications by the International Standard Organization (ISO) with the intent of describing in an abstract way and standardizing the computer networks [13–15]. The resulting standard was called Open Systems Interconnection architecture (ISO-OSI).

The abstract methodology followed by ISO was later also adopted by international telecommunication union-telecommunication (ITU-T) (the international standard organization for telecommunications) that applied it to TDM transport networks extending it and introducing a set of ad hoc definitions and procedures.

What came out was a general synthesis technique that has a wide use, much wider than the original intention of its creators. As abstract methodology, it is useful every time a complex function has to be realized by designing a system starting from the building blocks performing much simpler functionalities.

Other examples of use of this model are software design [16], study of brain functionalities, and of artificial neural networks [17,18]. Naturally, the extension of the layering methodology to all these different applications frequently requires customizations of the method and introduction of ad hoc elements.

To introduce the layering methodology, let us imagine to have a complex function $F(\alpha)$ operating over some input α (in general, constituted by a collection of elements) in order to produce an output β (also a set of elements). Just to be specific, it can be the function of a WDM transmission system whose input are the data emitted by the signal sources feeding the system and the output are the data delivered at the destination equipment. In this case, α is the set of bit streams representing the input data, one for each wavelength, and β is the set of bit streams representing the data delivered at the destination.

Let us image also that the function has to be implemented via an equipment built out of a few building blocks (let us indicate them as $b_1, b_2, ..., b_n$), all operating on a subset of the elements constituting α.

A layered representation of the function $F(\alpha)$ can be achieved as follows.

Let us imagine to identify a first set of functions $g_1, g_2, ...$ simpler than $F(\alpha)$, but still more complex than $b_1, b_2, ..., b_n$. Let us also assume that $F(\alpha)$ can be implemented by a finite number of applications of the functions $g_1, g_2, ...$ to the input α. In the language of the layering model, it is said that there is a client server relationship between $F(\alpha)$ and $g_1, g_2, ...$ or that the functions $g_1, g_2, ...$ are at the lower layer with respect to $F(\alpha)$. This relationship can be also graphically represented with a graph like those in Figure 3.17.

Since there is no physical device implementing directly the functions $g_1, g_2, ...,$ each of them has to be further decomposed into simpler elements and this procedure has to be repeated as shown in Figure 3.10 up to a layer in which the considered functions are directly performed by the available physical building blocks.

At that point, the layered decomposition is terminated and every function in a layer can be built by using the functions of the bottom layer, up to achieving the target function $F(\alpha)$.

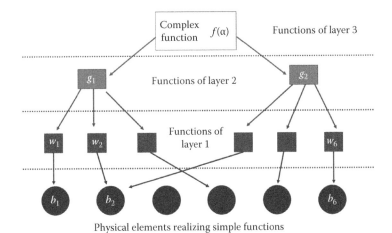

FIGURE 3.17
Graphical scheme of the layered model of a complex system.

Thus, $F(\alpha)$ can be synthesized by traveling the layered model bottom up and at the end it will result in its implementation by the available simple physical elements.

Up to now no news, the technique of dividing a complex problem into simpler elements is diffused in all the engineering fields. The strength of the layering technique is that it gives a structured procedure to carry the synthesis of complex functions through its formalism, which has to be somehow adapted to the particular field of application.

Here we will not introduce in a completely formal way the layering technique, since it is a task requiring a dedicated book. The scope of this section is to give an idea of the layering technique and a few details of the graphic language associated to it so as to be able to understand the layered structure of telecommunication networks.

Analyzing the layered structure, two interactions among functions can be individuated: interactions among functions in the same layer and interactions among functions in different layers. The first interaction type is needed to coordinate the work of the functions belonging to a layer; the second type of interactions is needed to realize the client–server relationship on which the model is based.

Functions of the same layer communicate via messages coded using specific rules that are called layer protocols. Thus, every layer has a set of specific layer protocols. Of course, the fact that the functions have to exchange messages means that they also have to accept control signals as input. Moreover, the functions output will be constituted by processed input data plus a set of control signals.

The communication of management information between different functions is carried out via a service communication network that can be either embedded in the system under design or completely separated.

As far as the interlayer communication is concerned, the first point is that while functions belonging to one layer generally exploit a logical layer network to communicate, this does not exist if functions belonging to different layers have to be connected.

To realize interlayer connectivity there are the so-called access points that are simply points in which the output of a function of the upper layer can be connected with the input of a function of the lower layer or vice versa, operation that cannot be done without the presence of this particular interface.

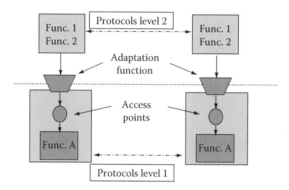

FIGURE 3.18
Graphical definition of protocols, adaptation functions, and access points.

Moreover, if interlayer communications have to exist, a communication interface has to be foreseen, which solves the problem that functions in different layers will have completely different protocols.

In order to realize the interlayer data exchange, a special function is needed, which translates the signals produced by the source layer in a form that the destination layer can process.

This is the so-called adaptation function and in general it is represented between two layers.

The scheme of intralayer and interlayer communication is shown in Figure 3.18 where layer protocols, access points, and adaptation functions are represented in the layering graphical language.

Using the concepts of access points and adaptation functions, we can go further in detail in the layered structure of a telecommunication network.

In general, a layer receives an input in correspondence to an access point from the output of an adaptation function and, since we are dealing with communication networks, its role is to deliver the output of its processing carried out from the layer functions to another access point that transmits it to the upper layer via another adaptation function.

The layer operation can be represented like a path through a logical layer network that connects a suitable number of functions so that the output of the first is the input of the second and so on. The layer network is simply the virtual network constituted by the layer functions located in correspondence to the locations of the physical network nodes.

As will be clear describing specific examples, not all the network nodes are equipped to implement all the functions belonging to a layer. There are network nodes whose equipment implements only a subset of the functions of a certain level so that in the corresponding location only those functions are available.

On the other hand, every network element has its own layered structure, thus providing functions at several layers.

The path through the network layer connecting the input access point to the output access point is called "trail" (or "link" in Internet engineering task force [IETF] documents) and in order to mark the trail start and the trail end the "trail termination points" (TTP) are introduced. TTPs are not only markers of trail start and end, but are functions in their own right.

As a matter of fact, at the trail ends a set of operations have to be performed, generally related to layer monitoring, and those operations are carried out by the TTPs.

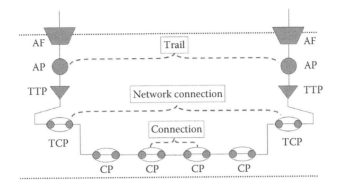

FIGURE 3.19
Graphical representation of a trail and its components within a single layer. AF, adaptation functions; AP, access points; TTP, trail termination points; TCP, trail connection points; CP, connection points.

TTPs are connected to the first and to the last layer function by the "terminal connection points" (TCPs)* while the layer path itself is created by connecting the functions via the "connection points" (CPs). The role of TCPs and CPs is somehow similar to the role of the access points. They represent the points performing every adaptation that is needed to allow different functions to communicate.

The structure of the trail through a generic network layer is represented in Figure 3.19 where all the trail elements are evidenced.

In practice different communication network models exist, built over the abstract method of layering and using different layers to model and design the network.

In order to adopt a common language, the layers are generally numbered following as much as possible the OSI definitions. This is an informal way to proceed, since there is no complete correspondence between the layer functions as coded in the OSI model and the layer functions as coded in the model used in practice (see, e.g., the Internet model or the MPLS technology) and care has to be adopted not to create confusion and errors due to the application of OSI characteristics that are not reflected in the model under consideration. However, this gives a quick reference and a way for a first comparison among different models.

Thus, it is useful to briefly describe the functions of the seven layers of the OSI model, keeping in mind that it was standardized assuming a computer network while present days public networks, even if based on packet switching, convey a much more complex set of services [13].

1. *Layer 7, application layer*: This layer supports application and end-user processes. Communication partners are identified, quality of service is defined, user authentication and privacy are considered, and any constraints on data syntax are identified. Everything at this layer is application specific. This layer provides application services for file transfers, e-mail, and other network software services.

2. *Layer 6, presentation layer*: This layer provides independence from differences in data representation (e.g., encryption) by translating from application to network format, and vice versa.

---

* It is to be noted that the acronym TCP in this context is used as "Terminal Connection Point" is also used with the meaning of "Transmission Control Protocol." describing the internet protocol suite. The two meanings indicate very different concepts so that no difficulty should arise to identify the correct meaning from the context.

3. *Layer 5, session layer*: This layer establishes, manages, and terminates connections between specific applications. The session layer sets up, coordinates, and terminates conversations, data exchanges, and content broadcast between the applications at each end.

4. *Layer 4, transport layer*: This layer provides end-to-end data transfer between the hosts; end-to-end message integrity and flow control are delegated to this layer.

5. *Layer 3, network layer*: This layer provides routing technologies, creating logical paths in case virtual circuits are used; otherwise the layer functions process individual datagrams at each network node. Routing and forwarding are functions of this layer, as well as addressing, error handling, and congestion control.

6. *Layer 2, data link layer*: This layer furnishes transmission protocol knowledge and management and handles errors in the physical layer, flow control, and frame synchronization. The data link layer is divided into two sub layers: the media access control (MAC) layer and the logical link control (LLC) layer. The MAC sublayer controls how a computer on the network gains access to the data and permission to transmit it. The LLC layer controls frame synchronization, physical flow control, and error checking.

7. *Layer 1, physical layer*: This layer conveys the bit stream—electrical impulse, light or radio signal—through the network at the electrical (or optical) and mechanical level. It provides the hardware means of sending and receiving data.

3.2.3 Internet

The Internet model is a complete model for the packet network and the IP suite is the set of protocols used at the various levels of the pure Internet network model [19,20].

In its pure version, the Internet is a datagram switched network: that is, a network where the data stream from one termination to the other is divided into packets traveling independently through the network

The fact that packets belonging to the same message can be routed through different paths in the network potentially causes a disordered arrival of packets at the destination and a difficulty to recognize packet lost.

The layer model of the Internet is shown in Figure 3.20 [21,22].

The layered model is composed of four layers plus the physical layer that collects all the equipment used for transmission and that is not part of the Internet stack.

Due to the datagram nature of the Internet, control signals travels through trails in the various layers either as headers of the data packets or as dedicated protocol packets, depending on the protocol.

This means that when a layer passes a packet to the lower layer, the packet is enveloped with a header and at link layer also a tail so that the whole packet of the upper layer becomes the payload of the packet of the lower layer.

This enveloping process [22] creates a packet structure where the headers relative to different layers follow one after the other starting with the lower layer header, as shown in Figure 3.21 [21].

In Figure 3.20, it is also shown that, from an equipment point of view, the end terminations of an Internet trail are hosts (i.e., computers with the suitable software to implement Internet connectivity) while transit machines are called routers. Routers implement only the first two layers of the Internet model, since their role is to store arriving packets at the input, calculate the output direction for each packet on the ground of routing tables and send each packet to the correct output.

Network physical link

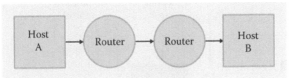

Layered model of the link

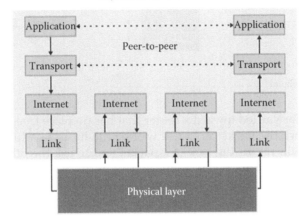

FIGURE 3.20
Network and layered model of a trail through an IP-based network.

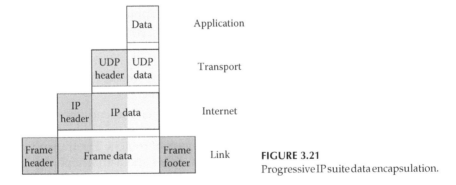

FIGURE 3.21
Progressive IP suite data encapsulation.

The high-level block scheme of a router is represented in Figure 3.22. When a datagram arrives at the router, the link header and tail are seprated and used to verify the correct packet transmission, while the IP layer packet is stored in an input queue. After storing, the IP address is read and used to access the routing tables to determine the exit direction of the packet [23,24].

The output of the routing table is the output port destination. In the absence of packet conflicts, the router configures the switch fabric in order to send the packet to the correct output queue and transfer the packet there. In the output queue, the link envelope is recreated and the packet is launched toward the next router (forwarding process).

In the presence of conflicts on an output port, a router uses a conflict solution criterion to decide the order in which the packets have to be transmitted, criterion that can be as simple as a first in first out (FIFO) or as complex as a complete QoS algorithm.

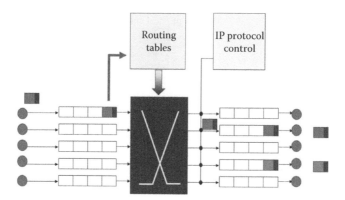

FIGURE 3.22
Router functional block scheme.

When a packet is forwarded toward a router that has no place free in the queue on the designated input port, the packet is lost. It will be the host via a higher layer protocol (e.g., the TCP) that will report the lack of a packet and to ask the other communication end the packet retransmission.

From this discussion, it results that routing through an IP network is performed via a distributed algorithm running on all the network routers; no router has a general control role and in order to reconstruct packet route there is the need to monitor contemporarily all the routers of its trail [24].

Summarizing, the four levels of the Internet model can be described as follows:

- *Application layer*: This is the layer allowing different applications to communicate for exchanging users' data or control signals; in this layer, the fact that the connected applications are in the same host or not is irrelevant. The communication partners are often called *peers*. This is where the "higher level" protocols such as Send Mail Transfer Protocol (SMTP), File Transfer Protocol (FTP), Secure Shell (SSH), Hypertext Transfer Protocol (HTTP), etc. operate.

- *Transport layer*: The transport layer manages the end-to-end communication between two network hosts, either directly on a local network or on a geographical network where the hosts are connected via a set of routers. The transport layer provides a uniform networking interface that hides the actual topology of the underlying network connections. This is where flow-control, error-correction, and connection protocols, such as TCP, are located.

  With reference to equivalent OSI layers, transport layer protocols are considered layer 4 protocols.

- *Internet layer*: The Internet layer has the task of exchanging data packets across network boundaries. This layer defines the addressing and routing of the packets to conduct them through the network from the origin to the destination. The primary protocol in this scope is the IP, which defines IP addresses. Protocols belonging to the Internet layer are layer 3 protocols in the OSI language.

- *Link layer*: It is used to move packets between the Internet layer interfaces of two different hosts or routers on the same trail. The processes of transmitting and receiving packets on a given trail can be controlled both in the software device driver of the network card, as well as on specialized chipsets. These will perform data link functions such as adding a packet header to prepare it for transmission, then actually transmit the frame over a physical medium. The Internet model includes specifications to translate the network addressing methods used in the IP to data link addressing, such as MAC; however, all other aspects below that level are implicitly assumed to exist, but are not explicitly defined.

The list of the main protocols belonging to the IP suite is reported in Table 3.1 besides a very short description of the protocol role.

In order to better understand the working of the Internet model, we will analyze in some more detail a few of the principal protocols listed in Table 3.1.

3.2.3.1 Transport Layer: Transmission Control Protocol

The TCP is one of the older and most used protocols of the IP suite; it is optimized for accurate rather than timely delivery, and therefore, it sometimes incurs relatively long delays (in the order of seconds) while waiting for out-of-order messages or retransmissions of lost messages.

Thus, TCP is not particularly suitable for real-time applications such as Voice over IP. For such applications, protocols like the Real-Time Transport Protocol (RTP) running over the User Datagram Protocol (UDP) are usually used (see Table 3.1b).

A TCP segment consists of a segment header and a data section. The header contains 10 mandatory fields, and an optional extension field, as shown in Figure 3.23.

The main fields of the header are the source port and destination port identifiers and two fields for the segment progressive number and the segment reception acknowledge. A checksum for error detection and an end of transmission flag are also included.

TCP protocol operations may be divided into three phases:

1. Connections establishment
2. Data transfer
3. Connection termination

3.2.3.1.1 Connection Establishment

To establish a connection, TCP uses a three-way handshake. Before a client attempts to connect with a server, the server must first prepare a port for connection binding it to a connection request: this is called a passive open. Once the passive open is established, a client may initiate an active open.

The active open is performed by the client sending a segment with the suitable flag on to the server. It sets the segment's sequence number to a random value, let us call it A.

In response, the server replies with an acknowledgment segment whose sequence number is another random number, let us call it B.

Finally, the client sends its own acknowledgment back to the server.

At this point, both the client and server have received an acknowledgment of the connection with coherent acknowledgment and sequence numbers and the connection is set up.

3.2.3.1.2 Data Transfer

The main features of TCP during the data transfer phase can be described as follows:

Reliable transmission: TCP uses a sequence number to identify each byte of data. The sequence number identifies the order of the bytes sent from each computer so that the data can be reconstructed in order, regardless of any fragmentation, disordering, or packet loss that may occur during transmission.

TABLE 3.1

Main Protocols of the IP Suite: (a) Application Layer, (b) Transport Layer, and (c) Internet Layer

Protocol Acronym	Protocol Function	IETF Standard
(a) Application layer		
DHCP	Protocol used by hosts (DHCP clients) to retrieve IP address assignments and other configuration information	RFC 2131
HTTP	Protocol for distributed, collaborative, hypermedia information systems	RFC 2616
HTTPS	Combination of the Hypertext Transfer Protocol with the SSL/TLS protocol to provide encryption and secure	RFC 2818
SMTP	Internet standard for electronic mail (e-mail) transmission	RFC 5321
POP3	Protocol used by local e-mail clients to retrieve e-mail from a remote server over a TCP/IP connection	RFC 1939
IMAP	Protocol for e-mail retrieval	RFC 3501
FTP	Network protocol used to exchange and manipulate files	RFC 959
SNMP	UDP-based network protocol. It is used mostly in network management systems	See Ref. [56,57]
SIP	Signaling protocol, widely used for controlling multimedia communication sessions such as voice and video calls	RFC 3261
RTSP	Control protocol for use in entertainment and communications systems to control streaming servers	RFC 2326
Telnet	Protocol used to provide a interactive communications facility. Typically, telnet provides access to a remote host	RFC 854
VRRP	Redundancy protocol designed to increase the availability of the default gateway servicing hosts on the same subnet	RFC 3768
RTP	Protocol that defines a standardized packet format for delivering audio and video	RFC 3550
BGP	Core routing protocol of the Internet	RFC 4271
RIP	Dynamic routing protocol used in local and wide area networks	RFC 2453
(b) Transport layer		
TCP	Protocol that provides reliable, ordered delivery of a stream of bytes from an host to another	RFC 1323 RFC 2818 RFC 2018, RFC 1323, RFC 1146,
UDP	Connectionless transport protocol	RFC 368
SCTP	Protocol, that provides reliable, in-sequence transport of messages with congestion control	RFC 4960
RSVP	Protocol designed to reserve resources across a network	RFC 2205
ECN	Protocol that allows end-to-end notification of network congestion without dropping packets	RFC 3168
OSPF	Dynamic routing protocol	RFC 2328
DCCP	Protocol that implements reliable connection setup, teardown, congestion control, and feature negotiation	RFC 4340
(c) Internet layer		
IP	Datagram oriented protocol	See Ref. [20,21]
ICMP	Protocols used by computers to send error messages	RFC 792
IGMP	Communications protocol used to manage the membership of IP multicast groups	RFC 3376

Bit offset	0	1	2	3	4	5	6	7	8	9	10	11	12	13	14	15	16	17	18	19	20	21	22	23	24	25	26	27	28	29	30	31
0	Source port																Destination port															
32	Sequence number																															
64	Acknowledgment number																															
96	Data offset				Reserved				CWR	ECE	URG	ACK	PSH	RST	SYN	FIN	Window size															
128	Checksum																Urgent pointer															
160	Options (if data offset >5)																															
…	…																															

FIGURE 3.23
TCP header content.

Error-free data transfer: The TCP checksum is a weak check by modern standards. The weak checksum is partially compensated for by the common use of more powerful error correcting codes (ECCs) or error detecting codes (EDCs) at layer 2 or even at the physical layer. However, this does not mean that the 16 bit TCP checksum is redundant: introduction of errors in packets between code-protected hops is common, but the end-to-end 16 bit TCP checksum catches most of them.

Flow control limits the rate a sender transfers data to guarantee reliable delivery. The receiver continually hints the sender on how much data can be received (controlled by the sliding window). When the receiving host's buffer fills, the next acknowledgment contains a 0 in the window size, to stop transfer and allow the data in the buffer to be processed.

Congestion control TCP also performs congestion control. It uses a number of mechanisms to achieve high performance and avoid "congestion collapse," where network performance falls. These mechanisms control the rate of data entering the network, keeping the data flow below a rate that would trigger collapse.

3.2.3.1.3 Connection Termination

The connection termination phase uses, at most, a four-way handshake, with each side of the connection terminating independently.

It is also possible to terminate the connection by a three-way handshake, when first host sends a flag of end of transmission and the second host replies with a segment containing the same flag and a suitable acknowledgment. At this point the first host replies with an acknowledgment. This is perhaps the most common method to tear down a connection.

It is possible for both hosts to send end of transmission flags simultaneously then both just have to acknowledge.

3.2.3.2 Transport Layer: User Datagram Protocol

UDP is the second used transport protocol in the Internet; UDP applications use datagram sockets to establish host-to-host communications. Sockets bind the application to service ports that function as the endpoints of data transmission. A port is a software structure that is identified by the port number, a 16 bit integer value, allowing for port numbers between 0 and 65,535.

Bits	0–15	16–31
0	Source port	Destination port
32	Length	Checksum
64	Data	

FIGURE 3.24
UDP header content.

UDP provides no guarantees to the upper layer for message delivery and retains no state of messages once sent.

UDP provides application multiplexing (via port numbers) and integrity verification (via checksum) of the header and payload. If transmission reliability is desired, it must be implemented in the user's application.

The scheme of the UDP header is shown in Figure 3.24. The UDP header consists of four fields.

1. *Source port*: This field identifies the sending port when meaningful and should be assumed to be the port to reply to if needed. If not used, then it should be zero.
2. *Destination port*: This field identifies the destination port.
3. *Length*: A 16 bit field that specifies the length in bytes of the entire datagram: header and data. The minimum length is 8 bytes since that is the length of the header. The field size sets a theoretical limit of 65,535 bytes (8 byte header + 65,527 bytes of data) for a UDP datagram.
4. *Checksum*: The 16 bit checksum field is used for error-checking of the header and data.

Due to its extreme original simplicity, UDP has been enriched in subsequent versions of IP; in particular, IPv6 introduces more complexity in the protocol, always maintaining its nature of pure datagram-oriented transport layer protocol.

3.2.3.3 Internet Layer: Internet Protocol

The IP is the main protocol for data transmission in the Internet layer. The principle of IP working relies on the segmentation and encapsulation of segments coming from the upper layer.

In particular, each segment is divided, if necessary, in datagrams, whose maximum length is 65,535 bytes and encapsulated into an IP envelop; the IP header contains, among other information, the sender and the destination IP address, and in case of IPv4, also a checksum is calculated to provide protection from errors. When IPv6 was defined, the checksum was eliminated in favor of faster routing, since the checksum is calculated every time an IP datagram is processed by a network router.

The encapsulation provides complete abstraction from the content of datagrams: every service related issue has to be solved by other layers.

From the moment in which they are created, datagrams travel independently through the network, routed on the ground of the destination address.

Since IP does not establish any connection between the source and the destination of a message (it is a connectionless protocol), different datagrams, even constructed starting from parts of the same upper layer segments, can arrive in the wrong order at destination, and datagram loss or duplication can happen without any IP control. The message integrity is all responsibility of upper layer protocols.

The key of IP addressing is constituted by the fact that all addressable network entities are labeled with an IP address that is used by the different routing algorithms to route datagrams through the network.

As a matter of fact, when an IP datagram is received and stored in the inner queue of a router input port, its IP address is read and constitutes the input to the routing table, whose output is the output port to which the datagram has to be transmitted.

The routing table can be constructed using simple or sophisticated algorithms. The simpler routing algorithm works as follows: every router of the network is associated to a subnetwork and to a reference gateway that is external to the subnetwork. Each time the router receives a datagram, it verifies if the IP address is comprised in its subnetwork. If this is true, the router sends the datagram to its destination, otherwise it sends it to the reference gateway. This is a simple and effective algorithm that is often used in data network, but does not take into account elements like QoS and it is not designed to optimize the computation capability of the router. If more control is desired on the routing, more complex algorithms have to be used.

Using IP routing is a powerful method, but in the Internet history a proliferation of IP addresses has been created close to the maximum number that were possible to code with the original IP address format [25].

This was the main reason for the evolution of the IP through a series of subsequent versions that were able to accommodate larger and larger IP addresses.

Up to now mainly two versions of the IP are used in the network: IPv4 and IPv6.

The IPv4 address format is shown in Figure 3.25 both in binary and in decimal dotted notation.

Even if the possible number of globally unique IPv4 addresses is huge, the fast evolution of the Internet rapidly created the risk of IPv4 addresses exhaustion. Several ways were devised to avoid it, all of them are now obsolete, but for the concept of Private Network address.

With the diffusion of the IP Suite, several private networks like companies local area networks (LANs) adopted the same protocol suite. Even if all the computers in a private LAN are labeled with an IPv4 address, these addresses have not to be globally unique, as far as the computers have no need of connecting to the Internet.

If a user from inside a private LAN wants to connect with an external Internet server, the network translation address (NAT) is used [21]. The LAN communicates to the Internet through a gateway that has a globally unique IPv4 address. Outgoing IP datagrams are coded by the gateway to be correctly routed by the Internet while incoming datagrams are routed by the gateway to reach the suitable LAN destination.

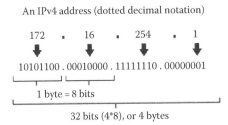

FIGURE 3.25
IPv4 address in dotted decimal notation and in binary notation.

The rapid exhaustion of IPv4 address space, despite conservation techniques, prompted IETF to explore new technologies to expand the Internet's addressing capability. The permanent solution was a redesign of the IP itself. This next generation of the IP was named IP Version 6 (IPv6) in 1995 [26]. The address size was increased from 32 to 128 bits or 16 octets, which, even with a generous assignment of network blocks, is deemed sufficient for the foreseeable future. Mathematically, the new address space provides the potential for a maximum of about 3.4×10^{38} unique addresses.

The new design is not only based on the goal to provide a sufficient quantity of addresses, but also targets the efficient aggregation of subnet routing prefixes to occur at routing nodes. As a result, routing tables are smaller, which is an important result for faster working of routers.

The new design also provides the opportunity to separate the whole Internet into subnetworks with an independent local administration. As a matter of fact, the IPv6 address can be divided into a prefix indicating the address of the subnetwork within the whole Internet and a suffix that indicates the position of individual machines inside the subnetwork.

If the global routing policy changes, IPv6 has facilities that automatically change the routing prefix of entire networks without requiring renumbering of the subnetwork suffix.

The partition of the Internet in subnetworks managed by gateways also transformed the routing process. As described earlier, the gateway has an IPv6 address whose prefix (the first part of the address) is common with all the entities in its reference subnetwork. Thus, when searching routing tables, routers need not search a complete correspondence between the destination IP address and the table entry, but a partial correspondence between a prefix of the address and a table entry, that is the gateway address.

Since different gateways can have similar addresses, it is possible that more than one prefix correspondence is found. Among them, the longest corresponding prefix has to be chosen since it indicates the correct gateways.

Although this procedure allows a better management of IPv6 addresses, it also complicates routing table searching. In fact, while the search of a complete correspondence ends when it is found, the fact that a prefix correspondence is found does not stop the prefix match search, since a longer match is always possible.

We will come back on this search mechanism and on its implementations in Chapter 7, when analyzing routers architectures and performances.

The large number of IPv6 addresses allows also large blocks to be assigned for specific purposes and, where appropriate, to be aggregated for efficient routing.

All modern desktop and enterprise server operating systems include native support for the IPv6 protocol, but it is not yet widely deployed in other devices, such as home networking routers, multimedia equipment, and network peripherals.

3.2.4 Carrier Class Ethernet

The adoption of the architecture of the packet network of the IP model leaves open the choice for the implementation of the packet network layer performing the interface with the transport network, that is, in line with OSI nomenclature, the layer 2 architecture.

For the extremely wide diffusion of its LAN version, the Ethernet is a natural candidate to this role, assuring wide compatibility with services (almost all service machines have Ethernet interfaces) and low prices due to the extremely high volumes in which Ethernet components are produced.

If the Internet model is defined in a more operative and informal way with respect to the network models developed by the ITU-T, Ethernet in its LAN-oriented implementation can be considered extremely simple, although sufficiently rich to manage complex and extended private networks [27,28].

The core of Ethernet is constituted by the collision detection/carrier sense multiple access (CD-CSMA) protocol that manages the sharing between several clients of a common transmission medium.

Each Ethernet station in a local network is given a 48 bit MAC address and the access to a share medium, in the simplest case a bus. When the upper layer (e.g., the Internet layer of a TCP/IP LAN) wants to transmit a packet, the packet is sent to the IP-Ethernet adaptation and then enveloping it into the Ethernet frame that contains the MAC address of the destination machine.

The Ethernet transmission procedure can be described as a simple algorithm that is based on the fact that each station can detect if the shared medium in a certain instant is idle and when it receives a frame it is able to send an acknowledgment at the sending station.

The algorithm can be written as follows:

Main Transmission Procedure—Carried Out by the Transmitting Node

    Frame ready for transmission, transmission abort flag FALSE.

    label AAA

    Is Transmission abort flag TRUE?

        YES: Abort transmission

        NO: Continue transmission procedure

    Is medium idle?

        NO, wait until it becomes ready

        wait the interframe gap period ($9.6\,\mu s$ in 10 Mbit/s Ethernet) and go ahead

        YES: Go ahead

    Start transmitting.

    Did a collision occur (i.e. no acknowledge from the receiving node)?

        YES: Go to collision detected procedure.

        NO: Reset retransmission counters

    end frame transmission.

Collision Detection Procedure

Continue transmission until minimum frame time is reached (jam signal) to ensure that all involved receivers detect the collision.

Increment retransmission counter.

Was the maximum number of transmission attempts reached?

        YES: Transmission abort flag TRUE.

        NO: Calculate and wait random backoff period based on number of collisions

            Transmission abort flag FALSE.

Re-enter main procedure at label AAA.

This is in brief the description of the Ethernet CSMA protocol [29]. We will return on the CSMA protocol in much more detail in Chapter 7 dealing with time division dynamic switches; thus, the reader that is interested to go more in detail is encouraged to go there both for a longer discussion and for a detailed bibliography.

On top of the simple bus LAN using CD-CSMA, several elements have been added to Ethernet to allow more complex LAN topologies to be managed: different Ethernet busses can be connected via bridges and hubs to extend the network and switches can be used to provide switching in more complex topologies [27].

It is out of the scope of this book that deals with the public network, to detail all the LAN-related Ethernet technologies.

The only thing we have to add regarding the LAN-oriented Ethernet is to describe the working of an Ethernet switch, which will be the base for the development of important equipment that can be used in the public networks.

Different Ethernet networks can be connected via bridges or hubs. When such connection is done however, the bandwidth of the transmission medium continues to be shared among all the nodes of the network: thus, bridging and using hubs solve the problem of creating complex Ethernet topologies, but not the problem that by increasing the number of nodes the average bandwidth per node decreases.

In order to solve this problem switches can be used.

When two Ethernet networks are connected via a switch, it detects and memorizes the location of all the MAC addresses in one network or in the other. When an Ethernet frame arrives, the switches memorize it in an input queue, read the MAC address, and recognize the network it belongs to. Thus, it sends the frame to the correct output port [30].

It is evident that such a switch separates the two networks and both of them have all the transmission bandwidth available. When more than one switch exists in an Ethernet network, the spanning tree algorithm is used for frame forwarding through the switches network up to the correct destination.

The spanning tree algorithm, besides CSMA, is a base characteristic of Ethernet networking and many discussions on the scalability of the Ethernet to the public network are centered on the potentialities and weakness of the spanning tree protocol.

For this reason, this protocol is described with some details in Appendix B.

At a first look, it could seem that a switch is somehow similar to a router, but it is not true, a part for the fact that both machines implement routing at their own layer.

A router deals with IP datagrams: at the router input, an IP envelope is open, the IP address read and, in case of IPv4, the checksum calculated.

Moreover, in public networks where real-time transmission protocols are used, the router manages QoS by controlling packets delays and transmission quality.

Last, but not least, since it terminates the Internet link layer, a router can be used in the home of a customer to separate the customer private home network from the public network, so separating what is the responsibility of the carrier from what is the responsibility of the customer.

If the customer has a private LAN, the router implements the NAT procedure to allow machines in the private LAN to enter the external Internet and it is able to connect with the remote authentication machine (the broadband remote access server [BRAS], that is another router, see Chapter 7) to authenticate the user when entering into the public Internet and, in case, to perform billing and other user account related functions.

A switch is a much simpler machine that is not able to perform all the operations we have listed for a router: the only function of a LAN switch is to separate, from the point of view of the transmission bandwidth, different segments of a LAN and to suitably route the Ethernet frames.

These differences also correspond to a major difference in the routing and forwarding engines of routers and switches, which will be detailed in Chapter 7, and that are caused by the fact that while an Ethernet domain is a limited network with a limited number of MAC addresses, the worldwide IP network has a huge number of IP addresses that cannot be managed by a single forwarding table forcing to adopt a forwarding policy based on prefixes and not on complete addresses.

Due to its effectiveness, the Ethernet is today practically the only layer 2 protocol used in LANs.

Since a huge quantity of data is produced in the Ethernet format and since the volumes of Ethernet subsystems is so high that prices are relatively low with respect to other subsystems used in public network equipment, it is quite natural to try to understand if Ethernet can be used at layer 2 (e.g., in between the Internet layer and the optical transport) in the public network also.

For sure this cannot be done without somehow adapting the LAN-oriented protocol, since it cannot be scaled to the public network dimensions and does not provide all the monitor capabilities that a carrier needs to support its services.

Recognizing this, the IEEE and successively a new industrial group called Metro Ethernet Forum (MEF) started to elaborate new Ethernet standards that are suitable for use in a wide area, public network, thus creating the Carrier Class Ethernet technology [31,32].

Requirements of this Ethernet evolution are as follows:

- Presence of functional support for a network management with complete capability
- Scalability to the public network scale
- Carrier level resilience
- Capability of assuring QoS to support services with SLA
- Capability of being integrated with legacy transport machines

3.2.4.1 Protocols to Support Management Functionalities

Even if all management functionalities are in principle requested, a few functionalities are the key to present a carrier Ethernet version that can be deployed in public networks. The basic functions that are necessary include the following [33]:

- Continuous connectivity check (CC), that is, the ability to monitor continuously the existence of connectivity along the links that are present in the network topology
- Tunnels loopback (LB), that is, the ability to monitor Ethernet tunnels (i.e., connections transiting through higher level machines like routers without being processed at higher layer, but simply passing through)
- Trace route (TR), that is, the ability to trace and report to the central management the route of individual Ethernet frames or of all the frames belonging to a higher layer connection
- Alarms filtering and suppression, that is, the capacity of discerning the origin of a failure when it generates alarms from a great number of different network entities due to its consequences, for example, lack of signal

- Equipment discovery
- Performance monitoring
- Survivability

The first three functions are needed to assure a correct reporting of the network status at the central management system. In order to implement them, new protocols are needed that are not included in LAN Ethernet.

These protocols have been introduced in MEF recommendation Y1731 and, in line with the Ethernet frame use, are based on the transmission of suitable control frames that are used both to monitor the existence of the connectivity also when no frame is transmitted between a certain couple of nodes and to realize the loopback of the connections.

As far as the trace or the frame routes is concerned, the passage of these control frames is registered into the switches with the frame identifiers. When the route of a certain set of frames have to be reconstructed, it is sufficient to interrogate the switches and to detect those having in memory the right route identifier.

3.2.4.2 QoS and Resilience

The key idea to reach this specification is the use of the virtual (V)LAN technology to separate data streams relative to different connections to transform Ethernet in a connection-oriented frame technology.

The VLAN was a concept introduced in the Ethernet protocol set to assure security and isolation to sensible data sent via a shared medium LAN.

Every VLAN that is created into a switched Ethernet network is identified by a VLAN identifier (VID) that is an integer number from 1 to 4094 (since 0 and 4095 are reserved).

Assuming that the network contains more than one switch, in order to use VLAN as virtual circuits, it is necessary that each packet traveling through a switch can be assigned to its own VLAN. In order to associate a packet to a VLAN, a two-byte header is added at the beginning of the Ethernet frame by the sender, which is also the VLAN creator. A single sender naturally can transmit packets on different VLAN if it is requested to do so, exactly as it could transmit packets on two different networks if it would be connected to both.

This header is called VLAN TAG and contains the VID of the packet. Every traversed switch can recognize that the packet belongs to a certain VLAN and switch it exactly as if it was on a physically separated network.

VLAN can be stacked and in the extreme case all the traffic can be ordered in a stacked set of VLANs.

This is exactly what carrier Ethernet does: Carrier Class Ethernet is a layer 2 architecture where the traffic is switched using virtual circuits (often called tunnels) each of which is constituted by a different VLAN. In a Carrier Class Ethernet network, a VLAN is created, managed, and tiered down exactly as a circuit in a circuit-oriented switched network.

A VLAN creation protocol exists that assures that a VLAN created by a network user is recognized by all the network switches and terminates at the right destination; a set of monitoring procedures exists for the management of a running VLAN; finally, a VLAN tier down procedure exists that eliminates the VLAN when the corresponding connection is ended and informs all the switches of the network that the VLAN has been tiered down.

The adoption of the virtual circuit concept through the VLAN technology opens the possibility to empower the carrier class Ethernet with a set of monitoring and QoS capabilities that justify the name "Carrier Class."

As a matter of fact, the concept of virtual circuit binds the packet belonging to the same communication to travel through the same path in the network. If we consider that the reception of each Ethernet frame is acknowledged there are all the elements to make it possible to control frames delay, loss, and duplication, while eliminating frames out of order arrival.

Once virtual circuits are used, survivability can be implemented via fast techniques, operating on the virtual circuits themselves.

The fast protection mechanism introduced in the standards is the 1:1 VLAN protection. This mechanism is based on the fact that, when it is needed to provision a VLAN, two of them are provisioned by the control plane. The two VLANs have the same endpoints but no other network element in common.

The traffic can be either divided among the two routes (implementing the so-called traffic balancing) or sent over one of them (the traditional protection). Moreover, the protection capacity can be also used to transmit best effort frames.

When a failure happens on a traffic-loaded path, the protocol managing protection starts. First of all, a control frame is sent by the terminal detecting the failure to the other end terminal of the involved VLAN. If no acknowledgment is detected, the terminal sets itself in the bidirectional failure mode, switches on the protection route after stopping low-priority traffic, and looks for the contact with the other end terminal via a control frame.

If the other end terminal has done the same thing, both the extremes of the affected VLAN are now on the protection route, they can verify the good performance of the route and their readiness to switch with a handshake, and after that, they deviate the high-priority traffic on the protection route.

In case of unidirectional failure, the terminals first enter in contact with the remaining working route, then switch on the protection one, and the process goes on as described in the bidirectional case.

During the failure, the terminals go on with a continuous probing of the failed link via control messages; when the failure is restored, the terminals can conclude a handshake on the working line and, after that, they can switch the traffic on the working path and recover the low-priority flux.

These protection mechanisms achieving protection times smaller than 50 ms can be applied both to rings and to point-to-point architectures.

3.2.4.3 Scalability

We have seen that in Carrier Class Ethernet, the traditional broadcast of frames on a shared medium supported by the CDMA algorithm to manage collisions is substituted by a virtual circuit frames management, where frames belonging to a connection are all routed along the same VLAN whose route represents the virtual circuit.

In order to gain efficiency and scalability, Carrier Class Ethernet also foresees the possibility of disabling the spanning tree to introduce more efficient frame routing. When this is done, generally dynamic MAC learning is also disabled in favor of a static MAC address configuration and recognition, more suitable for a network of great dimensions.

At this point, only the Ethernet frame and the concept of switching of the LAN-oriented Ethernet remain. Nevertheless, Carrier Class Ethernet revealed scalability problems due to its origin that had to be solved.

The first problem to face was linked to the use of VLAN itself. As a matter of fact, if carriers can use the VLAN concept to establish virtual circuits, customers using Ethernet

LANs and asking for public network Ethernet services want to use VLANs for their own original task.

If the customer frames are tagged with a VID, such VID cannot be confused with VIDs used from the carrier, requiring coordination not only between the carrier and the customer, but also among customers themselves not to repeat the same VID causing unwanted confusions in the public network.

This coordination cannot be done, also because it violates the principle of customers' signals isolation; thus, the problem was solved another way.

IEEE introduced the 801.ad standard [31] separating the public VID from the private VID. Following this standard, when a frame enters the public network, a new header field is added by the edge switch containing the public VID that refers to the so-called service VLAN. In this way, the carriers VLAN can be distinguished from the "customer VLAN" whose VID is coded in the inner frame header.

The frame is switched in the public network using the service VID, at the destination edge switch, the service VID is removed and the frame is sent again to the private network.

This mechanism is generally called provider bridge (PB) since it allows the service provider to separate customers' signals and to uniquely identify service VLANs and customers VLANs.

Although the introduction of PB solves the problem of isolating service VLANs from private VLANs, this still does not guarantee a real Carrier Ethernet Scalability.

As a matter of fact, MAC addresses have to be learned and used by the carrier even if they belong to the customers LAN. The problem is similar to those arising with IP addresses when a customer private network uses the TCP/IP stack as the public Internet.

While however IP domains can be separated by routers using the NAT principle, nothing similar exists in Ethernet since switches cannot perform MAC translations and MAC domains does not exist. Moreover, if the number of MAC addresses increases too much, switches forwarding tables become difficult to manage and private switches cannot be asked to have suitable dimensions.

On top of the MAC proliferation and the consequent copenetration of public and private networks, another problem also arose from the definition of service VLANs.

Following standard VID definition, the maximum number of service VLANs resulted to be 4094, really too small even for a small competitive local exchange carriers (CLEX) that has generally hundred thousand customers.

These problems were solved by IEEE with the so-called providers backbone bridges (PBB—IEEE 802-Iah) standard [31] that introduces three new fields in the Ethernet frame reserved to the carrier.

Using these fields, the edge router that accepts the frame at the boundary between the private and the public network creates an envelope containing carriers' network information that in this way is completely separated from the customer network one.

Customers MAC and VID are completely hidden for the carrier network and vice versa, carriers address and VLAN tag does not reach the customer LAN.

The four fields are as follows:

1. A backbone destination address (B-DA, 48 bits)
2. A backbone source address (B-SA, 48 bits)
3. A backbone VLAN tag (B-VLAN, 12 bits)
4. A service instance tag (I-SID, 24 bits)

The B-VLAN is used by the carrier to partition their PBB network in different broadcast domains that is useful not only for scalability but also in case broadcast applications like IP-TV have to be supported. Since the number of partitions is high (4096), the number of users collected in each partition is generally small (e.g., if the carrier has two millions customers, each partition collects only 500 customers). In this condition, the spanning tree protocol can be used for routing.

The I-SID is used to tag the service instances that is substituted in practice by the service VID of the PBB in individuating carriers' virtual circuits. Differently from the service VID, the I-SID scale over more than 16 millions, assuring scalability up to a number of users typical of a large continental carrier. I-SIDs are seen only by edge switches that provide, assign, and cancel it, and are ignored by all the intermediate switches of the carrier network.

The last two fields, B-DA and B-SA, take the role of destination and origin of MAC in the frame routing through the carrier network.

Finally, it is to be noticed that the new fields are introduced in such a way that passing from PB to PBB only requires the upgrade of edge switches, while intermediate switches can go on implementing PB, since they will automatically use the new fields.

3.2.5 Multi-Protocol Label Switching

If it is possible that carrier class Ethernet will be the layer 2 technology of choice in the metro and regional area, it seems that there is no possibility to use it in extended core network covering a country or a continent.

Thus, in the case of core network, the interface between the packet and the transport network has to be realized via some other technique. One of the most interesting candidates is the so-called multi-protocol label switching (MPLS).

MPLS is a link layer technology devised for networks adopting a plurality of protocols at network and transport layers [34,35]. This is a standard situation in the public networks, where different services have different requirements in terms of QoS and often are supported by different technologies.

Originally MPLS was designed to simplify the routers operation: as a matter of fact, it relies on the idea that all the datagrams belonging to the same connection and thus directed to the same destination are tagged with a label and are switched on the same route.

Once the label has been created and the route established, the intermediate routers do not have to process the IP address of every packet independently to calculate the route, but simply associate the incoming datagram to an existing tunnel via its label and switch it in the tunnel direction. In a data network implementing MPLS, the edge routers, where the tunnels originate and end, are called label edge routers (LER) while the intermediate routers are called label switching routers (LSRs).

This original intention however is no more the most important reason for pushing the adoption of MPLS especially in backbone IP networks.

In order to really simplify routers operations, the label processing should be much simpler than the IP address effective processing and this is not always the situation.

The presence of fixed tunnels conveying datagrams belonging to the same communication is instead precious to perform network monitoring and QoS management.

As a matter of fact, the transit of datagrams along a tunnel can be easily monitored producing statistics of errors in transmission, passing through time, loss of packets due to the transmission toward a full queue, and so on.

Moreover, a service related tag can be applied to each MPLS label that can be used to run priority algorithms in the input queues of the routers and so manage real-time services.

In its nature MPLS would be a layer 2 technology, since it is located in between the Internet layer and the pure physical layer. As we have seen, in practice it also performs operations that would be proper of the Internet layer (i.e., layer 3). This is due to the fact that Internet layer protocols lack the needed characteristics.

For this reason, often the MPLS is informally called a layer 2.5 technology.

As in the case of all connection oriented protocols, the MPLS manages tunnels through three phases:

1. Tunnel setup
2. Tunnel management
3. Tunnel tier down

Once a tunnel is created, the MPLS header is put in front of the Internet layer datagram; it is composed of 32 bits: 20 bits are reserved for the label, 3 bits code the priority, 1 bit indicates if there is a label stack, and finally 8 bits contain the time to live.

All the operations of setup, management, and tear down of a tunnel, besides other functionalities, are performed via the Label Distribution Protocol (LDP).

The LSR uses this protocol to establish tunnels through the network; this is done by mapping the network layer routing information that determines the end-to-end connection request directly to data-link layer switched paths, implemented via MPLS labels. These LSPs may have an endpoint at a directly attached neighbor (like IP hop-by-hop forwarding), or may have an endpoint at a network egress node, enabling switching via all intermediary nodes.

Two LSRs that use LDP to exchange label mapping information are known as LDP peers and they have an LDP session between them. Since in a single session each peer is able to learn about the others label mapping established by the LSR with which it is connected in the section framework, the protocol is called bidirectional.

There are four sorts of LDP messages:

1. Discovery messages
2. Notification messages
3. Session messages
4. Advertisement messages

Discovery messages announce and maintain the presence of a router in a network. Routers indicate their presence in a network by sending the hello messages periodically. The hello messages are transmitted as a UDP packet to the LDP port at the group multicast address for all routers on the subnet.

Notification messages provide advisory information and signal error information. LDP sends notification messages to report errors and other events of interest.

Session messages establish, maintain, and terminate sessions between LDP peers. When a router establishes a session with another router learned through the hello message, it uses the LDP initialization procedure over TCP transport. When the initialization procedure completes successfully, the two routers are LDP peers, and can exchange advertisement messages.

Advertisement messages create, change, and delete label mappings. Requesting a label or advertising a label mapping to a peer is a decision made by the local router. This is the local mechanism creating and deleting tunnels.

When a label request arrives at an LSR via an LDP packet, the router accesses routing tables, individuates the output port and the associated output label, and add the new tunnel to the tunnels list. The tunnel is stored memorizing the association of an input label (tagged to the port from which the label request was received) to the output label tagged to the corresponding output port. Moreover, a QoS tag is also associated with the tunnel [36] that will set the priority of packets belonging to it when contentions will present in the router queues.

As the last act of the tunnel setup, the router forwards along the tunnel route an LDP packet requesting a label association to the next router of the tunnel.

This mechanism goes on up to reaching the destination, where the tunnel ends. Here the edge router receives the LDP label association request with the label imposed by the last LSR, and proceeds to the correct termination of the tunnel associating the input label to an output drop port. In this way, when a packet will arrive at the LER along the tunnel, the tunnel label will be eliminated and the datagram will be delivered to the destination. If a bidirectional tunnel is to be set up, the LER also creates another LDP packet suitable to set up a tunnel where destination and source are exchanged and the process starts again.

Summarizing, an MPLS tunnel created by LDP can be viewed as a series of labels, one for each network link, so that the output label of a router is the input label of the following router along the tunnel path. The tunnel is identified by a router via the association of the input label and output label since each label is uniquely associated to a router port.

During the tunnel working, when a packet arrives at a router, the MPLS header is eliminated and the input label processed to determine the output port. The datagram is switched to the output port where the datagram is enveloped in the new MPLS header and it is sent to the next router. Tunnel monitoring is also performed by each router in terms of statistics on the QoS variables.

When the tunnel has to be deleted, a suitable LDP packet is generated by one of the LER and is transmitted all along the tunnel route; every router forwards it, and after forwarding cancels the tunnel from its tunnel list. Also in this case, in case of bidirectional tunnels, a double passage is needed, one in each direction.

If LDP is the protocol for tunnels management, tunnels routing is performed by Border Gateway Protocol (BGP) [37]. As we have seen, every LSR decides in an autonomous way (it is said an autonomous system) what is the suitable output port to route a new tunnel.

The primary function of a BGP is to exchange information about the presence of network connections linking a BGP system with other BGP systems. This network connectivity information includes information on the list of autonomous systems that the BGP packet traverses. This is sufficient to build inside each LSR a local graph of the network that is the base to perform the routing loops needed to construct the routing table of the LSR.

Due to its fundamental function for the network optimization, the BGP was updated several times so that several versions exist. The current version, BGP4, among other improvements, introduces mechanisms that allow aggregation of routes in subnetworks and aggregation of tunnels in tunnel classes. These features are important for the evolution of MPLS toward a generalized network control plane that will be discussed in detail in Chapter 8.

In order to increase the efficiency in network resources exploitation and to faster network survivability, the MPLS standard defines both centralized traffic engineering and fast reroute tunnels (FRR tunnels).

The first feature assumes that an off-line global traffic engineering tool is available to the carrier. If it is true, routing the network tunnels via this tool will require an off-line operation, much slower with respect to standard network working, but it will also give a more efficient network resources allocation. Thus, the MPLS also implements a feature allowing the central network management to route manually a few paths following the indications of the off-line traffic engineering tool in order to improve network efficiency [37].

In order to implement fast survivability, MPLS needs to route a couple of FRR tunnels [38,39]. They are tunnels having the same root and the same destination, but no other network resources in common. They can be used either exploiting one of them to route traffic them and leaving the other as a backup or using the so-called traffic balancing, consisting in sharing among the two tunnels the traffic load of a single tunnel.

When one of the FRR tunnels fails, all the traffic is automatically switched on the other tunnel. Due to the initial association between the FRR, this operation is very fast, generally below the 50 ms target typical of SDH/SONET protection mechanisms.

Finally, MPLS can implement virtual private networks (VPN). This is a key point for a carrier that wants to offer business-related services with a packet network [40].

In order to correctly implement VPNs, the address spaces of the MPLS core and all VPNs on the same shared network must be independent so that each customer can use the same address space without interfering with other (the same requirement already considered for carrier Class Ethernet carrying customers LANs). That is, every VPN customer and the core itself must be able to use the entire IPv4 address range completely independently.

Similarly, data traffic from each VPN must remain separate, never flowing to another VPN. A related requirement is that routing information for one VPN instance must be independent from any other VPN instance and from the core.

To achieve this separation, each VPN is assigned to a virtual routing and forwarding (VRF) instance. Every core LSR maintains a separate VRF instance for each connected VPN. Each VRF on the LER is populated with routes from one VPN, either through statically configured routes or through routing protocols that run between the carrier LER and the customer LER. Because every VPN is associated with a separate VRF, there is no interference among the VPNs on the carrier LER.

To maintain routing separation across the core unique VPN identifiers are added to the MPLS header and route distinguishers are added to the BGP header. VPN routes are exchanged across the core only by multiprotocol BGP and this information is not distributed again to the core network, but only to the associated carrier LER, which retains the information in VPN-specific VRFs. Thus, routing across a MPLS network remains separate for each VPN.

3.2.6 Synchronous Optical Network (SDH/SONET)

The synchronous optical network is a very important example of transport network model [41].

The introduction of SDH/SONET in substitution of the PDH, which was the previous transport network technology, was a revolution in terms of network manageability and transmission quality control. The modern idea of the management plane was born

TABLE 3.2

SONET SDH Acronym Correspondence

SONET	SDH
SPE	VC
STS-SPE	Higher order VC (VC-3/4/4-Nc)
STS-1 frame	STM-0 frame (rarely used)
STS-1-SPE	VC-3
STS-1 payload	C-3
STS-3c frame	STM-1 frame AU-4
STS-3c-SPE	VC-4
STS-3c payload	C-4
STS-12c/48c/192c frame	STM-4/16/64 frame AU-4-4c/16c/64c
STS-12c/48c/192c-SPE	VC-4-4c/16c/64c
STS-12c/48c/192c payload	C-4-4c/16c/64c

in practice with the SDH/SONET and till now the resilience and management of SDH/SONET transport network are the standard that carriers call "carrier class."

While IP was developed by IETF, a quite informal standardization institute born from the union of the main routers vendors, the SDH/SONET standard was generated by the official standardization institutes for telecommunications. That is European SDH standard was generated by ITU-T and European Telecommunications Standards Institute (ETSI) and the U.S. centric SONET by ITU-T and American National Standards Institute (ANSI) [42,43]. As far as Far Eastern countries are considered, generally they adopt the SDH standard.

We will not dwell on differences existing from SDH and SONET, since they are sufficiently similar to be dealt with the same network model. For a detailed analysis of differences and cross-compatibility the reader can start from Refs. [41,44], besides of course the official standardization documents. From now on, for simplicity, we will talk about SDH, but all will be valid also for SONET opportunely changing the names. A correspondence between SDH and SONET terminology is reported in Table 3.2 and a comparison between line rates is reported in Table 3.3.

TABLE 3.3

SONET and SDH Data Rates

Line Signal SONET	Line Signal SDH	Data Rate (Mbit/s)	Overhead Rate (Mbit/s)	Payload (Mbit/s)	User Data Rate (Mbit/s)
OC-1		51.84	1.728	50.112	49.536
OC-3	STM-1	155.52	5.184	150.336	148.608
OC-9	STM-3	466.56		451.044	445.824
OC-12	STM-4	622.08	20.736	601.344	594.824
OC-18	STM-6	933.12		902.088	891.648
OC-24	STM-8	1,244.16		1,202.784	1,188.864
OC-36	STM-12	1,866.24		1,804.176	1,783.296
OC-48	STM-16	2,488.32	82.944	2,400	2,377.728
OC-192	STM-64	9,953.28	331.776	9,600	9,510.912
OC-768	STM-256	40,000	1,327.104	38,500	—

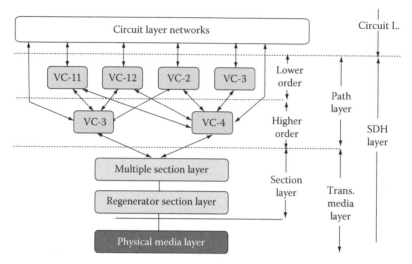

FIGURE 3.26
SDH layered model.

An SDH network is a transport network adopting deterministic TDM multiplexing to optimize the use of transport trunks. The high-level-layered model of an SDH/SONET network is reported in Figure 3.26. The model assumes to receive from the above service layer a set of circuit requests. A circuit is a permanent or semi-permanent connection between two network clients.

The first step is path layer multiplexing: the edge SDH machine consolidates all the circuits requests directed toward the same destination in SDH virtual containers (VCs, virtual tributaries [VTs] in the SONET standard), which will be the elements composing the payload of the SDH TDM frame.

The SDH standard defines a set of VCs, depending on the capacity needed, whose name is VC followed by a number, where the number is related to the VC capacity.

The path layer is divided into two sublayers, since a double stage multiplexing is foreseen. This means that low-capacity VCs can be enveloped in a high-capacity VC if needed.

The VCs are routed through the path layer network from the origin to the destination in the so-called SDH paths. This is realized by sending the VCs to the multiplex section layer. The adaptation function connecting the path layer with the multiplex section layer envelops the VCs adding header and footer. The synchronous transport signals (STSs) are multiplexed by the SDH multiplexer into the SDH Multiplex Section Frame, called STM-number, where the number is again related to the frame capacity. All the Multiplex Section Frames defined in the SDH and in the SONET standards are reported in Table 3.3.

It is to be underlined that STSs of different capacity can be multiplexed into a single STM so that so generating a flexible multiplexing scheme. This is the so-called matrix multiplexing, whose scheme is represented in Figure 3.27. This is an important example of deterministic TDM protocol, for the huge diffusion of the SDH/SONET transport in data networks also (see POS, e.g.). For this reason the SDH frame is described in detail in Appendix A. Here we will underline mainly its functional characteristics.

Different consecutive frames are arranged in a matrix and the multiplex section header is defined inside the matrix as shown in the figure. When a set of STSs have to be multiplexed, they are located into the payload part of the matrix in such a way to have a

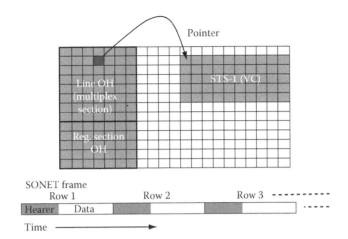

FIGURE 3.27
Principle of tributary multiplexing in an SDH/SONET frame.

smaller possible unusable area. The effect of this strategy is that the structure of the payload of a given STM can change its structure depending on the client STSs that was to be multiplexed.

To take trace of the STSs position inside the payload, a zone of the multiplex section header is dedicated to a set of pointers. Each pointer is associated to an STS, as reported in the header itself, and points in the payload area to the STS start. In this way the demultiplexer can read in the header what are the STSs multiplexed into the payload area and, following the pointers, can separate them by demultiplexing the frame.

It is evident that such a mechanism is very effective in giving a high multiplexing efficiency but, especially at high transmission speeds, is prone to errors in the presence of jitter. In particular, the low-frequency component of jitter (called generally wander) accumulates while the signal traverses the network, generates a random move of the pointers in the attempt to track this random change of the bit synchronism.

When the signal is demultiplexed in the adaptation layer before passing the information to the client, this pointer movement can cause abrupt changes of phase generating demodulation errors.

In order to minimize jitter accumulation, the SDH network does not rely on local clocks to synchronize the multiplex and demultiplexing operations, but it distributes a centralized clock via the so-called synchronization network. This is a conceptually separated overlay network distributing a clock signal to all the network elements. Due to this architecture, the overall jitter is order of magnitude smaller that in a network based on local clock generation allowing high transmission speed, at the expense of network architecture complexity.

Below the multiplex section layer there is the regeneration layer, dealing with the propagation of the signal between couples of in-line regenerators and, further below, the physical layer representing physical propagation of the signal along the transmission medium.

Summarizing, the physical scheme of a path in a synchronous optical network is represented in Figure 3.28, where both SONET and SDH nomenclatures are used for comparison.

In Figure 3.28, the main SDH network elements are also represented; they are the terminal equipment (TE), the digital cross connect (DXC), and the regenerator (REG).

The TE simply receives as the input a line signal and a set of connection requests from the network manager through the element manager. It demultiplexes the incoming line

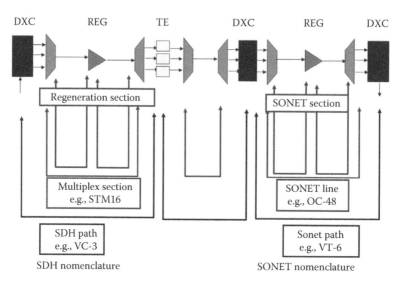

FIGURE 3.28
Sections in a trail through an SDH/SONET network. DXC, digital cross connect; TE, line terminal; REG, regenerator.

signal, delivers it to the client the VCs associated to the path ending in its node, multiplexes again all the incoming VCs that are not destined to its node with the client generated VCs and sends it to the output line. Naturally, since all the operations are performed on the digital signal, a TE also performs complete regeneration of the signal.

The DXC is a quite complex machine. It has several line inputs and several line outputs and it is also connected with several clients. The DXC function is to route the VCs toward the correct destination; in order to do that while efficiently exploiting the available transmission capacity, the DXC operates in the following way.

First, the client circuits requests are transformed in VCs, while the signals coming from the input lines are demultiplexed and the STSs are reported via adaptation to the path layer, extracting the contained VCs. All these operations are of course related to a set of performance and correct working controls that are carried out every time a signal transits through an adaptation function to be processed to a different layer.

When all the VCs are enucleated, the DXC read on the VCs the final destination and routes each VC toward the corresponding output port.

Now it is important to underline differences between the operation of a router and the operation of a DXC. First of all, a router works on statistically multiplexed signal; thus, conflicts can happen between different packets both at the input and at the output ports, thus generating packet loss. A DXC works on deterministic TDM frames that are statically configured by a central management software. This means that all the possible conflicts are solved a priori and the DXC does not lose any VC due to conflicts.

Moreover, as single SDH paths are statically configured, the DXC relies on an internal mapping table derived from the knowledge of all the network paths. If it is needed to determine what is the path traveled by a certain VC, it is simply stored in the central network control.

On the contrary, in the packet-based IP network, a router is subject to a high dynamic traffic [45] and it is practically impossible to exactly foresee the packet flux incoming in a future period of time.

Thus, the routing table used by a router is locally generated by using a local routing algorithm and in general the router does not know the whole trail traversed by a packet. If this trail has to be reconstructed, it is needed to start from the trail origin and to interrogate all the traversed routers, each of which knows from where the packet came in and from where it went out. At the end the packet trail can be reconstructed.

In synthesis, the SDH network is based on a static and centralized traffic engineering that is realized by configuring the DXC in the network in a suitable way from the central network manager. On the contrary, the IP network performs routing via distributed algorithms running in the routers and the centralized control system is used for reporting and traffic engineering.

The main automatic operation performed by an SDH network is protection switching. In general, traffic survivability is assured in an SDH network by two mechanisms: protection and restoration.

Protection is automatic and fast, relying on control plane protocols: in particular the SDH standard prescribes a protection time smaller than 50 ms.

If protection cannot recover the traffic, for example, in the case of multiple failure, restoration starts. It is a slow traffic rerouting carried out by the network manager that uses, if possible, free network resources to find alternative routes for the traffic interrupted by the failure.

Generally, restoration is implemented via DXCs only in meshed networks, because interconnected rings networks have a good resilience also to multiple failures, with the only condition that every failure happens in a different ring.

Protection mechanisms exist both at path and at multiplex section layers. When both the layers implement protection, a hierarchy is established setting a hold-off time retarding path protection. If it is possible to recover the failure at the lowest possible layer, it is done, otherwise higher layer protection is activated.

The simple kind of protection is 1+1 dedicated multiplex section protection (1+1 MSP), whose mechanism is shown in Figure 3.29a. An SDH multiplex section is routed along two links on different routes connecting the same endpoints. At destination, the signal on one route is detected, while the other signal is discarded. In case of failure, the receiving terminal detects a performance degradation below the threshold (or even a lack of signal) and switch to detect the signal coming from the other link.

This protection mechanism is very fast, since the switch is performed by the same equipment detecting the failure. However, it is worth noting that in SDH networks

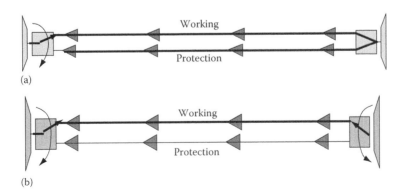

(a)

(b)

FIGURE 3.29
Functional schemes of an SDH/SONET 1+1 dedicated linear multiplex section protection (a) and of an SDH/SONET 1:1 dedicated linear multiplex section protection (b).

multiplex sections are always bidirectional. Thus, once a terminal switches, it sends a signaling to the other terminal on the return route to inform it of the failure. If the failure is also bidirectional, the other terminal will not receive the message, but it has already switched due to the detection of the failure. If the failure is unidirectional, the signal is received by the other terminal. In this case, the other terminal can leave the signal on the working route, implementing the so-called unidirectional protection, or it can switch too, if the protection is bidirectional. Generally, bidirectional protection is preferred, due to the need of avoiding big delay differences between the two directions of the same bidirectional multiplex section.

A better use of the protection capacity can be achieved in 1:1 MSP, whose scheme is represented in Figure 3.29b. In this case, the signal is sent only on one of the two paths connecting the same ends, called the working path. The other path is either left empty, or occupied with low-priority traffic, whose SLA accept more frequent interruptions. If one terminal detects a failure it sends a message to the other terminal, since in this case it is the source, not the destination to switch. In case of bidirectional failure, the terminals cannot communicate; thus, after a time in which each terminal does not detect a signal arriving on the protection route it identifies the bidirectional route and switches. In case of unidirectional route, the source terminal of the link interested by the failure detects a failure indication coming on the return route and switches the transmitted signal on the protection route. Also in this case, the protection can be unidirectional or bidirectional.

In both cases, when the failure is repaired, the standard describes two modes of working. One is the so-called nonrevertive protection, when the signal remains on the protection route up to a manual switch via the management system. The other is the revertive protection. In this case, every terminal whose transmitted signal goes through the protection link sends a test message periodically on the working link. When the terminal at the detection side sees the test message sends an acknowledgment on the returning working link.

After a double handshake, the terminals identify the failure as repaired and automatically switch the signals on the working link.

Besides MPS, there is also $1+1$ dedicated path layer protection ($1+1$ PLP) and 1:1 path layer protection (1:1 PLP). The principle is the same of the analogous mechanisms in the lower layer, but the paths and not the multiplex sections are involved. In order to underline the difference, both $1+1$ MSP and $1+1$ PLP are shown in Figure 3.30. In the figure, the MSP protection involves only one link and terminates on contiguous DXCs, due to the fact that the DXC terminates the multiplex section in each network node. On the contrary, the PSP starts and terminates on the DXCs at the path end, while the working and protection paths travel independently through the network. This is due to the fact that the path layer is accessed only at the path ends, while intermediate DXCs only pass through the paths.

Passing from meshed to ring networks, the concepts of $1+1$ and 1:1 protection can be applied at rings at both path and multiplex section layer, designing the so-called multiplex section–dedicated protection ring (MSDP)ring (ULSR: Unidirectional Line Switched Ring in SONET language) and the PLPRing (UPSR: in SONET language). An example of working of ULSR is reported in Figure 3.31. In these cases, in a two-fiber ring, one fiber is devoted to working traffic and the other to protection.

In the case of ring topology, also another protection mechanism can be implemented, which uses the protection capacity more efficiently. Let us imagine a four-fiber ring, as represented in Figure 3.32a. At the ring nodes there are add drop multiplexers (ADMs), essentially small DXC designed for rings.

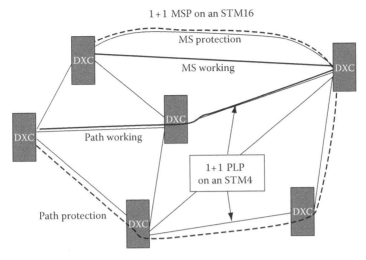

FIGURE 3.30
End-to-end path linear protection (PLP) versus multiple section protection (MSP) in a mesh SDH/SONET network.

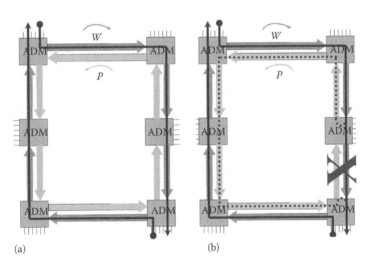

FIGURE 3.31
Scheme of an SDH/SONET path traveling through a 1+1 path protection SONET/SDH ring (PDPRing) in normal conditions (a) and in protection (b).

The first difference between the multiplex section–dedicated protection rings (MSDPring) and the multiplex section–shared protection ring (MSSPring) is in path routing. In particular, comparing Figure 3.31a with Figure 3.32a it is clear that in MSDPring the two-path directions (from A to B and from B to A if A and B are the connected ADMs) are routed both on the same fiber and in the same direction (from which the name unidirectional is coined in the SONET acronyms). In MSSpring, the two-path directions are routed on different fibers and in opposite directions (from which the name bidirectional is coined in the SONET acronyms). This routing algorithm can also be applied on a two-fiber MSSpring, exploiting the fact that the time slots of each fiber are logically divided into two parts: one

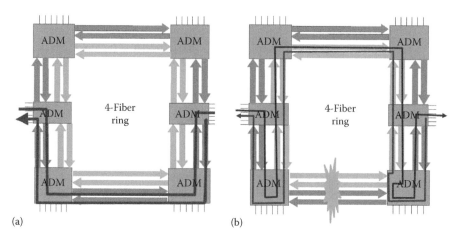

FIGURE 3.32
Scheme of an SDH/SONET path traveling through a multiple section share protection SONET/SDH ring (MSSpring) in normal conditions (a) and in protection (b).

simulating a virtual working fiber and the other a virtual protection fiber, to create a logical situation similar to a four-fiber ring.

Due to this particular routing, while in MSDPring, a multiplex section (MS) occupies all the capacity of the fibers (on one for working on the other for protections), in the MSSPring more than one MS can be routed on the ring with the condition that they are not topologically superimposed. In the extreme case when only the adjacent nodes have to be connected, up to n MS connections can be set up, where n is the number of nodes. All these MS connections share the same protection capacity, even if when one of them is switched in protection, the other cannot be switched too, thus avoiding collisions.

In this sense, the capacity is better exploited in the MSSPring with respect to the MSDPring.

Let us now imagine that the outer couple of fibers are used for working traffic and the inner one for protection. In case of failure, the protection switching occurs as in Figure 3.32b. Since the protection mechanism is located in the multiplex section, the nodes adjacent to the failure are involved in the protection procedure. In particular, each node detects the failure in the received MS and sends a multiplex section protocol to the other node to signal the failure. Since in the figure example the failure is bidirectional, no control signal is received by the nodes; thus, in the absence of an acknowledgment, both the nodes assume to be in the presence of a bidirectional failure and perform an automatic protection switch. After the switch, the nodes perform a handshake on the newly setup connection to verify its integrity and, after that, they send the traffic on the new route. Once the protection switch is performed, the two switching nodes also send a protocol along the ring that reaches all the other nodes signaling that the protection capacity has been occupied in certain ring sections. As a matter of fact, further protection switching forced by other failures that will send different signals on the same protection fibers segment is to be avoided.

If a unidirectional failure would be present, one of the nodes adjacent to the failed multiplex section will receive the failure protocol and will start a handshake with the sending node using the protection capacity for the acknowledgment. After that, both the nodes perform switching, verify the new connection integrity, and then proceed as in the case of bidirectional switch.

A similar procedure is performed to restore the working route when the failure is repaired. Also in this case, all the nodes have to be alerted through a specific protocol that the protection fibers are again available.

3.2.7 Optical Transport Network (OTN)

With the market success of optical WDM transmission, WDM (i.e., the "optical" name for the traditional frequency division multiplexing) became the standard of transmission systems.

The first step to introduce these systems in the network was to consider every wavelength channel as belonging to a separated SDH network. These brought to structure a network made of superimposed SDH rings; for example, if the transmission systems convey 16 wavelength channels, the network will be constituted by 16 superimposed SDH networks and in each network node up to 16 independent SDH equipment will be deployed.

This architecture, despite its intrinsic back compatibility and scalability, rapidly required a high CAPEX to multiply the number of SDH elements in every network node, so that a different solution had to be found.

As a reaction to this situation, the ITU-T standardized an independent model of the transport network based on optical WDM technology, following the same guidelines that emerged from SDH standardization, but for the obvious differences between the functionalities of equipment working on wavelengths with respect to equipment working on timeframes [46,47].

The definitions of the three layers composing the OTN are given in Figure 3.33 with the nomenclature defined in this standard.

In particular, the layers have the functions described in this section.

3.2.7.1 Optical Channel Layer

This layer receives requests of transport from the client layer in the form of a TDM frame to transmit through the network. Starting from these requests, the layer creates the optical channels (OChs) that are the elementary structures transmitted by the optical network. In creating OChs starting from client requests, the OCh layer performs multiplexing and grooming of client signals and enveloping them into the OCh frame.

Once the frame is created, the layer controls the OCh's transmission through the network performing routing and restoration, monitoring the transmission quality and reporting warnings and alarms to the central management system.

The OCh layer has several adaptation functions to accept as input various possible clients such as SDH/SONET, Ethernet, IP, and so on.

3.2.7.2 Optical Multiplex Section

The optical multiplex section (OMS) is the layer below the OCh. Its task is to perform wavelength division multiplexing and demultiplexing and to transmit the WDM signal.

Thus, its functions are as follows: multiplexing and demultiplexing, monitoring of the end-to-end WDM comb transmission quality, and a few networking functions related to the survivability of the whole DWDM signal and to its routing as a single element.

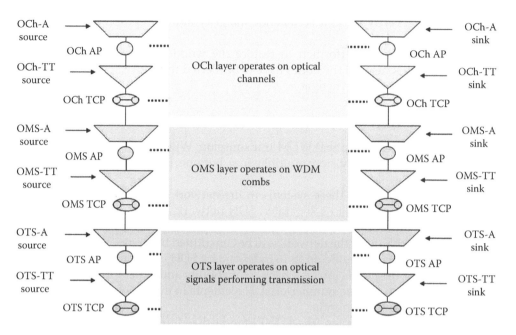

FIGURE 3.33
Scheme of the standard layers of the OTN.

Two frequency combs have been standardized by ITU-T for the OMS: the DWDM comb and the coarse wavelength division multiplexing (CWDM) comb: the standard wavelengths are shown in Figures 3.34 and 3.35, respectively.

In these figures also, the standard names the ITU-T has assigned to the different bands on the near infrared where fiber transmission can be realized are reported.

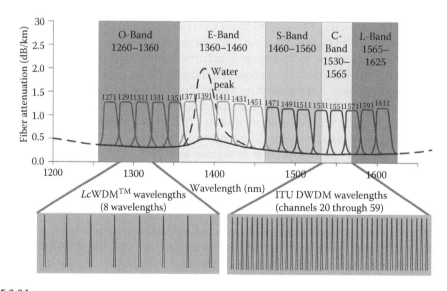

FIGURE 3.34
ITU-T standard wavelengths for DWDM transmission systems. In the figure, the standardized names of the optical bandwidths and the reference attenuation curve of a standard transmission fiber are also reported.

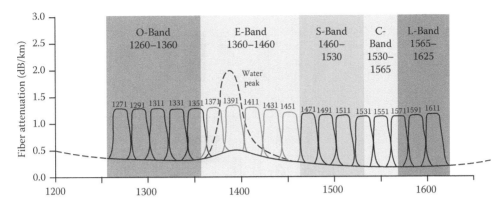

FIGURE 3.35
ITU-T standard wavelengths for CWDM transmission systems.

DWDM systems have wavelength in the C and L bands, often in extended versions of these bands depending on the needs of the transmission system.

The DWDM standard comb is constituted by 132 wavelengths with a spacing of 50 GHz. This comb can be used in different ways:

- Selecting a subset of wavelengths 50 GHz spaced
- Selecting a subset constituted only by even or odd wavelengths, thus obtaining a 100 GHz spaced comb
- Selecting a subset of 200 GHz spaced wavelengths

In all these cases, the available bandwidth around the central wavelength is also standardized, being smaller in the former case and larger in the last.

The CWDM comb is defined with the scope to create a standard suitable for the use of uncooled lasers. The channel spacing of 21 nm is derived from the analysis of frequency fluctuations of an uncooled distributed feedback (DFB) laser.

It is evident that the fiber attenuation changes greatly from channel to channel (the attenuation curve is reported in the figure) and since the signals at different wavelengths travel together in the same fiber link, the higher attenuation channel practically determines the link reach.

This is the reason why several times CWDM systems are realized selecting only eight channels to be multiplexed on the first eight frequencies of the standard.

The situation completely changes if no water peak fibers are used (whose attenuation curve is also reported in the figure). In this case, all the standard CWDM channels can be used.

Last, but not least, the optical transmission section (OTS), carries out physical transmission of signal through optical fibers.

The structure of a trail along the optical layer and its model are represented in detail in Figure 3.36.

Each layer has a signaling protocol; OCh protocol is embedded in the frame created by the adaptation function that create the OCh from the client signal, the OMS Protocol and the Optical Transmission Section Protocol are carried by the so-called service channel, a dedicated communication channel on a frequency on the edge of the useful spectrum.

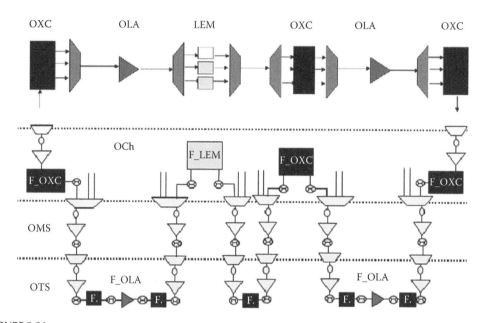

FIGURE 3.36
Layered model of a trail through an OTN: OXC, optical cross connects; OLA, optical line amplifiers; LEM, line extension module, that is, the electronic regenerator.

The structure of progressive client signal encapsulation needed to create the signal to be transmitted in the OTN is represented in Figure 3.37.

From Figure 3.37, it is clear that the process of producing the OCh frame from the client signal is quite complicated, taking into account all that is needed both to envelope different types of signals and to transmit all the needed network protocols. As a matter of fact,

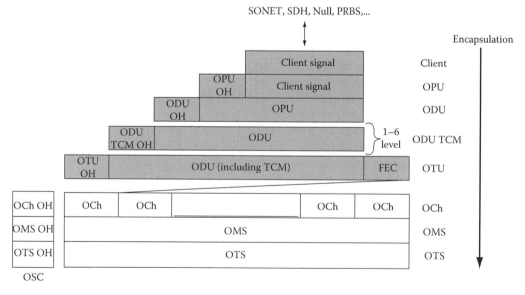

FIGURE 3.37
Progressive encapsulation of a client signal into the OTN frame.

the OTN foresees both protocols to control the control plane circuit switching of OChs and protocols to control automatic protection. Moreover, synchronous encapsulation of asynchronous signals is also implemented, in order to allocate into the OTN every possible client.

The OCh encapsulation is divided into several steps, each having its own envelope.

In the first step, the client signal is mapped onto the optical path unit (OPU) with the addition of an OPU overhead. The OPU is then enveloped into the optical data unit (ODU). The ODU is equivalent to the VC in SDH or the VT in SONET. It is the elementary frame of the optical path that connects OXC client ports after the OCh header is terminated thus managing electrical connection between clients via the OTN.

All information relative to optical path end-to-end routing between OXC client ports and, in case it is applied, optical path protection having as end ports the client ports is coded into the ODU header.

Different ODUs are multiplexed into the ODU TMC. This is essentially a multiplexing stage, allowing different tributaries to be transmitted together into the same OCh. As a matter of fact, a single OCh can have a transmission capacity much higher than the capacity of an ODU (lets think, e.g., of a 40 Gbit/s OCh) and in order to better exploit an OCh it is convenient to define a multiplex stage allowing multiplexing of different client channels onto the OCh.

Finally, the ODU TMC is enveloped into the OCh frame adding a header with all the information on the end-to-end OCh transmission between the optical terminations of the OCh inside the OTN and a tail with the forward error correcting code (FEC) redundancy.

It is important not to confuse the OCh with the optical path. The OCh is an all optical channel, transmitted on a chain of wavelengths (or on a specific wavelength if wavelength conversion is not present in the network) between optical interfaces into the OTN. The optical path is terminated, possibly in an electrical manner, on the client interfaces of the OXCs or the optical add drop multiplexers (OADMs). Just to make an example, an optical path could be correspondent to a single MPLS tunnel after enveloping with the ODU header to start and terminate on the ports connecting the label switching routers with the OXCs. The OCh could multiplex several MPLS tunnels with SDH MSs and ATM virtual circuits (each entity enveloped in a different ODU) and start and terminates on the end OXCs optical ports before the demultiplexing of the OCh envelop.

After the OCh envelope is built, different OChs are transmitted on different wavelengths; when all the wavelengths are multiplexed and the service channel added, we have the OMS and when the OTS header is also added in the service channel, the set of OTN envelopes is complete.

A few of the key network elements of the OTN are represented in Figure 3.36, where the difference between the equipment (represented in the upper part of the figure) and the key function of this equipment that is represented in the lower part is underlined and is indicated with the equipment acronym with an F_ in front to indicate that it is a function in the layer model.

In particular, in the trail represented in the figure, the following network elements are present:

- Channel optical cross connects (OXCs)
- OCh regenerators (LEM)
- Erbium doped optical amplifiers (OLA)

Other important network elements of the optical layer that are not present in the figure are

- Multiplex section optical cross connects (MOXCs)
- Optical add drop multiplexers (OADMs)
- Optical line terminals (OCLT)
- Raman amplifiers (ORA)

The OXC [47,48] is somehow similar to the DXC, if WDM is substituted to TDM. At the OXC line input, there is a number of WDM channels; these channels are demultiplexed, channels directed to the local clients are sent to the drop ports while the others are routed through a switch toward their destination. At the switch output, all the channels having the same destination are multiplexed together and sent to the output transmission systems.

As far as local clients generated channels are considered, they enter the OXC from the add ports and are sent through the switch at the correct output port. At the output, they are multiplexed in the suitable WDM comb and sent to the transmission systems.

As in the case of the DXC, the optical Channels Routing is performed by a centralized management system and the OXCs are statically configured by this central software.

Since the OXC main function is located in the channel layer, the service channel conveying the Multiplex Layers Protocols is terminated at the OXC input and recreated at the OXC output.

The availability of a suitable technology has heavily influenced the way in which OXCs are realized. After a huge amount of studies on OXCs realized with optical switching techniques, today, the commercial OXCs are almost all based on an electronic core switch. In these OXCs, after WDM demultiplexing, the optical signal is translated in the electrical domain and processed by an electronic core. At the output the signals are converted again to optical signals and WDM multiplexed.

These machines are OXCs and not DXCs since they work on the signal carried on a single wavelength as a whole, without performing TDM operations. To better clarify the logical functionalities of an OXC, the detailed layer model of the OXC is shown in Figure 3.38.

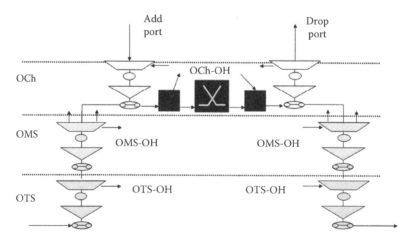

FIGURE 3.38
Layered functional scheme of an optical cross connect. The OTN layers are indicated with OTS, OMS, and OCh.

Besides the OXC, also MOXC can be devised in the optical layer; they manage the OMS as a whole, without WDM Mux/Demux. A MOXC has a certain number of input line ports, where the incoming WDM signals are not demultiplexed, but for passive extraction of the information carried out by the service channel. However, the service channel is not terminated, but it follows the signal in the switch.

As in the case of the OXC, the whole WDM combs are switched as a single entity and sent to the output ports. Clients of a MOXC are OXCs and the idea of the MOXC was born in consequence of the bubble traffic forecasts, envisioning a high traffic in the backbone to require stratification of the optical cross connect to avoid huge and expensive OXCs.

These forecasts are not realized in practice and the idea of the MOXC is almost abandoned since currently there is no commercial MOXC.

A particular discussion has to be done regarding the OADM. As a matter of fact, OADMs are the only wavelength routing equipment that are realized with pure optical techniques. While ADMs are a sort of small DXCs this is not true for OADM versus the OXC.

Even if OADMs are mainly OCh layer elements, generally they are not realized by demultiplexing the WDM channels and switching them individually, but exploit specific optical technologies to extract/reinsert one or more channels without demultiplexing the whole comb.

The OADMs can be divided into categories depending on the flexibility they have in choosing the channel to extract/reinsert. These categories are

- *Fixed OADM*: In this case the set of add/dropped channels is fixed and cannot be changed.
- *Switchable OADMs*: In this case the OADM has a certain number of Add/Drop ports, each of which is associated to one wavelength channel; the OADM can be configured to Add/Drop or to pass-through any of these wavelength channels while the channels that are not associated to one Add/Drop port can be only passed through.
- *Tunable OADM*: In this case the OADM has a certain number of Add/Drop ports and any wavelength channel can be Add/Dropped from any of the Add/Drop ports if the OADM is suitably configured.

Any further discussion on OADMs cannot ignore the technology used to build them. Thus, we delay it to Chapter 7, in which the architecture and performances of network elements devoted to route and switch information bearing frames or packets will be explicitly dealt with.

Due to the fact that OTN is built on concepts similar to those applied for SDH/SONET, it is not surprising that automatic protection is embedded in OTN.

Much like SDH/SONET, linear OCh protection and linear OMS protection can be implemented in point-to-point systems and in mesh networks. The only thing to note is that, while all SDH/SONET protection schemes also protect the optical transmitter, OMS protection is implemented after the transmitter and thus it is not protected. On the other hand, OCh protection is implemented before the transmitter, but requires an individual switch for each wavelength, while a single switch is needed for OMS-p.

These characteristics are shown in Figures 3.39 and 3.40 where the OCh layer and the OMS layer 1+1 dedicated protections are depicted decomposing the optical terminals in building blocks representing their main functionalities. Both directions are considered in the figures, and to underline OCh and OMS terminations, an add/drop card (ADC) has

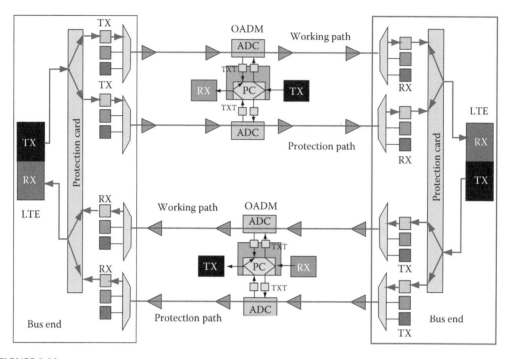

FIGURE 3.39
Scheme of a 1+1 OCh dedicated protection on a point-to-point WDM system with an intermediate branching point: LTE, line terminals; OADM, optical add drop multiplexers; ADC, add drop card; TX and RX, optical transmitter and receiver, respectively; PC, channel protection card.

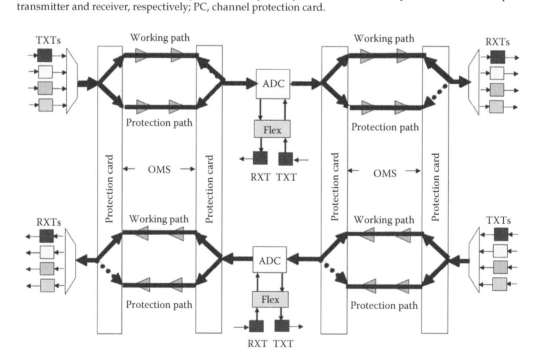

FIGURE 3.40
Scheme of a 1+1 optical multiplex section protection on a point-to-point WDM system with an intermediate branching point: ADC, add drop card; TXT and RXT, optical transmitter and receiver, respectively, flex a switch.

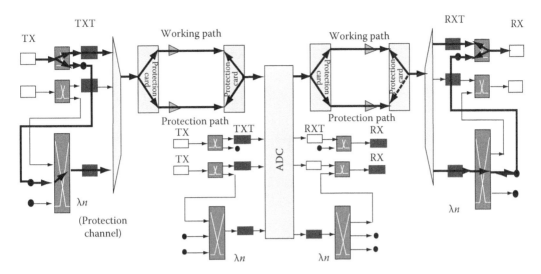

FIGURE 3.41
Scheme of the combination of a 1+1 optical multiplex section protection and an 1:*N* transmitters/receivers protection on a point-to-point WDM system with an intermediate branching point: ADC, add drop card; TXT and RXT, optical transmitter and receiver, respectively; TX and RX, electrical transmitter and receiver, respectively.

also been inserted along the protected link so that the OMS is terminated in between, while OChs passes transparently. The consequence is that, while in the OCh protection the ADC is to be doubled, in the case of OMS protection, the protection trail ends on the ADC and only one ADC is needed.

When OMS protection is used, in order to provide some resilience against transmitters failure, it is possible to adopt 1:*N* transmitter redundancy and correspondingly 1:*N* receiver redundancy. This kind of local protection that is shown in Figure 3.41, consists in reserving one wavelength to redundancy and in providing a set of switches that are able to direct any client signal either toward its working wavelength transmitter or toward the redundant wavelength transmitter.

When a local alarm detects a failure in a transmitter, automatically the signal is locally switched on the backup wavelength after a successful handshaking between the two OCh terminals and all the other switches are blocked to avoid conflict of more than one client signal on the backup wavelength.

At the receiver end, when the terminal sees no signal on the working wavelength it tries to receive the signal from the backup. If a successful handshake procedure is carried out through the backup wavelength with the sending terminal, the receiving terminal sets the receiver on the backup wavelength. Moreover, in case of bidirectional protection, that is almost the only used mechanism in OTN, the receiving terminal starts another handshaking on the backup wavelength of the return path and, if it is successful, the return signal is also switched on the backup wavelength.

It is to underline that 1:*N* transmitters protection could be implemented in SDH/SONET also when the MSP is adopted. The difference is that in the case of SDH/SONET, electrical signal sources are unprotected in case of MSP, while optical transmitters are always protected. Since electrical signal sources have a very high reliability, devising an ad hoc protection for them is often not a key point.

In case of OMS, the optical transmitters remain unprotected; after the fiber cable, optical transmitters are the building blocks with the lower availability, due to the presence

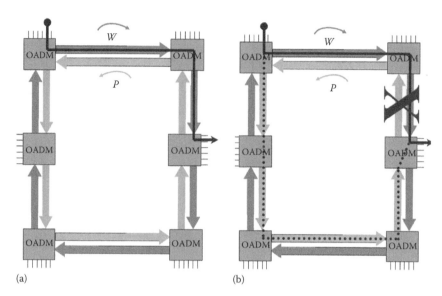

FIGURE 3.42
Scheme of an OChDPring: (a) path route in working mode; (b) path route after switching in protection mode due to the indicated failure (in both parts only one direction of the bidirectional path is shown, the return path run along the same fiber in the same direction).

of semiconductor lasers. Thus, devising a system to locally protect them is much more important.

As far as OTN rings are concerned [49,50], the situation is different from SDH/SONET since here a single wavelength represents a channel and not an MS.

In Figure 3.42 an OCh dedicated protection ring (DPRing) is shown, where dedicated 1+1 protection at OCh level is implemented. From the figure, the different routes followed in the protection state with respect to the SDH/SONET MSDPring are clear, while in both cases a whole wavelength is switched immediately before the optical transmitter. The cause of the difference is that, while in the case of MSDPring, the SDH/SONET MS is ended at each ADMs and the ADMs adjacent to the failure react when the failure is detected, whereas in the case of OChDPring, the OCh is terminated only at the origin and terminating ADMs and thus the protection works end-to-end.

In order to implement an optical multiple section protection, all the wavelengths have to be switched together by an optical switch immediately after the optical multiplexing, as shown in Figure 3.43. In this case, the OADMs adjacent to the failure react to the failure detection switching the entire OMS on the protection path.

Shared protection can be implemented also in OTN rings and it is more common to see it at the OCh layer.

The working of an OCh shared protection ring (SPring) is shown in Figure 3.44. Since generally a two-fiber ring is used in the OTN, the wavelength on each fiber is divided into two groups: a working group and a protection group, where the working wavelengths on the outer fiber are devoted to protection in the inner fiber and vice versa. Working channels, as shown in the figure, are routed bidirectionally using both the fibers on working wavelengths.

When a failure occurs, switching happens at the OCh end and the channels affected by the failure are rerouted in the opposite ring direction by using the other fiber and the set of protection wavelengths.

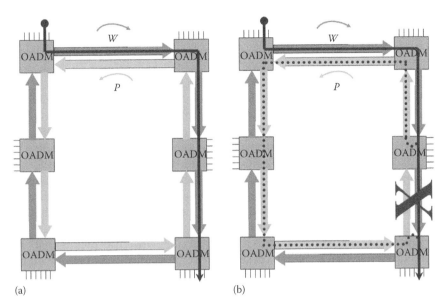

FIGURE 3.43
Scheme of an OMDPring: (a) path route in working mode; (b) path route after switching in protection mode due to the indicated failure (in both parts only one direction of the bidirectional path is shown, the return path run along the same fiber in the same direction).

Comparing OChDPring with OChSPring several differences come out. First, in channel dedicated protection ring (ChDPring), a channel is routed unidirectionally and, considering both working and protection wavelengths, occupies a single wavelength on all the ring extent and on both the fibers. Thus, in this kind of ring, whatever the traffic, the maximum number of channels that can be routed is equal to the number of available wavelengths in the WDM comb.

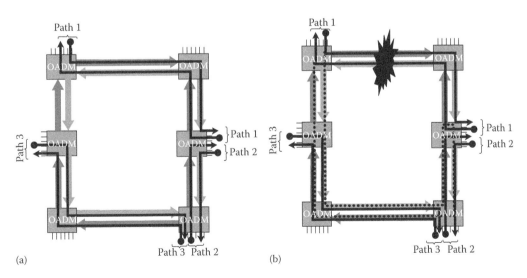

FIGURE 3.44
Scheme of an OChSPring: (a) paths route in working mode; (b) paths route after switching in protection mode due to the indicated failure.

In the case of the OChSPring, the bidirectional routing algorithm allows both working and protection wavelengths to be shared among different channels. As a matter of fact, as is shown in the figure, channels whose routes do not superimpose topologically can be routed using the same couple of working wavelengths and, in case of need, can be protected using the same couple of protection wavelengths. Naturally, they cannot be simultaneously protected; otherwise, a conflict on the protection wavelength happens.

This sharing mechanism is the cause of the fact that the number of channels that can be routed in an OChSPring depends on the traffic pattern: the more the sharing can be done the more the channels can be routed. In particular, the worst case happens when pure hub traffic is routed, where the situation coincides with that of an optical channel dedicated protection ring (OChDPring): the number of routed channels is equal to the number of wavelengths. This is due to the fact that, in case of hub traffic, all the channels are terminated on the same node (the hub) and all are topologically superimposed at least for a single link, so that sharing is impossible.

When a uniform traffic is loaded onto the OChSPring, since this means that OChs connects only contiguous nodes, the maximum sharing is possible since all the channels are either topologically or spectrally separated. This is the case in which the maximum number of channels can be routed in the ring: this number is simply evaluated to be equal to $(nm)/2$ where n is the number of nodes (at least 3 to be a ring) and m the number of wavelengths.

In all other cases, generic traffic patterns determine a maximum number of routable channels that is comprised between m (the number of wavelengths) and $(nm)/2$.

Also shared OMS protection can be applied to optical ring, in strict analogy with what is done in SDH/SONET simply substituting wavelengths to time slots. However, in the OTN rings, OMS protection is less used with respect to OCh protection, due to the need of protecting the optical transmitters. Thus, we will not detail optical multiplex section shared protection ring (OMSSPring) architecture being a simple extension of the SDH/SONET case.

Also the coordination protocols that bring all the information needed to perform correctly automatic protection switching are analogous in principles and implementation to the protocols for SDH/SONET and we will not repeat here in detail the analysis of their working. The only important point to consider is that both the OMS and the OTS headers with the relative layer protocols travel on the service channel that is a sort of header of the WDM comb.

3.2.8 Telecommunication Management Network

Up to now we have described different possible design of the network that generates sets of data plane plus control plane located at various levels of the reference open systems interconnection (OSI) system.

Now we want to describe the management plane, starting from the standard prescribed by ITU-T and that is potentially applicable at any type of communication network: the telecommunication management network (TMN) [51–53].

The most effective and complete implementation of this standard has been realized for the TMN controlling SDH/SONET networks and the same criteria was translated in the management of the OTN.

In such networks, a small number of operations, like fault protection, are automatically carried out by the network control plane via autonomous protocols, whereas almost all the management and maintenance operations are carried out by the TMN.

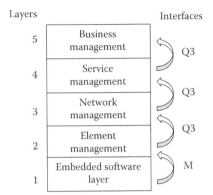

FIGURE 3.45
Layered architecture of the TMN.

In synthesis, network management software functionalities can be described with the acronyms FCAPS [52], which means

- Failure management
- Configuration management
- Accounting management
- Performance management
- Security management

The TMN is a very complex software, and in order to describe its functions and services [54] and to provide a design direction, the layered model was used to describe it.

The five layers of the TMN are represented in Figure 3.45.

3.2.8.1 Embedded Software Layer

This layer aggregates all the management software that is resident in the network elements: that is, the firmware that is present in the cards and the agent that runs in the network element control card.

The firmware is the part of software generally implemented on a customized hardware that directly operates on the hardware of an equipment particular card.

The firmware that is part of the TMN essentially performs two operations: generates alarm seeds and operates the card configuration.

To exploit the first function, the firmware reads the measurements that the physical sensors present on the card do to assess the performance of the various hardware parts of the card and compare it with thresholds. Usually, two thresholds are set. One to separate the working within specifications from the working out of specifications, the second to separate working from not working.

When one of the sensors goes beyond a threshold, the firmware produces an alarm seed that is transmitted to the agent in the control unit of the network element. In the example with two thresholds, the first threshold produces simple alarms and the second critical alarms.

To perform the second function, the firmware receives from the agent configuration instructions and implements them on the card hardware.

The agent is the software performing the management of the single network element as a whole. It receives from the upper layer the configuration instructions for the network element, declines them in configuration changes for the individual cards and sends commands to the firmware of the cards involved in the equipment reconfiguration.

Moreover, the agent receives alarm seeds from the firmware of various cards and filters them in order to recognize what are dependent from real failures and what are consequences of failures in subsystems different from that generating the seed. To make an example, let us assume that in a system the radiation emitted by a transmission laser is filtered by an optical filter and sent to the output fiber. A sensor measures the emitted radiation to verify laser performances, another sensor measures the optical power at the output of the filter to verify the filter integrity.

If the laser goes down, both the sensors go over the threshold and the firmware of the transmitter produces two alarm seeds. One of them, however, has to be filtered, since it is an indirect consequence of the other, so that the agent produces only one alarm that is sent to the upper TMN layer.

3.2.8.2 Element Management Layer

This layer is constituted by a software running on a workstation located in the central network management room.

The role of the element management layer is essentially to monitor and configure the network elements of the network and to identify failures through alarms collection and filtering.

The element manager can access all the network elements in the network and via this software, every card that is loaded in a network element can be configured. Moreover, the element manager has functions to load in a network element new cards (after the physical insertion at the network element location), to configure and test them and, finally, to free their resources to be assigned to live traffic paths.

Thus, it is the element manager that physically implements the paths routed and reserved by the upper layer by suitably configuring the network hardware.

The element manager has also the role of receiving alarms from the agents and of filtering them to avoid the alarms sequence that happen when a failure inhibits the working of all the network elements in a chain after the failed one. This operation is the fundamental instrument that the operators of network management use to locate failures after a protection switches and to plan field interventions.

Besides configuration and alarms management, the element management layer has a set of other functions. Among them, it is important to remember the following:

- Collection of statistical data on the utilization of network resources
- Measurement of data regarding the network element working like the consumed power
- Manage the software upgrade of the network elements both via firmware and software download and via the control of the local upgrade procedure

3.2.8.3 Network Management Layer

This layer of the TMN is dedicated to manage the logical configuration of the network: that is network paths. The main functionality of the network manager is to perform the operations that characterize the life of a path:

- Path setup
- Path monitoring
- Path tier down

During path setup, the network manager receives the request of a capacity between two endpoints and analyzes the network with an off-line tool to determine the optimum path routing taking into account the available network resources.

Once the path route has been identified and the resources needed to set it up are defined, the network manager asks the element manager for a resource reservation procedure. It is the element manager that matches the abstract resources considered by the network manager with the hardware present in the network and that sends a configuration command to the involved network elements to mark the resources as used by the considered path.

When the network manager receives from the element manager the resources reservation acknowledgment, it sets up the path in its logical path list and starts a path verification.

The path verification is first a logical operation, carried out by high-level simulation, and then it becomes a physical test of the new path carried out with test signals. Of course, in order to perform such a test, the network manager has again to interact with the element manager for the hardware operation.

If all the tests are passed, the new path can be loaded with real traffic.

During path working, the network manager monitors all the performances of the path and characterizes the traffic in statistical terms. Moreover, in case of fault, the network manager verifies the good working of the protection mechanism. If the traffic has to return on the working path when the failure is repaired and the protection mechanism does not implement it automatically, it is the network manager that supervises the traffic switching after a suitable test of the repaired path.

At the end of a path life, the network manager verifies that no traffic is on the path and switches it off by sending suitable commands to the element manager to release the corresponding network resources.

Due to the fact that we are considering either SDH/SONET or OTNs, TMN paths are semi-static. This means that, once a path is set up, generally it remains up until some great network restructuring: it is quite rare that paths are tear down just because the traffic request generating them is ended.

Moreover, the path setup via the TMN is not a very fast operation, due to the complexity of the procedure and to the fact that sometimes even manual intervention of the operators is required. On the other hand, due to the centralized algorithm used (Figure 3.46), the operation is generally optimized.

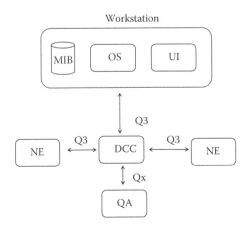

FIGURE 3.46
TMN functional scheme.

Besides paths-related functions, the network manager supervises all the "abstract" characteristics of the network: for example, MS routing if it exists and so on.

3.2.8.4 Service Management Layer

The service layer manages the services delivered to the end customer, providing customers authentication and billing, storing in its database the customer profile in terms of service fruition history, and providing a software platform to implement in the network the commercial offering of the carrier both in terms of service bundle and in terms of paying solutions.

Also security is implemented in this layer, by access keys and customers information separations.

Last but not least, this layer is the master of the procedure for the introduction of new services in the network.

3.2.8.5 Business Management Layer

This is the most abstract layer, whose role would be to take under control the business aspects of the network: costs and service related revenues, up to the evaluation of a sort of network profit-and-loss sheet.

From the earlier discussion, it is clear that the TMN is a very complex software, composed by several logically separated elements and also by physically separated parts.

Thus, for the TMN to work, it has to be provided with a communication facility to connect different parts (that is called the DCC: Data Communication Channel) and of a specialized control plane to use this facility correctly.

Frequently, the capacity designed for the overhead of the various layers is also used to implement the DCC. The most common solution is to exploit the Dx bytes embedded into the SDH/SONET MS overhead.

Another, less used, possibility to implement the DCC if WDM is present, is using a part of the capacity of the service channel that WDM systems provide on a wavelength on the edge of the standard WDM comb.

However, these are not the only possible solutions, and even a completely separated data network can function as DCC.

The DCC is managed by an OSI-like control plane that is called Q3 and that is optimized to interface TMN objects.

Taking into account all the elements we have mentioned, a possible scheme of a TMN is represented in Figure 3.46 where the operating system (OS), the user interface (UI), and management information base (MIB) represents respectively the operating system of the TMN workstation, its User interface and the TMN database, while NE represents the network element, Qx a simpler protocol stack with respect to Q3, and Qa is the adaptation function for the connection of elements that does not present a TMN standard interface [55,56].

In this very brief description, we have considered the standard ITU-T architecture of the TMN, but this is not the only used technology.

Different vendors can apply other technology solutions, deviating from the standard. Examples of this type are the Simple Network Management Protocol (SNMP)–based managers that are constructed around the SNMP that have been quite a success in the United States.

3.2.9 Central Management in IP Networks

In its intention, the TMN standard is defined to be applicable to every kind of network. If TMN is to be applied to a data network based on the Internet model, a few of its functionalities have to be disabled, since they are carried out via the IP suite in the control plane. For example, the TMN could implement off-line traffic engineering, but in normal condition the routing is performed by the control plane.

In practice however, due to the much simpler set of functionalities needed in IP-based networks, it is quite rare that TMN is chosen to implement the central management.

The IETF, the Internet standardization consortium, has defined a much simpler standard for the network management, based on the SNMP [56,57].

Three versions of SNMP have been standardized, and to some extent all are in use in deployed networks. It is interesting to note the existence of the so-called bilingual SNMP that support both SNMPv1 and SNMPv2 features.

As far as SNMPv3 is concerned, even if the IETF considers it the present version and calls the other versions "Historical," in practice it has not substituted SNMPv1 and SNMPv2.

An SNMP-based management system is constituted by the following elements:

- *Network elements.*
- *Agents.*
- *Managed object:* A managed object is a characteristic of something that can be managed. For example, a list of currently active TCP circuits in a particular host computer is a managed object. Managed objects differ from variables, which are particular object instances. Using our example, an object instance is a single active TCP circuit in a particular host computer. Managed objects can be scalar (defining a single object instance) or tabular (defining multiple, related instances).
- *MIB:* An MIB is a collection of managed objects residing in a virtual information store. Collections of related managed objects are defined in specific MIB modules.
- *Syntax notation:* A syntax notation is a language used to describe a MIB's managed objects in a machine-independent format. Consistent use of a syntax notation allows different types of computers to share information. Internet management systems use a subset of the International Organization for Standardization's (ISO's) Open System Interconnection (OSI) Abstract Syntax Notation (ASN.1) to define both the packets exchanged by the management protocol and the objects that are to be managed.
- *Structure of management information (SMI):* The SMI defines the rules for describing management information. The SMI is defined using ASN.1.
- *Network management stations (NMSs):* NMSs are usually engineering workstations. At least one NMS must be present in each managed environment.
- *Parties:* Newly defined in SNMPv2, a party is a logical SNMPv2 entity that can initiate or receive SNMPv2 communication.
- *Management protocol:* A management protocol is used to convey management information between agents and NMSs. SNMP is the Internet community's de facto standard management protocol.

SNMP standards do not define which variables a managed system should offer. Rather, they use an extensible design, where the available information is defined by MIBs [58].

MIBs describe the structure of the management data of a device subsystem and in its real nature, SNMP is simply a protocol for collecting and organizing information.

This approach is coherent with the design of the Internet-based network, where all the network operations are automatically carried out by the control plane exploiting the routers distributed intelligence and the main role of the centralized network management is to report the network status and statistics to the central control, to allow failed subsystems to be identified and substituted, and to monitor customers parameters like authentication, billing, and so on.

3.3 Network Convergence over IP

In Chapter 2, we have derived from the evolving market analysis a set of requirements for the telecommunication network, the first of which is the need of reducing the operational expenditure (OPEX).

In order to reduce the OPEX, the number of different technologies used in the network has to be reduced so that evolving toward convergence of all the services over the same network platform is a primary task for carriers.

Convergence of network platforms also allows convergence of the network control platform.

As noted in Chapter 2, the data network as it is realized at present is controlled and managed using different technologies.

In particular, the physical layer, either SDH/SONET or OTN, has a very thin control plane, essentially used for automatic protection, and a powerful management plane, often TMN based, which centralizes almost all the management and configuration functions.

On the contrary, the data network, either based on TCP/IP or on other technologies, has a powerful control plane, carrying out almost all the functions related to network configuration and survivability, and a slim management, generally based on SMNP.

When a coordinated operation has to be carried out, it is manually implemented by accessing to the two systems in parallel.

This way of managing the network is surely not optimized from the OPEX point of view, but a common control plane and management plane cannot be realized if there is no platform convergence.

Data being the prominent type of messages to be transmitted, because also real-time services like TV and telephone can be reduced to data flows with an appropriate QoS, it is quite reasonable to converge at transport and network layers (i.e., layers four and three) on the Internet model.

The IP suite has had such a huge success in the era of the Internet revolution that it is proved on very large networks and the corresponding equipment is produced in large volumes. This condition assures both solid standards and a trend of price decrease that probably will last over a long period.

Moreover, the flexibility of the Internet model allows a continuous protocol suite adaptation to new exigencies, that is, the key to introduce it into carriers' networks. Carriers have very specific requirements in terms of network monitoring and service quality and the evolution of telecommunication services in the near future is not easy to foresee; thus, the network reference architecture has to be flexible and scalable to adapt to future situations.

On the other side of the layered network architecture, optical transmission in almost every part of the network is out of discussion. No other technology allows such a large capacity to be transmitted at such a long distance with an affordable CAPEX. Only in the last mile, copper pairs are generally still used, but a slow trend exists also in this part of the network to substitute copper cables with fiber.

If selecting the technology for the upper and the lower layers of the network is quite natural and there is a large agreement on the choice, the situation is not the same for the intermediate layers.

As a matter of fact, even if conceptually it is not impossible to directly carry IP datagrams on the OTN frame, this solution has several issues if realized in the simpler way.

A very-poor layer-two functionality is not easy to accept in the design of a carrier network. In this case, for example, since all the QoS would have to be managed by Internet layer protocols, IP could not be used to convey the greater part of carrier services and some other Internet layer protocol has to be devised, partially loosing the advantage of the convergence on a widely used network layer.

Moreover, the typical content source, for example, a VoD server, does not format its output as a steam of pure IP datagrams (or the equivalent in some other Internet layer protocol), but frequently it is designed to format the output as a stream of Ethernet frames. With this choice, it adopts the most diffused format for local networks (even the big local networks adopted, e.g., by TV producers) and leverages on the low cost of Ethernet subsystems due to the high volume production.

Also data-oriented services, like network storage or network servers mirroring are based on transmission of Ethernet frames.

Last, but not least, the network is generally evolved starting from a legacy situation in which, in the vast majority of cases, a telephone circuit switched network was served by a SDH/SONET transport.

As a consequence, a lot of SDH/SONET machines are deployed in the field, some of them out of the last generation of TDM machines, thus with a long operating lifetime in front of them.

If the telephone switches cannot be reused, but have to be substituted, it is desirable to reuse, at least for an important part of their lifetime, the most modern SDH/SONET machines.

All these considerations bring to the conclusion that a more complex layering is needed in between the IP and the WDM layer.

3.3.1 Packet over SDH/SONET Model

Packet over SDH/SONET (POS) [41,59] is a layer 2 technology that uses two protocols: Point-to-Point Protocol (PPP) [60,61] and High-Level Data Link Control (HDLC) [20,61] to encapsulate IP packets into a frame. It is generally used to adapt IP packets to be transported by an SDH/SONET network, but this is not the only possible use.

POS enabled machines can be also directly connected by a bidirectional fiber link, thus generating the simpler example of direct IP over fiber.

Since this solution has its wider application in the United States, from now on we will adopt the SONET language, even if, all we will describe is also valid for SDH with the needed names and acronyms adaptation. When this will not be true for some specific aspect it will be remarked clearly.

Since a POS-based IP network uses SONET transport, if underlying WDM transmission systems exist, then the interface is well defined, since it is the interface between SONET and WDM transmission.

Adopting POS, the reuse of last generation SONET machines and the reuse at physical layer of SONET management and survivability mechanisms that are well known and appreciated by carriers, is immediate.

Moreover, if a legacy data layer has to be integrated into the network besides the Internet layer, it can be done simply using SONET multiplexing capability. As an example, if an overlay Asynchronous Transfer Mode (ATM) network has to survive besides the IP one, it can be inserted via the ATM/SONET adaptation function in a suitable VT that can be multiplexed using the SONET frame with the data from the IP network.

The disadvantage is the physical layer redundancy when WDM is used under the SONET layer. In this case, the POS advantages are paid with a larger and larger CAPEX while the number of SONET machines increases following the increase of traffic and WDM wavelengths.

The current POS specifications [59] define the requirements that are needed to transport data packets through POS across a SONET network. These requirements are summarized as follows:

- *High-order containment:* POS frames must be placed in the required synchronous transport signals used in SONET. For example, an optical carrier-12 (OC-12) concatenated POS interface requires a Synchronous Transport Signal-12 (STS-12) circuit to contain the payload. The client-generated IP packet has to be encapsulated into a layer 2 format to be adapted to the SONET frame. The POS standards specify PPP encapsulation for POS interfaces. The layer 2 PPP information is encapsulated into a generic HDLC header and placed into the appropriate field of the SONET frame. This can be a confusing concept at first. Although HDLC and PPP are different, mutually exclusive layer 2 protocols, HDLC is used as delimiter in the SONET frame. The encapsulation process of an IP packet to a SONET frame is illustrated in Figure 3.47.

- *Bytes alignment:* This refers to the alignment of the data packet byte boundaries to the STS byte boundaries.

- *Payload scrambling:* The ANSI standard for T1 transmission requires a density of 1s of 12.5% with no more than 14 consecutive 0s for unframed signals and no more than 15 consecutive 0s for framed signals. The primary reason for enforcing a 1s density requirement is for timing recovery or network synchronization. However, other factors such as equalization and power usage are affected by 1s density.

SONET has a default scrambler that was designed for voice transport. The 7 bit SONET scrambler is not well suited for data transport. Unlike voice signals, data transmissions

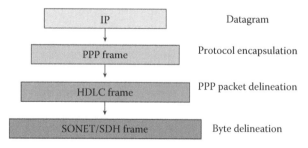

FIGURE 3.47
Packet over SONET (POS) encapsulation architecture.

might contain long streams of 0s or 1s. If the change from 0 to 1 is not frequent enough, the receiving device can lose synchronization to the incoming signal. This can also cause signal-quality degradation resulting from distributed bit errors. The solution to this synchronization and bit error problem is to add an additional payload scrambler to the one normally found within SONET environments. This scrambler is applied only to the payload section, which contains the data. The scrambling function is performed in hardware and does not affect performance in any way.

An important point is constituted by network reliability when POS is used.

Transport is carried out through a SONET network; thus, all SONET protection and restoration mechanisms apply. A possible point of weakness is the connection of the transport network with the router, where the IP datagrams are enveloped in the POS frame.

In order to improve reliability of this connection, the POS standard foresees different possibilities.

Since POS implements the SONET linear 1+1 MSP, if the router has an output card implementing the same protection mechanism, this solution can be used to protect this network segment against cable cut or other failures affecting one of the connections. Naturally, this means that a double cable has to be provisioned between the router and the POS enabled SONET equipment and the two cables have to run along different paths, so that it is improbable that both are cut at the same moment.

In this protection scheme, the router is a single point of failure. This problem can be mitigated by terminating the working and the protection links on different routers cards, or even by using two different routers.

These solutions improve network survivability at the expenses of resource allocation. The problem can be mitigated in the case of the use of two routers by applying load balancing.

Since POS uses the so-called concatenated MS frame, the slower possible channel is at 155 Mbit/s and if the data flux of the router toward that port is sensibly smaller, bit stuffing allows the OC-1 channel to be filled and utilized for transmission.

From a physical and topological point of view, a POS network will appear as schematically depicted in Figure 3.48 where the network topology is assumed to be a logical IP mesh over a physical SONET interconnected ring network.

Since the hardware that performs the POS encapsulation is in the router line cards, transit ADMs that are not connected to a router must only have the capability to recognize concatenated container created by POS and to manage them as a whole without demultiplexing, so that it is possible to use the most modern among the legacy machines.

POS is a powerful and relatively easy to implement solution to transport IP services over a TDM transport platform—it over performs the solution requiring ATM at level two, but it has some rigidities that causes some low efficiency.

The main point is the need of using concatenated high-speed SONET paths; that means that the lowest capacity POS path is a concatenated 155 Mbit/s STS. Concatenated means that the STS cannot be further demultiplexed by a SONET machine, being completely filled by client data: IP payload and bit stuffing up to the full available capacity.

Frequently, all this capacity is not needed for a single POS path and stuffing is abundantly used, loading the SONET layer with a void load that has to be transported in any case.

Moreover, this strategy makes it impossible to use at best SONET grooming capabilities that are embedded into SONET DXC and ADM.

For these reasons, a more effective adaptation layer has been defined between IP and a lower TDM physical layer that could be SONET or even OTN: the General Framing Protocol.

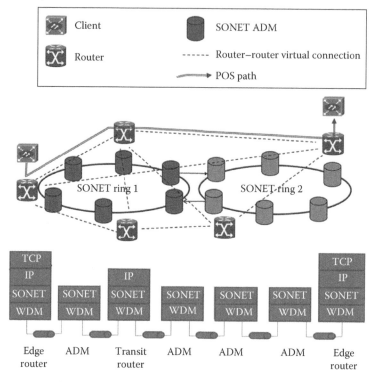

FIGURE 3.48
Representative topology and layering of an IP over POS network.

3.3.2 IP over Next Generation SONET/SDH

To overcome the limitations related to the use of Packet over SONET while maintaining the familiar SONET performances, an important group of carriers mainly in North America are pushing toward a new generation of SONET/SDH equipment, more suitable for delivering data-based services [62,63].

These equipment should be based on two main technologies: a new and very efficient packet encapsulation procedure (the general framing procedure [GFP]) and an embedded mechanism to manage the transmission bandwidth in a dynamic way to increase the exploitation efficiency of the physical layer resources.

3.3.2.1 General Framing Procedure

GFP is a powerful way to multiplex different packet-oriented formats on the same TDM frame.

Differently from POS, GFP does not rely only on the use of monolithic SONET frames, but defines a complex adaptation that is sufficiently flexible to allow both efficient multiplexing and effective grooming to exploit at best the transport platform.

GFP is conceived for application in high-capacity networks and for transport of a plurality of data formats, some of them datagram based such as IP, others frame based such as Ethernet or other datacom oriented formats such as the Fiber Channel. A first consequence is that WDM is assumed to exist as transmission facility.

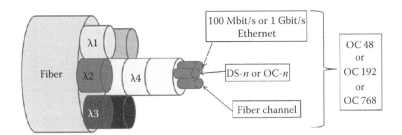

FIGURE 3.49
Tributary aggregation scheme of a GFP signal.

Relying on WDM, GFP considers the transmission trunk (the optical transmission channel) divided into wavelengths, forming a sort of frequency frame.

Each wavelength supports a TDM frame composed by a GFP "core header" and a "payload area" where different clients can be multiplexed. This structure of the optical transmission channel is shown in Figure 3.49. This figure is quite important, since we will see that the structure shown in the figure will recur also in the design of next generation control plane that we will consider in Chapter 8.

The core header supports all the client independent functions related to layer 2 link management, such as payload length indication and delimitation of the frame. The core header also has an independent error control, called core header error control (CHEC).

The payload incorporates not only the client data to be transported, but also a set of control data that are related to the specific client. These data are contained in the payload header, which also provides a payload-specific error control. Moreover, in the payload area there is also a frame alignment sequence.

The separation of payload and common header error control is done to also allow the system to send to the end user frames corrupted in the payload if the SLA so requires, while frames that are corrupted in the header have to be either repaired or somehow substituted by the layer 2.

In considering this error control policy two things have to be taken into account: first it is well possible that the upper layer has a powerful error correcting procedure that is capable of correcting corrupted frames in its end-to-end working. Moreover, in real-time services like TV or telephone, a corrupted frame is often better than no frame from the point of view of individual user perception. This is a point that layer 2 cannot cover: it will be the upper layer to judge if passing through corrupted frames, correcting them or even discard them.

Among possible payloads, the GFP distinguishes between client data frames (CDF) and client management frames (CMF). A CDF transports application data; a CMF transports data exchanged among management units on the management plane. This distinction is quite useful since it allows the GFP to manage differently the QoS for the different classes of payload.

The high-level structure of the GFP frame is reported in Figure 3.50.

In order to perform such a flexible multiplexing, GFP distinguishes two client types, assigning to each of them a corresponding transport mode: the frame mapped GFP (GFP-F) and the transparent mapped GFP (GFP-T). Even if these transport modes are quite different, they can coexist in the same frame, so that they provide more efficient and client-oriented adaptation without any rigidity.

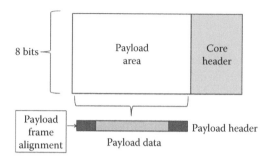

FIGURE 3.50
Schematic representation of the GFP envelop.

The GFP-F has been optimized for packet-based applications, primarily transport of IP datagrams, but also transport of Ethernet frames and even support to an MPLS upper layer.

In order to adapt to these types of clients in a better way, the GFP-F has, among the other specific features, the following characteristics:

- Variable frames
- Data multiplexing at packet level
- Variable data rate
- Can work at VT level
- It is MAC aware
- Requires buffering and thus introduces latency

The GFP-T is optimized for applications that require sensitivity to delay and real-time operation including all Storage Area Network Protocols.

GFP-T has, among the others, the following features:

- Constant length frames
- No buffering, thus no latency
- No MAC is required
- Retain idle frames having a lower bandwidth efficiency

The topological architecture of a network integrating different data clients over the same SONET transport using GFP is shown in Figure 3.51.

From Figure 3.51, it is clear that GFP allows different virtual networks to be supported by the same SONET network without losing efficiency in the data transport: thus, it is not exact to say that GFP allows convergence over IP. We can say that GFP is much more flexible: it allows convergence on TDM maintaining, if needed, different packet modes for different services.

3.3.2.2 Virtual Concatenation

Virtual concatenation (VCAT) is a technique that allows SONET channels to be multiplexed together in arbitrary arrangements. This permits custom-sized SONET pipes to be created that are any multiple of the basic rates. VCAT is valid for STS-1 rates as well as for VT rates.

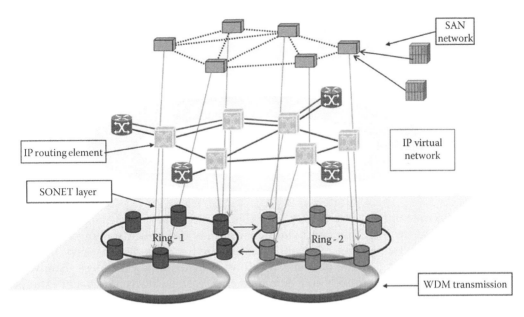

FIGURE 3.51
Representative topology of a network sustaining over a SONET multiring physical layer both an IP virtual network and a storage area network (SAN) using GFP encapsulation.

All the intelligence to handle VCAT is located at the endpoints of the connections, so each SONET channel may be routed independently through the network without requiring any knowledge of the VCAT. In this manner, virtually concatenated channels may be deployed on the existing SONET/SDH network with a simple endpoint upgrade.

All the equipment currently in the center of the network need not be aware of the VCAT.

In contrast, arbitrary contiguous concatenation that is used in POS architectures also allows for custom-sized pipes to be created, but requires that the concatenated pipe be treated as single entity through the network. This requirement makes deployment of this service virtually impossible over the legacy SONET networks without a significant upgrade.

For example, using VCAT, the SONET transport links may be "right-sized" for Ethernet transport. In effect, the SONET pipe size may be any multiple of 50 Mbit/s for high-order VCAT (STS-1), or 1.6 Mbit/s (VT1.5)/2.176 Mbit/s (VT2) for low-order VCAT.

The matching between virtually concatenated SONET channels and Ethernet frames is shown in Table 3.4 besides the very high frame efficiency so achieved.

In addition to sizing the transport paths to handle the peak bandwidth expected, VCAT may be used to create an arbitrary-sized transport pipe. The pipe may be sized for the average bandwidth consumed for a single connection, or may be sized in order to provide a statistically multiplexed transport pipe.

3.3.2.3 Dynamic Bandwidth Allocation

Along with VCAT, the capability to dynamically change the amount a bandwidth used for a virtual concatenated channel has been developed. This capability is commonly referred to as Link Capacity Adjustment Scheme (LCAS). Signaling messages are exchanged within the SONET overhead in order to change the number of tributaries being used by

TABLE 3.4

Efficiency of VCAT in Mapping Ethernet Signals

Ethernet Rate	VCAT Channel	SONET Rate	Frame Efficiency (%)
10 Mbit/s Ethernet	VT-1.5-7v	~11.2 Mbit/s	89
10 Mbit/s Ethernet	VT-2.0-5v	~10.88 Mbit/s	92
100 Mbit/s Fast Ethernet	STS-1 2v	~96.77 Mbit/s	100
1 Gbit/s Ethernet	STS-1 21v	~1.02 Gbit/s	98
1 Gbit/s Ethernet	STS-3c-7v	~1.05 Gbit/s	95

a Virtually Concatenated Group (VCG). The number of tributaries may be either reduced or increased, and the resulting bandwidth change may be applied without loss of data in the absence of network errors. The ability to change the amount of bandwidth allows for further engineering of the data network and providing new services. Bandwidth can be adjusted based on time-of-day demands and seasonal fluctuations. For example, businesses can subscribe to higher bandwidth connections (for backup, etc.) when the demand for bandwidth is low and hence the cost is lower. LCAS can further provide "tuning" of the allocated bandwidth. If the initial bandwidth allocation is only for the average amount of traffic rather than the full peak bandwidth, and the average bandwidth usage changes over time, the allocation can be modified to reflect this change. This tunability can then be used to provide (and charge for) only as much bandwidth as the customer requires.

3.3.3 IP over MPLS over OTN

GFP is a powerful tool for integrating packet switched networks over a SONET transport; it is so attractive for a carrier to rely on well-known and assessed manageability and reliability of SONET network that several vendors, mainly in North America, are designing a new generation SONET as a tool for network evolution.

However, relying upon SONET or SDH equipment does not allow reducing the number of equipment that it is necessary to deploy in a core network node. Increasing the number of wavelength in DWDM systems, the number of SONET/SDH network elements increases with two detrimental effects.

First of all, the CAPEX needed to expand the network increases almost linearly with the capacity due to the fact that if SONET machines process a group of wavelengths, for every new group that is activated a new SONET/SDH network element has to be added.

Second, the SONET/SDH network becomes rapidly composed of a large number of superimposed and potentially independent networks, one for each layer composed by a group of wavelengths (e.g., the wavelength entering the same set of DXCs). This is not the best way to efficiently exploit the transmission capacity, since grooming among different wavelengths is not possible and free capacity on one wavelength plane cannot be used to relieve congestion on another plane.

This last problem could be solved by a network design based on big DXCs and a mesh topology instead of ADMs and a ring topology, so that the DXC can terminate every wavelength of the WDM systems arriving at its node. In this case, the needed DXCs would have huge dimensions and the related cost would be huge too. As an example, to interface five in and five out DWDM systems at 10 Gbit/s using 64 wavelengths each, a DXC with a bidirectional capacity of 3.2 Tbit/s would be needed. Since a DXC performs SDH/SONET path layer operations, it should decompose the incoming signals into the client VCs (or VTs) in order to perform grooming optimization at the path layer.

Assuming, for example, that all the tributaries would be VC12 (2 Mbit/s payload) the DXC should manage up to 1.6 million channels, which is an impossible task. Of course, the situation gets better and better if the tributary channel speed increases. For example, working with VC3 tributaries, the number of paths to manage decreases up to about 66,000; that is a huge number but completely of another order of magnitude.

On the other hand, increasing the tributary speed, the efficiency in the exploitation of the transmission lines decreases as far as the grooming effectiveness is decreasing so that the choice of using DXCs is less and less justified.

On the survivability side, using DXCs changes completely the scenario since fast ring survivability is no more possible. Thus, either link is protected one by one using, for example, 1+1 DMSP or only restoration is adopted leveraging on DXCs rerouting capabilities.

In the first case, the number of DWDM systems has to be doubled with a clear CAPEX problem, in the second, there is no more a fast protection procedure and real-time services could experiment problems in case of a failure.

Thus, sooner or later IP over SDH/SONET over WDM architecture has to be evolved in order to face carriers CAPEX problems in pace with the required network evolution.

On one side, there is a certain number of carriers, especially in North America, that push for the evolution of SDH/SONET toward the so-called next generation SONET starting from the generalized introduction of GFP.

On the other side, other carriers simply call for the elimination of the SDH/SONET layer to build the packet virtual network directly on the optical layer.

Of course, this means that layer 2 functionalities have to be demanded to the packet network, and the most immediate solution is to design an MPLS-based packet network to directly put over the optical transmission layer.

This solution requires some adaptation function that matches the MPLS packets with the TDM frame at the optical level. This is not a difficult task and essentially two solutions exists.

The first relies on the adoption of the SDH/SONET frame only as an adaptation between packets and the transport layer. In this case, the GFP runs directly in the WDM equipment client interfaces just to create a TDM frame and to put MPLS packets inside. Then the frame is transmitted through the optical transmission systems up to the edge optical devices where GFP restores the packets and returns them to the MPLS layer.

The second solution is of course the introduction of the OTN frame with its own framing mechanism that is generally called digital wrapper. In this case, optical network elements works with the OTN frame (see Figure 3.37).

From a conceptual point of view, these are similar solutions, even if differences do exist in the efficiency of the frame and in the electronics that is needed to implement the adaptation functionalities at the adaptation points.

The resulting network architecture is shown in Figure 3.52. From the figure, it is clear that in this kind of network architecture, all the routing of data is performed by the label switching routers (LSR) and optical equipment is simply used for transmission.

In order to push this solution, a vendor has even proposed to directly integrate optical DWDM interfaces into the router equipment so that an optical line terminal is no more needed and the DWDM system is a sort of transmission extension of the LSR.

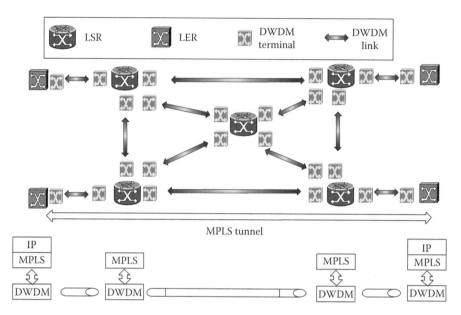

FIGURE 3.52
Representative topology and schematic layering of an MPLS over DWDM network.

Since this idea seems really attractive, we will discuss it in detail while dealing with routing equipment architectures in Chapter 7, since it hides more than what seems at first glance.

The architecture of Figure 3.52 is remarkably simple, but has an issue related to scaling to core networks. In a network as large as a nation or a continent, several times, a wavelength travels through a node without the need of processing but for regeneration at the transmission level if needed.

In the architecture of Figure 3.52, since there is no switching available at the optical layer, every wavelength has to be processed by an LSR at every crossed node. This will cause a quite inefficient LSR working since several label switching operations are not really necessary due to the fact that the input wavelength is simply reconstructed at the output.

It is possible to conceive a solution in which pass-through wavelength is not terminated on an LSR, but via a simple fiber patch panel sent to the output DWDM system. This architecture is surely CAPEX effective, but causes an embedded rigidity in wavelength assignment that results to be unacceptable for most carriers.

As a matter of fact, when a wavelength route has to be changed in such a network, manual intervention on the patch panels is needed.

A most effective solution consists in enriching the optical layer of a real wavelength routing capability exploiting the full OTN definition.

In order to distinguish the two solutions we will call MPLS over DWDM the solution where LSR and LER are connected directly via DWDM links and MPLS over OTN the solution in which either OXCs or OADMs (or both) realize a complete OTN layer on top of which the MPLS is built.

In MPLS over OTN, wavelengths are routed into the optical layer exploiting either an optical ring or an optical mesh topology architecture. Only when the tunnels contained into a wavelength have to be resolved the wavelength is sent to the LSR for MPLS label-related operations.

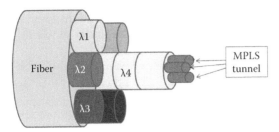

FIGURE 3.53
Tunnels aggregation in the signal traveling through a fiber of an MPLS over DWDM network.

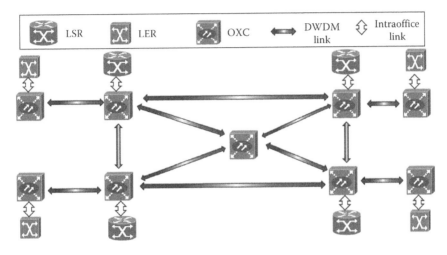

FIGURE 3.54
Representative topology of an MPLS over OTN.

The model of the multiplexed traffic is shown in Figure 3.53, where the traffic traveling through an optical fiber is decomposed in wavelengths that are routed into the OTN layer. Every wavelength multiplexes several MPLS tunnels that are managed by the LSR when the wavelength is passed to the MPLS layer.

The network architecture is schematically shown in Figure 3.54. In the figure, the OTN has a mesh architecture and OXCs are used to route whole wavelengths; however, a ring-based OTN layer can also be used where the OXCs are substituted by OADMs.

The adoption of OXCs, either all optical or with an electronic core switch, is quite different from the adoption of DXCs in the IP over SONET architecture. As a matter of fact, OXCs do not demultiplex the signal traveling on a single wavelength, but manage the wavelengths as a whole. Just to look at what happens in the aforementioned DXC example, a suitable OXC would have a switch fabric 320×320, able to send every input wavelength to every output port. Naturally, this implies the ability of the OXC to change the carrier wavelength during the operation, a property that is trivial for electrical OXCs, but not obvious for the all optical one as discussed in Chapter 10.

3.3.4 IP over Ethernet over OTN

The alternative to MPLS to provide the network with a carrier class packet-based layer 2 is the adoption of carrier class Ethernet that brings about the substitution of LSR with carrier class Ethernet switches while at the network edges the LER is substituted by a stack of a switch and a router.

In principle, this architecture can be deployed in both the schemes shown in Figures 3.52 and 3.54, considering a possible mesh and ring architecture of the underlying optical layer.

In practice, even if carrier class switches could have very good performances, the application of this architecture is conditioned to the scalability limits of Ethernet, even in its carrier class version.

There is a strong debate in the engineering community between Ethernet vendors and vendors of other technologies on the possible scalability of Ethernets networks and it is not the scope of this book to answer such a question, which among other things, is strongly influenced by future developments of Ethernet technology.

Looking to deployed networks, carrier class Ethernet is currently considered as essentially a metro network technology, but in the metro area it is experimenting a strong push especially in Europe and in a few Far Eastern countries.

3.4 Comparison among Different Core Architectures

In this section, we will try to compare the core network architectures we have considered both under a functional and from a CAPEX point of view.

As all the similar exercises numerical results have to be considered in their qualitative meaning, depending on technology and also on the even more unstable market situation that strongly influences the economic part of the comparison.

We will try to reduce these dependencies by using a set of countermeasures.

First of all, we will work with network costs for the network vendor, thus eliminating the impact of the vendor margin that is strongly customer dependant.

Moreover, all the results will be reported in a relative form, so that in the absence of technology breakthroughs, the market-related variation is partially compensated. As a matter of fact, in the absence of relevant technology changes, the price of products used for the same application varies almost in the same way.

The next step is to recognize that in order to carry out a significant comparison, single equipment cannot be compared out of a network environment, but networks built from different technologies have to be compared looking at both their cost and their functionalities.

3.4.1 Architectures Functional Comparison

Before the economic comparison, it is important to set clearly the relationship among the different architectures we have considered in terms of performances.

A too-low-performing architecture would force the carrier to ask for the lacking performances from other segments of the network (e.g., the upper layers) so that the CAPEX saving could be illusory if the whole network is considered.

3.4.1.1 Framing Efficiency

The original reason why POS was introduced to substitute IP over ATM over SONET was the frame efficiency. As a matter of fact, due to ATM segmentation of the layer 3 packets

TABLE 3.5

Packet over SONET Frame Efficiency
for "Good" Packet Sources

Client Packet Size (Bytes)	POS Efficiency (% of SPE Carrying Client Data)
64	86.8
128	94
256	97
512	98.6
1024	99.3
1518	99.5

TABLE 3.6

Packet over SONET Frame Efficiency for "Bad" Packet Sources

Data Bit Rate		SONET Rate	Effective Payload Rate	Bandwidth Efficiency (%)
10 Mbit/s	Ethernet	STS-1	~48.4 Mbit/s	21
100 Mbit/s	Fast Ethernet	STS-3c	~150 Mbit/s	67
1 Gbit/s	GbE Ethernet	STS-48c	~2.4 Gbit/s	42

in ATM cells and the need of adapting ATM cells to the SONET frame, the bandwidth efficiency of this solution is quite insufficient.

POS does much better, especially when the client rate is well matched with the used concatenated SONET frame. As an example, the POS framing efficiency is reported in Table 3.5 [64] for different "good" dimensions of the client packets.

The situation can be completely different in case of "bad" dimensions of client packets, as shown in Table 3.6 [65] where the POS framing efficiency is shown in case of Ethernet encapsulation.

It is to be noted that the situation of Table 3.6 is bad, but not uncommon, especially in the metropolitan area.

The situation gets better with the encapsulation of MPLS or Ethernet frames into the OTN frame—the so-called digital wrapper (compare Figure 3.37) when the client signal is high speed—starting from 2.5 Gbit/s. On the contrary, the OTN framing is quite inefficient for a large number of low-speed client signals.

This behavior is due to the way in which the OTN was designed: it was conceived as a physical layer that is able to transport huge traffic quantities, while processing of low-speed tributaries is left to the upper layer.

This effectively happens in the Ethernet over OTN scheme and even more in MPLS over OTN architectures, where switches and LERs, respectively, can offer to the OTN high-speed tributaries (e.g. 10 GbitE or 10 Gbit/s MPLS packets) for which the OTN frame efficiency is very high.

This qualifies the OTN as a core technology that can work with high frame efficiency but only in tandem with a suitable upper layer and in conditions of high traffic.

The greater framing efficiency for every client speed compatible with a core network is realized by GFP with VCAT. In this case, there is the greater frame flexibility and high

TABLE 3.7

Comparison among the Frame Efficiency of Different Adaptation Layers between the IP Network and the Physical Layer

		POS		Next Generation SONET		OTN	
		Mapping	Efficiency (%)	Mapping	Efficiency (%)	ODU1 (%)	ODU2 (%)
Ethernet	10 Mbit/s	VC-3	~30	VC-12-5v	~89	—	—
Fast Ethernet	100 Mbit/s	VC-4	~65	VC-3-2v	~100	—	—
GbE	1.25 Gbit/s	VC-4-16c	~40	VC-4-7v	~96	~40	~11
10 GbE (WAN PHY)	9.953 Gbit/s	VC-4-64c	~100	VC-4-64v	~100	—	~100
Digital video	270 Mbit/s	VC-4-4c	~43	VC-4-2v	~90	~9	~2.7
Serial digital HDTV	1.485 Gbit/s	VC-4-16c	~60	VC-4-10v	~100	~60	~15
SDH STM-16	2.488 Gbit/s	—	—	VC-4-17v	~94	~100	~25
SDH STM-64	9.953 Gbit/s	—	—	VC-4-68v	<100	—	~100

efficiency is reached even in the absence of efficient multiplexing and grooming in the upper layer.

The efficiency of POS, next generation SONET, and OTN are summarized in Table 3.7 [66] in different conditions.

3.4.1.2 Network Scalability: Core Network

The ability of a network to expand its capacity in pace with an increasing demand without the need of drastically changing its architecture is called network scalability.

There are essentially two ways of scaling a network: adding traffic on the same topology or adding nodes and/or transmission systems, thus modifying the topology [67].

From the scaling point of view, SONET-based architectures have a not so good performance. As a matter of fact, adding new traffic on existing links in SONET-based architectures can easily bring the need of installing new SONET equipment due to the saturation of the elements already present. This could be mandatory, for example, when a new wavelength is to be turned on.

As far as the addition of new topology elements is concerned, while the increase of traffic can cause the need of using unused fibers present in cables to deploy new transmission systems, the addition of a completely new core node is not a frequent event. Exception in this field are some Fast Eastern countries like China and India, where the development is so fast that the telecommunication network is also growing at a fast pace.

If a new transmission system has to be added to the network to relieve a condition of traffic saturation, it is important to take into account that next generation SONET is almost always deployed using multiring topology due to resilience and high control degree of this architecture. This is due both to the fact that implementing the new functionalities in ADMs is easier than implementing it in DXCs, and to the better survivability of a multiring network where independent protection is present on each ring and shared protection schemes can be adopted.

Moreover, integrating the next generation functionalities in such complex machines like a DXC is quite difficult and currently, the next generation DXCs are not diffused.

Adding a new link in a multiring architecture is difficult and this is another issue if a SONET-based network has to be scaled [67].

As far as the Ethernet-based architecture is concerned, we have already mentioned the discussion that exists around carrier Ethernet scalability. In this case, it would be the spanning tree algorithm to limit the scalability and the presence of switches that segment the network in separated spanning tree domains provides only a partial solution to this problem.

For sure, a better scalability is realized via MPLS based architectures. In MPLS over OTN architecture, if the OTN layer is realized via a multiring topology, all the difficulties related to rings exist as in the SONET case. The OTN layer however is more probably realized via OXC using a mesh topology. In this case, assuming that OXCs are scalable, the overall architecture is quite scalable both at the OTN layer and at the MPLS layer.

Even more scalable is the OTN over DWDM architecture, where only the transmission systems and the LER and LSR have to be scaled either when the traffic increases or when new topology elements are added.

This last operation is particularly simple if no switching/routing machine is installed at the physical layer. As a matter of fact, if either SONET or OTN are present, new network elements have to be added to the network manually not only at the hardware level, where installation of new equipment is needed, but also at the management system level; as a matter of fact, TMN has to be configured manually to recognize the new network configuration.

On the other hand, MPLS has the automatic network discovery feature that makes the network aware of the presence of new hardware even without a manual intervention on the central management system.

3.4.1.3 Network Scalability: Metro Network

If a metro network is considered instead of a core one, the issues related to network scalability change a bit.

The topology of the network in a highly populated metro area is practically standard: a two-level hierarchical ring topology that is represented in Figure 3.8.

In this context, scaling the network has two possible meanings: increasing traffic and adding a new access ring having as a hub an already existing metro core node. Apart from exceptional situations like catastrophes or an abrupt development of the metro area as that happened in some area of China like Shenzhen in the last 10 years, the addition of a new metro core node is a quite improbable event.

Adding a new access ring, provided that the hub node is not saturated and that fibers are present in the field, is quite easy. It requires the individuation of the central offices in which to install the access equipment and the deployment of new metro access systems; after that the network has to be configured either through the management system (in case of next generation SONET) or through the control plane (in case of carrier class Ethernet).

For this reason, many scaling issues that are present in a core network are not applicable in the metro area, and both next generation SONET and carrier class Ethernet seems more than suitable to be applied in this area also if scaling is considered a main requirement.

On the other hand, the fact that both next generation SONET and carrier class Ethernet present a direct Ethernet interface is quite important.

As a matter of fact, another scaling operation that can be done in the metro area (and has less meaning in the core network) is the move of some service sources from core nodes to metro nodes.

This network evolution can happen for services that requires a great number of content servers, like VoD. At the beginning of the service penetration, it is possible that the servers are placed in core nodes where, for example, the BRAS are placed.

However, if the service has really a great diffusion, the number of servers becomes huge and their installation in an area nearer to the end customer becomes advantageous.

Since almost all the content servers have Ethernet interfaces, having native Ethernet interfaces in the metro transport equipment facilitates a lot the installation of content servers in this area.

3.4.1.4 Network Survivability

All the proposed architectures implement fast protection at the physical layer. If no physical layer switching is possible since LSR and LER are directly connected via DWDM systems, only link-by-link protection will be possible, either 1 + 1 or 1:1 if low-priority traffic exists.

This kind of protection requires a duplication of all the DWDM systems with quite an increase in the cost of the transmission infrastructure.

Protection at the physical layer can be eliminated, relying only on MPLS fast restoration to prevent service interruption when a failure occurs. MPLS can contemporarily manage fast protection (i.e., realized by provisioning two disjoined tunnels when answering to the request of a new connection) and restoration (realized considering the tunnels hit by the failure as it were new connection requests and routing them through the network from which the failed element has been eliminated) [67,68].

When SONET or OTN exists at the physical layer, they allow both protection and restoration (at least in the case of mesh topology) to be implemented at the physical layer.

The physical layer and the layer 2 survivability mechanisms have of course to be coordinated in order to avoid either network instability or network freezing when a failure occurs [34], but can be done simply with hold off times that delay the intervention of layer 2 survivability mechanisms up to a moment when layer one procedures are surely ended.

In this way, layer 2 survivability becomes a backup with respect to layer 1 protection and restoration. Similar mechanisms will be analyzed in Chapter 8 when dealing with multilayer convergent network.

When dealing with a network survivability strategy, it is needed to consider the overprovisioning that is required to realize it.

As a matter of fact, implementing multilayer survivability causes an increase of the overprovisioned bandwidth due to the fact that the protection capacity in the upper layer is routed as high-quality client data in the lower layer since it is necessary to assure its integrity if the layer above switches in protection.

This situation is only partially mitigated by the adoption at the physical layer of shared protection rings, where the number of working channel is generally greater than the potential number of protection due to the sharing of protection capability.

In Table 3.8, the capacity over-provisioning is reported for different architectures and different protection schemes. Concerning MPLS over-provisioning, we have used data reported in Refs. [34,68,69]. Obviously the numerical values of the table have to be considered indicative and their value depends, among the other parameters, on the specific algorithms used to dimension the restoration paths both at the physical layer and at layer 2.

Moreover, the cases that are compared in the table are not characterized by the same network availability: it is quite intuitive that a mesh network with multilayer restoration and

TABLE 3.8

Qualitative Comparison among Different IP Convergent Network Architectures

Layered Architecture	Physical Layer Topology	Physical Layer		Layer 2		Global Over-Provisioning (%)
		Protection	Restoration	Protection	Restoration	
POS	Mesh	1+1	Yes	No	No	56
POS	Rings	Shared	No	No	No	50
Next Gen SONET	Mesh	1+1	Yes	No	No	56
Next Gen SONET	Rings	Shared	No	No	No	50
MPLS over DWDM	Mesh	No	Yes	No	Yes	39
MPLS over DWDM	—	1+1 (link based)	No	No	Yes	61
MPLS over DWDM	Mesh	—	No	Yes	Yes	53
MPLS over OTN	Mesh	1+1	No	No	Yes	61
MPLS over OTN	Rings	Shared	No	No	Yes	61

physical layer dedicated protection has higher availability with respect to mesh network built with the MPLS over DWDM technology where only MPLS restoration is implemented.

Also this aspect will be analyzed in a quantitative way in Chapter 8.

From the table, the effect of multilayer survivability emerges, for example, comparing line one with line six. In both cases, protection and restoration are both present, but in the first case they are at the same layer, in the second on different layers. The effect is that, while in the POS case the network manager of the physical layer can coordinate the two strategies, for example, avoiding 1+1 protection for restoration dedicated paths, in the second case this is not possible since restoration is on the MPLS layer while protection is on the physical one.

Due to this, when the capacity allocation proposals pass from the MPLS layer to the OTN layer, OTN cannot distinguish restoration paths from working paths and provides dedicated protection for both.

This is a side effect of the absence of a common control plane and is also an important cause of inefficiency of the network.

3.4.2 Network Dimensioning and Cost Estimation

After a performance qualitative comparison, we will carry out in this section a more quantitative comparison based on network cost as defined at the beginning of Section 3.4.

Since we are dealing with multilayer networks, a synthesis algorithm for these networks is needed.

In general, carriers have their own dimensioning algorithms that are often technology specific; network optimization means decreasing CAPEX and OPEX, so it has a great importance for carriers.

For a general comparison as that we are targeting now, using computationally heavy algorithms specific for a particular technology would not be the right choice. Different algorithms can introduce different levels of errors in the final solution and comparing results coming from network simulations adopting different algorithms is always complex requiring a preliminary assessment of the algorithms performance working on cases whose solution is known.

Thus, we will base our analysis on a single simulator sufficiently rich to simulate with a good accuracy complex multilayer networks.

The principles of the multilayer simulator working are detailed in Ref. [70].

The algorithm is based on the bare definition of a multilayer network, which is on the client server relationship between adjacent layers.

In particular, the highest layer is assumed to receive a traffic demand from its client, the algorithm determines the resources that this layer needs to carry out to satisfy these requests in terms of needed switching/routing hardware and transmission capacity.

Since the traffic request will be accompanied by a QoS requirement, it has to be taken into account in dimensioning the layer. Moreover, every layer specific equipment has to be represented by a detailed physical model that, among other things, represents the scalability of the equipment via the addition of new cards or similar procedures.

While switching/routing requests drive the design of the specific layer equipment, the traffic requests are passed at the lower layer that see them as client requests and start again the dimensioning.

The iterative procedure stops at the bottom layer, where transmission resources drive the design of the transmission WDM systems.

In order to design the adjacent layer, the layer topology has to be known (nodes cannot be added or subtracted during simulation).

When the physical layer is SDH/SONET or OTN, it can be designed either as a mesh or as a multiring network.

In the mesh case, there is no topology problem, since the mesh topology at the physical level is simply given from the set of nodes and WDM links that are deployed.

For a mesh network, in the case of static traffic, the optimization algorithm that is used to design the layer is based on integer linear programming (ILP). This method has been widely used in several network optimization problems [69,71] and results to be a very good compromise between effective optimization and computational complexity. We will return on that in Chapter 9 where we will describe in detail a dimensioning method for Automatically Switched Optical Networks/Generalized Multi-Protocol Label Switching (ASON/GMPLS) networks based on ILP.

Moreover, widely proved libraries exist for ILP optimization so that software development is also less prone to unresolved bugs.

In order to set up the network with dynamic traffic, first we dimension the network as if the traffic would be static using a traffic forecast derived from the dynamic traffic model. In order to route the capacity requests while they are dynamically presented to the considered layer, we have used a simple derivate of the Open Shortest Path First (OSPF) algorithm. In particular, the weight of a path is set equal to the cost of the resources used for routing the trail along that path. Periodically, a network optimization is carried out on the working network simulating a sort of centralized traffic engineering.

As a matter of fact, also in this simulated case, the continuous use of OSPF risks taking the network further and further from the optimum, due to the fact that local and not global optimization is performed.

Traffic engineering is carried out with an ILP algorithm very similar to that used for static traffic. The main modification is the addition of another weight to the value of the final solution, that is, the number of trails to reroute with respect to the field situation; this is done so to select, among those solutions with a similar cost, the solution rerouting the smallest possible number of trails.

In the case of multiring network, the first step is to optimally identify the rings starting from the given nodes and links. It is done only one time both in the case of dynamic and static traffic with the optimization algorithm reported in Ref. [67].

Once the rings are identified, the traffic routing is performed with the same algorithms used for the mesh, but on the ring topology.

As far as survivability is concerned, in the static traffic case and in the preliminary dimensioning of the dynamic case, we suppose that restoration is present at SDH/SONET when using DXCs, at MPLS layers and at the DWDM layer only if OXCs are used.

If a multiring SONET/SDH or OTN is considered, no restoration is present in the physical layer [72].

The coexistence of multilayer restoration, for example, in the MPLS over mesh OTN model, is simply managed via hold-off times.

In the case of a mesh network, fast protection is implemented at the physical layer for real-time traffic, while best effort traffic relies only on restoration. The protection is 1:1 MSP in case of SDH/SONET or 1:1 OCh in case of OTN.

This is not the better solution especially for a mesh architecture; however, it is not the scope of this study to optimize protection resources, it is enough not to insert a bias favoring one or the other solution and we achieve this goal by using for all architectures the same protection scheme.

To dimension the spare capacity needed in a layer for restoration we use a simple algorithm reported in Ref. [67].

3.4.3 Test Networks and Traffic Model

We consider as a test the network topology reported in Figure 3.2. In order to have sensitivity to the impact of network topology on our results, we will repeat a few runs using also the topologies reported in Figures 3.55 and 3.56. The first is the core German network as reported in Ref. [70], the second, a proposal for the IP-based core Italian network.

The architectures we are going to compare are as follows:

- Packet over SONET using the full GFP in a mesh network
- Packet over SONET using the full GFP in a ring network
- No switching below the MPLS router, routers are connected to the other routers with bare open WDM systems
- MPLS over OTN: physical mesh with electrical core OXCs
- MPLS over OTN: physical multiring 1:1 OChDPring

As far as the traffic is concerned, since the IP layer is not simulated, the higher layer that is dimensioned in the program is SDH/SONET or MPLS.

In both the cases, the traffic is represented via requests of connectivity.

In the static case, the requested traffic is organized in a traffic matrix, in the dynamic case, the requests of connectivity are generated through a mechanism in the IP layer that is not reproduced in the model. As far as layer 2 is concerned, the requests present themselves as a Poison process.

In both the cases, two QoS classes are considered: real-time services that have limitation in the maximum delay and need fast protection and best effort services that have no maximum delay requirement and face network failures only via restoration.

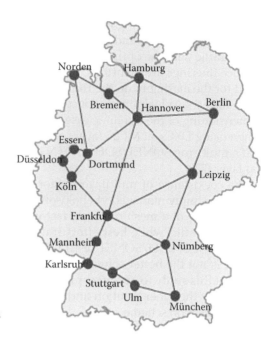

FIGURE 3.55
Topology of the German long haul network used in
some examples of network cost evaluation.

In all the cases that we are going to show here, the two classes of QoS have the same
number of trails requested from the IP.

3.4.4 Cost Comparison

Starting with static traffic, it is important to study first of all the behavior of the total
CAPEX needed to deploy the network.

In Figure 3.57, the total CAPEX is shown versus the traffic for the three solutions with a
mesh physical architecture.

Both the traffic intensity and the CAPEX are relative. Moreover, let us remember that the
CAPEX is calculated as costs for the vendors; thus, it does not include the vendor margin.

The solution interfacing directly the LSR and LER with the DWDM systems is the more
expensive, as it was easy to imagine, since for a fixed capacity, an MPLS router is a much
more complex and thus a more expensive machine with respect to an OXC that simply
switches wavelengths and also with respect to a DXC, at least as far as the traffic is not
huge.

Thus, using smaller LERs and smaller and perhaps less LSR by managing the pass-
through traffic through a physical layer switch like a DXC or an OXC is generally
advantageous.

Naturally, this is also due to the fact that in our model the router has optical grey interfaces (i.e., interfaces that are not suitable to enter directly into a DWDM system due to the
poor wavelength stability) and it is connected to the DWDM systems via transponders (see
Chapter 5). Directly mounting the DWDM interfaces in the router chassis for sure causes
a CAPEX decrease, but not as much as expected. We do not carry out in detail this discussion here since the router and the DWDM architectures have to be detailed; we will go
through this point in Chapter 8.

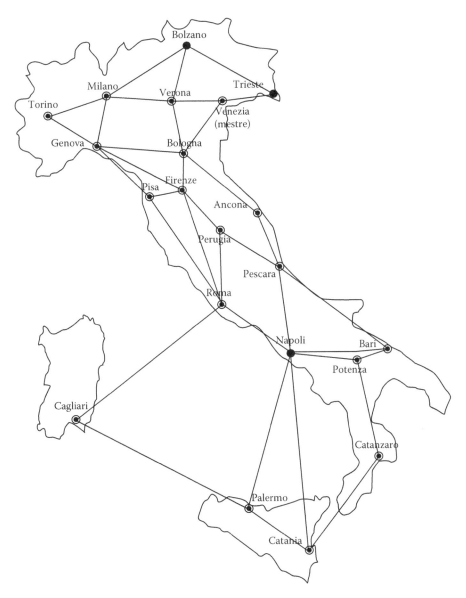

FIGURE 3.56
Topology of the Italian long haul network used in some examples of network cost evaluation.

For a low traffic volume, SDH mesh is advantageous with respect to OTN mesh. This result also is quite intuitive: if the traffic to process is not huge, the DXC core does not cost so much more with respect to an OXC core. On the other hand, for low traffic, the DXC is much more effective due to the ability to perform grooming at much lower capacity (the OXC operates on the wavelengths as a whole). Increasing traffic changes the situation, and the OTN version of the network gains more and more advantages with the need of the DXC to groom at greater and greater capacity level, so as not to grow to an impractical dimension.

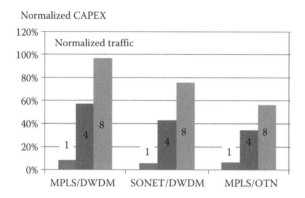

FIGURE 3.57
Normalized cost of three network architectures versus the normalized traffic level. The architectures assume a
mesh physical topology. The normalized traffic, evaluated considering only 10 Gbit/s channels in the network,
corresponds at the physical layer to an average number of wavelength per DWDM link equal to 4 (traffic = 1), 16
(traffic = 4), and 32 (traffic = 8), respectively. The corresponding average DWDM utilizations are 6% (40 Gbit/s),
25% (160 Gbit/s), and 50% (320 Gbit/s), respectively.

In the limit of very high traffic, the DXC manages paths that are concatenated 10 Gbit/s
so that grooming is no more possible and the DXC architecture is useless.

The partition of the CAPEX among the different network elements is also interesting
and it is reported in Figure 3.58 for the same architectures of Figure 3.57.

The first factor that results is the huge CAPEX that is used for pure transmission (fiber
cables, optical amplifiers, and regenerators essentially). This is due to the great distances
covered by the considered network that spans all North America: several links are near or
beyond 1000 km and the transmission has to be suitably assisted. If a similar evaluation is
done with the Italian network and the German one, the contribution of the pure transmission goes down to below 40%.

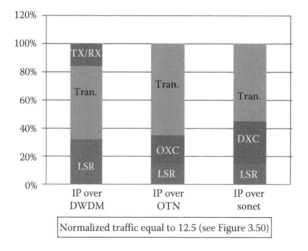

FIGURE 3.58
Relative cost of the different network elements in the same cases of Figure 3.50. Tran, DWDM transmission
systems; TX/RX, DWDM transponders; LSR, MPLS routers (both LR and LER are included).

Looking at the cost of the routers (indicated as LSR, but also LERs are included), it is obviously more important in the case of the absence of a real intelligent physical layer, while it is substituted partially by the cost of OXCs and DXCs in the case in which these equipment are present.

In order to compare networks with mesh and the multiring physical architecture, some topology parameters can be usefully introduced. The different topologies can be classified on the ground of these parameters to group topologies that are similar from some point of view. This approach follows completely [67].

A first possible parameter is the network connectivity α that is defined as follows:

$$\alpha = \frac{2M}{N(N-1)} \tag{3.1}$$

where
  M is the number of network links
  N is the number of nodes

This is a well-known parameter, obtained from the ratio between the number of links of the considered network and the number of links for a full meshed network with the same number of nodes.

Networks with the same connectivity are really "topologically similar" if they have almost the same number of nodes. However, the connectivity is not useful to compare networks with quite different number of nodes.

To introduce a network classification that is independent from the number of nodes, a new topological parameter has been introduced: the average connection degree X_m. The connection degree is defined as follows:

$$X_m = \frac{1}{\binom{N}{2}} \sum_{k=1}^{N} \sum_{j=1}^{N} x_{k,j} \tag{3.2}$$

where $x_{k,j}$ is the number of disjoined paths between the kth and the jth node.

Thus, X_m represents the average number of disjoint paths between two network nodes.

The connection degree is always greater than 2 in the considered networks, due to the fact that the minimum number of links cuts that divide the network into two unconnected parts is two. Moreover, in realistic networks, it is normally smaller than 2.5. For this reason, it is convenient to introduce a normalized connection degree $C_m = 100 \, (X_m - 2)$.

In Figure 3.59, the normalized CAPEX is reported versus the normalized connection degree for the different architectures at a normalized traffic level of 12.5.

The traffic level is quite high, reporting it to normal measuring units and looking at the network dimensioning it means that an average occupation of the DWDM link equals to about 50 wavelength at 10 Gbit/s per link. The fact that the solutions based on SONET are not quite competitive is due to this high traffic level; at a lower traffic the difference decreases up to a threshold. Below the threshold, IP over SONET (using GFP in our hypothesis) is cheaper with respect to IP over OTN (compare with Figure 3.56).

Comparing the most effective mesh and ring solutions, the presence of a threshold connectivity degree (C_m^*) can be noted. Above C_m^*, the mesh solution is convenient, while the multiring solution is more effective below C_m^*. This is due to the fact that the mesh

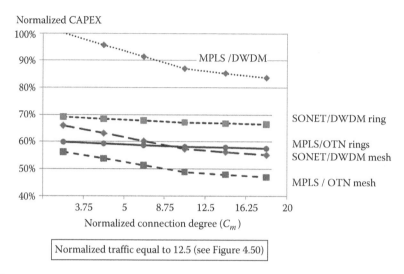

FIGURE 3.59

Normalized network cost versus C_m for the different architectures at a normalized traffic level of 12.5 (average number of wavelengths per fiber link 50, network load 78%, about 10% saturated links, with 64 channels out of a maximum of 64).

architecture receives better advantage from a well-connected network due to its capability to exploit all the possible routes. On the other hand, for low connectivity networks, the fact that the node equipment is less expensive for a multiring network is the most important element. This interpretation can be also confirmed by comparing the curves slope when increasing the network connection degree. In particular, the curves related to a mesh network decrease more rapidly increasing C_m for $C_m > 5$.

All the aforementioned results are evaluated starting from the North American network. It is interesting to evaluate the impact of the network topology on the results.

When comparing results relative to different networks we have to take into account the different impacts due to transmission in networks covering different geographical areas. For example, focusing on the MPLS over OTN architecture, the average length of the transmission links in the Italian network, Figure 3.56, is only 14% of the average length of links in the U.S. network, while the German network is in between with a relative average link length of 23%. Due to this difference, transmission cost is about 70% of the overall cost for the U.S. network, about 40% for the German one, and about 29% for the Italian one.

In order to make a meaningful comparison, we will normalize the geographical dimension of the network so that all the considered networks will have an average link length equal to the German one.

Considering C_m, the Italian and the German networks are quite meshed, having $C_m = 27$ and $C_m > 30$, respectively.

In Figure 3.60, the normalized network costs are reported for the three normalized networks with normalized traffic equal to 4 and in the presence of a mesh physical layer. It is evident that a topology dependence exists but it is not so important to cancel the general validity of the results.

A similar conclusion can be derived from Figure 3.61, where the network cost is evaluated versus C_m for an IP over next generation SONET architecture with a multiring and a mesh physical layers. The results are relative to a set of randomly generated networks with 12 nodes and the same average link length; circles refer to the ring topology, square to the

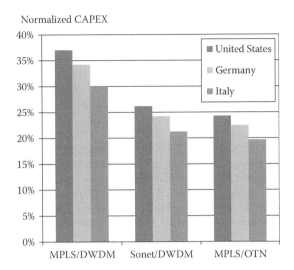

FIGURE 3.60
Normalized network cost for United States, Italian, and German normalized networks; normalized traffic equal to 4; mesh physical layer.

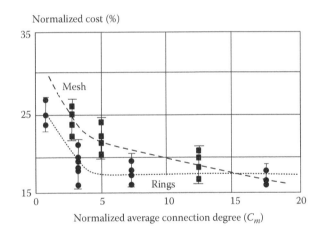

FIGURE 3.61
Normalized network cost versus C_m for an IP over next generation SONET with a multiring and mesh physical layers. Random networks: 12 nodes and the same average link length, circles refers to the ring topology, squares to the mesh one; normalized traffic equal to 3 (average 12 channels at 10 Gbit/s for DWDM link).

mesh one, and the normalized traffic is 3 (low traffic region, justifying the convenience of the ring topology at the physical layer).

References

1. Building Industry Consulting Service International (BICSI), *Telecommunications Cabling Installation*, 2nd edn., McGraw-Hill Professional, New York, s.l., 2002, ISBN-13: 978–0071409797.

2. Shimizu, M., Next-generation Optical Fiber Cable System Technology for FTTH Expansion Period. Reported on NTT Review *NTT Forum—Workshop Lecture 3*, NTT, Tsukuba, Japan, 2008, https://www.ntt-review.jp/archive/ntttechnical.php?contents=ntr200906sf3.html

3. Takai, H., Yamauchia, O., Optical fiber cable and wiring techniques for fiber to the home, *Optical Fiber Technology*, 15(4), 380–387 (2009), http://www.sciencedirect.com/science?_ob=ArticleURL&_udi=B6WP0-4W9S2XD-1&_user=10&_coverDate=08/31/2009&_rdoc=1&_fmt=high&_orig=search&_sort=d&_docanchor=&view=c&_acct=C000050221&_version=1&_urlVersion=0&_userid=10&md5=e223fcf06a3dc8ff18f35

4. Berger, M. et al. (from OPEN and PHOTON EU project teams), Pan-European optical networking using wavelength division multiplexing, *Communication Magazine, IEEE*, 35(4), 82–88 (April 2007), https://docs.google.com/viewer?url=http://www.ibcn.intec.ugent.be/papers/537.pdf

5. Aidarous S., Pleviak T., (eds.,) Quality of service in IP networks. In *Managing IP Networks: Challenges and Opportunities*, IEEE Press, Piscataway, NJ, s.l., 2003.

6. Iannone, E., Next generation wired access: Architectures and technologies. In K. Iniewski (Ed.) *Internet Networks: Wired, Wireless, and Optical Technologies (Devices, Circuits, and Systems)*, Taylor & Francis, Boca Raton, FL, s.l., 2009.

7. Golden, P., Dedieu, H., and Jacobsen, K. S., *Implementation and Applications of DSL Technology*, Auerbach, Boca Raton, FL, s.l., 2007, ISBN-13: 978-0849334238.

8. Bentivoglio, F., Iannone, E., New architectures and applications for integrated packet and optical routing equipment in metro area networks. *Optical Fiber Conference (OFC) Proceedings*, Los Angeles, CA, s.n., February 26, 2006.

9. Bentivoglio, F., Iannone, E., Cisco Systems Inc., Global Pricing List [Online], www.cisco.com (accessed: March 19, 2010).

10. Zhou, D., Subramaniam, S., Survivability in optical networks, *IEEE Networks*, 14, 16–23 (December 2000), https://docs.google.com/viewer?url=http:/users.encs.concordia.ca/~assi/courses/encs6811/surv1.pdf

11. Iannone, E., Solutions for the metro bottleneck: Evolution or revolution? *European Conference on Optical Communications (ECOC)—Market Forum*, Cannes, France, s.n., September 24–28, 2006.

12. Perros, H. G., *Connection-Oriented Networks: SONET/SDH, ATM, MPLS and Optical Networks*, Wiley, Chichester, England, s.l., 2005, ISBN-13: 978-0470021637.

13. ITU-T, X-Series Recommendations (accessed: June 30, 2011).

14. ITU-T, *Recommendation X.224*, November 1995, also ISO/IEC 8073.

15. ITU-T, *Recommendation Q.1400*, March 1993 (accessed: June 30, 2011).

16. Curtis, B., Krasner, H., Iscoe, N., A field study of the software design process for large systems, *Communications of the ACM*, 31(11), 1268–1287 (1988).

17. Davalo, E., Naim, P., Rawsthorne, A., *Neural Networks*, Macmillan (Computer Science Series), New York, s.l., 1991, ISBN-13: 978-0333549964.

18. Stergiou, C., Dimitrios, S., Neural networks. *Cambridge University Reports* [Online], November 20, 2004. http://www.doc.ic.ac.uk/~nd/surprise_96/journal/vol4/cs11/report.html (accessed: March 18, 2010).

19. IETF basic recommendations about IP suite.
   IP; Internet Protocol; RFC 791; STD 5
   ICMP; Internet Control Message Protocol; 792; STD 5
   ICMP; Broadcasting Internet Datagrams; 919; STD 5
   ICMP; Broadcasting Internet Datagrams in the presence of subnets; 922; STD 5
   ICMP; Internet Standard Subnetting Procedure; RFC 950; STD 5
   IGMP; Host extensions for IP multicasting; RFC 1112; STD 5
   UDP; User Datagram Protocol; RFC 768; STD 6
   TCP; Transmission Control Protocol; RFC 793; STD 7
   TELNET; Telnet Protocol Specification; RFC 854; STD 8
   TELNET; Telnet Option Specifications; RFC 855; STD 8
   FTP; File Transfer Protocol; RFC STD 9 (accessed: June 30, 2011).

20. Comer, D., *Internetworking with TCP/IP: Principles, Protocols, and Architecture*, Pearson Prentice Hall, Upper Saddle River, NJ, s.l., 2006, ISBN 0131876716.
21. Goralski, W., *The Illustrated Network: How TCP/IP Works in a Modern Network*, The Morgan Kaufmann Series in Networking, Burlington, MA, s.l., 2008, ISBN-13: 978-0123745415.
22. Davies, G., *Designing and Developing Scalable IP Networks*, Wiley, Chichester, England, s.l., 2004, ISBN-13: 978-0470867396.
23. Doyle, J., Carroll, J., *Routing TCP/IP*, Cisco Press, Indianapolis, IN, Vol. 1, 2005, ISBN-13: 978-1587052026.
24. Aweya, J., On the design of IP routers. Part 1: Router architectures, *Journal of Systems Architecture*, 46, 483–511 (2000).
25. Aweya, J., IPv4 Address exhaustion, mitigation strategies and implications for the US, a IEEE-UA White Paper, Editions IEEE, 2009.
26. 6net, *IPv6 Deployment Guide*, Javvin Press, Saratoga, CA, 2008, ISBN-13: 978-1602670051.
27. Spurgeon, C. E., *Ethernet: The Definitive Guide*, O'Reilly Media, Sebastopol, CA, s.l., 2000, ISBN-13: 978-15659266086.
28. IEEE basic recommendations defining the Ethernet standard
     802.3 (1985) Base standard (10B5)
     802.3a (1992) 10B2 Ethernet over thin coaxial cable
     802.3b (1985) Broadband Ethernet (using coaxial TV cable, now seldom used)
     802.3c (1985) Improved Definition of a Repeater
     802.3d (1987) Definition of Ethernet for Fibre (10BFOIRL) (now seldom used)
     802.3e (1987) 1Base5 or StarLAN (now seldom used)
     802.3h (1991) Layer Management
     802.3i (1990) 10BaseT, Ethernet over CAT-5 Unshielded Twisted Pair (UTP)
     802.3j (1993) defines Ethernet over Fibre (10BF)
     802.3p/q (1993) Definition of Managed Objects
     802.3u (1995) Definition of Fast Ethernet (100BTX, 100BT4, 100BFX)
     802.3x (1998) Definition of Full Duplex Operation in a Switched LAN
     802.3y (1998) Definition of Fast Ethernet (100BT2 over low-quality UTP)
     802.3z Definition of Gigabit Ethernet (over Fibre)
     802.3aa Definition of Gigabit Ethernet Maintenance
     802.3ab Definition of Gigabit Ethernet (over UTP CAT-5)
     802.3ac Definition of Ethernet VLANs
     802.3ad Definition of Ethernet VLAN Trunking (accessed: June 30, 2011).
29. IEEE /ISO CSMA/CD Standard, ISO/IEEE 802/3 (this integrates also old IEC TC83).
30. Seifert, R., Edwards, J., *The All-New Switch Book: The Complete Guide to LAN Switching Technology*, Wiley, New York, s.l., 2008, ISBN-13: 978-0470287156.
31. Raymond, X., Ethernet optical transport network. In K. Iniewski (Ed.) *Internet Networks: Wired, Wireless, and Optical Technologies (Devices, Circuits, and Systems)*, Taylor & Francis, Boca Raton, FL, s.l., 2009.
32. IEEE, MEF and ITU-T main standards regarding Carrier Class Ethernet
     Operation Administration and Management—IEEE M 802.3ah
     Connectivity Fault—IEEE 802.1ag
     Management—IEEE 802.1ag
     Grooming and Aggregation—IEEE 802.1ad
     Grooming and Aggregation—IEEE 802.1ah
     Discovery and Registration—IEEE 802.1ab
     Discovery and Registration—IEEE 802.1ac
     Discovery and Registration—IEEE 802.1ak
     Routing—IEEE 802.1aq
     Security—IEEE 802.1ar
     Demarcation—IEEE 802.1aj
     MEF 1 Ethernet Services Model—Phase 1 (obsoleted by MEF 10)

MEF 2 Requirements and Framework for Ethernet Service Protection
MEF 3 Circuit Emulation Requirements
MEF 4 MEN Architecture Framework—Part 1: Generic Framework
MEF 5 Traffic Management Specification—Phase 1 (obsoleted by MEF 10)
MEF 6 Metro Ethernet Services Definitions—Phase I
MEF 7 EMS-NMS Information Model
MEF 8 PDH over MEN Implementation Agreement (CESoETH)
MEF 9 Abstract Test Suite for Ethernet Services at the UNI
MEF 10 Ethernet Services Attributes—Phase I
MEF 11 User Network Interface (UNI) Requirements and Framework
MEF 12 MAN Architecture Framework—Part 2: Ethernet Services Layer
MEF 13 User Network Interface (UNI) Type 1 Implementation Agreement
MEF 14 Abstract Test Suite for Ethernet Services at the UNI
MEF 15 MEN Management Requirements—Phase 1 Network Elements
MEF 16 Ethernet Local Management Interface
MEF 17 Service OAM Framework and Requirements
MEF 18 Abstract Test Suite for Circuit Emulation Services
MEF 19 Abstract Test Suite for UNI Type 1
MEF 20 UNI Type 2 Implementation Agreement
MEF 21 Abstract Test Suite for UNI Type 2—Part 1 Link OAM
MEF 22 Mobile Backhaul Implementation Agreement
MEF 23 Class of Service Phase 1 Implementation Agreement
MEF 24 Abstract Test Suite for UNI Type 2—Part 2 E-LMI
Common Language ITU-T G.8001/Y.1354
EoT Definitions—ITU-T G.8001
Ethernet layer network architecture—ITU-T G.8010
Ethernet over Transport services framework—ITU-T G.8011
Ethernet private line service—ITU-T G.8011.1
Ethernet virtual private line service—ITU-T G.8011.2
Ethernet UNI and NNI—ITU-T G.8012
Ethernet transport equipment characteristics—ITU-T G.8021
Ethernet linear protection switching—ITU-T G.8031
Ethernet ring protection switching—ITU-T G.8032
Ethernet OAM—requirements—ITU-T Y.1730
Ethernet OAM—ITU-T Y.1731 (accessed: June 30, 2011).

33. Autenrieth, A. et al., Carrier grade metro Ethernet networks, Report of the fact group 5.3.3 on optical networks of ITC, https://docs.google.com/viewer?url=http://www.lkn.ei.tum. de/~akirstaedter/papers/2007_ITG_Eth.pdf
34. De Ghein, L., *MPLS Fundamentals*, Cisco Press, Indianapolis, IN, 2006, ISBN-13: 978-1587051975.
35. Main IETF recommendations defining MPLS
Multiprotocol Label Switching Architecture RFC 3031
MPLS Label Stack Encoding RFC 3032, updated by RFC 3443, 4182
LDP Specification RFC 5036, obsoletes RFC 3036
Carrying Label Information in BGP-4 RFC 3107
MPLS Support of Differentiated Services RFC 3270
RSVP-TE: Extensions to RSVP for LSP Tunnels RFC 3209, updated by RFC 3936, 4420, 4874
Fast Reroute Extensions to RSVP-TE for LSP Tunnels RFC 4090
BGP/MPLS IP Virtual Private Networks RFC 4364 (accessed: June 30, 2011).
36. Balakrishnan, R., *Advanced QoS for Multi-Service IP/MPLS Networks*, Wiley, New York, 2008, ISBN-13: 978-0470293690.
37. Zhang, R., Bartell, M., *BGP Design and Implementation*, Cisco Press, Indianapolis, IN, 2003, ISBN-13: 978-1587051098.

38. Black, U. N., *Traffic Engineering in MPLS Networks*, Prentice Hall, Upper Saddle River, NJ, 2006, ISBN-13: 978-0130358158.

39. Vasseur, J. P., Pickavet, M., Demeester, P., *Network Recovery: Protection and Restoration of Optical, SONET-SDH, IP, and MPLS*, Morgan Kaufmann Series in Networking, San Francisco, CA, s.l., 2004, ISBN-13: 978-0127150512.

40. Xu, Z., *Designing and Implementing IP/MPLS-Based Ethernet Layer 2 VPN Services: An Advanced Guide for VPLS and VLL*, Wiley, New York, 2009, ISBN-13: 978-0470456569.

41. Goralski, W., *Sonet/SDH Third Edition*, McGraw-Hill/OsborneMedia, New York, 2002, ISBN-13: 978-0072225242.

42. Main ITU-T Recommendations defining SDH
    Network Node Interface for the Synchronous Digital Hierarchy (SDH): ITU-T G.707
    Structure of Recommendations on Equipment for the Synchronous Digital Hierarchy (SDH): ITU-T G.781
    Types and Characteristics of Synchronous Digital Hierarchy (SDH) Equipment: ITU-T G.782
    Characteristics of Synchronous Digital Hierarchy (SDH) Equipment Functional Blocks ITU-T: G.783
    Architecture of Transport Networks Based on the Synchronous Digital Hierarchy (SDH): ITU-T G.803 (accessed: June 30, 2011).

43. Main ANSI Recommendations defining SONET
    SONET: Basic Description including Multiplex Structure, Rates and Format: s ANSI T1.105
    SONET: Automatic Protection Switching: ANSI T1.105.01
    SONET: Payload Mappings: ANSI T1.105.02
    SONET: Jitter at Network Interfaces: ANSI T1.105.03
    SONET: Jitter at Network Interfaces—DS1 Supplement: ANSI T1.105.03a
    SONET: Jitter at Network Interfaces—DS3 Wander Supplement: ANSI T1.105.03b
    SONET: Data Communication Channel Protocol and Architectures: ANSI T1.105.04
    SONET: Tandem Connection Maintenance: ANSI T1.105.05
    SONET: Physical Layer Specifications: ANSI T1.105.06
    SONET: Sub-STS-1 Interface Rates and Formats Specification: ANSI T1.105.07
    SONET: Network Element Timing and Synchronization: ANSI T1.105.09
    SONET: Operations, Administration, Maintenance, and Provisioning (OAM&P)—Communications: ANSI T1.119

44. Cisco, *Differences between SONET and SDH Framing in Optical Networks*, Cisco internal document ID 16180, last update and put in public domain November 2, 2006, https://docs.google.com/viewer?url=http://www.cisco.com/application/pdf/paws/16180/sonet_sdh.pdf

45. Baker, R. G. V., *Modeling Internet Traffic through Geographic Time and Space*, Springer, New York, 2010, ISBN-13: 978-9048189748.

46. Main ITU-T recommendations defining the OTN
    Optical Interfaces for Multichannel Systems with Optical Amplifiers: G.692
    Interface for Optical Transport Network (OTN): G.709
    Characteristics for the OTN Equipment Functional Blocks: G.798
    Framework for Reccommandations: G.871
    Architectures of the OTNs: G.872
    Management Aspects of OTN Elements: G.874
    OTN Management Information Model for the Network Element View: G.875
    Optical Interfaces for Equipments and Systems Related to SDH: G.957
    Optical Networking Physical Layer Interfaces: G959 (accessed: June 30, 2011).

47. Ramaswami, R., Sivarajan, K. N., Sasaki, G. H., *Optical Networks*, 3rd edn., Morgan Kaufmann, San Francisco, CA (Elsevier inprint), 2009, ISBN-13: 978-0123740922

48. Neilson, D. T., Frahm, R., Kolodner, P., Bolle, C. A., Ryf, R., Kim, J., Papazian, A. R. et al., 256×256 Port optical cross-connect subsystem, *IEEE Journal of Lightwave Technology*, 22(6), 1499 (2004).

49. Arijs, P., Meersman, R., Parys, W. V., Iannone, E., Tanzi, A., Pierpaoli, M., Bentivoglio, F., Demeester, P., Architecture and design of optical channel protected ring networks, *IEEE Journal of Lightwave Technology*, 19(1), 11 (2001).

50. Li, M. J., Soulliere, M. J., Tebben, D. J., Nederlof, L., Vaughn, M. D., Wagner, R. E., Transparent optical protection ring architectures and applications, *IEEE Journal of Lightwave Technology*, 23(10), 3388 (2005).

51. Main ITU-T recommendations defining the TMN
    Management Framework: X.700
    System Management Overview: X.701
    Common Management Information Service (CMIS) Definition: X.710
    Common Management Information Protocol (CMIP) Specification: X.711
    CMIP Protocol Implementation Conformance Statement (PICS): X.712
    Management Information Model: X.720
    Definition of Management Information: X.721
    Guidelines for the Definition of Managed Objects (GDMO): X.722
    Object Management Function: X.730
    State Management Function: X.731
    Attributes for Representing Relationships: X.732
    Alarm Reporting Function: X.733
    Event Report Management Function: X.734
    Log Control Function: X.735
    Security Alarm Reporting Function: X.736
    Summarization Function: X.738
    Workload Monitoring Function: X.739
    Security Audit Trail Function: X.740
    Test Management Function: X.745 (accessed: June 30, 2011).

52. Clemm, A., *Network Management Fundamentals*, Cisco Press, Indianapolis, IN, 2006, ISBN-13: 978-1587201370.

53. Plevyak, T., Sahin, V., *Next Generation Telecommunications Networks, Services, and Management*, IEEE Press Series on Network Management, IEEE Press, New York, 2010, ISBN-13: 978-0470575284.

54. Main ITU-T recommendations regarding ITU-T services
    Tutorial Introduction to TMN: M.3000
    Principles for a TMN: M.3010
    TMN Interface Specification Methodology: M.3020
    Generic Network Information Model for TMN: M.3100
    TMN Management Services Overview: M.3200
    TMN Management Capabilities at the F Interface: M.3300 (accessed: June 30, 2011).

55. Main ITU-T recommendations regarding Q3 and Qx interfaces
    Q.811 describes lower layer (1–3) protocols
    Q.812 describes higher layer (4–7) protocols suitable for Q3
    G.784 provides information about the protocol stack to be used for SDH-DCC

56. Main IETF recommendations regarding SMNP Network management
    (Standard 16)—Structure and Identification of Management Information for the TCP/IP-based Internets: RFC 1155
    (Historic)—Management Information Base for Network Management of TCP/IP-based internets: RFC 1156
    (Historic)—A Simple Network Management Protocol (SNMP): RFC 1157
    (Standard 17)—Management Information Base for Network Management of TCP/IP-based internets: MIB-II: RFC 1213
    (Informational)—Introduction and Applicability Statements for Internet Standard Management Framework: RFC 3410

(Standard 62)—An Architecture for Describing Simple Network Management Protocol (SNMP) Management Frameworks: RFC 3411

(Standard 62)—Message Processing and Dispatching for the Simple Network Management Protocol (SNMP): RFC 3412

(Standard 62)—Simple Network Management Protocol (SNMP) Application: RFC 3413

(Standard 62)—User-based Security Model (USM) for version 3 of the Simple Network Management Protocol (SNMPv3): RFC 3414

(Standard 62)—View-based Access Control Model (VACM) for the Simple Network Management Protocol (SNMP): RFC 3415

(Standard 62)—Version 2 of the Protocol Operations for the Simple Network Management Protocol (SNMP): RFC 3416

(Standard 62)—Transport Mappings for the Simple Network Management Protocol (SNMP): RFC 3417

(Standard 62)—Management Information Base (MIB) for the Simple Network Management Protocol (SNMP): RFC 3418

(Best Current Practice)—Coexistence between Version 1, Version 2, and Version 3 of the Internet-standard Network Management Framework: RFC 3584

(Proposed)—The Advanced Encryption Standard (AES) Cipher Algorithm in the SNMP User-based Security Model: RFC 3826

(Proposed)—Simple Network Management Protocol (SNMP) Context Engine ID Discovery: RFC 5343

(Proposed)—Transport Subsystem for the Simple Network Management Protocol (SNMP): RFC 5590

(Proposed)—Transport Security Model for the Simple Network Management Protocol (SNMP): RFC 5591

(Proposed)—Secure Shell Transport Model for the Simple Network Management Protocol (SNMP): RFC 5592

(Proposed)—Remote Authentication Dial-In User Service (RADIUS) Usage for Simple Network Management Protocol (SNMP) Transport Models: RFC 5608 (accessed: June 30, 2011).

57. Saperia, J., *SNMP at the Edge: Building Effective Service Management Systems*, McGraw Hill Professional, New York, 2002, ISBN-13: 978-0071396899.

58. Walsh, L., *SNMP MIB Handbook*, Wyndham Press, Stanwood, WA, 2008, ISBN-13: 978-0981492209.

59. Main recommendation defining POS:
    IETF: PPP over SONET, RCF 2615
    IETF: PPP in HDLC-like Framing, RFC 1662
    Bellcore: HDLC-over-SONET Mapping, GR-253, Issue 2, Revision 2, January 99
    ITU-T: SDH specification, G.707, G.783, and G.957

60. IETF recommendation defining PPP: RFC 1661 The Point-to-Point Protocol (PPP), RFC 1661

61. Freeman, R. L., *Reference Manual for Telecommunications Engineering*, Wiley, New York, 1993, ISBN-13: 978-0471579601.

62. Kartalopoulos, S. V., *Next Generation SONET/SDH: Voice and Data*, Wiley-IEEE Publishing, New York, 2004, ISBN: 978-0-471-61530-9.

63. Main ITU-T recommendation defining Next Generation SONT/SDH
    Generic Framing Procedure (GFP) G.7041
    Virtual Concatenation (VCAT) G.707/783
    Link Capacity Adjustment Scheme (LCAS) G.7042 (accessed: June 30, 2011).

64. Warren, D., Hartmann, D., *Cisco Self-Study: Building Cisco Metro Optical Networks*, Cisco Press, Indianapolis, IN, 2003, ISBN-13: 978-1-58705-070-1.

65. Dannhardt, M., *Ethernet over SONET*, Technology White Paper, PMC Sierra Inc., Sunnyvale, CA, 2002, PMC-202029669.

66. Berger, M. et al., *Optical Transport Networks (OTN) Technical Trends and Assessment*, ITG-positionspapier, VDE/ITG study group Photonic Networks, VDE Verband der Elektrotechnik, Frankfurt, 2006.
67. Binetti, S., Bragheri, A., Iannone, E., Bentivoglio, F., Mesh and multi-ring optical networks for long-haul applications, *IEEE Journal of Lightwave Technology*, 18(12), 1677 (2000).
68. Minei, I., Lucek, J., *MPLS-Enabled Applications: Emerging Developments and New Technologies*, Wiley, New York, 2008, ISBN-13: 978-0470986448.
69. Bigos, W., Cousin, B., Gosselin, S., Le Foll, M., Nakajima, H., Survivable MPLS over optical transport networks: Cost and resource usage analysis, *IEEE Journal of Selected Areas in Communications*, 18(5), 949–962 (2007).
70. Köhn, M., Bodamer, S., Gauger, C. M., Gunreben, S., Hu, G., Sass, D., Comparison of IP/WDM transport network architectures for dynamic data traffic, *Proceedings of ECOC 2006*, Berlin, Germany, September, 2006.
71. Tornatore, M., Maier, G., Pattavina, A., WDM network optimization by ILP based on source formulation, *Proceedings of IEEE INFOCOM*, New York, pp. 1813–1821, 2002.
72. Maier, G., Pattavina, A., De Patre, S., Martinelli, M., Optical network survivability: Protection techniques in the WDM layer, *Photonic Network Communications*, 4(3/4), 251–269 (2002).

4

Technology for Telecommunications: Optical Fibers, Amplifiers, and Passive Devices

4.1 Introduction

The ultimate role of the telecommunication network is to deliver services based on the network capability of transmitting the information generated in one place to faraway places.

Two basic functionalities are involved in this goal: transmission and switching.

Since the introduction of low-loss optical fibers and high-power semiconductor lasers, optics has been the prominent technology for transmission, even if microelectronics have also played a main role in the design of transmission systems.

The great success of optical fiber transmission is due to the coincidence of the low-loss wavelength region of optical fibers with the emission wavelength range of Indium Gallium Arsenide Phosphide (InGaAsP) semiconductor lasers.

Attenuations as low as 0.2 dB/km can be achieved with standard transmission fibers in the near infrared region (around 1.5 μm) where a single-mode semiconductor laser can emit a power as high as 16 dBm.

Using optical amplifiers to boost the signal power after a fiber propagation of 60–80 km, the signal can reach essentially unaltered distances of 1000 km.

Moreover, a large bandwidth exists around the minimum attenuation wavelength where the attenuation remains very low. Thus, several channels can be transmitted at different wavelengths, realizing wavelength division multiplexing (WDM) systems.

A commercial WDM system for long-haul applications can convey, just to give an order of magnitude, up to 150 channels at different wavelengths around 1.55 μm each of which transmits a signal at 10 Gbit/s up to a distance of 3000 km.

The main limitations to the reachable distance are the amplifiers' induced noise, single-mode propagation dispersion, and the nonlinear propagation effects.

Optical amplifiers work generally in a quantum noise limit regime; thus, there is no means of avoiding noise generation. Noise power accumulates at each amplification and at the end overcomes the signal.

Since the signal is modulated, different frequencies of the signal spectrum experience a different group velocity so that further signal distortion is generated.

Moreover, due to the small core area of fibers in which the main part of the optical power is confined, the transmission cannot be considered linear in several practical cases.

Nonlinear effects create signal distortions, coupling among signals at different wavelengths, and signal-noise coupling constituting another limitation to the reachable distance and transmission speed.

In this chapter, we will present a review of fiber optics technology in the two main applications that are relevant for telecommunication networks: transmission fibers and optical fiber amplifiers.

We will add to the bulk of the chapter devoted to fiber technology the review of a set of filtering and wavelength management devices that are of wide use in fiber-based telecommunication equipment.

4.2 Optical Fibers for Transmission

An optical fiber is a circular dielectric glass waveguide that has a wide number of applications in telecommunications, in sensor systems, in medicine, and so on.

Each application has different requirements and in this section we will only analyze optical fibers used in telecommunications.

4.2.1 Single-Mode Transmission Fibers

A single-mode fiber used for transmission is characterized, from the point of view of electromagnetism, by the condition of being a weakly guiding waveguide, that is, the refraction index variation along the fiber section radius is very small (generally less than 1%).

The fiber section of the simpler transmission fiber, the so-called step index fiber, is reported in Figure 4.1: it is constituted by a central cylindrical core with a refraction index n_c and a coaxial cladding with a refraction index n_e. Examples of indexes values used in telecommunications are $n_c = 1.4600$ and $n_e = 1.4592$, so that $\Delta n/n = 2(n_c - n_e)/(n_c + n_e) \approx 0.55\%$.

The field propagation in the fiber can be studied by solving the Maxwell equations in the structure [1,2].

Exploiting the weakly guiding approximation [3] it is possible to solve the propagation problem by individuating propagation modes that constitute a complete set for the description of the guided field. In long transmission systems, the optical power that is not guided by the fiber is irradiated in the first fiber segment; thus, a description of the guided field is sufficient.

The generic propagation mode in a perfectly cylindrical step index fiber in the absence of field sources can be described with the following expression of the electrical field:

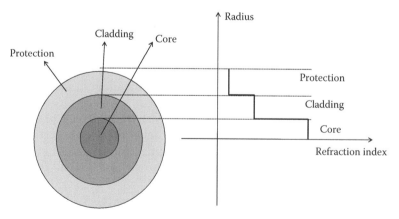

FIGURE 4.1
Section and refraction index distribution of an optical fiber.

$$\vec{E} = \vec{A}_{\zeta,l,m}(\rho)e^{i[\beta(\omega)z-\omega t]} \tag{4.1}$$

where
  ρ represents the radial coordinate in the fiber section
  $\beta(\omega)$ is the dispersion function, that is obviously mode dependent
  z is the axial coordinate

The individual propagation mode can be identified through one of the three mode classes, represented by the footer ζ in Equation 4.1 and by two progressive numbers: l and m.
  The mode classes are generally indicated as follows:

- *Transverse Electrical (TE) modes,* having the electrical field vector "transverse," thus orthogonal to the fiber section radius
- *Transverse magnetic (TM) modes,* having the magnetic field vector "transverse," thus orthogonal to the fiber section radius
- *HE modes,* where no field is "transverse"

The propagation description through the modes represented in Equation 4.1 is accurate, but difficult to manage for our purpose.
  In order to work with a simpler mode set, the solution of Maxwell equations in an orthogonal coordinate system can be attempted [3].
  By exploiting again the weakly guiding property of the fiber, two sets of modes can be derived that due to the adopted coordinate system are linearly polarized (LP).
  These modes (called LP) can be divided, looking at the electrical field, into x-polarized and y-polarized modes, where x and y are the orthogonal coordinates in the plane of the fiber section and due to the fiber cylindrical symmetry, every x-polarized mode is degenerate with a y-polarized mode.
  Naturally, as far as the adopted approximations are valid, LP modes can be represented as a sum of TE, TM, and HE modes.
  An LP mode, having, for example, linear polarization along the x axis, can be represented with the expression

$$\vec{E} = A_{l,m}(\rho)e^{i[\beta(\omega)z-\omega t]}\vec{x} \tag{4.2}$$

where
  $\vec{x}$ is the axis unitary vector
  l,m are the mode identification numbers

LP modes are much easier to use when the fiber propagation has to be considered for telecommunication purposes.
  The description of fiber propagation via LP modes is less accurate than via TE, TM, and HE modes with the effect that LP modes with the same mode index are not exactly orthogonal in real fibers, but a coupling exists.
  However, in studying fiber propagation, many other approximations are done (e.g., linear fiber axis, perfect circular section, abrupt index change between the core, and the cladding) that creates much more modes coupling both among LP and among TE, HE, and TM modes, so that the LP approximation can be adopted without any problem.

Different, nondegenerate modes propagate along the fiber with different group velocities due to the dependence of the function $\beta(\omega)$ appearing in Equation 4.2 from the mode numbers.

The dependence of $\beta(\omega)$ on the propagation mode (called mode dispersion [2]) generates signal distortion when transmitting a plurality of modes and greatly limits the possible span of a transmission system.

Even if a single LP mode is excited at the fiber input, the presence of material imperfections excites all the modes that can propagate through the fiber, thus causing multimodal propagation.

In order to solve this problem, a single-mode fiber has to be designed, that is, a fiber having only one mode that propagates at the minimum attenuation wavelength.

As in any kind of waveguide, modes propagating in an optical fiber are characterized by the so-called cutoff frequency, that is, the minimum frequency at which the mode can propagate. At lower frequencies, the mode propagation is impossible since the field rapidly vanishes advancing along the fiber, irradiating the greater part of the optical power out of the waveguide.

The mode cutoff is generally expressed in terms of normalized frequency V, defined as

$$V = \frac{2\pi a}{\lambda} \sqrt{n^2(0) - n^2(\omega)} - \frac{a}{c} \omega \sqrt{n^2(0) - n^2(\infty)} \tag{4.3}$$

where
  a is the core diameter
  c is the light speed in vacuum
  λ is the field wavelength
  ω is the field angular frequency

It is assumed that the diffraction index $n(\rho)$ is a decreasing function of the radial coordinate ρ so that $n(0)$ is the index at the core center and $n(\infty)$ is the index of the cladding far from the core.

At each fiber mode, a cutoff normalized frequency V_c can be associated, but to LP_{01} modes whose cutoff is zero. That means that these two polarization degenerate modes propagate at every frequency.

The cutoff frequency of the lower cutoff mode depends on the index profile; for a step index fiber with the indexes given in Figure 4.1, the mode with the lower cutoff is the LP_{11} and $V_c = 2.405$, corresponding to a wavelength of 1262 nm for $n_c = 1.4600$, $n_e = 1.4592$, and the core radius equal to 10 μm. This means that, at 1300 and 1500 nm (the wavelengths at which the fiber loss is minimum), the only modes that can propagate are the LP_{01}.

4.2.2 Fiber Losses

We have seen that the primary reason for the success of optical fiber as communication medium is the extremely low loss. Thus, studying loss mechanisms is very important.

When the light is injected into the fiber from an external light source and propagates along the fiber for a long span, the output light power is reduced with respect to the source emitted power by two kinds of losses: coupling losses and propagation losses.

Coupling losses are due to the fact that the field shape of the incoming light on the input fiber facet does not coincide with the fundamental mode profile, so that both radiation modes and below-threshold propagation modes are excited. After a very short fiber length only the fundamental mode remains, and all the power coupled with other modes is radiated out of the fiber core.

During propagation along the fiber, the fundamental mode power also decreases, due to propagation losses, to produce at the fiber output the attenuated output power.

4.2.2.1 Coupling Losses

Generally in a transmission system, light is injected into a single-mode fiber from a single-mode semiconductor laser source [4,5]. We will see dealing with lasers that the section of the waveguide composing the active region of a single-mode semiconductor laser is quite smaller than the section of a fiber (reference dimensions are $1\,\mu m \times 100\,nm$ for the square section active waveguide of a laser versus a diameter of $10\,\mu m$ for the fiber core). In this condition, a focusing system is needed to achieve good light injection into the fiber.

The focus system can be designed to focus the laser beam at the center of the input facet of the fiber. In this condition, light injection is performed as from a point source.

Since a point source emits a spherical wave, the coupling between the injected field and the suitably normalized fundamental modes is given by

$$\eta = \iint \vec{E}_s(\phi, \theta) \cdot \left[LP_{0,1}(\varphi - \phi, \vartheta - \theta)\vec{x} + LP_{1,0}(\varphi - \phi, \vartheta - \theta)\vec{y} \right] d\phi\, d\theta \qquad (4.4)$$

where
  $\vec{E}_s(\phi, \theta)$ is the electrical field of a spherical wave in spherical coordinates on the points of
    the fiber input facet
  $\cdot$ is the scalar product
  $\vec{x}$ and $\vec{y}$ are the unitary axes vectors
  integral is extended to the fiber input facet

This is the maximum possible coupling efficiency. In practice, this value is never reached, due to misalignments of the focusing system both along the z axis (the focus is not exactly on the fiber facet) and in the x–y plane (the focus is not at the center of the facet).

Moreover, fiber defects (like nonconcentricity of core and cladding and core deviation from a perfect cylinder) also contribute to increasing the coupling losses.

Considering coupling of a standard step index fiber with a distributed feedback laser, practical values from 1.5–3 dB are achieved, depending on the complexity and on the tolerance of the focusing lenses and of the assembly process.

4.2.2.2 Propagation Losses

Ideally, the fundamental mode does not attenuate during transmission along the fiber. In a real fiber however, there are several mechanisms causing power attenuation. All the loss mechanisms are wavelength dependent and the final fiber attenuation is obtained by their superposition.

The first loss mechanism is the photon absorption. The absorption can be divided into intrinsic absorption and extrinsic absorption.

Intrinsic absorption is due to the characteristics of the glass composing the fiber. If no impurities were present in the fiber material, all absorption would be intrinsic.

Since the glass is an amorphous and isotropic material, it can be described as a disordered ensemble of microcrystal structures. Thus, local phonons [6] can be introduced, which models with a good approximation microcrystals vibrations related to the glass temperature.

Local phonons can be excited by the photons of the propagating radiation, thus causing photons absorption that is efficient in the infrared region and increases increasing the wavelength.

Moreover, the propagating photons can be also absorbed by the external electrons of the glass molecules, promoting them to higher energy levels. This mechanism called electron absorption, is efficient in the ultraviolet region and decreases increasing the wavelength.

Besides intrinsic absorption, impurities in the glass composition cause extrinsic absorption. In the infrared region of interest for fiber optics communication, extrinsic absorption is mainly due to metal impurities and to hydroxyl ions. These last elements in particular cause two absorption peaks to appear in the near infrared that have a great importance in determining wavelength windows for fiber communications.

Besides absorption, another important loss element is light scattering, that is, the phenomenon causing a change of photons momentum and consequently their jump from the fundamental mode to a mode that does not propagate along the fiber.

At low propagating powers, linear scattering is prevalent. In this case, no wavelength change exists between the incident photon and the scattered photon for each scattering process and the scattered photons are lost generally for radiation.

Linear scattering can be studied introducing a parameter called scattering scale and defined as

$$\varsigma = \frac{\pi d_p}{\lambda} \tag{4.5}$$

where d_p is the diameter of the typical scattering particle.

If $\varsigma \ll 1$, then we are in the presence of Rayleigh scattering, if $\varsigma \approx 1$, we are in the presence of Meie scattering.

The prevalent form of scattering is the Rayleigh scattering, caused by particles and imperfections much smaller than the propagating field wavelength. Rayleigh scattering occurs with silica molecules composing the glass and it is greatly enhanced due to microdefects that are present in the fiber material [7].

Rayleigh scattering introduces an attenuation that decreases with the wavelength.

Finally, fiber imperfections such as imperfect concentricity of core and cladding introduce another loss factor in the field propagation.

In order to analytically express the behavior of attenuation versus the field wavelength, we will define the attenuation through the attenuation parameter α defined in such a way that, if L is the fiber length, P_0 the power coupled at the input with the fundamental mode, and P_1 the output power, it is

$$P_1 = P_0 e^{-\alpha L} \tag{4.6}$$

Frequently, the fiber loss is also characterized by the attenuation for a kilometer of fiber, measured in dB/km. The two parameters are directly related as can be easily derived.

A typical measure of the attenuation parameter of a step index fiber is reported in Figure 4.2, where the different contributions are also evidenced.

From the figure, it results that in a standard fiber, the low attenuation zone is divided into two parts by the hydroxyl absorption peak: a zone around 1.3 μm and a zone around

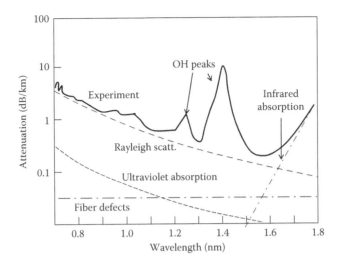

FIGURE 4.2
Loss contributions and total loss versus wavelength in an SSMF for signal transmission.

1.5 μm. For historical reasons, the two zones are called second and third transmission windows.

For a few applications, the hydroxyl (OH) absorption peaks are particularly problematic. In this case, the so-called zero water fibers can be used, where the hydroxyl ion content is so rigorously controlled that the correspondent absorption peaks are eliminated.

4.2.3 Linear Propagation in an Optical Fiber

In a real fiber, attenuation is not the only phenomenon causing the difference between the field coupled at the fiber input and the field emerging from the output.

In this section, we will assume that the power of the coupled field is sufficiently low that all propagation nonlinear effects are negligible. In this hypothesis, propagation along the fiber is purely linear and the fiber itself can be modeled as a linear distributed system.

In this condition, if the field coupled to the fiber at the input facet is a combination of the two main fiber modes, then

$$\vec{E}(0,\omega,\rho) = A(\rho)\left[a_x(0,\omega)\vec{x} + a_y(0,\omega)\vec{y} \right]e^{-i\omega t} \tag{4.7}$$

The field at the output will be obtained by the multiplication of the input field in the frequency domain with the frequency transfer function of the fiber.

In real fibers factors breaking fundamental modes degeneracy in terms of the transverse field shape are present, but are not important for telecommunication systems. Thus, we can assume that the transverse shape of the field remains equal to $A(\rho)$ in every fiber section and for every field polarization.

We have also seen that the field attenuation is a slowly varying function of the field frequency. For practical modulation, even at 100 Gbit/s, the attenuation variation in the signal bandwidth is completely negligible; thus, in writing the fiber transfer function we can assume a constant attenuation.

Fiber nonideality influences greatly the field phase evolution. As a matter of fact, in a real fiber it is not possible to assume that the propagation constants of the two LP fundamental modes are equal and that these modes do not couple during propagation.

On the contrary, the propagation constant depends on the field polarization and if a mix of the two fundamental modes is injected into the fiber, they will couple during propagation generating in every fiber section a different elliptical polarization.

Last but not least, microscopic imperfections cause coupling between the fundamental modes so that even if only one fundamental mode is launched into the fiber, that is the input field is perfectly LP, at the output, the field has a generic elliptic polarization revealing the excitation of both the fundamental modes.

Moreover, the output polarization will vary slowly in time, due to time variation of the coupling coefficient driven by phenomena like temperature and material stress variation.

These empirical observations can be summarized in the following expression of the field at a generic fiber section

$$\vec{E}(z,\omega,\rho) = A(\rho)e^{-((\alpha/2)z+i\omega t)}e^{i\beta(\omega)z} \begin{vmatrix} m_{xx}(z,\omega) & m_{xy}(z,\omega) \\ m_{yx}(z,\omega) & m_{yy}(z,\omega) \end{vmatrix} \begin{vmatrix} a_x(0,\omega)\vec{x} \\ a_y(0,\omega)\vec{y} \end{vmatrix} \tag{4.8}$$

where the matrix

$$[M] = \begin{vmatrix} m_{xx}(z,\omega) & m_{xy}(z,\omega) \\ m_{yx}(z,\omega) & m_{yy}(z,\omega) \end{vmatrix} \tag{4.9}$$

called Jones matrix, has to be unitary to fulfill energy conservation and represents both birefringence (i.e., the dependence of the propagation constant from the polarization) and mode coupling (through the off diagonal coefficients). It is to be noted that, since α represents the power attenuation, $\alpha/2$ appears in the field expression.

Equation 4.8 summarizes all the propagation effects characterizing the linear propagation regime in case of limited signal bandwidth.

4.2.4 Fiber Chromatic Dispersion

Chromatic dispersion in single-mode fibers is the phenomenon causing the broadening of a light pulse propagating along the fiber that is independent from the pulse polarization.

In Equation 4.8, chromatic dispersion is caused by the dependence on the angular frequency of the common mode propagation constant $\beta(\omega)$.

In particular, remembering the small signal bandwidth hypothesis and indicating with ω_0 the central angular frequency of the signal spectrum, we can write

$$\beta(\omega) \approx \beta(\omega_0) + \beta'(\omega_0)(\omega-\omega_0) + \frac{\beta''(\omega_0)}{2}(\omega-\omega_0)^2 + \frac{\beta'''(\omega_0)}{6}(\omega-\omega_0)^3 + \cdots \tag{4.10}$$

The terms of the Taylor series of $\beta(\omega)$ have different physical meanings.

The term $\beta'(\omega_0)z$ that is obtained substituting Equation 4.10 in Equation 4.8 represents a group delay, as can be easily shown from the property of the Fourier integral. This is the average delay a pulse experiments traversing the fiber span (as a matter of fact it has also the dimension of a time interval). To evidence this meaning, the time delay τ can be written as

$$\tau = \beta'(\omega_0)z = \frac{z}{v_g} \tag{4.11}$$

being $v_g = 1/\beta'(\omega_0)$ the group velocity.

In a first approximation, injecting in the fiber a pulse with bandwidth $\Delta\omega$, the pulse broadening Δt can be evaluated by the group velocity dispersion (GVD) between the frequencies at the spectrum borders. Using Equation 4.10 up to the second order, the following expression is obtained.

$$\Delta\tau = \frac{d\tau}{d\omega}\Delta\omega \approx \beta''(\omega_0)\Delta\omega z \tag{4.12}$$

Alternatively, considering wavelengths instead that angular frequencies

$$\Delta\tau = \frac{d\tau}{d\lambda}d\lambda \approx \frac{2\pi c}{\lambda^2}\beta''(\lambda_0)\Delta\lambda z = Dz\Delta\lambda \tag{4.13}$$

The parameter $D = 2\pi c/\lambda^2 \beta''(\lambda_0)$ is called the fiber dispersion parameter (measured in ps/nm/km).

There are fundamentally two phenomena contributing to the value of the parameter D: material dispersion and waveguide dispersion, so that D can be written as the sum of a material dispersion term D_M, a waveguide dispersion term D_W, and a mixed term D_{MW}, that in typical fibers is much smaller with respect to the other two.

Accurate expressions of the three terms are obtained as a result of the propagation modal analysis; here we will introduce approximated expressions for the material and guide terms in order to analyze their behavior with respect to the main fiber parameters.

The material term is related to the wavelength dependence of the silica refraction index; thus, it is the same term causing dispersion in bulk silica. It is positive, independently from the structure and design of the fiber.

In order to derive an approximated expression of this term, the propagating wave front can be approximated as a plane. The differential delay between two frequency components of the field can be written as a function of the frequency dependent group velocity as

$$\Delta\tau \approx L\left|\frac{1}{v_g(\omega)} - \frac{1}{v_g(\omega + \Delta\omega)}\right| = L\left[\left(\frac{d\beta}{d\omega}\right)_\omega - \left(\frac{d\beta}{d\omega}\right)_{\omega + \Delta\omega}\right] = \beta''(\omega)\Delta\omega L \tag{4.14}$$

The expression of the propagation constant in bulk material in the case of a plane wave is simply $\beta(\omega) = \omega\, n(\omega)/c$. Substituting this expression in (4.14) and considering that in the near infrared region that is of interest for telecommunication, the angular frequency is on the order of 10^{15}, it results as

$$\frac{\Delta\tau}{\Delta\omega} = L\left[\frac{\omega}{c}\frac{d^2 n}{d\omega^2} + \frac{2}{c}\frac{dn}{d\omega}\right] \approx L\frac{\omega}{c}\frac{d^2 n}{d\omega^2} \tag{4.15}$$

From (4.15) and the dispersion parameter definition, passing from ω to λ, the following expression is obtained.

$$D_M = -\frac{\lambda}{c}\frac{d^2n}{d\lambda^2} \tag{4.16}$$

The material dispersion, as it results from Equation 4.16 and the measured behavior of silica refraction index, is zero at a wavelength of 1300 nm, positive at longer wavelength, and negative at shorter once. In any case, D_M increases while increasing λ in all the near infrared region.

The waveguide term is due to the fact that the guided mode travels part in the core and part in the cladding. Considering the case of a step index fiber, if we analyze the distribution of optical power at different frequencies we found that it depends critically on the ratio a/λ due to the different effect of the core–cladding interface diffraction on waves with different wavelength.

This effect can also be represented by studying the dependence of the effective refraction index n_e (i.e., the ratio between the group velocity of a monochromatic guided wave and the velocity of light in vacuum) on the wavelength.

In particular, at short wavelengths, the effects of diffraction are smaller and the light is confined well within the core. In this condition, n_e is very close to the refractive index of the core.

As the wavelength increases, the effects of diffraction become more important and the light spreads slightly into the cladding. The effective refractive index decreases toward the refractive index of the cladding.

Finally, at long wavelengths, the effects of diffraction dominate and the light in the fiber spreads well into the cladding, n_e being very close to the refractive index of the cladding.

In order to obtain an approximate expression of the waveguide dispersion parameter, we will assume a zero material dispersion, a condition that can be realized only if the refraction index does not depend on wavelength. In this condition, it is useful to substitute the angular frequency with the normalized frequency V defined in Equation 4.3 that is directly related to the guiding properties of the fiber.

Thus, it is possible to write

$$\beta'(\omega_0) = \frac{1}{v_g} = \frac{d\beta}{dV}\frac{dV}{d\omega} = \frac{a}{c}\sqrt{n^2(0) - n^2(\infty)}\frac{d\beta}{dV} \tag{4.17}$$

Substituting (4.17) in the definition of the dispersion parameter it results, after expressing D_W as a function of λ

$$D_W = \frac{d}{d\lambda}\left(\frac{1}{v_g}\right) = -\frac{V^2}{2\pi c}\frac{d^2\beta}{dV^2} \tag{4.18}$$

The shape of the propagation constant in an ideal dielectric waveguide whose refraction index distribution does not depend on the wavelength can be varied widely by shaping the index profile being in particular either positive or negative. For example, in a step index fiber, there are two wavelength intervals: one in which the two dispersion terms are cumulative, the other in which they have different signs and one tends to compensate the other.

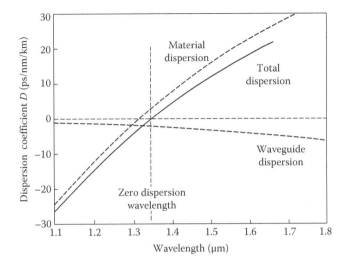

FIGURE 4.3
Dispersion contributions and total dispersion versus wavelength in a standard SSMF single-mode fiber for signal transmission.

This effect is shown in Figure 4.3 in the case of a step index fiber. In the figure, both the overall dispersion and the individual contributions are shown, evidencing the zero dispersion point (around 1.33 μm) where the two dispersion contributions are opposite.

The zero dispersion wavelength divides the wavelength axis in two intervals, that for historical reasons take the name of normal dispersion wavelengths ($\beta'' > 0$) and anomalous dispersion wavelengths ($\beta'' < 0$).

The dispersion parameter takes into account the greater contribution to fiber dispersion, but there are several cases in which the other terms of Equation 4.10 cannot be neglected. The most important of these cases is propagation around the zero dispersion frequency.

In this case, dispersion is almost all due to the third-order term, which at high transmission speed cannot be neglected.

As in the case of second-order dispersion, also in the third-order case, the pulse broadening can be approximately expressed as a linear function of the square signal bandwidth, thus defining a third-order dispersion coefficient.

In particular, the following equation is obtained:

$$D_3 = -\frac{2\pi c}{\lambda^2}\beta'''(\lambda_0)(\mathrm{ps/nm^2/km}) \tag{4.19}$$

The dispersion-related parameters of a standard step index fiber in the second and the third transmission window are reported in Table 4.1.

The fact that waveguide dispersion can be opposite in sign with respect to material dispersion can be used to shape the fiber dispersion curve. To control waveguide dispersion, the refraction index behavior along the fiber diameter is suitably designed.

Depending on the application, different types of dispersion-managed optical fibers can be used. Dispersion-shifted (DS) fibers are designed to move the zero dispersion point from the second to the third transmission window, dispersion flattened fibers are engineered to have a very low (but nonzero) dispersion in a wide wavelength interval, possibly

TABLE 4.1

Typical Values of the Parameters of a Step Index Single-Mode Transmission Fiber

Attenuation (dB/km)	α	0.25
Core refractive index (adim.)	n	1.58
Effective mode radius (μm)	ρ_m	4
Dispersion parameter (ps/nm/km)	D	15.6
Dispersion coefficient (ps²/km)	β_2	−20
Third-order dispersion coefficient (ps³/km)	β_3	0.05
Birefringence (m⁻¹)	$\Delta\beta$	0.1
PMD (ps/$\sqrt{km}$)	D_p	0.1
Random mode coupling characteristic length (m)	L_b	100
Real part of third-order susceptibility (W⁻¹)	$\chi^{(3)}_{xxxx}$	1.56×10^{-19}
Nonlinear coefficient (W⁻¹ km⁻¹)	γ	2
Brillouin gain coefficient (m⁻¹ W⁻¹)	g_B	6.02×10^{-12}
Brillouin gain bandwidth (MHz)	Δv_B	16
Raman gain coefficient (m⁻¹ W⁻¹)	g_R	7×10^{-15}

All values are at $\lambda = 1.55\,\mu$m.

including both second and third transmission window, nonzero dispersion fibers are generally realized to move the zero dispersion wavelength in between the two transmission windows.

Different index profiles corresponding to different categories of fibers for long-haul transmission manufactured by the main fiber vendors are represented in Figure 4.4.

In all the cases represented in the figure, the study of the structure-guided modes demonstrate that two degenerate fundamental modes of the LP type can propagate in the fiber and that the spatial profile of this mode is Gaussian with a good approximation, exactly as in a step index fiber. The variance of the field mode shape is different for different fibers causing a difference in the field confinement. As we will see, this will have an impact on the efficiency of nonlinear effects during field propagation.

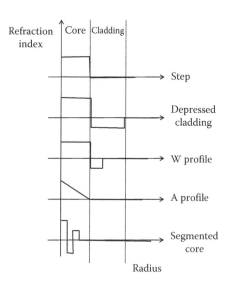

FIGURE 4.4
Different possible refraction index profiles for dispersion managed fibers.

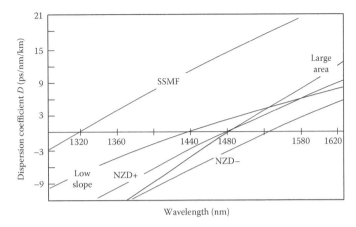

FIGURE 4.5
Dispersion characteristic versus wavelength for different dispersion-managed fibers.

The dispersion curve of a set of "managed dispersion" fibers is reported in Figure 4.5, and some important parameters of commercial fibers are summarized in Table 4.2.

The various performances are compared with the performances of a standard step index fiber.

The first dispersion-managed fibers to be produced were DS fibers, where the dispersion curve is shifted toward higher wavelengths so that the zero dispersion wavelength is moved around 1.5 μm.

These fibers were designed and deployed in the period in which long-haul single-channel systems were designed in order to decrease chromatic dispersion and increase the transmission reach.

With the introduction of dense wavelength division multiplexing (DWDM) systems, this class of fibers was practically abandoned, due to the fact that having the zero dispersion wavelength in the third transmission window causes the maximization of nonlinear channel crosstalk due to the channel phase matching.

In order to allow high channel count DWDM systems to be deployed and contemporary to decrease as much as possible the value of dispersion, nonzero dispersion fibers were designed.

TABLE 4.2

Comparison between the Properties of a Set of Fibers with Different Index Profiles

Name	Acronym	Index Profile	Dispersion Parameter (ps/nm/km)	Dispersion Slope (ps/nm²/km)	Attenuation (dB/km)	Effective Area (μm²)
Step index	SMF	Step	16 to 18	0.06	0.19	80
Dispersion shifted	DS	Segmented	<3.5	0.07	0.23	40
Nonzero dispersion +	NZD+	Segmented	3 to 5	0.045	0.21	50
Nonzero dispersion −	NZD−	Segmented	−3 to −2	0.07	0.21	50
Large effective area	LEAF	Segmented	3 to 6.5	0.08	0.21	70
Intermediate dispersion		Segmented	6 to 9	0.06	0.21	60

All values in the table that depend on wavelength are reported at λ = 1.55 μm.

In this kind of fibers, the dispersion curve is shifted to move the zero dispersion wavelength either in between the second and the third window or beyond the third window around 1570 nm.

In the first case, the fibers are called NZ+ and in the second NZ−.

Using these fibers allows avoiding zero dispersion in the transmission bandwidth while reducing the average dispersion.

Moreover, especially in the case of NZ−, the fiber design can be performed to achieve a greater effective mode ratio to decrease nonlinear effects. This point will be considered in more detail describing nonlinear propagation.

4.2.5 Polarization Mode Dispersion

Polarization mode dispersion (PMD) is the phenomenon causing strong coupling between the fundamental LP modes so that the propagating field polarization evolves during propagation [8,9].

The polarization not only changes from fiber section to fiber section, but also changes in time and wavelength. This is due to the fact that mode coupling is generated by fiber imperfection both at the material level and at a fiber form level. While temperature and mechanical conditions like micro-stresses evolve, mode coupling changes.

The fact that the field polarization changes with wavelength is particularly dangerous for modulated signals when the field polarization has to be recovered at the receiver. As a matter of fact, this becomes impossible if different frequencies of the signal spectrum have different polarizations.

Moreover, since the diagonal terms of the Jones matrix are not equal to one, different polarizations travel through the fiber at slightly different group velocities, thus causing the so-called polarization dispersion. This means that a pulse of finite duration sent into the fiber experiments a widening during propagation even in the absence of chromatic dispersion due to the PMD.

The base for the analysis of PMD impact is the definition of the principal states of polarization (PSPs), which for a fiber with small random birefringence somehow take the place of the birefringence axes in a high birefringence medium.

Given a fiber piece, the PSPs of that fiber in the considered deploying conditions are defined as the input states of polarization that generate at the fiber output a polarization state that does not depend on ω at the first order [8].

The PSPs can be derived directly from their definition and from the general expression Equation 4.8 of the field during linear propagation in the fiber. As a matter of fact, the definition of the PSPs implies that, at first order, their ω derivative has to be equal to zero.

Let us consider the slowly varying envelope of the field in Equation 4.8. To simplify the notation, it is useful to eliminate on both sides of Equation 4.8 the transversal field shape and to write

$$k(\omega) = -\frac{\alpha}{2}z + i\beta(\omega)z \tag{4.20}$$

$$\begin{bmatrix} a_x(0,\omega)\vec{x} \\ a_y(0,\omega)\vec{y} \end{bmatrix} = a(\omega)e^{i\phi(\omega)}\vec{p} \tag{4.21}$$

where $\vec{p}$ is the polarization vector.

If the input field is a PSP, by definition the first-order derivative of the output field has to be zero. Deriving Equation 4.8 and equaling the result to zero, the following equation is obtained:

$$\frac{d\lfloor M(\omega)\rfloor}{d\omega}\vec{p} = ih(\omega)[M(\omega)]\vec{p} \tag{4.22}$$

where

$$h(\omega) = \frac{d\phi}{d\omega} + i\left\{\frac{dk}{d\omega} - \frac{1}{a}\frac{da}{d\omega}\right\} \tag{4.23}$$

Taking into account that the Jones matrix is unitary, it is easy to recognize that Equation 4.22 always has two solutions, which are the fiber PSPs and whose coordinate expression is easy to evaluate solving Equation 4.22.

This procedure demonstrates that, for a given fiber piece, there exist two orthogonal input polarization states, called PSPs that characterize the polarization behavior of the fiber. In particular, if a field is launched into the fiber along one of the PSPs, the output polarization does not change, at least at first order, changing ω. This means that a finite duration pulse injected in the fiber along a PSPs does not experiment widening due to PMD.

Since a higher-order dependence of the PSP output polarization on the angular frequency exists, if the bandwidth of the input signal rises above a certain value, the PSP output polarization is no more ω independent and the concept of PSPs becomes quite useless.

The PSPs bandwidth is generally on the order of 100 GHz [10]; thus, it is quite larger than the bandwidth of normal modulated signals, but for modulation at 100 Gbit/s. In this case, if pure on–off modulation is used, PSPs probably cannot be used and some other means have to be adopted to describe polarization evolution.

Even if the PSP approximation can be used, the expression of the PSPs depends on the elements of the Jones matrix and thus slowly changes with time due to the change of the random birefringence [11].

Last but not least, macroscopically identical pieces of fiber can have completely different PSPs, due to the different distribution of micro-defects causing coupling between the fundamental modes.

Starting from the PSPs definition, the PMD characteristics can be derived.

Let us call $\vec{p}_+$ and $\vec{p}_-$ the unit vectors corresponding to the PSPs output polarizations of the considered fiber piece. A generic field injected into the fiber can be decomposed along the input PSPs so that, for the properties of the PSPs, the same decomposition holds for the output field with respect to the output PSPs.

This decomposition can be written in general as follows:

$$\vec{E}_{out} = E_{0+}\sin(\theta)e^{i\phi}\vec{p}_+ + E_{0-}\cos(\theta)e^{i\psi}\vec{p}_- \tag{4.24}$$

Starting from Equation 4.24, it is possible to evaluate the polarization dispersion contribution with a simple derivation exploiting the PSPs properties [12].

Squaring Equation 4.24 and remembering that the PSPs are orthogonal states, the following equation is obtained for the pulse envelope:

$$g(t) = g_+(t)\sin^2(\theta) + g_-(t)\cos^2(\theta) \qquad (4.25)$$

where the envelopes $g_+(t)$ and $g_-(t)$ are both replicas of the input pulse envelope that arrive at the output with different delays due to their travel along the two different PSPs.

Taking into account this consideration, we can imagine the propagation of two identical pulses launched through the fiber along the input linear polarization of the LP states.

The output envelope in the two cases can be written as in Equation 4.25 with the suitable coefficients obtaining

$$g_1(t) = g(t - \tau_+)\sin^2(\theta) + g(t - \tau_-)\cos^2(\theta)$$
$$g_2(t) = g(t - \tau_-)\sin^2(\theta) + g(t - \tau_+)\cos^2(\theta) \qquad (4.26)$$

where τ_- and τ_+ are the delays of the PSPs.

The differential delay due to PMD is by definition the time interval separating the instants in which the peaks of two identical pulses launched in the fiber along the LP fundamental states arrive at the fiber output.

Since Equation 4.25 simply gives the expressions of the output pulses envelopes, this can be the starting point for the PMD differential delay calculation.

For a finite duration pulse, it is possible to define the moment generating function and the general theory of the Fourier transform tells us that the abscissa of the pulse maximum (that in our model is the required arrival time) is given by $-iG'(0)$ if $G(\omega)$ is the moment generating function of its envelope, while the pulse variance is $iG''(0)$.

Thus, the differential delay can be written as

$$|\tau_1 - \tau_2| = |iG_2'(0) - iG_1'(0)| = |(\tau_+ - \tau_-)\cos(2\theta)| \qquad (4.27)$$

From Equation 4.27, it is evident that the delay is maximum for $\theta = 0$, that is, if the input pulses are launched along the PSPs, while it is minimum if the states along which the two input pulses are launched are inclined of $\pi/4$ with respect to the PSPs. In this case, both the input pulses are equally distributed among the PSPs and no differential delay exists.

The pulse spreading due to PMD can be evaluated starting from the definition

$$\sigma_{\Delta\tau} = \sqrt{\sigma_{out}^2 - \sigma_{input}^2} = \sqrt{i\left[G_{input}''(0) - G_{output}''(0)\right]} \qquad (4.28)$$

Exploiting the properties of the moment generating function and starting from (4.26), the following expression is found for the pulse spreading:

$$\sigma_{\Delta\tau} = \frac{1}{2}|(\tau_+ - \tau_-)\sin(2\theta)| \qquad (4.29)$$

Equation 4.29 confirms the fact that, if $\theta = 0$, that is, the input pulse is launched along a PSP, the pulse spreading is zero, while it is maximum if the pulse energy is equally divided between the PSPs at the fiber input.

Equation 4.29, even if it gives a very insightful expression of the pulse spreading, does not permit immediately to evaluate it starting from fiber parameters.

However, the evaluation of the average value of the pulse spread during propagation along a PMD-affected fiber can be carried out starting from Equation 4.29 with a further step.

We can imagine that the fiber link can be decomposed in several fiber pieces, each of which is completely independent from the others from a PSP point of view.

Since the output PSPs of the *j*th fiber piece are completely independent from the input PSPs of the $(j+1)$th, the final pulse broadening is the sum of a great number of independent contributions. We can consider the average pulse spreading given by Equation 4.29 also as the time delay standard deviation, due to the fact that in the presence of PMD the time delay experienced by a pulse during the propagation along a fiber piece is a random variable.

Thus, since the variance of a sum of independent random variables is the sum of the variances, the square of the pulse broadenings experienced by the pulse while traveling along each fiber piece can be added to derive the square of the final pulse spreading, obtaining

$$\sigma_{\Delta\tau} = \frac{1}{2}\sqrt{\Sigma_j(\tau_{+j}-\tau_{-j})\sin^2(2\theta_j)} \tag{4.30}$$

where *j* is the index running over the different cascaded pieces of fiber. Since the injection angle is completely independent from the delay difference, the average of the product can be calculated as the product of the average.

Assuming the injection angle is uniformly distributed, in the limit of infinite pieces of negligible length, the variance of the delay difference between the PSPs is proportional to $\sqrt{N}$, where *N* is the number of fiber pieces.

Thus, if we have to evaluate the pulse broadening for a certain fiber length *L*, we have to multiply and divide for the square of the length of the single piece l_j of fiber and make the limit $N \rightarrow \infty$ and $l_j \rightarrow 0$ in such a way that $Nl_j \rightarrow L$. By calculating the limit the following relation is obtained

$$\sigma_{\Delta\tau} = D_{PMD}\sqrt{L} \tag{4.31}$$

where the D_{PMD} is the PMD parameter and it is evident that, due to its statistical nature the pulse broadening due to PMD increases with the square root of the fiber length.

The same result can be derived from a statistical approach on the Jones matrix (or on the equivalent representation in the Stokes space).

A statistical polarization dispersion theory has been carried out both in the Jones and in the Stokes space [13,14] and a very elegant relationship has been developed between the two approaches based on the use of the σ spin matrixes [15].

The rigorous statistical theory arrives at the conclusion that, considering a very large ensemble of macroscopically identical fibers, the PSPs of every fiber piece are different. A statistical average of the delay can be done considering the link as a cascade of a large number of identical pieces and studying polarization evolution in this case.

The propagation along one piece of fiber generates a dispersion given by Equation 4.29, while the overall propagation, can be viewed as a random walk [14,15].

Due to the properties of the random walk, the standard deviation of the propagation time depends on the square root of the number of jumps. Since the jumps are equally probable in every fiber piece, depending on the probability on the fiber length *L*, the average pulse broadening $\sigma_{\Delta\tau}$, evaluated as time delay standard deviation, is given exactly by Equation 4.31.

4.2.6 Nonlinear Propagation in Optical Fibers

Even if the optical power launched in a fiber transmission system is not huge, the core radius is so small that it is not difficult to generate so high fields that they change the equilibrium of the dielectric composing the fiber [6,7].

In this case, the constants describing the material characteristics in the wave equation depend on the field, and nonlinear effects appear.

Nonlinear propagation effects can be divided in two categories: scattering effect and the Kerr effect.

Scattering effect includes the phenomena in which a propagating field photon is scattered by some material alteration caused by the field itself.

Kerr effect includes all the phenomena due to the dependence of the real part of dielectric susceptibility on the field.

Both the classes of effects are important in telecommunication systems, because they are either exploited to improve the system performance or avoided to avoid a too high system penalty.

The starting point for the propagation problem is always the wave equation with the boundary conditions corresponding to the fiber structure.

In particular, indicating with $\vec{P}(\vec{E})$ the polarization vector of the glass composing the fiber, the wave equation writes [16]

$$\nabla^2 \vec{E} - \frac{1}{c^2}\frac{\partial^2 \vec{E}}{\partial t^2} = -\mu_0 \frac{\partial^2 \vec{P}(\vec{E})}{\partial t^2} \tag{4.32}$$

where $\vec{P}(\vec{E})$ is the polarization vector.

In a nonlinear propagation regime, the polarization is not proportional to the field, but it depends on the field in a more complex way. The consequence of this more complex dependence is the presence of wavelengths in the output signal that were not present in the input signal.

If the more powerful of these nonlinear products is much smaller than the linear component (i.e., the signal power remaining along propagation in the input bandwidth) the propagation regime is told to be weakly nonlinear. In the weak nonlinear propagation regime a few approximations can be used that simplify the problem.

In normal propagation conditions, the impact of nonlinear propagation can be classified into three main effects:

1. *Brillouin effect:* It is caused by photons scattering from coherent phonons. This is a backscattering effect.

2. *Raman effect:* It is caused by photon scattering from glass molecules in an excited state.

3. *Kerr effect:* It is caused by incoherent interaction between molecules' external orbitals and the traveling field.

In order to set up a mathematical model of nonlinear propagation, an expression has to be adopted for the nonlinear polarization.

Assuming to be in the weak nonlinear propagation regime, series approximation of the nonlinear susceptibility can be adopted.

In general, the relationship between the susceptibility and the field is not a local and instantaneous relationship, due to the presence of distributed interactions as in the case of the Brillouin scattering. In similar cases, the susceptibility depends on the field not only in a single point and in a single instant, but incorporates a dependence extended to a finite area and to a finite time interval.

Formally, this property can be coded in mathematical terms considering the polarization vector as a convolution between the nonlinear susceptibility and the field [17], but the resulting integral–differential wave equation is really difficult to manage.

For this reason, we will assume that all fiber nonlinearities are characterized by a scalar nonlinear susceptibility. We know that in this way we will have a unified theory of Kerr and Raman effect, but we will be unable to describe Brillouin effect. For this last case, an ad hoc model will be developed.

In the local and instantaneous approximation, the Taylor series of the polarization reads

$$\vec{P}(\vec{E}) \approx \varepsilon_0 \left\{ \chi^{(1)}\vec{E} + \vec{E}^* [\chi^{(2)}]\vec{E} + \vec{E}^* [\chi^{(3)}]\vec{E}\vec{E} + \cdots \right\} \tag{4.33}$$

but for $\chi^{(1)}$ that represents the linear susceptibility, the generic term $[\chi^{(k)}]$ of the power expansion is an Euclidean tensor with rank $k+1$ so that the generic term results to be a vector, whose components are forms of degree k in the electric field components.

In particular, since the glass composing the fiber is an isotropic medium at molecular level, it can be demonstrated that $[\chi^{(2)}] = 0$; thus, the nonlinear propagation is mainly due to the third term of the expansion (4.33) [7].

The tensor nature of $\chi^{(3)}$ is responsible for nonlinear polarization evolution, that is, a quite weak phenomenon, but important in some kinds of long-haul systems [17] (compare Chapter 9, Ultra Long Haul 100 Gbit/s systems).

If nonlinear polarization evolution can be neglected, the third term can be simplified introducing a scalar third-order susceptibility, so that (4.33) rewrites

$$\vec{P}(\vec{E}) \approx \varepsilon_0 [\chi^{(1)} + \chi^{(3)} E^2]\vec{E} \tag{4.34}$$

$\chi^{(3)}$ is a complex term, whose real part $\chi_R^{(3)}$ causes the Kerr effect and whose imaginary part $\chi_I^{(3)}$ causes the Raman effect. A typical value of the real part of $\chi^{(3)}$ is reported in Table 4.1.

The presence of a nonlinear part of the susceptibility implies the dependence of the refraction index on the propagating field intensity $\mathcal{J}$. In particular, from the definition of the diffraction index, the following equation is derived:

$$n(\mathcal{J}) \approx n_0 + n_2\mathcal{J} = n_0 + \frac{3}{8n_0}\chi_R^{(3)}\mathcal{J} \tag{4.35}$$

From what we have seen up to now, it is evident that nonlinear effect is a distributed cause of modification of the input pulses and it is more and more effective increasing the transmission fiber length.

However, by increasing the transmission length, attenuation tends to decrease the nonlinear contribution to the index, which is proportional to the local field intensity. On the other hand, increasing the dimension of the mode, the peak intensity decreases and the nonlinear effect should be less effective.

It is thus intuitive that some characteristic length has to exist, depending on fiber attenuation, mode area, and launched power, such that when the transmitted signal goes beyond this characteristic length, it is so attenuated that the nonlinear effect is no more effective.

Let us define an effective length L_e such that the overall optical power present along the link is equal to the injected power multiplied by the characteristic length. The definition can be converted in a simple equation for L_e that reads $P_0 L_e = \int_0^L P_0 e^{-\alpha z} dz$ and whose solution is

$$L_e = \frac{1-e^{-\alpha L}}{\alpha} \qquad (4.36)$$

where L is the fiber link length. If the fiber length is much greater than $1/\alpha$, L_e results to be of the order of $1/\alpha$, that is about 20 km in standard transmission fibers.

Equation 4.36 is a pure empirical definition, but provides a useful idea of what is the fiber length causing relevant nonlinear effects.

The definition does not include the mode area. In order to also include this element, It can be suitably modified starting from the definition of the mode effective area that, in the hypothesis of perfectly symmetric mode, reads:

$$A_e = \frac{\left[\iint |E(\rho,\varphi)|^2 \, \rho \, d\varphi \, d\rho \right]^2}{\iint |E(\rho,\varphi)|^4 \, \rho \, d\varphi \, d\rho} = \pi \frac{\left[\int_0^\infty |A(\rho)|^2 \, \rho \, d\rho \right]^2}{\int_0^\infty |A(\rho)|^4 \, \rho \, d\rho} \qquad (4.37)$$

At this point, Equation 4.36 could be modified as follows:

$$L_e' = \frac{A_c}{A_e} \frac{1-e^{-\alpha z}}{\alpha} \qquad (4.38)$$

where A_c is the geometrical core area. In the case of a step index fiber with a core radius of 5.2 μm, $A_c/A_e = 0.95$, so that the results of Equations 4.36 and 4.38 practically do not differ. In other cases however, there is quite a difference. For example, in the case of a DS fiber with the same geometrical core radius, $A_c/A_e = 0.6$ clearly indicating the greater efficiency of nonlinear effects in this kind of fiber due to the greater field confinement.

However, it is the length defined by Equation 4.36 that appears in several equations regarding nonlinear effects and generally it is the quantity called characteristic length of fiber nonlinear effects.

4.2.7 Kerr Effect

Under the hypotheses set in the previous section, to analyze the impact of Kerr effect, we can neglect in a first approximation the polarization evolution and imagine dealing with ideal LP modes.

Several effects induced by the Kerr nonlinearity are much more efficient if the interacting signals polarizations are aligned so that a scalar model results to enhance the nonlinear effect with respect to the real vector case. A discussion on the applicability limits of the scalar model for the analysis of transmission systems is reported in [18].

If transmission penalties have to be evaluated, this is a reasonable approach, providing a worst case system performance, if the nonlinear effect is exploited to design some fiber-based device, the correctness of the scalar assumption has to be explicitly verified due to the risk of overestimating the device performances.

In these conditions, it is possible to derive a propagation equation in the presence of Kerr effect by applying the slowly varying approximation and considering the Kerr effect response time about zero so to deal with an instantaneous effect (being the Kerr characteristic time of the order of 100 fs this, approximation is always valid in telecommunications). This equation, called nonlinear Schrödinger equation, writes [6,19]

$$i\frac{\partial E}{\partial z} = -i\frac{\alpha}{2}E + \frac{\beta''}{2}\frac{\partial^2 E}{\partial \tau^2} - \gamma \mid E \mid^2 E \qquad (4.39)$$

where τ is the time in the pulse reference frame, thus $\tau = t - z/v_g = t - \beta' z$, and the nonlinear coefficient given by $\gamma = n_2 \omega_k/(C\pi r_m^2)$ with r_m representing the modal radius.

Also within this approximation, the evolution in the presence of a wideband signal is complex and, in order to understand from a qualitative point of view the signal distortions introduced by Kerr effect, signal propagation is generally studied assuming a few simple expressions for the fiber input signal.

These particular signals are conceived to put in evidence particular aspects of the Kerr effect, aspects that are called, for historical reasons "effects." Thus, we speak about self phase modulation (SPM) effect, cross phase modulation (XPM) effect, four wave mixing (FWM) effect, and so on. However, all these phenomena are Kerr effect–induced evolutions of particular input signals.

If the input signal is complex, the evolution can be qualitatively understood by considering the input as a superposition of a certain number of simple signals and using the fact that the nonlinear effect is small to try to model the final result as the superimposition of the effects on each simple signal (compare in Chapter 6 the rule of penalties addition and Appendix C).

This approach is quite useful to understand complex phenomena also, but has clear limits either when the power is very high and the superposition of the individual effects cannot be assumed or when the bandwidth is very large, so that parameters dependence on wavelength cannot be neglected. These conditions are not so far from the operational conditions of long-haul high-capacity WDM systems; thus, the validity of all the approximations has always to be verified when in front of a real system.

To access more quantitative results, several methods have been devised to numerically solve Equation 4.39 and their study can start from bibliographies [18,20].

4.2.7.1 Kerr-Induced Self-Phase Modulation

If an amplitude-modulated signal is injected into the fiber, the presence of nonlinear refraction index n_2 causes the intensity modulation to be reflected into a phase modulation in such a way that the signal phase, in the approximation of very weak effect, writes

$$\Phi(t) \approx n_0 z + \Phi_0 + \frac{2\pi}{\lambda} n_2 \mathcal{J}(t) z \qquad (4.40)$$

where $\mathcal{J}(t)$ is the field intensity that is time dependent due to modulation.

In the simplest case, the presence of a sinusoidal modulation, the sidebands of the signal spectrum are unbalanced for the superposition of the amplitude modulation and phase modulation.

If a signal with a more complex amplitude modulation is injected into the fiber, SPM causes a chirp, which is a dependence of the signal phase on time, and causes a nonlinear pulse broadening.

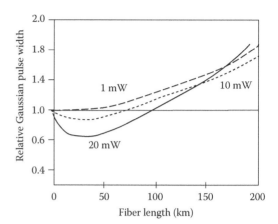

FIGURE 4.6
Broadening of a Gaussian pulse propagating in a DS fiber at 1.55 μm with zero chromatic dispersion versus the pulse power after the propagation through a variable length of fiber. (After Chraplyvy, A.R., *J. Lightwave Technol.*, 8, 1548, 1990.)

The broadening of a Gaussian pulse propagating in a DS fiber at 1.55 μm with zero chromatic dispersion is shown in Figure 4.6 [21] versus the pulse power after the propagation through a variable length of fiber.

It is interesting to observe that, even if for long distances the pulse is always broadened, at high transmitted power, there is a distance range where the pulse width decreases. This is due to the fact that SPM creates a phase modulation frequency distribution that on one side of the central carrier is out-of-phase with respect to the original frequency distribution that is due to the amplitude modulation. In the long run and at high powers, the nonlinear phase distribution prevails and determines the spectrum width, but a range of fiber lengths and field powers exist where the left modulation bands almost cancels, generating a pulse width decrease. This interpretation of the pulse shrink is confirmed by the fact that the minimum relative pulse width arrives near 0.5 and never goes below this value.

The SPM-induced broadening superimposes with the broadening due to fiber dispersion. In the normal dispersion regime, the two effects have the same sign and the observed pulse broadening is greater than that due to dispersion alone; on the contrary, in the anomalous regime the two effects have different signs and the overall broadening is smaller than that due to dispersion.

This effect is very important in transmission systems.

When the Kerr effect theory was consolidated, it was also discovered that, giving to the intensity of the field a particular shape, the linear and nonlinear effects completely cancel and the pulse propagates without broadening through thousands of kilometers of fiber. These particular pulses are known as solitons and in a first moment they were regarded as the key to design very high-capacity and ultra-long-haul systems [18,19,22].

The propagation dynamics of solitons however is quite complex: since a single pulse travels unchanged but for attenuation, and since the zero broadening condition is reached only for a precise value of the input power, a soliton can be realized only in a first approximation in a real fiber.

Even if a single soliton does not broaden, a random train of nearby solitons (representing a random string of bits) is affected by a sort of jitter due to the attraction between adjacent pulses. Moreover, if WDM is used, more jitter comes out due to the interaction between frequency adjacent channels.

Last but not least, perfect nonlinear and linear chirp compensation creates perfect phase matching among WDM channels belonging to the same comb, maximizing the dangerous nonlinear interchannel crosstalk.

A great amount of research has been devoted to solve these problems, producing brilliant solutions and a great insight on nonlinear propagation, but at the end, soliton systems never passed from laboratories to production, even if a great number of field tests were done.

The practical result of all this research was however the discovery that, even if mathematical solitons are too complex to be implemented in a practical system, the soliton principle can be exploited in any case.

In this way, the present day ultra-long-haul systems were born. In these systems, SPM is used to partially compensate chromatic dispersion to greatly reduce pulse broadening without creating perfect phase matching between spectrally adjacent channels.

4.2.7.2 Kerr-Induced Cross-Phase Modulation

When a comb of amplitude-modulated signals at different wavelengths is injected into the fiber, the expression of the phase of one of them becomes, in the approximation of weak nonlinear effect:

$$\Phi_j(t) \approx n_0 z + \Phi_{0j} + \frac{2\pi}{\lambda} n_2 z \mathcal{J}_j(t) + \frac{4\pi}{\lambda} n_2 z \sum_{k \neq j} \mathcal{J}_k(t) \qquad (4.41)$$

Besides the phase modulation imposed by SPM, another phase modulation term exists due to the presence of the channels at other wavelengths. In practical cases in which different channels are quite near (e.g., at 100 GHz spacing, that in wavelength units is a spacing of about 0.84 nm around 1.55 μm), n_2 is almost the same for all the channels and the weight of XPM is doubled with respect to the weight of SPM.

However, due to the time dependence of the intensity $\mathcal{J}_k(t)$, XPM is effective only when the intensity of the signal under consideration and the intensity of the interfering signal are superimposed in time. In a WDM transmission, due to the frequency difference of the different channels, there is a shift of one channel with respect to the other whose speed depends on the frequency distance of the considered channels. This shift, also called walk-off, reduces the efficiency of XPM and all the effects requiring phase matching.

Moreover, the effect of chromatic dispersion also causes a shift of one channel with respect to the other due to the different group velocity, implying that higher the dispersion less efficient is the XPM.

4.2.7.3 Kerr-Induced Four-Wave Mixing

Besides XPM, when several signals at a different wavelength propagate through an optical fiber, another important nonlinear interaction is caused by the Kerr effect: the so-called FWM.

Even if FWM is also generated by a couple of propagating waves, it is a four photons phenomenon, involving, in the case of two interacting wavelengths, two photons from one wavelength and one from the other, besides a photon at a different wavelength. This different wavelength photon is generated via absorption of photons of the traveling radiation and reemission through the transition at an intermediate molecular vibration level.

Thus, the nature of the FWM is more evident imagining three different radiations at angular wavelengths ω_i ($i = 1, 2, 3$) that travels through a fiber in the same direction.

The third-order nonlinear polarization can be obtained by substituting the expression of the overall field into Equation 4.34 and grouping the terms with the same frequency to determine the frequency components of the output spectrum.

Besides three components at the input frequencies, the output spectrum is composed of the four frequency components at the angular frequencies

$$\omega_j = \omega_1 \pm \omega_2 \pm \omega_3 \tag{4.42}$$

In spontaneous FWM, these new frequencies are created in a part of the spectrum that does not contain any radiation power at the fiber input, if on the other side a radiation is present at the FWM frequencies like in the case of WDM transmission, the FM component superimposes to the input field causing crosstalk.

At each FWM frequency, there is a corresponding term of the nonlinear glass polarization whose amplitude depends on the product of the amplitudes of the input fields and whose phase has the following expression

$$\theta_j = [\beta(\omega_1) \pm \beta(\omega_2) \pm \beta(\omega_3) - \beta(\omega_j)]z - [\pm\omega_1 \pm \omega_2 \pm \omega_3 - \omega_j]t \tag{4.43}$$

The term θ_j is called phase matching of the jth FWM frequency and the power of the jth FWM frequency is maximum when θ_j is equal to zero and minimum when θ_j is equal to $\pi/2$.

During the propagation through a real fiber, the phase matching for a certain FWM frequency depends on the axial coordinate z and effective generation of the new frequency happens only when the infinitesimal contributions generated in correspondence of each fiber section add together to form a sizeable optical power.

In order to have this superposition, the term depending on t in (4.43) has to vanish so that fixing selection rules for the frequencies where macroscopic FWM can be observed: they are all the frequencies where at least ω_j appears with the opposite sign with respect to the other angular frequencies.

In this case, the time dependent term in the phase matching expression vanishes with a suitable choice of ω_j and the phase matching condition becomes

$$\Delta\beta = [\beta(\omega_1) \pm \beta(\omega_2) \pm \beta(\omega_3) - \beta(\omega_j)] = 0 \tag{4.44}$$

In order to satisfy this condition, chromatic dispersion has to be negligible.

The evolution of the FWM generated waves, besides that of the waves injected at the fiber input, can be derived by solving the nonlinear wave equation. Following the method of the coupled waves equations it can be decomposed in a set of coupled wave equations, one for each wave, containing the coupling terms that derives from the nonlinear interaction.

In the so-called nondepleted pump condition, that is, when the nonlinear interaction is very low and the coupling terms in the equations describing the evolution of the injected waves can be neglected, the coupled waves equations can be analytically solved.

Assuming to have three waves at the fiber input, the power of the jth FWM wave is given by [23]

$$E_j^2(z) = 2\eta\gamma^2 d_e^2 L_e^2 E_1^2(z)E_2^2(z)E_3^2(z)e^{-\alpha z} \tag{4.45}$$

where
  L_e indicates the characteristic length defined in Equation 4.36
  d_e is the so-called degeneracy factor that is equal to 3 in case of degenerate FWM, when
    two interacting photons have the same frequency, and 6 in the other cases

γ is the so-called nonlinear coefficient

η is the FWM efficiency, that is, the parameter that represents the effectiveness of the nonlinear effect

The FWM efficiency η can be written as follows [23]:

$$\eta = \frac{\alpha^2}{\alpha^2 + \Delta\beta^2}\left[1 + \frac{4e^{-\alpha z}\sin^2(\Delta\beta\, z/2)}{(1 - e^{-\alpha z})^2}\right] \tag{4.46}$$

The dependence of the FWM efficiency on dispersion is clear from Equation 4.46, since $\Delta\beta$ cannot be equal to zero in the presence of sizable chromatic dispersion.

Approximating the propagation constants with their second-order approximation and assuming β_2 independent from the wavelength, an expression of the FWM efficiency as a function of the dispersion constant can be obtained.

The FWM efficiency in the higher-efficiency case ($\omega_1 \neq \omega_2$, $\omega_1 + \omega_2 - \omega_3 = \omega_j$) is shown in Figure 4.7 [21] versus the spacing between adjacent injected frequencies $\Delta\omega = \omega_1 - \omega_2 = \omega_2 - \omega_3$. All parameters are evaluated at 1.55 µm and two values of D are considered: 16 ps/nm/km (corresponding to a standard step index fiber) and 1 ps/nm/km (corresponding to a DS fiber).

The dependence of the efficiency from the input field spacing and on the dispersion is clear and similar plots can be used to understand the impact of FWM in WDM systems.

4.2.8 Raman Scattering

Raman scattering is the photon scattering caused by local optical phonons of the glass and in Equation 4.39 it is taken into account via the imaginary part of the third-order nonlinear susceptibility [6,7].

When a monochromatic light beam propagates in an optical fiber, spontaneous Raman scattering occurs. It transfers some of the photons to new frequencies. The scattered photons may lose energy (Stokes shift) or gain energy (anti-Stokes shift). If photons at other frequencies are already present then the probability of scattering to those frequencies is

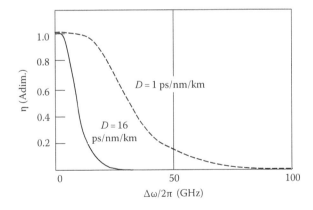

FIGURE 4.7
FWM efficiency in the higher-efficiency case ($\omega_1 \neq \omega_2$, $\omega_1 + \omega_2 - \omega_3 = \omega_j$) versus the spacing between adjacent frequencies $\Delta\omega = \omega_1 - \omega_2 = \omega_2 - \omega_3$. All parameters are evaluated at 1.55 µm and two values of D are considered: 16 ps/nm/km (corresponding to a standard step index fiber) and 1 ps/nm/km (corresponding to a DS fiber).

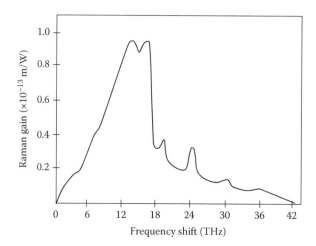

FIGURE 4.8
Raman gain spectrum in an SSMF versus the frequency shift from the pump wavelength. (After Shibate, N. et al., *J. Non Cryst. Solids*, 45, 115, 1981.)

enhanced. This process is known as stimulated Raman scattering and rely on stimulated emission due to the excited optical phonons.

In the Stokes scattering process, the only process happening in guided propagation due to the overall momentum conservation, the energy of incident photon is reduced to lower level and energy is transferred to the molecules of silica in the form of kinetic energy, inducing stretching, bending or rocking of the molecular bonds. The Raman shift $\omega_R = \omega_P - \omega_S$ (when P indicates the injected wavelength and S the Stokes wavelength produced by the nonlinearity) is dictated by the vibrational energy levels of silica.

The Stokes Raman process is also known as the forward Raman process since the Stokes wave propagates in the same direction of the input wave and the energy conservation for the process is

$$\mathcal{E}_g - \hbar\omega_p = \mathcal{E}_f - \hbar\omega_S \tag{4.47}$$

where $\mathcal{E}_g$ and $\mathcal{E}_f$ are ground state and final state energies, respectively, and in silica fiber $\omega_p - \omega_S \approx 12\,\text{THz}$ (that in the third windows are about 100 nm).

Since Raman scattering is not only a potential cause of channel crosstalk in WDM systems, but also the principle of operation of Raman amplifiers that are used in several transmission system architectures, we will delay the discussion of a detailed physical model of the Raman amplification to the next section.

Here it is interesting to note that the gain bandwidth of the Raman effect, whose experimental measure is reported in Figure 4.8 [24], is extremely wide and that, as all nonlinear effects generate new waves, the Stokes wave has an exponential growth in the limit of nondepleted pump, when the effect is thus very low.

In correspondence to the slope change of the exponential, a Raman threshold can be defined whose value is about 600 mW in conventional standard single-mode fibers (SSMFs).

4.2.9 Brillouin Scattering

Brillouin scattering is a nonlinear process that can occur in optical fibers at large field intensity. The large intensity produces compression in the material of the core of the fiber through the process known as electrostriction. This phenomenon produces density fluctuations in the fiber medium, which in turn modulates the refractive index of the medium and results in an electrostrictive nonlinearity [25].

The modulated refractive index behaves as an index grating, which is pump-induced and the scattered light is frequency shifted (Brillouin shift) by the frequency of the sound wave.

Quantum mechanically, the Brillouin shift originates from the coherent scattering between photon and local acoustic phonon, and the Brillouin shift is due to the Doppler displacement that is a consequence of the phonon momentum [26].

Brillouin scattering may be spontaneous or stimulated. In spontaneous Brillouin scattering, there is annihilation of a pump photon, which results in creation of Stokes photon and an acoustic phonon simultaneously. The conservation laws for energy and momentum must be followed in such scattering processes. For energy conservation, the Stokes shift ω_B must be equal to $(\omega_P - \omega_S)$, where ω_P and ω_S are frequencies of pump and Stokes waves.

The momentum conservation requires

$$\vec{k}_A = (\vec{k}_p - \vec{k}_S + j\,\vec{k}_{lattice}) \tag{4.48}$$

where
$\vec{k}_p$ and $\vec{k}_S$ are the pump photon and the phonon wave vectors
j is an integer
$\vec{k}_{lattice}$ takes into account the pseudo-particle nature of the phonon and accounts for other Stokes orders after the fundamental

In transmission systems, Brillouin effect has to be as small as possible; thus, only the first-order Stokes wave is to be taken into account, so that Equation 4.48 can be simplified as $\vec{k}_A = (\vec{k}_p - \vec{k}_S)$.

If v_A is the acoustic velocity, then the dispersion relation Equation 4.48 can be written as

$$\omega_B = v_A\,|\vec{k}_A| = v_A\,|\vec{k}_p - \vec{k}_S| = 2v_A\,|\vec{k}_p|\,\sin\!\left(\frac{\vartheta}{2}\right) \tag{4.49}$$

where ϑ is the angle between the pump and Stokes momentum vectors and the modules of the pump and the Stokes wave vectors are assumed almost equal due to Equation 4.48 and to the fact that the light speed is much greater than the sound speed. From the aforementioned expression, it is clear that the frequency shift depends on angle ϑ. For $\vartheta = 0$, the shift is zero, that is, there is no frequency shift in forward direction that means no Brillouin scattering. The only other possible direction in guided propagation is $\vartheta = \pi$ that represents backward direction and gives the maximum shift. The backward Stokes shift can be evaluated from Equation 4.48 obtaining

$$\omega_B = \frac{4\pi\,n\,v_A}{\lambda_P} \tag{4.50}$$

where n is the effective mode index, that is, the ratio between the light speed in vacuum and the mode group velocity.

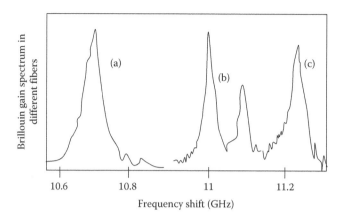

FIGURE 4.9
Brillouin gain spectra at pump wavelength 1.525 mm for a silica-core fiber (a), a depressed-cladding fiber (b), and a DS fiber (c). (After Nikles, M. et al., *J. Lightwave Technol.*, 15, 1842, 1997.)

The dependence of gain on frequency can be described evaluating the Brillouin gain spectrum. The finite life time T_B of acoustic phonons is the cause of the frequency dependence of the gain and of the small spectral width of the gain spectrum.

Solving the coupled wave equations of the input optical wave, the Stokes wave, and the acoustic wave, the following expression can be obtained for the Brillouin gain g_B

$$g_B(\omega) = \frac{g_{B0}}{1 + (\omega - \omega_B)^2 T_B^2} \tag{4.51}$$

The peak value of the Brillouin gain occurs at $\omega = \omega_B$. The gain g_{B0} depends on many parameters like concentration of dopants in the fiber, inhomogeneous distribution of dopants, and the electrostrictive coefficient. Figure 4.9 describes the Brillouin gain spectra at pump wavelength 1.525 mm for a silica-core fiber, a depressed-cladding fiber, and a DS fiber.

The inhomogeneous distribution of germania within the core of the depressed-cladding fiber used for the experiment is responsible for the double peak in the Brillouin gain spectrum of the figure, while difference in Brillouin shift among various fibers is due to the different percentage of germania in the core.

Exploiting the expression of the Brillouin gain, it is possible to develop a simple model of the evolution of the pump and Stokes wave intensities during propagation.

As a matter of fact, separating the equation for the pump and the equation for the Stokes from the general wave equation and applying the rotating wave and the slowly varying envelope approximations, the following coupled equations are obtained for the two optical waves intensities:

$$\frac{d\mathcal{J}_p}{dz} = -g_B \, \mathcal{J}_p \, \mathcal{J}_S - \alpha \mathcal{J}_p$$

$$\frac{d\mathcal{J}_S}{dz} = -g_B \, \mathcal{J}_p \, \mathcal{J}_S + \alpha \mathcal{J}_S \tag{4.52}$$

In the condition of very small nonlinear effect, the Stokes intensity can be neglected in the equation for the pump and the system (4.52) can be solved exactly.

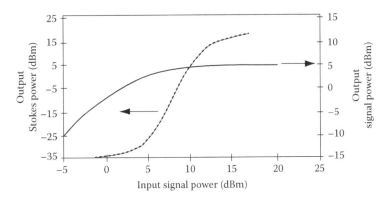

FIGURE 4.10
Power exiting from the far end (continuous line) and the near end (dotted line) of 13 km of DS fiber versus the input power at the near end of the fiber.

Since the optical power is proportional to the optical intensity via the simple relationship $JA_e = P$, where the proportionality coefficient is the effective area defined in (4.37), and since the Stokes is counter-propagating with respect to the pump, the solution for the Stokes power exiting from the fiber input facet is

$$P_S(0) = P_S(L)e^{-\alpha L}e^{g_B\, P_P(0)L_e/A_e} \tag{4.53}$$

From Equation 4.53, it is possible to evaluate the Brillouin threshold as the pump power at which the Stokes power at the input facet is equal to the pump power at the output facet.

Assuming $P_S(L)$ equal to the average spontaneous emission noise at the wavelength of 1.55 μm, thus representing the case of the absence of a coherent Stokes signal at the far end fiber facet and describing the pump evolution in the nondepleted pump approximation, the Brillouin threshold can be expressed as [27]

$$P_{th} = \frac{21\delta\, A_e}{g_B\, L_e} \tag{4.54}$$

The value of polarization factor δ lies between 1 and 2 depending on relative polarization of pump and Stokes waves. Typically, $A_e \approx 50\,\mu m^2$, $L_e \approx 20$ km, and $g_B = 4 \times 10^{-11}$ m/W for a fiber system at 1550 nm.

With these values and taking $\delta = 1$, $P_{th} \approx 1.3$ mW. The threshold power becomes just double if polarization factor is taken equal to 2. When threshold is reached, the effect of stimulated Brillouin scattering (SBS) on the signal power is described by the Figure 4.10, where the power exiting from the far and the near end of 13 km of DS fiber is plotted versus the input power at the near end of the fiber. Up to the threshold power, the transmitted power increases linearly. When scattered power attains the value equal to threshold power, the transmitted power becomes almost constant and independent of input signal power.

4.2.10 ITU-T Fiber Standards

The International Standardization Organization (the ITU-T) has standardized several types of fibers to allow system designers to rely on well-known set of fiber parameters.

TABLE 4.3

Summary of the Main ITU-T and IEC
Standards Regarding Transmission Fibers

	IEC	ITU-T
SSMF	B1.1	G.652 (A)
SSMF no water peak		G.652 (C)
Pure silica core	B1.2	G.654
DS fiber	B2	G.653
Dispersion flattened fiber	B3	—
NZD+	B4	G.655
NZD–	B4	G.655
LEAF	B4	G.655
Intermediate dispersion	—	G.566

Besides ITU-T, the most important organization that has done an extensive optical fiber standardization is International Electrotechnical Commission (IEC), whose influence is mainly present in North America.

A summary of the main recommendations of ITU-T and IEC for standardization is reported in Table 4.3.

The most common fiber in the world, deployed almost by all the carriers is the nondispersion-shifted fiber (NDSF) or G.652 fiber, also called SSMF.

This is a step index fiber optimized for low dispersion in the third windows. Even if SSMFs are the oldest among all the installed fibers, their characteristics are evolving on the push of the increasing speed of transmitted channels.

In particular, manufacturing processes are continuously improved to decrease PMD and to control the attenuation in the extreme parts of the spectrum.

Due to the relatively high value of the dispersion parameter, SSMFs allow a very good control of FWM and XPM interference in WDM systems.

Moreover, a particular class of G.652 fiber exists where the HO content in the fiber glass is so reduced that the HO attenuation peak almost disappears widening the available spectrum. This property is useful especially in coarse wavelength division multiplexing (CWDM) systems, where two of the standard channels are superimposed to the HO attenuation peak and cannot be used with SSMFs.

A few key specifications from ITU-T G.652 are reported in Table 4.4.

Second from the point of view of deployed length are the G.655 fibers. They are NZ+ and NZ– fibers, which trade a bit of attenuation with a substantially low dispersion in all the third window.

In reality, G.655 fibers have never demonstrated their capability to support 10 Gbit/s DWDM transmission systems with performance superior to that of G.652 fibers.

The continuous increase of the value of D at 1550 nm of the most recent G.655 fibers has demonstrated that the optimal trade-off between D, A_e and α has not been found yet.

DS fibers, standardized in ITU-T recommendation G.653, have on the contrary almost disappeared from the market today.

Avoiding dispersion in single-channel transmission at 1550 nm resulted in exacerbation of both XPM and FWM. Moreover, DS fibers have a higher PMD with respect to SSMF because of the core–cladding index difference two times higher than that of G.652 fibers.

TABLE 4.4

A Few Key Specifications from ITU-T G.652

	G.652 (A)	G.652 (C)
Cladding diameter (μm)	125 ± 0.7	125 ± 0.7
Core–cladding concentricity (μm)	≤0.50	≤0.50
Cladding on-circularity (%)	1	1
Cutoff wavelength in cable (nm)	≤1260	≤1260
Power attenuation in cable (dB/km)	≤0.40 (at 1310 nm)	≤0.40 (at 1310 nm)
	≤0.35 (at 1550 nm)	≤0.30 (at 1550 nm)
	≤0.40 (at 1625 nm)	≤0.40 (at 1625 nm)
Chromatic dispersion parameter (ps/nm/km)	≤3.5 at 1339 nm	≤3.5 at 1339 nm
	≤5.3 at 1360 nm	≤5.3 at 1360 nm
	≤18 at 1550 nm	≤18 at 1550 nm
Effective index (from LP group velocity)	1.466 at 1330 nm	1.466 at 1330 nm
	1.467 at 1550 nm	1.467 at 1550 nm
Mode field diameter at 1310 nm	(8.6–9.5) ± 0.7	(8.6–9.5) ± 0.7
D_3 (ps/nm^2/km)	≤0.093	≤0.093
Zero dispersion wavelength (nm)	$1300 \leq \lambda_0 \leq 1324$	$1300 \leq \lambda_0 \leq 1324$
PMD (ps/km$^{1/2}$)	≤0.20	≤0.50

A few key requirements from G.653 and from G.655 are reported in Tables 4.5 and 4.6, respectively.

4.2.11 Polarization Maintaining and Other Special Telecom Fibers

A polarization maintaining (PM) fiber is an optical fiber in which if the input field is injected along a predefined linear polarization (the polarization of the only LP fundamental mode of the fiber), this polarization is maintained during propagation [28].

Several different designs of PM fiber are used. Most work by inducing stress in the core via a noncircular cladding cross section, or via rods of another material included within

TABLE 4.5

A Few Key Requirements from G.653

	G.653
Cladding diameter (μm)	125 ± 1.0
Core–cladding concentricity (μm)	≤0.8
Cladding on-circularity (%)	2
Cutoff wavelength in cable (nm)	≤1270
Power attenuation maximum in cable (dB/km)	≤0.35 (at 1550 nm)
Chromatic dispersion parameter (ps/nm/km)	≤3.5 at 1550 nm
Mode field diameter at 1550 nm	(7.8–8.5) ± 0.8
D_3 (ps/nm^2/km)	≤0.093
Zero dispersion wavelength (nm)	$1525 \leq \lambda_0 \leq 1575$
PMD (ps/km$^{1/2}$)	≤0.5

TABLE 4.6

A Few Key Requirements from G.655

	G.655 (C)
Cladding diameter (μm)	125 ± 1
Core–cladding concentricity (μm)	≤0.80
Cladding on-circularity (%)	2
Cutoff wavelength in cable (nm)	≤1450
Power attenuation in cable (dB/km)	≤0.35 (at 1550 nm)
	≤0.40 (at 1625 nm)
Chromatic dispersion parameter (ps/nm/km)	≤1.0 at 1530 nm
	≤10 at 1565 nm
Chromatic dispersion coefficient sign	Positive or negative
Mode field diameter at 1550 nm	(8.0–11) ± 0.7
PMD (ps/km$^{1/2}$)	≤0.20

the cladding. Several different shapes of rods are used, and the resulting fiber design has the standardized name such as "PANDA" and "Bow-tie," as it is shown in Figure 4.11. There are small differences between different types of PM fiber design, more in the manufacturing process than in the final product performance.

It is to underline that a PM fiber does not polarize light as a polarizer does. Rather, PM fiber maintains the input linear polarization if it is launched into the fiber along the correct direction. If the polarization of the input light is not aligned with the stress direction in the fiber, the output will vary between linear and circular polarization (and generally will be elliptically polarized).

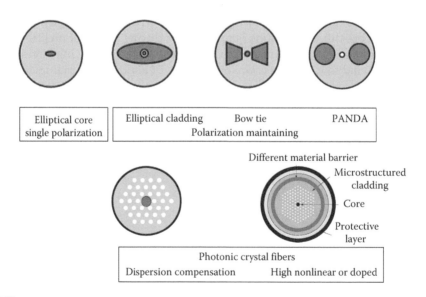

FIGURE 4.11

Schematic cross section of different special fibers for use in telecommunication systems. Polarization maintaining and single-polarization fibers are sketched in the upper part of the figure, examples of PCFs are reported in the lower part.

Polarization maintaining fibers are generally characterized by a minimum loss greater than 1 dB/km (typical for a PANDA fiber is 1.5 dB/km) that, besides the high cost, makes them unsuitable for transmission. However, there are several applications requiring small pieces of a PM fiber, such as optical components pigtailing, fiber optic sensing, interferometry, and quantum key distribution. They are also commonly used in telecommunications for the connection between a source laser and a modulator, since the modulator requires polarized light at the input.

Single-polarization fibers [28,29] are special optical fibers that can transmit light with a certain linear polarization direction, whereas light with the other polarization directions is either not guided or experiences strong optical losses. Such fibers should not be confused with PM fibers, which guide light with any polarization state, but can preserve a linear polarization state when the polarization direction is properly aligned with the birefringence axis.

In many cases, single-polarization guidance occurs in only a limited wavelength range. Outside that range, both polarization directions or no light at all may be guided. Also, some fibers exhibit a limited extinction ratio.

Different principles of operation can be utilized for single-polarization fibers. A common approach is the use an elliptical core, which introduces strong birefringence and also a polarization dependence of a cutoff wavelength, so that only light with one polarization direction is guided, whereas the fiber is a leaky waveguide for the other polarization.

After the fibers having particular polarization properties, a great importance in the design of DWDM systems is the dispersion compensating fibers (DCFs) [30,31].

This kind of fiber is designed to have a strong dispersion in the third transmission window so that a relatively short fiber piece can compensate an entire SSMF span from 70 to 100 km.

This result is achieved by suitably shaping the refraction index in the core and in the cladding of the fiber.

A typical index shape of a DCF is the segmented core profile represented in Figure 4.4. Two figures of merit can be given for DCFs:

1. *Efficiency (ε)* that is dispersion divided by the dispersion slope for a standard SSMF divided by the same ratio but for the DCF.

2. *Figure of merit (FM)* that is minimum dispersion in the C band divided by maximum attenuation in C band.

Efficiencies greater than 60% and FM as high as 150 can be obtained from commercial products that are completely sufficient for the field applications in DWDM systems. For example, a DCF length having 11.5 dB of loss can compensate 1640 ns dispersion in the C band (1545–1606 nm).

Photonic crystal fibers (PCFs) [32] are another type of special fibers that could find wide application in future telecommunication system mainly in active and passive fiber devices. They are an attempt to incorporate the bandgap ideas of photonic crystals into the fiber structure by stacking periodically a regular array of channels and drawing into fiber form (see Figure 4.11).

Two classes of PCFs exist [33,34], according to their mechanism for field confinement. Those with a solid core, or a core with a higher average index than the microstructured cladding, operates on the same index-guiding principle as conventional optical fiber.

However, they can have a much higher effective refractive index contrast between core and cladding, and therefore, they can have much stronger confinement for applications in nonlinear optical devices or, polarization-maintaining fibers. Alternatively, a "photonic bandgap" fiber can be created, in which the light is confined by a photonic bandgap created by the microstructured cladding—such a bandgap, properly designed, can confine light in a lower-index core and even a hollow (air) core [35].

In general, regular structured fibers such as PCFs, have a cross section (normally uniform along the fiber length) microstructured from one, two or more materials. Most commonly, the microstructure is periodically arranged over much of the cross section, usually as a "cladding" surrounding a core where light is confined. For example, the fibers demonstrated in [36] consisted of a hexagonal lattice of air holes in a silica fiber, with a solid or hollow core at the center where light is guided. Other arrangements include concentric rings of two or more materials.

Recently, it has been recognized that the periodic structure is actually not the best solution for many applications. This idea has been applied to reduce bend losses, and for achieving structured optical fibers with propagation losses below that of the SSMF.

PCFs with different designs of the hole pattern can have very remarkable properties, strongly depending on the design details:

- Single-mode guidance over very wide wavelength regions (endlessly single-mode fiber) is obtained for small ratios of hole size and hole spacing [37].

- Extremely small or extremely large mode areas are possible. These lead to very strong or very weak optical nonlinearities. PCFs can be made with a low sensitivity to bend losses even for large mode areas [38].

- Certain hole arrangements result in a photonic bandgap, where guidance is possible even in a hollow core, as a higher refractive index in the inner part is no longer required.

- Asymmetric hole patterns can lead to extremely strong birefringence for polarization-maintaining fibers [39]. This property can also be combined with large mode areas.

- Strongly polarization-dependent attenuation (polarizing fibers) [40] can be obtained in different ways. For example, there can be a polarization-dependent fundamental mode cutoff, so that the fiber guides only light with one polarization in a certain wavelength range.

- Similarly, it is possible to suppress Raman scattering [41] by strongly attenuating longer-wavelength light.

- Multicore designs are possible, for example, with a regular pattern of core structures in a single fiber, where there may or may not be some coupling between the cores.

- In the field of telecommunications however, probably the most important application of the holey fibers is in optical amplifiers.

Laser-active PCFs for fiber lasers and amplifiers can be fabricated by using, for example, a rare-earth-doped rod as the central element of the preform assembly. Rare earth dopants (e.g., ytterbium or erbium) tend to increase the refractive index, but this can be precisely compensated, for example, with additional fluorine doping, so that the guiding properties are determined by the photonic microstructure only and not by a conventional-type refractive index difference.

Generally, PCFs are constructed by the same methods as other optical fibers: first, one constructs a "preform" on the scale of centimeters in size, and then heats the preform and draws it down to the fiber diameter, shrinking the preform cross section but maintaining the same features. In this way, kilometers of fiber can be produced from a single preform.

Most PCFs have been fabricated in silica glass, but other glasses have also been used to obtain particular optical properties (such has high optical nonlinearity). Using this principle, for example, fibers optimized to have a very high Raman efficiency have been fabricated, which are conceived for the design of high-gain low-noise Raman amplifiers.

However, special care is required in various respects:

- Ends of PCFs may not be cleaned with liquid solvents, such as ethanol, as capillary forces may pull them into the holes.

- Pulling very long length of microstructured fibers is not easy due to the need of maintaining the cross section constant and in general there is a limit to the length of a piece of such fibers.

- Cleaving and fusion splicing PCFs is in principle possible, but can be more difficult, particularly for fibers with large air content. During fusion splicing, the air may expand and distort the fiber structure.

- Connections between fibers are also possible with a variety of mechanical splices, fiber connectors, protected patch cables, beam expansion units, etc.

- Even when the splicing process works well, there may be a substantial coupling loss due to a mismatch of mode areas, for example, when a small-core PCF is coupled to a standard single-mode fiber. There are special tapered single-mode fibers and tapered PCFs for enhancing the coupling efficiency, but these may not be easily available.

4.2.12 Fiber Cables

Simply covered by the polymer jacket, an optical fiber is impossible to deploy in the field. In order to be used, an optical fiber has to be packaged in a fiber cable [42].

The major benefits of fiber optic cabling are as follows:

- *Easy handling:* Some communication systems require tens or even hundreds of fibers (such as a metro backbone system). However, fibers in a cable make it very easy to install and maintain.

- *Protection from damaging forces:* Fiber optic cables have to be pulled into place through ducts (outdoor) or conduits (indoor). Pulling eyes are attached to the strength members or cable outer jackets. This is critical for isolating the fibers from the applied pulling forces. Glass fibers cannot endure more than 0.1%–0.2% elongation during installation.

- *Protection from harsh environment factors:* Cable structures protect fibers from moisture (outdoor cables), extreme temperature (aerial cables) and influx of hydrogen into the fiber (which causes light absorption peak at 1380 nm, which in turn impair fibers' transmission properties).

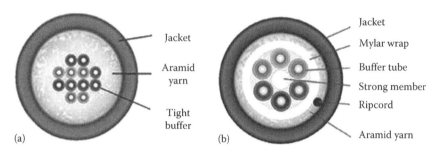

FIGURE 4.12
Scheme of the structure of a fiber cable: (a) tight-buffered cable and (b) loose tube cable.

For current telecommunication applications, practically only single-mode fibers are used, also in the access area. Thus, all the telecommunication cables include single-mode fibers, even if multimode fiber cables and even mixed fiber cables are still used for other applications or for compatibility with legacy systems generally in the access area.

The first element in determining the characteristics of a fiber cable is the choice of materials used to package the cable.

Depending on the environment in which the cable has to be deployed, different requirements are imposed to the design and different materials have to be chosen.

For indoor cables, used for LANs or internal cabling of carriers central offices, fire safety is the number one factor. Moreover, especially in the LAN application, protection of the inner fibers against excessive bending and easy installation in the typical LAN conduit is another key requirement.

In the case of outdoor cables, moisture resistance and temperature tolerance are the major factors when choosing materials. They also need to be ultraviolet resistant and capable of sustaining the stresses related to typical installation procedures without damage.

Finally, aerial self-supporting cables must endure extreme temperature ranges from sunlight heat to freezing snow. They also must survive high wind loading.

The structure of a fiber cable, schematically reported in Figure 4.12, is designed to allow these requirements to be fulfilled.

The fibers are sustained by an internal strength member made from steel, Aramid yarn [43], or some other material with high resistance to traction and high elasticity. The role of the strength member is to give mechanical resistance to the cable and to allow cable traction during installation. When a cable is pulled into a duct, the tension is applied to the strong member instead of the fibers.

Aramid yarn (Kevlar is its commercial name that is a product trademark) is a yellow color, fiber-looking material. It is strong and is used to bundle and protect the loose tubes or fibers in the cable.

Every fiber is individually protected with a polymer tube. These polymer tubes are also colored so that the fibers can be recognized at the two extremes of the cable. This is an important cable characteristic, since the circular symmetry make it impossible to recognize the single fiber from its position.

The set of fibers is protected as a whole by different concentric tubes, each of which has a different role in assuring either cable resistance to mechanical, chemical or thermal aggressions or other cable characteristics, for example, resistance to burning or nonpoisonous smoke if flaming.

A gel compound fills buffer tubes and cable interiors, making the cable impervious to water. It needs to be completely cleaned off when the cable end is stripped for termination.

The outer tube is called jacket and has a particular importance since it is the part of the cable that is exposed to the outside environment.

For this reason, the jacket material is quite important; in a typical fiber cable the jacket is composed of the following materials:

- Polyethylene (PE) is the standard jacket material for outdoor fiber optic cables. PE has excellent moisture and weather-resistance properties. It has very stable dielectric properties over a wide temperature range. It is also abrasion resistant.
- Polyvinyl chloride (PVC) is the most common material for indoor cables; however, it can also be used for outdoor cables. It is flexible and fire-retardant. PVC is more expensive than PE and that has an impact for very long cables.
- Polyvinyl difluoride (PVDF) is used for plenum cables because it has better fire-retardant properties than PE and produces little smoke.
- Low smoke zero halogen (LSZH) plastics are used for a special kind of cable called LSZH cables. They produce little smoke and no toxic halogen compounds. But they are the most expensive jacket material.

Embedded just below the cable jacket there is the ripcord: a thin but very strong thread. Its role is to split the cable easily without harming cable interiors.

Fiber optic cables are available in a wide variety of physical constructions. They can be anything from simple simplex or duplex (zipcord) cables used for jumpers to 144-fiber cable for intercity transmission.

Most of the fibers used in these cables come down to two basic configurations—900 μm tight-buffered fibers or 250 μm coated fibers (also called bare fibers). Actually, tight-buffered fibers cover a coated fiber (the coating is soft plastic) with a thick layer of harder plastic, making it easier to handle and providing physical protection.

There are two basic types of fiber optic cable constructions: Tight-buffered cable and loose tube cable. Their structure is shown in Figure 4.12a and b.

Multiple color-coded 900 μm tight-buffered fibers can be packed tightly together in a compact cable structure, an approach widely used indoors; these cables are called tight-buffered cables. Tight-buffered cables are used to connect outside plant cables to TE, and also for linking various devices in a premises network.

Multifiber, tight-buffered cables often are used for intrabuilding, risers, general building, and plenum applications.

On the other hand, multiple (up to 12) 250 μm coated fibers can be put inside a color coded, flexible plastic tube, which usually is filled with a gel compound that prevents moisture from seeping through the hollow tube. Buffer tubes are stranded around a dielectric or steel central member. Aramid yarn is used as primary strength member. Then an outer PE jacket is extruded over the core. These cables are called loose tube cables.

Loose tube structure isolates the fibers from the cable structure. This is a big advantage in handling thermal and other stresses encountered outdoors, which is why most loose tube fiber optic cables are built for outdoor applications.

Loose tube cables typically are used for outside-plant installation in aerial, duct, and direct-buried applications.

4.3 Optical Fiber Amplifiers

In this section, we will describe optical fiber–based amplifiers for telecommunication applications.

A specific attention to optical amplifiers is justified by the fact that design of single amplifiers and of an amplifiers cascade is quite different from what happens in the electronics domain. As a matter of fact, optical amplifiers behavior is dominated by the quantum mechanical properties of the light field and a quantum mechanical approach is required at least to set the main amplifier properties.

4.3.1 Basic Theory of Optical Amplifiers

Due to the specific properties of optical amplifiers, before dealing with particular devices and technologies it is important to fix a few characteristics that are common to all amplifiers working in a regime where thermal noise is negligible and the amplifier signal to noise ratio is determined by quantum noise.

These characteristics can be derived directly from the base principles of quantum mechanics applied to a system that has the characteristic to multiply the input signal by a constant greater than one, called gain.

Thus, in this section we will review the basic theory of quantum amplifiers [44], thus placing the bases for the analysis of specific amplifiers design.

4.3.1.1 Quantum Noise

An optical amplifier can be seen as a system designed to multiply the input optical power by a gain greater than one.

In order to transfer this definition in a quantum mechanical formulation, let us start from the expression of the quantized basic mode of the fiber.

The modal fiber analysis is not altered by the second quantization of the optical field [45] and we will assume in this so basic an analysis that propagation will be single polarization and that the fiber mode can be approximated as a plane wave.

In the Heisenberg representation [45], a single frequency component of the field at the amplifier input that is also a single frequency component of the fundamental mode of the fiber, can be written, starting from Equation 4.2 as

$$\hat{E}_{in}(t, \omega, z) = \sqrt{P_{in}}\ \hat{a}_{in}(\omega)\ e^{-i\theta(t, w, z)} \tag{4.55}$$

where
   $\wedge$ indicates operators
   $\hat{a}_{in}$ is the input creation operator and the field is measured, as it is generally advantageous in telecommunication related problems, in W^{-1}

The frequency dependent $\hat{a}_{in}$ contains all the amplitude fluctuations (e.g., the modulation), and the phase can be written as

$$\theta(t, \omega, z) = \omega\, t - k\, z \tag{4.56}$$

We will also assume that the field bandwidth, which we will call B measured in angular frequency, is much smaller than the amplifier gain bandwidth, hypothesis matching practical situations very well.

After traveling through the amplifier, following the amplifier definition and neglecting for the moment any noise contribution from temperature or spurious radiation, the output field frequency component can be written as a function of the amplifiers spectral gain $G(\omega)$ as

$$\hat{E}_{out}(t,\omega,z) = \sqrt{G(\omega)P_{in}}\ \hat{a}_{in}(\omega)e^{-i\theta(t-\tau,\omega,z)} + \sqrt{P_{in}}\ \hat{f}\ (\omega)e^{-i\theta(t-\tau,\omega,z)} \tag{4.57}$$

where τ is the delay due to the propagation through the amplifier, and with a redefinition of the time origin can be eliminated. The operator $\hat{f}(\omega)$ is added at the output to comply with the property that the output field has to be written as a function of the output creation operator as

$$\hat{E}_{out}(t,\omega,z) = \sqrt{P_{in}}\ \hat{a}_{out}(\omega)\ e^{-i\theta(t,w,z)} \tag{4.58}$$

and $\hat{a}_{out}(\omega)$ has to commute following the rule

$$[\hat{a}_{out}(\omega); \hat{a}_{out}^{+}(\omega')] = \delta(\omega - \omega') \tag{4.59}$$

where
+ indicates the Hermitian conjugate
$\delta(\omega - \omega')$ indicates the Dirac distribution

The relation (4.59) requires the presence in Equation 4.57 of the term containing $\hat{f}(\omega)$, that physically represents the contribution of the spontaneous emission that is unavoidable at the amplifier output.

Using Equations 4.57 and 4.58, the output creation operator expression can be obtained. Substituting this expression in (4.59), the following expression is obtained for the auto-commutation of the operator $\hat{f}(\omega)$:

$$[\hat{f}(\omega); \hat{f}^{+}(\omega')] = [1 - G(\omega)]\delta(\omega - \omega' \tag{4.60}$$

In order to provide a characterization of the quantum noise, the optical power flux through the amplifier can be evaluated.

Using the symmetric power flux expression [44], and assuming that the narrowband signal can be extracted from the wavelength integral with respect to the wideband $G(\omega)$, the expression of the optical power flux can be written as

$$P_{out}(t,z) = G(\omega_0)P_i(t,z) + Q(t,z) \tag{4.61}$$

where
ω_0 is the signal central angular frequency
$Q(t,z)$ is the quantum noise instantaneous power, whose average value is given by

$$\langle Q(t,z) \rangle = -\hbar\,\omega_0 \int [1 - G(\omega)]d\omega \approx \hbar\,\omega_0\,\frac{B}{2\pi}[G(\omega_0) - 1] \tag{4.62}$$

where $B/2\pi$ is the signal frequency bandwidth (remember that B was measured in angular frequency).

The presence of the quantum noise in a real amplifier is superimposed to the presence of the thermal noise, due to the black body emission of the fiber material and of the background. However, in almost all the optical applications, the dominant noise is the quantum noise and often the contribution of the thermal noise can be neglected.

4.3.1.2 Stationary Behavior of a Two-Level Amplifier

Once the role of quantum noise is clarified, we are ready to present a simple, but general theory of optical amplification. This will be based on the analysis of a two-level amplifier; thus, its results describe only qualitatively the behavior of real amplifiers. However, it will be useful to understand real amplifier behavior.

A two-level amplifier is a system where a material interacts with the light field only through one discrete transition between two stationary energy levels: the ground level at energy $\mathcal{E}_0$ and an excited state at energy $\mathcal{E}_1$. In order to simplify the model as far as possible, we will assume perfectly resonant interaction with a field of angular frequency $\hbar \omega = \mathcal{E}_1 - \mathcal{E}_0$.

In order to generate optical gain, a pump creates population inversion in the medium so that the incoming radiation is amplified through stimulated emission.

We will assume that, as in all practical cases in telecommunication, the amplification occurs during single-mode guided propagation (e.g., in an optical fiber).

In this case, neglecting the transversal mode shape and applying to the wave equation the usual rotating wave and slowly varying approximations [16,44], the stationary LP field that propagates along the amplifier medium follows this simple equation

$$\frac{dE}{dz} = \left[\frac{1}{2}g(\omega, P) + in(\omega)\right]E + \mu(t, z) \tag{4.63}$$

where $\mu(t, z)$ represents the quantum noise contribution and the local gain $g(\omega, P)$ can be written under quite general conditions as [44]

$$g(\omega, P) = \frac{g_0}{1 + (\omega - \omega_0)/\Delta\omega^2 + P/P_{sat}} \tag{4.64}$$

where
$\Delta\omega$ is the local amplification bandwidth
P_{sat} is the local saturation power

The noise process is generated by the amplification along the active medium of spontaneous emission; thus, in general, it should be expressed as a random superposition of coherent states generated by the amplification of each individual spontaneous emission photon. In conditions of high local gain, all these states are with a good approximation uncorrelated and populated by a high average number of photons; thus, the noise process can be assumed, within the amplification bandwidth, as a white Gaussian noise for the central limit theorem.

Under this approximation, taking into account imperfect population inversion,

$$\langle\mu(t, z)\mu(t + \Delta t, z + \Delta z)\rangle = \langle E_0^* | [\hat{f}(t, z); \hat{f}^+(t + \Delta t, z + \Delta z)] | E_0 \rangle = \hbar\,\omega\frac{\rho_1}{\rho_1 - \rho_0}g(\omega, P)\delta(\Delta t)\delta(\Delta z)$$

$$\tag{4.65}$$

where

$|E_0\rangle$ is the reference electrical field state in the Heisenberg representation

ρ_1 and ρ_0 represent the populations of the excited and ground states, respectively

and the population inversion term is inserted in Equation 4.65 phenomenologically to take into account a realistic condition (for a more rigorous derivation on the ground of the density matrix equations see [44,45]).

To describe the average evolution, Equation 4.63 can be averaged and added to the complex conjugate, obtaining an equation for the power density flux:

$$\frac{dP}{dz} = g(\omega, P)P \tag{4.66}$$

If the input power P_{in} is much smaller than the local saturation power P_{sat}, the solution of Equation 4.66 can be obtained in the following form, where the amplifier gain definition $G = P_{out}/P_{in}$ has been used to eliminate the output power in favor of the amplifier gain

$$(1-G)\frac{P_{in}}{P_{sat}} = \log\left(\frac{G}{G_0}\right) \tag{4.67}$$

where the unsaturated gain G_0 is given by $G_0 = \exp[g(\omega, P)L]$, L being the amplifier length.

The main phenomena related to the stationary working of an optical amplifier can be deduced from Equation 4.67.

The first point is gain saturation. As far as the input power increases, the amplifier passes from linear working, where the output power is simply proportional in average to the input one through the unsaturated gain G_0, to saturated working where the output power is constant and it is the gain that decreases while increasing the input power.

The gain behavior versus the input power is shown in Figure 4.13 for $G_0 = 18\,dB$ and $P_{sat} = 10\,mW$.

From Figure 4.13 it is also clear that the so-called macroscopic input saturation power P_S does not coincide with the microscopic saturation power P_{sat}.

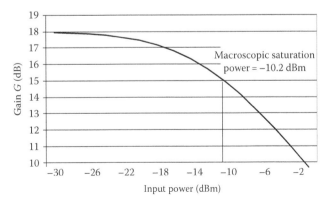

FIGURE 4.13
Gain saturation in an ideal two-level amplifier.

As a matter of fact, defining P_S as the input power at which the gain G is reduced to half the linear gain G_0, we get

$$P_s = \frac{2\log(2)P_{sat}}{(G_0 - 2)} \tag{4.68}$$

and the output saturation power is given by $P_S^{out} = G_S P_S = G_0 P_S / 2$.

This is essentially due to the fact that the macroscopic amplifier can be seen as the cascade of a huge number of microscopic sections with saturation power P_{sat}. The overall effect is integrated over the length and it is intuitive that the saturation is reached far sooner than in the single isolated section.

The same reason causes the overall gain bandwidth to be smaller than the local gain bandwidth. Defining B as the frequency bandwidth at which the gain G_0 is reduced to half (reduced by 3 dB), we get

$$B = \frac{\Delta\omega}{2\pi}\sqrt{\frac{\log(2)}{g_0\,L - \log(2)}} \tag{4.69}$$

The behavior of the amplifier bandwidth versus the linear gain G_0 is plotted in Figure 4.14 for a microscopic bandwidth of 2.5 THz.

Considering the amplifier noise performances, Equation 4.63 can be formally solved taking into account also the noise term and using the expression of the noise autocorrelation function, the following expression is obtained for the noise variance

$$\sigma_\mu^2 = \hbar\,\omega\,\frac{\rho_1}{\rho_1 - \rho_0}(G - 1)B_\mu \tag{4.70}$$

where B_μ is the noise bandwidth that is fixed either from a filter at the amplifier output or from the normalized integral of the amplified spontaneous emission (ASE) spectrum.

In practical communication systems, it is frequently important to evaluate the signal to noise ratio after the detection of the optical signal through a photodiode. For this reason, often optical amplifiers are characterized from a noise point of view through the so-called noise factor NF, defined as the ratio between the signal to noise ratio of the detected input signal (SN_0) and the signal to noise ratio of the same signal after amplification (SN).

Since in practical systems the signal incoming into the receiver is always with a very god approximation in a coherent state [44], in the absence of amplification, SN_0 is determined

FIGURE 4.14
Gain bandwidth of an ideal two-level amplifier versus the linear gain G_0 for a microscopic bandwidth of 2.5 THz.

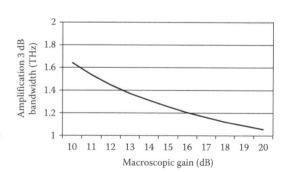

by the quantum noise related to a coherent state (i.e., the variance of the photon number in a coherent state).

This quantity can be thus expressed as

$$SN_0 = \frac{P_0}{2\hbar\,\omega_0\,B_e} \tag{4.71}$$

where B_e is the electrical signal bandwidth after detection by a photodiode.

After amplification, the signal has the expression

$$E_{out}(t,L) = \sqrt{G\,P_0}\,e^{-i\theta(t,L)} + \mu(t,L) \tag{4.72}$$

After detection, several random components are added to the term proportional to the optical power and that constitute the overall noise:

- The square of the quantum noise due to the input coherent state fluctuations (generally called shot noise)
- The square of the amplifier quantum noise (also called ASE)
- A beat term between the signal and the shot noise
- A beat term between the signal and the ASE noise
- A beat term between shot and ASE noise

indicating these current contributions with $c_i(t)$, ($i = 1,\ldots, 5$), we have that the overall current after the photodiode has the expression

$$c(t) = R_p\,G\,P_0 + \sum_{j=1}^{S} C_j(t) \tag{4.73}$$

where R_p is the photodiode quantum efficiency. As the noise terms are uncorrelated, their variances can be added. Evaluating all the variances and calculating the SN, the expression of NF can be easily found.

In telecommunication however, the used amplifiers are high-gain, low-noise components; thus, the signal arriving to the photodiode after amplification is orders of magnitude greater than the power of the shot noise.

On this ground, for practical cases, all the terms containing the shot noise can be neglected, besides the term containing the square of the ASE noise, which has quite a smaller power with respect to the term containing the beat between the ASE and the signal.

Retaining only the fourth term in the list given earlier, NF is immediately evaluated as

$$NF = 2\frac{\rho_1}{\rho_1 - \rho_0}\frac{(G-1)}{G} \tag{4.74}$$

It is apparent that in the ideal case, perfect population inversion and almost infinite gain, the minimum value of NF is 2. Of course, practical amplifiers are characterized by values of NF greater than 2 and generally, greater the gain, greater the value of NF since in the presence of great power inside the amplifiers the spurious phenomena causing a signal power decrease or a noise increase are more evident.

4.3.1.3 Dynamic Behavior of a Two-Level Amplifier

In practical systems, the optical signal to be amplified is modulated to transmit information. A stationary amplifier model is justified only if the relaxation time of the inverted medium is much longer than the typical signal fluctuation that generally can be expressed by means of the inverse of the bit rate R (i.e., the bit duration T).

This is not always the situation, and if the bit time is not much smaller than the relaxation time, a dynamic model has to be adopted since amplification changes the signal shape introducing distortions.

In these situations, the amplifier rate equations have to be considered, that in general are stochastic equations due to the ASE noise terms. Managing a system of two stochastic rate equations is quite difficult, but can be done [46] and it can be demonstrated that the noise properties of the amplifier are not affected in a relevant way by working in a dynamic regime.

Thus, here we will consider some dynamic properties of a two-level amplifier in a noiseless picture [7].

The rate equations write in this approximation

$$\frac{d\rho_1}{dt} = \mathcal{J} - \frac{\rho_1}{\tau} - g\frac{P(t,z)}{\hbar\omega}$$

$$\frac{dE}{dz} + \frac{1}{v_g}\frac{dE}{dt} = \left[\frac{1}{2}g(\omega,P) + in(\omega)\right]E \tag{4.75}$$

where $\mathcal{J}$ is the pumping rate and the local gain is related to the population inversion through the efficiency by $g = \gamma(\rho_1 - \rho_0)$.

From Equation 4.75, the population inversion can be eliminated by using the conservation of the number of molecules $\Gamma = (\rho_1 + \rho_0)$ and the expression $g = \gamma(\rho_1 - \rho_0)$ of the gain.

Moreover, the equation of the field can be separated into an equation for the power (obtained by adding it to the complex conjugate) and a separate equation for the phase.

At the end, defining the relative time $\theta = t - z/v_g$ the following system is obtained:

$$\frac{\partial g}{\partial \theta} = \frac{g_0 - g}{\tau} - \frac{g}{\tau}\frac{P(t,z)}{P_{sat}(t,z)}$$

$$\frac{\partial P(t,z)}{\partial z} = gP(t,z) \tag{4.76}$$

There is a standard procedure to arrive to a simplified form of these equations that can be solved by numerical algorithms [18].

The first effect to observe is dynamic gain saturation. Let us consider, as a simple example, a two-level amplifier with the parameters of Table 4.7. In Figure 4.15, the input pulse, the dynamic gain, and the output pulse are shown versus time.

The amplifier works in deep saturation, since the input pulse has a peak power of −3 dBm against a macroscopic saturation power of −5 dBm. Moreover, the lifetime of the excited molecules is on the order of the input pulse duration.

From the figure, the effect of dynamic gain saturation and the consequent distortion of the output pulse is evident. This distortion, as it can be appreciated from the figure, creates a much steeper growing front of the pulse with respect to the input pulse. If the pulse were

TABLE 4.7

Parameters for the Two-Level Amplifier Used in the Quantitative Examples of This Section

P_s	-5.21390228 dBm
P_{sat}	$5.00\text{E}{-}02$ W
G_0	20 dB
τ	$1.00\text{E}{-}09$ s
g	0.2
L	10 m

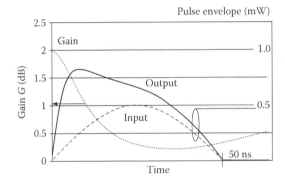

FIGURE 4.15

Distortion of the pulse at the output of an ideal two-level amplifier with the lifetime of the excited state of the same order of the pulse duration.

destined to be detected after filtering, part of the power will be rejected by the detection filter due to its high-frequency collocation.

Let us now imagine that a pulse train is modulated so that a pulse represents a bit "one" and the absence of a pulse a bit "zero." In the absence of a pulse, the signal level is assumed constant and very small.

In Figure 4.16, the effect of an amplification of such a signal with an amplifier with a lifetime smaller than the bit time is shown. In this case, the bit time is 5 ns (corresponding to a bit rate of 200 Mbit/s) and the amplifier lifetime is 1 ns. Moreover, the amplifier works again in deep saturation, with a pulse peak power of 50 mW.

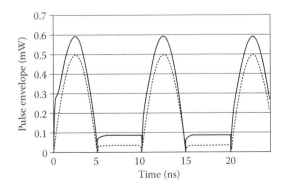

FIGURE 4.16

Dynamic range reduction of a return-to-zero (RZ) signal passing through an ideal amplifier with the lifetime of the excited state of the same order of the bit duration.

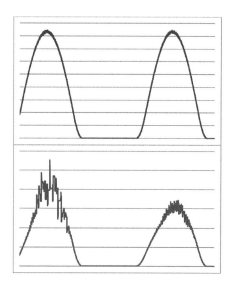

FIGURE 4.17
Instantaneous power of the optical wave after narrowband filtering around the shortest wavelength signal when two signals that are 10 GHz distant in frequency are injected into an ideal amplifier; both the signals are amplitude modulated with a train of pulses at 100 Mbit/s.

The effect of the amplifier dynamic nonlinearity is that, when the pulse is present, the gain is heavily saturated and the output pulse is practically equal to the input one, while when there is practically no signal, the gain is high and the small residue power present when a zero is transmitted is amplified.

In the case of the figure, that is quite extreme to better show the effect, the energy transmitted in the zero intervals is approximately multiplied by three, but the "one" energy increases only of 7%. The signal dynamic ratio (energy of a one divided energy of a zero) is thus reduced from 9.5 to 6.0 dB, causing a great degradation of the performances of a receiver that has to detect the signal and extract the binary sequence.

A final effect that we would like to observe is the interference between frequency multiplexed channels due to gain dynamics.

This phenomenon is illustrated in the example of Figure 4.17 where two signals at different frequencies that are 10 GHz distant are injected into the amplifier; both the signals are amplitude modulated with a train of pulses at 100 Mbit/s. The plot reported in the figure shows the instantaneous power of the optical wave after narrowband filtering around the shortest wavelength signal.

In Figure 4.17a, the result is reported without the second wavelength, and in Figure 4.17b, the second wavelength is present. The interference due to the dynamic amplifier gain is clear from the plots, and of course it would be a problem in case of DWDM systems.

Up to now we have neglected the field phase, whose equation is very simple

$$\frac{d\varphi}{dz} = -n(z) \tag{4.77}$$

In the ideal case, this equation hides no problems, but in some practical amplifiers, the same mechanism that causes population inversion also causes the refraction index to depend on local gain and through it on the input signal. In this case, a further nonlinear effect is added to the nonlinear gain dynamics that creates a phase modulation in correspondence to a traveling amplitude-modulated signal.

From a qualitative point of view, this is an effect similar to Kerr nonlinearity in fibers and the related effects take the same name: self and cross phase modulation, FWM, and so on.

Naturally, since the amplifier dynamics is different from the fiber Kerr effect, the quantitative description of these phenomena will be different. We will present the results of this kind of analysis talking about semiconductor optical amplifiers in Chapter 10.

4.3.1.4 Amplifiers Functional Classification and Multistage Amplifiers

In practice, an amplifier for use in telecommunication rarely is composed of the bare active medium and the pump: it is a complex system. An example of practical amplifier scheme is provided in Figure 4.18 [47].

Besides the pumped amplifying medium, the amplifier includes at least an isolator and a filter at the output; the isolator is needed to avoid the reflections from the line entering into the amplifier changing its behavior, the filter is needed to limit the ASE power propagating along the line and entering in the successive amplification stages.

Depending on the position they occupy in a transmission system, amplifiers are specialized as follows:

- *Buster amplifiers:* They are used generally after the receiver to enhance the power sent along the line; they are built to have a high saturation power and to amplify efficiently a relatively high input power signal coming out from the transmitter; in these amplifiers, the linear gain is not so high since the noise factor NF has to be maintained within low values and the key parameter is the saturation power.

- *Preamplifiers:* They are generally used in front of the receiver; thus, they have to be designed with a very low noise factor, moreover, having a low input power and working generally in high saturation to provide stabilization of the signal arriving to the receiver, the saturation power can be lower than in booster amplifiers. The gain should not be huge, but has to be sufficient to send to the receiver a signal power high enough to neglect shot noise and receiver thermal noise; summarizing for preamplifier amplifiers, the key parameter is the noise factor.

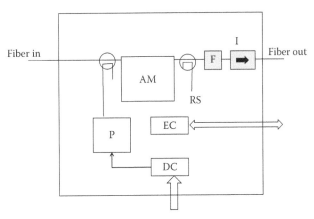

FIGURE 4.18
Block scheme of an amplifier system. AM, active medium; P, pump; EC, electronic control; DC, power supply; RS, residual pump; F, filter; I, isolator.

- *Line amplifiers:* They are used along the line to compensate propagation losses; moreover, they are also used to compensate losses introduced by in-line optical components like DCFs and optical add drop multiplexers; for these reasons, they must have a limited noise factor and, even more important, a high saturated gain; as a matter of fact, higher saturation gain allows longer spans to be realized, decreasing the transmission system CAPEX cost.

In order to achieve this specialized design, generally, the structure of Figure 4.18 is not sufficient and a more sophisticated design is needed.

Almost all line amplifiers and several boosters and preamplifiers are in fact multistage amplifiers with two or even three stages. Two-stage amplifiers can be designed in several ways, as an example, a possible scheme is provided in Figure 4.19.

From the figure it is clear that a two-stage amplifier is the cascade of two optical amplifiers divided by a group of passive components (an isolator, a filter, and, e.g., a DCF) providing an overall interstage loss.

In order to analyze the design of such an amplifier, two elements have to be added to the theory we have developed up to now.

First of all, as we have already underlined, even if the ideal ASE noise is simply almost proportional to the gain, in a practical amplifier, the noise factor NF depends, among the other things, on the efficiency of pump conversion; thus, it is higher if the gain is increased and if the output power is increased. Under this point of view, we can imagine that by tuning the pump power and the material characteristics (e.g., via doping) different amplifiers can be obtained, for example, achieving high gain and high output saturation power at the expense of a high NF or obtaining lower G and P_s, but with a much lower noise power.

Moreover, when we have a loss in an amplification chain, we have to take into account that the overall shot noise does not change, so that the traveling signal SN degrades.

From a quantum point of view, every loss is like a four-way splitter that takes away from the drop port a part of the incoming signal and adds from the add port a proportional amount of vacuum state [44]. This model is represented in Figure 4.20 and has to be taken into account when analyzing the noise in multistage amplifiers.

In order to optimize the amplifier with respect to the noise after detection, which is the important element in a communication system, we have to evaluate the SN at the optical

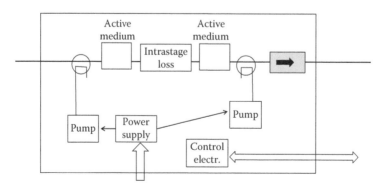

FIGURE 4.19
Block scheme of a two-stage amplifier system with an interstage loss.

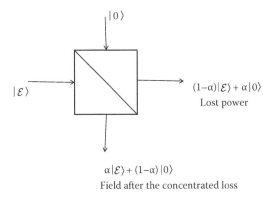

$|0\rangle$

$|\varepsilon\rangle$

$(1-\alpha)|\varepsilon\rangle + \alpha|0\rangle$
Lost power

$\alpha|\varepsilon\rangle + (1-\alpha)|0\rangle$
Field after the concentrated loss

FIGURE 4.20
Quantum scheme of a concentrated loss: it is equivalent to a quantum splitter eliminating part of the signal and adding an equivalent part from the vacuum state.

level. As a matter of fact, as we will see in Chapter 6, it is possible generally to express the error probability as a function of this parameter.

The LP electrical field can be written as

$$E_{out}(t, L) = \sqrt{GP_{in}}\, e^{-i\theta(t,L)} + \sqrt{G_2 \alpha}\mu_1(t, L) + \sqrt{G_2}\mu_2(t, L) + \mu_3(t, L) \tag{4.78}$$

where the random processes $\mu_j(t, L)$ represent the noise terms at the output of the three sections of our multisection amplifier and α the inter-stage loss.

Thus, evaluating the noise term power, we get

$$\sigma_\mu^2 = \frac{\hbar\omega}{2} B_\mu G \left[NF_1 + \frac{1}{\alpha G_1} + \frac{NF_2}{G} \right] \tag{4.79}$$

where B_μ is the bilateral noise bandwidth. This means that if the noise bandwidth is determined by a filter whose passband is W, it is $B_\mu = W$.

From Equation 4.79, it is clear that the expression of the overall noise factor of the two-stage amplifier is mainly dependent on the noise of the first stage, at least as far as the interstage loss is not so high to overcome the gain of one of the two stages. Thus, the result already known for electronic amplifiers is obtained also for optical amplifiers: the first stage has to be designed with a low noise factor, while the design of the second stage is mainly driven by the need of a high output power, since the noise factor contributes to the overall noise divided by the whole amplifier gain.

Starting from Equation 4.79, it is also possible to face another type of design problem related to the fact that the design of the whole transmission system dictates requirements for the subsystems and in particular for optical amplifiers.

Such requirements in general consists in the input and in the output power, with the condition that such input power pushes the amplifier sufficiently deep in saturation to be stable against the input signal or the environmental fluctuations. Moreover, the system design also dictates the value of the intrastage loss, since it depends on the dispersion compensation and on what are the other optical components that have to fit between the two stages.

In this condition, minimizing the noise means of course to reduce NF_1 as much as possible, increasing the population inversion and reducing the system nonidealities, but also means to distribute the overall gain between the two stages determining G_1 and G_2 so that σ_μ^2 is minimum.

This problem can be solved by expressing Equation 4.79 as a function of one of the unknown gains using $G = G_1 \alpha G_2$ and the expression (Equation 4.74) of NF.

Indicating with γ the population inversion of the two stages, the expression of the noise variance transforms in the following equation:

$$\sigma_\mu^2 = \frac{\hbar \omega}{2} B_\mu \left[\frac{(\gamma_1 G \alpha^2 + \gamma_2 - \gamma_1^2 \alpha^2) G_2^2 + \gamma_2 G_2}{G G_2} \right] \tag{4.80}$$

Equation 4.80 can be minimized with respect to G_2 to obtain the optimum gain distribution in a two-stage amplifier. The result can be written as $G_2 = \sqrt{(\gamma_1 G / (\gamma_1 G \alpha^2 + \gamma_2 - \gamma_1^2 \alpha^2))}$.

This equation, with realistic amplifier parameters, simply tells us that the gain of the second stage has to be very small with respect to the gain of the first and, in the limit of small interstage loss, a single-stage amplifier with the loss after the amplification performs better than a two-stage amplifier.

This solution however fails to represent a very important characteristic of real amplifiers that disappear when the quantum limit approximation is performed.

As a matter of fact, realistic amplifiers have a noise factor greater than what is represented by Equation 4.74 and, more important, the noise factor is generally bigger for amplifiers with a great gain. Thus, in our previous solution, if all the gain is concentrated in the first stage, also the noise factor of the first stage increases so that the overall noise is no more the minimum possible.

In order to represent this situation, let us do a very simple model of the amplification mechanism taking into account this phenomenon. In a real amplifier, the situation is much more complex, but this model allows us to concentrate on the main phenomena without the complexity of a real model of the real amplifier mechanism.

We will assume that the active medium is a passive matrix with the insertion of the active molecules, exactly as happens in erbium-doped fibers.

In this situation, the local unsaturated gain g_0 can be assumed proportional to the density of active molecules. Let us also image that the insertion of the active molecules somehow risks damaging locally the host matrix, thus creating microscattering locations whose density is proportional to the density of active molecules too.

Thus, eliminating the density of active molecules, we can say that the number of scattering imperfections is proportional to the local unsaturated gain.

Once a photon is scattered by an imperfection in the matrix, it can be considered lost for the amplified radiation; thus, this phenomenon reflects in a decrease of the local gain that is proportional to the gain itself.

Including this very simple model of active medium imperfection, Equation 4.63 can be rewritten as

$$\frac{dE}{dz} = \left[\frac{1}{2} g (1 - \delta g_0) + in \right] E + \mu \tag{4.81}$$

where the effects of the scattering points on the noise can be neglected if the loss parameter δ is very small. The impurity parameter has the dimension of a length (we can measure it in meters) and it can be considered as an impurity characteristic length.

In the approximation of small δ (i.e., $\delta g_0 \ll 1$), Equation 4.81 can be solved with the perturbation method with respect to δ, and the first perturbation term is sufficient to model the effect of the scattering imperfections.

The equation of the first-order perturbation is simple giving $P_{out}^{(1)} = -0.5\delta g_0^2 L P_{in}$.

Substituting the expression of the output power in the signal to noise ratio after amplification and detection and using again the definition (Equation 4.74) of NF, the expression of NF formally does not change, but for a multiplication factor that, in the hypothesis of first-order approximation in δ, is written as $(1 - 0.5\delta g_0^2 L)$.

As anticipated, the noise factor itself depends on the amplifier gain when amplifier imperfections are considered.

What we have obtained means that the factors γ_j in Equation 4.80 are no more constant, but depends on the gain of the corresponding stage. In particular, using the expression of the global unsaturated gain G_0, it comes out that

$$\gamma_j = 2\left(\frac{\rho_{1,j}}{\rho_{1,j} - \rho_{0,j}}\right)\left[1 + 0.5\frac{\delta}{L}(\log G_{0,j})^2\right] = 2\left(\frac{\rho_{1,j}}{\rho_{1,j} - \rho_{0,j}}\right)\left\{1 + 0.5\frac{\delta}{L}\left[\log G_j - (1 - G_j)\frac{P_{in,j}}{P_{sat,j}}\right]\right\}$$

(4.82)

where all has been expressed as a function of the saturated gain exploiting Equation 4.67.

Let us now try the amplifier optimization by minimizing the expression of the noise in a realistic parameters configuration described in Table 4.8.

The behavior of the noise curves (the normalized noise is represented) versus the gain of the second stage G_2 is reported in Figure 4.21; from the figure, it is evident that a minimum of the noise power for a value of G_2 is different from zero, thus confirming the experience that the optimum amplifier does not have only one stage.

The optimum gain G_2 in the case of an overall amplifier gain of 26 dB is shown in Figure 4.22 versus δ for different values of the interstage loss. From the figure, the presence of a sort of threshold value of δ is also seen that depends on the interstage loss and divides two regions where completely different criteria have to be adopted for the amplifier optimization: a region for low values of δ (below the threshold) where the optimum value of G_2 is quite low and the single-stage solution is competitive and a region (above threshold) where the two-stage solution is really needed. In the considered case for an interstage loss of −10 dB, the threshold value is about 0.002 m while for the interstage loss of −13 dB the threshold value is 0.0011 m.

TABLE 4.8

Parameters of the Two-Stage Amplifier Used
in the Optimization Example of This Section

Overall gain G (dB)	26
Interstage loss α (dB)	−10
γ_{10} (adim.)	0.9
γ_{20} (adim.)	0.65
Input power (at the stage 1) (dBm)	−13
Local saturation first stage (dBm)	0
Local saturation second stage (dBm)	10
Active medium length (m)	1

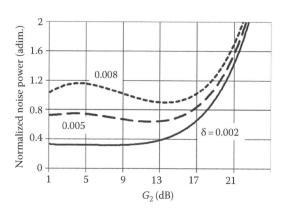

FIGURE 4.21
Noise power versus the saturated amplification of the second stage in the case of a realistic model of a two-stage amplifier where the noise factor depends on the amplification due to the characteristics of a simple material model. Different values of the material defect density parameter are considered. The presence of an optimum value of the gain for which the noise is minimum is evident.

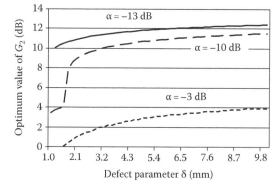

FIGURE 4.22
Optimum value of the gain of the second stage versus the material defect density parameter for different values of the interstage loss and an overall amplifier gain of 26 dB.

Looking at the amplifier application, three situations that have practical meanings are represented in the figure:

1. A low value of the interstage loss (−3 dB) that corresponds to the situation of a short metro system in which only a limited dispersion compensation is performed.

2. A interstage loss (−10 dB) corresponding to the length of DCF that is needed to compensate the dispersion of the span whose attenuation is compensated by the amplifier. As a matter of fact, if we consider a net loss of 0.3 dB/km for an SSMF in cable taking into account connectors, patch panels, bending, and margins for aging, the amplifier is able to compensate a span of 86 km with its 26 dB of overall gain. If we imagine a DCF with a dispersion parameter $D = -72.5$ ps/nm/km and a loss of 0.5 dB/km, we see that to compensate the dispersion of 86 km SSMF with $D = 17$ ps/nm/km, we need km 20 of DCF, whose loss is exactly −10 dB.

3. An interstage loss of −13 dB, allowing to allocate also an OADM in the amplifier site to construct a branching point.

In the first case, the optimum value of G_2 is really small and it is a very good solution to use a single-stage amplifier before the concentrated loss.

In both the other cases, the concentrated loss is sufficiently high to justify a second stage with a relevant gain. In condition of very high defect parameter (when the approximations we have done are at the working limit), the second stage has to gain more than the first, inverting what happens in the ideal case.

4.3.2 Erbium-Doped Fiber Amplifiers

The optical amplifier most used in telecommunication systems is a fiber amplifier: the erbium-doped fiber amplifier (EDFA) [48,49].

This amplifier uses as gain medium a fiber optics doped with erbium ions that are excited via optical pumping.

The use of a fiber optic as amplification medium allows the pump and the signal fields to be confined in a very small area (on the order of $60\,\mu m^2$), thus achieving high pump power density, which is a condition for a good quantum efficiency, and a very good matching between the area in which population inversion and the signal mode distribution happen. All this elements cooperate to increase the gain and reduce the noise.

In order to amplify signals in the third fibers transmission window (i.e., around 1550 nm), the dopant has to have a suitable transition and must be compatible with the glass matrix.

The first condition allows stimulated emission at the chosen wavelength, the second is key to manufacture doped fibers where the dopant is uniformly distributed (without the clustering effect that practically inhibits the interaction with light of the great part of dopant ions) and that are stable in time.

Several rare earth ions exhibit transitions in the useful wavelength range, and erbium ions, whose electronics energy levels are shown in Figure 4.23, can be inserted in the glass matrix in a uniform and stable way.

Thus, erbium-doped fibers can be used as active elements in architectures like that in Figure 4.24.

In particular, the transition between the levels $I_{15/2}$ and $I_{13/2}$ is around 1531 nm, thus about at the center of the third fibers transmission window.

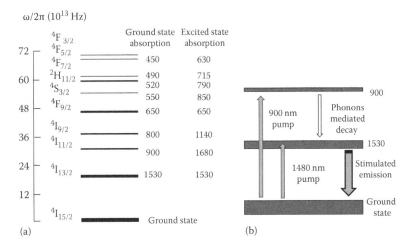

FIGURE 4.23
Energy levels of the erbium ion in a glass matrix (a). The density of states in the various states bands is indicated qualitatively with the line thickness. Quantum transitions used in EDFA amplifiers (b).

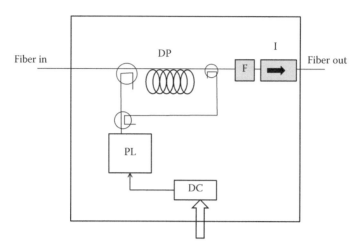

FIGURE 4.24
Block scheme of a single-stage EDFA. DP, doped fiber; PL, pump laser; DC, power supply; F, filter; I, isolator.

The photons–electrons cross section is sufficiently big to foresee a good stimulated emission efficiency and the excited level lifetime is sufficiently long to guarantee stationary amplification at all the interesting bit rates.

Once the erbium ion is inserted in the glass matrix, due to the interaction with the glass and with nearby ions, there is a split of the electronic levels, each of which becomes a thin level bandwidth. This is another important characteristic of erbium ions, allowing wide bandwidth amplifiers, suitable for DWDM systems.

The choice of erbium ions as dopants of a single-mode glass fiber to realize fiber amplifiers is then natural.

The first step to do to design a practical amplifier is to define an efficient and practical pumping scheme. From Figure 4.23, it immediately results that pumping has to be optical to excite some state at higher energy with respect to $I_{13/2}$ to achieve a three level working. On the other hand, the wavelengths giving absorption from the ground state to an excited state beyond $I_{13/2}$ are in the range typical of semiconductor lasers; thus, a suitable optical pump does exist.

In particular, 800 and 980 nm are particularly interesting as pump wavelengths, existing powerful semiconductor lasers based on AlGaAs technology. Moreover, the split of the resonant level at 1531 nm in a small energy band allows also semiresonant pumping at 1480 nm via InGaAsP semiconductor lasers. These lasers are in general less powerful than AlGaAs one, but due to the pumping state, the pump should be more efficient.

Among these three potential pumping frequencies, the 800 nm wavelength suffers the phenomenon of excited state absorption (ESA). This phenomenon consists in the absorption of a photon at 800 nm by means of electrons already excited at the $I_{13/2}$ state to jump to the $S_{3/2}$ state (see Figure 4.23). From $S_{3/2}$, the electrons then decays in a non-radiative way.

The ESA imposes a limit on the population inversion due to the progressive depletion of the lasing state when the pump populates it; thus, 800 nm is not a suitable pump wavelength.

Two possible pump wavelengths remain: 1480 and 980 nm. Using these wavelengths, different pumping schemes can be adopted for single- and double-stage amplifiers. In the single-stage case, the simpler schemes are

- Simple copropagating pumping either at 1480 or at 980 nm
- Simple counter-propagating pumping either at 1480 or at 980 nm
- Double pumping at the same wavelength (see Figure 4.24)
- Double counter-propagating pumping at different wavelengths

Comparing 980 and 1480 nm in similar configuration, the 1480 nm pump does not exhibit the expected efficiency advantage since resonant pumping is never reached due to the broadening of the pumping level that is also the stimulated emission level. In this condition, the intraband transition happening after the ion excitation is substituted by a similar interband transition and real two-level operation is never reached.

On the other hand, pumping at 980 nm attains lower noise factor and can be operated with more powerful lasers. Thus, in modern low-noise EDFAs almost only 980 nm pumping is used.

Once the pumping scheme is individuated, the overall stage architecture is fixed and it has to be optimized for the required characteristics.

The first parameters to be optimized are the doped fiber length and the pump power. In order to do this optimization, a complete model of the amplifier has to be developed, that in its simpler form is a three-level version of the model developed in Section 4.3.2; thus, the main parameters have been already introduced and the main qualitative phenomena are already known.

We do not report here a detailed model for the EDFA for sake of simplicity and also because practical EDFAs are really near to the ideal working. The interested reader is encouraged to start from the model reported in Appendix D and from the related bibliography.

In order to make an optimization example, we have also to detail a set of realistic parameters for a doped fiber. In Table 4.9, microscopic parameters of a doped fiber optimized for in-line amplifiers are reported. Using these parameters, the single-stage amplifier linear gain G_0 is reported versus the pump power in Figure 4.25 for single copropagating pumping at 980 nm. Different fiber lengths are considered.

The effect of pump saturation on the linear gain is clear, in this case, if this pumping scheme is chosen, going beyond 6 mW of injected pump is practically useless.

The situation is better if the pump is divided by two and a double counter-propagating pumping scheme is used: in this case, the result is the family of dashed curves of Figure 4.25. The reason for the improvement is clear: if the pump is splitted and injected in the two directions in the doped fiber, the pump distribution along the fiber is much more

TABLE 4.9

Parameters of an Erbium-Doped Fiber Optimized for In-Line Amplifiers

Core Area	$13\,\mu m^2$
Er^+ concentration	$5.4 \times 10^{24}\,m^3$
Signal-Er^+ overlap integral	0.4
Pump- Er^+ overlap integral	0.4
Signal emission cross section	$5.3 \times 10^{-25}\,m^2$
Signal absorption cross section	$3.5 \times 10^{-25}\,m^2$
Pump absorption cross section	$3.2 \times 10^{-25}\,m^2$
Signal local saturation power	1.3 mW
Pump local saturation power	1.6 mW

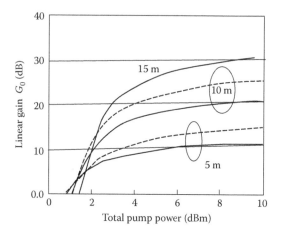

FIGURE 4.25
Gain saturation with pump power in EDFA amplifiers for different lengths of the active fiber.

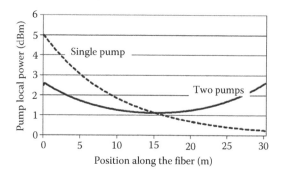

FIGURE 4.26
Pump power profile along the active fiber in a single-pump and double counter-propagating pump
schemes.

uniform and the pump local power smaller. This situation is shown in Figure 4.26, where
the pump local power is shown for the two pumping schemes versus the position along
the doped fiber.

Once a pumping scheme is chosen, it is easy to demonstrate that an optimum fiber length
exists, as is shown in Figure 4.27, where the linear gain behavior is represented versus the
fiber length for single copropagating pump at 980 nm.

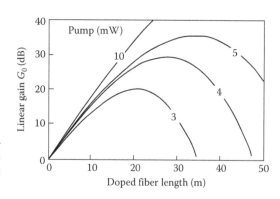

FIGURE 4.27
Gain saturation with the length of the active fiber
in EDFA amplifiers for different values of the pump
in a single-pump configuration. The presence of an
optimum value of the active fiber length is evident.

In deciding the doped fiber length however, it is to consider the fact that the noise factor also increases while increasing the fiber length, as it is quite intuitive, and this trend is fast while going near the optimum length. This makes the choice of the fiber length a delicate problem, and it is not rare that a shorter length is chosen with respect to the optimum gain value because having a lower noise factor is more important than having the maximum possible linear gain.

Fixed the pump wavelength and the doped fiber parameters, the number of stages and the pumping scheme for each stage and finally the doped fiber length for each stage, the EDFA amplifier is dimensioned for what regards the optical part.

Such amplifiers exhibit all the phenomena we have introduced in the previous section: gain saturation with signal power, ASE noise, and so on. Naturally, since the lifetime of the excited state is on the order of milliseconds, no dynamic distortion is created at every bit rate of interest in systems using EDFAs.

Another important characteristic of the EDFAs is their wide amplification bandwidth, which seems to be in contrast with the nature of the mechanism causing amplification, that is, stimulated emission in a discrete energy levels spectrum.

This is due to the fact that the electrons energy levels of the erbium ion population in glass matrix are composed more of small energy bands [50].

This splitting among electrons levels is due essentially to two effects: Stark splitting and homogeneous line broadening.

Stark splitting is due to the interaction of the electron orbitals of each erbium ion with the electrical field due to the surrounding glass molecules. Since the microscopic position of each ion is different, this influence generates a level splitting.

Homogeneous line broadening is due to the interaction with the ions electrons with local phonons of the glass. This mechanism, which is called homogeneous since it influences every ion in the same mode, causes the typical Lorenz type line widening. Inhomogeneous mechanisms exist too in the interaction between erbium ions and the host glass, but generally homogeneous broadening is dominant.

From the earlier discussion, it is clear that in general the amplification bandwidth of an EDFA cannot be uniform, but on the contrary EDFA spectral gain will have a complex behavior.

Just for an example, the gain spectrum of a practical single-stage amplifier is reported in Figure 4.28. Considering the nature of the phenomena causing bandwidth shaping, it is also evident that the gain flatness can be improved by working on the doped fiber

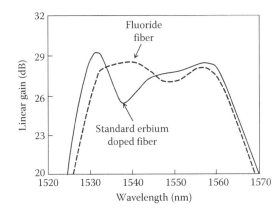

FIGURE 4.28
Gain bandwidth of a standard EDFA amplifier and an EDFA realized with a fluorine doped fiber. A single-pump scheme is considered. (After Ohishi, Y. et al., *Opt. Lett.*, 22(16), 1235, 1997.)

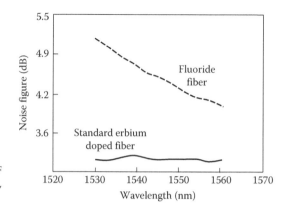

FIGURE 4.29
Noise figure versus wavelength for a standard EDF and a fluorine doped EDF. (After Ohishi, Y. et al., *Opt. Lett.*, 22(16), 1235, 1997.)

material. For example in Figure 4.28, the spectrum relative to a fluoride-doped fiber is also reported, whose uniformity is by far better with respect to that of an EDFA using a standard doped fiber. Several studies have been carried out along this direction, achieving good results [51–53].

However, changing the material composing the doped fiber also causes changes in the amplifier noise factor NF. In Figure 4.29, the spectral behavior of NF is shown for the same amplifier of Figure 2.28. The noise degradation due to the fluoride glass is evident from the figure. Better results have been obtained with improved fabrication techniques and materials, but in any case an NF degradation has to be taken into account.

Another technique to obtain an intrinsically gain-flattened erbium-doped fiber is to manage suitably the fiber index profile [54]. For example, using a staircase index profile [55], good flattening results can be achieved. Also in this case however, the price to pay is an increase of the noise factor due to the lower field confinement that reduces the superposition between pump and signal with the erbium ions distribution lowering the gain for a given pump power and increasing ASE noise generation.

Even if working on the doped fiber itself is interesting under several aspects, the most used method to obtain an almost flat gain bandwidth is to shape suitably the transfer function of the filter that always exists at the amplifier output to limit the amount of ASE power propagating along the transmission line. This is the so-called gain flattening filter [56,57].

A typical transfer function of a realistic gain flattening filter is represented in Figure 4.30. This transfer function can be realized by the thin films dielectric technology, a consolidated technology that is able to shape almost every wideband optical transfer function with a good approximation.

Another technology that is often used to manufacture gain flattening filters is fiber Bragg gratings technology.

Flattening filters allow very good performances to be achieved in terms of uniformity, as shown in Figure 4.31 where the flattened gain of a commercial single-stage EDFA is shown in all the amplifier useful bandwidth. Such results are achieved at the expense of a relevant reduction of the gain maximum value. As a matter of fact, a filter is a passive device and gain flattening is practically achieved through the introduction of the filter wavelength selective loss.

Performing a selective wavelength suppression of the peak gain causes the noise to be more dangerous, simply due to the smaller signal exiting from the amplifier end.

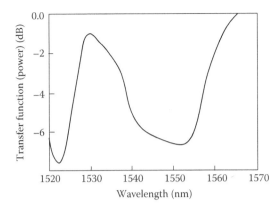

FIGURE 4.30
Shape of the transfer function of a gain flattening filter for a single-pump EDFA.

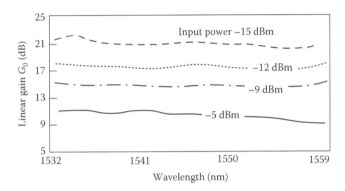

FIGURE 4.31
Spectral gain behavior of an EFDA using a gain flattening filter for different values of the signal input power.

The parameters of a few real EDFAs that have been optimized for use in communication systems are compared in Table 4.10.

The first thing that emerges from the table is the extreme variety of performance an EDFA can exhibit depending on the type of doped fiber, on the pump scheme, and on the inner architecture.

This variety demonstrates the flexibility of EDFA technology and somehow explains its great success.

All the considered amplifiers are designed to work in deep saturation and in stationary state mode. The first condition is assured by the prescription of a minimum input power: the amplifiers performances are not guaranteed below this power since deep saturation is no more verified.

The second condition is assured for all the signals of interest in telecommunications by the long lifetime of the $^4I_{13/2}$ state (compare Figure 4.21).

The amplifier design depends on the application and ranges from a simple single-stage single pump for single-channel amplification to a very complex three stages, two pumps design allowing to accommodate into the amplifier a great interstage loss (13 dB) while retaining a saturated output power as high as 25 dB. This kind of amplifier can be used as in-line amplifier to accommodate both a DCF and an optical add drop.

In this last case, the complex multistage design also allows a low noise factor to be maintained during standard saturated working, a fundamental condition for the use of the amplifier in long-haul and ultra-long-haul transmission systems.

TABLE 4.10

Characteristics of Practical EDFAs Optimized for Different Applications

Application	DWDM Preamplifier	From One to Four Channels Booster	DWDM Booster	DWDM In-Line Amplifier	DWDM In-Line and Preamplifier
Architecture	Single stage	Single stage	Two stages	Two stages	Three stages
Pump scheme	Two pump lasers at 980 nm in opposite directions	Copropagating at 980 nm	Pump sharing among the stages	Two pumps at 980 nm	Two pump lasers at 980 nm
Electronic control	AGC–APC–NM interface	AGC–APC	APC	AGC	AGC, NM interface
Gain bandwidth (nm)	22	3	65	34	35
Saturated output power (dBm)	20	10	17	21.5	20
Average input saturat. power (dBm)	−20	−5	6	—	−6
Minimum power in (dBm)	−35	−10	—	−30	−22
Maximum mid-stage loss (dB)	—	—	8	11	13
Maximum noise figure (dB)	6	7	7	6.9	6
Gain flatness (dB)	1	0.5	2	1.0	0.5

AGC, automatic gain control; APC, automatic power control; MN, separately manageable by network manager.

All the described amplifiers are naturally equipped with electronic control that allows the user to drive them via digital signals after the insertion into the transmission system.

Specific functions required in high-performance amplifiers are the automatic gain control and the automatic power control. They are control loops that stabilize the saturated gains or the output optical power, respectively, by acting on different amplifiers parameters (like the pump power or the position of an output variable attenuation).

One of the considered devices is also equipped with a network manager interface that means that it is also designed to work alone in a system amplification site so that it has to be recognized by the element manager as a network element.

4.3.3 Raman Fiber Amplifiers

Raman amplifiers are based on the stimulated Raman emission creating a Raman gain in a medium where an optical pump has created population inversion between the Raman levels [58].

Differently from EDFAs, Raman amplifiers exploit a nonresonant incoherent phenomenon; as a matter of fact the Raman gain curve is very wide (compare Figure 4.9).

For the same reason, while erbium-doped fibers gain is located in a specific wavelength range, depending on the erbium ion transitions, the Raman gain bandwidth only depends on the optical pump wavelength; thus, Raman gain can be obtained at whatever wavelength, only assuming that a suitable pump laser is available.

Also shaping the Raman gain spectrum is in principle an easy task. As a matter of fact, if the pump is not monochromatic, but is composed of an opportune set of lines, the resulting Raman gain bandwidth can be shaped as suitable for the application [59].

These characteristics of Raman amplifications make this technique quite useful in transmission systems, not to replace, but to complement EDFAs.

However, besides these advantages, Raman amplifiers also pose a set of important challenges that caused the fact that, even if researchers were working on Raman amplification far earlier than on erbium-doped fibers, the presence of Raman amplifiers installed in the field is quite recent.

The first issue is the low pumping efficiency characteristic of Raman scattering and due to the nonresonant nature of the Raman interaction.

This means that in order to obtain a certain gain, a Raman amplifier needs much more pump power with respect to an EDFA amplifier [60,61].

Besides pump efficiency, also the stimulated radiation buildup is less effective, due to the weakness of the nonlinear effect. This means that a concentrated Raman amplifier needs a much longer fiber with respect to a similar EDFA.

This last issue however has been overcome in two different ways, so that the characteristic of Raman amplifiers to need a long fiber is at present days nearly reverted to an advantage.

In concentrated Raman amplifiers, the Raman effect is obtained in the DCF [62,63], which in any case should be installed along the line to control chromatic dispersion.

If this solution is adopted, no interstage loss has to be designed into the Raman amplifier (differently from what happens in EDFAs), thus achieving a better gain and lower noise.

An alternative solution is to use as active Raman fiber the same transmission fiber, thus designing a distributed Raman amplifier [64].

This last solution is particularly effective when used either in systems where in-line amplifiers are not foreseen (e.g., short undersea systems) or in conjunction with EDFAs in ultra-long-haul systems.

A third concern that has to be taken into account when designing a Raman amplifier for application in telecommunication systems is that the Raman lifetime is very short: on the typical telecommunication time scale, Raman effect is almost instantaneous.

In this condition, distortion related to gain dynamics can affect a signal amplified by Raman effect (compare with Section 4.3.1.3) both as pulse dynamics reduction and interference among channels at different wavelengths [65].

A large effort has been made in the last 10 years to overcome these limitations and to device practical Raman amplifiers and currently there are Raman amplified DWDM systems installed in the field.

The first key development has been the availability of higher Raman gain fibers with relatively low loss. As an example, commercial DCF has a Raman efficiency almost 10 times greater compared to standard single-mode fiber. Moreover, new Raman gain fibers continue to be introduced commercially with different dispersion profiles and dispersion slopes.

A second key development for Raman amplifiers has been the availability of high pump power laser diodes. Commercial laser diodes are available with more than 800 mW output powers with the ability to inject more than 500 mW in an SSMF, and more than 1 W of fiber injected power has been realized in experimental systems.

Using sufficiently powerful pump lasers at the right wavelengths, in principle, it is possible to shape the Raman gain to achieve a sufficiently high gain over all C and L bands [66].

For example, with two pumps a sufficient equalization of the Raman gain over all the C band (35 nm) can be achieved and better results can be obtained with three pumps. Equalization of the gain over 60 nm (C and L bands) can be achieved using three or four pumps.

All the pumps adopted in this application are high-power semiconductor lasers. However, in using such devices, some issues could arise that does not exist when using the much less powerful EDFA pumps.

With a power of some Watt traveling through the fiber within 1400 and 1450 nm, connectors adjacent to the amplifier have to be ready to support a great stress and particular attention has to be devoted to their correct alignment.

Moreover, pump reflections can be dangerous and particular care has to be paid to reduce the reflection coefficients of the components along the pump path.

A third point in using semiconductor laser pumps is that Raman gain is polarization dependent; thus, in order to equalize the gain over all the incoming signal polarization, it is necessary to have an unpolarized pump. This can be obtained from a semiconductor laser, which emits a LP field either with a concentrated element or, more frequently, injecting the pump field into a piece of polarization maintaining fiber with the polarization directed at $\pi/4$ with respect to the polarization maintaining birefringence axis (see Section 4.2.6). In this way, at the output of the fiber, a depolarized pump is obtained.

In order to better understand advantages and issues of current Raman technology and to look at future evolutions, the two important cases of Raman amplifiers have to be dealt separately; as a matter of fact, concentrated and distributed Raman amplifiers have different characteristics and problems.

Conceptually, the block scheme of a concentrated Raman amplifier is given by Figure 4.18, where the active medium is a high Raman efficiency fiber. Frequently it is a DCF that is also used to compensate part of the transmission link dispersion.

Raman-concentrated amplifiers frequently use more than one pump, due to the fact that multiple pumps assure gain flatness.

The first point to analyze is the reduced pump efficiency that in a concentrated amplifier, designed to compete with EDFAs, is a particularly important point.

In standard conditions, this is smaller than the efficiency of erbium-doped fibers, but there are particular and interesting cases in which this is not true.

In Figure 4.32 [60], the efficiency of a Raman amplifier realized with a DCF is compared with the efficiency of a single-stage EDFA booster amplifier in a typical booster configuration, with an input power as high as 20 mW.

The input is a signal at 1529.5 nm with an input signal power of 20 mW. For the Raman amplifier curve, a Raman pump wavelength of 1433 nm was chosen, so that the signal falls on the gain peak at 13.2 THz away from the pump wavelength.

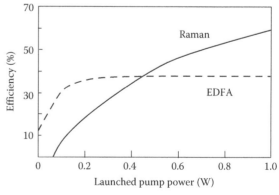

FIGURE 4.32
Efficiency of a Raman amplifier realized with a DCF versus the efficiency of a single-stage EDFA booster amplifier in a typical booster configuration, with an input signal power of 20 mW and different values of the pump power. (After Islam, M.N., *IEEE J. Sel. Top. Quantum Electron.*, 8(3), 548, 2002.)

The loss of the fiber at the signal wavelength is 0.472 dB/km, the length of the gain fiber is 5.64 km, and the Raman gain coefficient is 13.5 dB/(W km).

For the EDFA curve, it is assumed that the amplifier is forward pumped by a 1480 nm pump. The passive loss in the EDFA is assumed to be 0.15 dB/m and the length of the coil (20 m) is optimized to give maximum signal output power when pumped with 750 mW of 1480 nm pump for 20 mW of input power.

The definition of the amplifier efficiency η is given by the following equation:

$$\eta = \frac{P_{s,out} - P_{s,in}}{P_{P,in}} = \frac{(G-1)P_{s,in}}{P_{P,in}} \tag{4.83}$$

where
$P_{s,out}, P_{s,in}$ are the signal output and input powers, respectively
$P_{P,in}$ is the total pump input power (adding all the used pumps if the case)
G is the amplifier gain

It is clear that the efficiency of EDFA is limited by the pump saturation for high values of the pump, while the efficiency of the Raman amplifier does not have this limitation.

As a consequence, a threshold pump power exists above which, at least in the booster application, the Raman efficiency becomes higher than the EDFA one.

After verifying that the Raman efficiency allows designing functional amplifiers having their own advantages, as in the case of EDFAs, the pumping configuration has to be selected.

This point for Raman amplifiers is even more critical than for EDFAs since a certain degree of gain flatness can be achieved by an effective pumping scheme.

Even if it was proposed to pump Raman amplifiers with broadband pumps to achieve gain flatness [67], the most practical solution is to use several pumps, often propagating along the Raman fiber in both the directions.

Besides the issues related to the amplification efficiency, when this is the case, and to the optimization of the pumping scheme, Raman concentrated amplifiers are also limited by another important factor: the presence of spurious noise contributions besides ASE noise.

As a matter of fact, the global noise factor of a Raman amplifier cannot be foreseen correctly by considering only quantum noise, but it is needed to take into account other phenomena that can be as important as quantum noise.

The first is related to the fact that Raman scattering is practically an instantaneous phenomenon on the typical time scale of telecommunication systems.

This means that gain saturation due to signal is effective in reducing signal dynamics since gain is high when the input power is lower, that is, when a "zero" is transmitted, while when signal is high (i.e., a "one" is transmitted), the gain is smaller due to saturation. This phenomenon, however, is very fast and no other signal distortion is implied.

If the signal dynamic decrease can be managed by using high power suppression in correspondence of zeros, more dangerous is the fact that any amplitude fluctuation of the pump, also the intensity noise that is present in all the semiconductor lasers, reflects in a fluctuation of the Raman gain.

This creates the second important noise source that is called pump relative intensity noise (RIN) transfer [68].

A third noise source is the so-called double Rayleigh scattering constituted by multiple ASE reflections that create an incoherent ASE contribution to the overall noise. Naturally, the component due to two reflections, one backward and the other forward, is the greater and usually the only one that is needed to have a good description of the noise in the amplifier.

Finally, a fourth source of noise arises from the so-called phonon-stimulated optical noise. This effect is due to the population of thermally induced phonons in the glass fiber that spontaneously experience gain from the pumps, thereby creating additional noise for signals close to the pump wavelengths. It has been shown that this effect can lead to an increase in noise figure of up to 3 dB for signals near the pump wavelength [69].

In order to analyze the impact of different pumping schemes and to correctly foresee the amplifier noise properties, a model of a concentrated Raman amplifier has to be developed, taking into account the specific characteristics of this component.

A complete modeling of the amplifier brings about a set of coupled differential equations containing all the terms corresponding to every possible pump scheme, to a plurality of amplified channels, and to all the noise terms.

We include such equations with a brief comment in Appendix D since there are situations in which a more simplified model cannot be used. In such cases, a numerical solution of a more complete model has to be carried out.

Here, in order to find a simpler model allowing us to investigate lumped Raman amplifiers potentialities, we will introduce the following approximations:

- The equations for the noise terms will be separated by the equations for the signals eliminating the noise contribution from the signal equations (high signal to noise ratio condition).

- Stationary regime is assumed, that means that no time variation is induced in the amplifier parameters due to the signal modulation. We know that this is equivalent to neglecting the dynamic range compression due to the very short Raman response time.

- Backscattering of signal waves is neglected.

- All the RIN noise terms are neglected, thus eliminating also the RIN transfer from the pumps and the signal.

- Phonon-induced noise is neglected (that means that no useful signal is near the pump wavelengths).

- The undepleted pump approximation is used. It is worth noting that, under an architectural point of view, this means that the pumps are extracted from the line after the end of the active fiber.

- The useful signal is assumed to be a single narrowband channel.

Just to be practical, we will analyze the approximated model in the case of a two back-propagating pumps.

Eliminating all the relevant terms, the general equations of the Raman amplifier (see Appendix C) become

$$\frac{\partial P_s}{\partial z} = -\alpha P_s + \sum_{j=1}^{2} \frac{g(\omega_j)}{2AK} P_j P_s$$

$$\frac{\partial P_j}{\partial z} = +\alpha_p P_j \quad (j=1,2) \tag{4.84}$$

$$\frac{\partial N_A}{\partial z} = -\alpha\left(N_A - \frac{\hbar\omega}{2}\right) + \sum_{j=1}^{2}\frac{g(\omega_j)}{2AK}P_j\left(N_A + \frac{\hbar\omega}{2}\right)$$

where
the signal power flux is represented with P_s
the pump power flux is indicated with P_j $(j=1,2)$
the ASE power flux is indicated with N_A
the attenuations of the signal and of the pumps are α and α_p, respectively
$g(\omega_j)$ is the microscopic Raman gain, measured in m/W
A is the fiber effective core area
K is the polarization factor, 2 in our case of unpolarized pumps

The first two equations of the system (4.84) are independent from the noise and can be solved exactly obtaining simple attenuated propagation for the pumps and the following signal expression

$$P_s(z) = P_s(0)\, e^{-\alpha z} e^{\left\{\frac{P_{10}g_1}{2KA\alpha_p}\left[1-e^{-\alpha_p z}\right] + \frac{P_{20}g_2}{2KA\alpha p}\left[1-e^{-\alpha_p z}\right]\right\}} \tag{4.85}$$

where
$P_s(0)$ is the power of the signal at the amplifier input $(z=0)$
P_{10} and P_{20} are the powers of the two pumps injected in the fiber at the amplifier end $(z=L)$

From (4.85), the following simple expression of the gain is obtained

$$G = e^{-\alpha L}\exp[(P_{10}\, g_1' + P_{20}\, g_2')L_e] \tag{4.86}$$

where it is assumed $K=2$ and the effective local gain, measured in $1/W$, is given by $g_j' = (g_j/4A)$. This corresponds to the practical condition of unpolarized pumps that is needed in Raman amplifiers for telecommunications due to the random signal fluctuations.

Frequently, Raman amplifiers are characterized through the so-called on–off gain G_R, instead of the standard net gain G. The on–off gain is defined as the ratio between the output power with pumps on and the output power with pumps off while the net gain is defined as usual as the relationship between the signal at the output and at the input of the amplifier. The relation between the net gain G and the on–off gain can be written as follows, where all the gains and losses are expressed in dB:

$$G_R = G - \Lambda \approx 0.4343\,(P_{10}\, g_1' + P_{20}\, g_2')L_e \quad \text{(all in dB)} \tag{4.87}$$

where Λ is the (negative) loss of the Raman fiber with pumps off in dB.

Equation 4.87 clearly explains the utility of the concept of on–off gain: as a matter of fact, the Raman amplification depurated from the losses of the fiber where amplification occurs is exactly equal to the on–off gain. Thus, this parameter characterizes the effectiveness of the amplification process without taking into account the losses of the Raman fiber also.

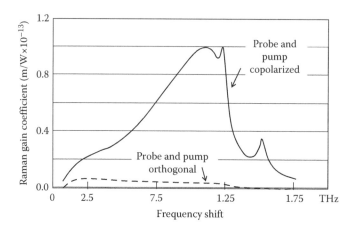

FIGURE 4.33
Polarization dependence of the Raman gain curve. (After Islam, M.N., *IEEE J. Sel. Top. Quantum Electron.*, 8(3), 548, 2002.)

Generally, a draft design of a Raman amplifier is carried out starting from (4.86) or a similar approximation, then it is verified by simulating the obtained structure with a numerical solution of a more complete model (Figure 4.33).

The first step to design a Raman amplifier is to optimize the pumps position. As far as the efficiency is concerned, configurations with counter-propagating pumps are more efficient with respect to configurations with copropagating pumps.

This effect is evident from Figure 4.34, where the pump power needed to reach a certain maximum saturated gain in a single-pump configuration is compared for a copropagating and counter-propagating pump injected at 1480 nm into the fiber. The greater efficiency of the counter-propagating configuration is evident.

If two pumps are used, as generally needed if a good gain is to be obtained in all the C or L band, the better configuration is bidirectional pumping, not only due to the higher global efficiency, but also to the opportunity to use the two pumps to flatten the gain.

In principle, due to the nonlinear interaction among the pumps, this could be done only by simulation.

In practice, the so-called pumps superposition rule holds: this means that it is possible with a very good approximation to superimpose the gains caused by different pumps to approximate the total gain with their products (their sum in dBs) [70].

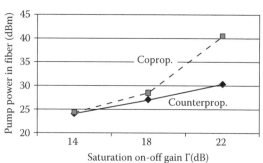

FIGURE 4.34
Pump power needed to reach a given maximum saturated gain in a single-pump configuration copropagating and counter-propagating pump injected at 1480 nm into the fiber. (After Islam, M.N., *IEEE J. Sel. Top. Quantum Electron.*, 8(3), 548, 2002.)

TABLE 4.11

Parameters of a Raman Amplifier Used in
the Numerical Examples of This Section

DCF fiber length	20 km
Dispersion parameter DCF	−70 ps/nm/km
signal attenuation in DCF	0.35 db/km
pumps attenuation in DCF	0.4 dB/km
Maximum local Raman gain $g(\omega_1,\omega_s)/A^2$ (DCF, at 1550 nm)	4.50E−03 1/m/W
Pumps insertion loss	3 dB

To have an idea of what is possible to achieve with a lumped Raman amplifier using a DCF as active fiber we will use the parameters of Table 4.11, which we will adopt also in Chapters 6 and 9 while analyzing WDM systems using Raman amplification.

The shape of the Raman local gain in the chosen fiber, a DCF optimized for Raman effect, is reported in Figure 4.35. It is different from the Raman spectral gain in an SSFM due to the different modal area and to the particular shape of the refraction index. The Raman shift is with a good approximation the same as an SSMF, being dominated by the microscopic glass properties.

The amplifier spectral gain in the C band is reported in Figure 4.36 for a two counter-propagating pumps scheme. From the figure, it results that with a specialized fiber a high gain can be achieved with a good flatness over all C band and with reasonable pumps power.

For all the curves in Figure 4.26, the pumps are located at 1430 and at 1400 nm. In the case of the higher-gain curve, realizing an average net gain of 20.5 dB, the pump powers are 450

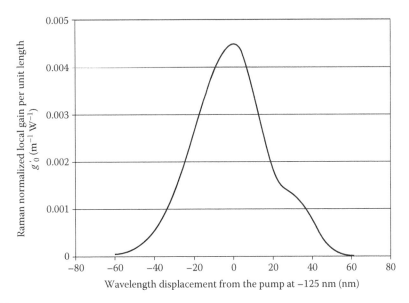

FIGURE 4.35
Normalized Raman gain of a highly nonlinear DCF manufactured for Raman amplification.

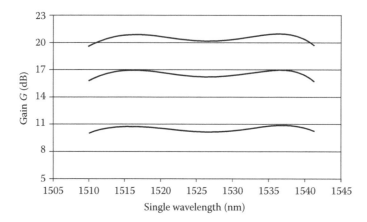

FIGURE 4.36
Gain of a Raman amplifier using a DCF and two pumps in positions designed to achieve the best possible uniformity within the fixed design gain and maximum pumps power constrains.

and 525 mW, respectively, with a pump insertion loss of 3 dB. The gain flatness in the C and is in this case 1.3 dB over all the C band.

In the intermediate curve, the net average gain is 16.5 dB and the pump powers are 390 and 450 mW, while the gain flatness is 1.1 dB.

Finally, the lower curve exhibits an average gain of 10.5 dB with a flatness of 0.7 dB using two pumps with a power of 340 and 300 mW.

Globally, it is clear that there is a sort of trade-off between the flatness and the gain, due to the fact that in the adopted approximation, the local gain due to a pump is proportional to the pump power.

It has to be noted also that the coupling loss between the pumps and the active fiber is a key in determining the Raman gain.

Considering very wide band amplifiers, a flat gain over C + L band, a band of about 60 nm, can be achieved by three or four pumps and gain in excess of 21 dB has been reported in literature as shown in Figure 4.37 [71]. Advantages can also be achieved by the so-called double pass architecture, where the signal is reflected at one amplifier length to pass through the amplifying fiber two times in opposite directions [72]. This design, quite

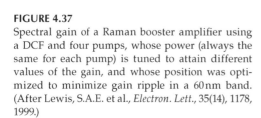

FIGURE 4.37
Spectral gain of a Raman booster amplifier using a DCF and four pumps, whose power (always the same for each pump) is tuned to attain different values of the gain, and whose position was optimized to minimize gain ripple in a 60 nm band. (After Lewis, S.A.E. et al., *Electron. Lett.*, 35(14), 1178, 1999.)

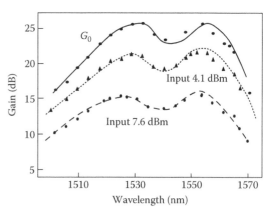

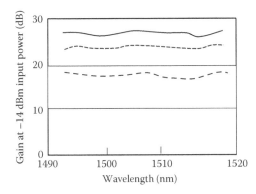

FIGURE 4.38
Gain of a lumped Raman amplifier designed to work in the S band.

interesting since suitable for low levels of input signals, has been experimentally implemented observing a maximum gain of 22 dB gain with a noise figure smaller than 5 dB over all the C band [72].

Finally, there is to consider another possible application of Raman amplifiers: amplification in the S band, where EDFAs cannot be used. Very good results have been achieved in this spectral area using high Raman efficiency fibers, as reported in [60] and shown in Figure 4.38 where a few experimental data reported in [60] are interpolated with the model presented in this section by adopting the parameters of a DCF and four pumps, whose power (always the same for each pump) is tuned up to a maximum of 330 mW, to attain different values of the gain and whose position was optimized to minimize gain ripple in the S band (from 1493 to 1523 nm).

Let us now consider the noise characterization of Raman amplifiers.

Starting from the ASE noise, we have to consider the noise equation and solve it to find the noise spectral density. Since the evolution of pumps and signal are known, it is not a difficult task and the following equation is obtained:

$$N_A = \hbar \omega \left(G - \frac{1}{2} \right) + \hbar \omega \alpha G \int_0^L \frac{dz}{G(z)} \qquad (4.88)$$

The values of the ASE spectral density obtained with this equation are plotted in Figure 4.39 versus wavelength considering a single-pump-concentrated amplifier made by a DCF 30 km long whose parameters are reported in [68]. The results are compared with the simulated ASE spectral density reported in [68]. The simulation has been carried out using a model similar to that reported in Appendix D, taking into account a wide variety of phenomena.

The spectral density in the signal bandwidth is very well modeled by Equation 4.88 in a very large part of the spectrum. Near the pumps, the ASE spectral density is in reality much higher than that foreseen by Equation 4.88 due to the fact that parametric amplification caused by the pumps, which is not included in the simple model we are considering, amplifies ASE up to quite high powers.

As a consequence, even if the gain should be sufficient, no signal has to be within a guard bandwidth of about 50 nm from the pumps in order to avoid a high ASE zone.

If this rule is respected, Equation 4.88 model the ASE spectral density very well, and it can be used to evaluate the Raman amplifier noise factor obtaining

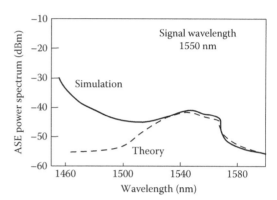

FIGURE 4.39
ASE noise spectrum for a single-pump Raman amplifier. (After Bristiel, B. et al., *IEEE Photonics Technol. Lett.*, 18(8), 980, 2006.)

$$NF = 1 + \frac{G(L)-1}{G(L)} + 2\alpha \int_0^L \frac{dz}{G(z)} \qquad (4.89)$$

A more accurate analytical solution of the noise equations can also be found taking into account time fluctuations (thus removing the stationary assumptions). In this way, a more accurate representation of the ASE noise can be achieved that is also valid for very high-gain amplifiers [68], where Equation 4.88 is less accurate. An example of noise figure calculation versus gain is reported in Figure 4.40 for an amplifier whose parameters are reported in [68] where it is also shown that similar curves agree very well with experimental results.

Among the noise term that we have neglected in simplifying the Raman equations, the RIN transfer from the pump to the signal is the most important; thus, a model for this noise source is useful when dealing with the design of systems including Raman amplifiers.

In order to study this phenomenon, following [68], it is needed to add to the system (4.84) the equations for the signal and the pumps RIN. These equations can be obtained from the complete model presented in Appendix D by applying to the complete RIN evolution equations the approximations already applied to the pumps and signal propagation equations.

The RIN equations can be solved analytically too, due to the knowledge of the signal and pumps evolution, so that the signal RIN at the output of the amplifier is given by

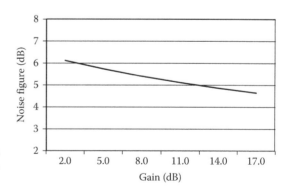

FIGURE 4.40
Raman amplifier noise factor versus net gain for an amplifier using 3 km of DCF fiber. (After Bristiel, B. et al., *IEEE Photonics Technol. Lett.*, 18(8), 980, 2006.)

$$RIN_{s,out} = RIN_{s,in} + \log(G_R)\left[RIN_{p,in} + \frac{\hbar\,\omega_p}{2P(0)}\alpha_p(L - L_e)\right] \qquad (4.90)$$

Equation 4.90 is in good agreement with experience, as shown in [68], but for the case in which the noise in the bandwidth near the pump is considered.

A last observation that can be important in the Raman lumped amplifiers design is that the effect of temperature can be important in degrading the amplifier noise performances. Just to make an example, an increase of the noise spectrum as great as 5 dB has been measured passing from 77 to 300 K at 1460 nm in a 9 km single-stage amplifier pumped by a pump of 850 mW at 1455 nm. In the same conditions, the peak gain at 1545 nm increases from 12 to 18 dB.

All the multistage techniques that are well established in the EDFA design can be adopted to design effective Raman amplifiers.

For example, multistage amplifiers can be devices with low-gain stages putting in between isolators that prevent the accumulation of double Rayleigh scattering and allows forward and backward pumps to be effectively separated.

Today Raman amplifiers are not considered as possible substitutes of EDFAs, even if this would be not impossible at least in high-performance systems, but are adopted as complementary amplification besides EDFAs. An exception of course is the case in which transmission in S band is needed, a case in which Raman amplifiers are the only solution.

4.3.4 Hybrid Raman-EDFA Amplifiers

Erbium-doped and Raman amplifiers can be usefully combined in very high-performance transmission systems. In this case, distributed Raman amplification can be used either to extend the bandwidth of EDFA amplifiers or to partially compensate the span loss via distributed amplification [60,73].

In the first case, the amplifier scheme is that shown in Figure 4.41. The optical signal is divided into bandwidth, C and L bands are amplified via specialized EDFAs, while the other bandwidths are amplified through concentrated Raman amplifiers.

This is a promising solution for ultra-high-capacity systems, even if balancing problems among amplifiers in the different bands can be difficult to solve.

The other application, however, is more than promising, and ultra-long-haul systems using this solution are already on the market.

The scheme of the hybrid amplifier is reported in Figure 4.42, while in Figure 4.43 the power profile along a link adopting hybrid amplifiers is compared with the power profile along a link adopting only concentrated EDFA amplifiers.

From Figure 4.42, some key advantages of using hybrid amplifiers can be seen immediately. From the fact that the Raman amplifier works as a low-noise preamplifier (it is distributed; thus, the noise figure is small) and from the general theory of cascaded amplifier, it is immediately deduced that the hybrid amplifier shall have a better noise figure with respect to an equivalent concentrated amplifier.

Thus, either a lower signal powers can be used or a higher loss can be tolerated. Alternatively, a longer transmission distance can be used between regenerators. This property explains why hybrid amplifiers are mostly used in ultra-long-haul systems, where the reach is key in the system value proposition.

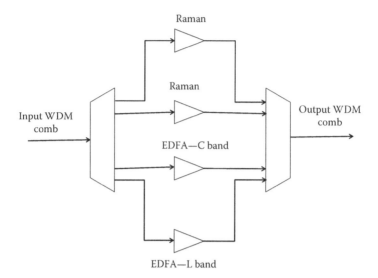

FIGURE 4.41
Block scheme of a wide band amplifier using EDFAs for C and L bands and Raman for S band.

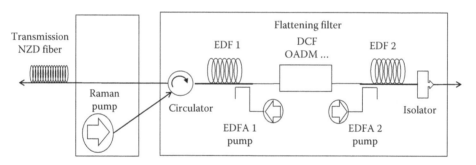

FIGURE 4.42
One of the possible schemes of hybrid EDFA and Raman amplifier.

A second advantage is a more uniform gain along the length of the fiber. This gives rise to better signal to noise ratio performance and reduced nonlinear penalty.

Finally, the gain equalization, gain level correction, Add–Drop multiplexers, and dispersion compensation can all be placed in between the two stages of the EDFA, thus requiring no modification to the standard architecture and design of such subsystems.

On the other hand, there are also challenges in using hybrid amplifiers. A first challenge is constituted by the high Raman pump powers propagating in the transmission fiber. Pumps on the order of 500 mW or even 1 W are common for Raman-distributed amplifiers if using an SSMF transmission fiber. At these power levels, connectors are highly vulnerable to damage. In addition, there are issues associated with higher sensitivity to spurious reflections and to environmental and mechanical changes.

Careful optimization and engineering are needed to design practical hybrid amplifiers; however, despite all these difficulties, many ultra-long-haul systems adopt them.

The potential performances of hybrid amplifiers are illustrated in Figure 4.44 where SN_0 and NF are shown for a high-performance hybrid amplifier [74]. Here SN_0 is the signal to

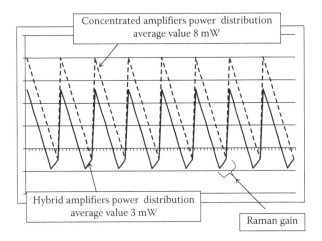

FIGURE 4.43
Signal power distribution along a line with EDFAs and with hybrid amplifiers.

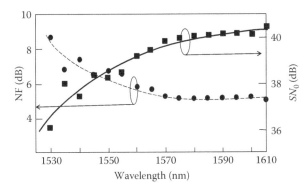

FIGURE 4.44
Spectral distribution of SN_0 and NF of a high-performance hybrid amplifier. (After Optiwave White Papers. Hybrid amplifiers [Online], 2009, http://www.optiwave.com/literature/article_whitepaper.html (Accessed: May 15, 2010).)

noise ratio at the amplifier output if the input signal is in a coherent state. It is not difficult to compare these figures with an equivalent two stages EDFA amplifier and recognize the advantage of the hybrid solution in having a small overall noise factor.

4.4 Optical Filters

The concept of filter in optics is almost the same as that in electronics: a passive and linear system used generally to admit into a system only the signals in a determined bandwidth.

There are a great number of optical filters used for a wide variety of applications [16]. Here we will briefly review the most important filters used in telecommunication equipment.

Optical filter can be classified in two classes: periodic and nonperiodic filters.

In some filters, the transfer function, or the shape of the filter passband, repeats itself after a certain period. The period of such devices is called free spectral range (FSR).

The FSR usually depends on various physical parameters in the device, such as cavity or waveguide lengths.

The *finesse* of a periodic filter is a measure of the width of the transfer function. It is the ratio of FSR to the filter 3 dB bandwidth.

For all optical filters that are placed on the signal path, the attenuation is a fundamental characteristic, besides dispersion and passband shape.

4.4.1 Fixed Wavelength Optical Filters

4.4.1.1 Grating Filters

One implementation of a fixed filter is the diffraction grating [16]. A bulk diffraction grating is essentially a flat layer of transparent material (e.g., glass or plastic) with a row of parallel grooves cut into it. The grating separates light into its wavelengths component by reflecting light incident with the grooves. The reflected spectral components interfere at the grating output and, at certain angles, only one wavelength interferes constructively; all others cancel by destructive interference.

This allows the desired wavelength to be selected by placing a spatial filter (essentially a hole) filter at the proper angle.

A grating can be also impressed onto a guiding structure, like a fiber or a monolithic waveguide. These gratings can be either transmissive or reflective.

4.4.1.2 Fiber Bragg Gratings

A fiber Bragg grating is a periodic or pseudo-periodic perturbation of the effective refractive index in the core of an optical fiber [16]. Typically, the perturbation is approximately periodic over a certain length from a few millimeters to centimeters for long gratings, and the period is on the order of hundreds of nanometers, or much longer for *long-period fiber gratings*.

The refractive index perturbation leads to the reflection of light in a narrow range of wavelengths, for which a *Bragg condition* is satisfied

$$\frac{2\pi}{\Lambda} = 2\frac{2\pi}{\lambda}n_{eff} \tag{4.91}$$

where
Λ is the grating period
λ is the vacuum wavelength
n_{eff} is the effective refractive index of light in the fiber

Essentially, the condition means that the wave number of the grating matches the difference of the (opposite) wave vectors of the incident and reflected waves.

In that case, the complex amplitudes corresponding to reflected field contributions from different parts of the grating are all in phase so that they can add up constructively. Even a weak modulation index (with an amplitude of, e.g., 10^{-4}) is sufficient for achieving nearly total reflection, if the grating is sufficiently long (e.g., a few millimeters).

Light at other wavelengths, not satisfying the Bragg condition, is nearly not affected by the Bragg grating, except for some side lobes that frequently occur in the reflection spectrum (but can be suppressed by apodization of the grating [47]).

The reflection bandwidth of a fiber grating, which is typically well below 1 nm, depends on both the length and the strength of the refractive index modulation. The narrowest bandwidth values are obtained for long gratings with weak index modulation. Large bandwidths may be achieved not only with short and strong gratings, but also with aperiodic designs.

4.4.1.3 Thin-Film Interference Filters

Thin-film filters offer another approach for fixed optical filtering [16]. These filters are similar to fiber Bragg grating devices with the exception that they are fabricated by depositing alternating layers of low-index and high-index materials onto a substrate layer. Thin-film filter technology allows the shape of the filter to be designed on the system needs since almost every pass-band shape can be realized via deposition of a sufficiently high number of layers. The main limitation is on the filter bandwidth that cannot be too tight. Typical values of the thin-film filter bandwidth are 1 nm at a wavelength of 1.5 µm.

4.4.2 Tunable Optical Filters

4.4.2.1 Etalon

The etalon [16], or Fabry–Perot filter, consists of a single cavity formed by two parallel mirrors. Light from an input fiber enters the cavity and reflects a number of times between the mirrors. By adjusting the distance between the mirrors, the cavity resonates on the desired wavelength that passes through the cavity, while the other wavelengths are reflected.

The varying transmission function of an etalon is caused by interference between the multiple reflections of light between the two reflecting surfaces. Constructive interference occurs if the transmitted beams are in phase, and this corresponds to a high-transmission peak of the etalon. If the transmitted beams are out-of-phase, destructive interference occurs and this corresponds to a transmission minimum. Whether the multiply-reflected beams are in-phase or not depends on the wavelength (λ) of the light (in vacuum), the angle the light travels through the etalon (θ), the thickness of the etalon (ℓ), and the refractive index of the material between the reflecting surfaces (n).

The phase difference between each succeeding reflection is given by δ:

$$\delta = \left(\frac{2\pi}{\lambda}\right) 2 n\ell \cos\theta = \left(\frac{2\omega}{c}\right) n\ell \cos\theta \tag{4.92}$$

If both surfaces have a reflectance $\mathcal{R}$, the transmittance function of the etalon is given by the following:

$$H(w) = \frac{(1-\mathcal{R})^2}{1-\mathcal{R}^2 - 2\mathcal{R}\cos\left[(2\omega/c)n\ell\cos\theta\right]} = \frac{1}{1+\mathcal{F}\sin^2[\delta(\omega)]} \tag{4.93}$$

where $\mathcal{F} = \mathcal{R}/(1-\mathcal{R})^2$ is the coefficient of finesse.

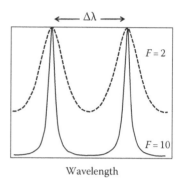

FIGURE 4.45
Periodic transmission of an etalon for different values of the finesse.

The transmission of an etalon as a function of wavelength is plotted in Figure 4.45 for different values of the finesse coefficient. A high-finesse etalon shows sharper peaks and lower transmission minima than a low-finesse etalon.

The etalon is probably the most used optical filter and many types of etalons exist for different applications.

Air-spaced etalons have fused silica substrates and an air gap between the cavity mirrors; the outside-facets are antireflection coatings to prevent extraneous interference patterns from forming. Spacers determine the parallelism of the mirrors and the etalon's FSR.

Solid state etalons typically have fused silica substrates and a solid material, transparent at the desired wavelengths, to form the etalon body. Dielectric (or, rarely, metallic) coatings provide the reflectivity necessary for the required finesse. The faces are ground, polished, and figured typically to better than 1/100 flatness to achieve the desired control of the transfer function. Solid state etalons cannot be tuned by moving the mirrors, but tuning can be achieved with other means, for example, changing the material refraction index.

Both air spaced and solid etalons exist in miniaturized versions for telecommunication applications.

To overcome the limit of the thinness of a solid etalon and yet retain its mechanical strength, Deposited etalons exist, realized by means of layers deposited on a substrate.

4.4.2.2 Mach Zehnder Interferometer

In an integrated Mach Zehnder interferometer (MZI) [16], a splitter splits the incoming wave into two waveguides and a combiner recombines the signals at the outputs of the waveguides as shown in Figure 4.46. In the figure, an example of filter working is also indicated, in particular the response to two signals entering from the same port.

An adjustable delay element controls the optical path length in one of the waveguides, resulting in a phase difference between the two signals when they are recombined. In order to widen the bandpass, sometimes the MZI branches are not straight, as also shown in the figure.

Wavelengths for which the phase difference is 180 are filtered out. By constructing a chain of these elements, a single desired optical wavelength can be selected [75].

While the MZI chain promises to be a low-cost device because it can be fabricated on semiconductor material [76], its tuning control is complex, requiring a delay element setting in each stage to be dependent on the settings in previous stages of the chain.

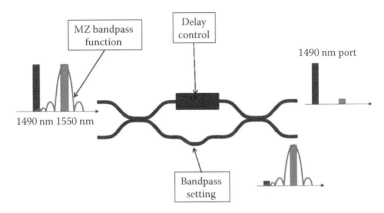

FIGURE 4.46
Block scheme of a wide band MZI filter.

4.4.2.3 Microrings Filters

Optical microring resonators (OMRs) are versatile elements for implementing filtering function in integrated photonic circuits [77].

A microring is generally constituted by a ring waveguide resonator with two tangent waveguides that, in the nearer point, passes at a very small distance from the ring. This distance, which is a critical parameter determining the coupling efficiency between the resonator and the input/output waveguides, is called characteristic gap.

Each ring, with its input and output ports, constitutes a single-pole add/drop periodic filter: resonant frequencies are sent from the input to the so-called drop port, while nonresonant frequencies are sent to the so-called line port. This behavior is represented in Figure 4.47.

The single-pole filter has a Lorentzian bandpass characteristic and can be used as building blocks for multipole add–drop filters (see Figure 4.47). The bandwidth of the single-pole filter is given by [78]

$$\Delta\omega = \frac{\omega_0}{\displaystyle\sum_{j=0}^{2} 1/Q_j} \qquad (4.94)$$

where
  ω_0 is the resonant angular frequency
  Q_0 is the circular cavity quality factor
  Q_1 and Q_2 are the quality factors of the coupling between the input and the output guides, respectively, and are given by

$$Q_j = \frac{\pi\omega_0\rho\, n_e}{c\,\kappa} \qquad (4.95)$$

where
  ρ is the ring radius
  n_e is the effective mode index
  c is the light speed in vacuum
  κ is the coupling coefficient

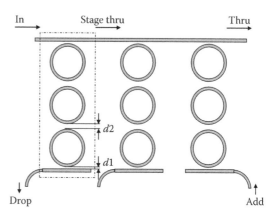

FIGURE 4.47
Scheme of a three-pole multiring filter.

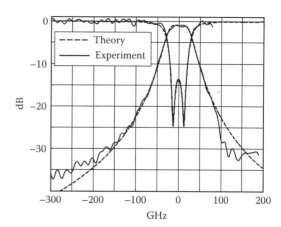

FIGURE 4.48
Notch and passband filtering characteristic of a two poles multiring filter: theoretical (dotted line) and measured (continuous line) characteristics. (After Romagnoli, M. et al., *Proc. SPIE*, 6996, 699611, 2008.)

As shown in Figure 4.47, microrings can be coupled together to build a complex filter with sharp transfer functions. In Figure 4.48 [79], the transfer function measured for a two-stage microring filter is compared with its theoretical estimation.

The weakness of this technique, which has as many advantages in terms of power consumption, dimension, and functionality, is the loss that today are still too high for the use in the signal line. However, microrings are more and more diffused in the field in which losses are less critical.

4.4.3 WDM Multiplexers and Demultiplexers

In principle, several types of optical filters are suitable to form arrays working as multiplexers and demultiplexers.

For example, several CWDM multiplexers are realized as shown in Figure 4.49 with an array of thin-film filters.

The way in which the filters are arranged implies that the different channels undergo a different loss. This problem is generally overcome by arranging the filters in the demultiplexer in the inverse order with respect to the multiplexer arrangement. Thus, the channel undergoing the greater loss at the demultiplexer experiment the smaller at the multiplexer and so on.

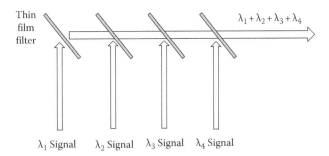

FIGURE 4.49
Example of WDM mux/demux realized with thin film filters.

Thin-film filters have also the quality of being low loss and very controlled in shape; thus, they constitute excellent elements for mux/demux.

Unfortunately, it is very difficult to obtain thin-film filters suitable for processing DWDM channels, whose distance is 100 or even 50 GHz with a modulation of 10 Gbit/s, while channels modulated at 40 Gbit/s are spaced 100 or 200 GHz.

To work at this small spacing, thin-film filters require a too high number of different coatings decreasing the repeatability and generating too great prices to be adopted in real products.

Practically all DWDM systems use array waveguide (AWG) multiplexers/demultiplexers.

Figure 4.50 shows a schematic layout of an AWG demultiplexer; a multiplexer is identical but for the fact that several signals enter from the waveguides that are the demultiplexer output and the WDM comb emerges from the demultiplexer input.

The device consists of three main parts which are as follows:

1. Multiple input and output waveguides
2. Two slab waveguide star couplers (also called free propagation region [FPR])
3. A dispersive waveguide array with the equal length difference between adjacent waveguides

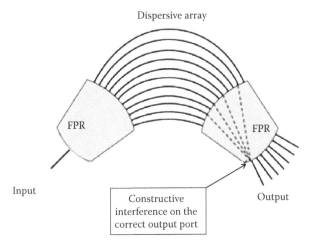

FIGURE 4.50
Scheme of an AWG demultiplexer.

TABLE 4.12

Characteristics of a Realistic AWG
Used for Multiplexing/Demultiplexing
DWDM Channels

Number of channel	40
Channel spacing	100 GHz
Band profile	Gaussian
Center wavelengths	ITU grid
Center wavelength accuracy	±5 GHz
−3 dB bandwidth	0.4 nm
Insertion loss	6.5 dB
Cross-talk in adjacent channel	−25 dB
PMD	0.5 ps
Dispersion	−10 ps/nm

The operation principle of the AWG multiplexer/demultiplexer is described as follows. A DWDM signal launched into one of the input waveguides will be diffracted in the first slab region and coupled into the arrayed waveguide by the first FPR. The length of the AWGs has been designed such that the optical path length difference (ΔL) between adjacent AWGs equals an integer (m) multiple of the central wavelength (λ_c) of the demultiplexer.

As a consequence, the field distribution at the input aperture will be reproduced at the output aperture. Therefore, at this center wavelength, the light focuses on the center of the image plane (provided that the input waveguide is centered in the input plane). If the input wavelength is detuned from this central wavelength, phase changes occur in the array branches.

Due to the constant path length difference between adjacent waveguides, this phase change increases linearly from the inner to outer AWGs, which causes the wavefront to be tilted at the output aperture. Consequently, the focal point in the image plane is shifted away from the center. By placing receiver waveguides at proper positions along the image plane, spatial separation of the different wavelength channels is obtained.

AWG are able to process a great number of nearby channels with acceptable loss and a great reproducibility and their structure allows even bigger mux/demux to be designed by cascaded of suitably designed AWG.

The performance of a realistic AWG used for multiplexing/demultiplexing DWDM channels is reported in Table 4.12.

References

1. Snider, A. W., Love, J. D., *Optical Waveguide Theory*, Chapman & Hall, London, U.K., 1983, ISBN-13: 978-0412099502.
2. Hunger, H. G., *Planar Optical Waveguides and Fibers*, Clarendon Press, Oxford, U.K., 1977.
3. Gloge, D., Weakly guiding fibres, *Applied Optics*, 11(10), 2252–2258 (1971).
4. He, Y., Shi, F., *Beam Propagation Method and Microlens Design for Optical Coupling: Finite-Difference Full-Vectorial Beam Propagation Method Development and Microlens Design for Fiber to Laser Diode Coupling*, VDM Verlag, Berlin, Dr. Müller, s.l., 2010, ISBN-13: 978-3639209723.

5. Skutnik, B. J., Moran, K. B., Spaniol, S., Optical fibers for improved low loss coupling of optical, *Proceedings of Progress in Biomedical Optics and Imaging*, San Jose, CA, January 27–28, s.l., 2004.

6. Boyd, R. W., *Nonlinear Optics*, Academic Press, San Diego, CA, 2008, ISBN-13: 978-0123694706.

7. Agrawal, G. P., *Nonlinear Fiber Optics*, 4th edn., Academic press, San Diego, CA, 2006, ISBN-13: 978-0123695161.

8. Pool, C. D., Wagner, R. E., Phenomenological approach to polarization dispersion in single mode fibers, *Electronics Letters*, 22, 1029–1030 (1986).

9. Curti, F., Daino, B., Mao, Q., Matera, F., Someda, C. G., Concatenation of polarization dispersion in single mode fibers, *Electronics Letters*, 25, 290–292 (1989).

10. Betti, S., Curt, F., Daino, B., De Marchis, G., Iannone, E., Matera, F., Evolution of the bandwidth of the principal states of polarization in single mode fibers, *Optics Letters*, 16, 467–469 (1991).

11. De Angelis, G., Galtarossa, A., Gianello, G., Matera, F., Schiano, M., Time evolution of polarization mode dispersion in a long terrestrial link, *IEEE Journal of Lightwave Technology*, 10, 552–555 (1992).

12. De Marchis, G., Iannone, E., Polarization dispersion in single mode optical fibers: A simpler formulation based on pulse envelope propagation, *Microwave and Optical Technology Letters*, 4, 75–77 (1991).

13. Bononi, A., Vannucci, A., Statistics of the Jones matrix of fibers affected by polarization mode dispersion, *Optics Letters*, 26, 675–677 (2001).

14. Curti, F., Daino, B., De Marchis, G., Matera, F., Statistical treatment of the evolution of the principal states of polarization in single mode fibers, *IEEE Journal of Lightwave Technology*, 8, 1162–1166 (1990).

15. Gordon, J. P., Kogelnik, H., PMD fundamentals: Polarization mode dispersion in optical fibers, *Proceedings of the National Academy of Sciences*, 97, 4541–4550 (2000).

16. Born, M., Wolf, E., *Principles of Optics: Electromagnetic Theory of Propagation, Interference and Diffraction of Light*, 7th edn., Cambridge University Press, Cambridge, U.K., s.l., 1999, ISBN-13: 978-0521642224.

17. Shen, Y. R., *The Principles of Nonlinear Optics*, Wiley Interscience, Hoboken, NJ, s.l., 2002, ISBN-13: 978-0471430803.

18. Iannone, E., Matera, F., Mecozzi, A., Settembre, M., *Nonlinear Optical Communication Networks— Appendix A1*, Wiley, New York, s.l., 1998, ISBN-13: 978-0471152705.

19. Gordon, J. P., Mollenauer, L. F., *Solitons in Optical Fibers: Fundamentals and Applications*, 1st edn., Academic Press (Kindle Edition), Burlington, MA, s.l., 2006, ASIN: B000PY3CWE.

20. Hamza, M. Y., Tariq, S., Split step Fourier method based pulse propagation model for nonlinear fiber optics, *International Conference on Electrical Engineering ICEE '07*, Hong Kong, IEEE, s.l., pp. 1–5, July 8–12, 2007.

21. Chraplyvy, A. R., Limitations on lightwave communications imposed by optical fiber nonlinearities, *Journal of Lightwave Technology*, 8, 1548–1557 (1990).

22. Kivshar, S. Y., Agrawal, G., *Optical Solitons: From Fibers to Photonic Crystals*, Academic Press, San Diego, CA, s.l., 2003, ISBN-13: 978-0124105904.

23. Tkach, R. W., Chraplyvy, A. R., Forghieri, F., Gnauck, A. H., Derosier, R. M., Four photon mixing and high speed WDM systems, *IEEE Journal of Lightwave Technology*, 13, 841–849 (1995).

24. Shibate, N., Horigudhi, M., Edahiro, T., Raman spectra of binary high-silica glasses and fibers containing GeO_2, P_2O_5 and B_2O_3, *Journal of Non-Crystalline Solids*, 45, 115–126 (1981).

25. Cotter, D., Observation of stimulated Brillouin scattering in low loss silica fiber at 1.3 µm, *Electronics Letters*, 18, 495–496 (1982).

26. Buckland, E. L., Boyd, R. W., Electrostrictive contribution to the intensity-dependent refractive index of optical fiber, *Optics Letters*, 21, 1117–1119 (1996).

27. Nikles, M., Thevenaz, L., Robert, P. A., Brillouin gain spectrum characterization in single-mode optical fiber, *IEEE Journal of Lightwave Technology*, 15, 1842–1851 (1997).

28. Mendez, A., Morse, T. F., *Specialty Optical Fibers Handbook*, Academic Press, Burlington, MA, s.l., 2007, ISBN-13: 978-0123694065.

29. Li, M. J., Chen, X., Nolan, D. A., Berkey, G. E., Wang, J., Wood, W. A., Zenteno, L. A., High bandwidth single polarization fiber with elliptical central air hole, *Proceedings of the 20th Annual Meeting of the Lasers and Electro-Optics Society*, Lake Buena Vista, FL, IEEE, pp. 3454–3460, December 19, 2007.

30. Ramachandran, S., *Fiber Based Dispersion Compensation*, Springer, New York, s.l., 2007, ISBN-13: 978-0387403472.

31. Wandel, M., Kristensen, P., Veng, T., Qian, Y., Le, Q., Gruner-Nielsen, L., Dispersion compensating fibers for non-zero dispersion fibers, *Optical Fiber Communication Conference and Exhibit, OFC 2002*, Anaheim, CA, IEEE, s.l., May 17–22, 2002.

32. Richardson, D. J., Belardi, W., Furusawa, K., Price, J. H. V., Malinowski, A., Monro, T. M., Holey fibers: Fundamentals and applications. Summaries of papers presented at the *Lasers and Electro-Optics, CLEO '02*, Long Beach, CA, IEEE, s.l., Vol. 1, pp. 453–454, 2002.

33. Poli, F., Cucinotta, A., Selleri, S., *Photonic Crystal Fibers: Properties and Applications*, Springer, New York, s.l., 2009, ISBN-13: 978-9048176090.

34. Borzycki, K., Holey fibers-application issues, *Proceedings of the 3rd International Conference on Transparent Optical Networks*, Cracow, Poland, s.n., pp. 92–95, June 18–21, 2001.

35. Bjarklev, A., Riishede, J., Photonic crystal fibers—A variety of applications, *Proceedings of the 4th International Conference on Transparent Optical Networks*, Warsaw, Poland, Vol. 2, p. 97, April 21–25, 2002.

36. Ming-Yang, C., Polarization-maintaining large-mode-area microstructured-core optical fibers, *IEEE Journal of Lightwave Technology*, 26, 13 (2008).

37. Hayes, J. et al., Advanced fibre designs for high power laser beam delivery and generation, *High Power Diode Lasers and Systems Conference*, Coventry, U.K., s.n., pp. 1–2, October 14–15, 2009.

38. Akowuah, E. K. et al., An endlessly single-mode photonic crystal fiber with low chromatic dispersion, and bend and rotational insensitivity, *IEEE Journal of Lightwave Technology*, 27(17), 3940–3947 (2009).

39. Ortigosa-Blanch, A. et al., Highly birefringent photonic crystal fibres, *Optics Letters*, 25(18), 1325–1327 (2000).

40. Nolan, D. A. et al., Single-polarization fiber with a high extinction ratio, *Optics Letters*, 29(16), 1855 (2004).

41. Zenteno, L. A. et al., Suppression of Raman gain in single-transverse-mode dual-hole-assisted fiber, *Optics Express*, 13(22), 8921–8926 (2005).

42. Hogari, K., Yamada, Y., Toge, K., Novel optical fiber cables with ultrahigh density, *Journal of Lightwave Technology*, 26(17), 3104–3109 (2008).

43. Hearle, J. W. S., *High-Performance Fibres*, Woodhead Publishing Ltd., Cambridge, England, s.l., 2001, ISBN: 1855735393.

44. Loudon, R., *The Quantum Theory of Light*, 3rd edn., Oxford Science Publications, Oxford, NY, s.l., 2000, June 2001. ISBN: 0198501765.

45. Itzykson, C., Zuber, J.-B., *Quantum Field Theory*, New Edition, Dover Publications, New York, s.l., 2000, ISBN: 0486445682.

46. Holden, H., Øksendal, B., Ubøe, J., Zhang, T., *Stochastic Partial Differential Equations: A Modeling, White Noise Functional Approach*, Springer, New York, s.l., 2009, ISBN-13: 978-0387894874.

47. Ramaswami, R., Sivarajan, K. R., Sasaki, G., *Optical Networks: A Practical Perspective*, 3rd edn., Morgan Kaufmann, San Francisco, CA, s.l., 2009, ISBN-13: 978-0123740922.

48. Desurvire, E., *Erbium-Doped Fiber Amplifiers: Principles and Applications*, Wiley, New York, s.l., 1994, ISBN-13: 978-0471589778.

49. Malin Premarante and Govind P. Agrawal *Light Propagation in Gain Media: Optical Amplifiers*, Cambridge University Press (March 14, 2011), Cambridge, ISBN-13: 978–0521493482.

50. Wyart, J. F., Blaise, J., Bidelman, W. P., Cowley, C. R., Energy levels and transition probabilities in doubly-ionized erbium (Er 111), *Physica Scripta*, 56, 446–458 (1997).

51. Tanabe, S., Development of rare-earth doped fiber amplifiers for broad band wavelength-division-multiplexing telecommunication. In *Photonics Based on Wavelength Integration and Manipulation*, The Institute of Pure and Applied Physics—IPAP Books, Tokyo, Japan, s.l., pp. 101–112, 2005.

52. Ohishi, Y., Yamada, M., Kanamori, T., Sudo, S., Shimizu, M., Low-noise operation of fluoride-based erbium-doped fiber amplifiers with 4I11/2-level pumping, *Optics Letters*, 22(16), 1235–1237 (1997).

53. Girarda, S., Marcandella, C., Origlio, G., Ouerdane, Y., Boukenter, A., Meunier, J.-P., Radiation-induced defects in fluorine-doped silica-based optical fibers: Influence of a pre-loading with H_2, *Journal of Non-Crystalline Solids*, 335(18–21), 1089–1091 (2009).

54. Pedersen, B., Bjarklev, A., Povlsen, J. H., Design of erbium doped fibre amplifiers for 980 nm or 1480 nm pumping, *Electronics Letters*, 27(3), 255–257 (1991).

55. Thyagarajan, K., Anand, J. K., Intrinsically gain-flattened staircase profile erbium doped fiber amplifier, *Optics Communications*, 222(1–6), 227–233 (2003).

56. Bae, J. K., Bae, J., Kim, S. H., Park, N., Lee, S. B., Dynamic EDFA gain-flattening filter using two LPFGs with divided coil heaters, *IEEE Photonics Technology Letters*, 17(6), 1226–1228 (2005).

57. Yu, A., O'Mahony, M. J. Analysis of dual-stage erbium-doped fibre amplifiers with passive equalisation filters, *IEE Proceedings on Optoelectronics*, 146(3), 153–158 (1999).

58. Headley, C., Agrawal, G. (Eds.), *Raman Amplification in Fiber Optical Communication Systems*, Prentice Hall Academic Press, Burlington, MA, s.l., 2004, ISBN-13: 978-0120445066.

59. He, J., Guo, T., Gu, W., Xu, D., Study on optimal design broadband and flat-gain multi-wavelength pumped fiber Raman amplifiers, *Journal of Optical Communications*, 24, online 42, 1–4 (2003).

60. Islam, M. N. Raman amplifiers for telecommunications, *IEEE Journal of Selected Topics in Quantum Electronics*, 8(3), 548–558 (2002).

61. Tang, M. Gong, Y. D., Shum, P., Design of double-pass dispersion-compensated Raman amplifiers for improved efficiency: Guidelines and optimizations, *IEEE Journal of Lightwave Technology*, 22(8), 1899–1908 (2004).

62. Hansen, P. B., Jacobovitz-Veselka, G., Gruner-Nielsen, L., Stentz, A. J., Raman amplification for loss compensation in dispersion compensating fibre modules, *Electronics Letters*, 34(11), 1136–1137 (1998).

63. Dung, J.-C., Sien, C., Characteristic of the reflective type Raman amplification in a dispersion compensating fiber, *Proceedings of the 5th Pacific Rim Conference on Lasers and Electro-Optics*, Taipei, Taiwan, Vol. 1, p. 4, December 2003.

64. Matsuda, T., Kotanigawa, T., Naka, A., Imai, T., 62×42.7 Gbit/s (2.5 Tbit/s) WDM signal transmission over 2200 km with broadband distributed Raman amplification, *Electronics Letters*, 38(15), 818–819 (2002).

65. Banoni, M., Fuochi, A., Transient gain dynamics in saturated Raman amplifiers with multiple counter-propagating pumps, *Optical Fiber Conference, OFC*, Atlanta, GA, IEEE, Vol. 1, p. 439, March 23–28, 2003.

66. Islam, M. N., Alam, M. S., Design of multiple pumps Raman amplifiers for optical communication systems, *Journal of Optical Communications*, 27(2), 75–78 (2006).

67. Pustovskikh, S. M., Kobtsev, A. A., Improvement of Raman amplifier gain flatness by broadband pumping sources, *Laser Physics*, 14(12), 1488–1491 (2004).

68. Bristiel, B., Jiang, S., Gallion, P., Pincemin, E., New model of noise figure and RIN transfer in fiber Raman amplifiers, *IEEE Photonics Technology Letters*, 18(8), 980–982 (2006).

69. Fludger, C., Handerek, V., Jolley, N., Mears, R. J., Fundamental noise limits in broadband Raman amplifiers, *Optical Fiber Communication Conference*, OSA, Anaheim, CA, March 17, 2001.

70. Namiki, S., Recent advances in Raman amplifiers, *European Conference on Optical Communications (ECOC)*, Amsterdam, the Netherlands, Vol. Tutorial, October 1–4, 2001.

71. Lewis, S. A. E., Chernikov, S. V., Taylor, J. R., Gain and saturation characteristics of dual wavelength-pumped silica-fibre Raman amplifier, *Electronics Letters*, 35(14), 1178–1179 (1999).

72. Tang, M., Gong, Y. D., Shum, P., Dynamic properties of double-pass discrete Raman amplifier with FBG-based all-optical gain clamping technique, *IEEE Photonics Technology Letters*, 16(3), 768–770 (2004).

73. Carena, A., Curri, V., Poggiolini, P., On the optimization of hybrid Raman/Erbium-doped fiber amplifiers, *IEEE Photonics Technology Letters*, 13(11), 1170–1172 (2001).

74. Optiwave White Papers. Hybrid amplifiers [Online], 2009, http://www.optiwave.com/literature/article_whitepaper.html (Riportato: May 15, 2010).

75. Kuznetsov, M., Cascaded coupler Mach-Zehnder channel dropping filters for wavelength-division-multiplexed optical systems, *IEEE Journal of Lightwave Technology*, 12(2), 226–230 (1994).

76. Okayama, H., Yaegashi, H., Ogawa, Y., Sub-micron Si waveguide design for polarization independent Mach-Zehnder filter, *Proceedings of the 6th International Conference on Group IV Photonics, GFP '09*, San Francisco, CA, IEEE, Vol. 1, September 9–11, 2009.

77. Popovic, M. A., Barwicz, T., Watts, M., Rakich, P. T., Dahlem, M. S., Gan, F., Holzwarth, C. W., Socci, L., Smith, H. I., Ippen, E. P., Kartner, F. X., Strong-confinement microring resonator photonic circuits, *Proceedings of the 20th Annual Meeting of the Lasers and Electro-Optics Society*, Lake Buena Vista, FL, IEEE, pp. 399–400, October 21–25, 2007.

78. Hossein-Zadeh, M., Vahala, K. J., Importance of intrinsic-Q in microring-based optical filters and dispersion-compensation devices, *IEEE Photonics Technology Letters*, 19(14), 1045–1047 (2007).

79. Romagnoli, M., Socci, L., Bolla, L., Ghidini, S., Galli, P., Rampinini, C., Mutinati, G., Nottola, A., Cabas, A., Doneda, S., Di Muri, M., Morson, R., Tomasi, T., Zuliani, G., Lorenzotti, S., Chacon, D., Marinoni, S., Corsini, R., Giacometti, F., Sardo, S., Gentili, M., Grasso, G., Silicon photonics in Pirelli, *Proceedings of SPIE*, 6996, 699611-699611-8 (2008), doi:10.1117/12.786539.

5

Technology for Telecommunications: Integrated Optics and Microelectronics

5.1 Introduction

In Chapter 4 we have presented a survey of fiber technology and of the technology needed to fabricate optical filter and wavelength multiplexer and demultiplexers for telecommunications.

In this chapter we will present a survey of the so-called integrated technology that is the technology leveraging planar processes to manufacture electronics and optical components.

The first part of the chapter is devoted to optical planar components built on the III–V platform.

The most important of those components is the semiconductor laser, both for the huge number of telecom and non-telecom applications and for the key role it plays in telecom equipments.

We will not try to make a survey of the semiconductor laser technology per se, but we will only summarize the characteristics and the main evolution potentialities of the devices used in telecommunications.

These devices are essentially signal sources or pump lasers for fiber amplifiers and we will consider both the applications, which require quite different devices.

We will add here also a brief description of the linear characteristics of semiconductor optical amplifiers (SOAs), which will be considered for the potentialities related to their nonlinear behavior in Chapter 10.

After a brief analysis of the modulators, both from III–V materials and from lithium niobate (LiNbO$_3$), the first part of the chapter devoted to planar optics comes to an end.

In the second part of the chapter we will analyze the impact of electronic evolution on telecommunication equipments.

Electronics is by far the most important technology used in telecommunication equipments and modern telecommunications would be unconceivable without very large-scale integrated circuits.

Naturally it is not possible to make a review of electronics technology from first principles, and we will assume that the reader has already a familiarity with basic electronics principles and potentialities.

Thus only the most recent electronic development that has an impact on telecommunications will be reviewed, excluding wireless technologies, both traditional radio bridges and wireless access in all its different standards.

After a brief review of the evolution of the complementary metal oxide semiconductor (CMOS) transistors, the base element of electronic circuitry, the discussion on electronics is divided into the analysis of error correction and compensators for transmission systems and the analysis of the base elements of electronics switching.

Only the analysis of the switch fabric itself, starting from the elementary cross-point, is delayed to Chapter 7, in the more suitable framework of the switching and routing machine architectures.

5.2 Semiconductor Lasers

5.2.1 Fixed-Wavelength Edge-Emitting Semiconductor Lasers

Semiconductor lasers are the most diffused laser category, whose application ranges from consumer products (like compact disc players) to highly specialized industrial applications like transmitters in optical fiber systems and sensor equipment [1].

Here we will limit our review to two categories of semiconductor lasers that are instrumental in developing optical transmission systems: transmission lasers and pump lasers.

Transmission lasers are used to provide the optical carrier over which the information is coded.

Different applications have different requirements for the optical carrier parameters like power, linewidth, wavelength stability, and so on, as well as different cost targets; thus there are a wide variety of transmission lasers.

Depending on the specific application, the modulation of the carrier can be performed by an external modulator (a component completely different from the source laser), by a modulator integrated into the same chip of the laser or by the laser itself, with the so-called direct laser modulation.

Something similar happens for pump lasers, whose application ranges from the cheap erbium-doped fiber amplifiers (EDFAs) that could be used in the optical access network to the high-performance distributed Raman amplifiers used in ultra-long haul dense wavelength division multiplexing (DWDM) systems.

5.2.1.1 Semiconductor Laser Principle

The principle at the base of the working of a semiconductor laser is to create population inversion between valence and conduction electrons (or holes, depending on the material), exciting electrons from the valence to the conduction band of a semiconductor via current injection.

In order to achieve photon-stimulated emission, the transition between valence and conduction bands corresponding to the desired laser frequency has to be a "direct transition."

This means that pseudo-electrons in the valence and conduction bands must have the same crystal momentum so that decay can happen with photon emission without the need of higher-order processes to compensate the momentum difference. In Figure 5.1 [2] the band structure of silicon, characterized by an indirect transition between the valence and the conduction band, is shown, while the band structure of an InGaAs alloy [2], characterized by a direct transition in the near-infrared region, is shown in Figure 5.2.

The requirement of direct transitions practically selects the materials that are suited to build semiconductor lasers: they are essentially the alloys of III–V semiconductors, starting from GaAs and GaAsP, used for pump lasers, to InGaAsP, which is generally the alloy in the active region of lasers used for transmission.

Once the material is selected, a structure suitable for electron pumping has to be realized. The best possible structure is an inversely polarized p–n junction. In a p–n junction

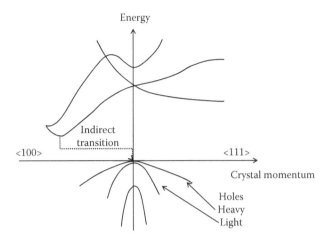

FIGURE 5.1
Schematic representation of the band structure of silicon.

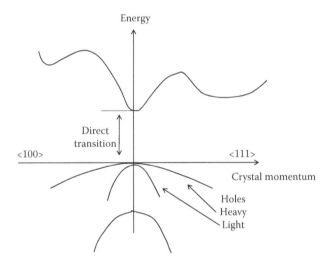

FIGURE 5.2
Schematic representation of the band structure of a II–V alloy having direct optical transition at the band gap.

under inverse polarization, the field in the junction zone is opposite to the electron flux; thus if a current is pumped in the junction, electrons pass through being promoted to higher-energy levels to overcome the junction potential barrier. In this condition the average population of the conduction band can become greater than the population of the valence band, thus causing population inversion.

However, if a simple junction is used, creating the conditions for lasing operation is quite difficult.

As a matter of fact, in a structure like that of Figure 5.3, there is no lateral confinement either for the injected current or for the optical field. Thus on the one hand, there is a constant electron loss due to the leak current flowing away from the inversion zone on the side of the device and, on the other hand, the optical mode is very large and the superposition of the mode with the inverted area of the junction where optical gain is concentrated is quite scarce.

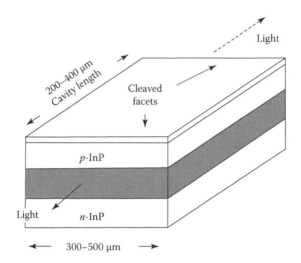

FIGURE 5.3
Simpler possible structure of a semiconductor laser.

In order to achieve effective lasing operation, both the current and the optical mode have to be effectively confined in the junction area.

Several architectures have been proposed to achieve vertical and lateral confinement, at present the most used solution is constituted by heterostructure index–guided architecture [1].

The heterostructure solution consists in realizing a p–i–n junction where the neutral zone is realized in a different alloy with a smaller band gap with respect to the doped zones. The band profile across the heterostructure is shaped approximately as shown in Figure 5.4 when an inverse polarization is applied.

The fact that the intrinsic material has a smaller energy gap with respect to the surrounding doped layers creates a sort of trap for the injected carriers: when the carriers arrive in the intrinsic zone, they fall in the potential energy hole and are confined to the intrinsic zone.

If the heterostructure provides confinement in the direction of the current injection, confinement in the lateral direction has to be achieved via some other system.

The most used structure for lateral confinement is the so-called index-confined architecture, where the optical mode is confined in the laser-active region by realizing an optical waveguide whose core is the active region itself. In this way the different materials provide confinement of the current and the diffraction index difference provides confinement of the field in the same region.

A transversal section of a buried heterostructure laser is represented in Figure 5.5. The gray area represents the active waveguide that is buried below the surface structure and

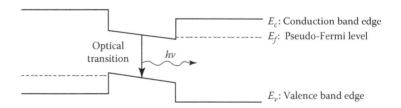

FIGURE 5.4
Schematic band profile of a heterojunction in inverse polarization.

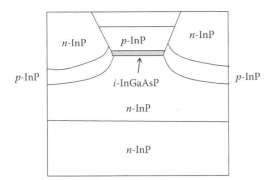

Figure 5.5
Transverse section of a buried heterostructure laser.

assures contemporary good lateral current confinement and good superposition between the optical mode and the inverted region.

Once inversion is effectively realized, in order to start the lasing operation, a positive counterreaction has to be introduced via an optical cavity to sustain spontaneous oscillations of the structure.

The easier way to do this is to cleave the front and rear facets of the laser to create a reflectivity.

The inner laser waveguide behaves like a Fabry–Perot (FP) cavity assuring the needed feedback to the laser.

The structure of an FP laser is schematized in Figure 5.3. With the typical dimensions of semiconductor lasers, the FP cavity has a free spectral range so small that several modes are enhanced by the optical gain.

Thus in general, FP semiconductor lasers oscillate in multimode regime even if, with some care in designing the laser, few modes can be selected. Even when single-mode operation is achieved by using strongly wavelength-dependent mirrors (that can be realized by multiple layer coating of the FP facets) this is not a stable condition and side modes can start to oscillate due to environmental fluctuations or laser modulation.

In order to achieve real single-mode operation a strongly wavelength-selective feedback has to be introduced.

This can be done in several ways, among which distributed feedback (DFB) is by far the most used.

5.2.1.2 Semiconductor Laser Modeling and Dynamic Behavior

Once confinement is achieved both for the optical field and for the injected charges, the laser can be considered as a waveguide inserted into a resonant cavity with the core subject to population inversion to generate optical gain.

A physical modeling of the semiconductor laser would require a quantum representation of both the optical field and the pseudo-particle population inside the crystal.

In this way the laser equations will appear as a couple of interacting equations, one for the density matrix elements representing the electron population and the other for the quadrature operators of the field [3].

Even though this analysis is possible, it is quite complicated and, from the point of view of the physical comprehension of laser dynamics, is not more effective with respect to a semiclassical model.

In a semiclassical model, the laser behavior is represented with a set of rate equations for the propagating field and for the carrier density.

The only element that cannot rise from a semiclassical analysis is the spontaneous emission noise, which is a pure quantum element (compare Section 4.3.1.1). Thus the noise terms have to be added phenomenologically to the semiclassical rate equations [4].

With this in mind, the rate equations can be derived by the coupled field wave equation and electrons balance equation and by applying the slowly varying and rotating wave approximation to the wave equation [5].

At the end of these procedures, calling $N(x,y,z,t)$ the carrier population in the conduction band, $S(x,y,z,t)$ the optical energy inside the cavity, which is proportional to the photon density, and $\phi(x,y,z,t)$ the field phase, the coupled equations of a semiconductor laser can be written as

$$\frac{\partial N}{\partial t} + D_N \nabla^2 N = \frac{J}{q d_a} - G_N(N - N_0)\frac{s}{(1 + \varepsilon s)} - \frac{N}{\tau_N} + F_N$$

$$\frac{\partial P}{\partial t} = \gamma\, G_N(N - N_0)\frac{s}{(1 + \varepsilon s)} - \frac{s}{\tau_P} + \hbar\,\omega_0\,R_{sp} + \hbar\,\omega_0 F_s$$

$$\frac{\partial \phi}{\partial t} = -(\omega_0 - \Omega) + \frac{\alpha}{2}\left[\gamma\, G_N(N - N_0) - \frac{1}{\tau_N}\right] + F_\phi \qquad (5.1)$$

The symbols that appear in the rate equations have the following meanings:

D_N is the carrier diffusion coefficient. In many problems the carrier density can be considered approximately constant in the active region and zero elsewhere, thus neglecting the diffusion term $D_N \nabla^2 N$ and the dependence of N from the spatial coordinates.

J is the pumping current.

d_a is the active layer thickness.

q is the electron charge.

τ_N is the carrier's lifetime.

G_N is the population inversion induced gain.

$(N - N_0)$ is the population inversion factor.

ε is the gain saturation constant.

τ_P is the photon's lifetime.

γ is the confinement factor, that is, the ratio between the active region volume and the modal volume. This parameter is evaluated by solving the mode propagation into the active region.

ω_0 is the angular frequency of the optical field.

Ω is the resonance frequency of the laser cavity.

R_{sp} is the spontaneous emission rate.

α is the so-called linewidth enhancement factor. The fundamental limit for the linewidth of a free-running laser is provided by the Schawlow Townes formula. Semiconductor lasers exhibit significantly higher linewidth values due to a coupling between intensity and phase noise, caused by the dependence of the refractive index on the carrier density. In the semiclassical model, the linewidth enhancement factor α quantifies this amplitude–phase coupling mechanism; essentially, α is a proportionality factor relating phase changes to changes of the amplitude gain [6].

F_j, $j = S, N, \phi$ are the noise terms that are added phenomenologically to the rate equations. These are independent random processes whose first moments have to be determined from the results of the quantum theory. In particular we have

$$\langle F_j(t, x, y, z)\rangle = 0 \quad j = S, N, \phi \qquad (5.2)$$

$$\langle F_j(t,x,y,z)F_k(t',x',y',z')\rangle = D_{jk}\delta_{j,k}\delta(t,t')\delta(x,x')\delta(y,y')\delta(z,z') \tag{5.3}$$

D_{jk} are the elements of the so-called diffusion matrix which, in the framework of the semi-classical approximation, can be written as

$$D = \begin{pmatrix} D_{SS} & D_{SN} & D_{S\phi} \\ D_{SN} & D_{NN} & D_{N\phi} \\ D_{S\phi} & D_{N\phi} & D_{\phi\phi} \end{pmatrix} = \begin{pmatrix} R_{sp}S & -R_{sp}S & 0 \\ -R_{sp}S & \dfrac{R_{sp}S}{\hbar\omega_0} + \dfrac{N}{\tau_N} & 0 \\ 0 & 0 & \dfrac{\hbar\omega_0 R_{sp}}{4S} \end{pmatrix} \tag{5.4}$$

Realistic values of the parameters of a standard DFB laser used as a source in DWDM systems are reported in Table 5.1.

To describe the laser behavior on the grounds of Equation 5.1 what we have assumed up to now is not sufficient.

As a matter of fact, these equations are stochastic equations that cannot be solved if the distribution of the noise term is not known.

The most correct distribution of the terms F_j ($j = S, N, \phi$) is not Gaussian: it is sufficient to consider that the term representing the shot noise (i.e., the fluctuation of the optical power) should be distributed following a Poisson distribution.

However, in the case of a stable lasing operation, it is possible to demonstrate that, due to the fact that the number of excited carriers and of the photons is very high, the Gaussian approximation implies a small error, in line with the overall accuracy of the semiclassical model.

Starting from Equations 5.1 and from the laser structure the main characteristics of a specific type of semiconductor laser can be derived. Despite the intrinsic complexity of their behavior, gain saturation happens also in semiconductor lasers.

In particular, due to the presence of the saturation constant, increasing the power in the cavity, the laser inner gain decreases so that the dependence of the emitted power from the pump is not exponential as foreseen in a laser without gain saturation.

Moreover, the dependence of the carrier density on a temperature typical of a p–i–n junction induces a sensible dependence of the curve relating the pumping current and the emitted power on temperature.

TABLE 5.1

Values of the Microscopic Parameters of a DFB Used at the Transmitter in DWDM Systems

Active layer thickness (d_a)	100–200 nm
Optical confinement factor (γ)	0.35
Carrier lifetime (τ_N)	3 ns
Photons lifetime (τ_P)	1 ps
Gain coefficient (G_N)	3×10^{-12} m^3/s
Carrier density at transparency (N_T)	10^{24} m^{-3}
Coherence time (s)	10^{-7}
Saturation carrier density (n_0) (m^{-3})	10^{-23}
Local saturation power (1/0) (mW)	1.3
Linewidth enhancement factor (α)	6 (1.5 MQW)

A typical emitted power versus current characteristic of a DFB laser is shown in Figure 5.6. From the figure, the laser threshold, the dependence of the laser behavior on temperature, and the saturation of the cavity gain are evident, so that at each temperature a maximum emitted power exists at the edge of saturation.

Above the maximum emitted power, the gain saturation drives a decrease of the emitted power while increasing the driving current.

When emitting in continuous wave (CW) mode, a semiconductor laser is affected both by phase and by intensity noise due to the amplified spontaneous emission into the laser cavity and to the fluctuations of the carrier density.

The noise characteristics can be obtained from the rate equations with a perturbation approach, at least as far as the noise is small [7]. In particular, substituting the variables of the rate equations the following expressions

$$N = N_0 + \delta N$$

$$S = S_0 + \delta S \tag{5.5}$$

$$\phi = \phi_0 + \delta\phi$$

where N_0, S_0, and ϕ_0 are the stationary solutions of the rate equations without noise, and linearizing with respect to the noise terms, the noise statistical characteristics can be derived.

In particular, the linearized equations assume a simpler form by introducing the vector $\overline{X} = (S, N, \phi)$ and the matrix

$$\Gamma = \begin{pmatrix} \Gamma_{SS} & \Gamma_{SN} & \Gamma_{S\phi} \\ \Gamma_{NS} & \Gamma_{NN} & \Gamma_{N\phi} \\ \Gamma_{\phi S} & \Gamma_{\phi N} & \Gamma_{\phi\phi} \end{pmatrix} = \begin{pmatrix} \dfrac{\epsilon\, S_0}{\tau_P(1+\epsilon\, S_0)} & \dfrac{\gamma G_N S_0}{\hbar\omega_0(1+\epsilon\, S_0)} & 0 \\ \dfrac{1}{\tau_P(1+\epsilon\, S_0)} & \dfrac{1}{\tau_N} + \dfrac{\gamma\, G_N\, S_0}{\hbar\omega_0(1+\epsilon\, S_0)} & 0 \\ 0 & \dfrac{\alpha\gamma G_N}{2} & 0 \end{pmatrix} \tag{5.6}$$

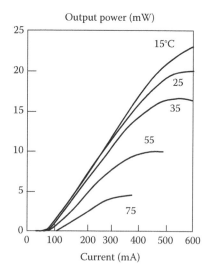

FIGURE 5.6
Typical emitted power versus current characteristic of a DFB laser.

In order to evidence the macroscopic noise terms in the expression of the emitted field, it can be written as

$$\vec{E} = \sqrt{P_0} a(x,y) r(t) e^{i\omega t + \varphi(t)} \vec{x} = \sqrt{\frac{\Im S_0}{\tau_c}} a(x,y) \sqrt{\left(1 + \frac{\delta S}{S_0}\right)} e^{i\omega t + [\phi(t) + \psi]} \vec{x} \qquad (5.7)$$

where
   $r(t)$ and $\varphi(t)$ are the relative intensity noise (RIN) and the phase noise
   $\Im$ is the end mirror transmittance
   τ_c is the cavity round trip time that allows us to pass from internal energy to emitted power
   ψ is the phase contribution of the extraction mirror

The power spectral density of the RIN is given by

$$\text{RIN}(\omega) = \frac{\hbar\omega}{S_0^2} \frac{2D_{NN}\Gamma_{SN}^2 + 2D_{SS}\left(\Gamma_{NN}^2 + \omega^2\right) + 4D_{SN}\Gamma_{SN}\Gamma_{NN}}{\left[\left(\Omega_R^2 - \omega^2\right)^2 + \omega^2(\Gamma_{NN} + \Gamma_{SS})^2\right]} \qquad (5.8)$$

where it is not difficult to recognize the typical behavior of a resonance, whose natural frequency is $\Omega_R^2 = (\Gamma_{SS}\Gamma_{NN} + \Gamma_{SN}\Gamma_{NS})$ and whose dumping rate is $(\Gamma_{NN} + \Gamma_{SS})$.

The observation that the laser behavior is characterized by a dumped resonance is very general, not limited exclusively to the RIN spectrum.

As a matter of fact, the same behavior is exhibited by the frequency noise (i.e., the process $\dot{\varphi} = \delta\varphi/\delta t$), whose power spectral density is given by [8]

$$F_\varphi(\omega) = D_{\varphi\varphi} + \frac{\Gamma_{\varphi N}^2 \left[D_{NN}\left(\Gamma_{SS}^2 + \omega^2\right) + \Gamma_{NS}^2 D_{SS} - D_{SN}\Gamma_{SS}\Gamma_{NS}\right]}{\left[\left(\Omega_R^2 - \omega^2\right)^2 + \omega^2(\Gamma_{NN} + \Gamma_{SS})^2\right]} \qquad (5.9)$$

Besides the presence of the laser characteristic resonance, the expression of the frequency noises spectrum allows also the other typical diode laser mechanism to be evidenced.

As a matter of fact, it is made by two terms: the first is due to the phase of random emitted photons, and the second is proportional through $\Gamma_{\varphi N}^2$ to α^2 and represents the additional phase noise due to the coupling between amplitude and phase through the carrier density [9].

The power spectral densities of the RIN and of the phase noise for a typical DFB laser emitting at 1550 nm are represented in Figures 5.7 [10] and 5.8 respectively.

From the theory we have presented, it results that the same phenomenon causes both the excess RIN and the excess phase noise with respect to the quantum limit. Thus a relationship must exist between the RIN and the phase noise spectra.

This relationship can be obtained by simplifying the denominator in the expressions of the two spectra and deriving an expression for the relation between them.

This expression can be experimentally verified by passing from the phase noise to the linewidth that is one of the most commonly measured quantities in a laser.

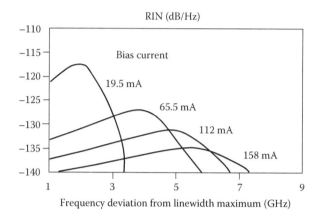

FIGURE 5.7

Power spectral density of the RIN for a typical DFB laser emitting at 1550 nm. (After Aragon Photonics Application Note, Characterization of the main parameters of DFB using the BOSE. 2002, www.aragonphotonics.com [accessed: May 25, 2010].)

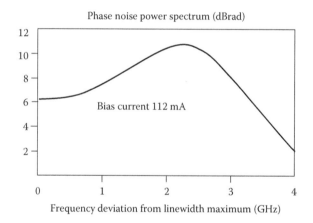

FIGURE 5.8

Power spectral density of the phase noise for a typical DFB laser emitting at 1550 nm. (After Travagnin, M., *J Opt. B. Quant. Semiclass. Opt.*, 2, L25, 2000.)

A rigorous derivation of the laser linewidth that is possible starting from (5.9) [8] would conclude that the laser linewidth is not Lorentian, but is composed by a Lorentian lobe plus two side peaks far from the central wavelength, exactly the frequency of relaxation oscillations. However, in high-quality lasers these lobes are so reduced that a Lorentian approximation of the linewidth is more than accurate.

Assuming a Lorentian linewidth, the expression of the full width at half maximum (FWHM) is given by

$$\Delta v = R_{sp} \frac{(1+\alpha^2)}{(4\pi r)} \tag{5.10}$$

where r is the cavity refraction index.

The curve relating the linewidth and the maximum of the RIN spectrum is shown in Figure 5.9 besides a set of experimental points from [11]. In particular the theoretical curve is obtained by best fitting the experimental data.

Another important characteristic of the semiconductor lasers is that they can be directly modulated acting on the bias current. However, when the bias current fluctuates, a complex dynamical behavior is observed, mainly due to the coupling between field amplitude and phase and to the resonance at Ω_R.

This phenomenon causes the contemporary presence of intensity and phase modulation, which always creates a frequency chirp in the direct modulated pulse train.

Besides this fundamental effect, a second phenomenon constraining the direct modulation bandwidth of semiconductor lasers is the presence of parasitic capacities in the structure. Reducing them is a key issue to increase the direct modulation bandwidth.

Considering again the rate equations with spatially constant variables and neglecting the noise terms, it is possible to study direct modulation by applying the substitutions in Equation 5.5 where now the fluctuations are due not to the noise but to the modulation.

In this condition the equations can be linearized again around the laser bias point. In this case Equation 5.7 can be used to pass from the field into the cavity to the emitted field so that the modulation index (i.e., the derivative of the emitted optical power with respect to the pump current fluctuations) results:

$$\frac{\delta P}{\delta J} = -\frac{\hbar\omega}{q d_a} \frac{\Im}{\tau_N} \frac{\Gamma_{NS}}{\omega_m^2 - \Omega_R^2 + i\omega_m(\Gamma_{NN} + \Gamma_{SS})} \tag{5.11}$$

where ω_m is the modulation frequency. The presence of the natural laser resonance is always evident and in this case it limits the modulation bandwidth.

If the expression of the coefficients Γ_{kj} is substituted in Equation 5.11 it is evident that, besides the presence of a resonance, gain saturation (represented by the factor ε also contributes to shrink the bandwidth with respect to the theoretical limit, and that higher the gain saturation coefficient, the smaller is the modulation bandwidth.

Similarly to the power modulation index, a frequency modulation index can be evaluated whose expression is

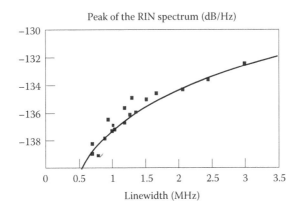

FIGURE 5.9
Linewidth versus RIN spectrum peak relationship for a DFB laser. (After JDSU Application Note, *Relative Intensity Noise, Phase Noise and Linewidth*, JDSU Corporate, s.l., 2006, www.jdsu.com [accessed: May 25, 2010].)

$$\frac{\delta\dot{\phi}}{\delta J} = \Gamma_{\phi N} \frac{\delta N}{\delta J} \qquad (5.12)$$

Equation 5.12 represents the theoretical frequency response of a semiconductor laser that is responsible for both the possibility of modulating the output field frequency by modulating the bias current (property exploited for example to implement the Brillouin dither) and the chirp that unavoidably is present when the laser is directly modulated.

Equation 5.12, however, due to all the approximations that are the base to derive the rate equations, does not take into account several phenomena that cause the deviation of the frequency response of real lasers from the simple proportionality to the carrier density variations.

Generally the frequency response is almost flat up to the resonance of the laser, but here it shows the resonance effects through a peak and then decreases rapidly.

5.2.1.3 Quantum Well Lasers

A quantum well laser is a laser diode in which the active region of the device is formed by one or more regions so narrow that quantum confinement occurs, alternated by wider regions of higher bandgap.

The carriers in the active region are generally divided into two populations that occupy a completely different set of energy levels: carriers in the wells and carriers out of the wells.

Carriers out of the wells are ordinary pseudo-particles obeying the rules of carrier motion in crystals, and in particular they cannot have any energy in the forbidden band of the semiconductor forming the active region.

Inside the quantum well, quantum confinement of the carriers creates energy sub-bands inside the forbidden energy band. The central energy of the sub-bands depends on the well thickness.

This is easily understood remembering that in the simple case of a one-dimensional infinite energy well the energy levels of a particle are equal to $E_n = n^2(\hbar^2\pi^2/8mL^2)$, where n is the quantum number and L is the width of the well.

Due to these characteristics the wavelength of the light emitted by a quantum well laser is determined by the width of the active region rather than just the bandgap of the material from which it is constructed. This means that much shorter wavelengths can be obtained from quantum well lasers than from conventional laser diodes using a particular semiconductor material. The efficiency of a quantum well laser is also greater than a conventional laser diode due to the stepwise form of its density of states function.

Beside the use of a quantum well, in order to shape the energy levels inside the active region, the so-called strain technique can be incorporated.

This consists in introducing in the quantum well zone a tensile or compressive strain, whose effect is to change the conduction sub-bands due to the quantum well.

One of the main effects of a controlled strain is to reduce the threshold current of the laser due to the increased separation between energy bands that renders recombination more difficult [12].

5.2.1.4 Source Fabry–Perot Lasers

Probably FP semiconductor lasers used as signal sources are the lasers produced in larger volumes on the market for a huge number of applications, from telecommunications to consumer electronics to sensors and so on.

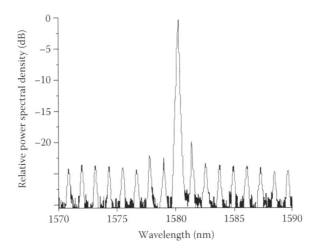

FIGURE 5.10
Spectrum of an FP InGaAsP laser emitting in a quasi-single-mode regime.

In telecommunications such lasers are adopted at transmitters either to generate the carrier or to generate the information-carrying signal through direct modulation when single-mode laser operation is not required but the cost of the source is a key issue. As a matter of fact, due to the simpler structure and the great volume production, FPs are cheaper with respect to single-mode lasers like DFB.

Thus, FPs are the standard optical sources in several types of access systems, from point-to-point to A and B class G-PON. Moreover, such lasers are used in the low-performance client cards of transmission systems, where it is needed to connect systems in the same central office and the connection speed is not so high.

The typical emission spectrum of an FP InGaAsP laser emitting in a quasi single-mode regime is plotted in Figure 5.10. The side modes suppression ratio (SMSR) is about 19 dB, which is much lower than the typical 35–40 dB of a DFB laser, but is enough for several applications, especially if the laser has not to be modulated.

The presence of a few lasing modes in an FP laser generates the growth of a particular noise phenomenon when it is directly modulated: the mode partition noise [13].

Mode partition noise is due to the random fluctuations of the lateral modes that causes a change in the spectrum and a fluctuation in the power of the main mode due to the fact that the steady-state power emitted from the laser in a first approximation is constant and depends only on the pump and on the laser quantum efficiency.

The effect is an enhancement of the RIN well above the limit given by the spectrum in a strict single mode operation. This effect is shown in Figure 5.11 where the fact that the sensible amount of RIN is mainly due to the mode partition noise is clear from the form of the RIN spectrum.

A selection of characteristics of practical FP diode lasers for different applications is reported in Table 5.2. From the table the wide span of applications covered by FP lasers corresponding to the possibility of matching different set of requirements is evident.

5.2.1.5 Source DFB Lasers

In a DFB laser the feedback is obtained by building a grating immediately on top of the active zone or on the side of it, influencing the propagation of the tail of the guided mode.

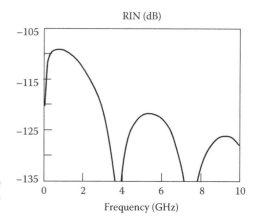

FIGURE 5.11
RIN spectrum induced by relevant mode partition noise in a multimode FP laser. (After Wentworth, R. H. et al., *J. Lightwave Technol.*, 10(1), 84, 1992.)

TABLE 5.2

Characteristics of Practical FP Diode Lasers for Different Applications

Application	Client Card Short Reach or CWDM	• Source for Free Space Wireless Systems • Transmitter for Remote Sensing • High-Power External Cavity Lasers	• Datacom • Client Interface Telecom
Fiber window	C band (L optional)	C band	Second window
Direct modulation	2.5 Gbit/s	NO	2.5 Gbit/s
CW optical power	5 mW	160 mW	0.3 mW
Cooling	NO	NO	
Package	TO 18	TO	
Threshold current	10 mA	50 mA	6 mA
Operating current	40 mA	450 mA	16 mA (CW)
			32 mA (pulse, duty rate 50%)
Peak wavelength	1530, 1550, 1570 nm	Custom in C band	1310 nm
Spectral width	2.5 nm	10 nm	1 nm
Operating temperature	−40°C + 85°C	−20°C + 70°C	20°C + 70°C
Notes	MQW active region	MQW active region	
	All data at 25°C	All data at 25°C	All data at 25°C
	Typical values	Typical values	Typical values

Via interaction with the mode tail, the grating blocks the amplification of all the wavelengths, save the one resonating with the grating itself, thus guaranteeing single-mode operation.

The scheme of a DFB semiconductor laser is shown in Figure 5.12 while its single-mode spectrum is shown in Figure 5.13.

Altering the temperature of the device causes the pitch of the grating to change due to the dependence of the refractive index on temperature. This dependence is caused by a change in the semiconductor laser's bandgap with temperature and thermal expansion.

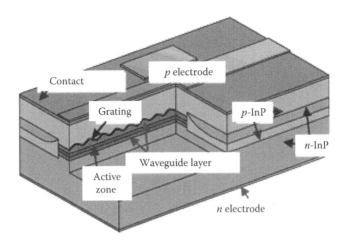

FIGURE 5.12
Schematic section of a high-performance DFB laser.

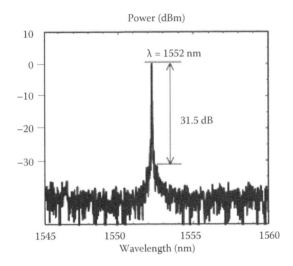

FIGURE 5.13
Emission spectrum of a typical DWDM DFB.

A change in the refractive index alters the wavelength selection of the grating structure and thus the wavelength of the laser output, producing a narrow band wavelength tunable laser.

The tuning range is usually on the order of 6 nm for an ~50 K (90°F) change in temperature, while the linewidth of a DFB laser is on the order of a few megahertz. Altering of the current powering the laser will also tune the device, as a current change causes a temperature change inside the device.

If this temperature sensitivity can be exploited to tune the laser on the opportune wavelength, it also implies that, in order to match the stability requirements of high-performance DWDM systems, the laser has to be temperature-stabilized. This is generally done through a Peltier temperature controller and a feedback control loop that assure sufficient temperature stability in the face of environment changes and the device heat production during working.

TABLE 5.3

Characteristics of Practical DFB Lasers for Use as Sources in Telecommunication System

	CW Source in Single-Channel Short-Reach Interfaces	High-Performance DWDM Carrier Source
Threshold current	10 mA	8 mA
Operating current	45 mA	150 mA
Modulation current	25 mA (Estintion ≈6 dB)	Only CW operation
Output power	2.5 dBm	13 dBm
Slope efficiency	0.025 mW/mA	—
Direct mod. extinction ratio	9 dB	—
Wavelength	1550–1560 nm	ITU-T grid—100 GHz
Spectral width	1 MHz	250 KHz
Side mode suppression	31 dB	45 dB
Relative intensity noise	−120 dB/Hz	−140 dB/Hz
Polarization extinction ratio	—	25 dB
Operating temperature	−20°C to +70°C	15°C–50°C

There are generally two distinct types of DFB lasers. Traditionally, DFBs are antireflection-coated on one side of the cavity and coated for high reflectivity on the other side (AR/HR). In this case the grating forms the distributed mirror on the antireflection-coated side, while the semiconductor facet on the high reflectivity side forms the other mirror. These lasers generally have higher output power since the light is taken from the AR side, and the HR side prevents power being lost from the back.

Unfortunately, during the manufacturing of the laser and the cleaving of the facets, it is virtually impossible to control at which point in the buried grating the laser cleaves to form the facet. So sometimes the laser HR facet forms at the crest of the buried grating, and sometimes on the slope. Depending on the phase of the grating and the optical mode, the laser output spectrum can vary. Frequently, the phase of the highly reflective side occurs at a point where two longitudinal modes have the same cavity gain, and thus the laser operates at two modes simultaneously. Thus such AR/HR lasers have to be screened at manufacturing and parts that are multimode or have poor SMSR have to be scrapped. Additionally, the phase of the cleaving affects the wavelength, and thus controlling the output wavelength of a batch of lasers in manufacturing can be a challenge.

An alternative approach is a phase-shifted DFB laser. In this case both facets are antireflection-coated and there is a phase shift in the cavity. This could be a single 1/4 wave shift at the center of the cavity, or multiple smaller shifts distributed in the cavity. Such devices have much better reproducibility in wavelength and theoretically they are all single mode, independent of the production process.

A selection of characteristics of practical DFB lasers for use as sources in telecommunication system is reported in Table 5.3.

5.2.2 High-Power Pump Lasers

If the ability to produce laser diodes with a strictly monochromatic output field that are stable in wavelength and power is instrumental to design optical transmitters for DWDM systems, high-power durable pump lasers are equally instrumental to realize optical amplifiers.

The success of fiber amplifiers, both based on erbium-doped fibers and Raman effect, is largely due to the availability of high-power laser diodes at suitable pump wavelengths.

Considering laser diodes designed to pump optical amplifiers, the key performances to care for are the emitted power, the stability of the wavelength, and the ability to emit a mode that can be focused easily on the small spot represented by a fiber-optic core.

Reliability is also a key property of these lasers, due to the fact that the reliability of the whole amplifier is largely determined by the reliability of the pumps.

The lifetime of semiconductor lasers decreases more than linearly, increasing the average emitted power, due to the fact that the more photons are present inside the laser cavity, the faster is the growth of fabrication micro-defects up to a state where they thwart the laser working.

Thus a major challenge for the manufacturers of pump lasers is to assure a long lifetime without scarifying the emitted power.

Another key design point for this type of lasers is the power consumption, which has to be limited as much as possible.

For this reason, it is important to have a high number of degrees of freedom in designing pump lasers, a condition that is assured by the use of not only a carefully selected alloy of III–V semiconductors for each laser section, but also a strained multi-quantum well (MQW) structure in the active zone.

In this way, the emitted wavelength is determined mainly by the MQW and the threshold current can be reduced allowing the emission of the required power at a lower bias current.

Moreover, in order to maintain the structure as simple as possible and avoid local micro-defects due to complex fabrication processes, almost all the pump lasers have an FP structure.

Since the emission wavelength is determined by the MQW structure, the base laser material can be selected to assure stability and reliability to the laser, leveraging on suitable fabrication processes.

Typical wavelengths for pump diode lasers used in telecommunication amplifiers are 800 nm, 980 nm, and a wide band around 1480 nm, where almost all the pump lasers for Raman amplification are located.

A key parameter for a power laser that has to work as an amplifier pump is the brightness. It is defined as the maximum power density that can flow through the output mirror per unit area and per unit solid angle.

Whatever the potential output power, the real laser performance is limited by the brightness limit at which the output mirror undergoes catastrophic damage that ruins the laser working.

In order to improve this threshold, it is important to coat the output facets of the laser with suitable coatings that tune the mirror reflectivity to the optimal value.

Just to give an example, an uncoated InGaAs/GaAs laser emitting at 980 nm can have a brightness limit of 10 MW/cm^2, while a similar uncoated laser realized in InGaAsP/InGaAs to pump at 1480 nm does not go over 5 MW/cm^2.

A suitable coating realized with a set of dielectric films can raise these figures up to 20 and 15 MW/cm^2 respectively, improving greatly the laser performances.

In Table 5.4 a set of characteristics of practical pump lasers for different applications is reported.

TABLE 5.4

Characteristics of Practical Pump Lasers for Different Applications

Application	EDFA Pump	EDFA Pump	Raman Pump	Raman Pump
Optical power	100–300 mW	160 mA	350 mW	420 mW
Threshold current	40 mA	50 mA	150 mA	
Operating drive current	250–600 mA	600 mA	1.2 A	1.7 A
Centre wavelength	974 nm	1480 nm	1420–1510 nm	1420–1510 nm
Spectral width	1 nm	10 nm	2 nm	1.5 nm
Operating temperature	−20°C to 75°C	0°C to 70°C	−20°C to 70°C	−20°C to 70°C
Footprint (w/pins, w/o pigtail)	50 × 37 mm	50 × 28 mm	50 × 40 mm	50 × 40 mm

5.2.3 Vertical Cavity Surface-Emitting Lasers

The vertical cavity surface-emitting laser (VCSEL) is a type of semiconductor laser diode with laser beam emission perpendicular from the top surface, contrary to conventional edge-emitting semiconductor lasers, which emit from surfaces formed by cleaving the individual chip out of a wafer [14].

There are several advantages to producing VCSELs when compared with the production process of edge-emitting lasers.

Edge-emitters cannot be tested until the end of the production process. If the edge-emitter does not work the production time, the processing materials and especially the test cost in term of manpower have been wasted.

On the contrary, VCSELs can be tested at several stages throughout the process to check for material quality and processing issues.

Additionally, because VCSELs emit the beam perpendicular to the active region of the laser as opposed to parallel as with an edge emitter, tens of thousands of VCSELs can be processed simultaneously on a 3-in. wafer.

Furthermore, even though the VCSEL production process is more labor- and material-intensive, the yield can be controlled to a more predictable outcome.

The VCSEL laser resonator consists generally of two distributed Bragg reflectors (DBRs) parallel to the wafer surface obtained alternating deposited layers of different materials.

The active region consists of one or more quantum wells allowing the control of both the threshold current, which in this kind of laser design can get very high if the laser is not designed carefully, and the emitted wavelength through the quantum well's structure.

The planar DBR mirrors consist of layers with alternating high and low refractive indices. Each layer has a thickness of a quarter of the laser wavelength in the material, yielding intensity reflectivity above 99%. High reflectivity mirrors are required in VCSELs to balance the short axial length of the gain region.

In common VCSELs the upper and lower mirrors are doped as p-type and n-type materials, forming a diode junction. In more complex structures, the p-type and n-type regions may be buried between the mirrors, requiring a more complex semiconductor process to make electrical contact with the active region, but eliminating electrical power loss in the DBR structure.

In Figure 5.14 three different VCSEL architectures are shown that essentially differ for the way in which the Bragg reflector is realized and, as a consequence, in the mirror's structure and position.

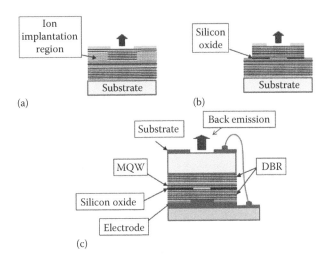

FIGURE 5.14
Scheme of three different VCSEL architecture: A VCSEL relying on ion implantation zones for current confinement is shown in (a), A VCSEL achieving current confinement via a passivation zone is shown in (b) while a structure using two confinement zones on both sides of the wafer is shown in (c).

The architecture shown in Figure 5.14a is the simplest to realize, but also the one with lower performance. One of the fundamental problems of this architecture and, in some way, of all VCSELs, is the fact that the optical mode and the current are parallel, traversing the laser-active zone from top to bottom.

To realize current confinement, generally in this type of structure the active region is surrounded by a region where the current cannot penetrate, for example, a zone where a proton implantation has been done.

The architecture shown in Figure 5.14b presents a passivation level immediately on top of the active zone that has the role of confining the injected current, as the proton-bombed zone in architecture Figure 5.14a.

Finally, architecture Figure 5.14c is more complex, requiring to process both sides of the wafer, but provides also better current confinement and in general better performances.

In their telecommunication application, where long wavelength lasers are needed to match the fiber transmission windows, VCSELs have also some challenge to overcome to be suitable substitutes of edge-emitting lasers in low-performance applications.

The first and perhaps more important point is the emitted power. While long-wavelength VCSELs emit several milliwatts, moving the wavelength toward the telecommunication region the emitted power decreases and common VCSELs at 1550 nm emit a power on the order of 0.1 mW [15], even if particular architectures have demonstrated larger emitted powers.

An example of high-performance VCSELs [16] emitting at 1551 nm, whose optical power versus current characteristic is plotted in Figure 5.15, is reported in [16,17], demonstrating that progress in this direction is rapid and products are almost ready to hit the market with suitable emitted power characteristics.

A second important challenge is constituted by the operating temperature [18]. Long-wavelength VCSELs are quite temperature-sensitive, while their application target requires uncooled sources.

An example is shown in Figure 5.16 [102], where the temperature sensitivity of the bias current of the VCSEL considered in Figure 5.15 is shown.

Also from this point of view progress seems to be rapid.

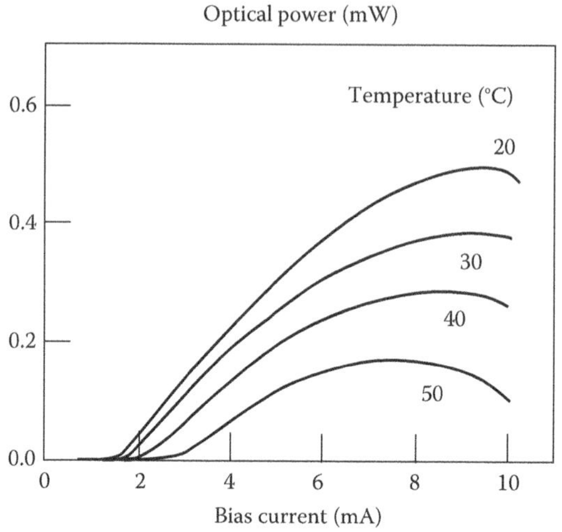

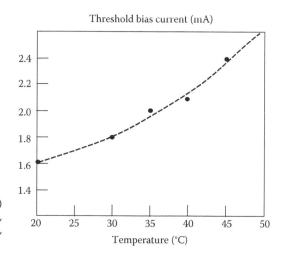

FIGURE 5.15
Long-wavelength VCSEL (emitting at 1550 nm) current optical power characteristic. (After RayCan, 1550 nm Vertical-cavity surface-emitting laser, 090115 Rev 4.0.)

FIGURE 5.16
Long-wavelength VCSEL (emitting at 1550 nm) current temperature characteristic. (After RayCan, 1550 nm Vertical-cavity surface-emitting laser, 090115 Rev 4.0.)

The third point is direct modulation, which is another key characteristic for low-end source lasers for telecommunication equipment. This point, which was a key issue for a certain time, seems to be solved with more recent VCSEL structures, so that direct modulation at 1.25 and 2.5 Gbit/s seems a consolidate feature.

Moreover, the frequency response of long-wavelength direct-modulated VCSELs has an interesting flat characteristic.

As a matter of fact, from the analysis of a high-confinement VCSEL it is possible to evaluate its response to frequency modulation by obtaining the following approximate equation that is valid for small modulation regimes [19]:

$$H(f) = H_0 \frac{\omega_R^2}{\omega_R^2 - \omega^2 + i\left(\mu\omega / 4\pi^2\right)} \frac{1}{1 + i\left(\omega / \omega_P\right)} \tag{5.13}$$

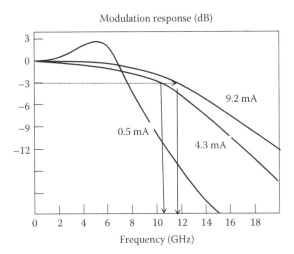

FIGURE 5.17

Spectral response to amplitude direct modulation for different values of the bias current of a long-wavelength VCSEL emitting at 1550 nm. (After Hofmann, W. et al., Uncooled high speed (>11 GHz) 1.55 μm VCSELs for CWDM access networks, *Proceedings of European Conference on Optical Communications—ECOC 2006*, s.n., Cannes, France, 2006.)

where

ω_R is the natural resonance frequency of the VCSEL

μ is the dumping factor of the relaxation oscillations

ω_p is a characteristic parameter whose expression can be derived from more fundamental characteristics of the VCSEL and that represents the intrinsic low-pass response of the structure in the absence of laser effect due to parasitic and thermal effects

The frequency response deriving from Equation 5.13 is plotted in Figure 5.17 [19] for an experimental VCSEL emitting at 1550 nm up to 2 mW. The resulting modulation bandwidth is never lower than 10.8 GHz if the bias current is greater than 4.3 mA, qualifying this experimental VCSEL as a very interesting source for high-capacity access networks like G-PON and 10 G-PON and for coarse-wavelength division multiplexing (CWDM) systems.

5.2.4 Tunable Lasers

Almost all semiconductor lasers are tunable by controlling the laser working temperature. Thermal tuning is generally limited to few nanometers, but for devices specifically designed for tuning where it can reach 5–6 nm.

Thermal tuning is generally used to maintain the laser on the correct wavelength through contrasting changes during an operation due to aging and environment.

Another way of tuning a standard semiconductor laser is to change the current injection. As a matter of fact, continuous current injection changes the carrier's density inside the laser cavity and causes a slight change of the emission wavelength.

Although this second method of tuning is even less effective than thermal tuning it indicates that there are two fundamental ways to tune a semiconductor laser: changing the index through a change in carrier concentration or changing the laser cavity length.

In all the cases, if tuning is achieved by index change, $\Delta\lambda/\lambda \sim \Delta r/r$; if tuning is achieved by changing the cavity length, $\Delta\lambda/\lambda \sim \Delta L/L$.

TABLE 5.5

Characteristics of Practical Tunable Lasers for Use as Transmitters in Telecommunication Systems

Technology	External Cavity with Liquid Crystals Mirror	Laser Array	Multisection DBR
Output power	13 dBm	15 dBm	13 dBm
Output power adjust	No	No	4 dBm
Tuning range	35 nm (C band)	40 nm (C band)	35 nm (C band)
Tuning mode	Discrete (ITU-T grid)	Discrete (ITU-T grid)	Discrete (ITU-T grid)
Channel number	80	96	80
Channel spacing	50 GHz	50 GHz	50 GHz
SMSR	50 dB	45 dB	40 dB
Linewidth	250 KHz	5 MHz	10 MHz
RIN	−155 dB/Hz	−145 dB/Hz	−140 dB/Hz
Laser gain current	250 mA	125 mA	100 mA
Reflector currents	NA	NA	25 mA
Phase current	NA	NA	7.5 mA
SOA Current	NA	NA	90–160 mA
Case operating temperature	−5°C to 70°C	0°C to 70°C	−5°C to 75°C

Network applications involving DWDM systems require a much wider tuning with respect to the potentiality of thermal or current injection tuning in order to use tunable laser effectively.

Generally tuning over an extended C band is required, that is, tuning over 40 nm, while retaining the characteristics of fixed-wavelength lasers.

Today technology offers essentially three types of widely tunable lasers: multisection lasers, external cavity lasers, and laser arrays [1]. In Table 5.5 the main parameters of practical tunable lasers for use as transmitter in telecommunication systems are summarized.

5.2.4.1 Multisection Widely Tunable Lasers

The simpler type of multisection widely tunable laser is a multisection DFB. Since a standard DFB can be tuned thermally only about 5 nm or less, wider tunability is achieved by creating a laser with multiple cavities. Nonuniform excitation along the cavity in DFB laser enables a change in the lasing condition.

Multi-electrode DFB lasers have been realized to achieve nonuniform excitation where the electrode is divided into two or three sections along the cavity [20].

The tuning function is provided by varying the injection current ratio into two sections. A typical geometry for this kind of laser is shown in Figure 5.18.

Tuning in these devices results from the combined effects of shifting the effective Bragg wavelength of the grating in one or several sections and the accompanying change of the optical path length for the change of the refractive index itself.

Combining these effects with thermal discontinuous tuning, a bandwidth on the order of 15 nm can be achieved, which is much more than pure thermal tuning, but it is not enough to cover the entire C band.

Much wider tuning is achieved using multisection DBR lasers.

A principal scheme of a tunable DBR is represented in Figure 5.19 [21]. The central region of the laser is an active region where a heterojunction is realized to allow population

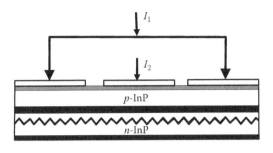

FIGURE 5.18
Lateral section of a multisection tunable DFB laser.

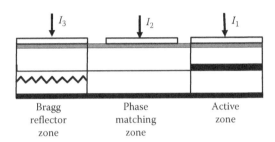

Bragg   Phase   Active
reflector   matching   zone   **FIGURE 5.19**
zone   zone   Lateral section of a multisection tunable DBR laser.

inversion and optical gain. The laser cavity is realized by cleaving one of the active region facets and providing, on the other hand, wavelength-selective feedback through a Bragg grating realized in a portion of the laser waveguide external to the active region, called the Bragg grating region.

Between the active region and the Bragg grating region there is an intermediate part of the waveguide that is used to adapt the field phase to the cavity length through the change of the refraction index caused by carrier injection. This zone is called phase adaptation zone.

In order to satisfy the resonance requirements, the laser mode has to satisfy the relation

$$\phi_1 - \phi_2 = 2k\pi$$

where
  ϕ_1 and ϕ_2 are phase change of the Bragg reflector in the active and the phase control regions, respectively
  k is an integer

Phase change ϕ_2 can be written as

$$\phi_2 = \beta_a L_a + \beta_p L_p = \left(\frac{2\pi}{\lambda} \right)(r_a L_a + r_p L_p) \tag{5.14}$$

where
  subscripts a and p denote the active and phase control regions, respectively
  β indicates a propagation constant
  λ is the wavelength of the lasing mode
  r is the refraction index

The phase ϕ_1 accumulated by the lasing mode passing through the Bragg reflector can be derived starting from the properties of Bragg reflectors.

The characteristic parameter γ of the Bragg reflector can be defined, following [22], as

$$\gamma^2 = \kappa^2 + (\alpha + i\Delta\beta)^2 = \kappa^2 + \left[\alpha + 2\pi i \left(\frac{r_b}{\lambda} - \frac{1}{2\Lambda}\right)\right]^2 \tag{5.15}$$

where Λ, κ, α, and r_b denote the corrugation period, the corrugation coupling coefficient, the loss of the corrugated region, and the equivalent refractive index for the Bragg reflector, respectively.

The phase change experimented by the field when reflected by the Bragg reflector can be expressed as a function of the characteristic parameter γ and the overall cavity length L as

$$\exp(i\phi_1) = -\frac{i\kappa}{r_b} \frac{1}{\gamma tgh(\gamma L) + (\alpha + i\Delta\beta)} = -\frac{i\kappa}{r_b} \frac{1}{\gamma tgh(\gamma L) + \left[\alpha + 2\pi i\left(r_b/\lambda\right) - (1/2\Lambda)\right]} \tag{5.16}$$

Substituting Equations 5.14 and 5.15 in Equation 5.16 the relation determining the possible lasing wavelength of the DBR laser can be obtained [1,21].

In particular, the laser will pass from one wavelength to the other when the index of the Bragg reflector section is changed by changing the carrier density with the injection of a current in the area; in general, however, the mode will start free oscillations only if the index of the phase matching section is suitably adjusted to satisfy the phase-matching condition (Equation 5.15).

It is clear that the phase-matching conditions can be attained contemporarily by a set of modes, due to the presence of the addendum $2k\pi$ in the phase-matching condition.

Among the phase-matched modes only one starts laser oscillation: the mode having the highest cavity gain.

The cavity gain of a mode in the DBR structure is given by the gain of the gain section minus the losses in the phase-matching and Bragg sections. But these losses are again controlled by the carrier densities; thus the specific mode to start laser oscillations among those permitted by the phase-matching condition can be selected by correctly polarizing the three laser regions.

The DBR laser is a good tunable laser, but the tuning range is just slightly wider than 5 nm and never exceeds 10 nm, since it is essentially due to the change of the index in the Bragg section (the other indexes are changed to satisfy the phase matching and to select the correct mode, and do not directly contribute to tuning); thus the relationship $\Delta\lambda/\lambda \sim \Delta r/r$ still holds and the limited changing range of the refraction index defines the changing range of the emitted wavelength. It is possible to combine index-driven tuning with thermal tuning and in this way a tuning range as wide as 22 nm has been reached with simple DBR structures.

In order to achieve much wider tuning ranges, the tuning cannot be based only on index changes. Now it is useful to remember that only three causes can generate the tuning of a semiconductor laser structure [23]:

1. Index changes
2. Cavity length changes
3. Mirror wavelength-selective reflectivity tuning

Somehow all these three techniques have to be used if a tuning range twice the DBR maximum of 22 nm has to be reached.

This is possible by using a structure with multiple cavities and a short grating as reflector for each cavity. Sampled-grating DBR (SG-DBR) and super-structure grating DBR (SSG-DBR) lasers implement this idea in two different manners [24].

The sampled-grating design uses two different multielement mirrors to create two reflection combs with different free spectral.

The laser operates at a wavelength where a reflection peak from each mirror coincides. Since the peak spacing is different for various mirrors, only one pair of the peaks can line up at a time, so that, when the lasing wavelength passes from a couple of peaks to the others a big change of emitted wavelength is obtained with a small change of the mirror peak positions (Vernier effect [25]).

The scheme of a discrete tuning SG-DBR and a plot showing the idea of the Vernier effect are shown in Figure 5.20.

Even if this simple explanation of the SG-DBR principle seems to limit this kind of laser to step tuning, a suitable design can achieve also semi-continuous tuning, at the expense of a more complex control algorithm.

Most of the features of the SG-DBR are shared by the SSG-DBR [26] design. In this case, the desired multi-peaked reflection spectrum of each mirror is created by using a phase modulation of grating rather than an amplitude modulation as in the SG-DBR.

Periodic bursts of a grating with chirped periods are typically used. This multielement mirror structure requires a smaller grating depth and can provide an arbitrary mirror peak amplitude distribution if the grating chirping is controlled.

Using SG-DBR or SSG-DBR, discontinuous tuning ranges as wide as 60 nm have been reached, arriving in this way to oversatisfy the requirements.

On the positive side, they are compact, they maintain the properties of fixed-wavelength lasers, and, perhaps more importantly, they are suitable to be integrated with other devices leveraging the InP platform.

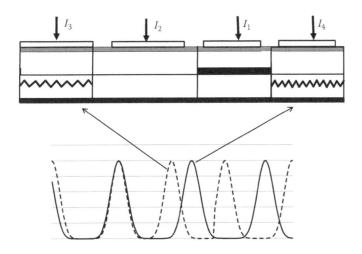

FIGURE 5.20
Lateral section of a multisection sampled grating DBR (SG-DBR) tunable laser. In the lower part of the figure the Vernier effect used to tune the laser by superimposing only one passband of the two gratings that create the laser cavity is also illustrated.

For example, if the output power has to be increased, it is possible to integrate immediately out of the laser, on the same waveguide, a semiconductor optical amplifier to achieve high output power, or if a compact component that can be directly modulated is needed, an InP modulator (either Mach–Zehnder or electro-absorption [EA]) can be integrated at the laser output.

Moreover, the typical reliability of integrated components is higher than that of components composed by different and possibly moving macroscopic parts.

On the negative side, there is the complex electronic control, which is more complex for more performing structures and the difficult processes needed to produce these lasers.

5.2.4.2 External Cavity Lasers

Limitation in tuning range experienced by integrated lasers when tuning is driven by the change of the optical cavity length is due to the fact that in integrated devices this is possible only by injecting carriers and changing the refraction index.

The situation is completely different in external cavity lasers, where the external part of the cavity can change its length with physical means in order to realize sufficient variations for very wide tuning.

The principal scheme of an external cavity laser is shown in Figure 5.21: after reducing the reflectivity of the cleaved active chip facet, part of the emitted radiation is reintroduced in the cavity of an FP semiconductor laser from an external etalon so as to realize a two-cavity system: the FP cavity inside the chip and the external cavity.

The effect of the feedback on the original laser emission can be analyzed by introducing two characteristic factors of the external cavity:

1. The field phase shift $\Delta\phi = \omega_0\tau$ experienced during a single external cavity round trip, where ω_0 is the field angular frequency and $\tau = 2L_c/c$ is the cavity round trip, L_c being the cavity length.

2. The feedback strength κ, defined as the amount of optical power reinjected into the laser.

The feedback strength can be evaluated starting from the reflectivity R_E of the external mirror and R_s of the facet of the chip obtaining

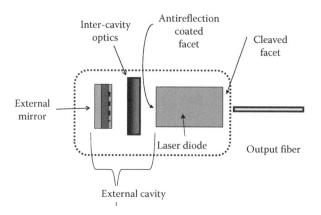

FIGURE 5.21
Schematic view of an external cavity laser.

$$\kappa = \frac{1 - R_s}{\tau_s} \sqrt{\frac{R_E}{R_s}} \qquad (5.17)$$

where τ_s is the laser cavity round trip time.

Naturally the external cavity will influence the laser behavior only if the feedback is not too weak. In order to state a qualitative condition that will help us to understand the order of magnitude of the parameters, let us state that the feedback is relevant for the laser dynamics only if it is more powerful than the spontaneous emission. This seems a reasonable condition since it requires that the feedback be visible by the laser above the noise plateau.

Since in a real laser the reflectivity of the external mirror will be on the order of 0.5 to allow sufficient feedback and sensible output optical power, it can be put at approximately $R_s/(1 - R_s) \approx 1$, so that the aforementioned condition becomes

$$R_E \gg \left(\frac{R_{sp}\tau_s}{r} \right)^2 \qquad (5.18)$$

Equation 5.18 states an important property of external cavity lasers: the shorter the cavity (i.e., the smaller the τ_s), the more sensitive is the laser to small amounts of feedback. Thus, if the feedback has to strongly influence the laser behavior it needs to design a short cavity laser.

Moreover, it is to be noted that R_{sp}/r is practically proportional to the laser linewidth $\Delta\nu$ (compare Equation 5.8), but for a factor on the order of 1. This factor, due to the condition $\gg$ appearing in Equation 5.18 can be neglected and the condition can also be written as $R_E \gg (\Delta\nu\tau_s)^2$ showing that the purer the laser emission (i.e., the smaller the $\Delta\nu$), the more the laser is sensible to reflections.

In order to study an external cavity semiconductor laser, the rate equations have to be written taking into account the electrical field composing the optical wave and not the optical energy, since the feedback has to be represented in terms of the electrical field.

Neglecting the space dependence of the variables and writing the linearly polarized electrical field as $E(t) = A(t) \exp[i\omega_0 t + \phi(t)]$, the external cavity rate equations may be written as follows:

$$\frac{dA}{dt} = \frac{1}{2} \left\{ g[N(t) - N_0] - \frac{1}{\tau_P} \right\} A(t) + \frac{\kappa}{\tau_i} \cos[\vartheta(t)] A(t - \tau)$$

$$\frac{d\phi}{dt} = \frac{\alpha}{2} \left\{ g[N(t) - N_0] - \frac{1}{\tau_P} \right\} - \frac{\kappa}{\tau_i} \sin[\vartheta(t)] \frac{A(t - \tau)}{A(t)} \qquad (5.19)$$

$$\frac{dN}{dt} = J - \frac{N(t)}{\tau_N} A(t) - g[N(t) - N_0] A^2(t)$$

where τ_i is the round trip time of the gain chip inside the laser and $A'(t) = (\kappa/\tau_i) \cos[\vartheta(t)] A(t - \tau)$ is the intensity of the delayed light injected from the external cavity to the active cavity. The overall ratio between the emitted and the reinjected intensity is indicated by κ and the phase coupling angle is given by the following formula:

$$\vartheta(t) = \omega_0 \tau + \phi(t) - \phi(t - \tau)$$ (5.20)

In Equation 5.19 also the noise terms have been neglected, since we are interested in the analysis of the deterministic behavior of the laser.

From the solution of the rate equations the dynamic of the external cavity laser can be studied.

In particular the following property is derived. The laser operation is strongly dependent on the so-called feedback parameter defined as $\xi = \sqrt{(1 + \alpha^2)}\kappa\tau$, where α is the linewidth enhancement factor of the gain chip.

If $\xi < 1$, the laser is in a regime of weak feedback, the laser cavity coincides with the chip cavity, and the feedback is a perturbation for the laser working. In this regime the rate equations can be solved using the perturbation theory where the "small" parameter is κ.

If $\xi < 1$ but $\kappa \ll 0.1$, different regimes alternate with stable and unstable regions where, depending on the cavity length and feedback intensity, there is a strong mode hopping or a single-mode operation. However, when single-mode operation is achieved, the laser behavior in this region is quite unstable and changes in environmental parameters or aging can completely change the laser emission.

Moreover, laser tuning is difficult due to this instability. Thus this region is difficult to exploit practically.

If $\xi \gg 1$ so that $\kappa > 0.1$, a stable operation regime of high injection is reached. The laser cavity is de facto the extended cavity (i.e., the chip plus the external cavity) and the presence of the chip internal facet can be considered a small perturbation to the extended laser working.

In this regime the laser is quite stable, its linewidth is sensibly reduced with respect to the value of the free running chip laser, and tuning can be operated by changing the cavity length without altering laser stability and without provoking mode hopping.

This is the regime in which practical external cavity tunable lasers work.

Considering from now on only the strong feedback regime, the laser output power can be evaluated once the exact laser structure is known. Considering in particular the structure in Figure 5.21, the output power is given by

$$P_o = \frac{\eta \hbar \omega}{\varepsilon} (J - J_{th}) \left[\frac{\ln\left(1/\sqrt{R_s}\right)}{\alpha L_i + \ln\left(1/TR_sR_i\right)} \right]$$ (5.21)

where
  η is the chip quantum efficiency
  ε is the dielectric constant
  L_i is the length of the gain chip
  T is the total loss internal to the external cavity (e.g., focusing lenses)
  R_i is the reflectivity of the cleaved chip facet

The aforementioned equations for power output indicate that, for a given injection current, the external-cavity laser generally has a somewhat lower power output than that of the solitary diode laser and, moreover, has higher threshold current, as shown in Figure 5.22.

As far as tunability is concerned, the emitted frequency is given, with a very good approximation, by the simple equation

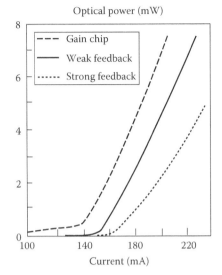

FIGURE 5.22
Modification of the current optical power characteristic of an FP laser as a consequence of the antireflection coating of a facet and the addition of an external cavity.

$$\omega = 2\pi \left(\frac{L_i r + L_c}{c} + k \right) \tag{5.22}$$

where k is an integer. Also in this case, among all the possible oscillating modes, generally only one experiences spontaneous oscillations, that is, the mode having the greater net cavity gain.

Different schemes have been proposed to change the external cavity length, but in practice two have been adopted in commercial products: the use of a wavelength-selective liquid crystal mirror that can be tuned by the applied voltage and a mechanically movable grating mounted on a micromachining electromechanical switch (MEMS) actuator.

In both cases the selectivity of the tunable mirror does not assure stable single-mode operation. On the other hand, a tunable laser for application in DWDM transmission does not need to be continuously tunable; on the contrary it has to be restricted to the International Association for Standardization in Telecommunications (ITU-T) DWDM wavelengths.

Taking into account such a restriction, sufficient wavelength selectivity can be achieved by inserting a solid-state etalon in the laser cavity, whose FSR coincides with the ITU-T grid spacing.

In order to have good control of the optical signal reinjection and to avoid the creation of spurious cavities, special gain chips are produced with a curve-active waveguide, so that the entire cavity appears at the end to be a curve.

The resulting more practical scheme of the external cavity laser is shown in Figure 5.23.

Beside tunability, one of the most interesting features of the high-feedback external-cavity laser is its narrow linewidth. By extending the optical cavity length, the spontaneous recombination phase fluctuation in the laser linewidth can be dramatically reduced. Since resonance oscillations are damped by the small reflectivity of the internal-gain chip facet, the linewidth can be considered Lorentian with an FWHM given by

$$\Delta v = \frac{\hbar \omega_0}{P_0} g R_{sp} (\Delta v_f)^2 \alpha_t (1 + \alpha^2) \tag{5.23}$$

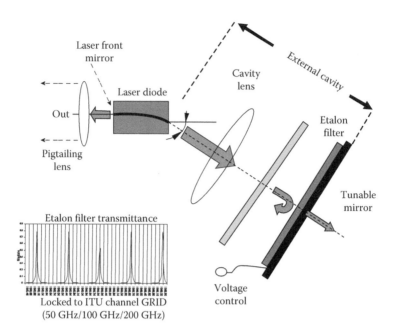

FIGURE 5.23
Practical implementation of an external cavity laser based on a liquid crystal mirror and on a curved cavity gain chip.

where

P_0 is the power in the mode

R_{sp} is the number of spontaneous emission photons in the mode

g is the local gain

Δv_f is the free running linewidth of the gain chip

The linewidth enhancement factor of the gain chip is indicated as usual with α.

The total cavity loss is indicated with $\alpha_t = a - \ln\left(T^2\sqrt{R_s R_i}\right)$, where a is the optical propagation loss.

It is not difficult to derive from Equation 5.23 substituting realistic values to the parameters that the external cavity reduces the linewidth of the gain chip of a factor that can be something like 5 or 7 so that external-cavity lasers with a linewidth of 250 or 500 kHz are used as sources in DWDM systems.

5.2.4.3 Laser Arrays

Laser arrays, where each laser in the array operates at a particular wavelength, are an alternative to tunable lasers. These arrays incorporate a combiner element, which makes it possible to couple the output to a single fiber.

If each laser in the array can be tuned by an amount exceeding the wavelength difference between the array elements, a very wide total tuning range can be achieved.

The DFB laser diode array might be the most promising configuration for wavelength division multiplexing (WDM) optical communication systems due to its stable and highly reliable single-mode operation.

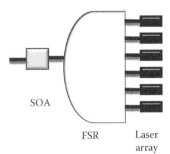

SOA

FSR   Laser **FIGURE 5.24**
               array   Block scheme of a laser array used as a tunable laser.

A possible scheme of a laser array used as tunable laser is sketched in Figure 5.24, where a free space propagation (FSP) region, similar to that used in array waveguides (AWGs), is used to collect the laser output and an SOA to compensate losses.

In principle, this is the easiest way of realizing a tunable laser since it is simply a collection in a single chip of a set of well-known devices: DFB lasers, passive waveguides, an MEMS deflector or an FSP, in case of an output SOA to reinforce the signal.

In practice, the real difficulty consists exactly in creating a chip where all these components are realized together.

In particular, two kinds of implementation problems arise, design and fabrication problems.

From a design point of view, thermal and electrical isolation of all the components is not a trivial problem, since the component footprint has to fit into a standard laser package. Moreover, the design has to be so accurate that all the individual lasers hit the correct wavelength region so that each laser can be tuned on the frequencies for which it is designed by simple thermal tuning.

In case an MEMS deflector is used to focus the light emitted from the laser that is currently switched on onto the output fiber, this element has to be realized with great care for reliability. As a matter of fact, laser interfaces are the elements with a smaller lifetime in a DWDM terminal even when fixed lasers are used. Further shortening of the laser life span is not acceptable from a system point of view.

On the other hand, if an FSP element is used it could be difficult to have a sufficiently high output power without an amplifier at the output of the FSP. In this case, besides the need of a very low-noise and wide-spectrum amplifier, further integration problems could arise.

However, laser arrays have been the first "tunable lasers" to hit the market and they still are a type of tunable lasers widely used in practical systems.

5.3 Semiconductor Amplifiers

Before closing the section on semiconductor lasers a brief presentation of an SOA is needed.

If the reflectivity of the chip facets is eliminated from the structure of an FP laser the chip cannot generate self-oscillation due to lack of feedback, but it can maintain an optical gain due to population inversion in the heterojunction depletion zone.

The result is a traveling-wave semiconductor amplifier.

The dynamics of an SOA is from several points of view similar to that of an ideal amplifier described in Section 4.3.1 with the advantage of having a broad gain curve and a small form factor. Two phenomena limit the use of SOAs in telecommunication systems as signal amplifiers:

1. As reported in Table 5.1 the typical carrier's lifetime is on the order of a few nanoseconds, that is, the order of magnitude of the bit time in 10 and 40 Gbit/s transmissions. Thus we have to expect that gain saturation happens during the amplification of signal pulses creating nonlinear pulse distortion as analyzed in Section 4.3.1.3.

2. The carrier dynamics in a semiconductor creates a link between the phase and the amplitude of the amplifier field so that pure amplitude gain cannot be achieved, but an induced chirp is always present.

Due to these elements the only practical use of SOAs as linear amplifiers is in InP-integrated components where an SOA section is often used to compensate the losses of attenuating components like multiplexers.

The nonlinear dynamics of SOAs is on the contrary an opportunity if they are used not as linear amplifiers but as elements in optical processing circuits.

This is the reason why a more detailed analysis of SOAs, with a particular attention to their nonlinear dynamics, will be delayed up to Chapter 10, where optical signal processing in new-generation networks will be discussed.

5.4 PIN and APD Photodiodes

A photodiode is in general terms a diode whose structure has been designed so as to allow the incoming light beam to reach a zone very near to the p–n junction [27].

If the junction is inversely polarized, all the carriers created by absorbed photons are immediately removed from the junction area due to the inverse polarization field and constitute the photocurrent.

Since the number of generated carriers is proportional to the incoming photon number, the intensity of the photocurrent is proportional to the optical power, so that a photodiode performs quadratic detection of the incoming field.

Since for telecom applications the interesting wavelength region is the near infrared, telecom photodiodes are built using materials with a high-absorption coefficient in this region of the spectrum.

Almost all the photodiodes used in telecom equipments are built using III–V alloy like InGaAsP or InP in order to optimize the absorption in the desired wavelength region [105], but the research on new materials has recently pointed out new alternatives that could give good results in the future [28,29].

Essentially two types of photodiodes are used: PIN photodiodes, where the junction is a p–i–n junction so as to optimize the absorption efficiency, and Avalanche photodiodes (APD), where internal gain is achieved by the avalanche multiplication of the photo-generated carriers.

Considering PIN photodiodes, the first important characteristic is the quantum efficiency η of the photodiode, defined as the probability that an incoming photon will generate a carrier couple that will contribute to the photocurrent.

The expression of the quantum efficiency is given by

$$\eta = (1-f)\Gamma(1-e^{-\alpha d}) \tag{5.24}$$

where

  f is the facet reflectivity, so that $(1 - f)$ is the probability that the photon is not reflected

  α is the absorption coefficient of the junction zone

  d is its thickness, so that $(1 - e^{-\alpha d})$ is the probability that the photon is absorbed in the junction zone creating free carriers

  Γ is the probability that the created carrier exits from the junction depletion zone without recombination

From Equation 5.24 the importance of the absorption coefficient is evident and it is this characteristic that selects the photodiode material once the wavelength region is determined.

The photocurrent created by a PIN photodiode as a consequence of the detection of a light beam is composed by four components:

The first term is constituted by the photon-generated carrier flux that is given by $c_{pc} = \eta q \Phi$, where q is the electron charge and Φ is the photon flux, that is, the number of photons per unit time arriving on the photodiode front area. It is important to underline that Φ is a random variable, corresponding to the measure of the quantum operator number ($\hat{n}$) relative to the incoming field quantum state.

In almost all the telecommunication applications, the incoming field can be considered a time sequence of quasi-stationary coherent states; thus Φ is a Poisson variable whose parameters slowly change as a function, for example, of the field modulation. Assuming a constant incoming field and separating the average and the fluctuations, the photo-generated carrier flux can be written as

$$c_{pc} = q\eta\langle\Phi\rangle + q\eta\big(\Phi - \langle\Phi\rangle\big) = R_p P + n_{sh}(t) \tag{5.25}$$

where P is the power of the incoming optical beam, $R_p = q\eta/\hbar\omega$ is the so-called photodiode responsivity, and the random term $n_{sh}(t)$, which is in all respects a noise term, is the so-called shot noise. The shot noise has a distribution determined by the fact that the photon flux is a Poisson variable and that an incoming photon has a probability η of generating a photocarrier.

Thus, using the probability composition law, the probability distribution of the photocurrent flux can be written as

$$P\left(\frac{c_{pc}}{q\eta} = M\right) = \sum_{j=0}^{\infty} \frac{\langle\Phi\rangle e^{-(\Phi)}}{M!(j-M)!} \eta^M (1-\eta)^{(j-M)} \tag{5.26}$$

However, if the number of photons arriving on the photodiode in the relevant time unit is much higher than 1, the distribution of the number of photocarriers can be assumed Gaussian, with a standard deviation equal to the incoming optical power.

The second current term is the dark current, which is the current that always pass through an inversely polarized p–i–n junction.

The third term is the current generated by thermally excited carriers and, finally, the fourth is the background noise, which is the current generated by the detection of environmentally generated photons.

Summarizing, the photocurrent is composed of four terms, represented in the following equation:

$$c(t) = R_p P(t) + n_{sh}(t) + n_{dark}(t) + n_{th}(t) + n_e(t) \tag{5.27}$$

The noise term can be considered in general as Gaussian white noise.

Equation 5.27 holds as far as either the incoming signal bandwidth does not exceed the photodiode bandwidth or the incoming power does not go above the photodiode saturation power.

In the first case, due to the finite time response of the photodiode essentially determined by spurious and junction capacities, the photodiode works as a lowpass filter, cutting too high frequencies.

In the second case, the linear relation between average current and optical power does not hold anymore and the photocurrent tends to saturate, increasing the incoming optical power.

The second type of photodetector is the so-called APD [30].

APDs are high-sensitivity, high-speed semiconductor light sensors. Compared to regular PIN photodiodes, APDs have a depletion junction region where electron multiplication occurs, by application of an external reverse voltage.

The resultant gain in the output signal means that low light levels can be measured and that the photodiode bandwidth is very large. Incident photons create electron–hole pairs in the depletion layer of a photodiode structure and these move toward the respective p–n junctions at a speed of up to 105 m/s, depending on the electric field strength.

If the external bias increases this localized electric field to about 105 V/cm, the carriers in the semiconductor collide with atoms in the crystal lattice with sufficient energy to have a nonnegligible ionization probability. Ionization creates more electron–hole pairs, some of which cause further ionization, giving a resultant gain in the number of electron–holes generated for a single incident photon.

Naturally the avalanche process is not a deterministic one and the generated gain also implies a multiplication noise that adds to the other noise terms typical of a PIN photodiode.

The excess noise is generally represented as a sort of shot noise amplification, so that the expression for the power of the shot noise term has to be modified as follows:

$$\sigma_{shot}^2 = 2G_m^2 e R_p P_o F_A \tag{5.28}$$

where
  G_m is the avalanche gain
  F_A is the excess noise factor, which is a function of the ionization ratio r_i, and can be written as [31,32]

$$F_A = r_i G_m + (1 - r_i)\left(2 - \frac{1}{G_m}\right) \tag{5.29}$$

Besides multiplication noise, when using an APD, it is to take into account that the dark current noise also is amplified by the avalanche mechanism.

In order to give a correct expression of the amplified dark current noise, one has to take into account that dark current can be divided into two components, depending on the physical phenomenon generating it.

There is a bulk dark current and a surface leakage dark current; the first is really amplified by the APD, and the second does not undergo amplification due to the fact that it is a surface phenomenon.

Thus, indicating with S_{bd} and S_{sd} the spectral densities of the two components, the final APD dark current power in the detector bandwidth will be [32]

$$\sigma_{dc}^2 = G_m S_{sd} B_e + G_m^2 S_{bd} F_A B_e \qquad (5.30)$$

where B_e is the electrical bandwidth of the front end immediately following the photodiode.

5.5 Optical Modulation Devices

5.5.1 Mach–Zehnder Modulators

As we have been discussing the modulation performances of semiconductor lasers at 10 Gbit/s and more it is very difficult to transmit at long distance using laser direct modulation due to the limited modulation bandwidth and to intrinsic chirp associated to it.

Among the external modulators those having better performances, if used in telecommunication equipments, are based on a Mach–Zehnder architecture where the interferometer arms are realized with a material exhibiting a strong electro-optic effect (such as LiNbO$_3$ or the III–V semiconductors such as GaAs and InP) [33,34].

For telecommunication applications, the ones most used are those based on LiNbO$_3$ and on InP.

By applying a voltage on the interferometer branches through suitable electrodes, the optical signal in each path is phase-modulated as the optical path length is altered by the variable microwave electric field.

Injecting an optical field in the interferometer and combining at its output the fields coming from the two branches of the phase modulation it is converted into amplitude modulation due to interference.

If the phase modulation is exactly equal in each path but different in sign, the modulator is chirp-free; this means that the output is only intensity-modulated without incidental phase (or frequency) modulation.

In Figure 5.25 the schematic of a Mach–Zehnder modulator is shown with a particular reference to a LiNbO$_3$ modulator.

Mach–Zehnder modulators can be either dual-drive or single-drive. In a single-drive modulator a single electrode driven by a signal proportional to the data induces phase modulation on one interferometer branch. This phase modulation is designed so that, at the output of the interferometer, the modulated field combines with the unmodulated one either in phase or in quadrature, so as to produce no chirp and the maximum dynamic range in the intensity-modulated optical field.

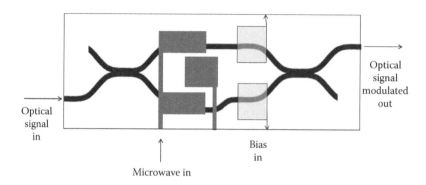

FIGURE 5.25
Scheme of a Mach–Zehnder modulator.

The dual-drive configuration uses a data-driven signal on one branch and a signal driven by the inverted data on the other branch to induce modulation on the two branches of the interferometer.

The dual-drive configuration is more stable and resistant to changes in the driving signal and in the environment.

The transfer function of a Mach–Zehnder modulator is shaped as an interferometer characteristic response like $0.5[\cos(x) + 1]$. This means that the modulator cannot be driven directly with the signal proportional to data, otherwise the nonlinear transfer function will completely distort the signal.

It is needed to bias the modulator with a continuous voltage so as to bring the working point at the center of the linear part of the characteristic, as shown in Figure 5.26. Once correctly biased, the microwave data-carrying signal is correctly modulated onto the optical field.

When digital modulation is adopted, which is almost always, the nonlinear nature of the modulator transfer function is also beneficial, since it helps to eliminate possible overshots in the driving signal.

Due to the need of a bias, every Mach–Zehnder modulator presents two types of electrodes: bias electrodes, to be used for the bias voltage, and signal electrodes, to be used for the data-carrying signal [35].

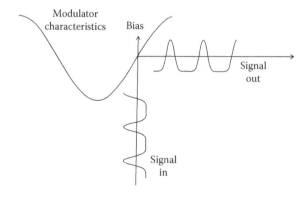

FIGURE 5.26
Instantaneous transfer function of a Mach–Zehnder modulator and ideal bias point.

Naturally the instantaneous relation between the modulating and the modulated signals is correct only as far as the signal bandwidth is within the flat frequency response of the modulator.

The modulator frequency response is limited both by the intrinsic bandwidth of the microwave circuit conveying the data and by the spurious capacities embedded in the structure.

Taking as a reference a dual-drive LiNbO$_3$ modulator and assuming an input field perfectly polarized parallel to the modulator waveguide mode, the real modulator response to the optical field is given by [36,37]

$$H(t) = \gamma L \cos\left[\frac{\varphi_1(t) - \varphi_2(t)}{2}\right] e^{i(\varphi_1(t) - \varphi_2(t))/2} + H_{XT}(t) \qquad (5.31)$$

where
γ is the modulator attenuation
L is the modulator length
$\varphi_1(t)$, $\varphi_2(t)$ are the phase changes on the two arms
$H_{XT}(t)$ represents the interference between the arms

Due to the finite frequency responses $h_1(t)$ and $h_2(t)$ of the microwave circuit, the phase responses can be written as

$$\varphi_j(t) = \varphi_0 + \frac{\pi}{2} \int h_j(\tau) \frac{V(t - \tau)}{V_\pi} d\tau \quad (j = 1, 2) \qquad (5.32)$$

where
φ_0 is the bias-induced constant phase that allows to work around the center of the linear part of the cos(x) characteristics
$V(t)$ is the signal carrying the data
V_π is the voltage that is needed to provoke a phase shift of $\pi/2$
$h_j(t)$ is the impulse response of the microwave circuit and ideally $h_1(t) = -h_2(t)$

From Equation 5.32 it is evident that the frequency response depends, at least in the absence of important defects of the chip, from the microwave circuit conveying the modulating signal.

In Figure 5.27 the frequency response of a dual-drive LiNbO$_3$ modulator is shown whose transfer function is quite flat in a bandwidth compatible with 10 Gbit/s modulation (requiring at least 12 GHz bandwidth due to overhead, see Chapter 3).

A second important effect limiting the response of Mach–Zehnder modulators, especially those built on LiNbO$_3$ for the greater length of the interferometer's branches, is the different propagation speeds of the microwave and optical fields.

As a matter of fact, this difference creates a misalignment between the modulating function and the carrier to be modulated that depends on the distance along the waveguide, thus inducing signal distortion.

This problem has to be solved with a very careful design of the microwave contacts that convey the microwave signal along the optical waveguide path.

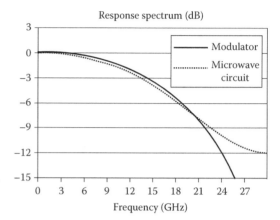

FIGURE 5.27
Frequency response of a Mach–Zehnder modulator and corresponding response of the microwave circuit conveying the signal.

The design can be conceived in such a way as to use the waveguide contribution to the microwave group velocity to compensate, at least partially, the group velocity mismatch.

Considering overall modulator performances, LiNbO$_3$ modulators have the better characteristics in terms of modulation bandwidth, chirp-free or chirped operation, and dynamic range.

Due to the values of the electro-optic constants of the LiNbO$_3$, modulators built out of this material are bigger than those built out of InP and need much higher microwave voltage.

Performances of InP-based modulators are improving fast and it is possible that in not so long a time the performance difference will almost vanish [38].

At present, LiNbO$_3$ modulators are used in high-quality DWDM interfaces, when compromises are not possible with the signal quality, while InP modulators are used in compact interfaces, like transceivers, where direct laser modulation is not possible. In Table 5.6 the main characteristics of a few practical Mach–Zehnder modulators are summarized.

TABLE 5.6

Characteristics of Practical InP and LiNbO3 Mach–Zehnder Modulators

Electro-Optic Material	LiNbO$_3$	LiNbO$_3$	InP
Bit rate	10 Gbit/s	40 Gbit/s	10 Gbit/s
Operating case temperature	0°C –70°C	0°C –70°C	0°C –70°C
Operating wavelength	1525–1605 nm	1530–1600 nm	1520–1600 nm
Optical insertion loss (connectorized)	5.0 dB	4.5 dB	6 dB
Insertion loss variation (EOL)	−0.5 to 0.5 dB	−0.5 to 0.5	
Modulator chirp parameter	−0.1 to 0.1	0.7	Selectable −0.2 to 0.2
Optical return loss	40 dB	35 dB	20 dB
Optical on/off extinction ratio (at DC)	20 dB	20 dB	10 dB
E/O bandwidth (−3 dB with linear fit)	12.0 GHz	42 GHz	10 GHz
Vpi RF port (at 1 GHz)	5. V	5 V	2 V
DC bias voltage range (EOL)	−8 to 8 V	9 V	3 V
Footprint (w/ electrical connections, w/o pigtail)	95 × 20 mm	99 × 14.5 mm	26 × 12 mm

EOL, end of life value.

5.5.2 Electro-Absorption Modulators

EA modulators are based on the change of the absorption coefficient when an electrical field is applied.

This effect is known as quantum-confined Stark effect when happening in quantum well structures, and almost all the practical EA modulators are realized using quantum well structures [31].

The physical description of the effect is quite complex [39], but the final result is that the absorption coefficient, and in particular the absorption edge, that is, the wavelength at which the material lowers abruptly its absorption coefficient and behaves like a transparent material, depends on the applied optical field.

Since the transition between strong absorption and good transparency is quite abrupt and the Stark effect is very fast, this dependence can be used to design a modulator [40].

This effect is shown in Figure 5.28, where the attenuation of a packaged EA modulator is shown at different wavelengths in the fiber third transmission window versus the applied voltage.

The structure of an EA modulator is based on a p–i–n junction built using suitable III–V alloy (e.g., InGaAsP) in order to work at the required frequencies.

Since the modulation mechanism is essentially based on absorption coefficient modulation, the longer the modulator, the greater is the extinction ratio, since the loss in the off state is greater and the signal is better suppressed.

On the other hand, a longer modulator has two issues to solve: the loss is high also in the on state, increasing the modulator insertion loss. Moreover, the longer the modulator, the greater are both the junction and the parasitic capacities, and thus the smaller the modulator bandwidth.

It is clear that there is an unavoidable trade-off between the extinction ratio on the one hand and the insertion loss and bandwidth on the other.

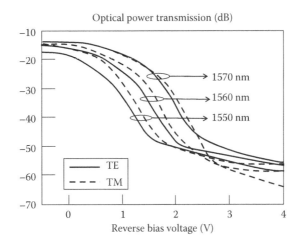

FIGURE 5.28
Characteristic curve (transmission versus applied voltage) of an EA modulator realized in InGaAsP for different wavelength and polarization of the input carrier. (After Yamanaka, T. and Yokoyama, K., Design and analysis of low-chirp electroabsorption modulators using bandstructure engineering, 2nd Edn., *Second International Workshop on Physics and Modeling of Devices Based on Low-Dimensional Structures*, 1998.)

In order to try to remove this limitation, which also constrains the selection of the material with which the modulator is fabricated, traveling-wave EA modulators have been proposed.

Figures 5.29 and 5.30 show the different structures between a lumped and a traveling-wave EA modulator.

In the lump configuration, the microwave signal is applied to the center of the waveguide. At high frequency, due to the strong microwave reflections at both ends, the device speed is strongly dependent on the total resistance capacity (RC) circuit time constant. As shown in the figure, the lumped EA modulator can be modeled as p–i–n junction capacitance C_i and the differential resistance R_d, which represents the resistivity of p- and n-semiconductor materials and metal contact.

In the traveling-wave configuration (Figure 5.30) the device is designed to have microwave co-propagation with optical waves. By matching the output termination, the electro-optic conversion is dominated by the distributed interaction between microwave and optical wave.

The electrical equivalent model of the traveling-wave modulator, as shown in the figure, is a transmission line. The microwave experiences distributed junction capacitance and inductance along the waveguide; thus the limitation is not the total capacitance, but the microwave characteristics, termination and the velocity matching, somehow similar to what happens in the Mach–Zehnder modulator.

Due to the large amount of studies done on Mach–Zehnder modulators and many other microwave circuits, there are several possible solutions to adapt microwave and optical propagation; thus it is possible to design the traveling-wave structure so as to have better performance with respect to the lumped architecture [41].

However, in both the configurations, the more important property of EA modulators is that they can be integrated with other InP-based components.

There are several commercial devices that are built as a laser plus an integrated EA modulator or even a set of DFBs, an EA modulator, and an SOA.

These integrated components are instrumental in the design of small and low-consumption optical interfaces like high-performance XFP transceivers.

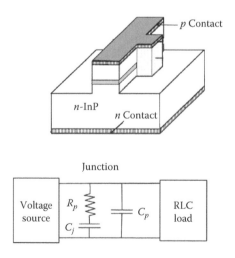

FIGURE 5.29
Scheme and equivalent circuit of a lumped EA modulator. R_p, junction resistance; C_j, junction capacity; C_p, parasitic capacity.

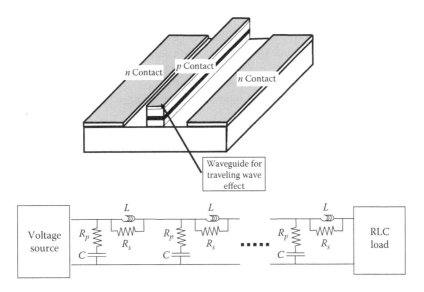

FIGURE 5.30
Scheme and equivalent circuit of a traveling-wave EA modulator. (After Zhang, S. Z. et al., *IEEE Photon. Technol. Lett.*, 11(2), 191, 1999.)

EA modulators suitable for modulation up to 40 Gbit/s have been introduced on the market and a summary of the characteristics of practical modulators is reported in Table 5.7.

5.5.3 Integrated Optical Components

It is quite spontaneous to think that, since several optical devices can be fabricated with planar technologies, the evolution of optical devices should somehow repeat the evolution of electronics through multiple function integration and large-scale integration circuits.

In the optical field there are even two possible integration platforms: silicon and III–V alloys, both compatible at first glance with integration of optical devices with electronics.

In practice, there have been several attempts to develop a technology platform for integrated optics and the results of these attempts always have been inferior to expectations.

TABLE 5.7

Characteristics of Practical Lumped and Traveling-Wave EA Modulators

Architecture	Lumped	Traveling Wave
Small signal modulation bandwidth	45 GHz	60 GHz
Operating bit rate	10–40 Gbit/s	10–40 Gbit/s
Operating wavelength	1550 nm	1550 nm
Optical insertion loss (connectorized)	6.0 dB	5 dB
Polarization-dependent loss	−0.5 dB	−0.7 dB
Maximum input optical power	10 dBm	8 dBm
OPTICAL RETURN LOSS	25 dB	35 dB
Optical on/off extinction ratio (at DC)	12 dB	15 dB
DC reverse bias voltage	6 V	7 V
Footprint (w/electrical connections, w/o pigtail)	30 × 32 mm	40 × 25 mm

As we will see in Chapter 10 a similar attempt leveraging the III–V platform is ongoing and it is possible that it will have success. But also in this case, the foundation of this possible success will be different from the causes of the success of microelectronics, since it is founded more on the performances allowed by optical integration than on the diffusion of a pervasive generation of low-cost new optical devices.

There are essentially two causes of this evolution: one more fundamental and the other related to the use of optical components.

5.5.3.1 Electrons and Photons in Planar Integrated Circuits

In a CMOS integrated circuit, electrons move into very small wires and experiment several quantum effects due to their associated wave nature, like a tunnel effect.

Consider an electron moving inside the channel of a CMOS transistor at a nonrelativistic energy of $2\,eV$ and with an effective conduction mass of $1.08m_e$, m_e being the electron rest mass. In these conditions, which are realistic in a standard CMOS working, the electron wavelength is around $6\,\text{Å}$. With this small wavelength it never happens that the electron behaves like a wavelength into the CMOS circuits and the entire design can be performed considering particle properties.

On the contrary, the wavelengths used in optical integrated circuits are generally on the same order or even smaller than the characteristic dimensions of optical integrated circuits.

Thus the photon should always be considered mainly like a wave.

This difference deeply influences the way in which integrated optical and electronic circuits are realized.

The shape of the waveguide core that is used in optical integrated circuits cannot be scaled below a certain dimension, not to risk losing any guiding property.

On the contrary, scaling is always possible with present-day technology in the case of electronics.

5.5.3.2 Digital and Analog Planar Integrated Circuits

Another element that distinguishes planar integration in optics and electronics is the fact that using CMOS circuits it is always possible to realize the required functionalities with a digital circuit, even if they are analogical in nature.

In the optical case, only analogical functionalities exist that have to be integrated together.

It is well known that analog functionalities are much more difficult to insulate in an integrated circuit, especially in a waveguide-based circuit, where the behavior of the photons reflects oscillatory field characteristics.

Just to give an example, it is enough to consider the evanescence-wave directional coupler that is based on the principle that a field can completely pass from one waveguide to another simply because they are near, even without any contact between the waveguide cores.

On the other hand, a sufficiently simple digital optics to push monolithic integration seems faraway, requiring substantial steps ahead before being sufficiently mature for the integration step.

5.5.3.3 Role of Packaging

Unless very high-speed circuits are considered, packaging of very large-scale CMOS circuits is a commodity that can be automated for large volumes through suitable packaging machines.

The situation for optics is completely different at least from two points of view.

The first is related to the way in which the optical field enters into and exits from the package. Unless the component is designed to work in free space, fiber optics is the most natural way to interface the optical chip with the field.

A standard monomode fiber has a core radius around 10 μm, and coupling it with a waveguide is more and more difficult while the waveguide gets smaller and smaller.

In order to achieve a good coupling, either a mode adaptor has to be realized, increasing both the package Bill of Materials and the amount of work to be done serially chip by chip, or complex structures (like multilevel tapers) have to be realized on the planar circuit so that a great part of the circuit area is occupied by coupling structures more than by the component for which the circuit has been realized.

In any case, active alignment is needed in both the cases to arrive at a coupling on the order of magnitude of 3 dB, an operation that is quite expensive.

Besides the coupling problem, high-performance integrated optics has also a temperature stabilization problem.

Both in the III–V platform and in the silicon one, the main materials used to realize planar optical devices are quite sensitive to temperature; moreover, fabrication tolerance creates quite a spread on the characteristics of similar devices (another consequence of the analog nature of optical circuits).

This condition creates the need for a thermal control of the optical components, generally carried out via a Termistor hosted inside the package.

The final effect of these characteristics of optical packaging is that in several cases the cost of the package is higher than the cost of the integrated circuitry inside it. In the case of a DWDM DFB laser, for example, the package and the related elements are worth about 70% of the component Bill of Materials.

5.5.3.4 Integrated Optics Cost Scaling with Volumes

The aforementioned reasons cause the cost model of integrated optics to be similar but not equal to that of microelectronics.

This fact has been studied in particular in the field of optical access regarding the access line home termination. In this field, potential volumes on the order of millions are present on the market, apparently stimulating the development of an integrated optics platform.

The variation of the cost of a component devoted to application in the optical access network is shown in Figure 5.31 versus the volumes for a microelectronic gigabit passive optical

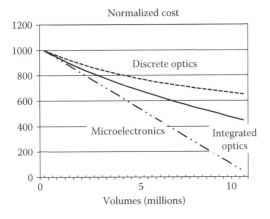

FIGURE 5.31
Price decrease versus volumes for a microelectronic GPON processor, a discrete optics GPON interface, and a monolithic integrated GPON interface.

network (GPON) processor, an integrated optics GPON interface, and a discrete optics GPON interface. The prices are normalized so as to appreciate better the different trends.

From the figure it is evident how the cost of integrated optoelectronic interface scales with volumes a bit faster with respect to the discrete optics components, but quite slowly with respect to microelectronic chips, due to the importance of the package Bill of Materials and the presence of important manufacturing processes serially executed component by component.

5.5.3.5 Integrated Planar III–V Components

Even if in the absence of a radical technology breakthrough there is no possibility to reply with integrated optics the success of microelectronics, it is evident that there are specific situations in which monolithic integration is useful.

The most important is the case of lasers with an integrated modulator (either EA or Mach–Zehnder) and in the case of an output semiconductor optical amplifier.

Devices of this kind are regularly used in telecommunication systems and have the advantage of putting into a single package three elements, sparing on package and connections and reducing interface loss.

This solution is particularly appreciated when a tunable multisection laser is used in conjunction with a Mach–Zehnder modulator.

5.6 Optical Switches

Under the category of optical switches all the optical components can be collected that deflect the light either in free space or in guided propagation from one direction to the other.

In telecommunication, optical switches are used almost only in conjunction with fiber propagation, but for the free space application, which is sometimes used as temporary connection either between nearby units or between parts of a broken cable [42].

Switches used in fiber-optic systems can be classified into two categories: semi-static switches and routing switches.

Semi-static switches are those subsystems that are installed instead of a patch panel to gain flexibility: from a functional point of view they are not routing elements in the network, instead they are part of the cable infrastructure like patch panels, ducts access points, and so on. These switches are generally micromechanical switches and change status quite rarely, sometimes never in their life.

Thus, it is very important that they maintain the characteristics of attenuation and polarization insensitiveness for a long time and that they are able to work after a long time of inactivity.

On the other hand, no important requirements on speed, power consumption, or dimension are generally imposed on these components, but for the fact that without power supply they cannot change state, whatever the state they are occupying (it is said that they have to be latching).

On the contrary, routing switches are designed to provide signal routing, either to find the right path in the network or to protect the signal from a failure.

Here we will talk mainly about routing switches technologies, even if a few of them are applied also to semi-static switches.

There are a plethora of different technologies that have been used to design optical switches, which can be classified as follows [43]:

- MEMS: involving arrays of micromirrors that can deflect an optical signal to the appropriate direction
- Piezoelectric beam steering involving piezoelectric ceramics providing enhanced optical switching characteristics
- Microfluidic methods involving the intersection of two waveguides so that light is deflected from one to the other when a liquid micro-bubble is created
- Liquid crystals that rotate polarized light either 0° or 90° depending on the applied electric field to support switching
- Thermal methods that vary the index of refraction in one leg of an interferometer to switch the signal on or off
- Acousto-optic methods that change the refractive index as a result of strain induced by an acoustic field to deflect light
- SOA amplifiers used as gates in an integrated III–V waveguide matrix

Naturally the only way to review in detail all the aforementioned technologies is to devote an entire book to optical switches, and it is not the scope here.

Thus, we will concentrate on a few technologies, those allowing us to better underline achievements and challenges in designing optical switches for telecommunications.

5.6.1 Micromachining Electromechanical Switches (MEMS)

MEMS fabrication techniques utilize the mature fabrication technology of CMOS-integrated circuits (see next section). The fact that silicon is the primary substrate material used in the integrated circuit and that it also exhibits excellent mechanical properties make it the most popular micromachining material.

Limiting our analysis to optical switching applications, there are essentially two techniques that can be used to realize optical switches with silicon-based MEMs: bulk micromachining and surface micromachining [44,45].

Bulk micromachining is the most mature and simple micromachining technology; it involves the removal of silicon from the bulk silicon substrate by etchants. Depending on the type of etchants that are used, etching can be isotropic, that is, oriented along one of the silicon crystal planes, or anisotropic, that is, oriented along a plane different from the crystal main planes.

Anisotropic etching is generally used to expose the silicon crystal planes in order to build mirrors or similar other structures.

Using bulk micromachining the need of complex planar processes is reduced to a minimum, but only the simplest structures can be created.

Surface micromachining is a more advanced fabrication technique. It is based on the use, near the layer of structural materials that are destined to compose the final structure, of layers of sacrificial materials, which are destined to be etched out so as to leave empty spaces between structural layers.

An example is shown in Figure 5.32, which demonstrates how a sacrificial material layer is removed in order to create a suspended mirror.

In addition, from the architecture point of view there are essentially two possible ways of constructing an MEMS optical switch: 2D MEMS and 3D MEMS.

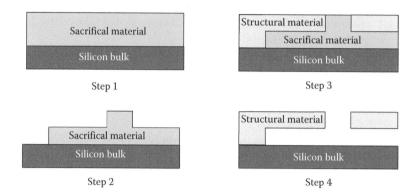

FIGURE 5.32
Schematic process flow for the creation of a suspended mirror in a planar MEMS. Step 1: sacrifical material deposition; step 2: sacrifical material shaping etchant 1; step 3: structural material deposition; step 4: sacrifical material elimination etchant 2.

In a 2D architecture the movable mirrors are positioned at the intersections of a grid of light paths, either as waveguides or as free space beams as is shown in Figure 5.33. In this case the mirrors act as 2×2 switches of a so-called cross-bar matrix (see Section 7.2.1). As an example this can be realized via a mirror that can be lifted and released to deflect or not the light beam, as is shown in Figure 5.34. As evident from the switch architecture, this 2D approach requires N^2 mirrors to form an $N \times N$ switch. Alternative approaches will be discussed in Section 7.2.1; however, in general, decreasing the number of mirrors causes a more complex interconnection layout that in waveguide circuits is a price that is often hard to pay.

A 3D or analog MEMS has mirrors that can rotate about two axes. Light can be redirected precisely in space to multiple angles—at least as many as the number of inputs.

Since each elementary mirror of this switch has much more than two states, it is expected that a much smaller number of mirrors is needed. As a matter of fact, depending on the exact switch design the number of mirrors varies between N and $2N$.

The majority of the commercial 3D MEMS use two sets of N mirrors (total of $2N$ mirrors) to minimize insertion loss as is shown in Figure 5.35; if only N mirrors were used, the port count will be limited by insertion loss that results from a finite acceptance angle of fibers/lens on the input or on the output side.

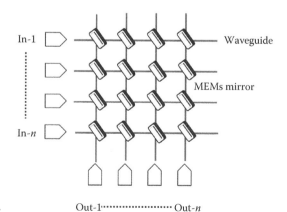

FIGURE 5.33
Scheme of an MEMS-based cross-bar optical switch.

FIGURE 5.34
Pictorial representation of an MEMS movable mirror for integration in a cross-bar optical switch.

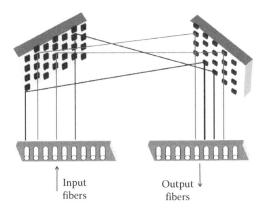

Input
fibers

Output
fibers

FIGURE 5.35
Scheme of a 3D MEMS-based optical switch.

This architecture can be scaled to thousands by thousands of ports with high uniformity in losses. Inevitably the control of the mirrors, being analog and not digital as in the 2D case, is much more complex than in the 2D case and a continuous analog feedback is needed on each mirror to put it in place at the correct angle.

Last thing to consider is the actuating mechanism, which can be either electrostatic or electromagnetic. Electrostatic forces involve the attraction forces of two oppositely charged plates. The key advantage of electrostatic actuation is that it is a very well-consolidated technology, applied in several fields in which MEMS are used like sensors or automotive and very robust and repeatable mechanisms. The disadvantages include nonlinearity in force versus voltage relationship, and requirement of high driving voltages to compensate for the low force potential.

Electromagnetic actuation involves attraction between electromagnets with different polarity. The advantage of electromagnetic actuation is that it requires low driving voltages because it can generate large forces with high linearity. However, the isolation of the different actuators inside the same switch is not easy and it requires a very accurate design.

Since MEMS technology is applied in several fields and MEMS are produced in billions in the world, there are several commercial optical switches based on this technology.

A few reference characteristics are reported in Table 5.8 to get a feeling of what can be done in production devices.

TABLE 5.8

Characteristics of Practical MEMS-Based Optical Switches with Different Architectures

Dimension	16×16	1×8	56×56
Architecture	2D	2D	3D
Wavelength Range	Third Fiber Window	1240~1640 nm	1290–1330 nm 1530–1570 nm
Insertion loss	2 dB	2.0 dB	2.2 dB
Return loss	45 dB	45 dB	50 dB
Polarization-dependent loss	0.2 dB	0.1 dB	0.15 dB
Cross-talk	−70 dB	−75 dB	−50 dB
Switch speed	40 ms	1 ms	30 ms
Operating voltage	12 VDC	5 VDC	12 VDC
Power consumption	700 mW	50 mW	700 mW
Operation temperature	−5°C to 70°C	0°C~70°C	−5°C to 70°C
Size (L×W×H)	142×254×25 mm	76×93×9.5 mm	100×80×15 mm

5.6.2 Liquid Crystals Optical Switches

Another consolidated technology that finds application in large areas of the consumer market is that of liquid crystals, which has been used to design optical switches for routing applications [43].

The scheme of Figure 5.36 shows the operational principle of an elementary 2×2 liquid crystal switch.

An input polarization diversity system divides the x and y linear polarizations among two branches of the device. On each branch the light passes through a liquid crystal cell and a polarization splitter, which is deflected or not depending on the orientation of the liquid crystal that is fixed by the applied voltage [46].

After the switching the two polarizations are put together again by a polarization splitter.

When liquid crystals started to be used in optical switches their performances were hardly limited by a highly temperature-dependent loss and a very slow switch time. From

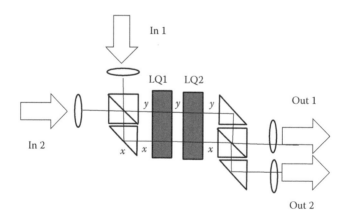

FIGURE 5.36

Scheme of a polarization diversity 2×2 optical switch based on two liquid crystal cells (LQ1 and LQ2), each of which work on one of the polarizations (x and y) after their separation via a polarization beam splitter.

that date, however, liquid crystal cells have gone several steps ahead and now we can realize cells with a switch time on the order of 300 μs and a loss on the order of 1.2 dB per cell.

These cells also retain the advantage of liquid crystals to consume a low power and to incorporate easily other functionalities inside the switch like broadcasting (obtained by switching polarization of 45°), selective attenuation used to equalize the switch output, and other possible features.

On the other hand, liquid crystal–based devices are necessarily built on micro-optics; thus they scale with difficulty to include thousands of elementary cells, like MEMS do.

For this reason, while liquid crystals are for sure competitive when small switches are considered, it is difficult for large switches to be fabricated with this technology.

5.6.3 Wavelength-Selective Switches

An important application of both LC (liquid crystals) and MEMS is in wavelength selective switches (WSSs), a type of optical switches quite diffused to realize tunable optical add drop multiplexers (OADMs) (see Chapter 7).

The scope of a WSS is essentially to eliminate from the input DWDM comb whatever set of channels is indicated by the network element control.

The functional scheme of a WSS is shown in Figure 5.48: it is essentially a set of components integrating together the demultiplexer and the drop channel separation so that the architecture of the single-side OADM is simplified. The implementation technology is similar to that of 3D MEMs, as shown in Figure 5.37.

In the figure the incoming WDM comb is directed through a discrete 3D optics to a space diffraction grating that divides the different wavelength channels. The channels are successively deflected by the same spherical mirror that has sent them on the grating to a 3D MEMS that directs any channel on the desired direction. In particular, channels to be dropped are sent to one fiber among the array of WSS drop fibers, while channels that are to pass through are sent again on the grating to be multiplexed and deflected toward the output line fiber. This is not the only possible architecture for a WSS, but tunability is generally a characteristic of all WSSs, so that any channel can be send to any drop port.

At the beginning of their introduction there was a lot of suspicion on WSSs due to their characteristics of being completely built through discrete 3D optics: it was believed that they were unstable and unreliable, at least in the long run.

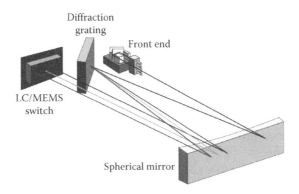

FIGURE 5.37
Block scheme of a quite diffused WSS architecture.

This prediction, at least up to now, has been demonstrated to be unfounded and WSSs are at present the most affirmed technology for tunable OADMs.

5.7 Electronic Components

When optical transmission was born, in the early 1990s, the industry was completely committed to the development of optical components and there was a diffused idea that optics was going to replace almost completely the electronics in telecommunication equipments.

At present this idea has been demonstrated to be completely incorrect.

Electronic components have evolved so fast in terms of cost and functionalities that electronic signal processing plays a fundamental role in the design of modern transmission systems and switching/routing machines are built using electronic switch fabrics and the situation seems not to change in the medium term.

As a matter of fact, signal processing CMOS circuits based on 45 and 32 nm technology [25] are so fast that real-time signal processing is possible also at the very high-speed characteristics of optical transmission. Thus all the equalization and compensation techniques that have been developed for other applications can now be applied to optical transmission.

As far as switch fabrics are concerned, electronic-based fabrics are convenient both in terms of footprint and of cost, simultaneously offering an unmatched reliability and a great richness of complementary functions.

5.7.1 Development of CMOS Silicon Technology

Beyond any architectural property, a baseline element determining the performance of an electronic equalizer or a switch fabric is the technology with which it is realized.

In this section we will try to understand achievements and challenges for microelectronics when used in telecommunication applications and in particular in electronic equalization of transmission distortions and in switch fabrics.

CMOS circuits are almost the only type of electronic circuits that are used out of analog and special applications.

Their diffusion is due to the combination of very large-scale integration, very low cost, and high performances granted by the great miniaturization.

CMOS require both n and p metal oxide field effect transistors (MOSFETs); hence the name complementary. The base advantage of CMOS versus other types of transistors is that a CMOS circuit only dissipates power when transistors are switched, not when they stay in a definite state, apart from any leakage currents in the system.

Figure 5.38 shows a schematic cross section of a typical advanced CMOS device with low-doped drain contacts implanted through Si_3N_4 spacers on either side of the gate to reduce the electric field between source and drain.

All gates and ohmic contacts are treated with $TiSi_2$, $CoSi_2$, or NiSi to reduce the resistances of gates and contacts.

The value chain of microelectronics is one of the most complex from wafer vendors to fabrication machine suppliers to foundries and chip producers.

Such a great number of players have to be coordinated to hit common goals and allow an ordinate development of the technology. This is the role of the International Technology

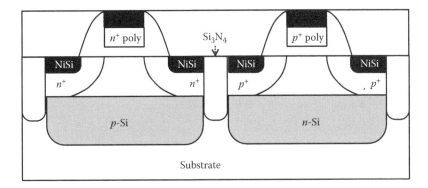

FIGURE 5.38
Section of an advanced CMOS system.

Roadmap for Semiconductors (ITRS), an industry road mapping group coordinating the efforts of the players in the microelectronics industry.

In order to fix clear targets for the major technology advancement, the ITRS has defined the concept of *technology node*, the moment at which the entire cycle that is needed to introduce a significant technology improvement through research, experimentation, development, and start of production starts to have an impact on products presented on the market.

Over the last 15 years, there has been a new CMOS technology node approximately every 2 years. The key feature of every node has been a doubling of the density of chips on the wafer and an increase of ~35% in switching speed and power consumption.

This improvement history is summarized in Figure 5.39 where the sequence of the different technology nodes is shown representing the decrease of the CMOS transistor gate length with time [47,48].

Different reasons have pushed the decrease of the dimensions of the CMOS, all oriented toward the improvement of CMOS circuit performances.

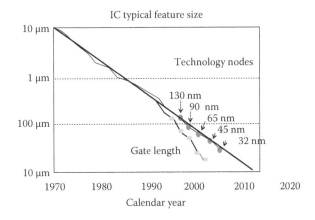

FIGURE 5.39
Shrink of the characteristic dimensions of CMOS in time and indication of the main technology nodes of the last 10 years. (After Parthasarathy, S. E. and Thompson, S., *Mater. Today*, 9(8), 20, 2006.)

- The gate length reduction is expected to decrease the time constant associated to the gate of each MOS increasing the circuit speed.
- The electronic charge associated to the gate is smaller, thus a switch operation is expected to burn less power.
- More circuits fit on the same wafer area, thus a circuit cost is expected to decrease.

5.7.1.1 CMOS Speed Evolution up and beyond the 32 nm Node

As far as the intrinsic RC circuit represented by the transistor gate is dominant over all the spurious RC contributions, the decrease of the gate length is directly related both to the switching speed of the transistor and to the power consumption.

Just to carry out an order of magnitude evaluation, both the capacitance and the equivalent resistance of the CMOS depend on the linear dimensions of the components. Assuming that the material does not change and that all the dimensions but the gate depth scale proportionally (both are in reality not true) in a very first approximation we can write

$$C = C_p + \varepsilon_0 \kappa \frac{WL}{d} \tag{5.33}$$

$$R = R_p + \rho \frac{\ell}{A} \tag{5.34}$$

where

C_p and R_p are the parasitic capacitance and resistance

ε_0 is the dielectric constant of vacuum

κ is the permittivity of the material filling the gate

ρ is the resistivity of the doped silicon electrode

W, L, d are the spatial gate dimensions, d being the depth, ℓ the electrode equivalent length, W the electrode width, and A its equivalent section

If all dimensions scale with the same constant the second term of both Equations 5.33 and 5.34 scale with L, while it is expected that the parasitic term increases, decreasing the characteristic dimensions of the CMOS.

Thus, as far as the parasitic elements can be neglected, it is

$$B \approx \frac{1}{\tau} \approx \frac{1}{\left(C_p + \varepsilon_0 \kappa \left(WL/d\right)\right)\left(R_p + \rho(\ell/A)\right)} \approx \frac{1}{\varepsilon_0 \rho \kappa} \frac{dA}{WL\ell} \tag{5.35}$$

Since the depth of the gate does not scale and A is also proportional to d, the second term of Equation 5.35 would increase almost as the inverse of the square of the gate length.

Unfortunately, the parasitic capacity and resistance cannot be neglected in a high-performance CMOS after the 90 nm node, so that the behavior given by Equation 5.35 is no more even approximated.

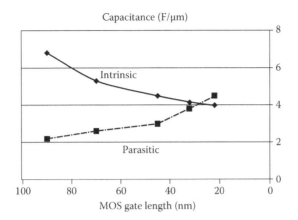

FIGURE 5.40
Intrinsic and parasitic CMOS gate capacitance versus the different component generations. (After Parthasarathy, S. E. and Thompson, S., *Mater. Today*, 9(8), 20, 2006.)

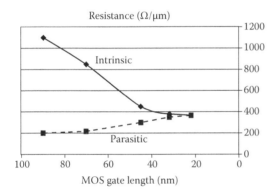

FIGURE 5.41
Intrinsic and parasitic CMOS gate resistance versus the CMOS generations. (After Parthasarathy, S.E. and Thompson, S., *Mater. Today*, 9(8), 20, 2006.)

The decrease of the intrinsic capacitance of the CMOS and the behavior of the parasitic one is demonstrated in Figure 5.40 and a similar plot is shown in Figure 5.41 for the intrinsic resistance and for the parasitic resistance [47].

Several phenomena imply a growth of the parasitic parameters.

- As MOSFET geometries shrink, the voltage that can be applied to the gate must be reduced to maintain reliability. To maintain performance, the threshold voltage of the MOSFET has to be reduced as well. As threshold voltage is reduced, the transistor cannot be switched from complete turn-off to complete turn-on with the limited voltage swing available.

- The final MOS design is a compromise between strong current in the "on" case and low current in the "off" case. Sub-threshold leakage which was ignored in the past can now consume upward of half of the total power consumption of modern high-performance very large-scale integration (VLSI) chips [49,50].

- The gate oxide, which serves as insulator between the gate and the channel, should be made as thin as possible to increase the channel conductivity and performance when the transistor is on and to reduce subthreshold leakage when the transistor is off.

However, with current gate oxides with a thickness of around 1.2 nm (which in silicon is ~5 atoms thick) the quantum mechanical phenomenon of electron tunneling occurs between the gate and the channel, leading to increased power consumption.

Insulators (referred to as high-*k* dielectrics) that have a larger dielectric constant than silicon dioxide, such as group IV metal silicates, for example, hafnium and zirconium silicates and oxides, are being used to reduce the gate leakage from the 45 nm technology node onward. Increasing the dielectric constant of the gate insulator allows a thicker layer while maintaining a high capacitance (capacitance is proportional to dielectric constant and inversely proportional to dielectric thickness).

- To make devices smaller, junction design has become more complex, leading to higher doping levels, shallower junctions, and so forth, all to decrease drain-induced barrier lowering. To keep these complex junctions in place, the annealing steps formerly used to remove damage and electrically active defects must be curtailed [51], increasing junction leakage. Heavier doping is also associated with thinner depletion layers and more recombination centers that result in increased leakage current, even without lattice damage.

- Traditionally, switching time was roughly proportional to the gate capacitance. However, with transistors becoming smaller and more transistors being placed on the chip, interconnect capacitance (the capacitance of the wires connecting different parts of the chip) is becoming a large percentage of capacitance [50]. Signals have to travel through the interconnect, which leads to increased delay and lower performance.

Figures 5.40 and 5.41 individuate a threshold technology node with a dimension of the gate on the order of 22 nm, where the parasitic resistance and capacitance will become almost equal to the intrinsic one.

Before that point, the strategy to decrease the dimensions is effective in increasing the CMOS switching speed; beyond that point, it seems to become less and less effective up to a point at which the component will become slower while shrinking.

This evolution is individuated also in Figure 5.42, where the inverse of the response time is plotted versus the technology node.

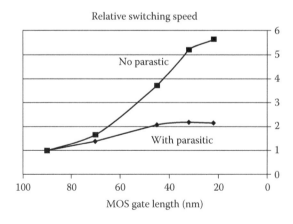

FIGURE 5.42
Impact of parasitic capacitance and resistance on CMOS commutation speed. (After Parthasarathy, S.E. and Thompson, S., *Mater. Today*, 9(8), 20, 2006.)

TABLE 5.9

Characteristic Speed and Power Consumption of
65 nm Gate CMOS Elements Optimized for
Different Applications

	T_{PD} (ns)	Total Power Consumption at Maximum Speed (μW)
Standard gate	75	750
High speed	0.9	760
Advanced	0.5	738
Low voltage	9	720

To manage the increase in parasitic R_p and C_p a number of solutions are under consideration, among them using novel silicides different from Si_3N_4 and reduced polysilicon height [10] that will bring to a vertical the scaling of the structure.

Another option is that along critical paths of a circuit, the devices are not placed as close as possible to the nearby devices in the wafer. In this case there is some area penalty, but there may be a path for reducing the device degradation caused by the nominal device pitch.

Concluding the section on CMOS switching speed, Table 5.9 provides figures for last-generation CMOS gates in terms of characteristic propagation time [52,53]. This is expressed through the propagation delay, which is the time for an input instantaneous switch to manifest at the gate output.

Considering the digital circuits speed, last generation CMOS are currently used to design processors and other digital circuits dedicated to very high volume markets with a clock up to 20 GHz, while special applications ASICS have been developed up to a clock as high as 50 GHz, reached not only via the choice of the most advanced transistor, but also through a careful circuit design that uses all the structures needed to achieve a so high speed.

So high speed CMOS can be used, and in practice have been used, to design also analog microwave circuits, having performances that are comparable with the more traditional III–V transistors used in these applications.

Analog microwave circuits have been realized with 45 nm and 32 nm CMOS having a passband as high as 50–60 GHz confirming the exceptional speed of these components.

5.7.1.2 CMOS Single-Switch Power Consumption

An ideal CMOS dissipates power only when switching from one state to the other. In reality, the presence of a leakage current, due to several reasons, induces a power consumption also in steady state.

As far as the switching power consumption is concerned, it is related to the dissipation of the electron charge accumulated into the dielectric part of the charged MOS.

Thus the power consumption during switching depends on the switching frequency (number of switching operations per second) and on the dimensions of the gate.

It is interesting to analyze the power consumption at the maximum switching speed, for which indicative values are reported in Table 5.9.

When the CMOS is used to build a switch fabric it is interesting to analyze the energy needed to switch an elementary CMOS gate.

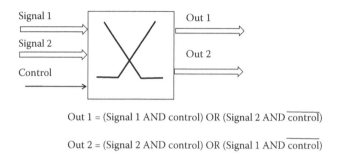

Out 1 = (Signal 1 AND control) OR (Signal 2 AND $\overline{\text{control}}$)

Out 2 = (Signal 2 AND control) OR (Signal 1 AND $\overline{\text{control}}$)

FIGURE 5.43
Realization of a 2×2 switch with logical CMOS gates.

A simple digital 2×2 switch can be constructed as a circuit with three inputs and two outputs: the three inputs are the two signals and the control; if the control is *on* the switch is in the bar state, if the control is *off* the switch is in the cross state (see Figure 5.43). Realizing the circuit with simple gates and counting the number of elementary CMOS it is possible to evaluate the power consumption per bit of the switch.

Assuming a clock at 3 Gbit/s and a total power consumption of 750 μW, we obtained a value of 3 pJ/bit for the energy consumption per bit.

As underlined in the caption of Table 5.9, data we are using are relative to the 65 nm technology node since this component generation is widely diffused in high-volume products and reliable average data are available.

It is possible to imagine that passing to 35 and 22 nm the power consumption will further decrease and it will be possible to get near the value of 1 pJ/bit, which is considered the target for the consumption of electronic switches.

5.7.1.3 CMOS Circuit Cost Trends

A powerful driver both for the shrink of CMOS dimensions and for the increase of wafer radius is the continuous push to decrease prices of CMOS circuits even if functionalities grow.

The price to pay for this evolution is the use of more and more expensive planar processes, whose cost is partially due to the requirement to fabricate smaller and smaller structures and partly to the need of maintaining the process uniformity over greater and greater wafers.

This evolution is represented in Figure 5.44, where the time evolution of the cost in dollars of a single CMOS (estimated starting from the average cost of an integrated circuit) is compared with the time evolution of the cost of a lithography machine [47].

The cost of the CMOS is decreased as much as seven orders of magnitude, even if the lithography tooling has become continuously more expensive.

At the end of the story, this is really the propulsive strength of microelectronics: it is able to conjugate better performances and lower prices.

Before illustrating specific telecommunication applications of electronics, we will briefly illustrate the structure and the performances of three base electronics-integrated circuits: the application-specific integrated circuit (ASIC), the field-programmable gate array (FPGA), and the digital signal processor (DSP). The evolution of these circuits is not mainly driven by telecommunications, since they find application in a much wider market, comprising consumer appliances and almost all the industrial machinery.

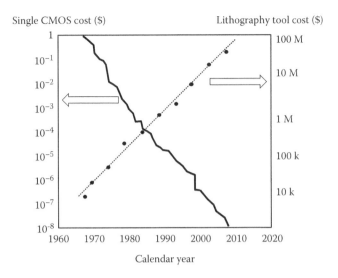

FIGURE 5.44
Trend of the CMOS cost and of the cost of the lithography tooling. (After Parthasarathy, S.E. and Thompson, S., *Mater. Today*, 9(8), 20, 2006.)

It is a fact, however, that the continuous enhancement in the capabilities of such circuits has conditioned strongly the design of telecommunication equipments. Electronic equalization would not be possible without suitable DSPs and FPGAs, while the evolution of router capacity has been strongly driven by the evolution of the ASIC performances.

ASICs and FPGAs have been largely used to fasten the complex functionalities of next-generation SDH/SONET ADMs and of Carrier Class Ethernet switches, up to the level that the design of an interface card of a packet machine (would be a router or a switch) partially coincides with the design of ASICs or with a programming of FPGAs with suitable characteristics.

5.7.2 Application-Specific Integrated Circuits

ASICs are, as the name indicates, nonstandard integrated circuits that have been designed for a specific use or application. Generally an ASIC design will be undertaken for a product that will have a large production run, and the ASIC may contain a very large part of the electronics needed on a single integrated circuit.

Despite the cost of an ASIC design, ASICs can be very cost-effective for many applications. It is possible to tailor the ASIC design to meet the exact requirement for the product and using an ASIC can mean that much of the overall design can be contained in one integrated circuit and the number of additional components can be significantly reduced.

This is why ASICs have such a big success in the line cards of large switching machines. These are quite complex subsystems, including a large processing capacity besides the physical interface with the transmission systems and besides memories, and all that is needed to complement a fast dedicated processor.

In Chapter 7 we will see that complex operations in a switching machine are carried out directly in the line card, such as QoS management.

Concentrating a large number of functionalities in a single ASIC is in these cases a key tool to ease the design and maintain, despite the complexity, high performances.

The development and manufacture of an ASIC design including the ASIC layout is a very expensive process. In order to reduce the costs, there are different levels of customization that can be used. These can enable costs to be reduced for designs where large levels of customization of the ASIC are not required. Essentially there are three levels of ASIC that can be used [54]:

1. *Gate array*: This type of ASIC is the least customizable. Here the silicon layers are standard but the metallization layers allowing the interconnections between different areas on the chip are customizable. This type of ASIC is ideal where a large number of standard functions are required which can be connected in a particular manner to meet the given requirement.

2. *Standard cell*: For this type of ASIC, the mask is a custom design, but the silicon is made up from library components. This gives a high degree of flexibility, provided that standard functions are able to meet the requirements.

3. *Full custom design*: This type of ASIC is the most flexible because it involves the design of the ASIC down to transistor level. The ASIC layout can be tailored to the exact requirements of the circuit. While it gives the highest degree of flexibility, the costs are much higher and it takes much longer to develop. The risks are also higher as the whole design is untested and not built up from library elements that have been used before.

In any case, the ASIC design is a complex task and several studies have been carried out on the process for ASIC design and on the ways to decrease the risk.

As far as the performances are concerned, a good design allows an ASIC to have a better possible performance for the specified task, due to its specialization.

Still at 90 nm, ASICs with a clock rate on the order of 20 GHz were produced for a variety of applications [55] and this limit decreases continuously, up to the ASICs with a clock rate exceeding 50 GHz that were produced using 45 nm technology [56].

It is to be expected that 32 and 22 nm ASICs will be still faster, and it is not unbelievable that 22 nm ICs will reach 80 GHz clock, which is a threshold for some telecommunication functionalities, like nonlinear equalization for 100 Gbit/s multilevel transmission (see Chapter 9).

As far as the throughput is concerned, experimental ASICs for security applications have been constructed with a single line serial throughput as high as 61 Gbit/s for an aggregate throughput on two parallel ports in excess of 121 GHz [57].

Another key feature of ASICs is the maximum area available. As a matter of fact, this is directly related to the number of transistors that can be integrated and thus with how functional each of the ASICs can be.

Measuring the ASIC area in a number of equivalent gates, ASICs with an area of more than 10^5 equivalent gates have been already built and this number will continuously increase while 32 nm technology will become more consolidated and the gates and functional libraries will enlarge, allowing easier ASICs design. It is to be taken into account that high-performance 45 nm ASICs are in production with an equivalent area of more than 7×10^5 gates [58]; thus it is expected that 32 and 22 nm technology will do better due both to smaller CMOS and to the continuous effort to increase the process planar uniformity.

5.7.3 Field Programmable Gate Array

FPGA chips are the most powerful and versatile programmable logic devices. They combine the idea of programmability and the architecture of uncommitted gate array which is one of the ways to develop ASICs [59].

The main elements of an FPGA chip are a matrix of programmable logic blocks (these blocks are named differently, depending upon a vendor) and a programmable routing matrix.

The interconnections consist of electrically programmable switches which is why FPGA differs from Custom ICs (see Section 5.7.2), as these are programmed using integrated circuit fabrication technology to form metal interconnections between logic blocks (see Section 5.7.2).

An abstract FPGA block scheme is shown in Figure 5.45, where these fundamental building blocks are shown in a common FPGA architecture, in which the logic blocks are arranged as a square matrix. However, different vendors use different internal architectures, functional to their particular logic block and routing technology and to the performances [60].

FPGA configuration memory is usually based on volatile SRAM (static random access memory), so these FPGAs must be connected to a dedicated Flash chip if stable memory has to be implemented. However, there are also both SRAM-based FPGAs with integrated flash module and pure Flash-based FPGAs that don't use SRAM at all.

5.7.3.1 Programmable Connection Network

In order to build the programmable connection network, present-day FPGAs adopt essentially two different technologies: SRAM-controlled switches and antifuse.

An example of the use of an SRAM-controlled switch is shown in Figure 5.46 where the connection between two logic blocks through two switches constituted by one pass transistor and then a multiplexer is depicted. The whole subsystem is controlled by SRAM cells [61].

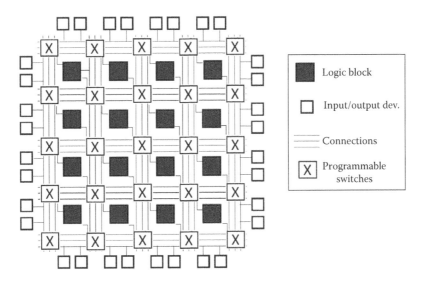

FIGURE 5.45
Functional block scheme of an FPGA.

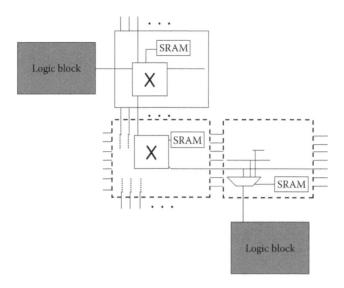

FIGURE 5.46
Example of interconnection between two logic blocks within the reconfigurable network inside an FPGA based on elementary switches and SRAM controllers. (After Actel, *Technical Reference for Actel Antifuse FPGAS*, 2010, www. Actel.com [accessed: October 18, 2010].)

The other type of programmable switch used in FPGAs is the antifuse. Antifuses are originally open circuits and take on low resistance only when programmed. Antifuses are suitable for FPGAs because they can be built using modified CMOS technology [62].

Generally antifuses rely on saturable insulators, that is, materials that have a very high resistance without an applied voltage or if the voltage is below a threshold, but if the voltage is brought above the threshold abruptly, the resistance goes down to values typical of a conductor.

Amorphous silicon has a similar behavior and is effectively used in a few antifuses, but much more specialized materials exist that are compatible with the CMOS technology.

5.7.3.2 Logic Block

Logic blocks of an FPGA can be constituted by a large variety of sub-circuits, depending on the use for which the FPGA is designed.

Logic blocks simply as a couple of transistors or a flip-flop are used in some FPGAs, like in other complex structures using a look-up table and a set of logic ports are adopted.

An example of a complex-structure logic block is shown in Figure 5.47 [61]. Such a logic block includes look-up tables and two flip-flops, besides a certain number of logical gates.

Logic blocks of this type are also called configurable, since their functionality can be configured through the content of the look-up tables.

5.7.3.3 FPGA Performances

Since FPGAs are modular elements whose inner connectivity depends on the task to be realized; the performances of these elements depends both on the particular FPGA inner architecture and on the type of function to be carried out.

In telecommunication applications FPGAs are used, for example, to implement digital signal processing in equalization engines, to manage header information in simple communication systems or to carry out protocols in switching machines.

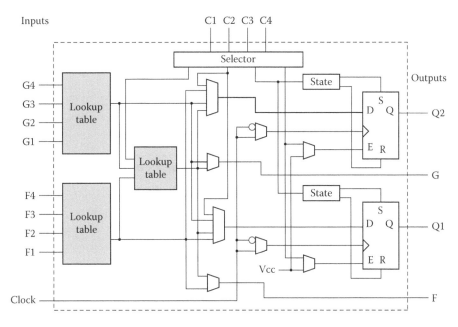

FIGURE 5.47
Example of reconfigurable logical block of a high-performance FPGA. (After Actel, *Technical Reference for Actel Antifuse FPGAs*, 2010, www.Actel.com [accessed: October 18, 2010].)

We will select a particular application that is quite studied in literature and that is a key for several equalization algorithms: basic linear algebra subroutines (BLAS).

We will essentially follow [63] in comparing the performances of a typical high-performance FPGA with C programs to execute linear algebra operations with both a standard high-speed central processing unit (CPU), and with a massively parallel graphic processing unit (GPU) running a standard mathematic kernel for high-performance and high-precision applications.

The time elapsed to perform a standard operation on an $N \times M$ matrix (like the multiplication for a $1 \times M$ vector) is shown in Figure 5.48. The disadvantage of the FPGA versus a fast CPU with an advanced mathematical kernel is not so significant, and even the parallel GPU with the mathematical kernel gains only about a factor of 3 in speed, at the cost of an expensive and complex parallel processor.

Moreover, a correctly programmed FPGA is superior to a standard C programming of about a factor of 1.8, qualifying as a real-time processor.

On the other hand, the FPGA advantages are clear in terms of dissipated power, which is one of the main parameters to judge the electronic circuitry for signal processing.

In all the considered examples the FPGA is able to go over the 3000 iteration per Joule of burned energy, while the GPU can carry out an average number of 800 iterations for burned Joule and the CPU less than 600.

This is an important advantage that qualifies FPGA as a competitive alternative for many telecommunication applications, where the consumed power is a key issue.

Another useful comparison is between FPGAs and ASICs. In this field it is clear that the FPA will always have worse performances due to the specific optimization of a well-designed ASIC. On the other hand, carrying out an FPGA project can be 10 times less expensive with respect to an ASIC project and the final ASIC, once in production, has a

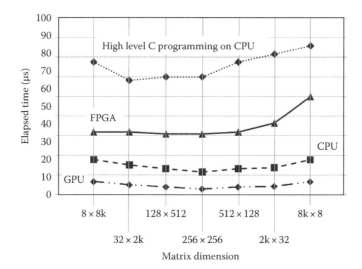

FIGURE 5.48
Example of performance evaluation of a programmed FPGA in the task of executing linear operations on arrays and vectors. (After Kestur, S. et al., *BLAS comparison on FPGA, CPU and GPU* [a cura di], IEEE Computer Society, *Annual Symposium on VLSI (ISVLSI), 2010*, IEEE, s.l., pp. 288–293, 2010.)

competitive price only if produced in very large volumes, which is not the case generally for telecommunications.

This consideration often leads designers to use FPGAs instead of ASICS every time the required performances allow it.

5.7.4 Digital Signal Processor

A DSP is a particular type of CPU optimized to run real-time digital processing algorithms like fast Fourier transform (FFT), digital filtering, digital correlation, and so on.

In order to do that, a DSP is realized with a set of hardware-implemented functions that are suitable for the target of the processor, while generally the configuration interfaces are not so powerful as those of a standard CPU, since a DSP is not built to change continuously the program in order to carry out a large variety of tasks (as a standard computer does), but to execute on a high-speed transit signal a well-defined set of operations that perhaps will be programmed at the beginning of the DSP's life and will never change.

Due to their design target, DSPs are used almost always as local controllers or equalizers, and the telecommunication application is not an exception.

For convenience, DSP processors can be divided into two broad categories:

1. *General-purpose DSP processors*: These are basically high-speed microprocessors with hardware and instruction sets optimized for DSP operations.
2. *Special-purpose DSP processors*: These include both hardware designed for efficient execution of specific DSP algorithms like FFT or feed-forward equalizer (FFE)–decision feedback equalizer (DFE) filtering and hardware designed for specific applications like forward error correction (FEC) coding–decoding.

5.7.4.1 DSP Hardware Architecture

In most of the applications requiring the use of a DSP, digital algorithms like digital filtering and FFT involve repetitive arithmetic operations such as multiplication, additions, memory access, and heavy data flow through the CPU.

The Von Newman architecture of standard microprocessors is not suited to this type of activity due to the fact that heavy mathematics without any access to the memory or to the input/output interfaces of the system practically nullify the ability of the processor to share the computation power among different processes exploiting the time each process takes to access the memory or the peripheral devices of the system.

DSPs overcome this difficulty by a hardware architecture that exploits a sufficient degree of parallelism so as to avoid the intense mathematics that will destroy the processor performance.

An example of basic DSP architecture is shown in Figure 5.49; this is a particular implementation of the so-called Harvard architecture [64] and exploits different levels of parallelism to speed the real-time processor operations:

- Program memory is separated by the data memory, thus massive data transfer does not collapse the transfer of program instructions to the arithmetic–logical unit (ALU). Moreover, in a standard microprocessor, where the program codes and the data are held in one memory space, the fetching of the next instruction while the current one is executing is not allowed, because the fetch and execution phases each require memory access. On the other hand, since the execution of an instruction requires fetching and decoding before the effective execution phase, it would contribute to speed up operations if the fetching phase would be carried out contemporary to the execution of a previous instruction. This is possible if data memory and program memory are separated and exploit separate busses. Thus, in a DSP with the architecture of Figure 5.49 two instructions are contemporary under processing in the ALU, one is executed, the other fetched.

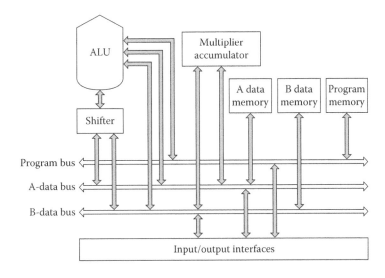

FIGURE 5.49
Architecture of a basic DSP.

In many DSPs the pipeline process, which is the process constituted by putting in parallel different operations with a different state of completion, is even more developed. In these cases, three instructions are active at the same time: one is in the fetching state, the second in the decoding state, and the third in the execution state.

- The data memory is separated into two independent memory units, each of which has its own bus; in this way parallel data transfer is possible, easing operation on real-time data coming from the environment, for example, on data arriving from a transmission line.

- The ALU is supported by the parallel operation of a multiplier with an accumulator. The basic numerical operations in DSP are multiplication and addition. Multiplication in software is time-consuming, and additions are even worse if floating-point arithmetic is used.

  To make real-time DSPs possible, a fast dedicated hardware Multiplicator and Accumulator Circuit (MAC), using either fixed-point or floating-point arithmetic, is mandatory.

  This device also generally uses a pipeline process, since the operations can be represented in three steps: data read, execution and accumulation. Thus generally also the MAC has three active, contemporary operations, each of which is in a different state of its life cycle in the MAC.

- The ALU is also supported by a shifter that allows recurring operations to be carried out efficiently.

Beyond the basic architecture of Figure 5.49, a much higher degree of parallelization can be realized in high-performance DSPs, as shown in Figure 5.50. This architecture is inspired by what is presented in [65]; however, it is not intended to be the real product architecture, as it is simplified and schematized for explanation purposes and to present a more general example. Here, up to four MACs can be put in parallel to support the ALU in executing recurrent operations. The MACs are not connected to the ALU via the program bus, but via dedicated connections that allow a much more effective pipelining.

Moreover, different DSPs can be built, where one or two MACs can be substituted by other kinds of support units that operate in parallel, like shifter or more complex logical units that can perform also complex operations using hardware-coded instructions.

The pipeline operation and the division of the computation load among the ALU and the other parallel processing units are cared for by a specific part of the chip, a program control unit. This is supported by a sequencer that constructs the sequence of instructions for each parallel element.

The program control unit has also as support an emulation unit that allows more complex operations that have to be emulated via the inner language of the machine to be solved without charging the ALU. As a matter of fact, emulation is a complex operation that would slow the pipeline operation if it would be in charge of the ALU.

Finally different input/output units exist for data and program that are not shown in the figure.

5.7.4.2 DSP-Embedded Instruction Set

Besides architecture that facilitates intense real-time mathematics, the DSPs are equipped with an internal set of machine language instructions that are very different from those of general-purpose CPUs.

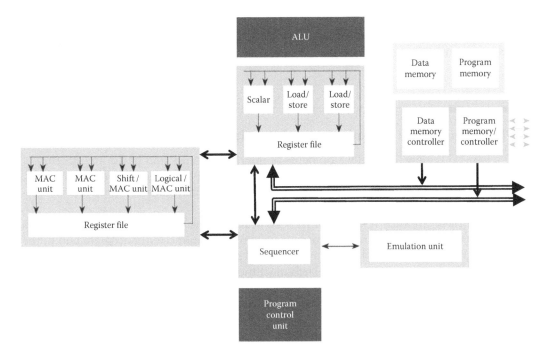

FIGURE 5.50
Architecture of a more advanced DSP. (After Ceva, *The Architecture of Ceva DSPs—Product Note*, 2010, www.ceva-dsp.com [accessed: October 18, 2010].)

As a matter of fact, depending on the DSP-specific architecture, several recurring operations in digital processing are carried out by hardware parts of the ALU or by parallel elaboration circuitry, like MACs.

For example, key parts of an FFT algorithm can be encoded into a single machine language instruction in a DSP optimized for this kind of application.

Moreover, a DSP-embedded instruction set is almost always a reduced instruction set computer (RISC) set, where no micro-language is present to elaborate complex instructions.

The characteristics of a RISC set are as follows:

- The instructions are short and the execution time does not vary so much from instruction to instruction.
- The instructions have a fixed length.
- Every instruction is executed in a clock cycle.
- Memory is addressed with a small number of simple instructions.
- In order to face the reduced number of instructions and the absence of micro-language, registries are used much more than in a traditional CPU.
- Pipelining is easier.

5.7.4.3 DSP Performances

As in the case of FPGAs, it makes little sense to talk in detail about DSP performances, as they vary depending on the application and on the inner DSP architecture.

In the following part of this chapter we will see a lot of applications requiring DSPs or FPGAs in order to have an idea of what can be achieved with these electronic circuits in the telecommunication environment.

5.8 Electronics for Transmission and Routing

After a general review of the developments of microelectronics and of the architecture of a few general-purpose circuits that have a relevant impact on telecommunication systems, we discuss in this section a set of specific applications of microelectronics to telecommunication equipments.

Naturally these are only a subset of the use of ICs in telecommunications and cover a set of recent evolutions both in transmission system performance enhancement and in the solution of the addressing problem in switches and routers, which has been for a long time the bottleneck of the performances of such machines.

There are a lot of other applications of microelectronics, like integrated switches and single cross-point implementation, control processors for element and network management and CPUs for protocol running.

A few of these applications are in the field of general-purpose electronic systems, like the control board of the telecommunication equipments; others are consolidated technology, even if the continuous CMOS platform development leads to a continuous performance enhancement.

However, the field of electronics application is so large that for sure it can happen that the reader is interested in something that is not even cited here. More specific books on electronics for telecommunications surely will help the interested reader to go more in deep in this subject. [64,66,67].

5.8.1 Low-Noise Receiver Front End

Before reviewing digital signal processing applied to optical transmission, we will consider an important analog device that is present in all transmission systems: the low-noise amplifier that has the role of amplifying the photocurrent coming out from the photodiode up to a level suitable to be processed by the electronic receiver [32,68,69].

This amplifier noise characteristic determine almost completely the thermal noise performances of the receiver, since it is the first element of the electronic amplifier chain.

In order to have a low noise figure, the input impedance of the amplifier has to be high, so a high-impedance amplifier is suitable for this application.

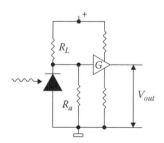

FIGURE 5.51
Electrical scheme of a high-impedance receiver front end.

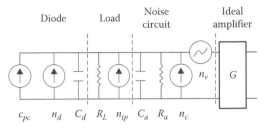

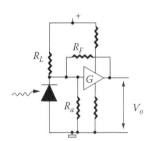

FIGURE 5.52
Equivalent circuit of a high-impedance receiver front end.

FIGURE 5.53
Electrical scheme of a transimpedance receiver front end.

The scheme of a high-impedance amplifier following a photodiode is plotted in Figure 5.51, while its equivalent noise circuit is shown in Figures 5.52 and 5.53.

In the equivalent circuit, the photodiode is represented as a capacitor in parallel with the photocurrent generator and the dark current generator while the amplifier is divided into three sections: the load resistance, the noise circuit representing the thermal noise contribution, and an ideal amplifier providing the gain G and the lowpass characteristics with a bandwidth equal to B_e.

This structure is aimed to minimize the noise contribution with a very high input resistance. This design implies a small passband and in order to achieve a large electronic bandwidth the amplifier has to be accompanied by an equalizer.

From the equivalent circuit it is clear that the noise results from two main sources: the photodiode noise, which can be modeled as a noise current, and the amplifier noise process, which can be modeled as a current noise and a voltage noise.

Thus the overall circuit noise can be modeled as follows:

$$n_t(t) = G[n_{tp}(t) + n_c(t) + n_{cv}(t)] \tag{5.36}$$

where the noise contributions are defined in Figure 5.51 and $n_{cv}(t)$ is the noise current generated by the voltage noise at the amplifier input.

The power of the photodiode thermal noise can be written as

$$\sigma_{tp}^2 = \frac{4k_B\Im}{R_L} B_e \tag{5.37}$$

where
k_B is the Boltzmann constant
$\Im$ is the absolute temperature
R_L represents the photodiode load resistance

The power of the process $n_{cv}(t)$ can be evaluated by evaluating the input admittance of the amplifier. In particular, calling the admittance $Y(\omega)$, the power spectral density of the voltage process $n_v(t)$ with $S_v(\omega)$ and the ideal amplifier transfer function $H_G(\omega)$, we obtain

$$\sigma_{cv}^2 = \int |Y(\omega)|^2 |H_G(\omega)|^2 S_v(\omega)d\omega \approx S_v \left[\frac{B}{R_{in}^2} + \frac{4}{3}\pi^2(C_a + C_d)^2 B^3 \right] \tag{5.38}$$

Combining Equations 5.36 with 5.37 and 5.38, the overall power of the thermal noise at the receiver is given by

$$\sigma_{th}^2 = G^2 \left\{ \frac{4k_B\Im}{R_L}B_e + S_cB_e + S_v \left[\frac{B_e}{R_{in}^2} + \frac{4}{3}\pi^2(C_a + C_d)^2 B_e^3 \right] \right\} = F_a \frac{4k_B\Im}{R_L}B_e \tag{5.39}$$

where the noise factor of the receiver F_a is given by

$$F_a = G^2 + \frac{G^2 S_c R_L}{4k_B\Im} + \frac{G^2 R_L S_v}{4k_B\Im} \left[\frac{1}{R_{in}^2} + \frac{4}{3}\pi^2(C_a + C_d)^2 B_e^2 \right] \tag{5.40}$$

From Equation 5.39 it results that the minimum noise is achieved by increasing R_L as much as possible. The limit to the increase of R_L is given by the bandwidth of the front end which is approximately proportional to $1/R_L$. Even if an equalizer is used to improve the bandwidth, the relationship between noise and bandwidth constrain the design of low-noise high-impedance front ends.

This limitation can be overcome by means of the so-called transimpedance front end, whose scheme is shown in Figures 5.52 and 5.53 while the equivalent circuit is sketched in Figure 5.54.

The transimpedance of the front end is defined as the ratio between the amplifier input current (that is the photocurrent) and the output voltage and can be written as

$$Z_G(\omega) = \frac{R_F}{1 + (1/G) + (R_F/R_LG) + (i\omega C_{in}R_F/G)} \tag{5.41}$$

If the circuit after the front end read the voltage at the front end output as a useful signal, the bandwidth of the front end can be written as

$$B_e = \frac{G}{2\pi R_F(C_a + C_d)} \tag{5.42}$$

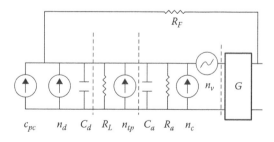

FIGURE 5.54
Equivalent circuit of a transimpedance receiver front end.

while the noise at the amplifier output can be evaluated with a procedure similar to that carried out for the high-impedance front end, obtaining

$$\sigma_{th}^2 = G^2 \left\{ \frac{4k_B \Im}{R_F} B_e + S_c B_e + S_v \left[\frac{B_e}{R_F^2} + \frac{4}{3} \pi^2 (C_a + C_d)^2 B_e^3 \right] \right\} = F_a \frac{4k_B \Im}{R_L} B_e \qquad (5.43)$$

The principle on which the transimpedance amplifier is based is clear from Equations 5.42 and 5.43: the noise is limited by the high feedback resistance while the bandwidth can be wide if the open loop gain is sufficiently high.

5.8.2 Distortion Compensation Filters

During propagation through an optical fiber the signal undergoes both linear and nonlinear distortion and the performance of an optical transmission system largely depends on how these distortions are managed.

The main deterministic linear distortion is chromatic dispersion and dispersion compensation is mandatory in high-performance transmission systems.

If dispersion is optically compensated, through a distortion compensation filter (DCF), for example, it is possible to compensate many wavelength channels together; moreover, since optical compensation is applied before the receiver, the only nonlinearities that are present are those related to fiber propagation, which generally can be considered small. Thus in a first approximation optical dispersion compensation can work on a linear channel.

On the other hand, if dispersion is to be compensated after the detection via an electronic circuit, one compensator per channel is needed; moreover, the photodiode square detection law creates a strong nonlinearity in the channel before the electronic compensator.

Nevertheless, electronic dispersion compensation has several attractive characteristics, so that it is nowadays almost always present in high-performance systems.

The cost of electronic circuits is not so high, so that from a cost point of view, having one compensator for each channel is not a big problem. On the other hand, per channel compensation allows accurate management of the dispersion slope and of the polarization mode dispersion (PMD), which is not possible using a compensator that processes all the wavelength channels due to the fact that the bandwidth of the principal states of polarization (PSPs) is generally smaller than the wavelength extension of the multiplexed comb.

Moreover, using more complex algorithms, electronic compensation can cope also with nonlinear distortion [68,70].

Electronic compensators can be divided into three categories [56]: electronic post-compensators (EPCs), electronic pre-compensators (EBCs), and electronic pre-distortion circuits (EPDs):

EPCs are circuits located at the receiver that try to eliminate intersymbol interference caused by dispersion and, just in case, nonlinear effects by processing the photocurrent.

EBCs on the contrary are located at the transmitter and distort the signal before modulation onto the carrier to simulate the transit through a virtual dispersion compensation device. Once the modulated signal is built and transmitted through the fiber, the predistortion is compensated by the fiber dispersion and ideally the signal arrives undistorted at the receiver.

The case of EPDs is more extreme: generally an EPD is designed to compensate not only linear effects but also nonlinearities due to fiber propagation. For this reason, the signal distortion caused by an EPD is much more complex than that caused by an EBC.

5.8.3 Electronic Dispersion Post-Compensation

Among all the electronic distortion compensators, EDCs are the most diffused. Essentially there are three types of EDCs:

1. Continuous-time filters (CTFs)
2. FFE/DFE
3. Maximum likelihood sequence estimator (MLSE) equalization

The working principles of CTFs are quite simple: the bandwidth of the receiver is sliced in a certain number of sub-bands and every sub-band is multiplied by a complex number. A feedback circuit regulates these multiplication factors so as to maximize the eye opening or (in case an FEC is mounted in the receiver) the estimated error probability.

This is a simpler type of EDC, but also the lowest performing one, and is going to be substituted in new systems by more performing algorithms.

The most diffused EDC is the FFE/DFE.

5.8.3.1 Feed-Forward/Decision Feedback Equalizer

The idea at the basis of the FFE/DFE is to compensate the dispersion via digital filtering of the received signal.

The electrical current after detection can be in general written as

$$c(t) = R_p \sum b_j g_j(t - jT) + n(t) \tag{5.44}$$

where
$b_j = 0.1$ is the transmitted bit
$g_j(t - jT)$ is the received pulse, whose shape depends on the position in the bit stream due to the presence of dispersion
T is the bit interval

Assuming that the pulse received in the absence of distortion would be $f(t)$ and neglecting the pulse shape dependence on the position in the bit stream, the distortion equivalent filter $D_\omega(\omega)$ is a filter whose characteristic is to transform the ideal pulse in the real one, that is

$$g(t) = \int_{-\infty}^{\infty} f(\tau)D(t - \tau)d\tau \tag{5.45}$$

where
$D(t)$ is the inverse Fourier transform of $D_\omega(\omega)$

The distortion equivalent filter is naturally only an approximation of the real channel behavior, which is nonlinear at least due to the photodiode. However, once $D_\omega(\omega)$ is introduced, the distortion can be in principle compensated by inverting $D_\omega(\omega)$. If this operation is well conditioned, which is the case in the optical channel, the current after the ideal compensator would be given by

$$c_c(t) = R_p \sum b_j g_j(t - jT) \otimes D^{-1}(t) + n(t) \otimes D^{-1}(t) \tag{5.46}$$

where $\otimes$ indicates convolution in the time domain.

Sampling at the Nyquist rate and truncating the resulting series to a finite number of terms results in the FFE compensation algorithm

$$c_c(v) = R_p \sum_j b_j \left\{ \sum_{k=1}^m g_j(v-k-jT)d(k) + \sum_{k=1}^m n(v-k)d(k) \right\} \qquad (5.47)$$

where $d(k)$ are the coefficients of the FFE filter.

The algorithm of Equation 5.47 can be easily implemented by a digital circuit shown in Figure 5.54 that is essentially a set of taps and delay lines.

However, Equation 5.47 differs a lot from Equation 5.46 due to the finite number of taps and all the other approximations, thus a simple FFE does not assure a good chromatic dispersion correction. In order to improve the performance of the compensator generally the FFE is complemented with a DFE.

The DFE is essentially another set of delay lines and taps that is put as a feedback to the FFE in order to correct FFE operation.

Once the circuit is designed, its effectiveness depends on the choice of the FFE and DFE coefficients and on the number of taps. In practice, five to nine is a feasible number of taps; thus it is required to set up 10 or 18 coefficients to optimize the working of the compensator [71,72].

After initial setting, high-performance FFE–DFE relies on an adaptive coefficient setting algorithm that continuously adjusts the coefficients to optimize the receiver performance.

The determination of the FFE–DFE coefficient is in general a process of filter parameters determination in the presence of noise and many FFE–DFE use a least mean square (LMS) algorithm to adapt the coefficient on the grounds of the receiver performances. This solution allows a well-established processing to be used [73].

However, if shot noise cannot be neglected or if amplified optical noise is the dominant noise contribution (see Chapter 6) the noise affecting the photocurrent is not Gaussian so that the LMS algorithm is not the theoretically optimum solution. Other algorithms have been proposed, suitable for the coefficient determination in the presence of non-Gaussian noise that are more complex than LMS, but LMS seems to perform better especially in conditions of high dispersion and the presence of PMD [74]. As a matter of fact, while chromatic dispersion is in principle a deterministic phenomenon and presents only very slow changes due to aging and temperature fluctuations, PMD is a random fluctuating phenomenon. For this reason using an optimum algorithm for the equalizer coefficient identification is much more critical in the presence of a sizeable PMD.

Moreover, it is also demonstrated that, when using an electronic dispersion compensator based on the FFE–DFE the availability of an automatic threshold control (ATC) algorithm is very beneficial for the overall system working. As a matter of fact the compensator, while equalizing the intersymbol interference, also filters the noise with a digital filter that is generally adapted for equalization and not for noise filtering (Figure 5.55).

Since the equalizer implements a feedback structure the statistical distribution of the noise is changed after the equalizer, and more adaptive the equalizer, the more the noise distribution is filtered differently at different moments. An example of noise distributions is shown in Figure 5.56, where the decision-variable distribution estimate is shown in correspondence to the transmission of a "1" and the transmission of a "0." Besides the distribution in the absence of equalizers, the distributions for different equalizers are shown.

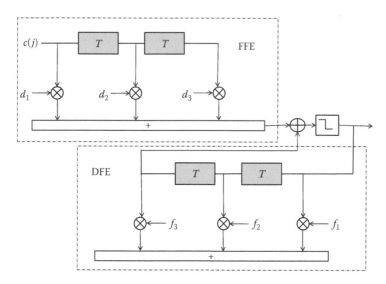

FIGURE 5.55
Block scheme of an FFE–DFE electronic dispersion compensator.

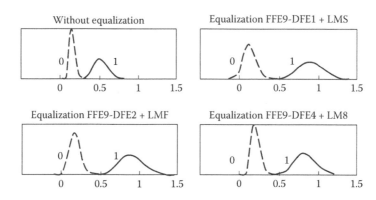

FIGURE 5.56
Effect of the electronic equalization at the receiver with an FFE–DFE on the decision sample probability distribution. (After Koc, U.-V., *J. Appl. Signal Process.*, 10, 1584, 2005.)

In particular, the equalizers are represented with the number of taps after the name of the filtering section (e.g., FFE3–DFE2 is an equalizer with three taps in the FFE and two taps in the DFE) and with the name of the adaptive algorithm used for the estimate of the coefficients. Here LMF [75] and LM8 [76] are two algorithms alternative to the LMS proposed in [74]. The effect of the equalizer in changing the decision variable distribution is clear from the figure, and the greater the equalizer complexity, the greater is the distribution change. Thus continuous optimization of the decision threshold is a key point for the optimum system working.

The fact that the equalizer changes the noise distribution has also another important impact on the overall working receiver.

In high-performance receivers an FEC is almost always implemented to enhance the system performance. The FEC gain in terms of SN_o needed to reach a certain post-FEC error probability is generally evaluated for a Gaussian noise at the FEC input. In case of optical systems there is a long experience that allows the gain of the most common FEC to be estimated also in the presence of the amplifier noise.

However, if the error statistics changes due to an equalizer, the FEC gain can also change. This is a problem that mainly hits nonlinear equalizers and can be important in the general design of the receiver.

An example of performance of the FFE–DFE is shown in Figures 5.57 and 5.58 [74]. Here the performance of a receiver in the presence of a signal carrying optical Gaussian noise (e.g., from optical amplification) is considered. In particular, the increment of the optical signal to noise ratio (ΔSN_o) needed to overcome the effect of dispersion is plotted versus the chromatic dispersion in Figure 5.57 and versus the PMD in Figure 5.58. An equalizer with nine taps in the FFE and four in the DFE is considered, with two different adaptation algorithms and with and without ATC.

From the figures it is evident that the insertion of an equalizer brings a performance degradation in the absence of dispersion, but increasing the dispersion increases the performance of the equalized receiver.

From an implementation point of view, the FFE–DFE is generally realized in digital CMOS technology. In order to digitalize the band-limited photocurrent, a fast ADC is required. After the ADC, depending on the complexity of the tracking algorithm and on the number of taps, both a high-performance DSP and an FPGA can be used. It is difficult that the volumes related to long-haul and ultra-long-haul systems justify the use of a specific ASIC, but FFE–DFE has been proposed also for interconnection on multimode fibers

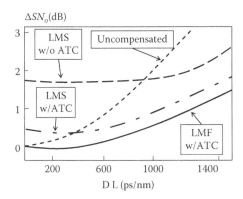

FIGURE 5.57
Performance of an FFE–DFE dispersion compensator on a dispersive fiber line. An equalizer with nine taps in the FFE and four in the DFE is considered, with two different adaptation algorithms and with and without ATC. (After Koc, U.-V., *J. Appl. Signal Process.*, 10, 1584, 2005.)

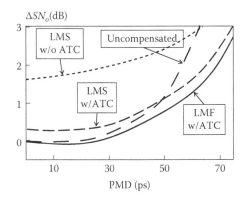

FIGURE 5.58
Performance of an FFE–DFE dispersion compensator on fiber line with PMD. An equalizer with nine taps in the FFE and four in the DFE is considered, with two different adaptation algorithms and with and without ATC. (After Koc, U.-V., *J. Appl. Signal Process.*, 10, 1584, 2005.)

and class C G-PON. In these cases, if the use of equalizers becomes diffused, the development of dedicated ASICs is more than probable.

From an implementation point of view, the ADC is the bottleneck of the FFE–DFE design, since the ADC is a power-consuming component at high speeds and cannot exploit the parallelization of the digital signal in the following components. Today, ADCs that realize a sampling at 5–10 GHz are available on the shelf, but arriving at 80 GHz requires specialized and power-consuming circuits, while 200 GHz is for the moment out of reach.

This is why the use of these circuits is a consolidated practice at 10 Gbit/s, and is a challenge at 40 Gbit/s, where the advantage of having a greater receiver tolerance to dispersion has to be balanced with the greater power consumption, so it is not feasible for serial transmission at 100 Gbit/s.

To close the discussion on FFE–DFE, it is worth mentioning the proposal of nonlinear FFE–DFE based on the Volterra algorithm [77]. These nonlinear equalizers can mitigate the effects of dispersion much better than the classical FFE/DFE, but are more complicated to build. In a practical system the Volterra nonlinearity is limited to second and third order. However, the number of filter coefficients for such a nonlinear FFE of appropriate order (e.g., five taps) is still very high even if methods to reduce the complexity separating the linear from the nonlinear parts have been proposed.

5.8.3.2 *Maximum Likelihood Sequence Estimation Equalizers*

Theoretically much more powerful with respect to FFE–DFE, the equalizers based on the MLSE algorithm, after a period of pure theoretical study, have nowadays started to become practical circuits while electronics becomes more and more fast and functionally rich [78].

An MLSE equalizer solves the problem of identifying any hidden sequence of initial states (in the transmission case the transmitted bits) starting from the observed sequence of the final states (in our case the received samples before decision).

The Viterbi algorithm [79] operates on the assumption that the system under observation is a state machine, that it can be characterized by a state map composed of a finite number of states and by a set of allowed transitions that connect each state to the states that can be reached starting from it.

The algorithm is based on a few key assumptions:

- If the system starts from one state and arrives at another state, among all the possible paths connecting the two states on the state graph, one and only one is the most likely, which is called the "survivor" path.

- A transition from a previous state to a new state is marked by an incremental metric, usually a number. Typical metrics are the transition probability or the transition cost.

- The metric is cumulative over a path, that is, the metric of a path is the sum of the metric of each transition belonging to the path.

On the grounds of these hypotheses the algorithm associates a number to each state. When an event occurs, in our case every time a new sample is received at the decision level, the algorithm knows the state in which the system is and the previous sequence of states. Starting from them it examines a new set of states that are reachable from the initial state and that are compatible with the observed event, for each state combines the weight associated to the starting state with the metric of the transition and chooses the transition that creates the better combined value. This transition results in a new starting state that enriches the state sequence.

The incremental metric associated with a transition depends on the transition probability from the old state to the new state conditioned to the observed event and from the state transition history for a length equal to the channel memory.

After computing the combinations of incremental metric and state metric, only the best survives and all other paths are discarded. It is to be underlined that, since it is the path weight that is at the center of the Viterbi algorithm, the path history must be stored in order to work the algorithm.

In the case of an equalizer for the distortion of the optical channel, whatever technique is used to identify a finite number of states within the continuous response of the optical channel, the core issue is that the state graph is too big to be completely represented into the memory of the signal processor.

Thus the key issue of this kind of equalizers, besides the sampling at very high rate, is the reduction of the Viterbi algorithm both in terms of number of states and in terms of number of transitions, so as to be able to manage it with a DSP or programming an FPGA [80,81].

One of the most used techniques is that leveraging the circuit scheme of Figure 5.58. Once a new sample is received the probability density function that a "1" is transmitted conditioned to the value of the received sample is estimated starting from the previous N samples. This is the transition metric for the transition to the two possible new states: a "1" is received or a "0" is received.

Two operations help in limiting the complexity of the MLSE [82]:

1. Limiting the number of bits per sample and the number of samples per received intervals; as a matter of fact, if n samples are taken each bit interval and each sample has N bits and the number of MLSE states to be considered in every received bit interval is $n2^N$, a number that easily becomes too large to be managed by a DSP.

2. Simplifying the probability estimation (i.e., the channel characterization) contributes to speed the algorithm.

These two elements have been drastically simplified in order to realize equalizers that work at 10 and even at 40 Gbit/s, and the surprising thing is that the Viterbi algorithm works well also with a very simplified state machine.

As an example let us consider the architecture presented in Figure 5.59, where a traceback of three bits is realized via a parallel processor working at 622 Mchips/s. The number of bits per sample is eight, but only one sample is extracted per bit period.

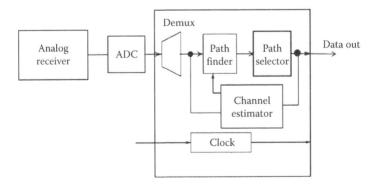

FIGURE 5.59
Block scheme of an electronic equalizer implementing MLSE equalization with the Viterbi algorithm.

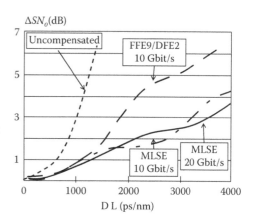

FIGURE 5.60
Performance of an electronic equalizer implementing MLSE equalization with the Viterbi algorithm at 10 and 20 Gbit/s on a fiber line with chromatic dispersion. The performances of an FFE–DFE with nine FFE taps and two DFE taps, LMS adaptation algorithm with ATC is also reported for comparison. (After Xia, C. and Rosenkranz, W., *IEEE Photon. Technol. Lett.*, 19(13), 1041, 2007.)

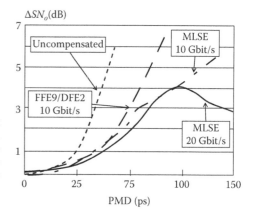

FIGURE 5.61
Performance of an electronic equalizer implementing MLSE equalization with the Viterbi algorithm at 10 and 20 Gbit/s on a fiber line with polarization dispersion. The performances of an FFE–DFE with nine FFE taps and two DFE taps, and LMS adaptation algorithm with ATC is also reported for comparison. (After Xia, C. and Rosenkranz, W., *IEEE Photon. Technol. Lett.*, 19(13), 1041, 2007.)

The performance of such an equalizer is shown in Figure 5.60 for a link affected by chromatic dispersion and in Figure 5.61 for a link affected by PMD [71]. As in the case of the FFE–DFE, the performances of the equalizer are measured through the increment of the optical signal to noise ratio (ΔSN_o) needed to overcome the effect of dispersion.

In both cases, and both at 10 and 20 Gbit/s, MLSE outperforms the considered FFE–DFE as soon as the distortion becomes relevant.

The MLSE has also been demonstrated to be effective in mitigating the impact of nonlinear distortion in DWDM systems. Both four-wave mixing (FWM) [83] and self-phase modulation/cross-phase modulation (SPM/XPM) [84,85] effects can be attenuated, contemporary to PMD and chromatic dispersion compensation.

As an example, let us consider the structure proposed in [84] where the MLSE digital equalizer comprises a 3 bit analog-to-digital converter operating at up to 25 Gsamples/s and a four-state (two bits per sample, one sample per transmitted bit) Viterbi decoder. This is a case of extreme reduction of the Viterbi algorithm, quite far from the unconstrained working.

Nevertheless, it is effective in mitigating both linear and nonlinear fiber distortions. This is demonstrated in Figure 5.62, where the performance of the MLSE EDC is compared with that of a similar uncompensated receiver in the case of 10 Gbit/s speed. The total fiber link length was 214 km of the G.652 fiber with a nominal dispersion value at 1550 nm of 17 ps/(nm km), for a total link dispersion of about 3638 ps/nm. The amount of transmitted power was varied to evaluate the performance of the MLSE EDC against SPM as a function of

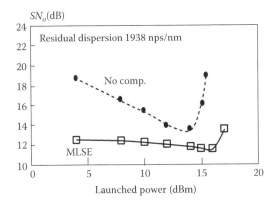

FIGURE 5.62

Performance of the MLSE EDC compared with that of a similar uncompensated receiver in the case of 10 Gbit/s speed. The total fiber link length was 214 km with a nominal dispersion value at 1550 nm of 17 ps/(nm km) and part of the dispersion is optically compensated at the receiver. The amount of transmitted power was varied to evaluate the performance of the MLSE EDC against SPM as a function of residual chromatic dispersion. The baseline SN_0 is about 12 dB due to the use of the FEC at the receiver. (After Downie, J.D. et al., *IEEE Photon. Technol. Lett.*, 19(13), 1017, 2007.)

residual chromatic dispersion. The baseline SN_0 is about 12 dB due to the use of the FEC at the receiver.

The EDC figure of merit is always the increment of the optical signal to noise ratio (ΔSN_0) needed to overcome the effect of dispersion in a receiver equipped with a standard FEC.

Even if MLSE algorithms perform better even in their most reduced form, the circuitry implementing MLSE is in any case much more complex with respect to the circuitry implementing an FFE–DFE and it requires a much more complex design. This is going to influence prices of MLSE equalizers, at least up to the moment in which they will be diffused in large volumes.

Moreover, if an FFE–DFE introduces slight modifications to the signal statistical distribution, the MLSE changes it more radically.

As a matter of fact, MLSE is a sort of convolution decoding and the surviving errors at the MLSE output have characteristic configurations depending on the MLSE structure [86]. This affects the performances of a traditional FEC. Optimum performances are achieved by a joined design of the MLSE and the FEC so that the interaction between the two algorithms takes into account the correct noise statistics and optimizes the performances of both FEC and MLSE [87]. At present this design is greatly constrained by memory requirements both in terms of memory quantity and memory-processor communication speed. However, it is possible that further progress in electronics will make this technology more accessible.

5.8.4 Pre-Equalization and Pre-Distortion Equalizers

A simple technique to avoid the intrinsic limitation of post equalization due to the presence of the nonlinear photodiode response is the implementation of electronic precompensation, that is, signal distortion in the transmitter before propagation.

This technique can be implemented, for example, by using an I/Q modulator, or a dual-drive Mach–Zehnder modulator.

The simpler scheme of signal pre-distortion that is meant for compensation of chromatic dispersion impairments is shown in Figure 5.63.

The principle of operation of this circuit is simple: the modulator is driven by two signals that are built in such a way that the modulator output coincides with the signal to be transmitted filtered by a filter whose spectral response is the inverse of the fiber dispersion spectral characteristics.

In general, this way of compensating linear distortion is prone to several mathematical difficulties related to the fact that the inverse frequency function can be impossible

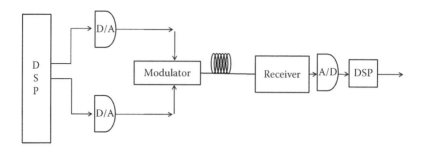

FIGURE 5.63
Block scheme of an electronic dispersion pre-compensator.

to define when the distorting filter has a big attenuation, so as to almost cancel the signal at those frequencies. Fortunately this is not the case with fiber dispersion, whose characteristic function is a pure phase filter, whose module is equal to 1 at all frequencies.

Thus, the driving currents for the modulator can be easily evaluated with simple calculations [88].

The pre-distortion filter implemented by the scheme of Figure 5.63 is quite effective in compensating large amounts of chromatic dispersion.

However, the pre-compensation technique allows the received pulse to be dispersion-free, but causes during propagation the broadening of the transmitted signals due to the combined effect of pre-distortion and dispersion.

This complex evolution, almost unavoidably, causes the apparition in some points along the link of complex structures with high maxima and deep depleted minima.

The high-power density maxima create a strong nonlinear effect, thus the pre-distorted signal is more prone to the nonlinear effect impairments with respect to a standard, non-compensated signal.

Nonlinear effects compensation can be operated in principle through the use of the so-called look-up table–based pre-distortion. This technique conceptually consists of the following steps:

- A physical model of propagation through the link is implemented in a DSP, based on a nonlinear propagation equation like Equation 4.39; this model depends on a set of parameters that are specific to the link under operation.

- At the beginning, in a set-up phase, the propagation model parameters are tuned by observing the link behavior when a set of known bit patterns is transmitted through the link. When a good estimate of the link parameters is reached they are used to fit the link behavior with the propagation algorithm.

- Once the algorithm is tuned on the real link characteristics it is used to construct the inverse of the propagation operator through virtual pulse back-propagation through the link.

- The inverse propagation operator is thus stored as a set of parameters in a look-up table that is used for fast pre-distortion during link operation.

- The link enters operation and the transmitted pulses are pre-distorted using the data stored in the look-up table.

- An adaptive algorithm adapts the look-up table parameters to the slow channel changes by repeating the link parameters identification periodically.

This method, based on back-propagation is also called electronic dynamically compensating transmission (EDCT).

A look-up table equalizer is quite effective, since its only limitation is the approximation implicit both in the propagation model and in the representation of the inverse propagation operator with a discrete set of numbers in a look-up table. Unfortunately, its direct application is at present difficult mainly due to the memory requirements that are difficult to satisfy with signal processors.

Thus, in order to exploit the idea of the look-up table algorithm, a mixed method is often proposed.

This scheme, represented in Figure 5.64, combines the look-up table technique with a further compensation through digital filtering [89]. The input bit sequence to be transmitted is used to address a look-up table, whose outputs values $S_I(j)$ and $S_q(j)$ correspond to the real and imaginary parts of the waveform suitably distorted to compensate nonlinear effects. These values are input to a set of digital filters with transfer functions $h_{Re}(j)$ and $h_{Im}(j)$. The resulting in-phase and quadrature drive voltages d_1 and d_2 are then obtained by complex-summing the outputs of the filters as shown in the figure.

The equalizer resulting from the composition of digital filters and look-up table is potentially very efficient, requiring a much smaller memory table with respect to a direct application of the look-up table method.

Nevertheless, the great potentialities of this equalization method are not free from challenges that have to be solved for a practical implementation.

The first challenge is the complexity of the electronic circuitry with respect to a simple FFE–DFE post-processing. In the mixed solution represented in Figure 5.64, the set of linear filters consume alone about two times a traditional FFE–DFE filter. In addition, there is the look-up table. Finally there is the real-time adaptive algorithm that is needed to adapt equalization to the channel fluctuations.

The implementation of this algorithm is another challenge: if all the equalizer parameters have to be adapted to channel variations, the adaptive algorithm results in being quite complicated and a lot of design work is needed to implement it in a practical circuit.

Last, but not least, the adaptive algorithm needs a return channel from the receiver to the transmitter to drive transmitter-adaptive evolution with the received pulse shape. Thus, the receiver also must sample at a very high speed the incoming pulses to feed the transmission-adaptive procedure.

A complete study of the back-propagation algorithm in case of the adoption of a scalar propagation model and of the efficiency and complexity of this method is reported in [90].

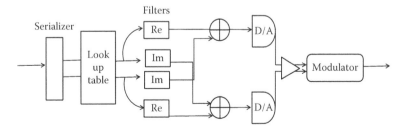

FIGURE 5.64
Scheme of an electronic pre-compensator that combines the look-up table technique with a further compensation through digital filtering. (After Killey, R.I. et al., Electronic dispersion compensation by signal predistortion, *Proceedings of Optical Fiber Conference,* OSA, s.l., 2006.)

Starting from the propagation equation (Equation 4.39) it is demonstrated that the reverse propagation is obtained just from the forward propagation equation with negative parameters.

In this way, the signal attenuation is transformed into a gain and the operators working on the phase of the signal induce a rotation into the opposite direction.

Thus, to obtain the input signal corresponding to a certain output (i.e., the unperturbed signal) it is sufficient to "invert" the parameters of the propagation equation and to apply the split-step Fourier method, the same that is used to simulate fiber propagation in normal conditions [91].

From a processing complexity point of view, the signal processing block at the transmitter is one of the critical components.

In a practical system this block can be either implemented as an ASIC or by using a DSP or an FPGA. It is worth noting that in this case the ASIC solution could prevail even if it is more expensive and time-consuming due to its optimization, which could allow the consumed power to be reduced reentering within the system limits while achieving higher processing speed.

If the pulse distortion is mainly caused by linear propagation and nonlinear effects are negligible, the complexity of this device is mainly driven by the number of complex taps necessary for the FIR filter to compensate for the chromatic dispersion.

Several studies have been done to optimize the number of taps that can be adopted in this case [92] and it is demonstrated that the complexity scales linearly with the amount of accumulated dispersion to be compensated. It was shown that the complexity grows quadratically with an increasing bit rate [90].

In the presence of substantial nonlinear effects, nonlinear filtering has to be used. When a look-up table is used for filtering, the number of entries is determined by the number ζ of symbols that significantly interfere with each other. In particular, the number of look-up table entries grows exponentially according to Θ^ζ, where Θ is the number of bits per symbol used in the signal sampling by the algorithm.

Regarding the channel memory, this is mainly due to dispersion; when dispersion is compensated optically and only intra-channel nonlinearity is compensated by EDCT, the number of entries remains relatively small.

However, for the joint compensation of chromatic dispersion and intra-channel nonlinearity the number of states can get rapidly impractical, especially beyond 10 Gbit/s. This does not only lead to increased storage requirements but it also increases the effort that needs to be spent on computation of pre-distorted waveforms.

In spite of all these challenges, the evolution of electronics technology for signal processing is rapidly eliminating the gap between the algorithm needs and the processing capabilities and similar methods are starting to be implemented in real products [93].

5.8.5 Forward Error Correction

The use of FEC is really instrumental in designing high-performance WDM systems, but has a much wider application field, comprising consumer electronics like PCs (in the transmission between the hard disc and the motherboard).

5.8.5.1 FEC Definition and Functionalities

An FEC is from a functional point of view an extension of the well-known error-detecting code that is able to correct the channel-induced errors if the error rate is within the correction ability of the code [94].

We will consider in this paragraph systematic block group codes without memory.

- A code is a block code if the uncoded message is divided at the encoder in words of the same length and every word is processed by the encoder so as to obtain a longer word (called the code word).

  Let us assume that the encoder receives a message constituted by a string of symbols extracted from the message alphabet $\mathcal{M}$, a collection of M symbols, that is, $\mathcal{M} \equiv \{\chi_j, j = 1, \ldots, M\}$. We will assume that the code word is composed of the same symbols of the message word. Thus the encoder transforms message words of m symbols into code words of d symbols all from the same alphabet $\mathcal{M}$ the difference $s = d - m$ is called code redundancy.

  Due to the encoding the line rate R_{Line} results in being greater than the message rate R_{mess}, being $R_{line} = (1 + (s/d)R_{mess})$.

- A block code is systematic when the code word is obtained by adding to the message word a string of s symbols calculated by the encoder, but the message word is transmitted unchanged.

- A block code is said to be without memory when every code word is created at the encoder independently from the previous words and is processed at the decoder also independently from the previous words. Every possible channel memory is considered contained in the length of the message word.

- Finally a block code is called a group code when its coding and decoding algorithms are based on the fact that the alphabet $\mathcal{M}$ is structured as a group by defining suitable operations among its members. We will always consider codes built on alphabets that are at least rings (which have the structure of the integer numbers). In the most common cases in optical communications the alphabet structure is even richer, that is, it has the Galois field structure, which we will call $\mathcal{G}(M)$. This is the structure of the real numbers, but for the fact that the alphabet is constituted by a finite number of elements, differently from the real numbers.

Practically all the FECs used in optical communications have the properties considered in the aforementioned list, due to the fact that the encoding and decoding algorithms have to be performed in real time in pace with the arriving bit stream, whose speed is a minimum of 10 Gbit/s if not 40 or 100 [95–98]. This trend, however, is going to change with the need of more efficient codes for 100 Gbit/s transmission; at the end of this section we will look at one of the possible solutions for the next-generation codes: the turbo block codes.

A block code can correct errors since not all the words of d symbols are possible code words, but only a subset of M^m words are acceptable. Thus, when an unacceptable word is detected, errors are revealed and, starting from the received word, an algorithm exists to estimate the transmitted word.

The number of nonzero symbols in the difference between two words of d symbols in $\mathcal{G}(M)$ is called distance between the words and it is possible to verify that it has all the properties of a well-defined distance. The minimum distance between two code words is called Hamming distance of the code and it is quite clear that the upper bound of the number of errors that the code can correct is exactly the Hamming distance. As a matter of fact, changing a number of symbols equal to the Hamming distance is possible to change a code word into another code word, creating an error that is impossible to correct.

The Galois fields that are used in the block codes commonly used in optical transmission are the fields constituted by the integer numbers module for a given field-generating number $\wp$ which has to be a prime or the power of a prime; such fields are called $\mathcal{G}(\wp)$. As an example, $\mathcal{G}(5) = \{0, 1, 2, 3, 4\}$, where all the numbers have to be an intended module 5, so that, within $\mathcal{G}(5)$, $4 + 3 = 2 \bmod(5)$.

The symbols of the alphabet $\mathcal{M}$ are considered elements of $\mathcal{G}(M)$ so as to be able to perform operations with them. On the grounds of this definition, every word is represented with a polynomial in $\mathcal{G}(M)$, associating to the finite sequence of symbols $\{\chi_j, j = 1, \ldots, d\}$ the polynomial $\mathcal{P}(y) = \sum_{j=0}^{d-1} \chi_{j+1} y^j$.

With this definition, encoding and decoding algorithms can be structured as an operation with polynomials on $\mathcal{G}(M)$, the type of operation that a dedicated logic like that of an ASIC or of a DSP executes very fast.

Different codes can be defined depending on the encoding and decoding algorithms; each code is characterized by a correction capability (number of independent errors the code can correct in a received word), a detection capability (number of errors the code can detect, but not correct, in a received code word), and a degree of complexity of the encode and decode algorithms.

5.8.5.2 BCH and the Reed–Solomon Codes

The category of codes that are more effective in providing a good error correction capability with simple implementation algorithms are the Bose–Chaudhuri–Hocquenghem (BCH) codes (named after their inventors). Reed–Solomon (RS) codes, even if historically invented independently, are a category of BCH codes and are the most used in high-speed transmission.

The encoding algorithm of BCH codes is based on the multiplication of the mth degree polynomial representing the message word by a generating degree δ polynomial $g(x)$ so as to derive the *d*th degree polynomial representing the code word.

The encoder can thus be implemented as a very simple linear feedback shift register (LFSR) as shown in Figure 5.65.

The characteristics of a particular BCH code depend on the choice of the generating polynomial.

To describe the way in which the generating polynomial is constructed, let us fix a generating prime number $\wp$, which is introduced to construct the code in $\mathcal{G}(\wp^m)$. Naturally in many cases of interest $\wp = 2$.

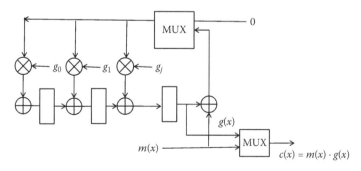

FIGURE 5.65
Bock scheme of a BCH encoder.

To search a code with a given Hamming distance, start to choose two positive integers d and c such that d and $\wp$ are reciprocally prime in $G(\wp^m)$ (their unique common divisor is 1) and m is the multiplicative order of $\wp$ modulo d.

To construct the generating polynomial it is needed to find the elements $(a^{c+j}, j=1,\ldots,\delta)$ where a is a primitive dth root of unit in $G(\wp^m)$. In correspondence of each of the a^{c+j} it is possible to find a minimal polynomial over $G(\wp)$, that is the minimum degree polynomial in the considered field with leading coefficient equal to 1 and having a^{c+j} as radix; let us call it $\mu_j(\chi)$.

The BCH code-generating polynomial is given by $g(x) = \text{lcm}(\mu_j(\chi), j=1,\ldots,\delta)$, where $\text{lcm}()$ indicates the lowest common multiple.

From the definitions, we can also say that a given word of d symbols χ_j in $\Gamma(\wp^m)$ is a code word if and only if the polynomial $P(x) = \sum_{j=0}^{d-1} \chi_{j+1} x^j$ has all the elements a^{c+j} as radix.

The construction we have described assures to the BCH code a minimum Hamming distance equal to δ and the ability to correct up to δ independent errors for each code word.

RS codes are particular BCH codes, where $c=0$ and the generating polynomial is given by

$$g(x) = \prod_{(j=1)}^{\delta} (x - a^j)\grave{u}$$

(5.48)

As far as decoding is concerned, there are many algorithms to decode BCH codes; we will describe the most used here, due to its easy implementation at high speed.

At the decoder the received word composed by d symbols is considered as a polynomial generated by the channel, adding to the sent polynomial $t(\chi)$ an error polynomial $e(\chi)$ so that correcting errors means determining $e(\chi)$ and subtracting it.

The algorithm steps are essentially three:

1. Compute the δ syndromes ξ_j of the received word, which are defined as $\xi_j = t(a^{c+j})$. It is evident that if all the syndromes are zero the word is a code word and it is considered correct. If not, there are errors to correct.

2. Starting from the syndromes find the error polynomial $e(\chi)$.

3. Reconstruct the correct code word as $c(x) = t(\chi) - e(\chi)$.

The second step, that is the core of the algorithm, is not obvious, and it is not within the scope of this section to detail a procedure to determine the error polynomial starting from the syndromes.

There are two main algorithms to do that: the Peterson–Gorenstein–Zierler algorithm and the Berlekamp–Massey algorithm, both based on a set of linear operations on matrixes and vectors in the space $G(\wp^m)$.

Both the algorithms for their nature are well adapted to be implemented in a dedicated signal-processing circuit, and this is the reason for the wide adoption especially of the RS codes, for which the algorithm can be further simplified.

In general, the BCH decoder has the bock scheme represented in Figure 5.59.

5.8.5.3 Turbo Codes

The BCH codes, and the RS that are a particular class of BCH, guarantee a simple coding/decoding algorithm that is suitable for application at very high speed, but are far from the optimality since the capacity of a channel using BCH codes is far from the Shannon limit [94], even if very long code words are selected.

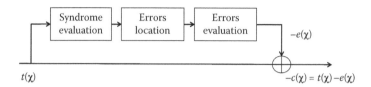

FIGURE 5.66
Principle scheme of a BCH decoder.

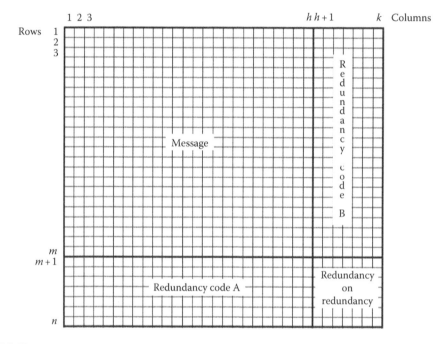

FIGURE 5.67
Composition of the code word of a systematic turbo block code.

This indicates that much better codes must exist, guaranteeing a higher number of corrected errors for a certain redundancy and code word length.

A class of better codes that is also suitable for application at high speed is constituted by the so-called turbo codes [99,100].

Turbo codes are a class of codes that construct the code word by concatenating two codes separated by an interleaver (see Figure 5.66).

The turbo code word, when the component codes are systematic, can be arranged in a matrix, whose rows are transmitted in a determined order, as shown in Figure 5.67, where it is clear that redundancy is added not only to check errors in the message, but also to check errors in the redundancy introduced by the composing codes.

As a result of the aforementioned definition, the encoder can be built from the encoders of the constituting codes so that the encoder of a systematic turbo code will appear as in Figure 5.68.

Starting from its very general definition, there are several classes of turbo codes: with and without memory (convolution or block codes), systematic or not (i.e., such that the redundancy is separated by the original message at the output or not).

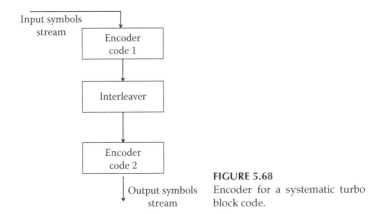

FIGURE 5.68
Encoder for a systematic turbo block code.

Since we target codes that can be applied in optical systems at a very high speed, we limit our attention to block turbo codes based on two equal BCH components, which results naturally as systematic codes.

The choice of the interleaver is a crucial part in the turbo code design. The task of the interleaver is to scramble the symbols in a predetermined way. The presence of the interleaver has two beneficial effects on the code.

The presence of code words with a high number of null symbols (in case of the BCH codes the zero of the Galois field) weakens the code, since in general they are separated by a smaller distance. From this it derives the general property that a good code has the highest possible average weight, where the code weight is defined as the average of the weight of the code words, that is, the average of nonzero symbols in the code words.

Now, in a turbo code, if the first encoder produces a low-weight word, it is improbable that the second also will do the same after scrambling. Generally it can be seen that scrambling increases the average weight of the code word, while decreasing at the same time its variability.

A second advantage of scrambling is constituted by the fact that decreasing the correlation between the two coded words that are transmitted increases the information that is sent to the decoder, thus allowing more effective decoders to be designed.

Ideally both these advantages are maximized by using a random scrambler, but it is difficult to realize at the high speeds typical of optical systems.

In Figure 5.69 the workings of a few possible interleaver designs are shown for the cases in which the sequences to be interleaved are binary [101].

Row–column interleaver: In this type of interleaver the data are written row-wise and read column-wise. This is a very simple design, adapted for use at very high speed, but it also provides little randomness.

Helical interleaver: In this type of interleaver, data are written row-wise and read diagonally.

Odd–even interleaver: In this encoder architecture, encoding happens before interleaving, and after encoding only the odd-positioned coded bits are stored. Then, the bits are scrambled and encoded again, but now only the even-positioned coded bits are stored.

The idea at the base of the decoder of turbo codes is that the signal arriving at the receiver is composed of two words encoded with the same code where the symbols are in a different order.

It is intuitive, even if it is not so easy to demonstrate, that the quantity of information the decoder has on the sent word is much greater than what it would have if only one code were used.

The input word is organized like a matrix	x_1	x_2	x_3	x_4	x_5
	x_6	x_7	x_8	x_9	x_{10}
	x_{11}	x_{12}	x_{13}	x_{14}	x_{15}

Output of the line-row interleaver														
x_1	x_6	x_{11}	x_2	x_7	x_{12}	x_3	x_8	x_{13}	x_4	x_9	x_{14}	x_5	x_{10}	x_{15}

Output of the helical interleaver														
x_{11}	x_7	x_3	x_{14}	x_{10}	x_1	x_{12}	x_8	x_4	x_{15}	x_6	x_2	x_{13}	x_9	x_5

Even-odd interleaver working														
Encoded word without interleaver														
x_1	x_2	x_3	x_4	x_5	x_6	x_7	x_8	x_9	x_{10}	x_{11}	x_{12}	x_{13}	x_{14}	x_{15}
y_1	—	y_3	—	y_5	—	y_7	—	y_9	—	y_{11}	—	y_{13}	—	y_{15}
Encoded word with row–column interleaver														
x_1	x_6	x_{11}	x_2	x_7	x_{12}	x_3	x_8	x_{13}	x_4	x_9	x_{14}	x_5	x_{10}	x_{15}
—	z_6	—	z_2	—	z_{12}	—	z_8	—	z_4	—	z_{14}	—	z_{10}	—
Final code word after encoding and interleaving														
y_1	z_6	y_3	z_2	y	z_{12}	y_7	z_8	y_9	z_4	y_{11}	z_{14}	y_{13}	z_{10}	y_{15}

FIGURE 5.69
Scheme of the operation of a few possible interleaver algorithms.

In this situation, the generic decoder of a turbo code is a sort of soft decoder having at the input the row samples of the received data and constituted by two different decoders, one for each component code.

The decoders exchange information about the received word and at the end, if no errors happen, both produce the same output symbol stream.

Naturally, if the two decoders produce in a certain step two different outputs, they detect it by exchanging data, and interpret it as the presence of errors starting a joined procedure to determine the nature of the error and to correct it.

All this procedure is much more complicated than the decoding algorithm of BCH codes, but nevertheless can be carried out fast using dedicated logic, like a very fast DSP or FPGA or more probably an ASIC.

There is a strong research effort to study properties and implementation of turbo code decoders and the interested reader can start from the following bibliographies [102,103].

5.8.5.4 ITU-T OTN Standard and Advanced FEC

The use of a FEC implies that besides the signal to be transmitted, a redundancy is sent to be used by the decoder to correct decision errors or to perform optimum soft decision.

Since this redundancy has to be placed within the transmitted frame, it is natural to compare the requirement of the most performing codes that are suitable for optical transmission application with the ITU-T frame standard for optical transmission.

Currently, ITU-T has standardized the use of codes with 7% redundancy (exactly 16 bit redundancy and 255 bits overall length), as recommended in ITU-T G.975 and its successor ITU-T G.975.1. A 7% code is applied directly to blocks of binary information bits, producing an output sequence, whose length is expanded by a factor of 255/239, and the resulting bit stream is transmitted using a binary modulation format. That is, redundancy is accommodated by sending more symbols.

The research on transmission systems, however, is going toward a different direction.

As we will see in Chapter 9, in order to transmit speeds higher than 40 Gbit/s binary modulation is not suitable and multilevel coding is a very attractive solution. When multilevel coding is used, redundancy can be either transmitted by increasing the symbol rate or by increasing the number of symbols. For example, if a 16-level transmission is used, redundancy can be accommodated by passing to 32 levels. This means that one redundant bit is transmitted every three message bits.

As far as specific codes are concerned, the RS (255, 239) code was the first to see widespread application in fiber-optic communication systems, due to its high rate, good performance, and efficient hard-decision decoding algorithm.

As the speed of electronics increased, it became practical to consider more powerful error-correcting codes. The ITU-T G.975.1 standard describes several candidate codes providing better correction performances, all with a 16/255 structure in order to be accommodated into the Optical Transport Network (OTN) frame in place of the RS (255, 239).

In all the considered cases, hard-decision decoding is applied, and improved performance can be partially attributed to the significantly longer block lengths considered.

Due to the period in which this standard has been developed, no soft decision and no block turbo codes have been considered.

Even if the optical community has, so far, standardized FEC at 7% redundancy, this choice is not at all fundamental. The 7% value certainly reflects a reasonable choice of code rate, given the manner in which optical communication systems have been operated to date.

Due to the high SN of optical channels, the uncoded symbol error rate for binary signaling schemes is small. For the RS (255, 239) code, communication with bit error rates less than 10^{-12} requires an uncoded error rate (or "FEC threshold") of less than 1.5×10^{-4}.

Since optical systems have traditionally been operated to achieve such low pre-FEC error rates, a high-rate code was sufficient to provide sizeable coding gains, and 7% redundancy provided a reasonable balance between bit rate and error-correcting capability.

From an information theoretic perspective, however, mandating 7% redundancy unnecessarily restricts the pre-FEC symbol error rate to be small.

When multilevel transmission will come into play, that seems a must for transmission at 100 Gbit/s; this implies that the used multilevel constellation has to include a small number of points so as not to cause too big an SN penalty (see Chapter 9).

To increase the spectral efficiency of fiber-optic communication systems (e.g., via multilevel modulation), it is necessary to design the coded modulation scheme using information theoretic considerations. Maximizing spectral efficiency may, in fact, lead to operating points in which the "raw" error rate requires the use of codes with greater than 7% redundancy.

In view of the next-generation systems, this discussion will be opened inside the ITU-T and for sure, if effective proposals will emerge, standardization will include them into the standard.

The opportunity to do so will be standardization of multilevel optical systems, which are all to be developed since only binary systems are covered by today's ITU-T recommendations.

Due to these considerations, in all our examples, we will not limit our self to consider 7% redundancy codes, even when binary systems are considered.

5.8.5.5 FEC Performances

A great effort of research has been devoted to determine the performances of various FECs in terms of SN gain. In the case of optical systems, the results obtained in the FEC-related literature have to be elaborated due to the particular form of the optical receiver.

As a matter of fact, we will see in Chapter 6 that the most important application of FEC is in long-haul and ultra-long-haul systems. In this case the error probability depends on the optical signal to noise ratio (SN_o), which is in turn proportional to the optical received power. In general the FEC performances are provided in terms of gain on the SN immediately before decision, which in the optical case is the electrical signal to noise ratio (SN_e).

In order to evaluate the performances in terms of optical gain, we have to link the SN_o with the SN_e using the receiver parameters.

Even if this relationship can be derived in a general way, this is not so significant. The most important cases are the two extreme cases in which there is no optical noise and the receiver noise plus shot noise dominates and the case in which the optical noise dominates.

As detailed in Chapter 6, these are the cases of unamplified systems and optically amplified systems.

In the first case, assuming as always in practice that the received optical signal can be modeled as a time sequence of coherent states, the overall noise after detection is equal to the variance of the shot noise plus the electronic noise (see Section 5.4).

Thus the electrical SN can be written as

$$SN_e = \frac{R_p^2 P_o^2}{2\hbar\omega B_e + F_a\left(4k_B\Im/R_L\right)B_e} \tag{5.49}$$

as obviously the SN_e depends on the square of the optical power; thus a gain of X dB in SN_e results in a gain of $X/2$ dB in terms of optical received power.

In the case of dominant optical noise, the SN_o can be written in general as

$$SN_o = \frac{P_o}{P_{ASE}} \tag{5.50}$$

where P_{ASE} is the ASE power in the optical bandwidth.

On the other hand, the SN_e has to be evaluated taking into account the quadratic detection law of the photodiode and neglecting shot noise and receiver noise.

In evaluating SN_e we do not neglect the quadratic term in the noise, due to the fact that in the presence of FEC sometimes it is possible to work with a small SN_e and the quadratic term in the noise can be relevant in situations in which errors are possible.

The quadratic term power, after electrical filtering, is evaluated assuming a flat optical noise spectral density in the band B_o; thus after square law detection the power spectral density of the term containing the square of the noise is a triangle with base extending from $-2B_o$ to $2B_o$. This triangular power spectral density is filtered by an almost ideal electrical filter of unilateral bandwidth B_e.

The receiver model with the definition of the relevant wavelength and frequency bandwidths is represented for clarity in Figure 5.70. The conventions in terms of bandwidths and SN definition that are shown in the figure will be maintained for the remaining part of the book.

The ASE noise is considered a white Gaussian noise with power spectral density Σ_0 so that the ASE power in the optical bandwidth, with the definitions of Figure 5.61, could be defined as $P_{ASE} = 2\Sigma_0 B_o$. Since, in order to have a parameter tailored for the application we will study in Chapter 6, we have defined B_o as unilateral bandwidth, it is convenient to define a unilateral ASE spectrum; let us call it $S_0 = 2\Sigma_0$ so that the ASE power can be written as $P_{ASE} = S_0 B_o$.

The fact that this definition is useful to write simply the equations immediately results from the case of a receiver using a single pre-amplifier.

In the case of a single amplifier, considering the expression of the ASE spectral density deduced in Chapter 4, we have, B_o being unilateral,

$$P_{ASE} = \hbar\omega_c(G-1)B_oNF \xrightarrow{\text{yields}} S_0 = \hbar\omega_c(G-1)NF \tag{5.51}$$

where
  ω_c is the carrier angular frequency
  NF is the amplifier noise factor

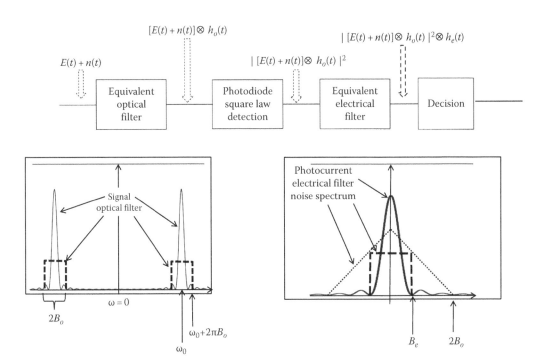

FIGURE 5.70
Definition of the formal receiver model used for the FEC performance evaluation and of the related bandwidths.

and the unilateral power spectral density S_0 does not present any factor beyond the terms related to amplifier noise

On these grounds, the power of the electrical noise current corresponding to the optical noise is easily evaluated as

$$\sigma^2 = \left\langle \left[n^*(t) + E_r^*(t)n(t) + n(t)n^*(t) \right]^2 \right\rangle = 2P_oS_0B_e + \frac{S_0^2}{2}B_e(4B_o - B_e) \qquad (5.52)$$

where

$E_r(t)$ is the received field average

$n(t)$ is the optical noise process whose average is supposed to be zero

P_o, S_0 are the signal power and the noise power spectral density, respectively

The electrical bandwidth is called B_e and it is often approximated with the symbol rate R_s and the optical bandwidth (unilateral as defined in Figure 5.69) is called B_o.

Let us introduce the parameter $M_{oe} = B_o/B_e > 0.5$. It is the ratio between the electrical and the optical bandwidths, thus it is also the number of samples at the Nyquist sampling rate of the optical signal that fit into the Nyquist sampling interval. This parameter will be quite useful in the analysis of transmission systems.

As a function of M_{oe} the electrical SN may be written as

$$SN_e = \frac{P_o^2}{2P_oS_0B_e + (S_0^2/2)\left(4B_oB_e - B_e^2\right)} = \frac{SN_o^2M_{oe}}{2SN_o + \left(2 - 1/2M_{oe}\right)} \approx \frac{SN_oM_{oe}}{2} \qquad (5.53)$$

where the approximation is valid in conditions of large SN, so that it can be written as $SN_o \gg 1$. In this case the SNs are equal if the electrical bandwidth is equal to half the unilateral optical bandwidth.

This relationship is plotted in Figure 5.71 for different values of M_{oe} and it is the relation that has to be used to evaluate the code gain in systems with Gaussian noise.

It is to be noted that Figure 5.71 is plotted with SN_o as an independent variable; decreasing M_{oe} at constant SN_o means that, for a constant value of the optical bandwidth, the electrical bandwidth shrinks so that less optical noise power is detected at the decision level. This explains why in the plot the SN_e increases, decreasing the electrical bandwidth.

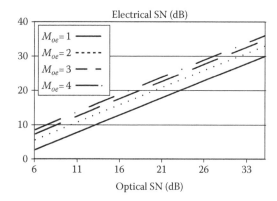

FIGURE 5.71

Relation between the optical signal to noise ratio (SN_o) and the electrical signal to noise ratio (SN_e) for different values of the ratio M_{oe} between the optical bandwidth and the electrical unilateral bandwidth (B_o/B_e).

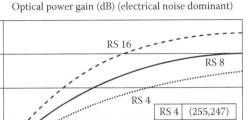

Optical power gain (dB) (electrical noise dominant)

RS 4	(255,247)
RS 8	(255,239)
RS 16	(255,223)

Log$_{10}$(BER) after FEC operation

FIGURE 5.72
SN_o gain of three different RS codes in an unamplified system.

In Figure 5.72 the gain of a few RS codes is plotted versus the target value of the BER (logarithm of the target BER after FEC operation) when no optical noise is present. The greater the code overheard, the greater is the code gain, even if in this case there is a modest gain due to the fact that SN_e depends on the square of the optical power while the noise is added after square law detection.

The FEC gain curve is shown in Figure 5.73 for the case in which the optical noise is dominant. In this case the FEC allows higher gains to be achieved. It is to be underlined that the FEC gain decreases at the increase of M_{oe}. This is due to the fact that in this case the plot is made maintaining constant the BER, and thus the SN_e.

Inverting (5.53) we obtain

$$SN_o = \frac{SN_e}{M_{oe}}\left[1 + \sqrt{1 + \frac{1}{SN_e}\left(2 - \frac{1}{2M_{oe}}\right)}\right]$$

(5.54)

Thus it is evident that increasing M_{oe} while SN_e is constant means decreasing SN_o, making clear the behavior of the curves of Figure 5.72.

From Figure 5.72 it is also derived that in the reference case $M_{oe}=2$, when the standard code RS (233, 249) (indicated with RS8 in this Figure) is used to reach a BER of 10^{-12}, a coding gain of 6 dB is attained.

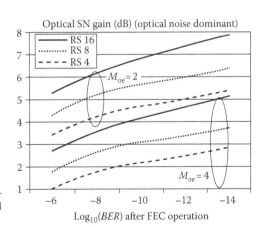

Optical SN gain (dB) (optical noise dominant)

——	RS 16
·········	RS 8
‐ ‐ ‐	RS 4

$M_{oe}= 2$

$M_{oe}= 4$

Log$_{10}$(BER) after FEC operation

FIGURE 5.73
SN_o gain of three different RS codes in a system with in-line amplifiers so that the performances are dominated by the optical noise. RS (233, 249) is indicated with RS8.

Another useful manner to evaluate the effectiveness of an FEC that does not need to take into account the receiver parameters is to relate the error probability that would characterize the system without FEC with the error probability of the system with FEC. Generally these error probabilities are called pre-FEC and post-FEC BER.

Once the pre-FEC BER is determined the entire system is designed so as to obtain the wanted pre-FEC BER. After the addition of the FEC, the real error probability is determined by the pre-FEC BER and by the FEC performances.

The relation between the pre-FEC and the post-FEC BER is shown in Figure 5.73 for a few BCH codes.

As far as turbo codes are concerned, several codes have been constructed with encoder and decoder and analyzed.

An example is given in Figure 5.74, where the code gain in terms of SN_e is shown for three block turbo codes built on RS codes [104]. In particular, the following component codes are selected:

- Code A: built on RS (15, 12), redundancy 36%, Hamming distance 4
- Code B: built on RS (31, 26), redundancy 30%, Hamming distance 6
- Code C: built on RS (63, 58), redundancy 15%, Hamming distance 6

From the figure it is evident that the code gain is very high, also for the high redundancy that these codes exhibit, and in particular a code gain a little below 10 dB can be achieved at BER $= 10^{-12}$.

The last observation to be made on the FEC performances is that we have assumed up to now a Gaussian distribution of the decision samples.

In reality we will see in Chapter 6 that this is not true: due to the square law detector, when the dominant noise is the ASE noise, the decision samples do not have a Gaussian distribution.

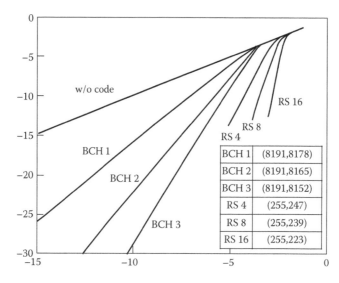

FIGURE 5.74
Relationship between pre-FEC BER and post-FEC BER for various BCH and RS codes.

It can be demonstrated that, taking into account the correct distribution of the electrical equivalent noise, the performances of the FEC gets better, less for BCH codes, while sometimes a more significant difference occurs in turbo codes [100].

This observation allows us to set the target for an advanced FEC to use in ultra-high-speed DWDM systems, to 11 dB of code gain.

5.8.6 Content Addressable Memories

As equalization and error correction is instrumental for high-capacity transmission, fast packet routing is fundamental to build the IP convergent network.

As detailed in Chapter 7, IP routers and carrier-class Ethernet switches have to compare the address of each incoming packet with the entries of a routing table, whose content indicates the output port where the packet has to be sent.

In the IP case, the operation is worsened by the fact that the so-called forwarding table does not contain whole addresses, but only prefixes, and routing is performed by searching the longest prefix matching the incoming IP address (see Chapter 3).

Since this operation has to be performed several million times a second, it is natural that in large packet switches it is carried out by specialized hardware whose core is generally a content-addressable memory (CAM) [105].

CAM is a derivation of the random access memory (RAM). In addition to the conventional READ and WRITE operations typical of RAMs, CAMs also support the SEARCH operations. A CAM stores a number of data words and compares a search key with all the stored entries in parallel. If a match is found, the corresponding memory location is retrieved. In the presence of multiple matches, either several addresses are returned or a priority encoder determines the highest priority match on the grounds of a matching priority policy (Figure 5.75).

CAM-based table look-up is very fast due to the parallel nature of the SEARCH operation. A schematic illustration of the CAM working is shown in Figure 5.76.

CAMs can be divided into two categories: binary CAMs and ternary content-addressable memories (TCAMs). A binary CAM can store and search binary words (made of "0"s and "1"s). Thus, binary CAMs are suitable for applications that require only exact-match searches. A more powerful and feature-rich TCAM can store and search ternary states ("1," "0," and "X"). The state "X", also called "mask" or "don't care," can be used as a wildcard entry to perform partial matching.

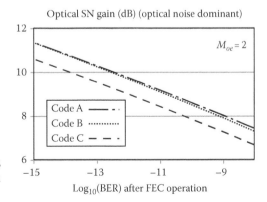

FIGURE 5.75
Coding gain for a few block turbo codes based on RS code components. The code characteristics are listed in the text.

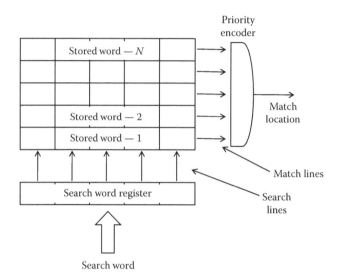

FIGURE 5.76
CAM functional scheme.

A typical CAM cell utilizing 10 transistors all in CMOS configuration (a 10 T cell) is shown in Figure 5.77. The bottom part of the figure (transistors from T1 to T6) replicates exactly the structure of an SRAM cell and is used to perform READ and WRITE operations.

The transistors from T7 to T10 implement the XNOR logic that is needed during the SEARCH to compare the table entry with the search key.

As in a SRAM, The WRITE operation is performed by placing the data on the bit lines (*Bs*) and enabling the word line. This turns on the access CMOS (T1–T2), and the internal nodes of the cross-coupled inverters are written by the data. The READ operation is performed by pre-charging the *Bs* to the power supply voltage V_D and enabling the word line.

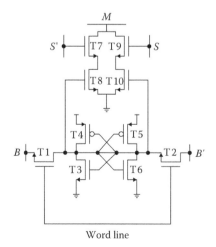

FIGURE 5.77
Circuit representation of a CAM cell.

The conventional SEARCH operation is performed in steps:

- Search lines (S and S') are reset to ground.
- M is pre-charged to V_D.
- The search key bit and its complementary value are placed on S and S'.

If the search key bit is identical to the stored value ($S = B$, $S' = B'$), both M-to-ground pull-down paths remain "OFF," and the M remains at V_D indicating a "match."

Otherwise, if the search key bit is different from the stored value, one of the pull-down paths conducts and discharges M-to-ground indicating a "mismatch."

A TCAM cell is conceptually similar to the CAM cell, but it needs two SRAM cells to be able to store ternary data.

Several variations have been proposed for the base cell architecture that is shown in Figure 5.77, all with the intent to lower the voltage and power consumption and/or to increase the cell speed [106–108], and at present every practical CAM and TCAM implementation, even if it relies on the same base principle, has its own design particularity.

A CAM or a TCAM is realized by a matrix of cells, not so differently from an SRAM. The matrix lines represent the stored words, sharing the M voltage terminal that is connected with the so-called M sense amplifier, which detects the presence or the absence of a match with the search word that is stored in a suitable registry. This architecture is shown in Figure 5.78.

Besides the pure dedicated hardware implementation that we have briefly described, the progress in terms of density of CMOS circuits have recently allowed the CAM and TCAM implementation via FPGAs using suitably optimized libraries [109].

In terms of performances, the main CAM parameters are

- Search speed (measured in terms of msps, or mega searches per second)
- The power consumption during the search operation

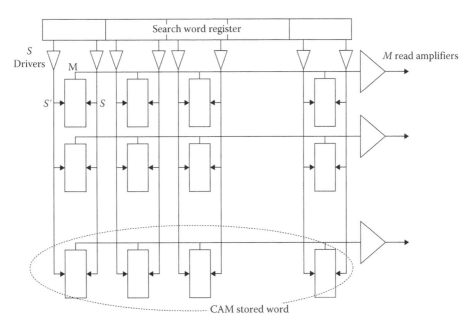

FIGURE 5.78
A complete CAM composed of an array of CAM cells.

- The dimension of the searchable area (traditionally the length of the CAM word is called width and the number of the word depth so that generally the capacity is called CAM area)
- Flexibility

As far as the speed is concerned, it is the search speed that is to be reported since in the main CAM and TCAM application, that is, router-forwarding engines and MAC-routing switches, the memory performs almost only search operations, being read and written only when the routing table is updated.

Flexibility indicates the possibility to use the CAM to perform different types of operations. Commercial CAMs frequently can be configured either as real CAM or as TCAM depending on the activation of a mask flag. Another configurable feature is the priority policy, whose parameters can be set via a CAM control. Sometimes the priority mechanism can also be disabled so that all the found matches are reported at the output.

A CAM/TCAM realized with dedicated hardware and CMOS 65 nm technology can reach 100 Mbit capacity with a width comprising between 40 and 640 bits. With a suitable clock speed (around 360 MHz) the search speed can arrive at 360 Msearch/s, all in a processor-like package 30×30 mm.

Perhaps the most critical performance of large CAM/TCAM is the power consumption and much industrial effort is directed to reduce it. A TCAM having the performances summarized in the earlier paragraph can consume about 15 W. Taking into account that in order to design a core router packet-forwarding engine eight TCAMs could be needed, a power consumption up to 120 W can result.

A possible way to reduce power consumption is to cascade different pipelines in the CAM/TCAM architecture, as shown in [107]. In this case, maintaining a low latency, which means limiting the number of cascaded pipelines, an energy consumption as low as 7 W seems attainable.

Another possible approach is to use FPGAs to program the CAM/TCAM functionality. In principle, FPGAs are general-purpose chips, being thus less optimized with respect to a specific dedicated hardware.

In practice, the huge volumes and the large application domain of FPGAs have pushed a rapid improvement of performances via so good a design optimization that sometimes they are competitive with dedicated chips.

From a density and power consumption point of view, the last three generations of FPGAs have seen the density scale from 10 k gates to 300 k gates and the power consumption from 30 to 12 W (see Section 5.7.3) [110]. Therefore it is possible in some conditions to spare a few watts using FPGAs, with the advantage that they allow working with volumes smaller than that necessary to fully justify the ASIC development expenses.

5.9 Interface Modules and Transceivers

When optical transmission was born, optical systems were produced by vertically integrated companies, manufacturing in-house all the key components that should be integrated into the WDM system.

With the consolidation of the technology and the decrease of the margins due to the so-called bubble crisis, the companies in the optical communication market concentrated their effort on the core business, separating the activities in other areas.

Contemporary to this market evolution, technology was going on producing more and more classes of different devices for different applications.

Thus system vendors faced a double problem: first, it was difficult to develop in-house the electronic control of components that was manufactured outside; as a matter of fact, sometimes this is very hard without a direct knowledge of the mechanisms that are at the base of the component design (this is true, for example, for multi-section tunable lasers, several types of switches, and so on).

Moreover, in the absence of any standardization on the technology, every component has its own control and its own parameters; thus functionally equivalent components from different vendors cannot be exchanged, but each of them has to be driven by specific electronics.

This fact makes it impossible for system vendors to leverage on multiple suppliers to have better prices and manage the risk that one of them faces the impossibility to go on with the furniture.

In order to solve these two problems, system vendors started to ask components vendors for the so-called optical modules. These are subsystems that, in the simpler hypothesis, will contain a component with its own electronic control so that this subsystem can be directly mounted on the system card and be driven by the system controller through a digital interface with an appropriate language.

Even if single-component modules are quite popular as far as switches and similar components are concerned, in the case of the transmitting and receiving interfaces a further evolution has pushed the system and component vendors to define Multiple-Source Agreements (MSA) for more complex modules comprising both transmitter and receiver.

These optical interface modules can be divided into three classes:

1. MSA TX/RX modules
2. Fixed transceivers
3. Pluggable transceivers

The so-called MSA modules are generally DWDM-quality interfaces (even if gray and short-haul MSA also exist) that are designed to be directly integrated into the system card. They exit with two fiber pigtails and two connectors so that they can be connected with the transmission fibers.

In addition, fixed transceivers are modules that are mounted on the system card, but they are generally low-quality interfaces, designed more to occupy a small space and dissipate less power. Since a couple of years they have been completely substituted by pluggable transceivers.

Pluggable transceivers are interfaces, both WDM and gray, that are mounted on a package inserted into the hosting card from the front through a suitable housing while the system is running.

The transceiver is recognized and switched on by the system controller, all without interfering with the system working.

In this way, when a failure occurs, there is no need of changing the whole card, but it is possible to change only the transceiver.

Due to their small form factor, pluggable transceivers cannot host high-performance interfaces, which are still reserved for MSA modules, but miniaturization of lasers and electronics is going on and perhaps the day is not far when pluggable transceivers will substitute MSA modules as well.

5.9.1 MSA Transmitting–Receiving Modules

The MSA 300 PIN transponder is the standard interface for high-performance DWDM systems, especially if tunable lasers are required. It is standardized in two different dimensions: MMS (medium size) and SFF (small size).

The mechanical dimensions of the two standards with the positing on the package of the main features are depicted in Figures 5.79 and 5.80.

The name comes from the 300 PIN connector that is located on the lower part of the package and assures the contact of the transponder with the host card.

The functional scheme of a 300 PIN MSA transponder is sketched in Figure 5.81. From the figure it is evident that the scope of the transponder is to supply also the drivers to the system vendor, mainly for the laser and the modulator.

No signal processing is in general present inside the transponder: no FEC, no electronic compensator; the design of a few among these elements is perceived as a differentiator by the DWDM vendors so that it is not wanted inside a multivendor standard module.

However, it is possible that this will change at the moment in which the MSA 300 PIN at 10 Gbit/s will be in competition with the smaller XFP. At that moment it is possible that substantial signal processing will migrate inside the MSA 300 to differentiate it from the XFP.

The functional scheme evidences also the digital standard interface that MSA 300 PIN modules use to communicate with the motherboard: the I^2C interface [52]. This is a simple standard, comprising a layered description of the transmission system that allows the motherboard both to configure the module and to read alarms coming from it.

As far as the MSA 300 PIN at 40 Gbit/s is concerned, no alternative format is present, since transceivers as they are standardized at present are not able to host 40 Gbit/s components, essentially for a power consumption and for a mechanical dimension reason.

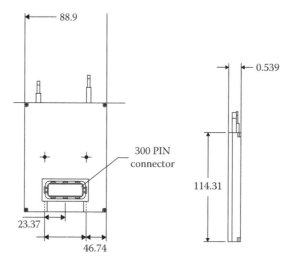

FIGURE 5.79
External dimensions of an MMF MSA 300 PIN transponder.

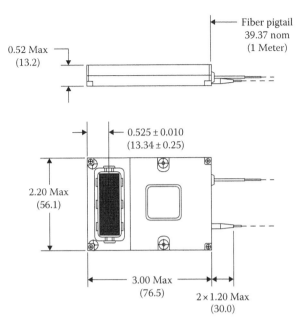

FIGURE 5.80
External dimensions of an SMF MSA 300 PIN transponder.

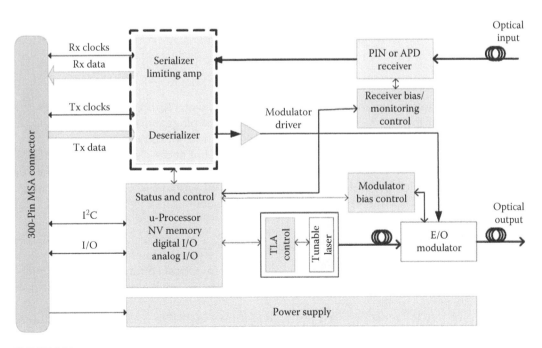

FIGURE 5.81
Functional scheme of an MSA 300 PIN transponder.

TABLE 5.10

Parameters of Practical MSA 300 PIN Transponders

	Transmitter		
Size	SMF	SMF	MMF
Type	Intermedium reach	Long reach	Very short reach
Tunability	No	Yes	No
Modulation	NRZ	NRZ	NRZ
Channel spacing	NA	50 GHz	NA
Embedded compensation mechanisms	No	Negative chirp	NO
Reach	40 km	80 km	2 km
Bit rate	10 Gbit/s	10 Gbit/s	40 Gbit/s
Output power	0.5 dBm	4.2 dBm	2 dBm
Extinction ratio	8.2 dB	9 dB	8 dB
Wavelength	1530–1570 nm (FS)	1530–1565 nm	1530–1570 nm (FS)
SMSR	30 dB	35 dB	30 dB
	Receiver		
Photodiode type	PIN	APD	PIN
Sensitivity	−16 dBm	−25 dBm	−5 dBm
Maximum received power	−1 dBm	−4 dBm	
Return loss	−37 dBm	−30 dBm	−27 dB
	General		
Operation temperature	−5°C to 70°C	−5°C to 70°C	−5°C to 70°C
Power consumption	6 W	8 W	17 W

In Table 5.10 an overview of practical performances of MSA 300 PIN modules is reported.

5.9.2 Transceivers for Carrier-Class Transmission

Pluggable transceivers were introduced initially in datacom equipment as an advanced version of fixed transceivers. However, the advantages of pluggability and the small form factor were so appealing that telecom transceivers were introduced very soon.

There are a plethora of standard transceiver formats, but at present only two are really used in the telecommunication environment, the others being either confined to datacom application or completely substituted by more advanced formats: small form-factor pluggable (SFPs) and XFPs. Thus we will talk in detail only about these two standards.

5.9.2.1 SFP Transceivers for Telecommunications

The SFP is the more compact optical transceiver used in optical communications. It interfaces a network equipment mother board to a fiber-optic or unshielded twisted pair networking cable.

This is probably the most diffused transceiver format available with a variety of different transmitter and receiver types, allowing users to select the appropriate transceiver for each link to provide the required optical reach over the available optical fiber (e.g., multimode fiber or single-mode fiber).

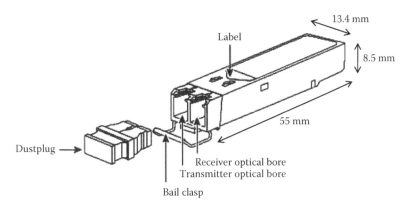

FIGURE 5.82
Pictorial representation and dimensions of a pluggable SFP transceiver.

A drawing of an SFP transceiver is presented in Figure 5.82, where the particular con-nector for the input and output fibers that, with different dimensions, is present in all the transceivers is evidenced. The way in which the SFP transceivers are hosted on the motherboard using a suitable cage allowing a hot plug is shown in Figure 5.83, where both the empty cages on the front of a system card and the cages with plugged SFPs are shown.

Optical SFP modules are commonly available in four different categories: 850 nm (SX), 1310 nm (LX), 1550 nm (ZX), and WDM, both DWDM and CWDM. SFP transceivers are also available with a "copper" cable interface, allowing a host device designed primarily for optical fiber communications to also communicate over unshielded twisted pair net-working cable.

Commercially available transceivers have a capability up to 2.5 Gbit/s for transmission applications; moreover, a version of the standard with a bit rate of 10 Gbit/s exists, but it can be used only to connect nearby equipment, and is very useful to spare space and power consumption as interface in the client cards of line equipments.

Modern optical SFP transceivers support digital optical monitoring functions accord-ing to the industry standard SFF-8472 MSA. This feature gives the end user the ability to monitor real-time parameters of the SFP, such as optical output power, optical input power, temperature, laser bias current, and transceiver supply voltage.

SFP transceivers are designed to support SONET, Gigabit Ethernet, Fiber Channel, and other communications standards.

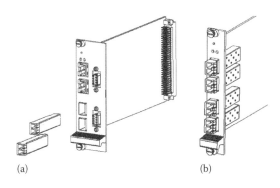

(a)   (b)

FIGURE 5.83
Hosting of SFP transceivers into a system card.

TABLE 5.11

Characteristics of Practical SFP Pluggable Transceivers for Different Applications

Bit rate	2.5 Gbit/s	2.5 Gbit/s	1.2 Gbit/s	10 Gbit/s
Type	CWDM	DWDM	GbE	10 GbE
Wavelength	ITU-T Standard	100 GHz spacing	Gray 850 nm	Gray 850 nm
Supply voltage	3.60 V	3.6 V	5.5 V	3.3 V
Supply current	290 mA	300 mA	50 mA	480 mA
Operating temperature	0°C to 70°C	−5°C to 70°C	−40°C to 75°C	0°C to 70°C
Transmitter				
Optical output power	3 dBm	2 dBm	−6 dBm	−1 dBm
Extinction ratio	8.2 dB	8.2 dB	15 dB	10 dB
Spectral width	4 nm	0.3 nm (−20 dB)	0.85 nm	Modal bandwidth 200 MHz/km
Side mode suppression	30 dB	30 dB		
RIN	−120 dB/Hz	−135 dB/Hz	−116 dB/Hz	−128 dB/Hz
Rise/fall time	180 ps	200 ps	260 ps	
Receiver				
Detector	PIN	PIN	PIN	PIN
Optical sensitivity	−28 dBm	−30 dBm	−17 dBm	−10.7 dBm
Maximum input optical power	−9 dBm	−6 dBm		

The standard is expanding to SFP+, which will be able to support data rates up to 10.0 Gbit/s (that will include the data rates for 8 Gbit Fiber Channel, and 10 GbE). Possible performances of different realistic SFP transceivers are reported in Table 5.11.

5.9.2.2 XFP Transceivers for Telecommunications

The MSA for the XFP transceiver was born after the start of the success of the SFP format to provide a transceiver with a form factor suitable to host 10 Gbit/s transmission components, but sufficiently compact to reproduce the advantages of the SFP.

In a short time it was evident that the XFP industrial standard was really tailored according to the system needs and at present this is the only type of transceiver used in telecom equipments whose evolution is targeted to high-performance interfaces at 10 Gbit/s.

At the beginning the target was a simple short-reach or medium-reach interface, but the evolution of the lasers and the Mach–Zehnder modulators integrated on InP platform is driving the development of a new generation of high-performance XFPs with tunable long-reach interfaces.

Thus, at present, there are a large number of different XFP transceivers designed for telecommunications: from the transceivers with gray short-reach interfaces for application in the client ports of optical equipments to short-, intermediate-, and long-reach DWDM interfaces, both with fixed and tunable lasers, to CWDM 10 Gbit/s transceivers.

The drawing of an XFP transceiver is presented in Figure 5.84 with the indication of the transceiver's main dimensions. XFP transceivers are slightly greater than SFPs, but they

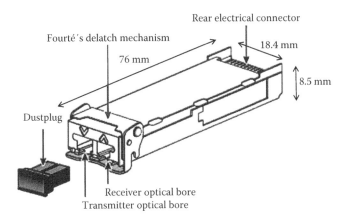

FIGURE 5.84
Pictorial representation and dimensions of a pluggable XFP transceiver.

are by far the smaller 10 Gbit/s interfaces suitable for DWDM transmission, and even if the transmission performances attainable with MSA 300 PIN are better, several optical systems, even when requiring long-haul transmission, adopt XFPs. As a matter of fact, the advantages in terms of space, power consumption, and failure management often overcompensate a certain transmission penalty.

The way an XFP is hosted on the motherboard is shown in Figure 5.85. As in the case of the SFP there is a suitable cage that has to be mounted on the motherboard in order to allow the XFP hot plug. Since high-performance XFPs have a high ratio between power consumption and area of contact with the cooling air flux, a heat sink is generally needed to increase the heat exchange area. This is mounted directly on the cage to minimize the thermal resistance (see Section 6.5).

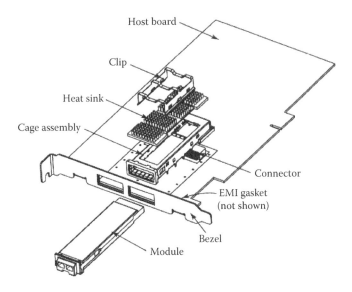

FIGURE 5.85
Hosting of XFP transceivers into a system card.

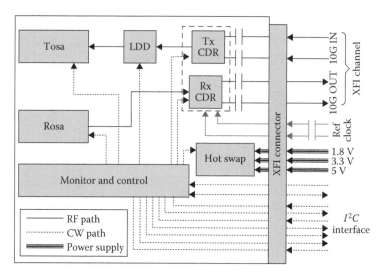

FIGURE 5.86
Functional scheme of a DWDM 10 Gbit/s XFP transceiver.

The functional diagram of a high-performance XFP is shown in Figure 5.86. From the figure it results that the module is controlled via an I²C interface, the same that is used also to control MSA 300 PIN modules. This clearly declares the fact that an XFP is not conceived as a low-performance module; on the contrary, it is equipped with a sufficiently powerful control interface to allow even the most complex features to be configured and managed.

TABLE 5.12

Characteristics of Practical DWDM XFP Transceivers Optimized for Different Uses

	XFP 40 km—Fixed	XFP 80 km—Fixed	XFP 80 km—Tunable
Optical			
Type	DWDM—100 GHz	DWDM—100 GHz	
Data rate	10 Gbit/s	10 Gbit/s	10 Gbit/s
Channel spacing	100 GHz	100 GHz	50 GHz
Wavelength stability	+/− 12.5 GHz	+/− 12.5 GHz	+/− 12.5 GHz
Output power	+1 dBm	+4 dBm	+4 dBm
Extinction ratio	8.2 dB	9 dB	8.2 dB
SMSR	30 dB	30 dB	30 dB
Relative intensity noise	−130 dB/Hz	−130 dB/Hz	−130 dB/Hz
Extinction ratio	8.2 dB	9 dB	10 dB
Maximum power on receiver	−2 dBm	−4 dBm	−4 dBm
Sensitivity	−16 dBm	−25 dBm	−25 dBm
Electrical			
Maximum dissipated power	3.5 W	3.5 W	3.5 W
Mechanical—Requirements			
Operating temperature	0°C–70°C	0°C–70°C	0°C–70°C

The input–output of the 10 Gbit/s channel is performed through the XFI parallel standard. This choice has been made to comply with the maximum possible number of system card design.

In Table 5.12 the main characteristics of three different DWDM XFPs are summarized.

References

1. Numai, T., *Fundamentals of Semiconductor Lasers*, Springer, New York, s.l., 2010, ISBN-13: 978-1441923516.
2. Tsidilkovskii, I. M., *Band Structure of Semiconductors*, Pergamon Press, Oxford, NY, s.l., 1982, ISBN-13: 978-0080216577.
3. Loudon, R., *The Quantum Theory of Light*, 3rd Edn., Oxford Science Publications, New York, s.l., 2000, ISBN 0198501765.
4. Vahala, K., Yariv, A., Semiclassical theory of noise in semiconductor lasers: Part II, *Journal of Quantum Electronics*, 19(6), 1102–1109 (1983).
5. Vey, J. L., Gallion, P., Semiclassical model of semiconductor laser noise and amplitude noise squeezing—Part I: Description and application to Fabry–Perot laser, *IEEE Journal of Quantum Electronics*, 33(11), 2097–2104 (1997).
6. Hui, R., Mecozzi, A., D'Ottavi, A., Spano, P., Novel measurement technique of alpha factor in DFB semiconductor lasers by injection locking, *Electronics Letters*, 26(14), 997–998 (1990).
7. Travagnin, M., Intensity and phase noise spectra in semiconductor lasers: A parallelism between the noise enhancing effect of pump blocking and nonlinear dispersion, *Journal of Optics B: Quantum Semiclassical Optics*, 2, L25–L29 (2000).
8. Piazzolla, S., Spano, P., Tamburrini, M., Characterization of phase noise in semiconductor lasers, *Applied Physics Letters*, 41(8), 695–696 (1982).
9. Mecozzi, A., Piazzolla, S., Sapia, A., Spano, P., Non-Gaussian statistics of frequency fluctuations in line-narrowed semiconductor lasers, *IEEE Journal of Quantum Electronics*, 24(10), 1985–1988 (1988).
10. Aragon Photonics Application Note. Characterization of the main parameters of DFB using the BOSE. 2002, www.aragonphotonics.com (accessed: May 25, 2010).
11. JDSU Application Note. *Relative Intensity Noise, Phase Noise and Linewidth*, JDSU Corporate, s.l., 2006, www.jdsu.com (accessed: May 25, 2010).
12. Lewis, G. M., Smowton, P. M., Blood, P., Chow, W. W., Effect of tensile strain/well-width combination on the measured gain-radiative current characteristics of 635 nm laser diodes, *Applied Physics Letters*, 82(10), 1524–1527 (2003), doi: 10.1063/1.1559658
13. Wentworth, R. H., Bodeep, G. E., Darcie, T. E., Laser mode partition noise in lightwave systems using dispersive optical fiber, *Journal of Lightwave Technology*, 10(1), 84–89 (1992).
14. Iga, K., Surface-emitting laser—Its birth and generation of new optoelectronics field, *IEEE Journal of Selected Areas in Quantum Electronics*, 6(6), 1201–1215 (2000).
15. Wistey, M. A., Bank, S. R., Bae, H. P., Yuen, H. B., Pickett, E. R., Goddard, L. L., Harris, J. S., GaInNAsSb = GaAs vertical cavity surface emitting lasers at 1534 nm, *Electronics Letters*, 42(5), 282–283 (2006).
16. RayCan, 1550 nm Vertical-cavity surface-emitting laser. 090115 Rev 4.0.
17. Jurcevic, P., Performance enhancement in CWDM systems, *Diploma Thesys—Technischen Universitat Wien*, s.n., Wien, Austria, May 2008.
18. Björlin, E. S., Geske, J., Mehta, M., Piprek, J., Temperature dependence of the relaxation resonance frequency of long-wavelength vertical-cavity lasers, *IEEE Photonics Technology Letters*, 17(5), 944–946 (2005).
19. Hofmann, W., Böhm, G., Ortsiefer, M., Wong, E., Amann, M.-C., Uncooled high speed (>11 GHz) 1.55 μm VCSELs for CWDM access networks, *Proceedings of European Conference on Optical Communications—ECOC 2006*, Cannes, France, September 24–28, s.n., 2006.

20. Kuznetsov, M., Theory of wavelength tuning in two-segment distributed, *IEEE Journal of Quantum Electronics*, 24(9), 1837–1844 (1988).
21. Kobayashi, K., Mito, I., Single frequency and tunable laser diodes, *IEEE Journal of Lightwave Technology*, 6(11), 1623–1633 (1988).
22. Yariv, A., *Optical Electronics*, 4th Edn., Saunders College, philadelphia, PA, s.l., 1991.
23. Coldren, L. A., Monolithic tunable didoe lasers, *IEEE Journal of Selected Areas in Quantum Electronics*, 6(6), 988–999 (2000).
24. Mason, B., Fish, G. A., Barton, J., Kaman, V., Coldren, L. A., Denbaars, S., Characteristics of sampling grating DBR lasers, *Optical Fiber Conference 2000*, Baltimore, MD, IEEE, s.l., March 2000.
25. Born, M., Wolf, E. *Principles of Optics: Electromagnetic Theory of Propagation, Interference and Diffraction of Light*, 7th Edn, Cambridge University Press, Cambridge, U.K., s.l., 1999, ISBN-13: 978-0521642224.
26. Ishii, H., Tanobe, H., Kano, F., Tohmori, Y., Kondo, Y., Yoshikuni, Y., Quasidiscontinuous wavelength tuning in a super-structure-grating, *IEEE Journal of Quantum Electronics*, 32(3), 433–441 (1966).
27. Piotrowski, J., Rogalski, A., *High-Operating-Temperature Infrared Photodetectors*, SPIE Publications, Bellingham, WA, s.l., 2007, ISBN-13: 978-0819465351.
28. Basu, P. K., Das, N. R., Mukhopadhyay, B., Sen, G., Das, M. K., Ge/Si photodetectors and group IV alloy based photodetector materials, *Proceedings of 9th International Conference on Numerical Simulation of Optoelectronic Devices*, NUSOD 2009, South Korea, IEEE, 2009, pp. 79–80, September 14–18, 2009.
29. Mathews, J., Radek R., Xie, J., Yu, S.-Q., Menendez, J., Kouvetakis, J., Extended performance GeSn/Si(100) p-i-n photodetectors for full spectral range telecommunication applications, *Applied Physics Letters*, 95(13), 133506–133506–3 (2009).
30. Campbell, J. C., Recent advances in telecommunications avalanche photodiodes, *IEEE Journal of Lightwave Technology*, 25, 109–121 (2007).
31. Agrawal, G. P., *Lightwave Technology: Components and Devices*, Wiley, Hoboken, NJ, s.l., 2004, ISBN 047121 5732.
32. Schneider, K., Zimmermann, H. K., *Highly Sensitive Optical Receivers*, Springer, Berlin, Germany, s.l., 2006, ISBN-13: 978-3540296133.
33. Hunsperger, R. G., *Integrated Optics: Theory and Technology*, 6th Edn., Springer, New York, s.l., 2009, ISBN-13: 978-0387897745.
34. Li, G. L., Yu, P. K. L., Optical intensity modulators for digital and analog applications, *IEEE Journal of Lightwave Technology*, 21(9), 2010–2030 (2003).
35. Jackson, M. K., Smith, V. M., Hallam, W. J., Maycock, J. C., Optically linearized modulators: Chirp control for low-distortion analog transmission, *IEEE Journal of Lightwave Technology*, 15(8), 1538–1545 (1997).
36. Muraro, A., Passaro, A., Abe, N. M., Preto, A. J., Stephany, S., Design of electrooptic modulators using a multiobjective optimization approach, *IEEE Journal of Lightwave Technology*, 26(16), 2969–2976 (2008).
37. André, P. S., Pinto, J. L., Optimising the operation characteristics of a LiNbO3 based Mach Zehnder modulator, *Journal of Optical Communications*, 22, 767–773 (2001).
38. Yasaka, H., Tsuzuki, K., Kikuchi, N., Ishikawa, N., Yasui, T., Shibata, Y., Advances in InP Mach-Zehnder modulators for large capacity photonic network systems, *Proceedings of 20th International Conference on Indium Phosphide and Related Materials, IPRM 2008*, Versailles, France, s.n., pp. 1–5, May 25–29, 2008.
39. Prasad, P. N., *Nanophotonics*, Wiley, Hoboken, NJ, s.l., 2004, ISBN 0741649880.
40. Yamanaka, T., Yokoyama, K., Design and analysis of low-chirp electroabsorption modulators using bandstructure engineering, 2nd Edn., *Proceedings of Second International Workshop on Physics and Modeling of Devices Based on Low-Dimensional Structures*, Aizu, Japan, 1998.
41. Zhang, S. Z., Chiu, Y.-J., Abrham, P., Bowers, E., 25 GHz polarization insensitive electroabsorption modulators with traveling wave electrodes, *IEEE Photonics Technology Letters*, 11(2), 191–193 (1999).

42. Kahn, X., Zhu, J. M., Free-space optical communication through atmospheric turbulence channels, *IEEE Transactions on Communications*, 50(8), 1293–1300 (2002).

43. El-Bawab, T. S., *Optical Switching*, Springer Verlag, New York, s.l., 2006, ISBN 0387261419.

44. Liu, A.-Q., *Photonic MEMS Devices: Design, Fabrication and Control*, CRC Press, Boca Raton, FL, s.l., 2009, ISBN 139781420045600.

45. Korvink, J. G., Paul, O., *MEMS: A Practical Guide to Design, Analysis, and Applications*, Springer Verlag & William Andrew, Heidelberg, Germany, s.l., 2006, ISBN 0815514972.

46. Chigrinov, V. G., Pasechnik, S. V., Shmeliova, D. V., *Liquid Crystals: Viscous and Elastic Properties in Theory and Applications*, Wiley, Weinheim, Germany, s.l., 2009, ISBN 9783527407200.

47. Parthasarathy, S. E., Thompson, S., Moors law, the future of Si microelectronics, *Materials Today*, 9(8), 20–25 (2006).

48. ITRS, *International Technology Roadmap for Semiconductors Industry*, 2009 Edn., 2009, www.itrs.com (consulted: July 5, 2010).

49. Roy, K., Yeo, K.-S., *Low Voltage, Low Power VLSI Subsystems*, McGraw Hill, New York, s.l., 2004, ISBN 007143786.

50. Soudris, D., Pirsch, P., Barke, E., *Integrated Circuit Design: Power and Timing Modeling, Optimization, and Simulation (10th Int. Workshop)*, Springer, Berlin, Germany, s.l., 2000, ISBN 3540410686.

51. Lindsay, R., Pawlak, B., Kittl, J., Henson, K., Torregiani, C., Giangrandi, S., Surdeanu, R., Vandervorst, W., Mayur, A., Ross, J., McCoy, S., Gelpey, J., Elliott, K., Pages, X., Satta, A., Lauwers, A., Stolk, P., Maex, K., *A Comparison of Spike, Flash, SPER and Laser Annealing for 45 nm CMOS. The Material Gateway*, The Material Research Society, s.l., 2000, http://www.mrs.org/s_mrs/sec_subscribe.asp?CID=2593&DID=109911&action=detail (accessed: July 5, 2010).

52. Storey, N., *Electronics: A Systems Approach*, 4th Edn., Pearson, Canada, s.l., 2009, ISBN 9780273719182.

53. Veendrick, H. J. M., *Nanometer CMOS ICs: From Basics to ASICs*, Springer, Heidelberg, Germany, s.l., 2008, ISBN 9781402083327.

54. Golshan, K., *Physical Design Essentials: An ASIC Design Implementation Perspective*, Springer, New York, s.l., 2010, ISBN-13: 978-1441942197.

55. Beggs, B., Sitch, J., Future perspectives of CMOS technology for coherent receivers, *European Conference on Optical Communications—ECOC 2009*, Vienna, Austria, Vol. Workshop 'DSP & FEC: towards the Shannon limit', September 20–24, 2009.

56. Papagiannakis, I., Klonidis, D., Curri, V., Poggiolini, P., Bosco, G., Killey, R. I., Omella, M., Prat, J., Fonseca, D., Teixeira, A., Cartaxo, A., Freund, R., Grivas, E., Bogris, A., Birbas, A. N., Tomkos, I., Electronic distortion compensation in the mitigation of optical transmission impairments: The view of joint project on mitigation of optical transmission impairments by electronic means ePhoton/ONe1 project, *Optoelectronics*, 3(2), 73–85, 2009.

57. Walker, J., Sheikh, F., Mathew, S. K., Krishnamurthy, R., *A Skein-512 Hardware Implementation*, Intel Public Report, s.l., 2010, http://csrc.nist.gov/groups/ST/hash/sha-3/Round2/Aug2010/documents/papers/WALKER_skein-intel-hwd.pdf (accessed: October 18, 2010).

58. Altera, *Leveraging the 40-nm Process Node to Deliver the World's Most*, 2009. White Paper, www.altera.com (accessed October 18, 2010).

59. Woods, R., McAllister, J., Turner, R., Yi, Y., Lightbody, G., *FPGA-Based Implementation of Signal Processing Systems*, Wiley, Chichester, U.K., s.l., 2008, ISBN-13: 978-0470030097.

60. Brown, S., Rose, J., *Architecture of FPGAs and CPLDs: A Tutorial*, Department of Electrical and Computer Engineering, University of Toronto, Toronto, Ontario, Canada, 2000, http://www.eecg.toronto.edu/~jayar/pubs/brown/survey.pdf (accessed: October 18, 2010).

61. Xlinx, *Vertex Family FPGA Technical Documentation*, 2010 version, www.xilinx.com (accessed: October 18, 2010).

62. Actel, *Technical Reference for Actel Antifuse FPGAs*, 2010, www.Actel.com (accessed: October 18, 2010).

63. Kestur, S., Davis, J. D., Williams, O., *BLAS comparison on FPGA, CPU and GPU* [a cura di], IEEE Computer Society, *Annual Symposium on VLSI (ISVLSI), 2010*, Lixouri Kefalonia, Greece, IEEE, s.l., pp. 288–293, 2010.

64. Chitode, J. S., *Digital Signal Processing*, Pune Technical Publications, Pune, India, s.l., 2008, ISBN 9788184311327.
65. Ceva, *The Architecture of Ceva DSPs—Product Note*, 2010, www.ceva-dsp.com (accessed: October 18, 2010).
66. Pederson, D. O., Mayaram, K., *Analog Integrated Circuits for Communication: Principles, Simulation and Design*, Springer, New York, s.l., 2010, ISBN-13: 978-1441943248.
67. Tomkos, I., Spyropoulou, M., Ennser, K., Köhn, M., Mikac, B., *Towards Digital Optical Networks: COST Action 291 Final Report*, Springer, Berlin, Germany, s.l., 2009, ISBN 13 9783642015236.
68. Säckinger, E., *Broadband Circuits for Optical Fiber Communication*, Wiley, Hoboken, NJ, s.l., 2005, ISBN: 978-0-471-71233-6.
69. Chen, W.-K. (ed.), *Fundamentals of Circuits and Filters*, CRC Press, Boca Raton, FL, s.l., 2009, ISBN-13: 978-1420058871.
70. Weber, C., Bunge, C.-A., Petermann, K., Fiber nonlinearities in systems using electronic pre-distortion of dispersion at 10 and 40 Gbit/s, *IEEE Journal of Lightwave Technology*, 27(16), 3654–3661 (2009).
71. Xia, C., Rosenkranz, W., Electrical mitigation of penalties caused by group delay ripples for different modulation formats, *IEEE Photonics Technology Letters*, 19(13), 1041–1135 (2007).
72. Azadet, K., Haratsch, E., Kim, H., Saibi, F., Saunders, J. H., Shaffer, M., Equalization and FEC techniques for optical transceivers, *IEEE Journal of Solid State Circuits*, 19(13), 954–956 (2007).
73. Hayes, M. H., *Statistical Digital Signal Processing and Modeling*, Wiley, New York, s.l., 1996, ISBN 0-471-59431-8.
74. Koc, U.-V., Adaptive electronic dispersion compensator for chromatic and polarization-mode dispersions in optical communication systems, *Journal on Applied Signal Processing*, 10, 1584–1592 (2005).
75. Zerguinea, A., Chanb, M. K., Al-Naffouria, T. Y., Moinuddina, M., Cowan, C. F. N., Convergence and tracking analysis of a variable normalised LMF (XE-NLMF) algorithm, *Signal Processing*, 89(5), 778–790 (2009).
76. Nocedal, J., Wright, S. J., *Numerical Optimization*, Springer, New York, s.l., 2006, ISBN 0-387-30303-0.
77. Fritzsche, D. et al., Volterra based nonlinear equalizer with reduced complexity, *Proceedings of Optical Transmission, Switching, and Subsystems V*, Wuhan, China, SPIE, pp. 67831R.1–67831R.8, November 2–5, 2007.
78. Bulow, H., *Tutorial Electronic Dispersion Compensation*, Bonn, Germany, s.n., 2006.
79. Madow, U., *Fundamentals of Digital Communication*, Cambridge University Press, New york, 2009, ISBN-13: 978-0521874144.
80. Fan, M. L., Zhangyuan, Z., Xu, C. A., Chromatic dispersion compensation by MLSE equalizer with diverse reception, *Proceedings of European Conference on Optical Communications*, Brussels, s.n., pp. 1–2, September 21–25, 2008.
81. Bae, H. M., Ashbrook, J., Park, J., Shanbhag, N., Singer, A., Chopra, S., An MLSE receiver for electronic-dispersion compensation of OC-192 fiber links, *Solid-State Circuits Conference, ISSCC. Digest of Technical Papers*, San Francisco, CA, IEEE International, s.l., pp. 874–883, February 6–9, 2006.
82. Kratochwil, K., Low complexity decoders for channels with intersymbol-interference, *MILCOM 97 Proceedings*, Monterey, CA, IEEE, Vol. 2, pp. 852–856, November 2–5, 1997.
83. Carrer, H. S., Crivelli, D. E., Hueda, M. R., Maximum likelihood sequence estimation receivers for DWDM lightwave systems, *Proceedings of Global Telecommunications Conference GLOBECOM '04*, Dallas, TX, IEEE, s.l., Vol. 2, pp. 1005–1010, November 29–December 3, 2004.
84. Downie, J. D., Hurley, J., Sauer, M., Behavior of MLSE-EDC with self-phase modulation, *IEEE Photonics Technology Letters*, 19(13), 1017–1019 (2007).
85. Rozen, O., Cohen, T., Kats, G., Sadot, D., Levy, A., Mahlab, U., Dispersion compensation in non-linear self phase modulation (SPM) and cross phase modulation (XPM) induced optical channel using vectorial MLSE equalizer, *Proceedings of 9th International Conference on Transparent Optical Networks, ICTON '07*, Roma, Italy, s.n., Vol. 1, pp. 302–304, July 1–5, 2007.

86. Hueda, M. R., Crivelli, D. E., Carrer, H. S., Agazzi, O. E., Parametric estimation of IM/DD optical channels using new closed-form approximations of the signal PDF, *Journal of Lightwave Technology*, 25(3), 957–975 (2007).

87. Haunstein, H. F., Schorr, T., Zottmann, A., Sauer-Greff, W., Urbansky, R., Performance comparison of MLSE and iterative equalization in FEC systems for PMD channels with respect to implementation complexity, *IEEE Journal of Lightwave Technology*, 22(11), 4047–4054 (2006).

88. Killey, R. I., Watts, P. M., Mikhailov, V., Glick, M., Bayvel, P., Electronic dispersion compensation by signal predistortion using digital processing and a dual-drive Mach–Zehnder modulator, *IEEE Photonics Technology Letters*, 17(3), 714–716 (2005).

89. Killey, R. I., Watts, P. M., Glick, M., Bayvel, P., Electronic dispersion compensation by signal predistortion, *Proceedings of Optical Fiber Conference*, Anaheim, CA, OSA, s.l., 2006.

90. Hanik, S., Hellerbrand, N., Electronic predistortion for compensation of fiber transmission impairments—Theory and complexity consideration, *Journal of Networks*, 5(2), 180–187 (2010).

91. Iannone, E., Matera, F., Mecozzi, A., Settembrem, M., *Nonlinear Optical Communication Networks—Appendix A1*, Wiley, New York, s.l., 1998, ISBN-13: 978-0471152705.

92. Watts, P., Waegemans, R., Glick, M., Bayvel, P., Killey, R., An FPGA-based optical transmitter design using real-time DSP for advanced signal formats and electronic predistortion, *IEEE Journal of Lightwave Technology*, 25(10), 3089–3099 (2007).

93. Watts, P., Waegemans, R., Glick, M., Bayvel, P., Killey, R., *Breaking the Physical Barriers with Electronic Dynamically*, Nortel Networks Whitepaper, Now under Ciena White paper, s.l., 2006, www.Ciena.com

94. Huffman, W. C., Pless, V., *Fundamentals of Error Correcting Codes*, Cambridge University Press, Cambridge, U.K., s.l., 2010, ISBN-13: 978-0521131704.

95. Hsu, H.-Y., Wu, A.-Y., Yeo, J.-C., Area-efficient VLSI design of Reed–Solomon decoder for 10 GBase-LX4 optical communication systems, *IEEE Transactions on Circuits and Systems*, 53(1), 1245–1249 (2006).

96. Trowbridge, S. J., FEC applicability to 40 GbE and 100 GbE, IEEE, Santa clara, CA, 2007, *Report to the Standardization Group IEEE-802*.

97. Yan, J., Chen, M., Xie, S., Zhou, B., Performance evaluation of standard FEC in 40 Gbit/s systems with high PMD and prechirped CS-RZ modulation format, *IEE Proceedings of Optoelectronics*, 151(1), 37–40 (2004).

98. Mizuochi, T., *Soft-Decision FEC for 100 Gb/s DSP Based Transmission*, Newport Beach, CA, s.n., pp. 107–108, 2009.

99. Kschischang, B. P., Smith, F. R., Future prospects for FEC in fiber-optic communications, *IEEE Journal on Selected Areas in Quantum Electronics*, 16(5), 1245–1257 (2010).

100. Cai, Y., Morris, J. M., Adali, T., Menyuk, C. R., On turbo code decoder performance in optical-fiber communication systems with dominating ASE noise, *IEEE Journal of Lightwave Technology*, 21(3), 727–734 (2003).

101. Kaasper, E., Turbo codes, Institute for Telecommunications Research, University of South Australia, Adelaide, Australia, 2005, Technical Report, http://users.tkk.fi/pat/coding/essays/turbo.pdf (accessed: October 20, 2010).

102. Abbasfar, A., *Turbo-like Codes: Design for High Speed Decoding*, Springer, Dordrecht, the Netherlands, s.l., 2010, ISBN-13: 978-9048176236.

103. Johnson, S. J., *Iterative Error Correction: Turbo, Low-Density Parity-Check and Repeat-Accumulate Codes*, Cambridge University Press, New York, s.l., 2009, ISBN-13: 978-0521871488.

104. Aitsab, O., Pyndiah, R., Performance of Reed-Solomon block turbo code, *Global Telecommunications Conference, GLOBECOM 96*, London, U.K., IEEE, s.l., Vol. 1, pp. 121–125, 1996.

105. Pagiamtzis, K., Sheikholeslami, A., Content-addressable memory (CAM) circuits and architectures: A tutorial and survey, *IEEE Journal of Solid States Circuits*, 41(3), 712–727 (2006).

106. Liu, S. C., Wu, F. A., Kuo, J. B., A novel low-voltage content-addressable-memory (CAM) cell with a fast tag-compare capability using partially depleted (PD) SOI CMOS dynamic-threshold (DTMOS) techniques, *IEEE Journal of Solid State Circuits*, 36(4), 712–716 (2001).

107. Lin, M., Luo, J., Ma, Y., A low-power monolithically stacked 3D-TCAM, *International Symposium on Circuits and Systems, ISCAS 2008*, Seattle, WA, IEEE, pp. 3318–3321, 2008.
108. Eshraghian, K., Cho, K.-R, Kavehei, O., Kang, S.-K., Abbott, D., Kang, S.-M., Memristor MOS content addressable memory (MCAM): Hybrid architecture for future high performance search engines, *IEEE Transactions on Very Large Scale Integration (VLSI) Systems*, 99(1), 1–11 (2010).
109. Ditmar, J., Torkelsson, K., Jantsch, A., A dynamically reconfigurable FPGA-based content addressable memory for internet protocol characterization, *Lecture Notes in Computer Science*, Springer, London, U.K., s.l., Vol. 1896/2000, pp. 19–28, 2000.
110. Altera, *Power-Optimized Solutions for Telecom Applications*, 2010, White Paper, http://www.altera.com/literature/wp/wp-01089-power-optimized-telecom.pdf (accessed: August 18, 2010).

6

Transmission Systems Architectures and Performances

6.1 Introduction

After a long period in which electronics switching was driving down prices while transmission via coaxial cables was costly and difficult, with the introduction of optical transmission in the early 1990s, the situation completely changed.

Although the end of the telecom market bubble (see Chapter 2) was followed by a big crisis influencing the market practically since 2001, it remains the fact that it is possible to transmit via optical fibers very high-capacity signals (like 100 channels at a speed of 10 Gbit/s each) at very long distances (in excess of 2000 km); transmission is today an abundant resource in the telecommunication networks while switching/routing remains more expensive.

This change of paradigm has driven the development of telecommunication networks in the last decade, and it does not seem to end.

Thus, the analysis of optical transmission is a key to observe the trends in the telecommunication market.

This chapter is devoted to the analysis of transmission. Since transmission is performed almost only via optical fibers, the chapter deals only with to optical transmission systems for metro and core networks. Access transmission systems will be analyzed in Chapter 11.

Radio bridges are not considered, even if they are still used in the network in particular situations.

Radio bridge architecture and performances cannot be analyzed without a suitable review of microwave techniques; this technology is not driven by telecommunications and we feel it would be impossible to condense it in a book on telecommunication networks.

The first part of the chapter is devoted to the most basic system: intensity modulation direct detection (IM-DD) systems without optical amplifiers.

Even if the use of optical amplifiers is widespread, unamplified systems are still used for particular applications, especially in rural areas, when a high-capacity trunk is needed to connect local central offices that are not so distant to justify optical amplification. The same systems are also used in the connection of mirroring data centers and in private security area networks.

However, the choice to consider this system first is due mainly to their greater simplicity that allows all the elements needed to analyze the performance of an optical system to be considered with minimum added complications.

In the second part, optical systems for long-haul and ultra-long-haul transmission are considered. Finally, in the third part of the chapter, the hardware architecture of a transmission system is analyzed, putting into evidence those issues, like power consumption and real estate occupation, that in practice are often more important than one more channel at 10 Gbit/s.

6.2 Intensity Modulation and Direct Detection Transmission

6.2.1 Fiber-Optic Transmission Systems

The block scheme of a generic single-channel transmission system is reported in Figure 6.1 [1–3]. A transmitter receives as the input a stream of symbols containing the message to be transmitted and the so-called baseband signal.

The transmitter encodes the symbol stream into the baseband signal and then uses the result to modulate the carrier; in optical systems the carrier is a continuous wave radiation emitted by a laser in the suitable transmission window of silica fibers. The component that performs this operation, as seen in Chapter 4, is called modulator.

The modulated signal is launched in the transmission line that is constituted by a transmission fiber interrupted, if it is the case, by online sites where the optical signal is processed, for example, amplified, filtered, regenerated, and so on, depending on transmission needs.

At the end of the transmission line, a detector transforms the incoming optical signal into an electrical current. Starting from the electrical current, the receiver reconstructs the carrier signal using, for example, an electronic phase locked loop and uses the reconstructed carrier to extract, via a demodulator followed by a sampler and a decision threshold device, the original symbol stream.

The method we have just described for estimating the bit stream is called hard decision [4], since it is carried out simply on the ground of the signal statistics, without any help from codes for error correction or channel estimation.

When an error-correcting code is used, it is located after the decision block to correct errors in the estimated bit stream.

Another procedure, could be possible where the line code is used not to correct errors after the bit stream estimation, but to help the estimation device not to make errors. This method is called soft decision and generally is few dB better than hard decision at the optical error probability levels.

Soft decision is not implemented generally in optical systems due to the speed required to the processing electronics by the high bit rate. This situation is changing with the big progress of complementary metal oxide transistor (CMOS) electronics, and we will see that soft decision is also entering into the optical transmission systems' designer resources.

Due to the improvements in the components technology and systems design, very high-speed transmissions are possible. At present, even if 2.5 Gbit/s systems are still installed in great numbers, the most used transmission speed in the core and metro areas is 10 Gbit/s, while 40 Gbit/s systems are starting to spread in the long-haul network, and system vendors are concentrating the research for a new generation of products on 100 Gbit/s transmission.

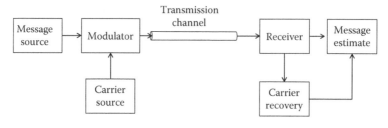

FIGURE 6.1
Block scheme of a generic single-channel transmission system.

6.2.1.1 Wavelength Division Multiplexing

Whatever design is adopted, the available bandwidth in the third transmission window of optical fibers is so wide that it is not possible to exploit it with a single time division multiplexing (TDM) channel.

Thus, almost all metro and core transmission systems use wavelength division multiplexing (WDM), as shown in Figure 6.2.

At the link near end, several transmitters generate different optical channels using carriers at different frequencies so that they can be multiplexed by an optical multiplexer and sent into the fiber. At the far end, the channels are demultiplexed and processed by an array of optical receivers.

When the optical channels are densely packed in the optical spectrum (e.g., 100 or 50 GHz apart from one another), the multiplexing is called dense WDM (DWDM).

DWDM systems are long-reach and high-capacity systems, require high-quality components, a great design and testing effort, and, therefore, are expensive.

In particular, DWDM transmitters have to be narrowband and frequency stabilized, requiring an accurate temperature control.

The WDM technology can also be adopted designing wide-spaced channels, so that thermal stabilization of transmitters is not needed. These are CWDM systems, designed mainly for application in the metro and access area.

6.2.1.2 Transmission System Performance Indicators

In all transmission systems, noise and other signal distortions are introduced at various points; thus, the extraction of the transmitted symbols from the received signal is a statistical operation prone to errors.

The point-to-point system performance is generally measured by the probability that detection fails in identifying a symbol, which is called symbol error probability. If the alphabet from which the symbols are taken is binary (i.e., there are only two symbols, "1" and "0"), generally this probability is simply called "error probability" or bit error rate (BER).

International standards fix a required error probability that depends on the application, so that systems have to be designed to reach such reference BER [5,6].

In the case of long-haul and ultra-long-haul transmission, the reference BER is 10^{-12}, and we will use this value every time this will be needed.

Frequently, it is interesting to know the degradation introduced in the system performances with respect to an ideal case by a certain phenomenon; for example, what is the degradation introduced by chromatic dispersion?

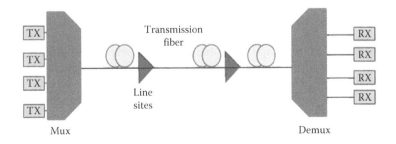

FIGURE 6.2
Block scheme of a WDM transmission system.

In this case, the degradation is measured through the increment with respect to the ideal case of the received power or of the ratio between the received power and noise (called signal-to-noise [SN] ratio) needed to reach the reference BER: this difference is called penalty.

In a regime of small effect of the nonideal elements, penalties are frequently added together like independent perturbations. This procedure is justified by the assumption that, in a regime of small perturbation, the relation between the BER and the SN ratio does not change and that the dependence of the SN on the perturbation parameters can be linearized.

Let us imagine that in ideal conditions, the BER has the following expression:

$$BER = e^{-SN_i} \tag{6.1}$$

SN_i being the ideal SN ratio both when a 1 is transmitted and when a 0 is transmitted.

Let us also assume that a practical system experiments two impairing phenomena characterized by two parameters A and B (e.g., chromatic dispersion and polarization mode dispersion [PMD]; in this case, the parameters would be D and D_{PMD}).

For a small impact of these phenomena, we can still approximate the BER with (6.1), where the ideal expression SN_i is substituted by an expression SN depending on A and B.

Generally, the logarithmic units are used for the SN ratio; thus, passing from linear to dB and developing at the first order in the small parameters representing the impairing phenomena, we can write

$$SN\,(\text{dB}) = SN_i(\text{dB}) + \frac{dSN}{dA}\,\Delta A\,(\text{dB}) + \frac{dSN}{dB}\,\Delta B\,(\text{dB}) \tag{6.2}$$

where ΔA and ΔB are the deviations of the parameter A and B from the values guaranteeing ideal system performances.

If such deviations are sufficiently low, they can be written as

$$\frac{dSN}{dA}\,\Delta A \approx \Delta SN\,|_A \quad \text{(all in dB)} \tag{6.3}$$

where $\Delta SN\,|_A$ is, expressed in logarithmic units, the variation of SN corresponding to a small variation of A. Thus, Equation 6.2 can be rewritten as

$$SN\,(\text{dB}) = SN_i\,(\text{dB}) + \Delta SN\,|_A\,(\text{dB}) + \Delta SN\,|_B\,(\text{dB}) \tag{6.4}$$

This is the simple rule of penalties addition that says that, for small deviations from the ideal behavior, the SN ratio in dB can be obtained by adding to its ideal value the sum of the penalties relative to all the impairing phenomena, all expressed in dB.

Equation 6.4 is the base for several design techniques, but it has always to be verified in its accuracy due to the approximation used to derive it.

A frequent reason for impossibility to use Equation 6.4 is the fact that the distribution of the noise changes greatly due to the presence of some transmission impairment so that the ideal expression of the error probability cannot be used. In these cases, Equation 6.4 can still be useful if an approximation of the relation between the BER and the SN is known or can be estimated via a suitable simulation.

Phenomena affecting system performances can be random or deterministic in nature. Deterministic phenomena often consist in unwanted distortions of the signal shape, like the effect of fiber nonlinearities or of the nonideal modulator transfer function.

In this case, an intuitive way of quantifying the distortion is the so-called eye opening penalty (EOP). If the receiver samples the signal in the instants t_k, indicating with $s(t,b)$ the signal immediately before sampling with its dependence on time and on the transmitted bit b, the EOP is defined as follows:

$$EOP = 10\log_{10} s_i(t_{i,k},1) - 10\log_{10} s(t_k,1) + 10\log_{10} s(t_k,0) \tag{6.5}$$

where $s_i(t_{i,k}, 1)$ is the ideal received signal sampled in the ideal instant $t_{i,k}$, while the ideal received signal for a zero is considered to be zero.

In the presence of random phenomena, like thermal or quantum noise, the EOP is a random variable, and often its average is considered to eliminate the noise and to identify the effect of distortions. In case more distortion effects are present, the EOP can also be expressed in a first approximation as the sum of the EOP relative to each single effect.

The procedure to demonstrate it is similar to that used for the SN and the carrier; it brings to write, in the case of two distortion causes with characteristic parameters A and B, respectively,

$$EOP\,(\mathrm{dB}) = EOP(A)(\mathrm{dB}) + EOP(B)(\mathrm{dB}) \tag{6.6}$$

where $EOP(A)$ and $EOP(B)$ indicate the EOP in the presence of only one of the distorting phenomena.

6.2.2 Ideal IM–DD Transmission

In this very preliminary analysis, we will assume the following simplifying assumptions:

- The transmitted message is binary.
- The system is point-to-point.
- Only one signal propagates along the transmission line.
- Information is coded in the signal amplitude.
- The fiber transmission is ideal, that is, without dispersion and nonlinear effects.
- All transmitter components and all receiver components are ideal; thus, only noise affects the receiver detection.
- No optical amplifier is used.

In this condition, the baseband transmitted signal can be written as

$$S(t) = \sum_k b_k m(t - kT) \quad (b_k = 0,1) \tag{6.7}$$

where
$m(t)$ is the pulse that has to shape the transmitted field
T is the period of the transmission: 1 bit transmission will take T seconds

The inverse of this period $R = 1/T$ is called bit rate.

Two possible modulation types exist depending on the shape of $m(t)$: if it is constant in the bit time and zero (or again constant but much smaller) in the adjacent bit times, the modulation is called non-return to zero (NRZ). As a matter of fact, the optical power does not return to zero between the transmission of two consecutive ones and a series of consecutive ones corresponds to a constant signal.

If the pulse $m(t)$ occupies a part of the bit time, leaving a guard time among consecutive pulses so that the signal returns to zero (or to a small value) between two consecutive ones, the modulation is called return to zero (RZ).

Logically, the modulation operation is divided into two steps: the binary message is coded on the electrical carrier (operation that in real systems is done by the remote signal originator) and the obtained current is coded by the modulator onto the CW field emitted by the transmitting laser.

The field is then coupled with the transmission line injecting it in an optical fiber. After coupling loss, the signal at the fiber input contains the information in the slow varying part of the optical field. In telecommunication systems, almost only single-mode fibers are used; thus, the propagating field is also single mode.

From Equation 4.2, we can write this signal as follows, where the mode transversal shape and the modulation pulses have been considered normalized to one and P_{in} is the average optical power injected into the fiber in the absence of modulation

$$\vec{E} = \sqrt{P_{in}}\, A_{1,0}(\rho) \sqrt{\sum_k b_k m(t - kT)}\; e^{i[\beta z - \omega t]} \vec{x} \tag{6.8}$$

After ideal propagation, the field that arrives in front of the receiver has the following expression:

$$\vec{E} = \sqrt{P_{in}}\, A_{1,0}(\rho) \sqrt{\sum_k b_k m(t - kT - \vartheta)}\; e^{-\alpha L/2} e^{i[\beta L - \omega \tau]} \mathbf{M} \begin{pmatrix} \vec{x} \\ \vec{y} \end{pmatrix} \tag{6.9}$$

where
 L is the fiber length
 α is the attenuation
 τ is the propagation time
 $\mathbf{M}$ is the Jones matrix that in the ideal case does not depend on frequency

After detection by the receiving photodiode, the electrical current is proportional to the optical power; thus, in the ideal case in which all the incoming power is detected by the photodiode, the electrical current writes

$$c(t) = R_p P_{out} \sum_k b_k m(t - kT - \tau) * \mathcal{H}(t) + c_{bias} + n_e(t) \tag{6.10}$$

where
 c_{bias} is the detection circuit bias current
 R_p is the photodiode sensitivity
 $P_{out} = P_{in} e^{-\alpha L}$
 $n_e(t)$ is the detector noise process

$\mathcal{H}(t)$ is the pulse response of the electrical front end following the photodiode
∗ represents convolution

In the absence of any optical amplification, the noise is fully due to the sum of two terms: the shot noise and the receiver thermal noise [3]. However, in practical systems without amplifiers, the signal arrives on the photodiode with a great attenuation due to fiber propagation; thus, the shot noise is generally negligible with respect to the thermal noise.

Assuming to have a photodiode based on a internal depletion zone (PIN), in Chapter 4 we have seen that, if the filter $\mathcal{H}(t)$ eliminates the low frequency $1/f$ noise contribution, $n_e(t)$ is a band-limited white Gaussian noise whose variance can be indicated with σ_n^2.

At this point, a small part of the signal is baseband filtered so to insulate only the carrier contribution and is sent to a phase lock loop (PLL) that maintains the sampler synchronized with the received signal.

The sampler extracts a sample of the signal (6.10) at the center of the bit interval and compares it with a threshold. If the sample is higher than the threshold, the received bit is assumed to be a one, otherwise a zero.

Calling c_{th} the threshold, two possible receivers do exist: fixed and adaptive threshold receivers.

In the case of fixed threshold, c_{th} is set at the beginning of operation to an optimum value, while in adaptive one it is adapted over times much longer than the bit time to follow slow random fluctuations in the power of the received signals.

If the received signal has a constant average power, the two receivers have the same performances, while this is not true in the presence of intensity noise.

In any case, the BER of our much simplified system is

$$BER = P(1/0)p_0 + P(0/1)p_1 \tag{6.11}$$

where
p_0 and p_1 are the probabilities that a zero or a one is transmitted
$P(a/a')$ is the probability that a is detected when a' is transmitted

Let us define the Q function as

$$Q(x) = \frac{1}{\sqrt{2\pi}} \int_x^\infty e^{-\xi^2/2}\, d\xi \approx \frac{e^{-x^2/2}}{\sqrt{2\pi x^2}} \tag{6.12}$$

where the asymptotic approximation holds for high values of x.

The error probability can be expressed in terms of the Q function like

$$BER = \frac{1}{2}\left[Q\left(\frac{c(1) - c_{th}}{\sigma_n} \right) + Q\left(\frac{c_{th} - c(0)}{\sigma_n} \right) \right] \tag{6.13}$$

where $c(1)$ and $c(0)$ are the average currents corresponding to the transmission of the two bits.

Equation 6.13 gives the expression of the BER in the very simplified case we have considered up to now, but it is an important reference. The optimum threshold is the value of c_{th} that minimizes the error probability. Deriving Equation 6.13 with respect to c_{th} and equating the derivative to zero it is obtained $2c_{th} = c(1) + c(0)$ and substituting in Equation (6.13) the BER with the optimum threshold becomes

$$BER = Q\left(\frac{c(1) - c(0)}{2\sigma_n}\right) = Q\left(\frac{1}{2}\sqrt{SN_e}\right) \tag{6.14}$$

where
    σ_n is the standard deviation of the noise term
    SN_e indicates the electrical signal to noise ratio

In this ideal system, $EOP = c(1) - c(0)$ so that it is clear why a phenomenon affecting EOP causes a BER reduction.

A realistic example of a phenomenon reducing EOP is a limited transmitter signal dynamic range. Ideally, for a given peak power, the BER is maximized for $c(0) = 0$. In practice, however, the signal zero level is not exactly zero, due to both the inability of the source laser to switch rapidly between the off and the on status and various phenomena like interference with other channels or pulse broadening during propagation that move energy into the bit periods in which a zero is transmitted.

Thus, if the ideal $SN_e = c^2(1)/\sigma_n^2$ in practice, it is $SN_e = [c(1) - c(0)]^2/\sigma_n^2$, with $c(0) > 0$.

The corresponding optical penalty due to the reduction of the dynamic range can be obtained by passing in logarithmic units so that

$$SN_e\,(\mathrm{dB}) = 20\log_{10}[c(1) - c(0)] - 20\log_{10}\sigma_n \tag{6.15}$$

taking into account that $c(1) \gg c(0)$ is needed to assure system working, it can be written

$$SN_e\,(\mathrm{dB}) = 20\log_{10} c(1) - 20\log_{10}\sigma_n + 20\log_{10}\left[1 + \frac{c(0)}{c(1)}\right] \approx SN_i\,(\mathrm{dB}) + \frac{20}{\ln(10)}\frac{c(0)}{c(1)} \tag{6.16}$$

Thus, the expression of the logarithmic power penalty due to the reduction of the transmitter dynamic range is

$$\Delta SN_e\,(\mathrm{dB}) = \frac{20}{\ln(10)}\frac{c(0)}{c(1)} \tag{6.17}$$

6.2.3 Analysis of a Realistic Single-Channel IM–DD System

The hypotheses that produce Equation 6.14 generally are not fulfilled by a realistic system, due to different reasons.

First of all, transmission through the line is not ideal; dispersion and nonlinear effects generate signal distortion that affects system performances. Moreover, the noise contribution at the receiver is not simply the same Gaussian process whatever bit is transmitted. Different noise processes contribute to the noise and some of them are signal dependent.

In this section, we will start to remove some ideal assumption and to confront ourselves with a realistic system performance evaluation.

In analyzing a realistic IM-DD system we will always assume that the system is functional; thus, all the negative effects have a small impact on system performances and the penalty addition rule can be applied.

6.2.3.1 Evaluation of the BER in the Presence of Channel Memory

In a real system, one of the major phenomenon to take into account is fiber propagation. Due to different effects (e.g., chromatic and polarization dispersion and self phase modulation [SPM]), there is a pulse spreading during propagation, moving energy between adjacent bit intervals.

This effect has a big impact when energy is moved from a "one" to a "zero," while there is almost no impact if a "one" is adjacent to another "one" or if there is a sequence of "zero."

This example shows that in a real system the probability that a bit is detected in the wrong way also depends on the nearby bits.

This is called channel memory and has to be taken into account when evaluating system performances.

Under the hypothesis of low distortion, we can limit the analysis to patterns of 3 bits, that we will tag with an index $k = 1, 2, ..., 8$. The bit in a pattern will be indicated with b_{kj} where $j = 1, 2, 3$ indicates the bit position and $k = 1, 2, ..., 8$ to which pattern the bit belongs.

The eight patterns are

- $\Phi_1 = 0\ 0\ 0$; $\Phi_2 = 1\ 0\ 0$; $\Phi_3 = 0\ 0\ 1$; $\Phi_4 = 1\ 0\ 1$
- $\Phi_5 = 1\ 1\ 1$; $\Phi_6 = 1\ 1\ 0$; $\Phi_7 = 0\ 1\ 1$; $\Phi_8 = 0\ 1\ 0$

Since errors will occur at the receivers, we will distinguish the transmitted pattern, ${}^t\Phi_k$, whose bits are ${}^t b_{kj}$ and the corresponding pattern estimated at the receiver Φ_k, whose bits are b_{kj}.

With these definitions, assuming the transmission of uncorrelated and equally probable bits, we have

$$BER = \frac{1}{8} \sum_{k=1}^{8} P\left(b_{k2} \neq {}^t b_{k2} \middle| {}^t b_{k2} \right) = \frac{1}{8} \sum_{k=1}^{8} P_e(k) \tag{6.18}$$

where $P(b_{k2} \neq {}^t b_{k2} / {}^t b_{k2})$ represents the probability to estimate the middle bit of the pattern equal to b_{k2} when ${}^t b_{k2}$ was transmitted and $b_{k2} \neq {}^t b_{k2}$.

The BER is evaluated on the middle bit of the pattern since this covers all the possible cases; the other bits of the pattern only serve to correctly represent the channel memory.

6.2.3.2 NRZ Signal after Propagation

In the NRZ case, we will assume the transmitted pulse perfectly squared.

We will assume also that both Raman and Brillouin scattering are negligible. Raman scattering has a high threshold, and we have to avoid a very high power that goes beyond it. As far as Brillouin is concerned, the threshold is quite low and the effect has to be eliminated somehow.

In order to increase the Brillouin threshold, the dependence of the Brillouin gain on the signal bandwidth (see Equation 4.51) is exploited. The spectrum of an amplitude modulated signal can be divided in the sidebands whose width is approximately equal to the bit rate and the central carrier (see Figure 6.3). Since optical systems transmit multigigabit signals, the sidebands' width is much greater than the Brillouin gain bandwidth and their effect can be neglected with respect to the central carrier.

In general, the central carrier power is above the threshold if effective transmission has to be carried out. However, it is sufficient to operate a direct phase modulation of the transmitting laser (often called dithering and operated by a modulation of the bias current) to increase the carrier spectral width well beyond the Brillouin linewidth $\approx$ 12 MHz [7].

In this condition the Brillouin gain becomes [8]

$$G_B(\omega) = G_{B0}(\omega) * \frac{S(\omega)}{P_s} \Rightarrow G_B(\omega_B) \approx \frac{B_B}{B_o} G_{B0}(\omega_B)$$ (6.19)

where
  B_B is the linewidth of the Brillouin gain curve
  B_o is the optical bandwidth of the signal pumping the Brillouin scattering
  $S(\omega)$ is its power spectrum
  P_s is its optical power
  $G_{B0}(\omega)$ is the Brillouin gain for monochromatic pump
  ω_B is the frequency of the maximum gain

Moreover it is assumed $B_o \gg B_B$.

From Equation 6.19 it is derived that a dithering of about 250 MHz causes the Brillouin threshold for an amplitude modulated signal to increase up to about 100 mW. Thus, as far as the transmitted power is below 50 mW Brillouin scattering can be neglected.

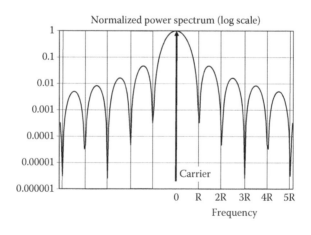

FIGURE 6.3
Baseband power spectrum of an intensity modulated signal.

Indicating with $(\vec{p}_+, \vec{p}_-)$ and $(\vec{p}_{i+}, \vec{p}_{i-})$ the output and the input principal states of polarization (PSPs) of the transmission fiber, the average field at the receiver input can be written as

$$\vec{E} = \sqrt{P_{in}}\, A_{1,0}(\rho) e^{-\alpha L/2}\, e^{i\beta_1 L}\, \hat{\wp}$$

$$\left\{ \sqrt{\sum_k b_k m(t - kT - \tau_+)}\, e^{-i\omega\tau_+} (\vec{x} \cdot \vec{p}_{i+})\vec{p}_+ + \sqrt{\sum_k b_k m(t - kT - \tau_-)}\, e^{-i\omega\tau_-} (\vec{x} \cdot \vec{p}_{i-})\vec{p}_- \right\} \qquad (6.20)$$

where
  $\hat{\wp}$ is the nonlinear propagation operator
  τ_+, τ_- are the PSPs' delays

In compliance with the hypothesis of small nonlinearity, the pulses are assumed to propagate independently so that $\hat{\wp}$ can be applied to every pulse individually [9].

It is to be noted that not only $\hat{\wp}$ is nonlinear in general, but also incorporates the channel memory so that it results to be signal dependent. However, this difficulty can be solved by considering the patterns Φ_k as independent signals instead of the single bits.

Considering NRZ modulation, if SPM is negligible, the pulse broadening due to chromatic dispersion and PMD can be analyzed separately and added at the end to obtain the overall pulse spread.

Chromatic dispersion effect is deterministic, and can be represented with the effect of a linear operator on the pulses. Due to the channel memory, it is necessary to define eight signals that represents the eight patterns Φ_k. Calling $\eta_T(t)$ the function equal to 1 if $0 < t < T$ and to 0 we can write

$$s_k(t) = \sum_{j=1}^{3} b_{kj}\, \eta_T(t - jT) \qquad (6.21)$$

It is noteworthy that $s_k(t) = \sqrt{s_k(t)}$ so that the signal received when one of the patterns Φ_k is sent can be rewritten, neglecting SPM and PMD, as

$$\vec{E} = \sqrt{P_{out}}\, A_{1,0}(\rho) e^{i\beta_1 L}\, \hat{\wp}_{DGD} \left\{ \sum_{j=1}^{3} b_{kj}\, \eta_T(t - \tau - jT) \right\} \vec{\vartheta} \qquad (6.22)$$

where
  P_{out} is the detected power
  $\hat{\wp}_{DGD}$ represents the dispersion operator
  τ is the average propagation delay
  $\vec{\vartheta}$ is a generic polarization vector

Now $\hat{\wp}_{DGD}$ is a well-known operator, and the dispersion related part of Equation 6.22 can be rewritten, passing from the Fourier domain, as

$$\hat{\wp}_{DGD} \left\{ \sum_{j=1}^{3} b_{kj}\, \eta_T(t - jT) \right\} = \frac{T}{2\pi} \sum_{j=1}^{3} b_{kj} \int_{-\infty}^{\infty} \frac{\sin(\omega T)}{(\omega T)}\, e^{i\frac{\beta''(\omega_0)}{2}(\omega - \omega_0)^2 L}\, d\omega \qquad (6.23)$$

The integral can be evaluated numerically via Fast Fourier transform (FFT) to have a correct shape of the received signal for each pattern.

However, a first idea of the broadening of the pulses composing the patterns Φ_k can be attained much simply by the definition of D itself (see Equation 4.13). In a first approximation the square pulse propagating along the fiber broadens of a factor $\Delta\tau = DL\Delta\lambda$. Naturally, during propagation the pulse shape does not remain squared (as resulting from Equation 6.23), but it is not a great error in regime of small distortion to imagine that the pulse broadens with triangular tails (a comparison between the real pulse shape and a triangular pulse is provided in Figure 6.4, in an extreme case, in which there is strong dispersion distortion).

In this case, just to give an example, in a very first approximation the energy translated from the central "one" to one of the nearby "zero" in the pattern $\Phi_8 = 0\ 1\ 0$ can be evaluated as $E_8 = 2PDLR\lambda^2/c$ where P is the relevant optical power [2]. If we want a rough estimate of the penalty due to pulse broadening in this case, we have to compare the signal power with the interference power which is the energy per unit time. Since E_8 is evaluated in the bit interval, this ratio is $Pen = E_8/PT = 2DLR^2\lambda^2/c$ so that, neglecting the noise, the penalty depends in a first approximation on the square of the bit rate and it is independent from the transmitted power.

If SPM cannot be neglected, the propagation problem cannot be solved analytically and a numerical integration of the propagation equation is needed.

In Figures 6.5 and 6.6 an example of the nonlinear evolution of an NRZ pattern is reported [8] comparing linear evolution (at a transmitted power of −10 dBm) and strongly nonlinear evolution (at a transmitted power of 13 dBm). From the figure, it is clear that the transfer of energy from the "one" to the nearby "zero" is much more pronounced in the linear case, even if the pulse shape is completely distorted in the case of nonlinear propagation. From this qualitative observation, justified by the opposite sign of the phase chirps imposed by chromatic dispersion and SPM, it is intuitive that in opportune conditions SPM can help in attenuating the effects of dispersion.

6.2.3.3 RZ Signal after Propagation

If RZ modulation is concerned exploiting again the hypothesis of small distortions, it is possible to demonstrate [10] that a Gaussian pulse propagating along a fiber remains

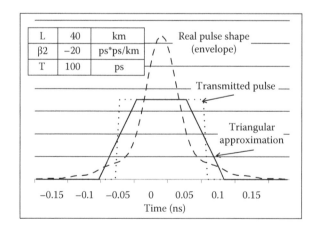

FIGURE 6.4
NRZ pulse after propagating through 40 km of SSMF with perfect linear propagation and absence of PMD. The triangular approximation of the distorted pulse is also shown.

Instantaneous power (AU)

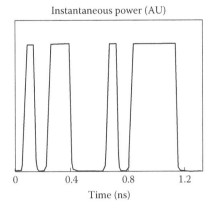

Time (ns)

FIGURE 6.5
NRZ pulse sequence transmitted to generate the example of Figure 6.6.

Instantaneous power (AU)

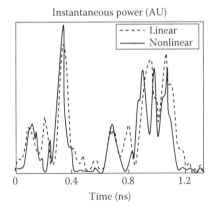

- - - - Linear
——— Nonlinear

Time (ns)

FIGURE 6.6
An example of the nonlinear evolution of an NRZ pattern comparing linear evolution (at a transmitted power of −10 dBm) and strongly nonlinear evolution (at a transmitted power of 13 dBm). An SSMF fiber is assumed with a length of 30 km.

Gaussian in shape, undergoing a broadening caused by interaction between chromatic dispersion, SPM, and polarization dispersion. While chromatic dispersion and SPM are deterministic phenomena, and their interaction cannot be neglected due to the fact that both produce a chirp that modifies the signal phase, PMD is a statistical phenomenon and the PMD induced broadenings can be considered additive.

In [10] the following approximated equation is found for the final pulse width T_1 relative to a Gaussian pulse propagating in the presence of chromatic dispersion and SPM:

$$T_1 = T_0 \sqrt{1 + \sqrt{2}\,\frac{LL_e}{L_D L_{NL}} + \left(\frac{L}{L_D}\right)^2 \left[1 + \frac{4}{3\sqrt{3}}\left(\frac{L_e}{L_{NL}}\right)^2\right]} \tag{6.24}$$

where
L is the link length
L_e is defined in Equation 4.36

The dispersion and nonlinear lengths are defined as follows:

$$L_D = \frac{T_0^2}{|\beta_2|}$$

(6.25)

$$L_{NL} = \frac{\lambda A_e}{2\pi n_2 P_0}$$

where both the effective area A_e and the nonlinear index n_2 are defined in Chapter 4.

Equation 6.64 even if obtained for the first time with a specific derivation, can be derived from the general pulse broadening equation presented in Appendix C.

The expression of the characteristic lengths related to the different phenomena is an important source of intuitive information on the system.

Due to their role in Equation 6.24 and many other equations related to nonlinear propagation in single-mode fibers, they represent the fiber length needed to have a relevant contribution of the considered phenomenon. Thus, longer the characteristic length, the weaker the phenomenon.

The average detected field expression can be rewritten as

$$\vec{E} = \sqrt{P_{out}}\, A_{1,0}(\rho) \sqrt{\sum_k b_k m_{out}(t - kT - \tau)}\, e^{-i\varphi}\vec{\vartheta}$$

(6.26)

where

P_{out} is the detected power, taking into account all the attenuation effects

τ is the average propagation delay

φ collects in a single constant all the constant phase terms

$\vec{\vartheta}$ is a slowly varying random unitary polarization vector

m_{out} is a normalized Gaussian pulse whose width T_{out} is given by $T_1 + \Delta\tau$, where $\Delta\tau$ is the PMD random delay (compare Equation 4.31)

Equation 6.26 allows us to determine the performance of RZ transmission.

6.2.3.4 Realistic Receiver Noise Model

We have seen that the field emitted by a laser is with an excellent approximation in a coherent quantum state [11]. Narrow band modulation does not change this situation since the coherence time of the modulated field remains very long with respect to the time T_ω needed for a complete revolution of the field phase ($\omega T_\omega = 2\pi$). Thus, the number of photons arriving on the photodiode has a poison statistics and generates the so-called shot noise or quantum noise.

If shot noise is considered, three noise terms appear in the expression of the current: squared shot noise, beat between shot noise and signal, and thermal noise.

6.2.3.4.1 Shot Noise Terms

The shot noise is a signal-dependent noise affecting mainly the transmission of a "one."

Neglecting the squared shot noise, since it is a second-order term in the noise, and considering that probability p that an incident photon generates an electron is related to the photodiode responsivity R_p by the formula $p = \hbar\omega R_p/\varepsilon$, where ε is the electron charge, the instantaneous power of the shot noise signal beat is

$$\sigma_{shot}^2 = R_p \varepsilon P_o B_e \tag{6.27}$$

where

B_e is the front-end, electrical bandwidth
P_o is the instantaneous received optical power

6.2.3.4.2 Detector Noise

The detector noise depends on the photodiode that is used.

Starting from a PIN photodiode, the electrical noise introduced by the photodiode with its front-end signal amplifier depends essentially on the electronic structure of the front-end amplifier (see Chapter 5). Here we will summarize all the parameters of the front end in the so-called noise factor, called F_a, so that we have the following expression for the total noise:

$$\sigma_{th}^2 = \frac{4\mathfrak{K}_B \mathfrak{T}}{\mathcal{R}_a} F_a B_e \tag{6.28}$$

where $\mathcal{R}_a$ is the front-end input resistance. The Boltzmann constant is indicated with $\mathfrak{K}_B$ and $\mathfrak{T}$ is the absolute temperature. In Equation 6.28, the contribution due to the spectral increase of the electronic noise at very low frequencies, the so-called 1/f noise, is neglected. Both for this reason and for coupling reason with the other electronic circuits of the receiver, practical front ends generally exhibit a minimum passband frequency, provided by an input inductive impedance. Below the minimum passband, the electrical signal is eliminated.

A minimum passband frequency around 100 kHz, for example, does affect neither the modulated signal nor the residual carrier (that has to be preserved for the sampling synchronization), and allows low frequency noise to be virtually eliminated.

Besides the thermal noise, practical PIN-based receivers exhibit another form of electrical noise that could give a relevant contribution in specific cases. It is the dark current noise (see Chapter 5) that depends on the fact that, even in the absence of detected radiation, a real PIN photodiode under inverse polarization is traversed by a small current, called dark current, whose amplitude is a random process so that it is seen as another noise source during signal detection.

In case of the use of an avalanche photodiode (APD), besides the thermal noise contribution (Equation 6.28), the APD introduces a so-called excess noise due to the intrinsic random nature of the avalanche gain (see Chapter 5).

The excess noise is generally represented as a sort of shot noise amplification, so that Equation 6.27 can be modified as follows:

$$\sigma_{shot}^2 = 2G_m^2 \varepsilon R_p P_o F_A \tag{6.29}$$

where

G_m is the avalanche gain
F_A is the excess noise factor, that is a function of the ionization ratio r_i (see Chapter 5) and
   can be written as

$$F_A = r_i G_m + (1 - r_i)\left(2 - \frac{1}{G_m}\right) \tag{6.30}$$

Besides multiplication noise, when using an APD, one has to take into account that also the dark current noise is amplified by the avalanche mechanism.

In order to give a correct expression of the amplified dark current noise, one has to take into account that dark current can be divided into two components, depending on the physical phenomenon generating it.

There is a bulk dark current and a surface leakage dark current; the first is really amplified by the APD, the second does not undergo amplification due to the fact that it is a surface phenomenon.

Thus, indicating with S_{bd} and S_{sd} the spectral densities of the two components, the final APD dark current power in the detector bandwidth will be

$$\sigma_{dc}^2 = G_m S_{sd} B_e + G_m^2 S_{bd} F_A B_e \tag{6.31}$$

6.2.3.5 Performance Evaluation of an Unrepeated IM-DD System

At this point, we have all the elements to evaluate the BER for an IM-DD system, at least in the hypothesis that the receiver signal processing (i.e., baseband filtering and clock recovery) can be assumed perfect.

Since dispersion involves nearby bits, we need to start from the consideration of the eight pattern of 3 bits that we have called Φ_k ($k = 1, 2, \ldots, 8$).

The current corresponding to a certain pattern after detection and front-end amplification can be written as

$$c_k(t) = R_p P_{out} G_m \sum_k b_k m_{k,out}(t - jT - \tau) * \mathcal{H}(t) + c_{bias} + n_{k,shot}(t) + n_{k,r}(t) \tag{6.32}$$

where $n_r(t)$ collects all the receiver related noise terms: thermal noise, excess thermal noise, and dark current noise and the diode gain G_m has to be set to 1 for a PIN.

Due to the presence of the shot noise term, the statistical distribution of the photocurrent now is not Gaussian. However, as we have noted in the previous section, in the case of the use of a PIN, the thermal noise is largely prevalent in practical systems; thus, it is not a great approximation to consider the noise distribution to be Gaussian.

In the case of an APD, the APD-generated noise can dominate the system performances, but the random behavior of the gain shapes the noise distribution so that a Gaussian approximation is very near to the reality.

Thus, after sampling and bias elimination, the current sample has the following expression:

$$C_{k,j} = R_p P_{out} G_m b_{kj} m_{k,out}(t_k - jT - \tau, \Delta\tau) + N_{kj} \tag{6.33}$$

where the noise sample N_{kj} is the sum of all the noise terms and can be considered a Gaussian variable and the dependence of the sample from the polarization induced delay difference is evidenced.

Once a value of $\Delta\tau$ is fixed, the BER can be evaluated with the technique that was shown in the previous section, only carefully taking into account that the noise variance is dependent on the signal sample.

Thus, the following conditional error probability expressions are obtained, the first for an adaptive threshold receiver, the second for a fixed threshold receiver with the threshold at half the detected signal dynamic range:

$$P\left(\frac{er}{\Delta\tau}\right) = Q\left(\frac{c(1,\Delta\tau) - c(0)}{\sigma_n(1) + \sigma_n(0)}\right) \quad \text{(adaptive threshold)} \tag{6.34}$$

$$P(er/\Delta\tau) = \frac{1}{2}\left[Q\left(\frac{c(1,\Delta\tau) - c(0)}{2\sigma_n(1)}\right) + Q\left(\frac{c(1,\Delta\tau) - c(0)}{2\sigma_n(0)}\right)\right] \quad \text{(fixed threshold)} \tag{6.35}$$

where "er" represents the error event.

Thus, applying the Bernoulli theorem, the BER is given by

$$P_e(k) = \int_{-\infty}^{\infty} p(\Delta\tau) Q\left(\frac{c(1,\Delta\tau) - c(0)}{\sigma_n(1) + \sigma_n(0)}\right) d\Delta\tau \quad \text{(adaptive threshold)} \tag{6.36}$$

$$P_e(k) = \frac{1}{2}\int_{-\infty}^{\infty} p(\Delta\tau) \left[Q\left(\frac{c(1,\Delta\tau) - c(0)}{2\sigma_n(1)}\right) + Q\left(\frac{c(1,\Delta\tau) - c(0)}{2\sigma_n(0)}\right)\right] d\Delta\tau \quad \text{(fixed threshold)} \tag{6.37}$$

where $p(\Delta\tau)$ is the distribution of $\Delta\tau$ (see Chapter 4).

This technique allows the random effect of the PMD to be incorporated in the evaluation of the error probability.

However, besides the average worsening of the BER, the PMD has also another effect. During the random variation of $\Delta\tau$ short periods can happen where the value of the PMD is very high. In this case the system working is completely destroyed even if for a short period.

The probability of this event called outage is small; therefore, it does not influence greatly the long-term error probability, but when it happens the system could even go down so its importance is beyond the pure influence on the long-term error probability.

This is the reason why generally also the so-called outage probability is evaluated to characterize the PMD influence on a system in situations of high PMD.

We do not detail the evaluation of this parameter in general, encouraging the interested reader to consult [12,13] and Section 9.2.3.2.

6.2.4 Performance of Non-Regenerated NRZ Systems

In order to analyze the effect of all the transmission impairments that are included into Equations 6.36 and 6.37, let us concentrate on the optimum receiver structure: that with the adaptive threshold.

When it will be necessary to evaluate numerical results in this section and in the following, we will use a set of representative parameters reported in Table 6.1, but for fibers' parameters, that are reported in the Table 4.1 of Chapter 4 for the SSMF and in Table 4.2 of the same chapter for NZ+ and NZ−, depending on the fiber type.

TABLE 6.1

Reference System Parameters for the Performance Evaluation Carried Out in the First Part of the Chapter

Transmitter		
NRZ		Perfectly squared pulse
Bit rate		10 Gbit/s (40 Gbit in some examples)
Dynamic range		20 dB
Transmitter loss (up to the booster input)		14 dB
Laser alternatives	Standard DFB	Nontunable	13–20 dBm emitted power
	Multisection DBR	Tunable	16 dBm emitted power (with SOA)
	External cavity	Tunable	13–16 dBm emitted power

Receiver	
Responsivity	0.9
Photodiode thermal noise current amplitude density $\sqrt{4 \Re_B \mathcal{J} F_a / \mathcal{R}a}$	2 pA/$\sqrt{\text{Hz}}$
Dark current power (PIN)	50 nA
Electrical bandwidth	10 GHz
INGaAsP APD ionization ratio (r_i)	0.7
Spectral densities of the APD dark current components (S_{bd} and S_{sd})	2.5×10^{-23} A^2/Hz
Receiver loss, from the preamplifier output to the APD	8 dB

The first impairment to analyze is chromatic dispersion since in standard single-mode fiber (SSMF) it generates a rapid widening of the transmitted pulses.

To analyze system performances we will require a BER of 10^{-12} and in agreement with ITU-T recommendations we will evaluate the dispersion limit reach for each system as the distance at which the dispersion induced penalty is 2 dB.

The performance of an NRZ system with no SPM and no PMD, are reported in Figure 6.7.

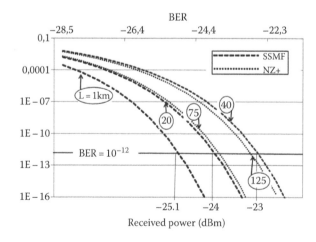

FIGURE 6.7
Performance of an NRZ system with no SPM and no PMD. System parameters are reported in Table 6.1 but for fibers parameters that are reported in Table 4.2.

As noted in the previous section, pulse broadening has the effect to transfer energy from one bit period to the adjacent one. If this causes energy to be transferred from a one to the adjacent zero, the signal dynamic range is reduced and the BER degrades. This phenomenon is also called intersymbol interference (ISI).

The presence of ISI has a different effect with respect to the noise on the BER curve. Increasing the noise power a greater value of the received power is needed to achieve the same SN_e and then the same BER.

However, whatever the required BER, it can be achieved, at least theoretically, transmitting a sufficiently high power.

In the case of ISI, both the energy detected on a "one" and the energy transferred to the adjacent time slots increase proportionally to the transmitted power. Thus, the ratio between the energy that remains in the right time slot and the disturbing energy is constant and no increase of the signal power can change it.

This causes a change of shape of the BER curves and it is expected that the penalty would rise much faster than linear increasing pulse broadening.

The chromatic dispersion penalty can be evaluated from the BER and it is plotted in Figure 6.8 where the foreseen rapid increase of the penalty can be verified.

From the figure, a limit distance of about 36 km is estimated for transmission on SSMF at 10 Gbit/s, while about 132 km are derived for transmission on NZ+.

To understand the penalty trend, when the bit rate changes, in the figure, also a curve relative to 40 Gbit/s on an NZ fiber is shown indicating for this case a dispersion limit of 11 km; thus, the dispersion-limited distance decreases approximately of a factor 16 passing from 10 to 40 Gbit/s.

This is a general trend that can be easily verified using the performance evaluation method detailed in this section: the dispersion limit decreases as the square of the bit rate, at least in a first approximation. This is a confirmation of the rough evaluation carried out in the previous section on the ground of the pulse widening.

It is clear from the aforementioned results that in order to obtain long-reach systems, dispersion has to be compensated.

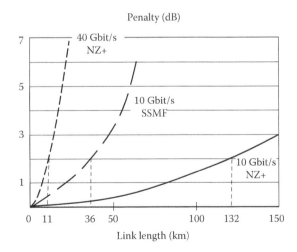

FIGURE 6.8
Chromatic dispersion penalty for an IM-DD system using NRZ modulation. System parameters are reported in Table 6.1 but for fibers parameters that are reported in the suitable table of Chapter 4.

Several methods have been conceived to compensate dispersion and the most used is the adoption of dispersion compensating fibers (DCFs). It is interesting however to analyze the effect of several different methods.

6.2.4.1 Dispersion-Compensated NRZ IM-DD Systems

Since dispersion is a linear and deterministic phenomenon, in line of principle perfect passive compensation with optical dispersion elements should be possible.

Two kinds of in-line dispersion elements exist: DCF and dispersion compensating gratings.

The DCF can be placed either at the transmitter or at the receiver, even if the receiver position seems advantageous since at the receiver it could be integrated with other devices.

The only real disadvantage in using DCFs is their high loss that in a system without optical amplifiers can be a limitation.

Besides the use of DCF or fiber gratings, electronic compensation or pulse pre-chirp can be used.

The basic idea of the pre-chirp method is to inject into the fiber a signal whose pulses have already a phase modulation exactly opposite to that caused by the fiber dispersion.

During propagation, dispersion compensates this pre-distortion and the pulse arrives at the receiver without any ISI [14–16]. This is a type of pre-distortion similar to that adopted through some electronic devices, but for the fact that it is performed optically.

An important advantage of pre-chirping is that no specific device is needed to apply this technique if an external modulator is used for signal modulation. In this case, pre-chirp can be achieved by simply modulating the transmitting laser bias current or using the pre-chirp embedded functionality that is present in many external modulators.

It is also to be noticed that pre-chirping does not conflict with dithering for Brillouin removal. This last modulation does not have requirements in terms of pulse shape, but only in terms of residual carriers' bandwidth width.

Pre-chirp has to be optimized for the dispersion of the particular link and generally this optimization is carried out by an automatic control loop driving the chirp parameters in order to maximize either the EOP that is detected at the receiver or, when available through an error correcting code, the estimated BER.

In real systems, the effectiveness of pre-chirp is limited essentially by nonperfect linearity of the phase modulation, by the spurious intensity noise that always comes with a semiconductor laser phase modulation, and by the jitter coming from imperfect synchronization of the phase modulation with the information bearing intensity modulation.

In Figure 6.9, the dispersion related power penalty of a 10 Gbit/s system using a SSMF fiber and optical pre-chirping is shown, evidencing both the effectiveness and the limits of pre-chirping techniques.

An advantage of the use of pre-chirping is that it can be combined with almost all the other dispersion compensating techniques with a good result and that it can be used also in the presence of SPM. Also in this last case, pre-distortion parameters have to be optimized for the specific case.

Another method to compensate for channel distortions is electronic compensation. As discussed in Chapter 5, electronic compensation has two important characteristics: it potentially compensates for every kind of channel distortion, independently from its causes, so that it can be useful also for management of nonlinear effects and, in the case of fast adaptive algorithms, also for PMD. On the other hand, since it is applied on the signal after detection, even if fiber propagation is perfectly linear, the channel as it is viewed by

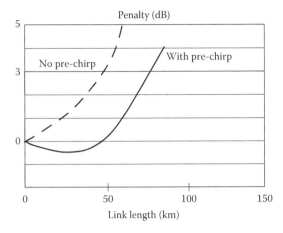

FIGURE 6.9
Dispersion related power penalty of a 10 Gbit/s system using NRZ modulation a SSMF fiber and pre-chirping technique. System parameters are reported in Table 6.1 but for fibers parameters that are reported in Table 4.1. (After Boyd, R.W., *Nonlinear Optics*, Academic Press, San Diego, CA, 2008.)

the electronic dispersion compensator (EDC) is a nonlinear channel due to the presence of square law detection carried out by the photodiode.

Since a large number of EDC are digital linear filters, we can expect only a partial compensation.

Of course, this is not true for nonlinear compensation compensator, that in principle can attain much better performances and that, due to the advance in microelectronics technology, is for the first time a practical possibility.

To have an idea of the possible performance of feed forward equalizer–decision driven equalizer (FFE–DFE) compensators, in Figure 6.10 the optical signal to noise ratio SN_o needed for a BER of 5×10^{-4} (i.e., the requirement for systems using a standard forward error correcting code [FEC], see Section 5.8.5.5) is plotted versus distance for propagation over an SSMF and detection using a PIN photodiode [17].

The FFE–DFE compensator allows the dispersion limit (about 36 km in this case without dispersion compensation) to be pushed up to about 70 km in this case, and slightly better results are reported in literature. In the figure also the performances of a Viterbi algorithm–based maximum-likelihood-sequence-estimator (MLSE) compensator are reported. As expected, it is more effective in compensating extra errors coming from signal distortion, pushing the dispersion limit in this case up to about 90 km.

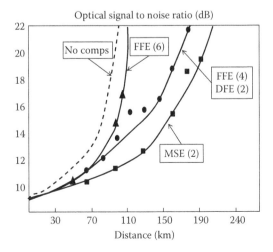

FIGURE 6.10
OSNR needed for a BER of 5×10^{-4} (i.e., the requirement for systems using FEC, see Section 5.8.5.5) versus distance for propagation over an SSMF and detection using a PIN photodiode of an NRZ IM-DD system using FFE–DFE electronic equalizer at the receiver. Also the performances of a Viterbi algorithm-based MLSE compensator are reported. System parameters are reported in Table 6.1 but for fibers parameters that are reported in Table 4.1. (After Xia, C. and Rosenkranz, W., *IEEE Photon. Tech. Lett.*, 19(13), 1041, 2007.)

As discussed in Section 5.8.3.1, Viterbi algorithm changes the statistics of errors after the compensator; thus, the effect of the FEC cannot be evaluated assuming independent errors. Several studies are ongoing to determine exactly the impact of this element and to find an optimum design for the overall MLSE + FEC system, and probably this difficulty will be soon overcome with a global design taking into account both the elements together.

Even more effective results can be obtained using pre-distortion of the transmitted signal through a look-up table designed to compensate both linear and nonlinear signal distortions.

This method is not easy to implement and to control during system working, but it is very high performing, as it is demonstrated, for example, in [18]. The performance of a mix look-up table and linear electronics equalizer are reported in Figure 6.11 [18], where a maximum distance of more than 500 km is foreseen with only signal pre-distortion in pure linear propagation.

Let us now assume that chromatic dispersion has been completely canceled. In this case limitations arise to the transmission reach from PMD and SPM.

Let us start to assume transmission at low power, so that SPM is negligible, and to operate with a fiber installed before the discovery of the importance of controlling the PMD (a lot of fiber installed before the years 2000 are manufactured with processes that can produce a high PMD).

Thus, let us assume the parameters of an SSMF fiber with a residue chromatic dispersion after compensation of $0.5 \, ps^2/km$ and a PMD parameter of $0.4 \, ps/km^{1/2}$.

The PMD induced penalty is shown in Figure 6.12 versus distance for 40 and 100 Gbit/s. It is clear that great areas of the figure do not correspond to links that can be realized with unamplified systems, but the graphic is quite important since, due to the linear nature of PMD, the effect is independent of the transmitted power. This means that Figure 6.12 approximately holds also for amplified systems (but for the fact that an accurate estimation of the penalty in that case has to take into account the different noise distribution).

Figure 6.12 demonstrates that PMD is an important impairment in long-reach, high bit rate systems, and in a few important cases, for example, when mixed erbium-doped optical

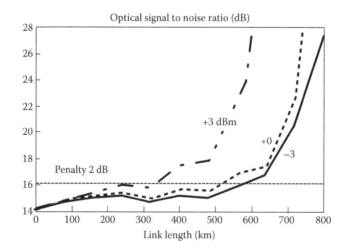

FIGURE 6.11
Performance of a mix look-up table and linear electronic equalizer: a maximum distance of more than 500 km is foreseen with only signal pre-distortion in pure linear propagation. System parameters are reported in Table 6.1 but for fibers parameters that are reported in Table 4.1. (After Killey, R.I. et al., *IEEE Photon. Tech. Lett.* 17(3), 714, 2005.)

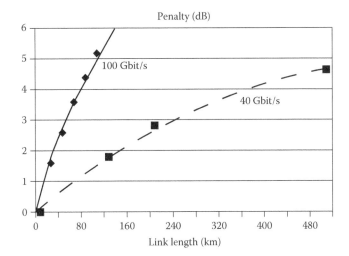

FIGURE 6.12
PMD induced penalty versus distance for 40 and 100 Gbit/s. The parameters of an SSMF fiber with a residue chromatic dispersion after compensation of $0.5\,ps^2/km$ and a PMD parameter of $0.4\,ps/km^{1/2}$ are considered. The system parameters are reported in Table 6.1.

fiber amplifier (EDFA)–Raman amplification is used to maintain the optical power low and avoid nonlinear effects, it can be the ultimate limit to the system reach.

PMD compensation is much harder with respect to chromatic dispersion compensation due to the fact that PMD not only is time varying, but it is also a random process. This means that the only possible way for compensation is to devise an adaptive compensator.

For their nature, all electronics adaptive dispersion compensators can be optimized so to face also PMD. The great majority of them does not distinguish pulse broadenings due to different physical causes so that, naturally, they try to compensate any distortion.

However, linear compensators like FFE–DFE suffer in terms of performances of the nonlinear nature of the communication channel.

Completely different is the situation for adaptive nonlinear compensator that reaches a very good level of compensation (see Figure 6.11).

Up to now, we have evidenced the penalty caused by linear and nonlinear propagation effects on the system performances.

Now we pass to consider another important element of the transmission chain: the detector. A PIN photodiode can be substituted with an APD, so to exploit the detector gain to increase the SN_e that is dominated by the receiver thermal noise.

From the previous section, we have all the elements to analytically estimate the BER in this case. There is only one important observation to do.

Looking at Equations 6.27 through 6.29 we can see that the overall noise power is composed of two terms. The thermal noise that does not depend on the APD gain and the sum of shot noise and dark current noise that depends on the gain.

For a great value of G_m the shot term prevails and increasing the gain the SN_e decreases; for small values of G_m the APD behaves like a PIN with a small gain so that increasing the gain the SN_e increases.

This behavior indicates the presence of an optimum gain that maximizes the SN_e. The optimum gain naturally depends on the parameters of the APD. Just for an example, the parameters of a medium performance InGaAs APDs are reported in Table 6.1. The

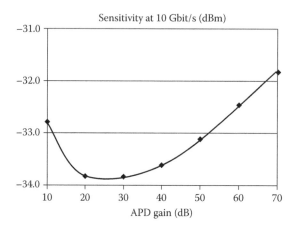

FIGURE 6.13
Sensitivity (input optical power needed to attain a BER of 10⁻¹² at 10 Gbit/s) versus APD gain in the absence of dispersion and nonlinearity. The parameters are reported in Table 6.1.

optimum gain at 10 Gbit/s is around 28. In Figure 6.13 the sensitivity (input optical power needed to attain a BER of 10^{-12}) is shown versus the APD gain in the absence of dispersion and nonlinearity.

In this ideal condition, the sensitivity gain due to the APD is around 9 dB, passing from a PIN sensitivity of about −24.5 dBm to an APD sensitivity of about −33.5 dBm.

The impact of the other performance decreasing factors on APD receivers are not different from what we have seen in the case of the PIN receiver, but for the sensitivity gain.

The only factor that is sensibly impacted from the presence of an APD is the role of SPM.

As a matter of fact, for a fixed fiber link, the presence of an APD at the receiver allows the optical power to be maintained much lower, so reducing the effect of SPM.

6.2.5 Performance of Non-Regenerated Return to Zero Systems

Considering RZ systems, the performances in condition of ideal propagation are the same with respect to NRZ, since they depend only on the signal and noise energy received in the bit interval.

The situation is completely different in the case of presence of dispersion. In the case of chromatic dispersion, two things happen when decreasing the pulse width.

On one side, the dispersion is more efficient, the relative pulse broadening being bigger, on the other side, the pulse does not occupy the whole bit interval; thus, some broadening can be accepted if no energy is transferred in the nearby intervals.

As usual, when two different phenomena tend to balance we have to expect an optimum value of the pulse width at which the penalty due to chromatic dispersion is minimum.

As we have mentioned in Chapter 4, the pulse propagation problem in the presence of chromatic dispersion can be solved exactly in the case of Gaussian pulse. In this case, it is possible to derive the following expression of the output pulse root square medium width (that is also half the width at $1/e$ below the maximum), obtaining [2]

$$T(z) = \sqrt{T^2(0) + \left(\frac{\beta_2 z}{T^2(0)}\right)^2} = T(0)\sqrt{1 + \left(\frac{\beta_2 z}{T^4(0)}\right)^2} \qquad (6.38)$$

that is Equation 6.17 in the limit of $L_{NL} \to \infty$. This expression is another example of the more general rule to analyze the widening of a pulse traveling through a cascade of filters that is derived in Appendix D. As a matter of fact, the factor that is added to the unit under the square root is exactly the square of the width of an ideal filter that has on the optical pulse the same effect of the dispersion.

Thus, $\Delta T = T(z) - T(0)$ is minimum at a distance L from the transmitter, if $T(0) = \sqrt{\beta_2 L}$. For an SSMF this means, at a distance of 40 km, $T(0) = 0.028$ ns that is a short pulse with respect to the duration of 0.1 ns of the bit interval at 10 Gbit/s, but it is not sufficiently short for 100 Gbit/s transmission, where the bit interval is 0.001 ns.

Moreover, increasing the distance the optimum pulse becomes longer. This can be understood by observing that, for very large distances, the term containing dispersion is so high that the other term under square root can be neglected and the final pulse width is inversely proportional to the square of the initial width so that shorter launched pulses produce longer output pulses.

For example, at 10 Gbit/s and with $\beta_2 = 20 \, ps^2/km$, the optimum pulse at 300 km is 77 ps. Since this is the half width at $1/e$, this means that we are quite near NRZ, with a pulse whose real width is 50% wider than the bit interval (if a real transmission would be set up, naturally only the part of the pulse within the bit interval would be transmitted).

In the following analysis of 10 Gbit/s systems, we will consider practical pulses never shorter than 10 ps and never longer than half the bit interval whatever the distance and the bit rate.

In order to define the pulse width we will use the so-called duty cycle of the RZ modulation that is defined as $(2T_0 R)$ so to be in practical cases always smaller than one so to be measured in percentage unit. In a 10 Gbit/s RZ modulation with a duty cycle of 30% the full width at $1/e$ of the used pulses is 30 ps out of the 100 ps of the bit interval.

The sensitivity penalty of an unrepeated and uncompensated RZ system using pulses of 25 ps is shown in Figure 6.14 for propagation on an SSMF and different values of the PMD parameter. The effect of PMD is evident also at short distances due to the short pulses.

For uncompensated systems on high dispersion fibers, some working zones can be individuates, where there is a partial compensation between the chromatic dispersion and the

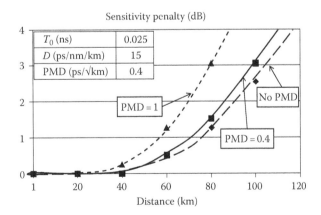

FIGURE 6.14
Sensitivity penalty of an unrepeated and uncompensated RZ system using pulses of 25 ps for propagation on an SSMF and for different values of the PMD parameter. System parameters are reported in Table 6.1 but for fibers parameters that are reported in the suitable table of Chapter 4.

SPM induced chirps. Although this phenomenon is not particularly useful in unrepeated system, it is important as the base on which the design of nonlinear aided WDM system is founded [8].

6.2.6 Unrepeated Wavelength Division Multiplexing Systems

The huge bandwidth available in the third transmission window of optical fibers naturally move the interest toward frequency multiplexing, that in the field of optical transmissions is called WDM.

In general, WDM systems are called DWDM if the channels are a few GHz far, one from the other. Typical channels' spacing for DWDM systems at 10 Gbit/s are 50 or 100 GHz, while for 40 Gbit/s a typical spacing is 200 GHz.

DWDM systems require frequency stabilized sources and suffers sizeable interference among adjacent channels due to different phenomena.

Differently, CWDM (coarse WDM) systems have 20 nm spaced channels, so that up to 18 channels can be used in the low attenuation bandwidth of an optical fiber, if the water absorption peak is not present (compare Chapter 4).

The 20 nm spacing is chosen on the ground of the standard characteristics of semiconductor lasers, so to be able to use uncooled sources to reduce the system cost.

CWDM systems are mainly used in the metro and metro access area, where distances are not so long and the cost is a key issue.

In addition to phenomena impacting the single channel, channel crosstalk is another relevant degradation source in WDM systems.

Assuming an NRZ signal, the received field can be written as follows:

$$\vec{E} = \sum_k \vec{E}_k = \sum_k \sqrt{P_k}\, e^{i(\beta_1 L + k\,\Delta\omega t)} \left\{ \sum_j b_{jk} g_{jk}(t - \tau_k - jT) \right\} \vec{\vartheta}_k \tag{6.39}$$

here k is the channel index that is assumed to run from $-N/2$ to $N/2$ being $N+1$ the channel number. Since the channels interact one with the other, the received pulse in the jth bit interval, not only depends on its position in the bit stream, but also on the considered channel; this is the reason why the pulse $g_{jk}(t)$ has both the channel index k and the slot index j. The channels will be asynchronous in time, since their group velocity is different, so that the propagation delay τ_k brings the channel index.

Similarly to the delay, the polarization also will be different from channel to channel, justifying the channel index on the polarization unitary vector $\vec{\vartheta}_k$.

Last but not least, the detected power also depends on the considered channel, due to the nonuniform wavelength response of optical components, and it is labeled as P_k.

In the usual hypothesis that the transmission impairments are small and can be treated as perturbation of the ideal system, we will go on analyzing the interference in form of a power penalty, to add to penalties coming from other sources.

Two kinds of interferences are potentially present in WDM systems: linear and nonlinear interference.

The term linear and nonlinear depends on the fact that the first type is always present, even when linear propagation can be assumed, while the second is related to nonlinear fiber propagation.

6.2.6.1 Linear Interference in Wavelength Division Multiplexing Systems

Linear interference arises due to the fact that there is an overlap between adjacent channel optical spectra, so that the power of a channel unavoidably interferes with the adjacent ones, as shown in Figure 6.15.

This effect has to be limited by optical filtering, since after quadratic detection, all the channel spectra superimpose in the electrical baseband. Indicating with $H_o(t)$ the optical pulse response of the optical filter, the electrical field relative to channel $k = 0$ after demultiplexing and, if needed, further optical filtering is given by

$$
\vec{E} = \sqrt{P_0}\, e^{i\beta_1 L}\left\{\sum_j b_{j0}\, g_{j0}(t - jT)\right\}\vec{\vartheta}_0 + \sum_{k \neq 0}\sqrt{P_k}\, e^{i\beta_1 L}\left\{\sum_j b_{jk} g_{jk}(t - \Delta\tau_k - jT)e^{ik\Delta\omega t} \otimes H_o(t)\right\}\vec{\vartheta}_k
$$

$$
= \sqrt{P_{0k}}\, e^{i\beta_1 L} G_{k0}(t,0)\vec{\vartheta}_{0k} + \sum_{k \neq 0}\sqrt{P_k}\, e^{i\beta_1 L} G_k(t, \Delta\tau_k) * H_o(t)\vec{\vartheta}_k
$$

(6.40)

where $G_k(t,\tau)$ represents the signal conveyed by the kth channel, whose expression can be easily deduced from Equation 6.33, in a first approximation the filtering effect of $H_o(t)$ on the selected channel is neglected and $\Delta\tau_k = \tau_k - \tau_0$.

Generally, for a well designed WDM system, only adjacent channels generate relevant linear interference; thus, the index k in Equation 6.40 can assume only the values -1 and 1.

From Equation 6.40, the current after detection can be derived obtaining in the jth time slot:

$$
c_j(t) = R_p P_0 G_j(t,0) + R_p \sum_{k \neq 0}\sqrt{P_{k0}P_k}\, G_{j0}(t,0)G_{kj}(t,\Delta\tau) \otimes H_o(t)\vec{\vartheta}_k \cdot \vec{\vartheta}_0 + \sum_{k \neq 0} R_p P_k\, G_{kj}(t,\Delta\tau) + n(t)
$$

(6.41)

where $n(t)$ represents the global noise term, and a PIN receiver is assumed.

From Figure 6.15 it is clear that the part of the spectrum of an adjacent channel entering the selection filter has to be small. In this condition, the 0–j beat term is quite bigger with respect to the j–j so that this last term can be neglected.

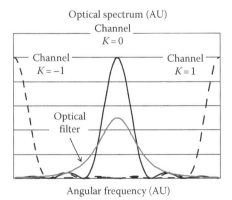

FIGURE 6.15
Linear interference due to the overlap between adjacent channel optical spectra.

Even with this simplification, the analysis of the linear interference impact is difficult due to the random nature of this effect. As a matter of fact, both the scalar product of the polarization vectors and the relative delay are slowly varying random variables.

However, due to the slow variation, calculating their distribution and the related average BER can be misleading. As a matter of fact, the random variables related to polarization and delay can assume, for a long time, values near the worst case, thus heavily influencing the transmission of many bits.

For these reasons, it makes sense to evaluate the worst case BER penalty.

The worst situation is obtained when interfering channels carrying a "one" signal are synchronous and co polarized with the useful channel (see Equation 6.41) [19].

With all these approximations the photocurrent writes

$$c_j(t) = R_p P_0 G_j(t,0) + R_p \sum_{k \neq 0} \sqrt{P_{k0} P_k} G_{j0}(t,0) G_{kj}(t,0) \otimes H_o(t) + n(t) \qquad (6.42)$$

and we can use the tools used in the previous sections to evaluate the BER.

In Figure 6.16, the worst case linear interference penalty is plotted versus the channel spacing at 10 Gbit/s and for different optical filters. Perfect dispersion compensation and absence of relevant PMD and SPM are assumed. The results are plotted searching the worst case with respect to the relative phase of the carrier of the considered and the interfering channel. This assumption is important for small channel spacing, where only few oscillations of the relative interfering carrier (with angular frequencies $\Delta\omega$) are comprised into the bit interval. In normal situations, where there is at least a factor 5 between the bit interval and the interfering carrier relative frequency this is not an important assumption.

Gaussian filters are considered in Figure 6.16a while ideally rectangular filters are used to plot Figure 6.16b.

Moreover, the optical bandwidth is unilateral and this convention will be always assumed in this book. Thus, the minimum optical bandwidth that catch the entire signal spectrum main lobe is $B_o = R$ and the Nyquist bandwidth is $B_o = R/2$.

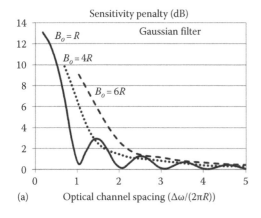
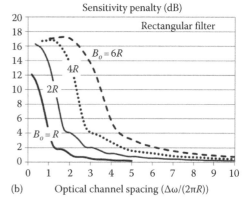

FIGURE 6.16

Worst case linear interference penalty versus the channel spacing for different optical filters: (a) shows the results for Gaussian filters (where the bandwidth is half width at $1/e$) and (b) for rectangular filters.

The impact of the optical filter shape is clear, since the amount of interference power depends on the filter shape.

From the figures, also the oscillating nature of the interference can be noted, that depends on the channels spectrum shape.

6.2.6.2 Nonlinear Interference in Wavelength Division Multiplexing Systems

Nonlinear interference depends on nonlinear fiber propagation. The effects that give the main contribution to this interference term are Kerr induced four wave mixing (FWM) and Kerr induced cross phase modulation (XPM), assuming as always that the Brillouin effect is neutralized with a sufficiently fast source dithering [10].

Equations 4.45 and 4.46 allow us to evaluate the FWM power on the various frequencies arising due to this effect. The main hypothesis here is that the channels can be considered monochromatic, that means the bit rate R is much smaller than the channel spacing.

This condition is verified several times, but not always. For example, it is not verified in the case of a DWDM system at 10 Gbit/s and a channel spacing of 25 GHz.

Fortunately, in the case of 10 Gbit/s systems with 100 or 50 GHz spacing, the monochromatic channel assumption is quite reasonable.

In principle, knowing the FWM power hitting the signal bandwidth does not allow a complete performance evaluation.

As a matter of fact, the FWM instantaneous power in a signal bandwidth fluctuates with time due to the channels modulation, walk-off, and polarization changes. The channel walk-off is due to the fact that different channels have different group velocities; thus, the alignment of the transmitted bit streams varies with time while one "slides" with respect to the other. Since an FWM term arises only when a "one" is transmitted on all the involved channels, walk-off causes the FWM power to change in time.

A similar variation is due to polarization fluctuations. Since we have seen in Chapter 4 that the bandwidth of the PSP is about 100 GHz, if the channel spacing is on the order of 100 GHz or greater, the polarization of different channels evolves independently. FWM happens only when the polarization of the involved channels is almost the same; thus, polarization fluctuations cause FWM power to change in time.

These FWM fluctuations have fast and slow components (as fast as twice the bit rate and as slow as polarization fluctuations) and their exact statistic is complex. In the realistic case, in which there are many WDM channels, several contributions are added on the same wavelength and in a first approximation the FWM can be considered a sort of narrowband colored Gaussian noise [2,8].

Once this approximation is made, the system performance in the presence of FWM can be evaluated by considering a central channel of the WDM comb (i.e., the channel in the worst situation regarding FWM products) and evaluating the number of products contained in the channel bandwidth and the overall FWM power as the sum of their individual powers [20].

As an example in Figure 6.17 [21], a system with nine channels, 30 km long and without dispersion compensation is considered. The transmitted power is 0 dBm per channel. In the figure, the power penalty due to FWM is shown versus the channel spacing.

The first thing to notice is the oscillatory behavior of the curves. This is due to the oscillatory behavior of the phase matching term and also the fact that only the main FWM terms are accounted for in the figure. Considering all the terms the oscillation results less deep in correspondence of the points in which the main terms nullify.

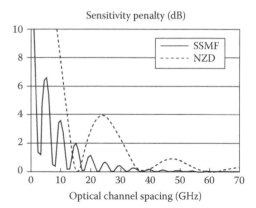

FIGURE 6.17
FWM power penalty versus channel spacing for a system with nine channels 30 km long without dispersion compensation. The other system parameters are reported in Table 6.1.

In designing a WDM system, it is not possible to exploit the fact that the FWM power is strongly reduced around some points for the aforementioned fluctuations of the phase matching conditions, and it is necessary to evaluate the FWM penalty using the envelope of the maximum values of the curve of Figure 6.17.

The strong effect of fiber dispersion is also evident from Figure 6.17, where the FWM impairments are much more severe in the case of an NZ fiber.

Last but not least, the impairment due to FWM has a step behavior typical of nonlinear phenomena. On this ground, an FWM induced capacity limit can be introduced, that is the limit to the product distance by bit rate imposed by the FWM effect.

In general, this is defined as the point on the envelope of the maxima of a figure similar to Figure 6.17 at which the FWM causes a penalty of 1 dB.

This limit can be evaluated starting from the FWM total power as follows. Let us define the FWM to noise ratio FN as the ratio between the FWM power and the optical noise power in the optical selection filter bandwidth: $FN = P_f/\sigma_o$.

If we consider a system limited by the optical noise (that is the most interesting case) and we assume the FWM power as an additional optical noise source, it results

$$SN_o = \frac{P_o}{P_f + \sigma_o}$$

(6.43)

Then, since ideal SN_o is given by $SN_i = P_o/\sigma_o$ the FWM induced penalty can be written as

$$Pen_{FWM} = \frac{SN_o}{SN_i} = FN + 1$$

(6.44)

The analysis of the XPM is a bit more complicated due to the fact that the expression of the XPM induced crosstalk power has to be carried out by dealing with the nonlinear propagation equation in a more complicated way with respect to FWM.

This study is carried out in detail in [22] while a simple but sufficiently accurate model in almost all the practical cases is reported in [23]. We will use this last model, since all the WDM designs we will do will always be verified by simulation.

The conclusion of the model presented in [23] is that also XPM can be considered like a power dependent noise due to the combined effect of modulation, walk-off, and polarization fluctuations.

In analogy with FWM, an XPM to optical noise ratio can be defined $XN = P_x/\sigma_o$, where P_x is the XPM equivalent noise power so that, the XPM induced penalty can be written as

$$Pen_{XPM} = XN + 1 \tag{6.45}$$

The XPM power affecting the kth channel of the WDM comb can be evaluated by the following equation

$$P_X = 4\gamma^2 \langle P \rangle^2 \sum_{\substack{j=1 \\ j \neq k}}^{N} \int_{-\infty}^{\infty} S_j(\omega) \, | \, H_{j,k}(\omega)|^2 \, d\omega \tag{6.46}$$

where

k is the index of the selected channel

$\langle P \rangle$ is the average optical power

$S_j(\omega)$ is the interfering channel normalized power spectral density so that $\int_{-\infty}^{\omega} S_j(\omega)d\omega = 1$

$H_{j,k}(\omega)$ is the representation of the interaction between the jth and the kth channels and it is given by

$$H_{j,k}(\omega) = i \left\{ \frac{1 - e^{[-\alpha L + i(\delta_{j,k}\omega - \Psi\omega^2)L]}}{\alpha - i(\delta_{j,k}\omega - \Psi\omega^2)} - \frac{1 - e^{[-\alpha L + i(\delta_{j,k}\omega + \Psi\omega^2)L]}}{\alpha - i(\delta_{j,k}\omega + \Psi\omega^2)} \right\} \tag{6.47}$$

where

α is the attenuation in m^{-1}

$\delta_{j,k}$ is the so-called walk-off factor, that is the inverse of the difference between the group velocities of the considered channels

Ψ is related to the fiber dispersion by the equation $\Psi = D\lambda^2/4\pi c$, λ being the central wavelength of the WDM comb and c the light speed in glass

L is the length of the link

Equation 6.46 has to be integrated numerically, but it is much easier than a simulation. We will use several times this approach for its simplicity.

6.2.6.3 Jitter, Unperfected Modulation, Laser Linewidth, and Other Impairments

Besides the phenomena we have dealt in the previous sections, there are several other potential performance impairments that have to be considered when evaluating the performance of an optical transmission system.

These effects are not so important as dispersion or FWM, but if not managed with a correct system design they can heavily affect system performances.

- *Timing jitter*: This phenomenon consists of the fluctuations of the receiver decision circuit sampling instant. It is intuitive that, if sampling does not occur at the center of the bit interval, a penalty is generated. Timing jitter generally depends on the imperfect working of the receiver PLL, causing the reconstructed clock to be affected by phase fluctuations. A correct design of the digital receiver PLL can make this phenomenon negligible [24].

- *Pulse jitter*: This phenomenon, affecting mainly RZ transmission, consists in the fact that the transmitted pulses are not located exactly at the center of the bit interval. The cause can be phase noise in the clock at the transmitter, a random component in the switch-on or switch-off time of the transmitting modulator, or even pulse attraction due to nonlinear propagation for quasi-soliton RZ pulses when very long, amplified systems are considered. Transmitted pulses' position fluctuations cause ISI and have to be carefully controlled during system design [25].

- *Limited transmitter dynamic range*: In this case the difference between the "one" level and the "zero" level is not sufficient to assure correct system working. This effect can be due to a wrong drive of the modulator or by the use of a modulator with unsuitable characteristics. Since a limited dynamic range reflects in a proportional increase of the EOP this has to be specified with great attention before accepting a certain transmitter in the system design.

- *Transmitting laser linewidth*: Even if semiconductor lasers are high performing and very stable sources, nevertheless, they have a finite linewidth due to homogeneous broadening of the emitted mode. Generally, lasers used for DWDM transmission have a linewidth on the order of 1 MHz or less and the impact on system performances is negligible. However, if the linewidth should increase it can disturb the receiver PLL causing timing jitter [2].

- *Polarization dependent losses*: Imperfect connectors positioning or other problems along the line (like transit through patch panels in the metro area) can cause polarization depending losses (PDL). PDL also affects the system performances increasing effective losses and creating unbalance between the output polarization principal states.

- *Excess losses due to passive optics*: In a realistic DWDM system there is a lot of passive optics (like connectors, patchcords, etc.) to bring the signal from one card to the other. All these elements introduce losses that can be important in case of bad mounting or of a damaged connector.

- *Amplifiers gain curve ripple*: Although gain flattening filters are usually adopted in optical amplifiers to smooth the gain curve, cascading several amplifiers could evidence also small imperfections. If a channel gain is globally greater it can depress the gain of nearby channels creating a relevant penalty.

6.3 Intensity Modulation and Direct Detection Systems Using Optical Amplifiers

In the previous section, we have analyzed optical transmission systems without optical amplifiers. Even if all the transmission impairments are somehow managed, the fiber attenuation and the receiver sensitivity set a limit to the product bit rate–distance that can be assumed as a sort of system quality factor.

With an APD receiver having a sensitivity of −34 dBm at 10 Gbit/s and a transmitter that injects about 1 dBm in the fiber, the pure power budget limits the transmission to a distance of 140 km due to the fiber attenuation assumed to be 0.25 dB/km.

Using optical amplifiers to boost the signal along the line this limit is completely removed. Not only the amplifier compensates for the fiber loss, but allows transmitting less power, thus limiting nonlinear effects.

Almost all the practical transmission systems that are on the market use optical amplifiers; thus, there is a wide variety of amplified systems depending on the application.

Here we will divide the systems in three categories:

1. Long-haul and ultra-long-haul systems are characterized by an amplifier chain constituting the transmission line; they can use any type of amplifier, depending on the needs, and their reach exceeds several thousand kilometers

2. Single span systems use only booster and preamplifiers and generally they are mixed EDFA–Raman amplifiers. They are used for applications in which the distances are not so big

3. Optical ring, characterized by the ring topology, they are used mainly in the metropolitan network

6.3.1 Long-Haul and Ultra-Long-Haul Transmission: Performance Evaluation

The principle scheme of the simpler long-haul DWDM system is shown in Figure 6.18. After the transmitter, a preamplifier brings the power to the desired level at the transmitter; the line is divided into spans and at each span end another amplifier is present. Since, due to the long reach of these systems, dispersion has to be compensated by some in-line device, in-line amplifiers will have the possibility to insert between the amplification stages a loss. This loss will be generally caused by a DCF, even if more complex situations arise when dynamic devices as some optical add drop multiplexers (OADMs) are used.

Thus, at the end both loss and dispersion are compensated, respectively, by optical amplifiers and DCFs.

Now we will apply all the methods introduced in the last sections to assess the performances of the system sketched in Figure 6.18 when EDFA amplifiers are used.

The noise contribution at the receiver will be mainly caused by amplifiers amplified spontaneous emission (ASE) accumulation, since the presence of a preamplifier will make the received signal quite more powerful than in unamplified cases. Moreover, a PIN is

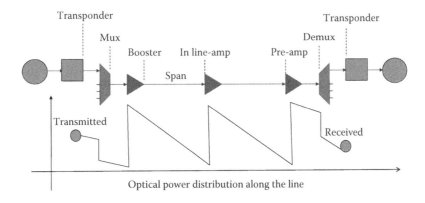

FIGURE 6.18
Block scheme of a long-haul DWDM system and optical power distribution along the line.

generally used at the receiver since further amplification after the preamplifier is not needed.

In order to determine the electrical noise, let us consider the transmission line like a cascade of amplifying and attenuating elements. In this simple model, let us imagine that all the in-line amplifiers are equal and that the net gain resulting from the two stages and the attenuation due to the interstage loss exactly compensate the loss of one span.

In real systems, the amplifiers are generally hosted in preexisting locations where the carrier has a point of presence. These locations could not be at the same distance, one from the other; thus, a more complicated situation arises. We will evaluate later the characteristics of a chain of amplifiers that are not at the same distance.

A single double stage amplifier with an interstage loss has a noise figure $NF_{total} = NF_1 + NF_2/(\alpha G_1)$ (compare Section 4.3.1.4) and it is optimized so to minimize the overall noise figure for the considered loss. Thus, we will assume that all the in-line amplifiers are optimized for the correct value of the DCF loss and have a noise figure equal to NF_{il}.

The system will have also a booster at the transmitter that we will assume a single-stage amplifier optimized to perform as booster and at the receiver a preamplifier will be present. The booster and preamplifiers noise figures will be NF_b and NF_P, respectively.

To provide an example of numeric performance evaluation, we will assume that the transmission fiber is an SSMF and that the values of the various system parameters are those reported in Table 6.2.

An exact evaluation of the noise at the amplifier chain is complex due to the fact that the electromagnetic field propagating in a generic system span is not in a minimum energy state due to the ASE contribution of the previous amplifiers [11]. The result of the exact quantum mechanical evaluation of the overall noise is reported in [19]. However, in a realistic telecommunication case, amplifiers are low-noise high-gain components, so that the only noise term that is relevant after detection among the noise processes caused by the amplifier chain is the so-called beat noise, that is the beat between the signal and the accumulated ASE.

TABLE 6.2

Parameters for the Design Examples of DWDM Amplified Systems

In-line amplifiers input saturation power	P_s (dBm)	−8
In-line amplifiers output saturation power	P_s^{out} (dBm)	24
Booster amplifiers input saturation power	$P_{s,booster}$ (dBm)	−8
Booster amplifiers output saturation power	$P_{s,booster}^{out}$ (dBm)	24
Preamplifiers input saturation power	$P_{s,preamp}$ (dBm)	−8
Preamplifiers output saturation power	$P_{s,preamp}^{out}$ (dBm)	24
Amplifiers noise factor (all the amplifiers)	NF (dB)	5
Multiplexer demultiplexer loss (128 channels)	α_M (dB)	9
Extra losses at the transmitter	α_{XT} (dB)	2
Extra losses at the receiver	α_{XR} (dB)	2
Amplifier insertion loss	α_c (dB)	0.3
Transmitter output power (injected in fiber)	P_L (dBm)	0
SSMF fiber loss	α (dB/km)	0.25
Receiver sensitivity @10 Gbit/s	S_{rec} (dBm)	−20
Receiver saturation power	P_{sRec} (dBm)	10

In this hypothesis, every amplifier introduces an optical noise whose power is given by

$$\sigma_j = \hbar\,\omega\,B_o G_j NF_j \quad (j = 1, \ldots, N_a) \tag{6.48}$$

where B_o is the unilateral optical bandwidth.

Here it is necessary to remember the convention we have already introduced in Chapter 5 to evaluate the noise power of amplified systems.

We define a unilateral power spectral density as $S(\omega) = \hbar\omega_c GNF$ where ω_c is the carrier angular frequency. Since the noise factor is involved in the expression of the noise power spectral density instead of the population inversion, the noise power is obtained by multiplying it by the unilateral optical bandwidth B_o. If the signal spectrum is mainly contained in the angular frequency interval $(\omega_c - \omega_M; \omega_c + \omega_M)$ and in the symmetric on the negative ω axis, B_o is defined as $B_o = \omega_M/2\pi$.

The signal at the end of the chain can be written as

$$\vec{E} = \sum_{k=1}^{N_c} E_k(t) e^{i\omega_k t}\vec{\vartheta}_k + \sum_{j=1}^{N_a+2} n_j(t)\vec{\xi}_j \tag{6.49}$$

where

N_a is the in-line amplifiers' number so that the overall number of amplifiers is N_{a+2}

$\vec{\vartheta}_k$ and $\vec{\xi}_j$ represent casual polarization vectors

The transmitted field intensity $E_k(t)$ contains the information. The indexes j and v run over the number N_a of amplifiers and the index k over the number N_c of WDM channels.

Due to the hypothesis that every amplifier compensates exactly the loss of the previous span, we can set

$$\prod_{j=1}^{N_a} G_j e^{-\alpha L_j} = \prod_{v=1}^{j-1} G_v e^{-\alpha L_v} = 1 \tag{6.50}$$

thus greatly simplifying expression (6.49). Moreover, from the consideration that all the ASE processes are independent since they come from different amplifiers, it results that the noise power is simply the sum of the ASE noise powers generated by each amplifier of the chain and the same holds for the overall noise figure, so that

$$NF_{ToT} = \sum_{j=1}^{N_a+2} NF_j = NF_b + NF_p + N_a NF_{il} \tag{6.51}$$

Starting from Equation 6.51, it is possible to optimize the amplifiers span, that is, to set the optimum value of N_a once given the amplifier noise characteristics and the link length L.

To do that, it is necessary to fix the bonds under which the optimization has to be performed. If amplifiers are used, periodically the signal returns to a high power level; thus, different from the case of unamplified systems, nonlinear effects happen over a long part of the link.

If the link has to work it is needed to limit the average power below a threshold, so to be sure that nonlinear effects remain under control. Since no solution in which the average power is higher than the threshold is acceptable, it is reasonable to optimize the overall link by maintaining constant the average power.

If this is the optimization bond it is useful to introduce in Equation 6.51 a different expression of the amplifiers number N_a.

From the equation $\prod_{j=1}^{N_a} G_j e^{-\alpha L_j} = G_{il}^{N_a} e^{-N_a \alpha L_s}$ where L_s is the span length, the following expression of N_a can be derived:

$$N_a = \frac{\alpha L_s}{\ln(G_{il})} \tag{6.52}$$

Substituting this expression of the number of amplifiers into the expression of the noise power generated by the line and using the noise factor to evaluate the overall ASE spectrum S_{ASE} is obtained

$$S_{ASE} = \left\{ \frac{\alpha L_s}{\ln(G_{il})} (G_{il} - 1)NF_{il} + (G_p - 1)NF_p + (G_b - 1)NF_b \right\} \hbar\omega \tag{6.53}$$

Since the average power in fiber and the bit rate are fixed, the SN_o and, thus, the error probability is optimized, if S_{ASE} is minimized. In order to minimize S_{ASE}, it is necessary to decrease L_s up to the limit of continuous amplification. In this limit, the amplifiers' gain is very small (tends to zero in the absence of DCF) and the noise figure is near to the quantum limit.

In reality, a long amplifying fiber is not feasible, but this result is nevertheless very important.

The penalty due to the use of concentrated amplifiers remains low if the span does not exceed 50–80 km and the use of lumped amplifiers is advantageous under many practical points of view.

Once the link is optimized from the point of view of attenuation compensation, it has to be optimized also under the dispersion compensation.

Chromatic dispersion is not the only component of the broadening of a pulse propagating through an optical fiber. SPM and PMD introduce phase distortion and thus pulse broadening.

On long-haul WDM systems the dispersion compensation is generally set so to partially balance XPM and SPM with residual dispersion, since propagation happens generally in the third transmission window of the optical fibers, where the dispersion coefficient β_2 is negative. This technique is called dispersion assisted transmission.

Last but not least, since this class of systems is explicitly designed to reach long distances, FEC is always used to increase the transmission distance. With standard FEC a post-FEC BER of 10^{-12} is achieved starting from a pre-FEC BER of about 10^{-4}, that we will take from now on as a second BER reference besides the value of 10^{-12}.

The received field after propagation and before the optical filtering that selects the wanted channel can be written in general as follows:

$$\vec{E} = \sum_{k=1}^{N_c} E_k(t) e^{i\omega_k t} \vec{\vartheta}_k + \sum_{v=1}^{N_{fwm}} n_v(t) e^{i\omega_v t} \vec{\varepsilon}_v + \sum_{h=1}^{N_{xpm}} n_h(t) e^{i\omega_h t} \vec{\mu}_h + \sum_{j=1}^{N_a+2} n_j(t) \vec{\xi}_j \tag{6.54}$$

where the four terms represent in the order information bringing terms (composed by N_c channels at different wavelengths), the FWM terms (whose number is N_{fwm}), the XPM terms (whose number is N_{xpm}), and the ASE noise terms (composed by one noise process for each of the $N_a + 2$ traversed amplifiers: N_a along the line plus the preamplifier and the booster).

At detection, the optical field is filtered to select the desired channel; this channel is detected by a PIN photodiode and processed by an electronic front end and by a decision circuit to produce an estimate of the transmitted bit stream.

The simpler model of the receiver processing chain is shown in Figure 6.19a.

As shown in the figure, thermal noise is also added at the receiver, with a process that can be represented like addiction of a white Gaussian noise after the post detection signal amplifier.

Carrying out system performance evaluation, using the analog model of the receiver presented in Figure 6.19 is quite difficult; moreover, it does not take into account that in almost all modern receivers, electronics functions are performed via digital circuits after a signal sampling operated immediately after amplification, when the electrical current has a sufficiently high power to correctly feed an A/D converter.

Thus, instead of the model of Figure 6.19a, we will assume for the receiver the model in Figure 6.19b.

In this model, sampling is performed immediately after the detector amplifier, the thermal noise is added as noise over discrete samples, and the analogical integrate and dump circuit that in the analogical model is placed before the threshold device is substituted by a discrete time integrator.

In this digital receiver model, electrical current sampling happens at a rate of $1/T_0$ samples a second where T_0 has to be equal or shorter than the inverse of the unilateral optical filter bandwidth B_o, if the circuit has to fulfill the Nyquist sampling condition.

Due to the Nyquist condition, no information related to the signal is lost due to sampling, so that, for the performance evaluation model, adding thermal noise after the sampler and not before it as physically happens makes no difference.

The optical signal incoming on the photodiode after optical filtering can be expressed in a useful form by introducing the so-called field quadratures.

Let us select a reference frame to express polarization vectors; if PMD has to be taken into account, it is useful to consider as a base the output PSP, otherwise the base of the linearly polarized modes can work; in any case, let us call $\vec{p}_+$ and $\vec{p}_-$ the two base unit vectors.

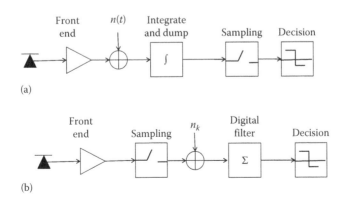

(a)

(b)

FIGURE 6.19
(a) Analog model of the receiver detection chain and (b) digital model of the receiver detection chain.

Naturally the field component along each of these vectors will be a complex function: the set of four real functions constituted by real and imaginary part of the field components are called field quadratures along the chosen base [19].

Setting

$$\varepsilon_k = \vec{\vartheta}_k \cdot \vec{p}_+ = \sqrt{1-(\vec{\vartheta}_k \cdot \vec{p}_-)^2}$$

(6.55)

the four field quadratures that we will indicate as $E_s(t)$ with $s = 1, 2, 3, 4$ can be written as

$$E_1(t) = \sum_{k=-N/2}^{N/2} \left[\sqrt{P_k} \cos(k\Delta\omega t + \varphi_k + \Delta\varphi) \left\{ \sum_u b_{uk} g_{uk}(t - \Delta\tau_{uk} - uT) \right\} \varepsilon_k \right] + n_1(t)$$

$$E_2(t) = \sum_{k=-N/2}^{N/2} \left[\sqrt{P_k} \sin(k\Delta\omega t + \varphi_k + \Delta\varphi) \left\{ \sum_u b_{uk} g_{uk}(t - \Delta\tau_{uk} - uT) \right\} \varepsilon_k \right] + n_2(t)$$

(6.56)

$$E_3(t) = \sum_{k=-N/2}^{N/2} \left[\sqrt{P_k} \cos(k\Delta\omega t + \varphi_k) \left\{ \sum_u b_{uk} g_{uk}(t - \Delta\tau_{uk} - uT) \right\} \sqrt{1-\varepsilon_k^2} \right] + n_3(t)$$

$$E_4(t) = \sum_{k=-N/2}^{N/2} \left[\sqrt{P_k} \sin(k\Delta\omega t + \varphi_k) \left\{ \sum_u b_{uk} g_{uk}(t - \Delta\tau_{uk} - uT) \right\} \sqrt{1-\varepsilon_k^2} \right] + n_4(t)$$

where the selected channel corresponds to $k = 0$ and $n_k(t)$ are band-limited Gaussian processes summarizing ASE accumulation and any other effect that can be described as a Gaussian noise—whose bandwidth is limited by the optical selection filter as FWM and XPM.

All the propagation effects causing signal distortion are summarized in the form of the received bit pulse $g_{jk}(t)$. The pulse shape depends both on the position of the pulse in the bit stream, due to the influence of the nearby bit through dispersion and similar effects introducing memory in the fiber response, and on the position of the considered channel in the WDM comb, due to the interference with nearby channels.

The photocurrent is easily evaluated on the ground of the expression of the field quadratures. After sampling, a sample of the photocurrent can be expressed as

$$C(t_h) = R_p \sum_{m=1}^{4} E_m^2(t_h) + n_r(t_h)$$

(6.57)

where $n_r(t_h)$ comprises all the receiver noise (i.e., thermal noise, dark current noise, and in case other noises introduced from the receiver).

The digital integrator performs the sum of a certain number of samples to produce the final value to compare with the threshold. Let us call M_{oe} the number of samples summed up by the integrator. The integration bandwidth, that is by definition the electrical bandwidth B_e, can be evaluated as the inverse of the integration time. Thus, $B_e = (M_{oe}T_0)^{-1}$ where T_0 is the sampling rate.

Since the bandwidth of the photocurrent before electrical filtering is B_o, due to the Nyquist condition we can assume a sampling rate $T_o = 1/B_0$.

It is to be noted that in real receivers, the Nyquist condition is often not fulfilled, due to the fact that it is not necessary to reproduce exactly the shape of the pulse, but it is enough to estimate the energy in the bit time.

It derives $M_{oe} = B_o/B_e$, where M_{oe} represents the ratio between the unilateral optical bandwidth and the unilateral electrical bandwidth (frequently approximated with the bit rate R) and it is the same parameter we introduced to relate the optical and electrical SN ratio in Chapter 5.

Let us also remember a convention we have done in Chapter 5 and that we will use throughout the book. The noise spectral density in the optical bandwidth S_o is defined so that multiplied for the unilateral optical bandwidth B_o gives the noise power. Thus, in the case of a single EDFA amplifier $S_o(\omega) = \hbar\omega_c NF(G - 1)$ being NF the amplifier noise factor (this is doubled with respect to the more common definition found in literature, where the bilateral optical bandwidth is used; see Figure 5.9 at page 307).

After discrete integration we have

$$C_w = R_p \sum_{j=1}^{M_{oe}} \sum_{m=1}^{4} E_m^2(t_j - wT) + n_r(t_j - wT) \tag{6.58}$$

It is now clear that the sample in front of the threshold comparison circuit is a quadratic form of Gaussian random variable plus a Gaussian independent term.

We can assume that the M_{oe} samples that constitute the decision variable C_w are statistically independent: this is the number of samplings at the Nyquist sampling rate of the optical signal that fit into the Nyquist sampling interval of the electrical signal (see Chapter 5). This is rigorously true if the sampling rate corresponds to the Nyquist rate and if the filters (both the optical and the electrical one) are perfect bandpass and low pass filters. Generally these conditions are only approximated by realistic systems; however, even in realistic conditions, assuming that the filters have spectral tails going to zero sufficiently fast and that the sampling is not very far from Nyquist condition, the statistical correlation among the samples can be neglected.

In this condition the characteristic function of the random variable C_w (i.e., the bilateral Laplace transform of its probability density) can be evaluated with a known method [26].

Let us individuate a set of bit patterns that allow us to manage the memory due to propagation phenomena (in general it will be a set of nine patterns of three bits as done in Section 6.1) and let us call $\Phi_j = \{b\}_j$ the jth pattern. The bit that is under decision is the central bit of the pattern, that will be indicated with b_{cj}.

The characteristic function $F(s)$ of the decision variable conditioned to the transmission of a certain pattern has the following expression, where $s = s_r + is_i$ is the Laplace variable:

$$F_c\left(s/1\Phi_j\right) = \frac{exp\left[-\left(SN_o(b_{cj}M_{oe}s/(1+s)) - (Q/4)s^2\right)\right]}{(1+s)^{2M_{oe}}} \tag{6.59}$$

where

- SN_o is the optical SN ratio that can be written starting from Equation 6.46 as

$$SN_o = \frac{P(b_{c_j})}{S_{ASE}B_o + P_{nli}(\Phi_j)} \tag{6.60}$$

with the power of the optical field in the assumed hypothesis indicated with $P(b_{c_j})$ and the sum of the FWM power and the XPM power indicated with $P_{nli}(\Phi_j)$

- $Q = \sigma_r^2 / [R_p S_{ASE}B_o + P_{nli}(\Phi_j)]$ is the ratio between the power of the receiver noise and the power of the optical noise

The BER, as in the unrepeated case, is expressed as

$$BER = \frac{1}{N_{pat}} \sum_{u=1}^{N_{pat}} P\left(\frac{\bar{b} - b_{uc} \neq 0}{\Phi_u}\right) \tag{6.61}$$

where
$\bar{b}$ is the estimate of the received bit
$N_{pat} = 8$ is the number of patterns

In order to estimate with a suitable approximation the BER, it seems that the specific form of the decision variable can be exploited to reduce the problem to a Gaussian probability.

As a matter of fact, it seems that, after the execution of the square of the field quadratures, the term with the ASE noise squared can be neglected with respect to the beat between the ASE noise and the signal.

With this approximation the problem is effectively reduced to a Gaussian problem, but the solution is affected by an error on the order of magnitude of 3 dB: that is, in order to reach a low BER (e.g., 10^{-12}), the needed power is estimated to be half of the correct value.

This is due to the fact that the beat term is much greater than the square noise term in average, but when an error occurs it means that the instantaneous noise power is on the order of magnitude of the signal power and in this condition it is not true that square noise is negligible. Neglecting it means more or less neglecting half the noise, when a wrong bit estimate occurs from which the magnitude of the error in the Gaussian approximation.

The terms of the sum (6.61) can be estimated correctly using the so-called saddle point approximation, obtaining a closed form expression for the error probability.

It is not within the scope of this section to demonstrate this approximation, the interested reader is encouraged to start from the bibliography regarding this point [27,28].

From an operative point of view, the saddle point approximation requires to find a real solution x_0 (if it exists) of the equation

$$\frac{1}{F_c(x)} \frac{\partial F_c}{\partial x} + C_{th} - \frac{1}{x} = 0 \tag{6.62}$$

where C_{th} is the threshold value. If this solution is found, defining the function

$$\Theta(x) = \ln[F_c(x)] + C_{th}x - \ln(x) \tag{6.63}$$

the error probability is approximated as

$$P\left(\bar{b} = 0/\Phi_u \cup b_{uc} = 1\right) = \int_{-\infty}^{C_{th}} p(c)\,dc \approx \frac{\exp[\Theta(x_0)]}{\sqrt{2\pi\Theta''(x_0)}} \tag{6.64}$$

and with a similar equation for $(P(\bar{b} = 1/\Phi_u \cup b_{uc} = 0)$, where it is only necessary to change c in $-c$ to use the approximation formula.

The saddle point approximation can be used to evaluate a very good approximation of the error probability in a very general set of cases.

It is interesting however to try to obtain a closed form approximation of the error probability at least when only the dominant ASE noise is present.

Assuming that no pattern effect is present and that thermal noise at the receiver is negligible with respect to the ASE noise, the error probability expression (6.64) relative to an intensity modulated systems with in line amplifiers simplifies as

$$P_e = \frac{1}{2}[P_e(0) + P_e(1)]$$

$$= \frac{1}{2}\frac{e^{-(\gamma SN_o - 2M_{oe})}}{\left(2M_{oe}/\gamma SN_o\right)^{2M_{oe}}} + \frac{1}{2}\frac{e^{-SN_o M_{oe}}\left[\left(X(SN_o)/(1 + X(SN_o))\right) - \gamma\left(X(SN_o)/M_{oe}\right)\right]}{[1 + X(SN_o)]^{2M_{oe}}} \tag{6.65a}$$

$$X(SN_o) = -\gamma SN_o + M_{oe} + \sqrt{M_{oe}}\sqrt{(\gamma SN_o + M_{oe})} \tag{6.65b}$$

where
   The value of the threshold in terms of a fraction of the signal to noise ratio at the optical level is given by $C_{th} = \gamma SN_o/M_{oe}$,
   M_{oe} is the ratio between the optical and the electrical bandwidth

Moreover, the saddle point approximation has been further simplified assuming a high SN ratio situation and $SN_o \gg M_{oe}$ (such a simplified approximation is also called Chernov bound).

This approximation is quite accurate for low values of M_{oe}, approximately up to $M_{oe} = 5$ if we look below a BER of 5×10^{-4} and up to $M_{oe} = 3$ if we look to a BER below 10^{-12}. Often in practical systems the target pre-FEC BER is on the order of 10^{-4} and the value of M_{oe} is below 3; thus, this is a good approximation.

Out of these intervals, in particular for systems with a large value of M_{oe} and a strong FEC so that the BER pre-FEC is very low, the complete saddle point approximation has to be used, that is much more accurate but produces an optimization equation whose solution has to be found numerically.

Naturally, to evaluate the BER, for each value of SN_o, the decision threshold (thus γ) has to be optimized.

The BER curves obtained by Equations 6.65a and 6.65b are plotted for reference in Figure 6.20 versus SN_e. If a plot versus SN_o is deduced, it can be noted that increasing M_{oe}, the BER seems to be better and better, against intuition. In reality this is not true, since it is not to forget that increasing M_{oe} a certain value of the optical SN ratio is more and more difficult to attain due to the increase of the optical bandwidth. If this effect is evaluated it is discovered that it largely overcomes the slight decrease of the BER with M_{oe}, so that if a plot versus the received power is done, increasing M_{oe} the power needed to attain a given BER would increase.

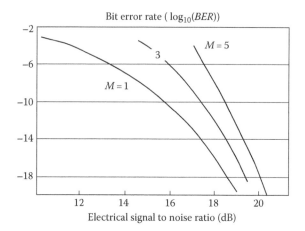

FIGURE 6.20
Ideal performance of a long-haul optical link for different values of the ratio M_{oe} between the bilateral optical bandwidth and two times the electrical bandwidth. The equation used to plot the BER curve is the approximation reported in the text and where the approximation is not valid the curves are interrupted.

From Figure 6.20 the value of SN_e when $BER = 10^{-12}$ and the optical bandwidth is minimum ($M_{oe} = 1$) is a bit less than 17 dB.

If a simple Gaussian model is used to evaluate the BER, but the noise variance is correctly taken into account like in [2], the required SN_e to achieve a BER of 10^{-12} in this condition is almost exactly 17 dB. Although this assumption underestimates a bit (about 0.5 dB) the system performance, the value of 17 dB is often assumed as a design reference and we will do the same in the examples of this chapter.

Sometimes, in order to express penalties due to different transmission impairments, the so-called Q factor is used, not to be confused with the Q function.

The Q factor is defined as

$$Q = Q^{-1}(BER) \tag{6.66}$$

This definition comes from Equation 6.14 and, if the BER would be given by the relation (Equation 6.14), there would be a direct relationship between the Q factor and the SN_o. In our case Equation 6.14 does not hold, and the Q factor has to be evaluated via its definition and the correct expression of the error probability.

If Equations 6.65a and 6.65b can be used, the relationship between the squared Q factor and the optical SN ratio is shown in Figure 6.21 for $M_{oe} = 2$.

Up to now, we have assumed that all the amplifiers that are deployed along the line are equal, meaning that all the spans are equal.

In real system installations this is not always true. Several times, DWDM systems are installed along the route of old systems where the points of presence where amplifiers can be installed are not equally spaced.

In the presence of nonuniform spans, Equation 6.43 becomes

$$S_{ASE} = \left\{ \sum_{v=1}^{N_a} (G_{ilv} - 1)NF_{ilv} + (G_p - 1)NF_p + (G_b - 1)NF_b \right\} \hbar \omega \tag{6.67}$$

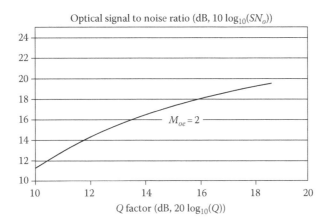

FIGURE 6.21
Relationship between optical SN ratio and Q factor for $M_{oe} = 2$.

Let us assume that every in-line amplifier compensates exactly the loss of the span after it and that the booster compensates the first span while the preamplifier compensates the end terminal losses.

For the sake of simplicity, in a first approximation, let us assume that the noise factors are all equal and independent from the gain of the amplifiers. Indicating with L_v the length of the vth span, Equation 6.67 can be rewritten as

$$S_{ASE} = \left\{ \sum_{v=1}^{N_a+1} (e^{\alpha L_v} - 1) + (G_p - 1) \right\} NF\hbar\omega \qquad (6.68)$$

If the span length is variable it is a relevant problem to find the optimum span distribution, that is, the span set that minimizes the overall ASE noise.

This problem can be solved easily with the method of Lagrange multipliers, finding the minimum of (6.68) with the condition

$$\sum_{v=1}^{N_a+1} L_v = L \qquad (6.69)$$

L being the overall link length.

Calling γ the Lagrange multiplier the Lagrangian function is

$$S_{ASE} = \left\{ \sum_{v=1}^{N_a+1} (e^{\alpha L_v} - 1) + (G_p - 1) \right\} NF\hbar\omega - \gamma \left(\sum_{v=1}^{N_a+1} L_v - L \right) \qquad (6.70)$$

and the derivative, with respect to the unknown span lengths are

$$\frac{\partial S_{ASE}}{\partial L_v} = NF\hbar\omega\alpha e^{\alpha L_v} - \gamma \qquad (6.71)$$

with the condition (6.69).

The only way in which all the N_a relations (6.71) can be equal to zero is that all the spans have the same length. Thus, as it is intuitive, the condition of equal length spans is an extreme of the ASE function.

The fact that it is a minimum and not a maximum is readily demonstrated by applying to the equal span condition the smallest possible perturbation, that is, assuming that one span is ΔL longer than the others and, to fulfill condition (6.71), another is ΔL shorter.

Calling S_{ASE} the noise with all the span equal and S'_{ASE} the noise after the modification it is easily calculated

$$\delta S_{ASE} = S'_{ASE} - S_{ASE} = 2NF\hbar\omega e^{\alpha L/N_a} \sinh^2\left(\frac{\Delta L\alpha}{2}\right) \tag{6.72}$$

This quantity is always positive, demonstrating that the condition of equally long span is the optimum span distribution in the sense that it minimizes the noise power.

The situation does not change if a dependence of the noise factor on the gain is considered, as can be readily demonstrated.

It is interesting to notice that minimizing the noise is not the only meaningful way to optimize the span distribution in a DWDM long-haul system.

Due to the importance of nonlinear effects in determining the system performances, it is also meaningful to search a condition of minimal nonlinearity for a given length of the link, span number, and transmitted power.

Such a condition can be analytically expressed through the minimization of the link equivalent length that depends on the amplifier spacing.

Thus, the function to minimize is now

$$L_e = \sum_{v=1}^{N_a} \frac{(1 - e^{-\alpha L_v})}{\alpha} \tag{6.73}$$

By repeating exactly the same procedure applied in the case of the ASE noise, it is possible to demonstrate that also in this case the minimum equivalent length is achieved when the spans have all the same length. This result is quite intuitive since if a span is longer the amplifier compensating that span has to launch more power in the fiber, causing more nonlinear effects and this is not compensated by the fact that the shorter span is less affected by nonlinearity due to the exponential growth of nonlinear efficiency with length.

From the previous discussion, it is clear that not only the situation in which all the system spans are equally long is optimum both under a noise and from a nonlinear effect point of view, but also that the performances rapidly degrade if this condition does not hold.

6.3.2 Design of Long-Haul Transmission Systems

The great part of the products offered on the market for long-haul transmission is based on DWDM and on EDFA amplifiers.

In this section, we will sketch the procedure to design from an optical point of view a system of this type. A few elements on the thermal and mechanical design will be covered at the end of this chapter.

Essentially the design of a long-haul DWDM system follows five steps:

1. *System draft design*: In the first phase the main system parameters are selected like the number of channels and their bit rate, the channel spacing, the link length, the span length, and so on. This means to position the DWDM product in a well defined category of DWDM systems.

2. *Power budget*: Starting from the first system specifications the optical power profile along the line is determined, thus selecting the components to use. At the end, the optical SN ratio is determined.

3. *Dispersion map*: Determining the dispersion map, that is, the distribution of dispersion compensating elements along the line and, in case, the presence at the transmitter and at the receiver of equalization components is a key step. It influences not only the system behavior versus chromatic dispersion, but also the impact of nonlinearities and of PMD, that in the majority of cases is equalized by electronic adaptive equalizers contemporary to chromatic dispersion.

4. *Penalties impact*: Using the penalty addition rule all the relevant penalties are added to the ideal SN_o (driven only by ASE noise) to arrive to a required SN_o that takes into account all the relevant phenomena. If the required SN_o is smaller or equal to that estimated performing the power budget the design can go on, otherwise some elements has to be changed to match the required performance.

5. *Simulation*: The design is confirmed via simulation to understand the impact of the numerous approximations.

6.3.2.1 Erbium-Doped Optical Fiber Amplifier Amplified Systems Design

6.3.2.1.1 Power Budget

In long-haul systems, EDFA amplifiers work in deep saturation, since this allows both power stabilization along the line and minimum signal distortion.

In this condition, in-line amplifiers are characterized by an output saturation power P_s^{out} that corresponds to an input saturation power P_s^{in}. This means that every in-line amplifier will inject into the line a power equal to $\alpha_c P_s^{out}$, where α_c is the coupling loss that in general is very small. At the amplifier input, the power would not be lower than P_s^{in}/α_c, due to the fact that the saturation condition has to be maintained.

Moreover, extra losses that are located before the booster and after the preamplifier have to be accounted for. These are due essentially to the multiplexer and the demultiplexer, to connectors and patchcords connecting the different cards, and to taps used to monitor the optical signal.

Let us indicate with α_M the loss of the multiplexer and the demultiplexer and α_{XT}, α_{XR} the extra losses at the transmitter and at the receiver respectively, with P_L the optical power exiting from a transmitter after modulation and with $P_{S\,PIN}$ the input saturation power of the receiver.

Realistic values for these parameters can be found in Chapter 4 and are summarized in Table 6.2.

The power budget equations for the system are as follows:

- The equation assuring that the booster amplifier is saturated so to inject into the line the correct power

$$10 \log_{10}(n_c) + P_L\,(\mathrm{dB}) - \alpha_M\,(\mathrm{dB}) - \alpha_{XT}\,(\mathrm{dB}) > P_{s,booster}\,(\mathrm{dB}) \tag{6.74}$$

where
    n_c is the number of channels
    P_L is the power emitted by the transmitting laser
    α_M is the multiplexer loss
    α_{XT} is the transmitter chain loss (containing the modulator, the passive optics, and so on)

Finally $P_{s,booster}$ is the booster input saturation power.

- The equation that determines the maximum span that can be covered by the in-line amplifiers (assuming that the output power of the booster is the same of that of in-line amplifiers)

$$L_s < \frac{P_s^{out}(\text{dB}) - P_s(\text{dB}) - 2\alpha_c(\text{dB})}{\alpha(\text{dB/km})} \tag{6.75}$$

where
    α is the fiber loss in dB/km
    α_c is the amplifier insertion loss
    P_s and P_s^{out} are the line amplifier input and output saturated power

The span length is indicated with L_s.

- The equation that assures that the received power is sufficient to neglect the thermal noise of the detector but does not saturate the receiver

$$S_{Rec}(\text{dB}) \ll P_{s,preamp}^{out}(\text{dB}) - \alpha_{DM}(\text{dB}) - \alpha_{RC}(\text{dB}) < P_{s,PIN}(\text{dB}) \tag{6.76}$$

where
    S_{Rec} is the receiver sensitivity
    $P_{s,preamp}^{out}$ is the saturated output power of the preamplifier
    α_{DM} is the demultiplexer loss
    α_{RC} is the whole receiving chain loss

In the case of the parameters reported in Table 6.2 Equation 6.74 is fulfilled with a channel number as high as 256 and as low as 16; thus, the transmitter power budget is almost automatically fulfilled.

Equation 6.76 is also satisfied, while the margin on the upper inequality is only 0.5 dB. This means that fluctuations in the values of the parameters risk to generate receiver saturation.

Probably, since the sensitivity is not a problem, a receiver with a higher saturation power has to be adopted, even with lower sensitivity.

Regarding the transmission line power budget, that is, Equation 6.75, spans as long as 136 km can be achieved. Increasing the span length generally the cost per Gbit/s per km of the system decreases, due to the decreasing of the number of amplifiers per unit length, but also decreases the maximum system reach. The relationship between the system length, the span length, and the amplifier number is shown in Figure 6.22.

The choice of the span length is thus a key choice to position a DWDM product on the market, being one of the main parameters governing the balance between cost per Gbit/s per km and system performances.

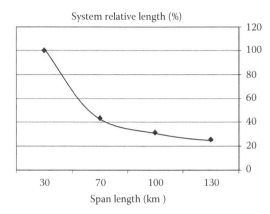

FIGURE 6.22

ASE limited system length versus the amplifier spacing. The system length is normalized to the maximum value attainable with distributed amplification. System parameters are reported in Table 6.1.

6.3.2.1.2 Erbium-Doped Optical Fiber Amplifier Amplified Systems Design: Dispersion Management

From the results shown in Section 6.2.3 it is clear that, in order to fully exploit the reach potential of DWDM amplified systems, chromatic dispersion has to be compensated.

Due to the close interaction between chromatic dispersion, SPM, XPM, and FWM, the way in which chromatic dispersion is compensated strongly influences also the way in which nonlinearity impacts transmission.

Different solutions are available to implement dispersion compensation:

- Use of in-line dispersion compensators (like DCF or fiber Bragg gratings [FBG])
- Pre-chirp of the transmitted pulses
- Pre- and post-electronic compensation

and quite often all solutions are used together in order to find the best mix to compensate dispersion and control nonlinear effects [29,30].

In any case, in-line dispersion compensators have to be used due to the presence of nonlinearities. In a very long link, not only dispersion, but also nonlinear effects accumulate with the distance. Once a strong nonlinear distortion has been accumulated, the pulse cannot be reshaped via linear compensation and also dispersion compensation is no more possible.

Thus, dispersion has to be compensated in-line so frequently that no sensible nonlinear degradation accumulates between two nearby compensators.

In order to evaluate the effectiveness of nonlinear effect accumulation, the effective length L_e is often used in unrepeated systems. The same can be done in the presence of amplifiers; since each amplifier rises the power of the optical signal, nonlinear effects could start after each amplifier so that the effective length of the amplified link is $L_e = n_a \cdot L_{e0}$, where L_{e0} is the effective length of a passive fiber link and n_a the span number.

The only case in which in-line dispersion compensation is not needed is the case in which a new technology called electronic dynamically compensated transmission (EDCT) is used. In this section we will first analyze the approach through in-line linear compensation and after that we will describe EDCT.

In practical systems in which in-line compensation through DCF or fiber Bragg grating (FBG) is used, the amplifiers distance is sufficiently short to allow dispersion compensating modules (DCMs) to be located in the amplifiers sites, generally in between the stages of a multistage optical amplifier.

TABLE 6.3

Parameters of a DCF

Attenuation (dB/km)	0.5
Dispersion (ps/nm/km)	−102
Dispersion slope (ps/nm²/km)	−0.2
Nonlinear index (m²/W)	2.6×10^{-20}
Effective area (μm²)	22

Three techniques can be used for in-line dispersion compensation:

1. *Pre-compensation*: The compensation module is at the beginning of the span; this means that a DCM is placed at the booster site to compensate the first span

2. *Post-compensation*: The compensation module is at the end of the span; this means that a DCM is placed at the preamplifier site to compensate the last span

3. *Mixed pre- and post-compensation*: In this case a DCM is preset both at the transmitter and at the receiver side

Several studies have been done to compare post-compensation and pre-compensation in different situations. The final result depends on the specific system setup.

In general, post-compensation performs better at the expenses of a more stringent tolerance on the compensator parameters choice. An example of this behavior is reported in Figure 6.23, where the Q factor is reported versus the compensation ratio (ratio between residual dispersion and total link dispersion) for a system transmitting a single 10 Gbit/s signal with 13 spans of 80 km of SSMF and a variable length of DCF, which is adjusted to define the compensation ratio.

The fiber parameters are given in Table 4.1, Chapter 4 for the SSMF and in Table 6.3 for the DCF. A two stage EDFA with 5 dB noise figure per stage is used in each section to regenerate losses and to host the DCF.

In Figure 6.23, experimental data reported in [31] are fitted with the performance evaluation model where the propagation of NRZ pulses through the fiber link is simulated with a program using the algorithm described in [8], that is based on the split step solution of the nonlinear propagation equation (see Section 4.39). Noise is taken into account, analytically, after the simulation of the deterministic part of the propagation.

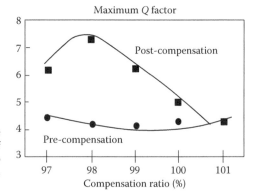

FIGURE 6.23

Q factor versus the post-compensation ratio for a single 10 Gbit/s channel in a system with 13 spans of 80 km of SSMF and a variable length of DCF, which is adjusted to define the compensation ratio. (After Peucheret, C. et al., *IEEE Photon. Tech. Lett.* 12(8), 992, 2000.)

Other situations exist where it is pre-compensation to be preferable, as in the examples reported in [32].

The situation changes completely if some pulse pre-distortion is implemented, either at the electrical or at the optical level.

Pre-distortion is intended to support in-line compensation in balancing the dispersion of the line. It consists in transmitting chirped pulses with a positive chirp in the case of transmission in the third window, where there is anomalous dispersion. The chirp, that can be generated by the transmitter electronics or by the optical modulator, is compensated by the first part of the fiber and provides an intrinsic robustness to dispersion.

In this case, in the great part of the systems the differences between pre- and post-dispersion compensation are much less pronounced.

Naturally, even better performances are attained when both pre- and post-compensation are used, if the overall link is suitably optimized.

As further refinement, also a combined pre- and post-compensation with individual channel compensators at the transmitter and at the receiver has been proposed [29], so attaining even better performances with respect to mixed pre- and post-compensation. The weakness of this solution is of course its cost, especially for systems with a high channel count.

From Figure 6.23, it also results that the higher value of the Q factor is realized with a certain degree of under compensation that is compensating only 98% of the line dispersion.

This result is reasonable, since a bit of residual dispersion allows SPM and XPM to be partially compensated, thus improving system performance.

Extensive study of under-compensation technique has been done [30,33,34]. The optimization technique is generally based on building a variable dispersion map, where the value of the DCF dispersion and dispersion slope are the same at each site and are the optimization parameters. Sometimes, when the requirements do not fix the amplifier sites a priori, also the span length is a third optimization parameter.

The scalar signal propagation is simulated via a numerical model and the Q factor or the SN_o are evaluated using the analytical expression of the noise variance.

At this point, an algorithm is available relating the parameters to be optimized, that is, the DCF dispersion and dispersion slope and, in case, the span length, with the link quality factor.

Thus, a numerical optimization procedure is used to find the optimum parameters combination. Among traditional nonlinear optimization methods [35] the Nelder–Mead optimization procedure, also called nonlinear simplex method, seems particularly suitable for this case in which the evaluation of the objective function is quite complex [36].

Naturally, due to the fact that any optimization will require several simulation executions, a key point is to use a computationally simple simulation procedure; thus, the experience in balancing accuracy and approximations intended to simplify the computation is a key point for this optimization procedure [37,38].

Generally, when optimizing the system, more than one local optimum exists and the choice is often guided more by the stability of the working point than by extreme optimization of the system length [31].

As a matter of fact, when implementing a design in a practical system, several considerations, besides pure performance optimization, lead the implementation choices. One of these elements is the stability of the chosen solution versus small variations of the design parameters, variations that can be caused both by tolerances of the components' characteristics and by aging.

Another important point is the comparison between NRZ and RZ modulation. In the case of unrepeated dispersion-limited systems, RZ performs better than NRZ. This behavior is even more evident in single channel optically amplified links, due to the robustness of RZ modulation to ISI due to pulse spreading. In this case, it is even possible to improve pure RZ performances by transmitting chirped pulses to achieve more robustness to residual dispersion.

To give an example, let us consider long-haul system with length of about 4600 km and 26.2 nm optical bandwidth over SSMF. The system can support 64 channels with 50 GHz spacing.

The Q factor is plotted versus the spectral position of the selected channel in Figure 6.24.

From the plot it is clear that chirped RZ is efficient in taking under control the interplay between dispersion an XPM/SPM.

On the other hand, when DWDM is considered, the wider optical spectrum of RZ modulation impacts on channel interference. An example assuming ideal modulation and demodulation with square filters can clarify the effect. The linear sensitivity penalty for the central channel of an NRZ 10 Gbit/s comb with a spacing of 50 GHz is 1 dB, while in the case of RZ transmission with 30% duty cycle it is as high as 4 dB and to reduce the penalty to 1 dB the channels have to be separated by a gap of about 120 GHz.

In-line dispersion management can also be combined with pre- or post-electronic dispersion compensation.

Careful system design can achieve good results with time domain electronic compensators like FFE–DFE, but better results can be attained especially at high speeds using a more advanced solution. It consists in a linear EDC working on the baseband electrical pulse before modulation. It consists of three steps: FFT, multiplication for the inverse fiber frequency response, and inverse FFT [39].

This solution requires electronic processing with a very high sampling rate, arriving up to 80 Gsamples/s if applied to channels at 40 Gbit/s and it is possible only with the more recent generation of CMOS circuits.

An example of comparison between a system exploiting standard pre-compensation via DCFs and a system substituting the DCF at the transmitter with the EDC is reported in Figure 6.25 [40]. At 10 Gbit/s the performances of the traditional compensation method is superior, but at 40 Gbit/s this is no more true, due to the better accuracy with which the electronic control manages very-high-speed signals compensating the dispersion channel by channel.

A different technique based on electronic pre-compensation of nonlinear propagation distortion has demonstrated very good results allowing very long reach and stabile EDFA-based systems to be realized without using in-line compensation [40] (see Chapter 4).

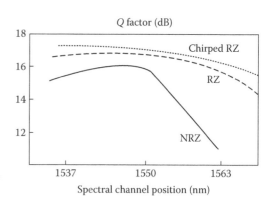

FIGURE 6.24
Q factor versus the spectral position of the considered channel in a long-haul system 4600 km long with 64 channels and 50 GHz spacing. System parameters are reported in Table 6.1.

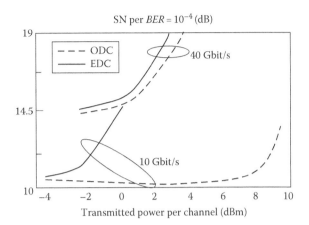

FIGURE 6.25
Comparison between standard pre-compensation via DCFs and substitution of the first DCF at the transmitter with electronic dispersion pre-compensation. System parameters are reported in Table 6.1. (After Weber, C. et al., *Optical Fiber Conference Proceedings, OFC 2009*, IEEE, s.l., 2009.)

The advantages of this technique are potentially very important: adaptive electronic pre-compensation can face effectively system parameters fluctuations due to tolerances and aging; it allows much simpler EDFAs to be deployed in-line, with an evident cost advantage.

On the other hand, implementing compensation of nonlinear propagation requires complex algorithms to be implemented in a real-time signal processor with a very high sampling rate (50 GHz in the specific case).

As an example of the performance of these classes of algorithms in Figure 6.26 [18], the SN_o penalty is shown versus the power launched at the transmitter into the fiber. The considered system is a single channel 1200 km long system with 15 span of 80 km each. The transmission fiber is a standard SSMF fiber with D 17 ps/nm/km, $\alpha = 0.21$ dB/km,

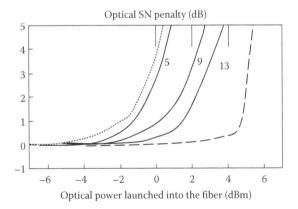

FIGURE 6.26
SN_o penalty versus launched power in a single-channel RZ IM-DD 1200 km long system with 15 span of 80 km each using SSMF fiber with D 17 ps/nm/km, $\alpha = 0.21$ dB/km, $\gamma = (1/(W \; km))$. Dispersion is electronically pre-compensated via an adaptive algorithm, no optical compensation is implemented. (After Killey, R.I. et al., *IEEE Photon. Tech. Lett.*, 17(3), 714, 2005.)

$\gamma = (1/(W \ km))$. No optical dispersion compensation is implemented along the line and conventional EDFAs are used. Information is coded and decoded through RZ IM-DD format without any line code and FEC.

The dotted line represents the system performances with perfect dispersion compensation and no compensation of nonlinear distortion. From this curve, it is evident that the system suffers a great performance degradation from nonlinear effects.

The solid curves represent penalty in the presence of nonlinear compensation with different orders of the look-up table addressing space (indicating the dimension of the look-up table). The performance becomes better and better with the progressive reduction of the difference between the ideal inverse propagation filter and the real processing implemented through the look-up table. The better curve is near the dashed line that represents the modulator limit.

6.3.2.1.3 Erbium-Doped Optical Fiber Amplifier Amplified Systems Design: Example of Draft Design

Here we want to apply the rules we have given in the previous sections to carry out an example of draft design of a long-haul DWDM system.

Starting from 10 Gbit/s, we state at the beginning not to use L band to avoid a too costly system.

Since L band channels are separated by C band in the fiber and require different amplifiers, the example can be easily extended to a C + L band system, only adding bands splitters when needed and all the components for L band processing.

Extended C band can allocate 90 channels at 10 Gbit/s with a spacing of 50 GHz on the ITU-T grid; thus, we will select this option as a good trade-off between cost and capacity.

We also select a span length of 70 km. With the amplifiers characteristics of Table 6.2 this is an intermediate span length (the maximum would be 136 km).

Good quality array waveguides (AWGs) are used to multiplex and demultiplex the optical channels so to arrive at a total unilateral optical bandwidth of $B_o = 20 \ GHz$ (0.16 nm at 1550 nm wavelength) with $M_{oe} = 2$.

The use of a FEC is almost obliged in long-haul systems and we choose the standard ITU-T FEC: the RS (255,239) that in the ideal conditions provides 6 dB of code gain.

Using Equations 6.65a and 6.65b it is easy to verify that the power budget equations for the transmitter and the receiver can be satisfied with standard DWDM quality components (see Chapter 4, the main parameters of transmitter and receiver are summarized in Table 6.1). The total ASE power is −16.4 dBm per channel, so that the ideal SN_o is 24 dB and the ideal $SN_e = 24$ dB (if $M_{oe} = 2 \ SN_o \approx SN_e$). This gives quite a margin with respect to the required 11 dB (17 from the receiver with $M_{oe} = 2$ minus 6 dB of code gain) that allows the allocation of more than 8.8 dB for extra penalties due to various impairments.

The last thing to do is to select the dispersion map. We have to design quite a standard link, following literature, we will adopt a post-compensation map with a certain degree of under compensation and a per channel FFE–DFE equalization of the residual dispersion at the receiver.

Following a simple map optimization, where the only parameter to setup is the under compensation, we will under compensate for 1500 ps/nm implying a differential delay of 493 ps in the signal bandwidth and will adopt a nine taps FFE-4 tap DFE adaptive equalizer. This allows the residual dispersion to be almost completely compensated at the receiver and the SPM effect to be attenuated.

6.3.2.1.4 *Erbium-Doped Optical Fiber Amplifier Amplified Systems Design: Penalties Impact*

Once the DWDM system has been specified fixing its main parameters, the system power budget and dispersion map have been designed, and all the tentative specifications for the main components have been set, it is time to start to verify if the design is sufficiently robust versus all the performance impairments to meet the performance requests, that, in general, are provided in terms of a maximum attainable BER.

The draft evaluation, essentially, consists in calculating all the relevant penalties in a situation in which the considered phenomenon is the only impairment affecting the ideal system and adding the obtained penalties with the system margin to arrive at a final requirement in terms of SN ratio.

Generally, it is easier to work at the optical level, specifying the SN_o, but sometimes the presence of electronic nonidealities at the receiver renders it even easier to evaluate the target SN_e.

The target SN is compared, at the end, with the SN coming out from the design to verify the system is working correctly.

Naturally, it is well possible that the system does not meet the requirement in terms of SN after the first design, in this case, the main causes of distortion and noise are identified and the design is reworked again to mitigate the corresponding degradation.

We have seen in the previous sections that a long-haul system with 90 channels at 50 GHz spacing satisfies the draft requirements.

A realistic penalty allocation is reported in Table 6.4. It is to be noted that, besides impairing effects, also engineering margins are added, 3 dB of system margin that takes into account fluctuations of components specifications and imperfect design implementation and 1 dB of aging margin.

As far as the aging penalty is concerned, there are two possible approaches in considering components degradation with time. Here we have decided to perform all the calculations using end of life parameters, that have to be evaluated by components suppliers during the components qualification stage. With this choice, a further aging penalty is only a sort of second margin on a wrong evaluation of the degradation of some component.

Another approach, followed, for example, in [2], is to perform design calculations using the new components parameters. In this case, all the aging degradation has to be added at this stage. Generally, in this case, an aging penalty of 3–4 dB is considered.

The required SN_o results to be 20 dB before FEC operation, smaller than the 24 dB resulting from the draft design. This gap generally exists in long-haul systems since it is used to allocate other components in the gap between the stages of in-line amplifiers, like OADMs.

Just to do an example, let us image that we have to place in four of the intermediate sites one OADM for branching that introduces a loss of 6 dB.

This loss has to be balanced by a greater gain of the EDFAs so generating more noise.

Carrying out the draft design, by introducing these new components, we see that the SN_o deriving from this new analysis is equal to 21.9 dB, thus getting much nearer to the bottom line for the system working.

6.3.2.1.5 *Erbium-Doped Optical Fiber Amplifier Amplified Systems Design: Simulation*

The system design coming out from the considerations of Table 6.4 has been simulated using two different simulation programs.

TABLE 6.4

Realistic Penalty Analysis for Long-Haul 10 Gbit/s Dense Wavelength Division Multiplexing System

Long-Haul 10 Gbit/s DWDM System		
System length	3500 km	
Channel speed	10 Gbit/s	Suitable EDFAs, span length 60 km, RS(255,239) FEC
Channel spacing	50 GHz	Extended C band, 38 nm in the third window, that is 4560 GHz, 90 channels @ 10 Gbit/s
Ideal SN_e (dB) ($B_e = 2R$)	17.2	NRZ modulation
Optical to electrical bandwidth ratio	$M_{oe} = 2$	Main contribution two Gaussian AWG with a unilateral passband of 20 GHz (bilateral 40 GHz $\approx$ 0.33 nm @ 1.55 μm)
Ideal SN_o (dB)	17.2	
	Penalty in dB	
Transmitter dynamic range	0.5	SFF MSA using tunable CW laser and LiNbO$_3$ modulator
APD diode noise + shot noise	0.3	
Linear crosstalk	0.8	Gaussian selection filter (AWG), $B_o = 4R$
		Channel spacing $\Delta f = 5R = 50$ GHz (see Figure 6.16)
Nonlinear crosstalk	1.4	Use of SSMF to thwart FWM, post-compensation with DCF
		From Equations 6.44 and 6.45 adapted to this case
SPM and chromatic dispersion interplay	1.3	Under compensation of chromatic dispersion at 98%
PMD	0.0	NRZ modulation
Aging	1	—
System margin	3	—
Polarization effects	0.5	PDL, connectors misalignments, etc.
Total SN post code	26 dB	—
Code gain (dB)	6 dB	Standard ITU-T code gain
Total SN_o precode	20 dB	To confirm with simulation
Total SN_e precode	20 dB	

The first is a Monte Carlo simulation that is able to simulate also the noise. This is a quite accurate program, but it is computationally very complex, especially when a great number of runs have to be carried out for statistical reasons.

Running on a pc equipped with an i7 Intel processor, it is possible to arrive to an estimate of the error probability with a precision around 10% with few days of running, only if the BER is not smaller than 10^{-4}, that means simulating the transmission of around 5×10^5 pseudorandom bits.

We have also used a semianalytical program that simulates the propagation of the deterministic part of the signal and, at the end, evaluates the error probability by the saddle point approximation on the ground of the theoretical properties of the noise (naturally taking into account the real shape of the filters, and so on).

Both the programs are scalar, so that all effects related to the vector nature of the field have to be inserted at the end in a phenomenological way.

The two programs give results in a good agreement when they can be superimposed.

The simulation result in terms of BER versus the SN_o are reported in Figure 6.27a for both programs, where the simulation points are interpolated with a BER function of the form (Equation 6.65) fitting the free parameters with the simulation points.

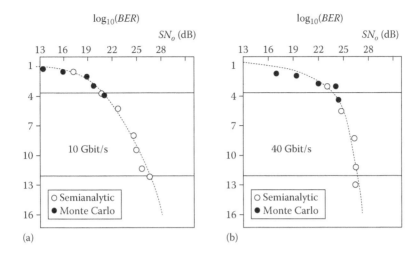

FIGURE 6.27
BER simulation of the system whose draft design reported in Table 6.4. Two types of simulation points are reported, Monte Carlo simulation and a semianalytic simulation where the signal propagation through the fiber and the different devices are simulated while the error probability is evaluated analytically on the ground of the theoretical noise distribution: (a) $R = 10$ Gbit/s and (b) $R = 40$ Gbit/s.

To attain the BER needed using the FEC an SN_o of 15.8 dB is needed. This number of course does not take into account system margin and aging, and since the simulation is scalar, also polarization penalty is not accounted for. Thus, 5 dB have to be added a posteriori, obtaining a simulation required SN of 20.8 dB that is not far from the 20 dB obtained in the last section, taking into account all the numerous approximations.

6.3.2.2 Long-Haul Transmission at 40 Gbit/s

6.3.2.2.1 Definition and Draft Design

It is interesting to repeat the design in the case of 40 Gbit/s channels.

Choosing a channel spacing of 200 GHz the overall bandwidth of the optical link, mainly determined by the cascade of the multiplexer and the demultiplexer, can be as tight as 60 GHz, attaining $M_{oe} = 1.5$ that gives a very good noise rejection.

In this case, we can allocate 22 channels at 40 GHz in the extended C band (see Table 6.5).

Reduced gain amplifiers have to be used at this speed due to the small number of channels and to the greater efficiency of nonlinear effects. The choice of the average power along the link is a strategic choice, since the right value comes out from a trade-off between the need of limiting nonlinear effects that call for a low power and the need of a high SN ratio that would bring to select a high power.

In our case, a fast look to the penalty plots relative to nonlinear effects reveals that at 40 Gbit/s a comb of 22 channels 200 GHz spaced suffers a penalty of 0.1 dB from FWM at a maximum power per channel of 0 dBm, the penalty rises at 2 dB at a maximum power per channel of 5 dBm and to 4 dB at a power per channel of 6 dBm.

Also XPM rises fast with the power per channel; it is smaller than 0.1 dB at a maximum power per channel of 5 dBm, but at 6 dBm is already slightly smaller than 2 dB.

It results that a power per channel of 5 dBm is a sort of threshold for nonlinear effects allowing to maintain the overall nonlinear crosstalk penalty at 2 dB that is a practical limit for penalty coming from a single effect for the working of our design procedure.

TABLE 6.5

Realistic Penalty Analysis for Long-Haul 40 Gbit/s Dense Wavelength Division
Multiplexing System

Long-Haul 40 Gbit/s DWDM System		
System length	2000 km	
Channel speed	40 Gbit/s	Suitable EDFAs, span length 50 km, RS(255,239) FEC
Channel spacing	200 GHz	Extended C band, 38 nm in the third window, that is 4560 GHz, 22 channels @ 40 Gbit/s
Ideal SN_e (dB) ($B_e=2R$)	17	
Optical to electrical bandwidth ratio	$M_{oe}=1.5$	
Ideal SN_o (dB)	18.2	
	Penalties in dB	
Transmitter dynamic range	0.5	SFF MSA using tunable CW laser and TiNbLi modulator
APD + shot noise	0.5	
Linear crosstalk	1.0	Gaussian selection filter (single stage AWG), $B_o=2R$
		Channel spacing $\Delta f=5R$ (see Figure 6.16)
Nonlinear crosstalk	1.5	Use of SSMF to thwart FWM and XPM, post-compensation with DCF. NRZ modulation
SPM	1	Under compensation of chromatic dispersion at 98%
		Unchirped NRZ modulation
Chromatic dispersion	0.5	Residual dispersion 500 ps/nm. Per channel electronic dispersion compensation at the receiver per 500 ps/nm. Used an FFE nine taps + DFE four taps (see Figure 5.56)
PMD	0.2	4 ps, NRZ modulation FFE–DFE at the receiver (see Figure 5.57)
Aging	1	—
System margin	3	—
Polarization effects	1	PDL, connectors misalignments, etc.
TOTAL SN post code	28.4 dB	—
Code gain (dB)	6	
Total SN_o precode	22.4	To confirm with simulation

Having a maximum power per channel equal to 5 dBm means to fix the output saturated power of our amplifiers at 18.5 dBm instead of 20 dBm. In practice, such a regulation can be done via software on every amplifier; thus, we can imagine using the same amplifier of Table 6.2 with a different software configuration.

We will need a high SN, since we want a total length of at least 2000 km; thus, we will chose a span length of 50 km so to reduce the ASE, as much as possible, within the constraint of a cost within market targets.

The transmitter and the receiver power budget equations are satisfied with a laser power of 13 dBm and a transmitter loss of 15 dBm, for the transmitter and a receiver loss of 10 dBm and an APD sensitivity at 40 Gbit/s of −20 dBm and an APD saturation power of −2 dBm.

The power per channel arriving on the APD is −5 dBm, sufficiently far from both sensitivity and saturation.

We can adopt at the receiver an electronic equalizer similar to that adopted in the 10 Gbit/s case.

The dispersion map, is also in this case, based on post-compensation and under compensation; the uncompensated differential group delay (DGD) is 500 ps/nm.

With the data, we derive from our draft design the ASE power in the optical bandwidth results to be −19.7 dBm, for a total SN_o of 23 dB.

The ideal SN_o is 20.0 dB; thus, taking into account the FEC gain, we have a great space available to accommodate all the relevant penalties.

6.3.2.2.2 *Long-Haul Transmission at 40 Gbit/s: Penalty Evaluation*

The next step of the system design is to evaluate all the penalties and obtain a draft estimate of the required, using the penalty addition method.

The result is summarized in Table 6.5. The required SN_o results to be 22.4 dB that is smaller than the 23 dB obtained from the draft design.

In this case, only semianalytical simulation has been carried out, from which the estimate of the required SN_o is essentially confirmed.

6.3.2.3 Long-Haul Transmission: Realistic Systems Characteristics

As a conclusion of this section, it is useful to review the characteristics of realistic long-haul systems as summarized in Table 6.6.

The characteristics of three different types of systems are shown, occupying different product segments in the category of long-haul DWDM systems.

The first is a system mainly designed for regional and metro–core applications, even if its characteristics allow a pure long-haul application to be feasible. It can convey a large number of different formats and data speed directly on a wavelength, and ring configurations are possible both in hub and in standard ring configuration via a flexible OADM. On the other hand, 40 Gbit/s channels are not supported.

The second system is a pure long-haul designed to be installed in the core network of a large carrier; it supports longer distances, 10 and 40 Gbit/s, and there is a possibility of supporting pure transparent branching via a fixed OADM.

Tunable OADM is also supported, in order either to drop a few channels in an intermediate site of a point-to-point link or to build optical transparent rings on long distances. Both optical channel and optical multiplex section (OMS) protections are supported.

Finally, the third system is a long-haul system also proposed for particularly high traffic regional areas.

6.3.3 Design of Ultra-Long-Haul Transmission Systems

If it is needed to design a DWDM system with a reach far greater than the number reported in Table 6.5, refining the design cannot help.

Nevertheless, the reach limitations of long-haul systems can be overcome by the combined use of EDFA and Raman amplification.

There are several possible architectures to use distributed or concentrated Raman amplification to design ultra-long-haul systems, each of which has its own design rules.

Raman amplification, can be used to provide distributed amplification as a complement to lumped EDFAs in the hybrid approach or all Raman amplification can be used, potentially both distributed and lumped.

TABLE 6.6

Characteristics of Realistic Long-Haul Systems Designed for Different Applications

Application	Channels Speed	Number of Channels	Channel Spacing	Reach (Only EDFAs)	Optical Bands	OADMs	Notes	Real Estate
Long haul Regional	100 Mbit/s to 10 Gbit/s	Up to 128	50 GHz	1200 km	C+L	From 4 to 40 wavelength per node	1+1 OChP	1 standard rack—64 channels
Long haul	10 or 40 Gbit/s	• Up to 96 @ 10 Gbit/s • Up to 80 @ 40 Gbit/s in the C-band • Any 10G/40G mix supported	50 GHz @10 Gbit/s 200 GHz @40 Gbit/s	• 2500 km @ 10 Gbit/s • 1600 km @ 40 Gbit/s	C+L	• Fixed OADM for passive branching configuration • Wavelength Selective Switch (WSS)-based technology	Line dispersion compensation per band dispersion compensation 1+1 OCH protection OMSP protection	Rack dimensions: • Height: 2200 mm • Width: 600 mm • Depth: 300 mm
Long haul	10 or 40 Gbit/s	• 192 @ 10 Gbit/s • 80 @ 40 Gbit/s	25 GHz @ 10 Gbit/s	• 5000 km @ 10 Gbit/s (with channel spacing 100 GHz) • 1000 km @ 40 Gbit/s	Extended C band	—	1+1 OChP 1:N transponder protection	ETSI standard cabinet • 2200 mm (height) × 600 mm (width) × 300 mm (depth)
Long haul Regional	2.5 or 10 or 40 Gbit/s	320 @ 10 Gbit/s	25 GHz @ 10 Gbit/s	• 4000 km @ 10 Gbit/s (with 100 GHz spacing) • 1000 km @ 10 Gbit/s Regional conf.	Extended C band		Electrical dispersion compensation (EDC)	Both ANSI and ETSI standard racks

Even if hybrid approach is used, co-propagating Raman pumps, counter-propagating Raman pumps, or both can be used.

In general, the design is more complicated with respect to what is described for long-haul systems.

We will introduce the main points of the design of such systems by analyzing one of the most common architectures.

This is the architecture based on hybrid EDFA–Raman amplifiers, where the Raman pumps are located in the EDFAs locations and generate signal gain by counter-propagation along the DCF.

We will assume that fiber dispersion is compensated by in-line DCF adopting under compensation to control XPM and SPM.

Due to the good Raman gain of the DCF, a single-stage EDFA is used, and the Raman pumps are propagated through the DCF to provide distributed amplification and blocked at the end of the DCF since propagating them through the SSMF used for transmission generates noise and creates limited gain due to the pumps attenuation experimented during the pass through the DCF.

The block scheme of the line site is reported in Figure 6.28.

Since the system will use extended C band, two Raman pumps are enough to achieve a flatness of the Raman gain on the order of 0.5 dB; overall gain flatness is achieved via a gain flatness filter at the output of the EDFA.

The advantages that are expected from the hybrid amplifier configuration are as follows:

- Once fixed, the received power that assures the required receiver sensitivity, the Raman gain allows the optical power at the EDFA output to be reduced so decreasing nonlinear effects.

- The overall optical gain is smaller and part of it is distributed; thus, less ASE is expected.

- A single-stage EDFA is sufficient, also implying less ASE at the receiver.

6.3.3.1 Ultra-Long-Haul Transmission at 10 Gbit/s: Draft Design

Our scope in this section is to determine the value of the optical SN ratio before detection, given the required system length via a power budget.

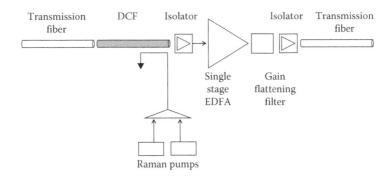

FIGURE 6.28
Block scheme of the line site of an ultra-long-haul system integrating a hybrid EDFA-Raman amplifier.

In this case, the system design is quite more complex, because hybrid amplifiers are to be optimized contemporary to the dispersion map, since Raman gain mainly occurs in the DCF.

We will assume that the required length of our system is around 10,000 km. In order to improve the control of dispersion and nonlinear phenomena, we will select RZ modulation with a short duty cycle: 25%. This means that the width of the transmitted Gaussian pulses at $1/e$ is as short as 25 ps.

Since the signal spectrum is 40 GHz large (unilateral), this choice forces to enlarge the transmission line bandwidth up to 40 GHz (unilateral, see Section 5.8.5.5) so that $M_{oe} = 4$. A channel distance of 50 GHz is not feasible with this optical bandwidth; the system will support in the extended C band (38 nm) 46 channels at 100 GHz of spacing.

In order to generate such short pulses, we select a high power external cavity laser with a linewidth of 300 kHz and an emitted power of 16 dBm.

The span length is fixed to 60 km to minimize as much as possible the ASE power while remaining within a reasonable number of required intermediate sites.

In order to carry out the power budget we have to design the hybrid amplifier.

Since we are interested in a draft design that will be confirmed by simulation, the hybrid amplifier will be optimized using the Raman amplifier model reported in Chapter 4, without resorting to the much more complex model of Appendix D.

Coherently, with this design approach we will consider the penalty, due to the various noise terms, that we are neglecting in a first time as penalties in the penalty analysis.

Following the approach of Chapter 4, the gain of the Raman pumped DCF is given by

$$G_R(\omega) = e^{-\alpha_D L_D} \exp\{[P_{10} g_1 + P_{20} g_2] L_{eD}\} \tag{6.77}$$

The parameters' meanings are as follows (see Equation 4.86):

- α_D is the DCF attenuation.
- L_D is the DCF length.
- P_{10} and P_{20} are the powers of the Raman pumps.
- g_1 and g_2 are the corresponding normalized spectral gains.
- L_{eD} is the DCF effective length.

The optimization consists in substituting in Equation 6.77 the expressions of the Raman local gain that depends on the wavelengths and power of the pumps and on the wavelength of the signal and evaluating the average and the standard deviation of the gain over the bandwidth occupied by the DWDM comb.

In these two expressions, the power and the wavelengths of the pumps are considered variables, and the optimization is carried out by minimizing the gain standard deviation with the condition to maintain the average gain constant and equal to the value that is dictated from the system design.

This optimization can be carried out both by numerical means and by the Lagrange multipliers if an approximated analytical expression is found for the local Raman gain curve (see references in Chapter 4 for several papers on pumps optimization).

The parameters we will use for the numeric optimization and for the following system performance evaluation are reported in Table 6.7, the parameters specific of Raman amplifiers are those reported in Table 4.11.

TABLE 6.7

Parameters of the Hybrid Amplified 10 Gbit/s Ultra-Long-Haul System

General	
Channel speed	10 Gbit/s
Optical bandwidth of the DWDM comb	Extended C band
Modulation	RZ—25% duty cycle
Number of channels	46
Channel spacing	100 GHz
Span length	50 km
Transmitting Chain	
Source laser emitted power	16 dBm
Mach Zehnder lithium niobate modulator loss	7 dB
Multiplexer (multistage AWG × 80) loss	7 dB
Patchcords, connectors, monitoring taps loss	1 dB
Detection Chain	
High-efficiency APD sensitivity @ 10 Gbit/s	−28 dBm
Demultiplexer (multistage AWG × 80) loss	7 dB
Patchcords, connectors, monitoring taps loss	1 dB
Electronic equalizer	FFE nine taps DFE four taps, LMF adaptation algorithm
Decision procedure	Hard decision, adaptive threshold, RS (255,223) code
Transmission Line (Assumed to be 10000 km Long)	
Fiber type	SSMF
Dispersion parameter SSMF	17 ps/nm/km
Dispersion parameter DCF	−70 ps/nm/km
DCF Length	12 km
Signal attenuation in SSMF	0.25 dB/km
Pumps attenuation in SSMF	0.35 dB/km
Signal attenuation in DCF	0.4 dB/km
Pumps attenuation in DCF	0.5 dB/km
Maximum peak signal	5 mW
Local Raman gain $g(\omega_1,\omega_s)/A^2$ (SSMF, @ 1550 nm)	8.00E−04 1/m/W
Local Raman gain $g(\omega_2,\omega_s)/A^2$ (SSMF, @ 1550 nm)	8.00E−04 1/m/W
Local Raman gain $g(\omega_1,\omega_s)/A^2$ (DCF, @ 1550 nm)	4.00E−03 1/m/W
Local Raman gain $g(\omega_2,\omega_s)/A^2$ (DCF, @ 1550 nm)	2.6E−03 1/m/W
Under compensation	1800 ps/nm

The Raman gain curve is the same DCF curve used in Chapter 4 (see Figure 4.35).

The pump wavelengths resulting from the optimization procedure are located at 1430 and at 1400 nm and the power are emitted by the pumps 330 and 290 mW.

Once the Raman distributed amplification is optimized, the design of the system can be done leveraging on two fixed numbers: the receiver sensitivity and the maximum peak power acceptable along the link to maintain nonlinear effects penalties within the limit compatible with the design.

In our case, we have assumed to have a high-efficiency APD at the receiver with a sensitivity of −28 dBm. Since we have the demux and a series of other small losses between the last EDFA and the APD, the power that has to exit from the last EDFA is equal to −20 dB (compare Table 6.7, the detection chain).

On the other hand, always assuming a link 10,000 km long, by considering under compensation for 2000 ps/nm; chirped RZ pulses with 25% duty cycle; and an electronic equalizer at the receiver composed by a nine taps FFE and a four taps DFE, the peak power along the line cannot overcome 5 mW per channel per span to maintain the SPM and the FWM related penalties within 2 dB each.

The SN_o has the following expression:

$$SN_o = \frac{P_{rec}}{\hbar\omega B_o a_{rec}[\xi(G_B - 1)NF_B + (N_a - 1)[G_E(G_R - 1)NF_R + (G_E - 1)NF_E]}$$

$$+ G_p(G_R - 1)NF_R + (G_p - 1)NF_p] \qquad (6.78)$$

where

- B_o is the optical noise bandwidth defined mainly by the cascade of the multiplexer and the demultiplexer
- P_{rec} is the power on the receiver; this is a design parameter that has to be chosen to satisfy the receiver power budget and to optimize the system performances. In our case we select −20 dBm, against a sensitivity of −28 dBm and a saturation power of −10 dBm
- a_{rec} represents the total optical loss between the output of the last EDFA amplifier and the APD, in our case the loss of the AWG and of all the monitoring taps and connectors
- G_B, G_R, G_E, G_p are respectively the gains of the EDFA booster, the Raman gain, the gain of the EDFA in-line amplifier, and the gain of the last EDFA, that is different from the others since the signal at the exit of the last span is not equal to the signal at the exit of the booster, but is equal to P_{rec}/a_{rec} in order to minimize the ASE power
- NF_B, NF_R, NF_E, NF_p are respectively the noise factors of the EDFA booster, of the Raman amplification, of the EDFA in-line amplifier and of the last EDFA
- ξ is the overall gain or loss of the transmission chain, from the exit of the booster to the exit of the last EDFA and can be expressed as

$$\xi = \frac{P_{rec}}{P_s(0)} = \frac{P_{rec}}{P_M} \qquad (6.79)$$

where $P_s(0)$ is the signal power at the output of the booster

Since, for all reasonable designs, it is the peak power of the link; it has a value P_M prescribed by the nonlinear effects that in our case we have tentatively set as 5 dBm

The condition on the maximum peak power can be written as

$$P_l \, a_T G_B = P_M \tag{6.80}$$

where
  P_l is the power emitted by the laser source
  a_T is the whole loss between the power source and the input of the EDFA booster; in our case, the losses of the multiplexer, the mux, the connectors, and the monitoring taps

The condition on the receiver sensitivity can be written as

$$G_p G_R P_M a_{rec} = P_{rec} \tag{6.81}$$

Solving (6.80) with respect to G_B and (6.81) with respect to G_p and remembering the definition of ξ, the expression of SN_o becomes a function of two unknown variables: G_E and G_R.

Generally, since Raman distributed amplification makes less noise and operates on a signal of lower level, the optimization consists in maximizing G_R, taking into account the limitations in the available power of the pumps and of the possibility of choosing different types of DCF, and then selecting consequently G_E.

At this point the dimensioning of the line is ended, and all the line parameters are summarized in Table 6.8.

TABLE 6.8

Line Parameters after the Draft Design of the
Ultra-Long-Haul 10 Gbit/s System

General	
EDFA noise factor	5 dB
Solution	
Power of the first pump	330 mW
Power of the second pump	290 mW
Wavelength of the first pump	1430 nm
Wavelength of the second pump	1400 nm
Average power in a span	−11.2 dBm
Booster EDFA gain	10 dB
In-line EDFA gain	5 dB
Raman gain	10 dB
Preamplifier EDFA gain	11 dB
Two Stages EDFA Solution	
Average power in a span	−6.61 dBm
EDFA booster gain	4 dB
First stage of in-line EDFA gain (also the preamplifier)	11.4 dB
Second stage of in-line EDFA gain	6 dB
Second stage EDFA preamplifier gain	11 dB

In order to carry out a comparison that should underline the advantages of the hybrid solution with respect to solutions based on EDFA alone, we want also design an ultra-long-haul link completely EDFA-based with the same parameters.

In this case the SN_o is given by

$$SN_o = \frac{p_{rec}}{\hbar\omega B_o a_{rec}[\xi(G_B - 1)NF_B + (N_a - 1)(G_E - 1)NF_E + (G_p - 1)NF_p]} \qquad (6.82)$$

where the booster amplifier is supposed to be a single-stage EDFA, post-compensation DCFs are used, as in the hybrid case, and the last EDFA is different to arrive with the correct power on the APD.

Also, in this case optimization is carried out with two constraints; the maximum power constraint is expressed by Equation 6.74, while the constrain relative to the power on the receiver is

$$G_p P_M a_{rec} = P_{rec} \qquad (6.83)$$

After that, the design procedure is identical to the used from the design with hybrid amplifiers.

All the results of the hybrid amplification design and of the two stages EDFA design we have achieved up to now are summarized in Table 6.8.

The characteristics of the selected components and the output of the draft design allow us to evaluate the SN_o at the link output that comes out 25.4 dB for the hybrid solution and 22.8 for the all EDFA solution. In both cases the FEC gain has not been accounted for, it will be considered in the penalties analysis.

A first advantage of the hybrid solution is revealed by these figures: the lower noise factor of the Raman amplifier and the automatic compensation of the DCF loss allow the ASE noise to be quite reduced.

6.3.3.2 Ultra-Long-Haul Transmission Systems: Penalties, Evaluation, and Simulation Results

Let us now define the target SN_o on the ground of the various transmission impairments in both the cases we have considered.

The detail of the penalty evaluation in the two cases is reported in Table 6.9, where 21.5 dB is obtained for the hybrid solution and 23.8 dB is obtained for the EDFA solution. Thus, the 10,000 km long link can be realized with the hybrid technique (the draft design SN_o was 23.5 dB) but not with a pure EDFA architecture (the design SN_o was 21.5 dB). This also taking into account that a margin has to exist between the draft design SN and the requirements derived from the penalties analysis, so to accommodate the approximation uncertainty that will become clear with the simulation results.

The difference in performances is mainly in the different role of nonlinear effects, as it can be clearly deduced by the table.

In order to better appreciate the reason of this difference, the signal power behavior in a single span of the hybrid system is shown in Figure 6.29.

It is clear how the average power along the transmitted line is much lower using distributed Raman amplifiers even if the optical power entering into the EDFA at the span end (i.e., into the preamplifier in front of the receiver) is the same in both cases.

TABLE 6.9

Penalty Analysis of the Ultra-Long-Haul 10 Gbit/s System Realized with Two Different Architectures: Hybrid Amplification and All Erbium-Doped Optical Fiber Amplifier Amplification

| | Ultra-Long-Haul 10 Gbit/s Transmission | | | |
| | EDFA Only | | Hybrid | |
Phenomenon	(dB)	Notes	(dB)	Notes
Ideal SN_e (dB) ($B_e = 2R$)	18.5		18.5	
Electrical to optical bandwidth ratio (M_{oe})	4	RZ with a duty cycle of 25%, $B_e \approx R$, $B_o \approx 4R$	4	RZ with a duty cycle of 25%, $B_e \approx R$, $B_o \approx 4R$
Ideal SN_o	15.5		15.5	
Signal dynamic range	1	Lithium niobate modulator	1.5	Lithium niobate modulator—Raman dynamic compression
Linear crosstalk	1.8	AWG multistage, spacing 2.5 signal bandwidth	1.8	AWG multistage, spacing 2.5 signal bandwidth
Nonlinear crosstalk	3	FWM and XPM (see Equation 4.45)	1.5	Smaller average power due to Raman
SPM–DGD interplay	3	RZ reduces relative penalty	1.2	Smaller average power due to Raman—RZ effect
PMD	0.5	10 ps, compensated by the FFE–DFE at the receiver (See Figure 5.57)	0.5	FFE–DFE at the receiver to cope with PMD (See Figure 5.57)
Polarization effects	1		1	
RIN transfer	—	Non-applicable	0.3	Compare Equation 4.90 with RIN parameters from Tables 5.4 for the pumps and 5.3 for the signal
Double Rayleigh	—	Non-applicable	0.2	
Photon induced noise	—	Non-applicable	0.0	
Aging	1		1	
System margin	3	Normal system margin	3	Normal system margin
Total w/o FEC	29.8		27.5	
FEC gain (RS 255,239)	6		6	
Required SN_o	23.8		21.5	

This fact alleviates greatly nonlinear effects allowing a longer link to be designed.

In both cases, however, nonlinear effects determine the maximum reach in the different situations.

An interesting element to further quantify is the reach difference between the two models' results.

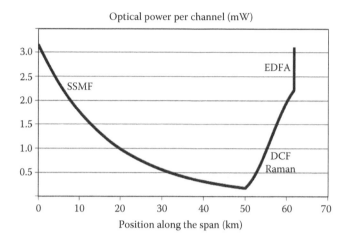

FIGURE 6.29
Single-channel power profile in a span of the ultra-long-haul system-based on hybrid amplification.

Doing again the dimensioning and the penalty evaluation, but maintaining the link length as a parameter, it is found that at a link length of 8000 km, the design SN_o of the EDFA alone system comes out to be 23.8 dB, while the requirements coming out from the penalties evaluation is 22.0 dB. In this condition the draft design can be accepted waiting for the simulation results.

Thus, with the longer link that can be realized with the EDFA alone technique and with our design choices is on the order of 8000 km long.

Naturally, the design can be further optimized, for example, using more effective electronic compensation, introducing pre-chirp to better manage nonlinear effects, optimizing both the duty cycle and the form of the transmitted pulse, besides an optimized dispersion map and better quality amplifiers.

All these steps help to improve the system performances above what we have achieved with our simple design, but it is true for both the system architectures and the difference always remains.

Moreover, the hybrid architecture presents possibilities that are not available if all the EDFA architecture is chosen.

Raman effect can be used in a more distributed way sending the pump also in the SSMF used for transmission, both co-propagating and counter-propagating pumps can be used in the Raman amplifiers and, finally, all Raman amplification can be adopted, eliminating completely EDFA amplifiers and going much nearer to the optimum scheme of distributed amplification.

All these steps, further improve the performance of the hybrid solution so increasing the difference with the all EDFA architecture.

6.3.3.3 Ultra-Long-Haul Transmission at 40 Gbit/s

At 40 Gbit/s, all the distortions depending on the signal speed are more severe.

The draft design is carried out with the same method used in the previous section starting from a system with 18 channels at 40 Gbit/s and spaced at 250 GHz.

The large spacing is due to the fact that, to control dispersion and the other sources of ISI, RZ is very useful.

Even if the duty cycle is reduced to 33% with respect to the 25% used at 10 Gbit/s, the optical signal spectrum is 120 GHz wide so that, in order to reduce linear interference, a very large channel spacing is needed.

A very simple inspection of the FWM and XPM penalty curves in this case shows that 10,000 km are too much. For this reason we will try to design a 6500 km long system.

The draft design can be carried out with the same procedure followed in the 10 Gbit/s case and the parameters are the same, but for the transmitter loss that, due to the 40 Gbit/s modulator, is 2 dB higher and the sensitivity of the PIN photodiode that substitutes the APD used for 10 Gbit/s is –12 dBm. The PIN photodiode is used to have a higher saturation power; as a matter of fact the power per channel arriving on the PIN is –2 dBm.

The final value of the SN_o derived from the draft design is 26.8 dB.

The summary of the penalty analysis for this system is reported in Table 6.10, from which a required SN_o equal to 25.6 dB is obtained. The system is thus within the required performances, if the last test via simulation does not differ too much from the draft penalty analysis.

In this case, the simulation has to take into account both nonlinear propagation and the complete model of Raman amplification. In order to simplify the simulation, we have simulated all the signal distortion, by coupling the nonlinear propagation in the SSMF with the complete pumps plus signal Raman equations, eliminating the noise terms.

Naturally, pump depletion is taken into account while analyzing Raman gain.

The noise power is evaluated analytically and the saddle point approximation is used to estimate the error probability.

The results of this simulation integrated with margins and aging confirm in all the considered cases the draft design with a good.

TABLE 6.10

Penalty Analysis of the Ultra-Long-Haul 40 Gbit/s System

Phenomenon	40 Gbit/s Ultra-Long-Haul Hybrid—40 Gbit/s	
	(dB)	**Notes**
Ideal SN_e (dB) ($B_e = 2R$)	17	
Electrical to optical bandwidth ratio	3	RZ modulation 33%
Ideal SN_o	18.7	
Signal dynamic range	1	Lithium niobate modulator
Linear crosstalk	1.5	AWG multistage—$B_o = 2R$
Nonlinear crosstalk	2	FWM and XPM
SPM and DGD interaction	2	Added electronic compensation
PMD	0.6	4.72 ps, compensated by the FFE–DFE at the receiver
Polarization effects	1	
RIN transfer	0.4	
Double Rayleigh	0.2	
Phonon induced noise	0	
Aging	1	
System margin	3	Normal system margin
Total w/o FEC	31.8 dB	
FEC gain (RS 255,223)	6 dB	
Required SN_o	*25.8 dB*	

Up to now, we have underlined all the advantages of the use of hybrid amplification. However, there are also a set of issues that explain the fact that pure EDFA DWDM systems are the most diffused products on the market till today.

The first and more important point is that using Raman amplification requires high power pumps with all the induced thermal management and high operational expense related to power consumption.

It is important to remember from Chapter 2 that, in general, the main goal of carriers is to decrease the operational cost (OPEX) of the network, more than increasing the performances of the transport layer.

When deploying an ultra-long-haul system, the greater cost related to operational expenses and to the higher system cost per km and per Gbit/s have to be compared with the costs implied by realizing the link with a cheaper long-haul system, using an intermediate electronic regenerator.

Also, from the security point of view, operating systems with such powerful lasers put some problems in the procedures for units' substitutions and maintenance in general.

6.3.3.4 Ultra-Long-Haul Systems with Electronic Pre-Compensation

The case studies we have carried out in the previous sections are intended to show the key points of long-haul and ultra-long-haul systems design.

In all the considered design variations, the compensation of the transmission line is carried out by optical means, but for the electronic compensation at the receiver.

A radically different way to solve the problem is to design a system leveraging on electronic pre-distortion via a combination of FFE–DFE and look-up table in order to deploy along the line only single-stage EDFA amplifiers to compensate losses (compare Chapter 5 for description and bibliography).

The potentialities of this strategy are all in the ability of the pre-distortion to correctly simulate and reproduce the signal propagation along the line.

From different experimental pre-distortion implementations, it results that at 10 Gbit/s the solutions based on optical compensations are quite more effective. Today, it seems that there is no viable implementation of pre-distortion circuit that at 10 Gbit/s allows about 10,000 km of SSMF to be traveled by a comb of 40 channels in extended C band.

On the other hand, the situation at 40 Gbit/s is quite different.

The reach of 40 Gbit/s systems is shorter, and nonlinear evolution is more evident.

In this situation, electronic pre-distortion is competitive from a performance point of view with optical compensation techniques [41].

Electronic pre-distortion can also be used not in alternative, but as a complement to optical compensation techniques. In this case the algorithm determining pre-distortion is a bit more complex, since all the optical compensation devices such as DCF and pre-chirping have to be simulated in back propagation, but the results can be significant.

Integrating pre-distortion with the contemporary use of look-up tables and FFE–DFE pre-compensation in the ultra-long-haul system at 10 Gbit/s analyzed in the previous section and assuming to use a nine taps FFE and a nine Taps DFE plus a 13 bit look-up table, the system can reach the impressive distance of about 13,000 km and adding also Raman co-propagating pumps in the SSMF in order to avoid excessive signal attenuation that request EDFA amplifiers, it is possible to go over 15,000 km.

However, such a complicated system, using a complex digital signal processor (DSP) per channel at the transmitter, four Raman pumps per site, and no EDFAs has quite a high

CAPEX and OPEX and could be applied only in a very small number of cases, at least in terrestrial networks.

6.3.4 Single-Span Systems

In practical networks, there are a certain number of systems that have to span a few hundred kilometers, too much for unamplified systems, but a short distance for the capabilities of long-haul systems with intermediate amplifier sites.

It is to remember that increasing the number of points of presence in the network increases the operation expenses due to maintenance and other activities related to the point of presence. Moreover, in submarine networks avoiding submerged amplifiers is even more important, if possible.

Thus, it is important to understand what is the potential performance of a system having amplification only at the terminals and no intermediate site.

Since these systems have to span the maximum possible length, there are several possible architectures that seem natural: using hybrid EDFA–Raman amplification and using the DCF also for enhancing Raman efficiency, using all Raman amplification to exploit also low Raman noise factor and more distributed amplification, or using complete link pre-compensation and amplifying through single-stage EDFAs.

Moreover, the use of differential phase encoding is appealing in this kind of application due to the better receiver sensitivity.

As examples we will design three 10 Gbit/s systems: two based on all Raman amplification, one using IM, the other differential phase shift keying (DPSK), the third based on IM and complete link electronic pre-compensation with single-stage EDFA amplification.

6.3.4.1 Single-Span Systems with Intensity Modulation and All Raman Amplification

A system based on all Raman amplification tries to leverage on the specific characteristics of Raman amplifiers: low global noise factor and distributed amplification in the DCF (Table 6.11).

This last characteristic, in particular, helps to maintain a low peak power along the link contributing to take nonlinear effects under control.

Due to the fact that nonlinear effects pose at the end the final limitation to the system reach, they need a careful consideration. In all the following reasoning we will assume to have a link around 250 km long. The attenuation of 250 km of SSMF is on the order of 63 dB, a formidable amount of attenuation to be compensated.

In order to achieve a high Raman amplification efficiency, we will divide the DCF part at the transmitter and part at the receiver.

The power injected per channel into the fiber at the end of the DCF located at the transmitter determines practically the amount of nonlinear effects, since Raman amplification from pumps located at the transmitter only happens into the DCF since the pumps are extracted at the DCF end.

Being SPM and XPM critical, RZ modulation is mandatory, even if this means having a big optical bandwidth accepting more ASE noise on the detector.

To make a trade-off between robustness to ISI and ASE noise power, we will select a duty cycle of 40% that means that the Gaussian pulse at the transmitter has a full width at $1/e$ equal to 40 ps.

We assume also to leave a residual dispersion of about 1900 ps/nm and to pre-chirp at the modulator the pulses with the same pre-chirp used to plot Figure 6.9.

TABLE 6.11

Parameters Values for the All Raman Single Span Intensity
Modulation and Direct Detection System

Channel Speed	10 Gbit/s
Optical bandwidth of the DWDM comb	Extended C band
Modulation	RZ
Duty cycle	0.4
Number of channels	40
Channel spacing	100 GHz
Transmitting chain	
Source laser emitted power	20 dBm
Mach Zehnder lithium niobate modulator loss	7 dB
Multiplexer (multistage AWG × 80) loss	7 dB
Patchcords, connectors, monitoring taps loss	1 dB
Detection chain	
High-efficiency APD sensitivity @ 10 Gbit/s	−28 dBm
Demultiplexer (multistage AWG × 80) loss	7 dB
Patchcords, connectors, monitoring taps loss	1 dB
Power per channel on the APD	−20 dBm
Transmission line	
Fiber type	SSMF
Dispersion parameter SSMF	17 ps/nm/km
Dispersion parameter DCF	−100 ps/nm/km
DCF length per Raman amplifier (total × 2)	20 km
Signal attenuation in SSMF	0.25 dB/km
Pumps attenuation in SSMF	0.35 dB/km
Signal attenuation in DCF	0.4 dB/km
Pumps attenuation in DCF	0.5 dB/km
Maximum peak signal (SPM and FWM pen. 2 B)	17.7 mW
Local Raman gain $g(\omega_1,\omega_s)/A^2$ (SSMF, @1550 nm)	0.0008 1/m/W
Local Raman gain $g(\omega_1,\omega_s)/A^2$ (DCF, @1550 nm)	0.004 1/m/W

Moreover, we assume to deploy at the receiver an MLSE as those whose performances are reported in Figure 5.55, that is a very simple circuit with respect to the potential complexity of an MLSE. It is to be noticed that in our example, the MLSE is used only to estimate the channel response and to compensate it inverting the ISI.

A much more efficient use of the MLSE could be done integrating it with the FEC and the decision in a soft decision system, but for simplicity we do not consider this alternative here.

With all these elements, the maximum peak power that is allowed along the link in order to maintain the SPM/XPM penalty within 1.5 dB is 17.5 mW per channel with a total power of 700 mW considering 40 channels in the extended C band with a spacing of 100 GHz. Such a big spacing is a consequence of the RZ modulation implying a unilateral optical bandwidth $B_o = 25$ GHz.

As far as FWM is concerned, Equation 6.37 suitably specified to our case tells that 1 dB penalty at 100 GHz spacing on 250 km of SSMF fiber is reached at about 60 mW per channel and that at 17.5 mW we have a negligible penalty (0.2 dB).

Also the interference between channels due to Raman effect is negligible, as far as the peak power per channel along the link is limited at 17.5 mW, as it can be demonstrated by evaluating it starting from Raman gain curves.

Thus, we will deploy two sections of DCF, one at the transmitter and one at the receiver, and we will optimize their length and dispersion coefficient to have the most efficient Raman gain.

The peak power per channel on the link, that is, the power at the output of the Raman amplifier at the transmitter, is maintained at 17.5 mW per channel, regulating the power of the Raman pumps.

A scheme of the resulting link architecture is plotted in Figure 6.30.

It is to be noted that in this case the DCF has a double role. It provides dispersion compensation and enhances the Raman gain.

The two Raman amplifiers can be dealt with independently.

Solving the Raman amplifier equation (see Chapter 4) for the amplifier at the receiver we obtain the following expression of the gain:

$$G_{R2} = e^{\Gamma_2(\omega)L_{eD} - \gamma_s L_2} \tag{6.84}$$

where

$$\Gamma_2(\omega) = g_2(\omega_1)P_1 + g_2(\omega_2)P_2 \tag{6.85}$$

and the symbols have the following meaning:

- $g_2(\omega)$ is the microscopic Raman gain in the DCF located at the receiver.
- γ_s is the attenuations of the signal in the DCF.
- P_1 and P_2 are the pumps powers injected at the DCF far end.
- L_2 is the length of the DCF located at the receiver.
- L_{eD} is the DCF effective length.

Analogously for the Raman amplifier at the transmitter the gain results to be

$$G_{R1} = e^{\Gamma_1(\omega)L_{eD} - \gamma_s L_1} \tag{6.86}$$

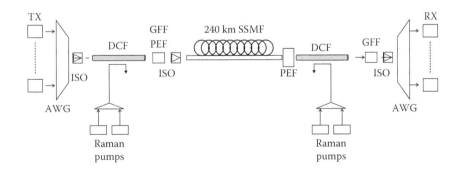

FIGURE 6.30
All Raman amplification single span IM-DD system architecture. TX, transmission transponders; RX, receiver transponders; ISO, isolator; DCF, dispersion compensating fiber; GFF, gain flattening filter (if needed); PEF, pump extraction filter; AWG, array waveguide; SSFM, standard single mode fiber.

with the corresponding meanings of the symbols.

The Raman gains have to be optimized both for flatness and for value.

Flatness is achieved by placing the pumps suitably as seen in Chapter 4. As far as the gain is concerned, it has to be set equal to the required value by setting the values of the pumps powers.

If the APD sensitivity is −28 dBm, it is reasonable to fix an amount of power P_{rec} on the APD greater than the sensitivity, to be sure that the electrical noise will be negligible.

Moreover, due to the performance required to the system, we will adopt a laser with an emitted power $P_l = 20$ dBm; it is a DBR tunable laser with an integrated semiconductor optical amplifier (SOA) providing a gain of 8 dB that brings the power from 12 to 20 dBm.

Calling $P_{max} = 17.7$ mW the maximum peak power per channel at the Raman amplifier at output of the transmitter, the power budget equations write

$$P_{rec} = G_{R1} + G_{R2} + P_l - A_T - A - A_{rec} \quad \text{(all dB)} \tag{6.87}$$

$$P_{max} = G_{R1} + P_l - A_T \quad \text{(all dB)} \tag{6.88}$$

where A_T, A, and A_{rec} are the losses of the transmitting chain, the transmission fiber, and the receiving chain. In the term A, there is the link length. A fast evaluation of the parameters versus the link length suggests fixing it on the order of 300 km.

Solving these equations, the target values of the gain of the two Raman amplifiers is determined and, from the gain expression, the needed pump powers are fixed. All the results of the draft design are summarized in Table 6.12.

In determining the gain of the amplifier, the DCF length has been optimized at both sides to achieve maximum amplification.

To understand the meaning of this optimization, the behavior of the single-channel optical power is reported in Figure 6.31. Observing Figure 6.31, emerges another important feature of Raman assisted transmission that is always present, but in this case is particularly evident.

During propagation of the signal in the DCF two phenomena occur: Raman gain and attenuation due to scattering and absorption.

TABLE 6.12

Results of the Draft Design of the All Raman Single Span Intensity Modulation and Direct Detection System

Pump 1 power-receiver amplifier	455 mW
Pump 2 power-receiver amplifier	350 mW
Pump 1 power-transmitter amplifier	525 mW
Pump 2 power-transmitter amplifier	600 mW
Raman pumps insertion loss	3 dB
DCF length transmitter	20 km
DCF length receiver	20 km
Raman gain transmitter	23.5 dB
Raman gain receiver	19.5 dB

Attenuation is driven by a constant coefficient that does not depend on the pumps and on the signal, while the gain depends on the pump. This means that regulating the DCF length while the pump powers are maintained constant the Raman gain can be maximized.

This fact is also illustrated in Figure 6.32, where the Raman gain in the DCF is shown versus the DCF length.

The Raman gain is built in the first part of the DCF, up to a maximum where the pumps have produced all the possible gain, from the maximum on, the fiber attenuation overcome the gain and the gain goes down up to a final zone in which the curve is almost linear (in dBm), demonstrating that the pumps have almost completely exhausted their gain capability and the signal evolution is governed by the attenuation only.

Once determined the optimum DCF length, the dispersion parameter of the two DCF elements can be derived from the equation

$$D_1 L_1 + D_2 L_2 - R_R + DL = 0 \tag{6.89}$$

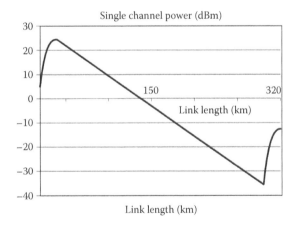

FIGURE 6.31
Single-channel optical power profile in the Raman amplified single span system depicted in Figure 6.30.

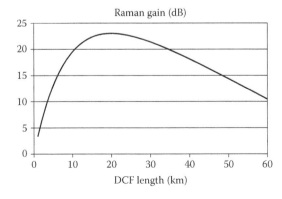

FIGURE 6.32
Raman gain in the DCF versus the DCF length: the pumps are maintained, fixed, and constant. The amplifier data are those relative to the system depicted in Figure 6.30.

where

D_1 and D_2 are the dispersion parameters of the DCF at the transmitter and at the receiver

R_R is the wanted residual dispersion

D is the SSMF dispersion coefficient

Since we have only one equation we can assume $D_1 = D_2 = D_D$ obtaining

$$D_D = \frac{DL - R_R}{L_1 + L_2} \tag{6.90}$$

The obtained DCF parameters are summarized in Table 6.12.

At this point the SN_o can be evaluated by assuming a standard noise factor for the Raman amplifiers (see previous section) obtaining $SN_o = 18.4\,dB$. This figure can seem very small but it has to be taken into account that the optical bilateral bandwidth is $2B_o = 75\,GHz$, five times greater than the bit rate; thus, when the signal will be filtered at the bit rate with an electrical filter after detection the SN_e will be larger.

After the draft design, it has to be verified by the penalties analysis.

The summary of this design phase is presented in Table 6.13. At the end, after the FEC gain, the required SN_o is 18.1 dB, smaller than the design value of 18.4 dB.

A final test has to be done by simulation, whose result completely confirms the draft design.

Thus, a link 320 km long (280 km SSMF + 40 km DCF) without intermediate sites can be realized using RZ modulation and Raman amplification even if it is short of margins out of the 3 dB standard added to the penalty analysis. It is to be noted that, since pumping of the DCF is always operated from the system end, the DCF itself can be cabled and used also for transmission.

As in all the examples of this chapter, the system is not completely optimized; thus, it is to expect that better performances can be attained with a careful optimization of all the design steps.

6.3.4.2 Single-Span Systems with Differential Phase Shift Keying Transmission and Raman Amplification

In this section, we will try to improve the design we performed in the last section by changing the modulation format. Essentially, we will use differential phase modulation.

The phase of the transmitted signal is modulated using a simple phase modulator and adopting differential encoding. Let us call b_j the jth transmitted bit, the signal in the jth bit interval writes

$$\bar{E}(t, k, w_j) = \sqrt{P_0}\, e^{i\theta(j)} \eta_T(t, j) \vec{p}_+ \tag{6.91}$$

where $\eta_T(t,j)$ is the Heaviside function that is equal to one for $t \in ((j-1)T, jT)$ and zero elsewhere.

The name differential comes from the fact that the transmitted bit is encoded into the phase difference between the two consecutive symbol intervals so that $\theta(j) - \theta(j-1) = b_j \pi$.

In terms of quadrature, components (6.91) writes

TABLE 6.13

Penalty Analysis of the All Raman Single Span Intensity Modulation and Direct Detection System

	All Raman Single Span System at 10 Gbit/s	
	10 Gbit/s	Notes
System reach	320 km	280 km SSMF, 40 km DCF
Channel number	30	Extended C band
Channel spacing	100	
Modulation format	RZ 40%	Gaussian pulses
Ideal SN_e (dB)	18	
B_o/B_e (due to RZ modulation)	4	$M_{oe}=4$ @ 10 Gbit/s, RZ modulation with 40% duty cycle
Required SN_o (dB)	15	
Impairment	Penalty (dB)	Notes
APD receiver noise	0.3	This is mainly the avalanche noise (see Equation 5.29), due to the low power on the APD (−20 dBm per channel)
Signal dynamic range	1	Lithium niobate modulator— Raman dynamic compression
Linear crosstalk	0.5	AWG multistage, signal spectrum Gaussian with unilateral spread of 25 GHz, Gaussian optical filter with 25 GHz width, see Figure 6.16
Nonlinear crosstalk (FWM XPM)	1	Wide spacing (100 GHz) and use of an SSMF
Residue chromatic dispersion and SPM interplay	1	Value due to the high input power. Managed with • Use of RZ modulation (paying a larger B_o) • MLSE as in Figure 5.59 • Residual dispersion of 1200 ps/ nm (300 km is 5100 ps/nm) • Pre-chirp at the modulator as in Figure 6.9
PMD	0.	MLSE at the receiver to cancel completely PMD that is very small (less than 2 ps per 400 km @ $PMD=0.1\,ps/km^{-1/2}$)
Polarization effects	0.5	
RIN transfer	0.3	Compare Equation 4.90 with RIN parameters from Table 5.4 for the pumps and 5.3 for the signal
Double Rayleigh	0.2	
Phonon induced noise	0.2	
Aging	1	
System margin	3	Normal system margin
Total w/o FEC	24	
FEC gain (RS 255,223)	6	
Required SN_o (dB)	*18*	

$$e_1(t) = \sqrt{P_0}\cos\left[\theta(j-1)+b_j\pi\right]\eta_T(t,j)$$

$$e_2(t) = \sqrt{P_0}\sin\left[\theta(j-1)+b_j\pi\right]\eta_T(t,j) \qquad (6.92)$$

$$e_3(t) = e_4(t) = 0$$

Differential encoding allows a very simple direct detection to be performed for DPSK signals through the receiver whose scheme is shown in Figure 6.33.

The incoming signal is split with a standard beam splitter; one of the outputs is delayed a bit time T and then the two fields are recombined with another beam splitter.

Starting from (6.91) it is easy to see that, after traveling through an amplified line, the photocurrent has the following expression:

$$c(t) = RP\cos(\pi b_j)\eta_T + 2R\sqrt{P}\,\eta_T Re\left\{[e^{i\theta(j)}+e^{-i\theta(j-1)}]\vec{\eta}'_{ASE}\right\} + R\,|\,\vec{\eta}'_{ASE}(t)\,|^2\,\eta_T + n_s(t) \quad (6.93)$$

where $\vec{\eta}'_{ASE}(t) = \vec{\eta}_{ASE}(t) - \vec{\eta}_{ASE}(t-T)$, being $\vec{\eta}_{ASE}$ the optical noise process, $\eta_T(t)$ is the Heaviside function as defined earlier, P is the received optical power, and R the photodiode responsivity. Finally, $n_s(t)$ is the electrical and shot noise contribution.

In order to evaluate the ideal error probability it is convenient to express (6.93) via the field quadratures. Assuming as reference a polarization unit vector parallel to the instantaneous received field polarization, the photocurrent can be written as

$$c(t) = R\left[\xi_1(t)\xi_1(t-T)+\xi_2(t)\xi_2(t-T)\right]+n_s(t) \qquad (6.94)$$

where $\xi_j(t) = e'_j(t) + \eta_j(t)(j=1,2,3,4)$ are the quadratures of the received field composed by a signal contribution and an ASE contribution. The signal part of the quadrature of the field, that is, $e'_j(t)$, is related to the transmitted quadratures via a rotation of the quadrature space induced by fiber propagation (see Chapter 4).

Since (6.94) is a quadratic form of Gaussian variables plus a Gaussian term, the method of the characteristic function can be used.

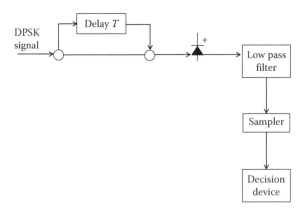

FIGURE 6.33
Block scheme of a DPSK receiver.

Neglecting as usual when amplifiers are used the receiver noise, the integral relating the characteristic function of the decision variable and the error probability can be solved exactly in this case providing

$$P_e = \frac{1}{2} e^{-SN_e} \tag{6.95}$$

This is a very simple expression of the error probability and the required SN_e for an error probability of 10^{-12} is 14.3 dB. This represents a gain of almost 3 dB with respect to IM, essentially, this gain is due to the fact that the average power along the line is doubled for the same peak power.

This advantage, however, is not without a price.

Not only phase modulation is affected by phase noise [8,42] that directly superimposes to the information bringing phase, but it is also more sensible to SPM, essentially for the same reason.

The draft design in case DPSK is used can be carried out with a procedure similar to that already used in the previous section.

The system parameters are reported in Table 6.14. Here it is a key to have a narrow linewidth; thus, a high emission power external cavity laser is selected as transmitter. Moreover, all the blocks constituted by splitters and delay line are supposed to be integrated into a Si–SiO$_2$ chip so that the additional loss due to interfacing is minimum.

The parameters of the line are not reported since they are identical to the IM case.

In order to try to achieve the maximum possible distance, we also add another couple of Raman pumps into the SSMF from the receiver with the intention of limiting the signal attenuation so to decrease the global noise amount.

Since the DCF is used also for transmission of the pumps into the SSMF fiber to increase the system length, the pumps for the SSMF have to be located in the far

TABLE 6.14

Parameter Values for the All Raman Single Span Differential Phase Shift Keying System

Channel speed	10 Gbit/s
Optical bandwidth of the DWDM comb	Extended C band
Modulation	DPSK NRZ
Number of channels	40
Channel spacing	100 GHz
Transmitting chain	
Source laser emitted power	16 dBm
Source laser linewidth	500 kHz
Mach Zehnder lithium niobate phase modulator loss	5 dB
Multiplexer (multistage AWG × 80) loss	7 dB
Patchcords, connectors, monitoring taps loss	1 dB
Detection chain	
High-efficiency APD sensitivity @ 10 Gbit/s	−28 dBm
Beam splitters + delay line + insertion in SiO$_2$	5 dB
Demultiplexer (multistage AWG × 80) loss	7 dB
Patchcords, connectors, monitoring taps loss	1 dB
Power per channel on the APD	−20 dBm

terminal of the system and propagated up to the insertion point, that is, for 20 km, into a dedicated fiber.

This system is called remote pumping. Since the loss of the SSMF at the pump wavelength is about 0.3 dB km; 20 km propagation is equivalent to 6 dB loss, that have to be added to the 3 dB insertion loss. Almost no loss there is at the insertion of the pump into the SSMF that transport them since we imagine that the pump lasers are sold packaged and the nominal power is the power in the fiber pigtail.

With these figures in mind, we have to add 9 dB to the remote pumps power to define the power of the pump lasers.

The scheme of the system is reported in Figure 6.34 and the single-channel power behavior in Figure 6.35. The Raman gain of this new amplifying section is again expressed by an equation like (6.92) with the parameters of the SSFM and the power budget equations becomes

$$P_{rec} = G_{R1} + G_{R2} + G_{R3} + P_l - A_T - A' - A_{rec} \quad \text{(all dB)} \tag{6.96}$$

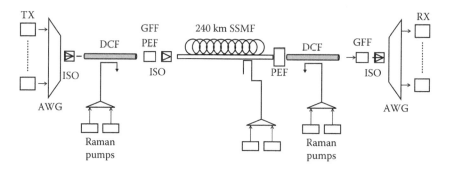

FIGURE 6.34
Architecture of a single span DPSK system using all Raman amplification both in the DCF and in the transmission SSMF. TX, transmission transponders; RX, receiver transponders; ISO, isolator; DCF, dispersion compensating fiber; GFF, gain flattening filter (if needed); PEF, pump extraction filter; AWG, array waveguide; SSFM, standard single mode fiber.

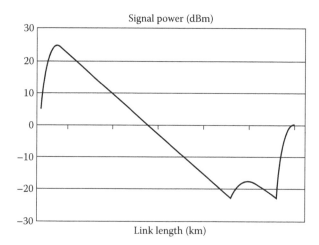

FIGURE 6.35
Single-channel optical power profile in the Raman amplified single span system depicted in Figure 6.34.

$$P_{max} = G_{R1} + P_l - A_T \quad \text{(all dB)} \tag{6.97}$$

where the symbols are those of the previous section but G_{R3} that is the gain of the new Raman section and A' that now refers to the passive part of the SSFM fiber, that is, $L-L_3$ if the Raman pumps are stopped at a distance L_3 from the end of the SSMF.

The gain in the SSMF is used differently from that in DCF. As a matter of fact, the pumps are stopped after a length when the power is the same it has to be at the output of the line (compare Figure 6.32 for the gain behavior with distance). The gain at that point is zero, but it is as if the length was reduced of L_3 since in this part of the link the effective loss is zero.

Naturally, ASE is generated, and in order to evaluate the generated ASE this piece of fiber has to be considered as a distributed amplifier adding the noise term to the Raman equations, linearizing the equations around the noiseless solution and estimating the linearized variance of the amplified field amplitude.

This is an easy but cumbersome procedure and we do not detail it here, even because the relative noise contribution is not very important.

In the case of the DPSK the maximum power per channel changes too, due to the greater sensitivity to nonlinear phase noise becoming 15.5 mW.

The design results are summarized in Table 6.15 for 315 km of SSMF fiber so that the total system length results to be 355 km.

For this system, the draft design estimates an SN_0 of 17.7 dB.

The summary of the penalties analysis is reported in Table 6.16 requiring an SN_0 of 16.7 dB, in good agreement with the simulation estimate.

At the end, it is possible to conclude that the system is feasible and its reach is 355 km, versus the 320 km achieved with IM-DD.

Two main observations can be done comparing this example with that of the previous section: the first is that the sensitivity advantage of the DPSK is almost all counterbalanced by the bigger losses at the receiver, while the use of NRZ instead of RZ has allowed reducing significantly the noise contribution.

TABLE 6.15

Results of the Draft Design of the All Raman Single Span Differential Phase Shift Keying System

Maximum transmitted power per channel	15 mW
Pump 1 power-receiver	480 mW
Pump 2 power-receiver	370 mW
Pump 1 power-transmitter	470 mW
Pump 2 power-transmitter	360 mW
Pump 1 power-SSMF	650 mW
Pump 2 power-transmitter	500 mW
DCF length transmitter	20 km
DCF length receiver	20 km
Total length traveled by pumps in SSMF	52 km
Equivalent amplifier length in SSMF	20 km
Raman gain transmitter	24 dB
Raman gain receiver	23 dB

TABLE 6.16

Penalty Analysis of the All Raman Single Span Differential Phase Shift
Keying System

	All Raman Single Span System at 10 Gbit/s with DPSK and SSMF Pumping	
	10 Gbit/s	Notes
Channel number	30	Extended C band
Channel spacing	100	
Modulation format	NRZ	DPSK modulation
Ideal SN_e (dB)	14.3	
B_o/B_e	2	
Required SN_o (dB)	14.3	
Impairment	Penalty (dB)	Notes
APD receiver noise	0.3	This is mainly the avalanche noise (see Equation 5.29), due to the low power on the APD (−20 dBm per channel)
Signal dynamic range	0.5	Lithium niobate modulator—Phase modulation eliminates Raman dynamic compression
Linear crosstalk	0.2	AWG multistage, signal spectrum Gaussian with spread of 10 GHz, Gaussian optical filter with 15 GHz width, see Figure 6.16
Nonlinear crosstalk (FWM XPM)	0.6	Wide spacing (100 GHz) and use of an SSMF
Residue chromatic dispersion and SPM interplay	1.3	Value due to the high input power. Managed with • MLSE as in Figure 5.59 • Residual dispersion of 800 ps/nm (300 km is 5100 ps/nm)
PMD	0.	MLSE at the receiver to cancel completely PMD that is very small (less than 2 ps per 400 km @ $PMD = 0.1\,\mathrm{ps/km^{-1/2}}$)
Polarization effects	0.5	
RIN transfer	0.2	Compare Equation 4.90 with RIN parameters from Tables 5.4 for the pumps and 5.3 for the signal. Smaller impact on DPSK
Phase noise	0.2	
Double Rayleigh	0.3	
Phonon induced noise	0.3	
Aging	1	
System margin	3	Normal system margin
Total w/o FEC	22.7	
FEC gain (RS 255,223)	6	
Required SN_o (dB)	*16.7*	

The second observation is that two additional technologies have been added in this dimensioning example: DPSK and pumping of the SSFM, both increasing the cost and the complexity of the system.

What has been gained is 35 km, which is not so much. However, in real applications, if it is very difficult or very expensive to plane in-line amplifiers (like in short submarine links) even 30 km can be the difference between realizing a project or not.

6.3.4.3 Single-Span Systems with Electronic Pre-Distortion at 10 Gbit/s

In the case in which electronic pre-distortion is used, as far as the link length is within a few hundred kilometers and no complex device is in line, it can be assumed that single-channel nonlinear effects and dispersion are completely compensated but for a fraction due to the imperfect working of the pre-distortion (finite number of bits, finite dimension of the look-up table, finite adaptation time, and so on; Table 6.17).

This uncompensated part of dispersion and nonlinear effects does not imply a penalty greater than 0.3 dB on the overall system performances at 40 Gbit/s, and an even smaller penalty at 10 Gbit/s.

Thus, the system architecture can be very simple: a single-stage EDFA booster at the transmitter and a single-stage EDFA preamplifier at the receiver, IM modulation and detection via a PIN diode due to the small loss between the preamplifier exit and the photodiode (essentially the AWG demultiplexer loss).

As usual the design starts from the power budget, considering that no dispersions have to be optically compensated.

The power budget equations are

$$P_{rec} = P_t + G_p + G_b - A_T - A - A_{rec} \quad \text{(all in dB)}$$

$$P_{max} = P_t + G_b - A_T \quad \text{(all in dB)}$$

(6.98)

TABLE 6.17

Parameters of the Single Span Systems with Electronic Pre-Distortion at 10 Gbit/s

Channel speed	10 Gbit/s
Optical bandwidth of the DWDM comb	Extended C band
Modulation	IM NRZ
Number of channels	40
Channel spacing	100 GHz
Transmitting chain	
Source laser emitted power	20 dBm
Source laser linewidth	5 MHz
Mach Zehnder lithium niobate IM modulator loss	7 dB
Multiplexer (multistage AWG × 80) loss	7 dB
Patchcords, connectors, monitoring taps loss	1 dB
Detection chain	
High-efficiency PIN sensitivity @ 10 Gbit/s	−20 dBm
Demultiplexer (multistage AWG × 80) loss	7 dB
Patchcords, connectors, monitoring taps loss	1 dB
Power per channel on the PIN	−15 dBm

where the symbols have the usual meanings. It is to underline that, also in this case, there exists a maximum peak power per channel that can be injected into the line due to the fact that the pre-distortion device has an upper limit in its equalizing capability and if the non-linear distortion is too pronounced or if it starts to involve nonlinear polarization rotation pre-compensation is no more possible.

The value of P_{max} is much higher than in the case of previous sections and, looking at the practical implementations of electronics pre-distortion and suitably scaling the presented results [43], it can be set at 34 mW (15.4 dBm) per channel if we target a system length much longer than the effective length, that is almost always the case.

From the second power budget equation it results $G_b = 16.77$ dBm, the total input power being 12.77 dBm and the output power 29.54 dBm. This is not an impossible amplifier, but it is for sure not a standard booster, especially for the very high output power. It is possible that a double stage amplifier is needed, but in any case, we will assume that such an amplifier will be available.

In the second power budget equation we have two unknown parameters: the system length and the preamplifier gain.

The relation in dB between these parameters is easy: they are proportional.

A second relationship can be obtained by the SN_o equation. This equation is

$$SN_o + P_{ASE} - P_t - G_p - G_b + A_T + A + A_{rec} = 0 \quad \text{(all in dB)} \tag{6.99}$$

where

$$P_{ASE}(\text{dBm}) = 10 \, \log_{10}(\hbar\omega B_o) + 10 \, \log_{10}\left[(G_b - 1)G_p \, e^{-\alpha L}NF_b + (G_p - 1)NF_p \right] \tag{6.100}$$

The required SN_o with the adopted optical bandwidth is 17 dB, while our dimensioning gives 21 dB for 220 km transmission, allowing the accommodation of the penalties.

The penalties analysis is summarized in Table 6.18 where the required SN_o is 17.8 dB, while the simulation estimates 18.2. In both cases, the value estimated in the draft design is higher, confirming the system feasibility.

Even if these results have to be considered more as an indication than as real limits, the characteristics of this category of transmission systems emerge clearly.

At 10 Gbit/s the overall performance of Raman assisted systems using mixed optical and electronic compensation performs a bit better.

However, the system based on electronic pre-distortion is much simpler, implying probably a smaller CAPEX in terms of cost per Gbit/s per km and a smaller OPEX, due to simpler maintenance, smaller cost of the spare parts warehouse, and globally better reliability.

Thus, in all the cases in which the two systems will enter in competition, the system using electronic pre-distortion is really competitive with respect to the system using Raman assisted transmission.

6.3.5 Metropolitan Optical Rings

As we have seen in Chapter 2, while in the core network the topology is generally a mesh and point-to-point transmission among complex core nodes is dominant, in the metro area ring topology is frequently used.

TABLE 6.18

Penalty Analysis of Single Span Systems with Electronic Pre-Distortion at 10 Gbit/s

	Electronical Pre-Compensation Single Span System at 10 Gbit/s	
	Notes	
System length	220 km	
Channel speed	10 Gbit/s	
Channel number	30	Extended C band
Channel spacing	100 GHz	
Modulation format	NRZ	Pre-compensation
Ideal SN_e (dB)	17	
B_o/B_e (due to NRZ modulation)	2	$M_{oe}=2$ @ 10 Gbit/s, NRZ modulation
Required SN_o (dB)	17	Due to the very high value of B_o/B_e
Impairment	Penalty (dB)	Notes
APD receiver noise	0.3	This is mainly the avalanche noise (see Equation 5.29), due to the low power on the APD (−20 dBm per channel)
Signal dynamic range	0.5	Lithium niobate modulator
Linear crosstalk	0.1	AWG multistage, NRZ signal spectrum with spread of 10 GHz (first zero), Gaussian optical filter with 40 GHz bilateral width, see Figure 6.16
Nonlinear crosstalk (FWM XPM)	0.4	Wide spacing (100 GHz) and use of an SSMF
Residue chromatic dispersion and SPM interplay	0.5	Value due to the high input power managed by electronic pre-compensation with a mixed FFE-DFE-Look-up-table system
PMD	0	Completely cancelled by pre-distortion
Polarization effects	1	
Aging	1	
System margin	3	Normal system margin
TOTAL w/o FEC	23.8	
FEC gain (RS 255,223)	6	
Required SN_o (dB)	17.8	

This difference reflects the different functionalities of the two networks, the first devoted to transport of signals among faraway places, the second to collect and distribute the signals generated or directed to users in the same area.

In this section, we will talk about optical rings only from a transmission point of view, while networking functionalities like wavelength routing and protection have been analyzed in Chapter 3 and the functional architecture of an OADM will be detailed in Chapter 7.

In the metro area there are two main technologies that coexist: CWDM and DWDM. As a rule of thumb, DWDM is used in the core of metro and regional network and CWDM in the metro access when single wavelength SDH/SONET systems are not sufficient to collect the traffic. However, exceptions are frequent, where, for example, CWDM is used in the core of small local networks or light DWDM systems are used to collect signals in big business centers where there is a huge amount of generated traffic.

6.3.5.1 Transmission in Dense Wavelength Division Multiplexing Metropolitan Ring

The main difference from a transmission point of view between a ring and a point-to-point link is that in a ring different optical channels, routed through the ring, generally have different sources and destinations. Moreover, the ring close topology allows spurious signal components and optical noise potentially to circulate along the ring.

The first point can represent a serious problem for amplified transmission. As a matter of fact, when designing a point-to-point system we have seen that the gain of EDFA amplifiers is flattened by the use of gain flattening filters so to exactly compensate the unbalance of the considered amplifier.

This is possible in a point-to-point system since the power of the channels traveling through a certain amplifier is fixed. The only thing that has to be managed is the progressive addition of other channels. Generally in a point-to-point system there is a rigid channel upgrade plan that prescribes the order in which the channels have to be added. In correspondence to the number of working channels, the element manager tunes the setting of the amplifiers so to maintain the required flatness.

A similar strategy is not possible in a ring network, due to the fact that the DWDM comb transiting through a generic node is formed by channels injected into the ring in different nodes and thus having very different powers. The lack of uniformity of optical power among the channels transiting in a node has an impact on the EDFAs' gain flatness.

As a matter of fact, the gain of an amplifier depends on the signal traveling through it. To make an example, let us consider our model of Chapter 4 of the two level amplifiers and let us consider an amplifier with center wavelength at 1550 nm, microscopic saturation power of 10 mW microscopic bandwidth 155 nm and microscopic gain 5 m^{-1}. With these parameters the linear gain is 21 dB and the macroscopic input saturation power is 0.1 mW.

This amplifier is in deep saturation if a signal of 0.5 mW power is injected at the wavelength of 1540 nm, as a matter of fact the gain of the signal is 16.7 dB.

In Figure 6.36, the gain of the signal at 1540 nm is shown versus the power of a second signal at 1460 nm, thus quite far from the first. It is evident that, increasing the second signal input power, the gain of the first signal decreases, up to compress greatly when the second is more powerful.

This behavior is due to the fact that, even if far from a spectral point of view, both the signals leverage on the same population of optically activated molecules; thus, they enter into competition inside the amplifier and the presence of the second signal de facto reduces the inverted population on which the gain of the first is built.

This phenomenon naturally also changes the spectral behavior of the macroscopic gain, depending on the power distribution of the input channels.

In Figure 6.37, the spectral gain of the amplifier is shown for different values of the two injected signals and it is clear that the percentage difference between the gain peak and the value, for example, at 1550 nm changes greatly, passing from 7.7 to 3 dB.

This behavior, even if obtained from a greatly simplified model, well represents what happens with EDFA amplifiers.

Since when a ring is deployed, it is not a priori known what will be the order in which the channels will be switched on and what will be the route of each channel; a static gain flattening cannot be used in optical rings.

Two strategies are available to face this problem: either the use of a dynamically adjustable flattening filter or a dynamic control of the channels power at the amplifier input.

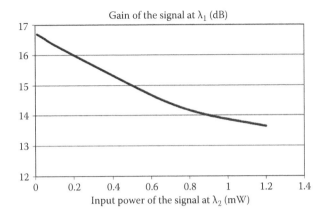

FIGURE 6.36
Decrease of the gain of an amplifier at a certain wavelength due to the presence of a strong signal at a different wavelength.

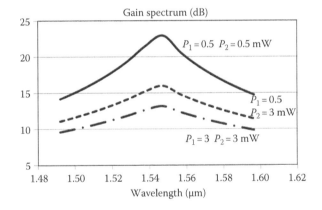

FIGURE 6.37
Dependence of the gain spectrum of an ideal amplifier on the power of a strong signal transiting in the active medium.

The first solution is always possible and relies, essentially, on a set of filters, one for each spectral slice, that can be automatically adjusted by the element manager when a new channel is inserted in the ring. Naturally, these filters introduce a greater loss with respect to fixed filters.

The other solution is viable only if the channels are demultiplexed at each node, as happens when the OADM is realized with a couple of back to back AWG with optical switches in between (as shown in Figure 6.38). Here a set of variable attenuators equalize the channels before entering in the second amplification stage so to achieve a globally flat amplification.

A second issue arising in DWDM optical rings is related to the fact that, in the absence of a node where the continuity of the optical signal is interrupted, both spurious signal contributions and the ASE out of the signal bandwidth can circulate all around the ring and, if the overall ring gain is positive, could create laser action.

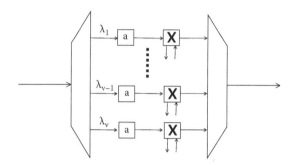

FIGURE 6.38
Block scheme of an OADM realized with AWG multiplexer/demultiplexer.

As far as ASE is considered, if no node interrupts the ring fibers continuity, it is possible to set up in one location along the ring a periodic filter that passes all the channels but stops the ASE out of the channel bandwidths.

The situation is completely different for the spurious signal components that arise from imperfect filtering of the drop OADM and continue circulation along the ring.

The situation is shown in Figure 6.39. In this figure a signal is added at node A and extracted at node C, but the extraction filter, being not ideal, allows a small part of the signal to pass through along the ring fiber until it arrives at node A again. At this point, if the suppression of the OADM is not sufficiently high, lasing action can start on the frequency λ_1, ruining completely the ring working.

Naturally, here it is not possible to stop the channel bandwidth since it should be possible to have the channel wavelength available over the entire ring.

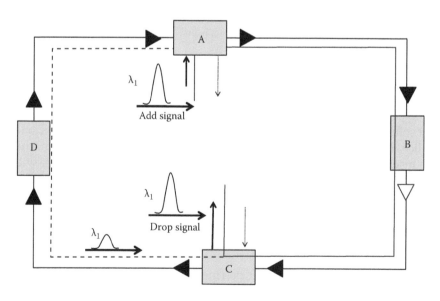

FIGURE 6.39
Recirculation mechanism due to imperfect optical filtering in optical transparent rings.

At this point, either a set of switches are deployed in the nodes to stop all the passing through bandwidths that are not occupied by a real channel or the ring is accurately dimensioned, taking into account aging and other potential parameter changes, to make sure that lasing condition never starts whatever is the channel load and routing.

Last but not least, in designing optical rings it has to be taken into account that potentially the signal could traverse a great number of filters, one for each traversed node.

The bandwidth of a cascade of several filters is smaller than the bandwidth of a single filter, for example, the cascade of n single pole filters has a bandwidth B_n given by $B_n = (\sqrt[n]{2} - 1)B_1$, where B_1 is the bandwidth of the single filter.

Thus, the design of the OADM filtering characteristic has to be made carefully.

This implies also that in optical rings ASE coming from amplifiers nearer to the receiver is filtered by a larger filter with respect to the ASE coming from far amplifier, and this effect has to be taken into account when evaluating the SN_o.

Just to make an example, let us imagine having a cascade of OADMs with the structure of Figure 6.38 and that the filtering characteristic of an AWG can be approximated with a Gaussian function.

Let us imagine that the signal traverses six nodes traveling from the source to the destination, that means traversing 12 AWGs, one at the source, one at the destination, and two for each intermediate node.

The overall filtering characteristic experienced by the signal is

$$H_s(\omega) = \left[H_{AWG}(\omega)\right]^{12} = H_0^{12} \exp\left[-\frac{6(\omega - \omega_s)^2}{B_{AWG}^2}\right] \tag{6.101}$$

apart from the attenuation that is compensated by the amplifiers the overall bandwidth (in this case intended as the value at $1/e$) now is $B_o = B_{AWG}/2\sqrt{3}$ so that if the design requires $B_o = 4R$ it should be $B_{AWG} \approx 13.85R$.

However, it would be wrong to assume that B_o is also the noise bandwidth. Let us imagine, for simplicity, that the ring node is constituted by OADMs with identical amplifiers before and after (see Figure 6.39), and that the first amplifier compensates exactly the OADM loss and the second the span loss. The overall noise arriving at the receiver, if we imagine that no additional filter is present in front of the photodiode, is (see Figure 6.39 to compute how many filters the ASE generated by each amplifier traverse)

$$S_{ASE} = \hbar\omega nGNF \sum_{j=1}^{n} \frac{1}{\sqrt{2j-1}} \frac{B_{AWG}}{n} = \hbar\omega nGNF \sum_{j=1}^{n} \frac{2\sqrt{3}}{\sqrt{2j-1}} \frac{B_o}{n} \tag{6.102}$$

where j is the number of traversed spans and the spans are numbered so that the number one is the nearer to the drop node. In the aforementioned example, the optical bandwidth is $4R$ for the signal, but the equivalent optical bandwidth for the noise is $B_n \approx 3.5B_o \approx 14R$.

This is a difference of more than 5 dB that has to be taken into account when designing the system.

Naturally, the problem can be faced by maintaining quite large the line filters and adding a noise selection filter in front of the receiver.

However, if this solution is selected, it is to take into account that in the filter bandwidth that is not occupied by the signal the ASE can circulate in the ring and the design has to carefully avoid the start of ASE lasing.

6.3.5.2 Transmission in Coarse Wavelength Division Multiplexing Metropolitan Ring

In the case of CWDM we are talking about rings without amplifiers, as a matter of fact, the CWDM occupies such a large optical bandwidth (compare Chapter 3) that it is unpractical to use amplifiers.

Moreover, in order to use uncooled sources, the channels are so spaced in frequency that there is no problem in using thin film filters to select them, avoiding any distortion due to filter cascading.

The challenge in designing CWDM systems is to satisfy the requirements in terms of power budget by maintaining the cost effectiveness that customers expect from this technology.

There exist two types of CWDM systems on the market, that are directed to different applications: pure CWDM and mixed DWDM and CWDM.

Pure CWDM systems are exclusively tailored on CWDM technology. Generally, they are simple machines whose main added value is the low cost.

These systems are generally completed with regenerators and fixed ADMs based on thin film technology and are used in the metro access area and in rural areas where the local exchanges are particularly far from the hub of the regional network.

Mixed CWDM and DWDM systems are instead capable to host on the same platform both CWDM and DWDM channels following the ITU-T standard channel allocation.

The multiplexing stage of mixed systems is generally composed by a first CWDM multiplexer, and a second DWDM multiplexer that fills with DWDM channels the CWDM passbands located in C band. This architecture is shown in Figure 6.40.

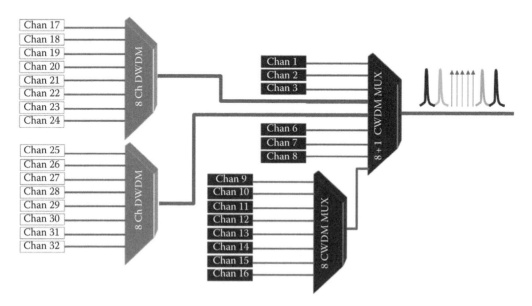

FIGURE 6.40
Multistage mixed CWDM/DWDM multiplexer.

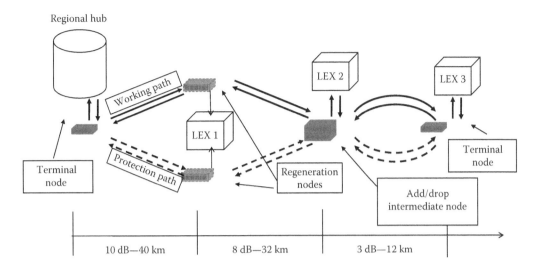

FIGURE 6.41
Example of CWDM system used to backhaul three small local offices to a hub.

These kinds of systems have higher cost with respect to pure CWDM, but allow slower growth after the installation using CWDM channels and faster when using DWDM configuration.

Even if CWDM interfaces at 10 Gbit/s exist and are used mainly for backhauling of remote local exchange in rural areas, more frequently CWDM wavelengths carry 2.5 Gbit/s SDH/SONET frame or directly application related formats line Fiber channel and FICON for data related applications or Gbit ethernet for small distance connections.

Typical power budgets of 2.5 Gbit/s CWDM systems are 8 dB for short-haul applications, 12 dB for long-haul, and up to 16 dB for extremely long connections.

An example of CWDM system used to backhaul three small local offices to a hub is shown in Figure 6.41. The adopted architecture is a protected bus that exploits the fibers already installed in the field. The first local exchange is reached by a complete path diversity, also using two diverse OADMs, since there were available two nearby but different sites on the ground. The second and the third local exchanges are reached with different paths but the OADM location is only one since two nearby locations were not available. In considering this architecture, it is to take into account that in rural areas, the points of presence not always are real central offices, but could also be big street cabinets where the local equipment is located.

6.4 Alternative Modulation Formats

Albeit almost all commercial equipment for optical transmission uses either NRZ or RZ modulation; a great number of other modulation formats have been devised, either to partially relieve the impact of different transmission impairments or to assure better receiver sensitivity or better wavelength selectivity.

The most promising set of different modulation formats is coupled with the use of the so-called coherent detection and is constituted by different kinds of phase modulations.

Even if in a few specific applications like single span systems, the greater sensitivity— provided by coherent detection of phase modulated signals could be useful, this kind of system never gave rise to commercial product due to their cost and to the success of IM-DD.

In facing the challenge of transmitting at 100 Gbit/s however, phase modulation and coherent detection could become a key technology and we will analyze a few applications in Chapter 9.

Among other alternative modulation formats we can remember single side band (SSB) modulation and duobinary modulation.

6.4.1 Single Side Band Modulation

The spectrum of an IM-DD signal is constituted by two sidebands and a carrier. Due to the real nature of the baseband signal, the sidebands are symmetric; thus, in line of principle, one of them is enough to reconstruct the signal at the receiver.

An SSB optical signal is generally obtained by generating a standard signal and filtering it before transmission so to eliminate one of the side bands.

An SSB optical signal ideally allows both a 3 dB SN_o to be gained due to the reduced bandwidth occupation and a more dense DWDM transmission [44].

In practice, this transmission technique has not found commercial application due to the fact that at 10 Gbit/s it is not easy to control the shape of optical filters about 10 GHz wide, so to eliminate one sideband completely without affecting the carrier or the other sideband.

Moreover, it is to be noted that an SSB signal is more sensible to distortion with respect to standard IM, due to the lack of sideband redundancy that gives robustness to the standard modulation format [45].

The situation could change at 100 Gbit/s and with the evolution of electronic equalization to control signal distortions.

6.4.2 Duobinary Modulation

Duobinary modulation has been proposed for optical communications but has not experimented wide diffusion. As a matter of fact, although as any proposed method, duobinary modulation has its own advantages, it is also true that it requires more complex transmitters and the greater global cost of the system has to be compared to the economic impact of the advantages in the network design.

The duobinary modulation is a type of amplitude modulation where the bit stream is manipulated to reduce the bandwidth of the modulated signal with respect to NRZ, so reducing the impact of chromatic dispersion and PMD.

The duobinary method was first applied to electrical digital signals. It combines two successive binary pulses in the digital stream to form a multilevel electrical signal.

To easily express the duobinary encoding, let us indicate the binary bit stream with $b_{h,k}$ where h is the progressive number that indicates the position of the bit in the bit stream, k is zero if the bit is a zero, otherwise it is a progressive number that indicates how many ones are passed in the bit stream.

Just to give an example in the bit stream 1001101, the bits are coded in this way: $b_{1,1} = 1$, $b_{2,0} = 0$, $b_{3,0} = 0$, $b_{4,2} = 1$, $b_{5,3} = 1$, ...

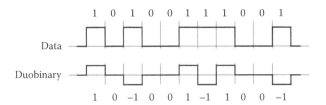

FIGURE 6.42
Example of duobinary coding.

Using this notation the duobinary signal d_h, where h is the progressive number of the duobinary symbols, can be expressed as follows:

$$d_h = (-1)^k b_{h,k} \qquad (6.103)$$

that is, if the bit is a zero, the code transmits a zero, if it is a one, the code transmits alternatively +1 and −1.

An example of duobinary coding is reported in Figure 6.42. Naturally in intensity modulation, duobinary is realized adding a bias to the modulated signal so that three different levels of power (one of them generally equal to zero) are used.

The spectrum of duobinary signal is slightly shrunk with respect to the spectrum of ordinary NRZ and much more compact than the spectrum of RZ, especially when small duty cycles are used to tolerate more pulse spreading without generating ISI.

Moreover, at the transmitter duobinary modulation can be performed by using a dual drive MZ modulator in conjunction with electrical low pass filters posed on the modulator's driving ports and the receiver is simply a standard IM-DD receiver, with the threshold in between the zero and the first of the two levels corresponding to a one.

The advantage in terms of dispersion tolerance is not so strong, and the application of these modulations' format is, at present, not as widespread as it was forecasted a few years ago.

6.5 Hardware Architecture of Optical Transmission Systems

In the previous section, we have analyzed a set of possible designs of optical transmission systems for application in the long-haul and metro network.

Typical quality parameters we have used to evaluate the performance have been capacity and BER.

However, it is important to recall what is derived from Chapter 2, where we arrived to the conclusion that a main target for carriers is to reduce network OPEX.

In order to reduce OPEX, it is surely important to optimize the network design using high-performing systems. As a matter of fact, decreasing the number and the complexity of central offices and points of presence is a very important task.

Once the network is optimized however, often the parameters that drive the carrier in the choice of a transmission system, among all the products that have more or less similar characteristics, are directly related to the decrease of OPEX.

Things like power consumption, real estate, easy maintenance, and safety are very often the parameters making the difference in the choice of a system among the market available products.

These considerations, strictly related to the engineering of the system, have driven strongly the evolution of DWDM systems especially after the telecom bubble end.

In this section, we will try to put the basis for a practical analysis of the engineering of an optical transmission system.

6.5.1 Mechanical Structure of a Dense Wavelength Division Multiplexing System

A typical DWDM system is mechanically hosted in one or more subracks, as shown in Figure 6.43. The same subrack is shown in Figure 6.44, without a few cards, to reveal the card positioning slides and the connectors that put the various cards in contact with the subrack backplane.

A rack supports a certain number of subracks (generally from three to five, but this is only an indication and there is a wide variety of solutions) and each subrack hosts a certain number of cards. The dimension of the rack follows detailed standards, due to the fact that they have to fit into the central office space in a better way. A diffused rack standard is the so-called 19′ rack, that was born in the United States, but is now diffused also in Europe and the far east [46–48].

A pictorial representation of a 19′ rack is presented, besides the rack standard measures, in Figure 6.45, while Figure 6.46 shows as the subracks being mounted one over the other into the rack.

Subracks on the contrary, as far as they fit into the standard rack, are largely different from system to system. Both European Standardization Institute (ETSI) in Europe and American Standardization Institute (ANSI) in the United States have defined standard subracks, but several nonstandard solutions are well accepted by the market as far as lower power and real estate sparing is demonstrated are demonstrated.

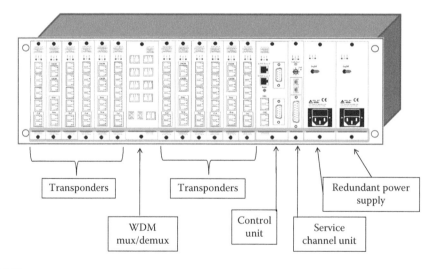

FIGURE 6.43

DWDM subrack: this is a metropolitan system with small cards, often long-haul and ultra-long-haul systems have higher subracks. (Published under permission of PGT Photonics s.p.a. from Arecco. F., *City8™ System Configuration Manual v 1.0.* Internal PGT doc, p. 25, 2003.)

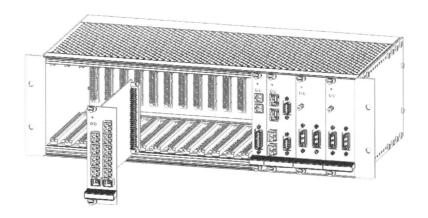

FIGURE 6.44
Cards allocation and backplane in a DWDM subrack. (Published under permission of PGT Photonics s.p.a. from Arecoo. F., *City8™ System Configuration Manual v 1.0*. Internal PGT doc, p. 25, 2003.)

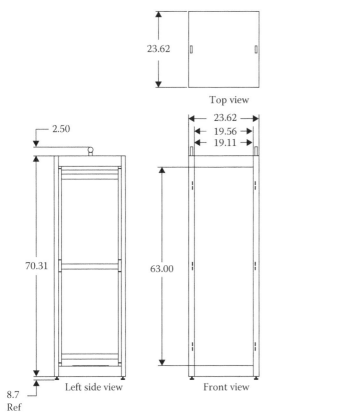

FIGURE 6.45
Standard measurements of the 19′ rack.

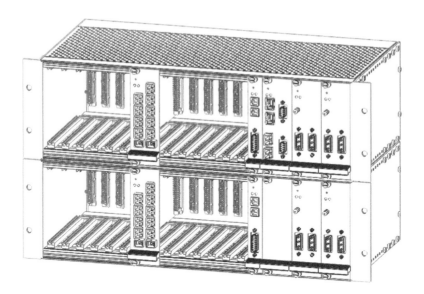

FIGURE 6.46
Two superimposed subracks ready to be mounted inside a 19′ rack. (Published under permission of PGT Photonics s.p.a. from Arecoo. F., *City8™ System Configuration Manual v 1.0*. Internal PGT doc, p. 25, 2003.)

The cards can fit into the subrack in different ways: vertically (like in Figure 6.44) is the most frequent solution since it allows better cooling of the system, but also subrack with horizontal cards exist.

The way in which the cards fit into the sub rack is important since it is determined by the cards dimension. Larger cards, fitting horizontally into the subrack, contain a more efficient electronics design (since more components are on the same card), but make the system less flexible.

Different cards, either belonging to the same subrack or to different subracks can exchange information though the back plane.

Let us assume that no switching capability is embedded into the system (e.g., no embedded ethernet switch or no SONET/SDH ADM). In this case, the system does not need the capability to demultiplex the TDM frame of the incoming signals in order to access to the tributaries in the deterministic case or to the packets in the statistical case.

On the other hand, the incoming signal is very high capacity. A standard ultra-long-haul system can manage more than 1.2 Tbit/s capacity shared among the wavelength multiplexed 10 Gbit/s channels. In this condition, almost all the DWDM systems are designed to transfer the 10 Gbit/s channels from a card to the other using fiber patchcords on the front panel. For this reason, almost all the racks designed to accommodate DWDM and CWDM systems comprise slides and other facilities to route the external fiber patchcords in an ordered way. This can be very important to facilitate the system maintenance, due to the fact that a long-haul DWDM system can have up to 600 patchcords in a single terminal to manage its 150 wavelength channels connecting transponders to the clients and to the mux–demux.

This means also that the backplane, differently from what happens in switching equipment, does not carry the signal from one card to the other, but only the commands and other information that are passed from a generic card to the system controller.

In the architecture of a generic DWDM system we can classify the cards in three classes:

1. Signal cards
2. Control cards
3. Support cards

6.5.1.1 Signal Cards

Signal cards perform all the functionalities needed to manage the incoming DWDM signal in the correct way. Essentially, we have

- *Transponders*: These are the cards where the incoming signal related to one or more channels is detected, regenerated, and retransmitted with the output power and the wavelength stability that is needed to have a correct propagation along the line. Transponders are used generally to feed the WDM system with client channels, but they can also be used (when suitable configuration is performed) to regenerate the channels in an intermediate site, when this is needed.

  Even if incoming and outgoing signals have essentially the same characteristics, transponders are needed in all the high-performing systems since a DWDM system has very specific requirements in terms of power and power stability, wavelength stability, and other characteristics that the client signal generally does not have.

  Naturally, open systems can exist, that is, systems that have to be used in conjunction of switching machines of the same vendor and that directly receive well-conditioned signals, since the DWDM interface is embedded in the client machine.

  In this case, the transponder is substituted by a front-end card that does not perform optical-to electrical-to-optical conversion, but simply sends the received signal online after performing measurements on the signal to make sure that it is compatible with the DWDM system.

  The open system solution however, up to now, does not have a great market success, especially in the incumbent carriers' networks. As a matter of fact, this technology obliges the carrier to buy a large amount of machines performing different network functionalities from the same vendor. Even if this brings the reduction of the amount of needed hardware, it also takes away the carrier possibility of leveraging on competition to lower the prices of different machines, since at that point the process of changing either DWDM or client vendors is very difficult and expensive.

  On the other hand, small competitive carriers that do not have and do not want to build a heavy company structure with test labs and research departments, frequently leave the network optimization to a specific vendor that becomes a sort of reference partner. In this case, integration of transmission and routing machines can be exploited at best. Transponders are also very important for the incumbent carriers providing high-speed transport wholesale services. As a matter of fact, in this case, the limit between the carrier network and the customer network is exactly the transponder, so that if an SLA has to be respected on both parts, the carrier needs sufficient monitoring capability to assess independently the incoming signal quality.

- *Multiplexers and demultiplexers*: These devices are generally hosted in dedicated cards, even if their functionality is not as complex as the transponder one. The reason is that, while a transponder processes one or few channels, a multiplexer and a demultiplexer have to process all the channels that have to arrive from connectors located on the card front end. Depending on the engineering choices regarding the mechanical system design, it could be even impossible to host all the needed connectors on the front of a single card. In this case, either the card is design so to occupy more than one slot into its subrack or the mux/demux is designed as a multistage component, locating different stages in different cards.

- *Amplifier cards*: These are the cards hosting optical amplifiers with all their structures, comprising pumps, filters, and so on. Different cards generally host different amplifiers (e.g., preamplifiers or in-line amplifiers).
- *OADM cards*: These are the cards hosting OADMs with all their structure, comprising in case, all the support needed for dynamic switching.
- *Protection cards*: If optical protection is adopted, generally optical protection switches are hosted in specific cards.

6.5.1.2 Support Cards

Support cards perform functionalities that are needed to assure correct system working and/or monitoring, but do not involve signal processing:

- The card detecting and terminating the service channel that brings control plan and central network manager information (e.g., the OMS and the ORS section headers) is a support card (see Section 3.2.7). Generally, some packet-oriented protocol is used on the management network, so that frequently this card has switching abilities that are used in the management network (e.g., it can include an ethernet switch).
- The cards allowing fans installation and scaling, while the system scales in terms of channels, is another support card. As a matter of fact, only in the small systems the fans are included in the system mechanics without the possibility to scale them while the system scales.
- The power supply is another support card. Depending on the system capacity, either a single power supply can be sufficient or power supply can exist in every subrack, all dependent on a central power supply. This solution, however, is adopted only when it is really needed, due to the additional real estate and complexity of the power distribution structure.

6.5.1.3 Control Cards

Control cards are essentially processors that run the network element agent and all the activities related to the control plane.

Generally, there is a single control card, having sufficient processing capability to manage all the needed software. Sometimes, due to different needs, every subrack has its own controller referring to the central controller via a local equipment network.

The central controller is also the card allowing the local craft to be connected to the system (the local craft is a light version of the element manager that is generally installed on a portable computer and allows to configure locally the network element).

Another control card is the subrack controlled. Depending on the control architecture of the system it can be a real controller or simply a switch for the equipment bus that allows different trunks of the equipment control network to be connected.

6.5.1.4 Redundancies

To reach the required network element availability, some cards are replicated twice in the system architecture and a 1+1 hardware protection is realized.

The network element processor is one of the cards that generally are doubled. One of the two processors works as mirror of the other, being in hot standby continuously updated so that, if the main processor fails, the second one can assure seamless network element management.

Also the power supply is generally protected by a hot standby.

6.5.2 Backplane Architecture

There are several possible solutions for the control architecture of a DWDM system, but the most common is that based on a bus communication structure. The block scheme of this architecture is shown in Figure 6.47.

The backplane is designed to create a bus for each subrack to allow all the cards that are part of the subrack to communicate each other. This bus terminates on the subrack controller.

All the subrack controllers are put in communication by the bus that connects all the subracks, that is also sustained by the backplane. In the hypothesis that more than one rack is needed, there is a bus extension that is able, from a backplane, to chain different racks.

Also in the multirack case, there is a hierarchical organization among the rack controllers so that one of them works as network element controller.

The first thing that has to be decided in designing a backplane implementing the architecture represented in Figure 6.47 is how to realize its connectivity.

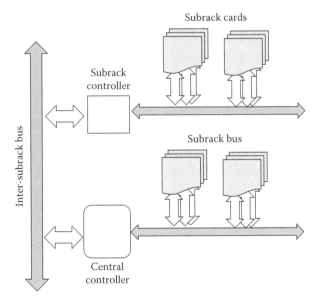

FIGURE 6.47
Bus-based communication architecture among the cards of a DWDM system.

Connectivity refers to how drivers and receivers are connected to the backplane and thus to the bus that allows them to communicate. In some systems it is necessary to connect multiple backplanes together. This can be done from a card to a backplane or from a card in one backplane to a card in another backplane.

After physical connection, the way in which the bus is physically implemented onto the backplane has to be defined.

In general, two types of transmission schemes are used in today's backplanes—serial and parallel. These can be differentiated as follows:

- In a parallel transmission scheme, the driver attached to the bus places bits of data in parallel signal lines onto the bus.

- In serial transmission, all data is sent point-to-point in a serial stream on a single data line. In a serial backplane, to achieve the same data rates of an analogous parallel backplane, data must be sent at much higher speeds. However, differently from other network elements, like routers or switches, in DWDM systems the signal does not travel along the backplane bus; thus, the backplane speed frequently has not to be very high. Often, the backplane speed is dictated by the need of performing protection switching in the required time of a few milliseconds and to do that a speed on the order of several Mbit/s is sufficient. Serial backplanes can be implemented using special serializer and deserializer devices that convert a parallel data stream to a high-speed serial data stream and then convert it back to parallel data at the backplane output.

Each design type has the advantages and disadvantages summarized in the following. When designing a backplane, it is important to select a scheme satisfying the needs of the system where it has to be installed.

The main advantages of the parallel design can be summarized as follows:

- Individual lines can be configured for control signals enabling a fast reaction time and a high degree of system and device control. This property could be very important when real-time protection switching commands have to be transmitted, especially in complex protection schemes like OChSPring.

- No time delay is required for serializing and deserializing data blocks. Besides the property described in the point aforementioned, this renders the parallel design suitable for machines driven by real-time commands. Besides protection, this could be a point if a GMPLS (generalized multiprotocol label switching) control plane should take the control of all the networks (see Chapter 8).

- Parallel design is common and widely used with a significant amount of industry design and implementation information available, including several standards.

While the main disadvantages of this design technique are as follows:

- Significant board and system space can be required, especially with wide or high card count design that is typical of DWDM systems, where, for example, more than one hundred transponder cards can be foreseen in a fully equipped system.

- Implementation costs can be high because of multiple terminations and devices, wide connectors, and high board trace counts on the backplane itself.
- Line-to-line signal skew must be matched and controlled.

As far as serial design is concerned, its main advantages are as follows:

- Serial data offer an adjustable approach to data rates, allowing longer cable lengths.
- Fewer ground lines and signal lines are required to transmit the data from one point to another.
- Serial design offers significant space savings over equivalent parallel designs. This point could be a key in highly compact DWDM designs, where real estate sparing is a system value proposition.

While the weaknesses of serial design can be summarized as follows:

- A time delay is incurred, when serializing and deserializing data blocks, that has to be controlled accurately if real-time protection switching is to be performed. Moreover, if design margins are not sufficient, it could be very difficult to upgrade the system to advanced network control requiring more and more a real-time response.
- Serial devices are generally more expensive than parallel backplane drivers.
- Higher speeds require greater attention to signal path layout and impedance matching.

Once the bus physical wiring is defined, there are still several ways allowing multiple clients to access a shared media to transmit data among them.

Two main technologies are specifically designed to address this issue in backplanes of complex equipment. For each of these techniques, both input/output electronics chips and specific design rules are available.

Single-ended or unbalanced communication is a technique characterized by the fact that one signal from a driver is carried on one wire or on one TDM frame slot to a receiver. The receiver compares the signal to a reference level with respect to ground. If differential or balanced communication is applied, both the signal and its complement are sent from the transmitted to the receiver.

The driver sends this differential signal pair on two matched wires to a receiver, and the receiver compares the two signals with each other. This topology offers noise immunity advantages over most single-ended technologies. Because the two differential signals are compared to each other and not to a ground or logical local reference, noise or ground shifts that affect both inputs equally will be ignored by the receiver. Also, most differential technology signal swings are offset from both grounds furthering input noise immunity.

Single-ended setups are much simpler than differential ones because they use half the signal lines, but noise can be a problem. For this reason, some systems benefit from differential signaling because of its superior noise immunity.

Table 6.19 summarize the advantages and disadvantages of single-ended and differential topologies.

Last key point in backplane design is the choice of the temporization scheme among synchronous, source-synchronous, or asynchronous signal interfaces.

TABLE 6.19

Advantages and Disadvantages of Single-Ended and Differential
Bus Topologies

Single-Ended	
Advantages	**Disadvantages**
Less expensive	Lower data transfer rates per bit
Easy implementation	Susceptibility to noise
More compact, less real estate	
Many technologies available	
Higher drive capability available	

Differential	
Advantages	**Disadvantages**
Typically higher data rates	Devices are more expensive
Lower output swing requirements	Board space
Generally lower electromagnetic emissions	Difficult to implement

Synchronous clock is the standard architecture used in many low and medium data rate systems. The design is implemented utilizing a single clock source that provides timing for all components requiring a clock. This means that all devices throughout the system receive the clock at the same time. This implies laying out clock lines of equal length to each device. Careful layout, system timing analysis, and minimization of clock skew are critical to this design.

Source-synchronous clock architecture uses a clock generated at every driver to toggle the receiving device. This is implemented by laying out a clock line between every transceiver and driver receiver pair that passes data. To ensure correct timing, the clock lines must be the same length and have the same loading as the data lines. A system master clock is normally used to retime received data coming out of backplane boards. Since timing is now "relative" from each driver to each receiver, much of the system latency is removed as compared to a synchronous system. Data can be passed between system devices without regard to clocking times of other components in the system. The benefits of this design are a significant increase in throughput and a potential for larger backplanes, because of the elimination of latency and speed issues associated with synchronous designs. The drawback is a significant increase in clock lines—one set for each pair of data-linked devices.

Asynchronous system architecture does not use a clock. Instead of clock-synchronized data, a local "handshake" protocol between the driver and receiver is implemented. There are many different design approaches for this protocol. One example is a request signal from the driver to send data, followed by an acknowledged signal from the receiver when the data is processed. Asynchronous design can be complex and is often unique for each design implementation. However, this architecture does have some unique benefits. For example, without a clock there are no clock timing or latency issues. This lack of clocking results in a potential increase in data-transfer speeds. In an asynchronous system, only the frequency and length of the data transmission limit data-transfer speeds. Clocked systems have a constantly running clock, regardless of whether data is being transferred or processed. An asynchronous design can yield a potential reduction in power usage due to the lack of a constantly running clock. Asynchronous architecture can be complex to implement and verify, since traditional timing verification methods do not work. In addition, data streams coming from a backplane of this type still need to be retimed for system processing. Despite these potential issues, asynchronous architecture has found favor with some designs.

6.5.3 Backplane Bus Protocols

Backplane bus protocols determine—electrically and mechanically—how signals are dropped off or picked up at every device attached to the line. Protocols determine how cards pass information, the speed of data transfer, the maximum number and type of peripheral devices allowed, and the physical size and layout of the system.

In general, the choice of the correct protocol for a backplane design is a complex decision that involves many design compromises. No one solution will give the best performance in each design category.

A few systems simply adopt ethernet, allowing asynchronous transmission along the backplane to be performed with a consolidated protocol. Frequently 1 Mbit/s or 10 Mbits ethernet is enough, otherwise there is 100 Mbit/s ethernet that for sure satisfy all the communication and configuration exigencies.

At present, chips performing ethernet switching and data multiplexing and demultiplexing in the ethernet format are cheap and diffused.

The issue in using ethernet is that the protocol implements several functionalities that are not required in a backplane bus. For example, in order to use ethernet in an architecture similar to that of Figure 6.47, every subrack controller must operate versus the bus like an ethernet switch.

Moreover, in order to operate correctly, an ethernet bus needs an appropriate, although potentially quite simple, physical layer that has to be constructed via the embedded cards electronics. This is for sure a powerful solution, but sometimes too complex for the requirements.

Specific bus protocols exist for backplane applications that are able to mange backplane busses fulfilling all their requirements.

The two most diffused among these protocols are the protocols of the PCI and of the VME family. Protocols in each family come from the family generator (e.g., original PCI) via tailoring for a specific requirement, as number of managed cards.

It is out of the scope of this chapter to detail the working principles of the protocols belonging to these two families; thus, we limit our analysis to a comparison of PCI and VME pros and cons whose key results are reported in Table 6.20, and to a brief comment on the set of protocols belonging to the two families.

TABLE 6.20

Comparison between Two Backplane Protocols: PCI and VME

Property	PCI	VME
Address space	32 bit	Up to 64 bit
Data path	Up to 64 bit	Up to 64 bit
Bus clock speed	33, 66, or 133 MHz	Asynchronous no standard clock
Data rate	Up to 1.06 Gbyte/s	Up to 512 Mbyte/s
Hot-swap capabilities	Yes	Yes
Operating voltages	3.3 or 5 V	3.3 or 5 V
Cost	Reduces cost	Can be high: licensing costs as well as lower volumes

- *Highest number of cards on a segment: VME64.* VME technology is the clear leader for the number of cards found in a segment (or continuous trace). VME is able to fit up to 21 cards on a standard 19′ backplane. It is important to note that, although PCI can have only four cards (eight for Compact PCI and two for PCI-X) on a segment, the technology can support up to 256 segments. However, this does have drawbacks because segments are arranged in a hierarchical tree that requires more logic bridges and longer propagation delay times, when adding additional segments to the bus.

- *Widest data path: VME64 (64 bit): PCI (64 bit).* The width of the data path has increased to 64 bits. This size is a standard option in VME64, VME64x, VME320, PCI, Compact PCI, and PCI-X.

- *Fastest data rate: PCI-X (133 MHz, 64 bit, 1.06 Gbyte/s).* The PCI protocol is, by far, the leader in speed. With the new PCI-X specification operating at 133 MHz and 64 bit data path, speeds of up to 1.06 Gbyte/s can be achieved. Current limitations allow only two cards on this high-speed segment and only one segment can run at 133 MHz, leaving the other segments to run at 66 MHz. The fastest competitor in the VME family is VME320, licensed by Arizona Digital, which has a bus bandwidth of over 320 Mbyte/s and a peak speed over 500 Mbyte/s.

- *Hot-swap capabilities: Compact PCI, PCI-X, VME64, VME64x.* To hot swap means to leave the bus interface component on, or in a known state while removing or inserting a card into the system. Hot swapping or live insertion is important for customers who require systems with zero downtime. All the major protocols are adopting specifications to support hot-swap capabilities. Compact PCI and VME64 both include special mechanical configurations in the connector to make and break the ground, prebias, control, and VCC connections before the data pins touch. This allows a card to be precharged and reduces insertion transients associated with noise and voltage spikes.

- *Low cost: PCI.* Devices manufactured for the PCI specification, including Compact PCI and PCI-X, are generally less expensive than VME technology. PCI devices are an inexpensive, mass-produced bus and board standard, with literally thousands of suppliers competing at low margins. If VME is used, it will have some PCI in the final product because compatibility issues force VME system designers to include PCI in their products. With over 800 companies supporting PCI and designing cards and accessories for the PCI open specification, VME customers demand that products they buy are compatible with their current and future PCI devices.

6.5.4 System Thermal Design

As we have seen in Section 6.5.1, a typical DWDM system is constituted by one or more racks having inside subracks where the units are packed once designed on a card. In a central office racks by different systems are aligned in order to gain access to power supply and external lines.

This means that the fan tries that are designed to cool the system have to be assisted by a careful system design if they have to provide optimized performances.

The thermal design generally starts from the requirements imposed by the system mechanical design, operational environment, and heath production of the various cards. For reference, the environment requirements standardized by ETSI in Europe and by Telcordia in North America are summarized in Table 6.21.

TABLE 6.21

Environment Requirements for Telecommunication Equipment Standardized by ETSI in Europe and by Telcordia in North America

Requirements Specifications		NEBS Level 3	ETSI Class 3.1
Normal	Environ. temp. (°C)	5–40	5–40
	Relative humidity (%)	5–85	5–85
Short	Environ. temp (°C)	5–50	5–45
term	Relative humidity (%)	5–90	5–90
Altitude		60–1800 m at 40°C	70–106 kPa
		1800–4000 m a 30°C (environ. temperature)	
Acoustic noise at attended room		78 dB at 27°C	7.2 Bells at 25°C
Rate of temp. changes (°C/min)		0.5	0.5
Maximum environ. temperature		40°C up to 96 h	NA
Maximum surrounding air flow (m/s)		NA	5

In particular inlets and exhaust air flows have to be carefully placed inside the system structure and correctly dimensioned.

If the rack is open, that is, no closed doors in front or in the rear, the best direction of inlets and exhaust air flows is front-bottom-in to rear-top-out, since this direction is privileged by the natural property that the air flow, hot after cooling the equipment, tends to move vertically up.

Since generally more than one subrack is stacked in a rack, the design of air flow has to take into account the cooling of each subrack.

The better solution should be independent cooling flux for each subrack.

In order to implement this kind of air flow design some space has to be foreseen between adjacent racks and this space should be organized so to render thermally independent adjacent subracks.

If a similar solution is not possible, the design can be carried out by foreseeing a single air flux traversing the whole rack, but in this case, the flux has to be sufficient to assure good cooling even taking into account the heat produced by every subrack.

This can be achieved by correctly optimizing the air flow path through the rack and the dimension and speed of each subrack fan set.

To comply with NEBS requirement the overall cooling system has to operate correctly in three situations:

1. Worst case environment conditions (taking into account to calculate temperature and humidity, the height on the see) with fans at full speed and a single dirty filter in all installation conditions.

2. One fan failure with the other fans at full speed and the environment at 40°C.

3. Standard conditions (environment at 40°C) with fans at minimum speed, assuring on all the components a temperature compliant with their foreseen live.

Thus, the thermal design has to be carried out taking into account effective compliance tests.

The approach to the analysis of the cooling capacity of a given air flux is in principle, easy starting from classical fluid motion theory and thermodynamic equations.

In particular, assuming laminar motion and calling z the vertical direction, we can write the following equations for the air flow motion and the heath exchange with the surrounding system:

$$\frac{D\vec{u}}{Dt} + \frac{1}{\rho}\nabla p = -g u\vec{z} \quad \text{(Navier–Stokes equation)}$$

$$\frac{D\rho}{Dt} + \rho\,\nabla\vec{u} = 0 \qquad \text{(fluid mass conservation)} \qquad (6.104)$$

$$\frac{DQ}{Dt} = c_p\frac{DT}{Dt} - \frac{1}{\rho}\frac{Dp}{Dt} \quad \text{(principle of energy conservation)}$$

where

$D/Dt = \partial/\partial t + \nabla.\vec{u}$ is the so-called Lagrangian derivative and represents the fact that the equations of motions are all written in the moving fluid particle reference frame

$\vec{u}$ is the fluid particle velocity in the absolute reference frame (as defined in the Navier–Stokes approach)

p is the pressure in the direction of the fluid motion (assumed laminar)

ρ is the fluid density

g is the gravity constant on the earth surface

Q is the heath density

T is the local temperature

c_p is the air specific heath at constant pressure

It is apparent that, even if the nature of the air motion in a well-designed cooling system allows several simplifications in the motion equations, solving the aforementioned equations system with suitable boarder conditions is a formidable mathematical problem.

The system designer has two possible options at his disposal.

The first consists in using a trial and error design procedure supported by a specific software tool that allows, once the air flow is designed and the power dissipation of each part of the system is fixed, to solve the aforementioned equations, numerically so deriving the regime temperature of all the system parts.

Commercial tools exist to do that and the probability of finding a suitable technical solution in a short time is completely dependent on the experience and physical insight of the designer that is able to start from a tentative solution sufficiently near to the optimum.

On the other hand, if some support in the first design is to be devised before analyzing the proposed structure with a software tool, either Equation 6.104 have to be drastically simplified or some heuristic rule has to be used.

A heuristic model that is largely used is that of concentrated constant heat circuit, that in electronic engineers groups is easily adopted due to its similarity with concentrated electronic circuit analysis.

In this model, every point in which a heath exchange is performed is modeled as a concentrated constant heat circuit, so talking in general of heat impedance and, in particular, of heat resistance, for example.

In this simple model, each part of the system that can be considered uniform from a thermal point of view (e.g., an interface card could be considered uniform in a first approximation, taking as a reference the worst card position, where the heat is generated by the transmitter) and that is in contact with a heat sink is characterized by a thermal resistance θ_T, so that the relationship between the temperature difference ΔT between the card and a static heat sink and the power P_d the card can dissipate can be written as

$$\Delta T = \theta_T P_d \tag{6.105}$$

Naturally, the thermal resistance depends not only on the system element, but also on the way it is connected to the heat sink. In our case, where there is a forced air flow that works as heat sink, there are technical graphics that, starting from measurements and analytical evaluations, give the thermal resistance existing between a subsystem and the air flow in the hypothesis of a complete immersion of the subsystem in the air flow, so that the air is in contact with every part of the subsystem. One of these graphics is reported in Figure 6.48.

At this point the procedure for the design is the following:

- Construct the thermal equivalent circuit remembering to insert all the heat resistances (also that related to a simple flow of the air along a space in the rack, since heat is exchanged with the metallic rack walls; Figure 6.49).
- For each subsystem evaluate the desired working temperature and the dissipated power.
- Construct the heat circuit model of the system also considering the incoming and outgoing air temperatures that are determined by the environmental conditions; the heat transfer model coming out are similar to the scheme of Figure 6.49, where a subrack with a few mounted cards is represented with a heat circuit.
- For each subsystem evaluate starting from Equation 6.90 and the heat circuit model the heat resistance that allows the power dissipation at the required temperature.

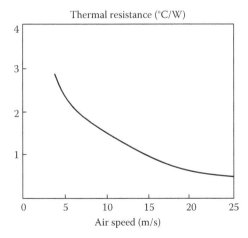

FIGURE 6.48

Example of plot providing the thermal resistance versus the air flow speed in conditions of laminar flow.

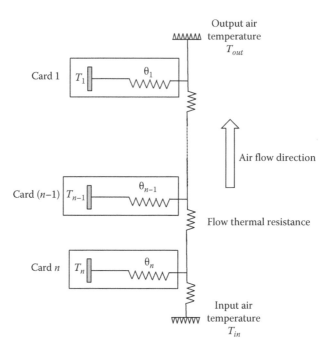

FIGURE 6.49
Example of simplified thermal equivalent circuit for a DWDM rack.

- From each heat resistance determine the corresponding, needed air flow from graphs similar to Figure 6.48 or from specific software.
- Design the air ducts, so to assure in every point of the system, at least the required flow, often the higher value of the air flow is the only considered and a worst case design is carried out trying to assure the same flow to all the system parts.

At the end of the design, the cooling system should be always analyzed with software solving numerically the fluid thermodynamic equations to verify with a less approximated model the expected performance.

References

1. Agrawal, G.P., *Fiber Optic Communication Systems*, Wiley, New York, s.l., 2002, ISBN-13: 978-0471215714.
2. Ramaswami, R., Sivarajan, K.N., Sasaki, G.H., *Optical Networks: A Practical Perspective*, 3rd edn., Morgan Kaufmann, San Francisco, CA, s.l., 2010, ISBN: 9780123740922.
3. Senior, J., *Optical Fiber Communications: Principles and Practice*, 3rd edn., Prentice Hall, Essex, U.K., s.l., 2008, ISBN-13: 978-0130326812.
4. Smith, D.R., *Digital Transmission Systems*, Kluwer Academic Publishers, Norwell, MA, s.l., 2004, ISBN: 1402075851.
5. Recommendations, ITU-T. M.21xx Series.
6. Recommendations, ITU-T. G.821, G.826, G.828, G.829.

7. Cotter, D., Observation of stimulated Brillouin scattering in low loss silica fiber at 1.3 μm, *Electronics Letters*, 18, 495–496 (1982).

8. Iannone, E., Matera, F., Mecozzi, A., Settembre M., *Nonlinear Optical Communication Networks*, Wiley, New York, s.l., 1998, ISBN-13: 978-0471152705.

9. Boyd, R.W., *Nonlinear Optics*, Academic Press, San Diego, CA, 2008, ISBN-13: 978-0123694706.

10. Chaplyvy, A.R., Limitations on lightwave communications imposed by optical fiber nonlinearities, *Journal of Lightwave Technology*, 8, 1548–1557 (1990).

11. Loudon, R., *The Quantum Theory of Light*, 3rd edn., Oxford Science Publications, Oxford, s.l., 2000, ISBN: 0198501765.

12. Bulow, H., System outage probability due to first- and second-order PMD, *IEEE Photonics Technology Letters*, 10(5), 696–698 (1998).

13. Winzer, P.J., Kogelnik, H., Ramanan, K., Precise outage specifications for first-order PMD, *IEEE Photonics Technology Letters*, 16(2), 449–451 (2004).

14. Watkins, L.R., Zhou, Y.R., Dispersion compensation for long-haul, high capacity NRZ systems by pre-chirping and optimum receiver signal processing. In *IEE Colloquium on High Capacity Optical Communications*, May 16, London, U.K., s.n., 1994.

15. Ito, T., Yano, Y., Ono, T., Emura, K., Pre-chirp assisted normal dispersion region transmission for highly marginal dense WDM transoceanic system. In *Proceedings of the Optical Fiber Conference*, February 22, 1998, San Jose, CA, s.n., 1998.

16. Nakazawa, M., Yamamoto, T., Tamura, K.R., 1.28 Tbit/s-70 km OTDM transmission using third- and fourth-order simultaneous dispersion compensation with a phase modulator, *Electronics Letters*, 36(24), 2027–2029 (2000).

17. Xia, C., Rosenkranz, W., Electrical mitigation of penalties caused by group delay ripples for different modulation formats, *IEEE Photonics Technology Letters*, 19(13), 1041–1135 (2007).

18. Killey, R.I., Watts, P.M., Mikhailov, V., Glick, M., Bayvel, P., Electronic dispersion compensation by signal predistortion using digital processing and a dual-drive Mach–Zehnder modulator, *IEEE Photonics Technology Letters*, 17(3), 714–716 (2005).

19. Betti, S., De Marchis, G., Iannone, E., *Coherent Optical Communications Systems*, Wiley Interscience, New York, s.l., 1995, ISBN: 0471575127.

20. Tkach, R.W., Chraplyvy, A.R., Forghieri, F., Gnauck, A.H., Derosier, R.M., Four-photon mixing and high speed WDM systems, *IEEE Journal of Lightwave Technology*, 13, 841–849 (1995).

21. Harboe, P.B., da Silva, E., Souza, J.R., Analysis of FWM penalties in DWDM systems based on G.652, G.653, and G.655 optical fibers, *World Academy of Science, Engineering and Technology*, 48, 77–83 (2008).

22. Xu, B., Brandt-Pearce, M., Comparison of FWM- and XPM-induced crosstalk using the Volterra series transfer function method, *IEEE Journal of Lightwave Technology*, 21(1), 40–53 (2003).

23. Bellotti, G. et al., XPM-induced intensity noise in WDM compensated transmission systems. In *European Conference on Optical Communications*, Madrid, Spain, Vol. 1, pp. 425–426, 1998.

24. Salehi, J.G., Proakis, M., *Digital Communications*, 6th edn., McGraw Hill, New York, s.l., 2008, ISBN-13: 978-0071263788.

25. Trischitta, P.R., Varma, E.L., *Jitter in Digital Transmission Systems*, Artech House, Norwood, MA, s.l., 1989, ISBN: 089006248X.

26. Morse, P.M., Feshbach, H., *Methods of Theoretical Physics—Part 1*, McGraw Hill, New York, 1935.

27. Mazo, J.E., Satz, J., Probability of errors for quadratic detectors, *Bell Systems Technical Journal*, 44(11), 2165–2187 (1965).

28. Helstrom, C.W., Performance analysis of optical receivers by saddle point approximation, *Transactions on Communications*, 27(1), 186–191 (1989).

29. Najto, T., Terahama, T., Shimojoh, M., Yorita, T., Chikama, T., Suyama, M., Pre- and post-dispersion compensation in long haul WDM transmission systems, *IEICE Transactions on Communications*, E83-b(7), 1409–1416 (2000).

30. Costa, N.M.S., Cartaxo, A.V.T., Influence of dispersion compensation granularity on the XPM-induced degradation in NRZ-IM-DD WDM links at 10 Gbit/s per channel with 50 GHz of channel spacing. In *Ninth International Conference on Transparent Optical Networks, ICTON '07*, IEEE, Roma, Italy, 2007.

31. Peucheret, C., Hanik, N., Freund, R., Molle, L., Jeppesen, P., Optimization of pre- and post-dispersion compensation schemes for 10-Gbits/s NRZ links using standard and dispersion compensating fibers, *IEEE Photonics Technology Letters*, 12(8), 992–994 (2000).

32. Frignac, Y., Antona, J.-C., Bigo, S., Hamaide, J.-P., Numerical optimization of pre- and in-line dispersion compensation in dispersion-managed systems at 40 Gbit/s, *Optical Fiber Communication Conference and Exhibit, OFC 2002*, IEEE, s.l., pp. 612–613, 2002.

33. Killey, R., Thiele, H.J., Mikhailov, V., Bayvel, P., Optimisation of the dispersion map of compensated standard-fibre WDM systems to minimise distortion due to fibre nonlinearity, In I.P. Kaminow and T. Li (Eds.) *Optical Fiber Telecommunications IV*, Academic Press, San Diego, CA, Invited Chapter, 2001.

34. Hayee, M.I., Willner, A.E., Pre- and post-compensation of dispersion and nonlinearities in 10-Gbps WDM systems, *IEEE Photonics Technology Letters*, 9(9), 1271–1273 (1997).

35. Sun, W., *Optimization Theory and Methods: Nonlinear Programming*, Springer, New York, s.l., 2009, ISBN-13: 978-1441937650.

36. Avriel, M., *Nonlinear Programming: Analysis and Methods*, Dover Publishing, Mincola, NY, s.l., 2003, ISBN: 0486432270.

37. Sinkin, O.V. et al., Optimization of the split-step Fourier method in modeling optical-fiber communications systems, *IEEE Journal of Lightwave Technology*, 21(1), 61–68 (2003).

38. Pachnicke, S., Chachaj, A., Helf, M., Krummrich, P.M., Fast parallel simulation of fiber optical communication systems accelerated by a graphics processing unit. In *International Conference on Transparent Optical Networks, ICTON 2010*, Munich, Germany, Paper Th.B1.5, IEEE, s.l., 2010.

39. Weber, C., Bunge, C.A., Winter, M., Fibre nonlinearities in 10 and 40Gb/s electronically dispersion precompensated WDM transmission. In *Optical Fiber Conference OFC/NFOEC 2009*, San Diego, CA, IEEE, s.l., 2009.

40. Yamazaki, E., Inuzuka, F., Yonenaga, K., Takada, A., Koga, M., Compensation of interchannel crosstalk induced by optical fiber nonlinearity in carrier phase-locked WDM system, *IEEE Photonics Technology Letters*, 19(1), 9–11 (2007).

41. Weber, C., Bunge, C.-A., Petermann, K., Fiber nonlinearities in systems using electronic predistortion of dispersion at 10 and 40 Gbit/s, *Journal of Lightwave Technology*, 27(16), 3654–3661 (2009).

42. Huang, J.-A. et al., Impact of nonlinear phase noise to DPSK signals: Experimental verification of a simplified theoretical model, *IEEE Photonics Technology Letters*, 17(10), 2236–2238 (2005).

43. Killey, R.I., Watts, P.M., Glick, M., Bayvel, P., Electronic dispersion compensation by signal predistortion. In *Proceedings of the Optical Fiber Communications Conference OFC 2006*, Paper OWB3, IEEE/OSA, s.l., 2006.

44. Laurencio, P., Medeiros, M.C.R., Analysis of optical single side band systems employing optical amplifiers. In *The International Conference on "Computer as a Tool", EUROCON*, 2007, IEEE, s.l., pp. 1017–1022, 2007.

45. Takano, K., Sakamoto, N., Nakagawa, K., SPM effect on carrier-suppressed optical SSB transmission with NRZ and RZ formats, *Electronics Letters*, 40(18), 1150–1151 (2004).

46. International Electrotechnical Commission (IEC), Basic dimensions of front panels, subracks, chassis, racks and cabinets, IEC 60297-3-101, 1st edn., 2004–2008.

47. International Electrotechnical Commission (IEC), Subracks and associated plug-in units, IEC 60297-3-101, 1st edn., 2004–2008.

48. International Electrotechnical Commission (IEC), Injector/extractor handle, Keying and alignment pin, Connector dependent interface dimensions of subracks and plug-in units, Dimensions and design aspects for 1 U chassis, IEC 60297-3-101, 1st edn., 2004–2008.

7

Switching Systems: Architecture and Performances

7.1 Introduction

In Chapter 6, we have described optical transmission systems-the type of equipment used in the telecommunication network to move the information bearing signal from one place to the other.

Besides transmission, the other key functionality of a telecommunication network is switching, that is, the ability of sending the signal from its origin to the correct destination selecting a suitable route in the network.

Differently from transmission, that essentially coincides with a physical function, switching exists almost at every network layer, since each layer needs to create routes for its proper signals.

This implies that while the base technology for transmission is well established and essentially unique, there is a wide variety of switching network elements, depending on the layer where they operate and on the technology with which they are realized (synchronous digital hierarchy [SDH], Ethernet, IP, etc.).

Moreover, it should be considered that almost all the signal processing that is needed in the network is carried out in the switching elements, since for their nature, they generally have powerful signal processors.

This contributes to differentiating further the switching systems, not only with respect to the technology and the layer where they are inserted, but also for the particular implementation of signal processing functionalities.

For all the aforementioned reasons, it is almost impossible to create an exhaustive review of all switching network elements.

As we have done in all the chapters of this book, we will try to give a clear idea of the main switching machines that will be the core of the IP convergent network; regarding other categories of equipment we will not be so accurate.

Dealing with the optical layer, we will classify the main architectures of optical cross connect (OXC) and optical add-drop multiplexers (OADM) introducing their properties and performances; moreover, we will also analyze new SDH/SONET ADMs.

As far as packet switching is concerned, besides a discussion about carrier class Ethernet switches, we will dedicate an important attention to Internet protocol (IP)/multiprotocol label switching (MPLS) routers.

All the aforementioned machines have in common a high-level architecture that is shown in Figure 7.1.

The switching machine is interfaced with transmission systems or with other machines, in case at a lower level, through input/output interfaces. The role of the interfaces is not only to regenerate the signal, but also to condition it to be processed by the system.

This conditioning comprises not only signal regeneration, but also, for example, the termination of the header at lower layers, possibly resynchronization of the incoming signals, error detection.

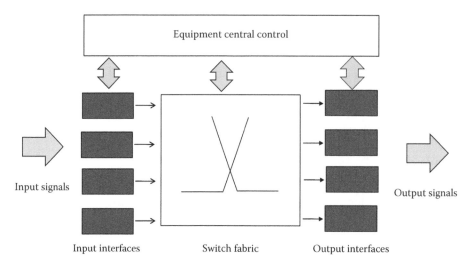

FIGURE 7.1
Functional scheme of a space diversity switch fabric.

In case of packet network machines, queues are located in the interface cards to perform the storage and forwarding of incoming packets.

The interfaces are connected with a switch fabric that has the role to create a correspondence among input and output ports to send the incoming signal in the correct direction.

Finally, a centralized or distributed processor manages the equipment.

The architecture of large switch fabrics is complex, due to the need of finding a trade-off between the ability of creating whatever permutation between input and output with fast switching times and using simple control algorithms and the need of avoiding a huge hardware structure with great costs and low reliability.

For this reason, we will devote a preliminary section to switching fabric architectures. This allows us to deal in a systematic way with this key point, reviewing a set of general results that will be useful several times dealing with different network equipment.

After the section dedicated to switch fabrics, we will also discuss the particular problems a platform for a switching machine has to face, which does not exist in platform for transmission systems. This is another general discussion that adapts to almost every kind of switching machine.

7.2 Space Division Switch Fabrics

In all the switching machines that are deployed in the telecommunication networks, there is a part of the machine devoted to transfer the signal from the input port to the correct output port. This subsystem is called switch fabrics [1,2].

The first kind of switch fabrics that comes under consideration is the class of the so-called space diversity switch fabrics, and we will concentrate for the moment our attention on them.

In this section, we will review a few schemes of space diversity switch fabrics that are both simple and important for applications. However, this is only a tiny part of the huge amount of different architectures that have been proposed and almost nothing will be reported of the important mathematical aspects of the interconnection networks topological study. A Few elements of switch fabric performance evaluation will be introduced in Appendix E. This topic has been delayed to an appendix since, even if it is fundamental for the assessment of switching machine performances, it is not needed to understand their architecture.

For a deeper understanding of these topics, a much wider review of different architectures for space division switch fabrics with a few elements of performance evaluation can be found in [3,4], while a mathematically oriented analysis of the interconnection theory can be found in [5,6].

A space diversity switch fabric is a subsystem having N input ports and N output ports, an internal interconnection network allowing the connection of the input ports with the output ports and a control processor that can configure the fabric to realize the required mapping among the input and the output ports.

A switch fabric can work in a static or in a dynamic fashion [7]. Static working is realized when the traffic is offered to the switch fabric in terms of semi-static connections, that is, connection with a very long duration and that has to be established without particular time constraint.

In the case of semi-static working, the configuration of the switch fabric is a part of a complex connection setup process, and live traffic travels through the established connection only after the completion of the whole setup process, generally closed by a functional verification of the newly established connection.

Examples of such semi-static switch fabrics are those inside equipment performing circuit switching, for example, at the optical transport network (OTN) layer.

Dynamic working is characterized by the fact that the switch fabric continuously follows a process of setup and teardown of internal connections acting in real time to satisfy streams of external requests.

A typical example of such a working is provided by the switch fabric of a router working in a datagram network.

In this case, the switch fabric works generally in tandem with a set of queues that can be located at the fabric input, at the fabric output, or in both the configurations. In the most general case of queues both at input and output, every time a packet arrives at an input, it is stored into a queue and its address is read. The processor identifies the output destination port, the switch fabric is configured to connect the input to the output, and the packet is transmitted at the output queue. After that, another packet is considered requiring another connection and so on.

The semi-static and the dynamic switch fabrics require different controllers [7,8].

In the semi-static case, the controller has the role of performing actions dictated by the overall connection setup procedure and to manage the hardware of the switch.

In the dynamic case, the controller, on top of the switch management and on the route finding inside the switch, has to manage conflicts arising when packets from different inputs either have to be sent to the same output or require the use of the same internal resource. This structure is represented schematically in Figure 7.2.

In this case, generally, either the controller runs two different processes or two different dedicated hardwares exist: one of them is the switch fabric manager; the other is the so-called arbiter.

The manager performs all the configuration operations on the switch fabric, while the arbiter performs conflicts solution on the ground of a suitable solution strategy [9].

Normally, the arbiter works on the ground of the priorities that are assigned to the input requests by an external entity: for example, by the request remote origin on the ground of its quality of service level.

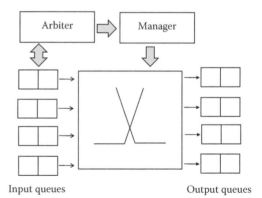

FIGURE 7.2
Block scheme of a dynamic space diversity switch with a switch fabric manager and a contention and queue policy arbiter.

However, to allow a fair allocation of resources and to achieve high-performance switch operation, the local fabric control system should be able to change locally the priority of the input requests on the ground of the local status of the switch [8].

If this would not be possible, it could happen that the arrival of a long sequence of high-priority requests on a specific input port will practically block the fabric, creating congestion on the other ports and increasing the average waiting time of lower priorities' requests out of the requirements of the quality of service.

In this way the priority changes is part of the policy employed by the switch fabric scheduling algorithm. For example, the round-robin policy, which is one of the most widely used priority update schemes, dictates that the request served in the current cycle gets the lowest priority for the next arbitration cycle.

It is out of the scope of this section to analyze the characteristics of the complete arbitration algorithms; one of the most used is the so-called wavefront arbiter [19] and the interested reader can start her study of arbiters from the indicated bibliography.

Different classifications of space diversity switch fabrics can be done, depending on what characteristic of the fabric is considered [1].

The first classification takes into consideration blocking events: a blocking event occurs in a switch fabric if a newly arrived request of connecting a free input port with a free output port cannot be satisfied due to the absence of a free route through the fabric, all the possible routes being blocked by ongoing connections.

The situation is called an absolute block if there is no possibility of rerouting the connections that are present in the fabric to free a route for the new connection, removable block if this is possible.

With respect to blocking events, switch fabrics can be classified as follows:

- *Strictly nonblocking*: No blocking event can occur in this type of switch fabrics, whatever pattern of requests is presented and whatever is the order of the requests. In other words, whatever the switch state is, a connection between a free input port and a free output port can always be established without the need of rerouting existing connections.

- *Rearrangeable nonblocking*: In this switch fabric, only removable block events are possible, whatever pattern of requests is presented to the switch fabric and in whatever order the requests are presented. In other words, whatever the switch state is, a connection between a free input port and a free output port can be established, but sometimes there is the need of rerouting existing connections.

- *Blocking*: In this switch fabric, absolute blocking events can occur; this means that there are statuses of the switch fabric in which a free input port cannot be connected to a free output port due to the absence of a free path connecting input and output. Moreover, there is no way to move the already established connections to free the resources to create the wanted route.

Considering traffic patterns, the ability to manage multicast connections is another characteristic of a switch fabric. In order to implement multicasting, the elementary node of the switch fabric must have the ability to send the incoming signal to more than one output port. This multicast ability can be present in all the switch fabric building blocks or only in a few of them.

A switch fabric can be defined strictly nonblocking or rearrangeable nonblocking versus multicast traffic if, in the corresponding definitions, a unicast connection request is substituted by multicast connection request, that is, a request to send the signal incoming from a free input port to a set of free output ports.

Another important property that a switch fabric could have is the property of maintaining the state if the power feed is switched off; in this case, the switch fabric is called latching.

This means that in case of power blackout, due to either a local fault or some other reason, transiting connections are not affected.

Moreover, and perhaps even more important, if the fabric does not need power feeding to remain in the current status, the power consumption in semi-static applications is due only to the switching events being smaller. On the contrary, if power feeding is needed to maintain the current status, then the overall power consumption of the fabric is an important parameter since it conditions all the designs of the equipment (see Chapters 2 and 3).

In dynamic switching applications, the switching fabric is continuously involved in a fast series of state changes, never remaining in the same state for a long time.

In this case the first performance parameter is the power consumption that can be on the order of several hundred watts or even kilowatts for large fabrics in datagram-oriented routers (e.g., IP routers).

The switch fabric real estate is an important parameter in any application, and several times this is one of the elements that suggest the adoption of an architecture instead of another.

Another quality element of a switch fabric, intimately related to its structure, is the fabric modularity. Generally, a switch fabric is designed to switch signals at a certain layer of the network; for example, an OXC is designed to switch optical channels, that is, signals at the OCh layer within the sublayering of the physical layer. On the contrary, an IP router is designed to switch IP packets, that is, layer 3 signals.

Let us imagine that our switch is designed to switch layer n signals: the fabric is said to be layer k modular, with $k \leq n$, if adding at the switch fabric input an aggregation of layer n signals representing a layer k signal, it can be scaled to accept the new inputs by only adding new elements, without the need of changing already existing elements.

For example, let us consider an OXC switch fabric: it can be fiber modular (that would be optical multiplex layer modular) and wavelength modular (which means optical channel layer modular) [10,11].

The fabric is fiber modular when adding at its input a new fiber, with its whole WDM comb, it can be scaled to accept the new input simply adding other parts, without the need of changing sometimes what is already deployed.

The same OXC fabric is wavelength modular if, adding a wavelength to a set of input WDM combs (from one to all), it can be scaled to accept the new input simply adding other parts, without the need of changing sometimes what is already deployed.

On the contrary, since an OXC switch fabric performs essentially an OCh layer function, asking if it is label switched path (LSP) modular has no meaning since LSPs are not discriminated by the OXC.

Requiring the LSP modularity is correct for the switch fabric of an IP router in the case that it works as a label switched router (LSR).

A last element that determines the performances of a switch fabric is its robustness in front of an internal failure [12].

Naturally, the possibility of rerouting connections affected by an internal failure is based on the availability of alternative disjoined routes, and it is well possible that the needed rearrangement of a switch fabric due to a failure will transform it from strictly nonblocking to blocking, for example.

To summarize, the main qualitative elements allowing the evaluation of the performance of a switch fabric are as follows:

- Blocking characteristics with unicast traffic
- Blocking characteristics with multicast traffic
- Static applications
    - Ability to switch correctly after a long inactive period
    - Ability to remain in a status without power feed
- Dynamic application
    - Power consumption
- Real estate
- Modularity at every lower layer
- Capability to preserve connections in front of internal failures

Besides these qualitative elements, a switch fabric can be characterized by its throughput characteristics. When a traffic level is offered at the switch fabric ports, for example, a certain input packets process characterized by its mean number of packets per second, the output traffic follows the input one only up to the moment in which congestion manifests itself inside the switch. At that point, the output growth is less than the input since part of the arrival traffic is rejected by the switch.

Moreover, in dynamic fabrics, the time delay from the arrival and the departure of a packet is approximately constant only within the uncongested working state, while it starts to grow more and more when congestion starts.

Evaluating throughput and fabric crossing delay is another type of quantitative performance evaluation of a switch fabric.

Here we will analyze all the qualitative elements, while some quantitative evaluation is covered to Appendix E.

7.2.1 Crossbar Switch Fabrics

The simpler switch fabric scheme is the so-called crossbar switch that is shown in Figure 7.3 [1,4].

The crossbar switch is constituted by a set of 2×2 switches placed at the crosspoints of a matrix of connections. The control of the switch fabric operates on each single switch independently by setting it to one of its four states, as shown in Figure 7.4.

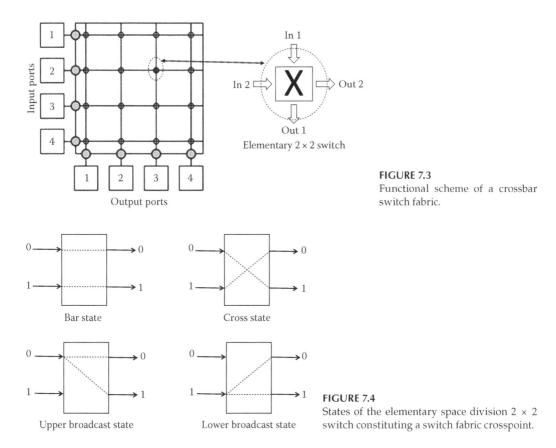

FIGURE 7.3
Functional scheme of a crossbar switch fabric.

FIGURE 7.4
States of the elementary space division 2 × 2 switch constituting a switch fabric crosspoint.

If the switch fabric is designed only for unicast traffic, only the bar and the cross state are used, otherwise all the states are exploited.

In this way, any input signal can be routed toward any free output port.

It is easy to see that a crossbar switch fabric is strictly nonblocking even if multicast connections are considered, if all the elementary 2×2 switches have multicast ability.

Moreover, a crossbar fabric is modular at every layer. As a matter of fact, greater crossbar fabrics can be built starting from a base crossbar with a recursive construction whose first step is the buildup of the basic crossbar from the 2×2 switches. In particular, starting from an $N \times N$ crossbar, a $2N \times 2N$ fabric can be built with four $N \times N$ elements, a $3N \times 3N$ fabric can be built using nine $N \times N$ building blocks, and so on. This construction is shown in Figure 7.5.

Since the path through the fabric from an input to an output port is not unique and there is no automatic routing algorithm, the algorithm used by the external controller to set up the signal path when a new connection between input and output is required is an important aspect of the crossbar working.

This is important in particular in dynamic applications, when the input of the switch fabric is generally constituted by the output of a set of queues, where the arriving switching requests are accumulated (see Figure 7.2) and fast switching based on the priorities of the incoming requests is the key to achieve high performances.

As shown in Figure 7.6, the route connecting an input port with a selected output is not unique, and in the worst case (when the signal enters from a port at the edge of the switch and has to exit from the port at the corresponding edge on the exit side) there are two completely disjoined paths between the input and the output.

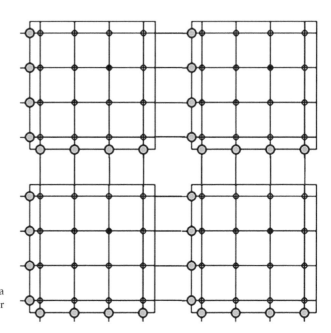

FIGURE 7.5
Example of the procedure to build a $2N \times 2N$ switch fabric starting from four $N \times N$ crossbar elements.

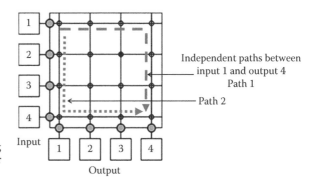

FIGURE 7.6
Example of the multiple routes connecting selected input and output ports in a crossbar switch fabric.

This means that the failure of a crosspoint in the switch fabric can create blocking patterns, that is correspondences between input and output that cannot be set, but not necessarily destroy the existing configuration, especially if a large number of ports is free.

From a performance point of view, the crossbar switch fabric is a good solution; unfortunately, it is not easy to scale it to a large switch dimension.

As a matter of fact, the number of elementary 2×2 switches required by a crossbar with $N = 2M$ input/output ports is M^2, so that the cost of the switch, the real estate and especially the dissipated power if it is used for dynamic switching growth with the squared number of input ports.

7.2.2 Clos Switch Fabric

Clos switch fabrics are characterized by three stages of architecture [13]:

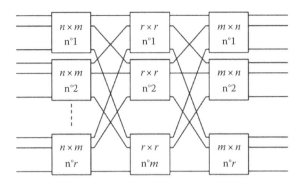

FIGURE 7.7
Block scheme of a Clos switch fabric.

1. The ingress stage
2. The middle stage
3. The egress stage

Each stage is composed by a set of crossbar switches. The switches of a stage are connected with those of the following stage but not among them.

All the characteristics of a Clos switch fabric are defined by three integers n, m, and r as shown in Figure 7.7 representing the general scheme of a Clos switch fabric.

The first stage of the switch is composed by r crossbar with n inputs and m outputs, so that the overall number of input lines of the switch fabric is $N = rn$.

The middle stage is composed by m crossbar switches with r input and r output ports. Thus, it can be done with exactly one connection between each output of the first stage and each input of the second.

Similar to the first, the third stage is composed by r crossbar with m inputs and n outputs so that each output of the middle stage is connected to one input of the egress stage.

The connection between the first and the second stage is realized by connecting the m ports of the first switch of the first stage to the first ports of all the switches of the second stage, the m ports of the second switch of the first stage with all the second ports of the switches of the second stage, and so on.

In general, the ports of the kth switch of the first stage are connected with all the kth ports of the switches of the second stage (compare Figure 7.7).

The relative values of m and n define the blocking characteristics of the Clos network [4], while the switch fabric dimension is fixed by r, once the dimensions of the crossbar has been chosen.

7.2.2.1 Strictly Nonblocking Clos Networks

If $m \geq 2n - 1$, the Clos network is strictly nonblocking.

Assume that there is a free input of an ingress switch, and this has to be connected to a free output on a particular egress switch. In the worst case, $n - 1$ other connections are active on the considered ingress switch, and $n - 1$ other connections are active on the considered egress switch.

Let us also assume, in the worst case, that each of these connections passes through a different middle stage switch. Hence, in the worst case, $2n - 2$ of the middle stage switches are unable to carry the new connection. Therefore, to ensure strictly nonblocking operation, another middle stage switch is required, making a total of at least $2n - 1$.

7.2.2.2 Rearrangeable Nonblocking Clos Networks

If $m \geq n$, the Clos network is rearrangeable nonblocking, which means that an unused input on an ingress switch can always be connected to an unused output on an egress switch, but to do that, existing connections may have to be rearranged by assigning them to different center stage switches in the Clos network.

To prove this property is out of this brief summary, but an example is shown in Figure 7.8 utilizing a Clos switch fabric with $r=n=m=4$, that results rearrangeable nonblocking.

Let us imagine receiving the set of connection requests that are shown in Figure 7.8a, where the input and output ports to be connected are identified with the same letter, and let us imagine that the connections arrive so that the letters are in alphabetical order. First arrives connection A and it is routed, then B, C, D, and all are routed, when connection E arrives, it is blocked. In order to route connection E, connection D has to be rerouted so to free the central stage switch. After D is rerouted (see Figure 7.8b) connection E can be routed, thus removing the situation of removable block that was created by the request of connection E.

7.2.2.3 Blocking Clos Networks

If $m < n$, the Clos network is blocking. For a long time, blocking Clos networks have found a large application in telephone switches. As a matter of fact, since telephone switches had a huge switch fabric, in order to spare money, a small blocking probability was accepted.

For this reason, it is important to have a means to evaluate the blocking probability of a Clos switch fabric. Two approximations have been developed for this reason: the Lee or the Jacobaeus [14] approximations.

In both cases, the starting assumption is that no rearrangements of existing connections is possible so that when a new connection arrives to an input switch, the potential number of other active connections on the ingress and egress switches that can block the new connection request is $u = n - 1$.

In the Lee approximation, it is assumed that each internal link between stages is already occupied by a connection with a certain probability p, and that this is completely independent between different links.

This overestimates the blocking probability, particularly for small r. The probability that a given internal link is busy is

$$p = (n-1)\frac{q}{m} \tag{7.1}$$

where q is the probability that an ingress or egress link is busy, that is related to the overall traffic offered to the switch fabric. The probability that the path connecting an ingress switch to an egress switch via a particular middle stage switch is free is the probability that both links are free, that is, $(1 - p)^2$. Hence, the probability P that this path is unavailable is

$$P = 1-(1-p)^2 \tag{7.2}$$

The probability of blocking P_B, that is, the probability that no path connecting the considered ingress and egress ports via whatever middle stage is available, is

$$P_B = P^m = \left\{1-\left[1-(n-1)\frac{q}{m}\right]^2\right\}^m \tag{7.3}$$

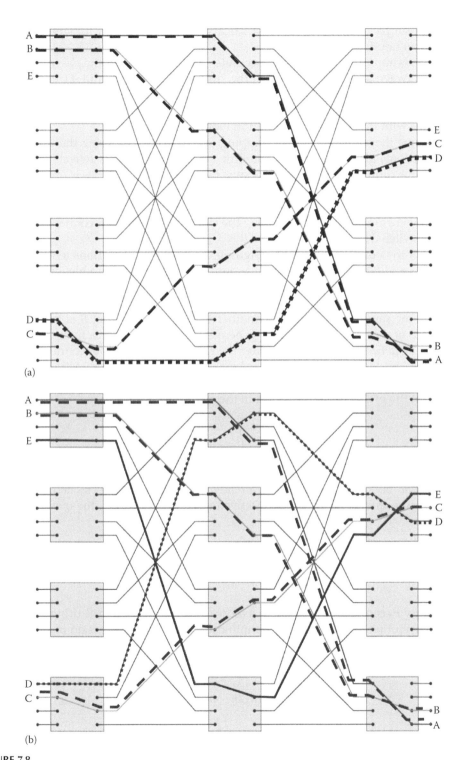

FIGURE 7.8
Example of rearrangement of a Clos switch fabric to eliminate a blocking situation: (a) the path between input and output ports labeled with E is blocked due to the configuration of the last element of the second stage; (b) the block is removed by rerouting the path between output and input D ports.

This is a very simple equation to use in practical cases, but especially for small values of r is a large upper bound.

The Jacobaeus approximation is more accurate since it is based on the observation that a connection between a free ingress port of the kth first stage switch and a free egress port of the hth third stage switch is blocked if and only if there is no middle stage available to accept the connection from the kth input port to transfer it to the hth output port (compare as an example with the routing performed in Figure 7.8).

Let us assume that some particular mapping of connections entering the Clos network (input connections) already exists onto middle stage switches. This reflects the fact that only the relative configurations of ingress switches and egress switches is of relevance.

Let us imagine that a new connection request between free input and output ports arrives. There are i input connections entering via the same ingress switch of the new request, and there are j connections leaving the Clos network (output connections) via the same egress switch. Hence, $0 \leq i \leq u$, and $0 \leq j \leq u$, being $u = n - 1$.

Let us call A the number of ways of assigning the j output connections to the m middle stage switches and B the number of these assignments which result in blocking. Then the number of cases in which the remaining $m - j$ middle stage switches coincide with $m - j$ of the i input connections to be inaccessible for the blocked connection is exactly B.

Looking at the same phenomenon in another way, we can see that B is the number of configurations of the i input connections generating $m - j$ input connections engaging the middle stage switches that remain free after the routing of the j output connections.

In this way, j middle stage switches are blocked by the output connections and $m - j$ are blocked by a subset of input connections and the newly arrived connection request is blocked. Then the probability of blocking in the considered situation is

$$p_{ij} = \frac{B}{A} = \frac{\binom{i}{m-j}}{\binom{m}{j}} = \frac{i!\,j!}{(i+j-m)!\,m!} \tag{7.4}$$

If f_i is the probability that i other calls are already active on the ingress switch, and g_j is the probability that j other calls are already active on the egress switch, the overall blocking probability is

$$P_B = \sum_{i=0}^{n-1} \sum_{j=0}^{n-1} g_j f_i p_{ij} \tag{7.5}$$

The probability P_B can be expressed with a more explicit formula substituting to g_j and f_i a binomial distribution. After a difficult simplification, the final expression of the blocking probability results to be

$$P_B = \frac{[(n-1)!]^2 (2-q)^{(2n-m-2)} q^m}{m!\,(2n-m-2)!} \tag{7.6}$$

7.2.2.4 Control of a Clos Switch

Since a Clos switch is composed by a collection of suitably connected crossbar switches, it is also composed of a set of independently driven 2×2 switches.

From the point of view of the switch control, it is somehow equivalent to the crossbar switch.

The route finding algorithm for a nonblocking Clos switch fabric can be formulated in several different ways. The most intuitive are probably the following:

- Individuate in the input stage the number of the switch from where the connection to be set up has to start; let us call it k.
- Staring from the first switch of the central stage, individuate the first central switch with the kth port free; let us call its number in the middle stage h.
- Switch the crossbar from where the signal comes in and the hth crossbar in the middle stage so as to transfer the connection request at the kth output port of the middle stage.
- Move the signal to the kth output port of the middle stage.
- Transfer the signal to the kth switch of the third stage.
- Transfer the signal to the correct output port.

The only thing to be noted is that, in the assumption of strictly nonblocking structure, there surely does exist a switch in the central stage having the kth input port free.

Much more complicated algorithms have been introduced both in order to reduce as much as possible the calculation complexity and to be able to manage rearrangeable nonblocking and blocking Clos networks. A good set of starting information is contained in [4,15,16].

7.2.2.5 Dimensions and Power Consumption

Let us imagine the need for a strictly nonblocking switch fabric with N input and output ports.

To implement this switch fabric with a Clos architecture, we have to satisfy the following equations: $nr = N$ and $m > 2n - 1$.

Working with the limit values, it results in $n = r = \sqrt{N}$, $m = 2\sqrt{N}$. Starting from these figures, the overall number $C(N)$ of crosspoints of an $N \times N$ Clos network is

$$C(N) = 6N\sqrt{N} \tag{7.7}$$

It results that, for a number of input/output ports greater than 36, a strictly nonblocking Clos switch fabric has less crosspoints than a crossbar switch of the same dimension. This difference becomes greater and greater while N increases, and for big fabrics it often makes the difference between a practical solution and an unpractical one.

For example, let us consider a dynamic switch fabric with input signals at 5 Gbit/s and realized with electronics crosspoints having a power consumption of 1 pJ/bit (i.e., a realistic number for 22 nm CMOS crosspoints).

Let us also imagine that connection between crosspoints are active elements (that again reflects the characteristics of CMOS technology) that have a power consumption of 0.5 pJ/bit.

Moreover, let us assume that the traffic load is $q = 0.5$, so that the average percentage of active connections among crosspoints is 50%.

Last, but not least, in order to make a realistic calculation of the overall power dissipated by the switch fabric, the power dissipated by the controller has also to be evaluated. It can be divided into two terms: the power dissipated by the processor, that is almost constant and we assume equal to 50 W, and the power dissipated by the unit commanding the

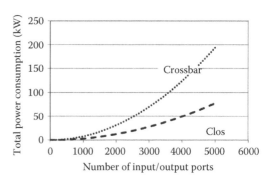

FIGURE 7.9
Power dissipated by a Clos switch fabric versus the power dissipated by a crossbar switch fabric with the same number of ports.

fabric, that is proportional to the sum of the number of crosspoints and the number of active connections through a coefficient of 0.2 pJ/bit.

In Figure 7.9, the power dissipated by a Clos switch fabric is compared with that dissipated by a crossbar switch fabric with the same number of ports.

Due to the different number of crosspoint of the two switch fabrics, the difference in the dissipated power increases thereby increasing the number of ports. A number of ports on the order of 1000 coincide with an overall switch capacity of 5 THz, on the order of magnitude of the capacity of a practical OXC switching 10 Gbit/s channels where the matrix has to be continuously fed with power to maintain the state. In this case, a Clos fabric dissipates about 1.4 kW, that is a great power, but still manageable from a standard optical platform. For example, such a switch fabric could fit in a standard ATCA sub-rack, since it can dissipate up to 200 W per slot and has a maximum of 12 slots. Considering that one slot will be occupied by the sub-rack controller and one from the card managing the connection of the local agent with the system central controller, 10 slots remain for a total dissipated power of 2 kW that is sufficient to host the Clos switch fabric.

A crossbar dissipates 8 kW, a power that would require nonstandard cooling methods and would fit with difficulties on a standard telecom platform, as demonstrated by the previous example.

Increasing the number of ports provides many more advantages for the Clos fabric. For example, if we consider an OXC switching 40 Gbit/s channels, a switch fabric with a capacity of 20 THz is needed, that is, a fabric with 4000 ports.

A Clos fabric dissipates 10.5 kW so that it is still possible to host it into five ATCA sub-racks occupying one rack and a half without any nonstandard power management.

A crossbar matrix dissipates 124 kW, which is simply too much for this kind of platforms.

7.2.2.6 Clos Switch Fabric Modularity

The price to pay for a small number of crosspoints with respect to a crossbar fabric is a more difficult scalability of a Clos fabric.

As a matter of fact, while similar crossbar elements can be combined easily to obtain a larger crossbar, this is not true for a Clos fabric.

Since it is impossible to increase the number of ports of a Clos switch fabric by maintaining the Clos architecture, other scaling strategies have to be found.

A trivial way of attaining a larger fabric is to use the same algorithm used for crossbars, since from a functional point of view, a Clos nonblocking switch is exactly equivalent to a crossbar switch.

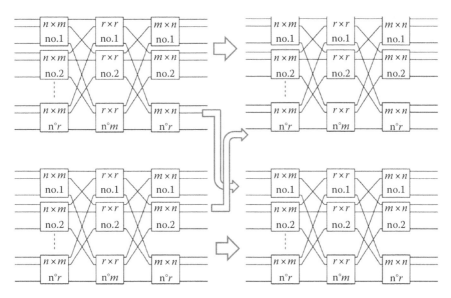

FIGURE 7.10
Four $N \times N$ Clos switches combined to obtain a $2N \times 2N$ nonblocking switch fabric.

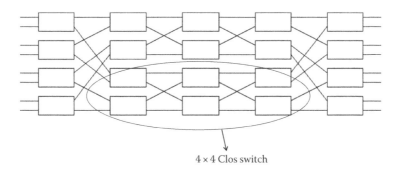

4×4 Clos switch

FIGURE 7.11
Example of the recursive way to extend a rearrangeable nonblocking Clos fabric building a rearrangeable non-blocking Benes fabric.

Thus, four $N \times N$ Clos switches can be combined as shown in Figure 7.10 to obtain a $2N \times 2N$ nonblocking switch. This technique, however, generates a quadratic growth rate of the switch crosspoints with the number of ports, gradually attenuating the advantage of the Clos fabric on the crossbar.

Another recursive way to extend a rearrangeable nonblocking Clos fabric is shown in Figure 7.11 for the case $n = 2$, $r = 2$, $m = 2$; what is obtained in the case is not another Clos fabric, but a rearrangeable nonblocking Benes fabric [17].

7.2.3 Banyan Switch Fabric

We have studied the Clos architecture in detail since this is widely used in practical switching equipment both for static and for dynamic applications.

We have also seen that, even if the number of crosspoints is smaller than that characterizing the crossbar switch, it is still on the order of $N^{3/2}$, increasing rapidly with the number of input/output ports N.

Moreover, the fact that the crosspoints have to be independently operated to create the required routes inside the switch requires the presence of a routing processing in order to satisfy connection requests across the switch.

The presence of a processor allows different priority strategies to be implemented and gives flexibility at the switch fabric, but also implies slower operation with respect to embedded function coded somehow in the hardware itself.

A structure that does not have these two limitations is that of Banyan switch [17].

The Banyan switch is based on the concept of perfect shuffle permutation. The perfect shuffle is a transformation that can be operated on a discrete set of ordered elements and its properties are deeply studied both for their intrinsic mathematical importance and for their large variety of applications [18,19].

In order to make a perfect shuffle permutation, the set of elements to be permutated has to be composed by p^d elements, with p a prime number and d a positive integer number.

Since d is finite, it is possible to enumerate these objects and to represent the corresponding number with a p-ary string (e.g., binary if $p=2$).

At this point, a perfect shuffle permutation can be defined as the permutation that exchanges the element whose representation is $b_1, b_2, b_3, \ldots, b_d$ with the element $b_2, b_3, \ldots, b_d, b_1$. In other words, this transformation operates an ordered left-side rotation of the bits in the element p-ary representation.

An example of perfect shuffle permutation is given in Figure 7.12.

The perfect shuffle permutation allows us to design an iterative construction mechanism that gives rise to the class of Banyan switch fabrics.

In general, the base element of a Banyan switch is a $p \times p$ crossbar matrix, where p is a prime number that is also assumed as the basis to code inside the switch fabric the addresses of the input and output ports. In this way, a connections request can be represented as two strings of elements in an alphabet of p elements: the addresses of the input and output ports to be connected. This representation, useful both in the dynamic and in the static case, also allows an automatic routing procedure to be devised that involves almost no computation from the local processor.

In order to be concrete, we will assume $p=2$, but the extension to other values of p is trivial.

With $p=2$, the building block of a Banyan network is a 2×2 crossbar switch that generally is assumed to have four states, as shown in Figure 7.4. Naturally, a similar building block is also suitable for multicasting into the switch fabric.

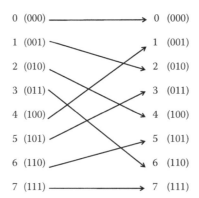

FIGURE 7.12
An example of perfect shuffle permutation.

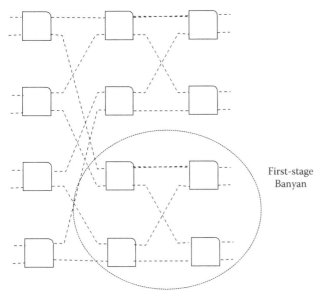

FIGURE 7.13
Scheme of the 8 × 8 Banyan network built starting from two 4 × 4 basic elements.

The input and output ports of the building blocks are tagged with the numbers from 0 to $p-1$ (0 and 1 in our case).

The first step of the Banyan switch creation generates a 4×4 Banyan switch by connecting a stack of two building blocks (the first stage) with a stack of two building blocks (the second stage) with a perfect shuffle connection. The scheme of the 4×4 Banyan network is reported in Figure 7.13 inside the circle.

After the first stage, the procedure can be iteratively repeated building at stage kth a $2^{(k+1)} \times 2^{(k+1)}$ Banyan switch by connecting with a perfect shuffle a first stage composed by $2^{(k-1)}$ building blocks and a second stage composed by two Banyan switches that have $2^k \times 2^k$ ports. The result of the second step of this procedure is also reported in Figure 7.13.

7.2.3.1 Routing through a Banyan Network

Automatic routing is one of the most useful characteristics of a Banyan network [20]. Let us imagine that the address of a connection request arriving at an input port is expressed as a number of k elements in base p (k bits in our particular case).

The routing through the switch fabric happens in this way: the building block of the stage ith reads only the ith bit of the address and switches the connection up if this is 0 and down if it is 1 (in case $p > 2$, the only thing to remember is that the building block would have p outputs).

The scheme of the obtained routing is reported in Figure 7.14.

In a dynamic application, the arbiter is always needed to solve conflicts and to assign local priorities, but no calculation is needed to solve the routing problem.

This property makes Banyan networks among the faster switch fabrics.

The automatically calculated route, however, is the only route existing between the input and the output port; thus, the switch has no mechanism helping to attenuate the impact of a building block failure.

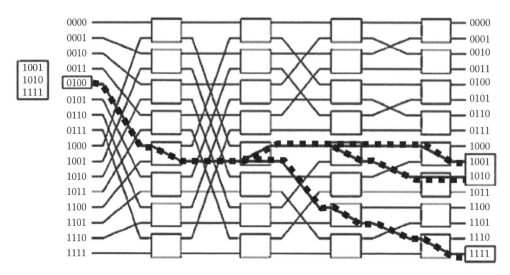

FIGURE 7.14
Example of routing in a Banyan network.

7.2.3.2 Modularity of a Banyan Network

A binary Banyan network has a base 2 modularity; thus, when the $2^k \times 2^k$ ports switch is no more enough, it is possible to step up only toward the $2^{(k+1)} \times 2^{(k+1)}$ switch core.

However, this operation can be performed simply adding new elements to the old switch without the need of changing the inner old switch structure.

7.2.3.3 Real Estate and Power Consumption of a Banyan Network

A $2^k \times 2^k$ Banyan switch fabric is composed a k stages, each of which comprises $2^{(k-1)}$ building blocks. Calling, in general, N the number of input/output ports and $C(N)$ the number of crosspoints and observing that every building block comprises four crosspoints results in

$$C(N) = 2k\, 2^k = 2N \log_2(N) \tag{7.8}$$

The growing pace of the number of crosspoints for a Banyan switch fabric is even slower than that of a Clos switch fabric. This is essentially due to the absence of redundancy in the switch: once an output port is individuated, the route from the input to the output is unique.

7.2.3.4 Variation on Basic Banyan Networks

Since Banyan networks have a very important set of characteristics, a large amount of work has been devoted to find even more performing interconnection networks starting from their base architecture.

The main targets for this research are as follows:

- Improving resilience by introducing more than one path between selected input and output without ruining the self-routing characteristic

- Distributing the memory among elementary switches inside the interconnects to improve dynamic performances
- Reducing the probability that the single path between selected input and output is blocked by other connections inside the switch

It is not within the scope of this brief review of Banyan networks' properties to analyze the main results of this research, which, however, have an impact on several aspects of real switch fabrics design.

Among the relevant references to start a detailed analysis there are [21–25].

7.3 Time Division Switch Fabrics

Up to now, we have considered space division switch fabrics. This means that in the switch fabrics we have considered the connections between a set of spatially divided input ports and a set of spatially divided output ports that were realized by configuring a switch fabric where signals coming from different inputs are routed through spatially separated routes.

This is not the only way to perform switching, that is, to establish and tear down connections among sets of input and output ports.

Another possible way is using time division multiplexing (TDM) and demultiplexing, that is, operating in the time domain [7].

7.3.1 Time Slot Interchange–Based Switch Fabrics

In order to introduce time domain switching, let us introduce with a bit of formality the simpler type of TDM multiplexing and demultiplexing. Much more complex methods are used when multiplexing is used to better exploit the transmission channel, and the reader is encouraged to look for the SDH/SONET description in Chapter 3 and Appendix A and to the related references.

In order to introduce time domain switching, our simple description is sufficient.

A TDM frame is a timely periodic sequence of N time slots, each of which is divided into a header field and a data field. We will assume that the header's field is composed of n_H bytes and that it contains the identifiers (IDs) of all the tributary channels multiplexed in the frame besides other information, for example, tributary channels nature (IP, Ethernet, …), FEC, and so on. We will also assume that the data field is composed by n_D bytes, divided into N blocks, each of which hosts data belonging to one of the N raw channels multiplexed in the TDM frame.

The frame efficiency is $\xi = n_D/(n_D + n_H)$, indicating the amount of bytes of the frame devoted to data among the overall frame capacity.

If each raw channel has a bit rate R_T, the TDM multiplexed signal has to have a bit rate equal to

$$R = \frac{N}{\xi} R_T \tag{7.9}$$

The operation of a TDM multiplexer is schematically shown in Figure 7.15. The working of the TDM multiplexer is divided into two phases: first the incoming tributaries are

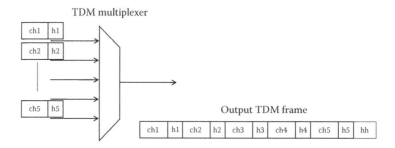

FIGURE 7.15
Scheme of the operation of a simple TDM multiplexer. In the figure, ch1, ch2, … are the payload and h1 … h2 the headers of the individual tributaries while hh is the header of the TDM frame.

synchronized with the multiplexer clock, which will be the multiplexed signal clock, then the tributary data are inserted into the suitable frame slot besides the header, whose content is produced by the multiplexer.

Generally, the incoming tributaries are not only asynchronous, but they are also affected by time jitter. The only way to synchronize different fluxes and compensate the time jitter is to operate with variable delays to align the bits.

Since it is not possible to anticipate a flux, delays increase more and more up to the moment in which an overall delay equal to a bit duration is accumulated.

At this point, we will assume that the multiplexer inserts a void bit, operating the so-called bit stuffing. The positions of void bits are indicated in the frame header so that they can be eliminated at the demultiplexer.

If the multiplexer creates the frame, the demultiplexer divides again the tributaries, sending them on separated output lines.

In general, TDM multiplexing has also to manage the jitter due to the presence in different equipment of different clocks. In our case, the multiplexer and the demultiplexer will belong to the same equipment; thus, this problem will not be present.

Simple TDM multiplexers and demultiplexers, like those considered in this chapter, can be realized with fast memories, where the memory is written in parallel and read in series for the multiplexer and the inverse for the demultiplexer.

Besides the multiplexers and demultiplexers, the base element of a time domain switch is the time slot interchanger (TSI) [26]. The scope of the TSI is to interchange two given time slots in the TDM frame without doing other changes.

The working of a single TSI is shown in Figure 7.16. A TSI can be realized with a simple high-speed memory that is written sequentially and read selectively so as to reproduce at the output the wanted order of the time slots [27].

The equivalent of an $N \times N$ crossbar matrix can be realized with a TDM multiplexer, a TDM demultiplexer, and N TSIs as shown in Figure 7.17.

This is a simple consequence of a more general equivalence between switch matrixes realized using different types of multiplexing as detailed in [28].

Since both the TSI and the TDM multiplexer and demultiplexer can be realized with memories, the required memory is a measure of the scaling pace of the time domain switch with the number of ports.

Assuming to switch channels at the same bit rate, the number B of memory bytes is

$$B = B_M + B_D + B_T = 3N(n_D + n_H) \tag{7.10}$$

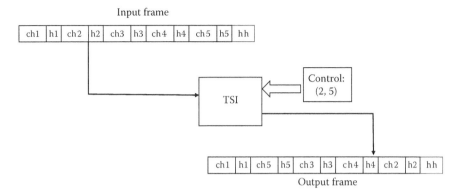

FIGURE 7.16
Scheme of the working principle of a TSI. In the figure, ch1, ch2, ... are the payload and h1 ... h2 the headers of the individual tributaries while hh is the header of the TDM frame.

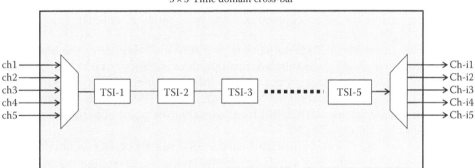

FIGURE 7.17
Time division switch equivalent to an $N \times N$ crossbar matrix realized with a TDM multiplexer, a TDM demultiplexer, and N TSIs.

where
   B_M indicates the memory used in the multiplexer
   B_D is the memory used in the demultiplexer
   B_T is the memory used in the TSIs

   Equation 7.10 shows that the complexity of the time domain crossbar is linear with the number of ports, contrary to the complexity of the space division crossbar whose number of crosspoints grows with the square of N.

   This advantage of the time domain crossbar is paid with the need of a much faster electronics. If the passband of the space division crossbar elements has to be at least equal to the bit rate of the signal, in the case of the time division switch the speed at which the memories have to work is the multiplexed frame bit rate that is given by Equation 7.9 and that scales linearly with the number of input ports.

   Thus, the limit of a time division crossbar based on TDIs is the speed of the used memories.

   Just to give an example, using CMOS circuits with a clock at 20 GHz, it is possible to realize a 16×16 switch where the tributary channels have a bit rate of 1.2 Gbit/s and the frame efficiency is 0.96.

7.3.2 Bus-Based Switch Fabrics

The use of TDIs is not the only way to realize time domain switching; it is a technique largely diffused in static switching applications, like SDH/SONET ADM, but practically unused in dynamic switching machines.

As a matter of fact, even if there is no general principle against the use of a switch based on TDIs in a dynamic switching application, the need of managing completely asynchronous incoming packets of different length creates a lot of additional complications in the switch design.

For this reason, in almost all dynamic switching applications, the bus-based time domain switch is used. This type of time domain switch fabric, reproducing the architecture of a bus network, is schematically represented in Figure 7.18 [29].

The input packets are stored in the switch fabric input queues, every queue is connected to a bus terminal that either in a slotted or in an asynchronous way transmits the packets in the queue through the bus after tagging them with the address of the output port where the output queue hosts the packet before sending it out of the switch fabric.

Naturally, the performance of such a time division switch is strongly dependent on the type of bus used inside the switch. However, a few key elements are independent from the particular bus implementation.

The most relevant of these characteristics is probably the behavior of the bus throughput. Let us indicate with $\eta(T_{in})$ the relative throughput of the adopted bus, that is, the ratio between the number T_{out} of packets per second that exit from the output ports and the number T_{in} of packets per second offered at the input ports. The dependence of $\eta(T_{in})$ on T_{in} depends on the bus implementation, but from a qualitative point of view has always the shape represented in Figure 7.19 [29].

The shape of this curve can be justified easily. As a matter of fact, at small values of offered traffic, the probability that a collision occurs on the bus is negligible and almost all the incoming packets go out of the bus without the need of retransmission.

Increasing the offered traffic, collisions are more frequent and the throughput is no more linear with the traffic, up to a point where a maximum throughput is reached, that is called saturation throughput.

For higher values of the traffic the throughput cannot increase and generally decreases, increasing the offered load due to the fast increase of the collision probability bringing the bus in a congestion state.

In order to present some examples that can give an idea of the effectiveness of the bus-based switch fabric, we will carry out a specific analysis of a few representative examples of bus-based switches.

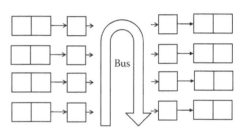

FIGURE 7.18
Dynamic time division switch fabric based on a bus architecture.

Input queues   Input terminals   Output terminals   Output queues

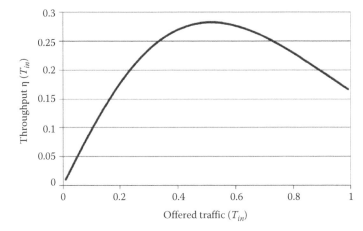

FIGURE 7.19
Qualitative behavior of the throughput of a generic bus versus the offered traffic.

In any case, we will do the approximation that the input queue overflow can be neglected, so that the entire traffic offered to the switch fabric is offered to the bus.

7.3.2.1 Switch Fabric Based on a Slotted Random Access Bus

In this case, we imagine that all the packets have the same duration, that is, all the packets are composed by n_D bytes and travel at a bit rate R_T.

All the packets in the switch fabric input queues are synchronized by inserting suitable delays and the TDM frame in the bus is divided into slots of suitable length so as to contain exactly a packet at the bus transmission speed R, taking into account also the tag added to the packet by the bus access point to direct it to the correct output and that we assume to be made by n_H bytes. The time slot length is thus $t_s = (n_D + n_H)/R$.

Each bus access point simply transmits every packet arriving at the head of the input queue as soon as the synchronization is realized. No communication exists between different access points.

If it happens that combining the arrival time of different packets and the time needed to synchronize and transmit them, they overlap in time, a collision happens. In this case both packets are completely ruined, condition that can be revealed at the receiving point via a suitable error detection code inserted in the packet tag.

Generally, the nth access point is integrated within the same chip with the nth detection point, so that all the information on the sending activity from the nth input is shared with the nth detection point.

Thus, when the nth detection point individuates a collision in the time slot where the corresponding access point has placed a packet, it can immediately send the information to the access point that selects a random waiting time and then retries the transmission. The random waiting time is used to decrease the probability of reiterated collisions between the same access points.

In order to evaluate the maximum throughput, let us now image that the offered traffic is so high that in any case there is always a packet waiting for transmission at each of the input switch fabric queues and that the probability that a bus access point transmits into a given time slot is p [29,30].

The probability that k access points among the N contemporary transmit in the same time slot is, thus, a binomial distribution that writes

$$P(k) = \binom{N}{k} p^k (1-p)^{(N-k)} \tag{7.11}$$

The relative throughput η coincides with the probability P_s that an access point transmits its packet successfully that is the probability that it is the only access point to transmit in the time slot. Thus,

$$\eta = Np(1-p)^{(N-1)} \tag{7.12}$$

A packet arriving at the input of the switch fabric has a duration $t_{in} = n_D/R_T$, containing $m = t_{in}/t_s$ time slots of the bus frame. Assuming that synchronization of a packet arriving at the head of an input queue happens in a random time slot within the following packet duration, since a packet is always present at each queue head, an approximated expression of p is

$$p = \frac{1}{m} = \frac{n_D + n_H}{n_D} \frac{R_T}{R} = \frac{R_T}{\xi R} \tag{7.13}$$

Moreover, in order to make possible a successful transmission attempt is has to be $N < m$. At the end, assuming exactly $N = m$, to reflect the high traffic regime (a packet is always present at a queue head) the saturation throughput is given by

$$\eta = N \frac{R_T}{\xi R} \left(\frac{\xi R - R_T}{\xi R} \right)^{(N-1)} \tag{7.14}$$

The throughput given by Equation 7.14 is plotted in Figure 7.20 versus the bus bit rate R. In order to remember the fact that changing R also the dimension of the switch fabric N changes, N is also reported on the abscissa. The input/output line cards bit rate is assumed 1.2 Gbit/s.

In order to evaluate the frame efficiency it is assumed $n_D = 512$ bytes, $n_H = \log_2(n_D) + 4$ bytes, where the two terms in which the expression of n_H is divided correspond to the address of the output port and to the error detecting code redundancy.

From Figure 7.20 it results that the dependency of the saturation throughput on the bus bit rate in the high traffic conditions is very low and the same happens with all the other parameters.

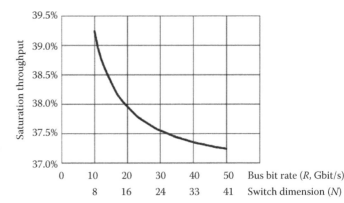

FIGURE 7.20
Saturation throughput of a time division switch fabric based on a random access slotted bus versus the switch fabric dimension N (or equivalently the bus internal bit rate R). The input/output line cards bit rate is assumed 1.2 Gbit/s.

The same conclusion can be obtained by noting that there exists a value of p maximizing Equation 7.12. Deriving with respect to p and substituting in Equation 7.12 the abscissa of the maximum, it is found that the maximum efficiency is attained for $p = 1/N$ and it is

$$\eta_{max} = \left(1 - \frac{1}{N}\right)^{(N-1)} \tag{7.15}$$

so that, for large values of N the throughput tends to $\eta_{max} = 1/e = 36.8\%$ as shown in Figure 7.20.

The information regarding the saturated throughput is quite important since in a functional system the offered load has to be smaller, and in general much smaller, than the saturated throughput. As a matter of fact, the queues of a realistic system are not infinite and an offered load greater than the saturated throughput will cause the average number of stored packets to increase continuously till it generates system congestion due to the inability of the switch fabric to accept new input packets.

In the example of Figure 7.20, fixing the bus bit rate at 20 Gbit/s and consequently N to 16, assuming that the realistic traffic that the switch fabric can manage is half the saturated throughput and that the traffic coming from different inputs is statistically similar and independent, it is obtained that the maximum offered load that the switch can manage with a minimal packet loss probability is on the order of magnitude of 445,000 packets per second per input port that is an average capacity of 216 Mbit/s per port, even if the line rate on the input line cards is 1.2 Gbit/s and the bus bit rate is as high as 20 Gbit/s.

7.3.2.2 Switch Fabric Based on an Unslotted Random Access Bus

In the case of the unslotted random access, the bus access policy is the same as in the precedent case, but no frame exists on the bus and every access point transmits the packets on the shared media in an asynchronous way.

We will maintain the hypothesis that the packets are all of the same length and are transmitted on input lines with the same speed. The hypothesis of heavy load is also maintained.

Let us fix an arbitrary time window of duration equal to t_s, that is equal to the packet duration at the bus bit rate, and let us call p the probability that a generic access point will start to transmit a packet within the interval.

If two or more access points will start to transmit within the interval, the packets will overlap generating collision, so that the probability of a successful transmission coincides with the probability that, starting one access point the transmission at a certain instant, no other access point will start to transmit in a time interval of duration t_s.

Summarizing, three conditions are needed to have a successful transmission [30]:

1. An access point starts to transmit in the instant t.
2. No access point is transmitting at the instant and no access point started transmission within the interval $(t - t_s, t)$.
3. No access point will start a transmission in the interval $(t, t + t_s)$.

The three conditions are independent; thus, the overall transmission success probability is the product of the three probabilities, that we will call P_i, $i = 1, 2, 3$.

The individual probabilities are easily evaluated using a model similar to that used in the previous section, since the bus access points are assumed independent. At the end, we obtain

$$\eta = P_1 P_2 P_3 = Np(1-p)^{(N-1)}(1-p)^{(N-1)} = Np(1-p)^{2(N-1)} \tag{7.16}$$

In this case, the probability p cannot be evaluated as in the slotted case, since no synchronization occurs. Optimization with respect to p can be done however in the same way, deriving (7.16) and finding the value of p that corresponds to the maximum value of η.

It is obtained that the maximum efficiency is attained for $p=0.5/N$ and it is

$$\eta_{max} = \frac{1}{2}\left(1 - \frac{1}{2N}\right)^{(2N-2)} \tag{7.17}$$

so that, for large values of N the throughput tends to $\eta_{max} = 1/2e = 18\%$.

7.3.2.3 Switch Fabric Based on a Carrier Sense Multiple Access Bus

Random access busses are very simple to implement and their performances are independent of the packet duration and the propagation delay; however, other possible bus protocols exist that are nearly as simple to implement and can significantly improve the throughput performances of the random access bus by exploiting the fact that the propagation delay along the shared medium is very short in the case of a switch fabric because the entire network is enclosed into a single equipment.

Carrier sense multiple access (CSMA) is one of these protocols that is extremely diffused for its application in Ethernet local area networks (LANs): we have informally introduced this protocol in Chapter 3 talking about Ethernet and now we want to analyze its working and performances in a more detailed manner. One of the aspects that renders CSMA so attractive for use in bus-based switches is that it is completely standardized [31] and widely used, so that adopting it hides a low risk and assures the possibility to rely on large volume electronics components.

A CSMA bus access point having a packet to transmit along the shared medium, instead of transmitting it as soon as possible, before transmission uses its receiving side to sense the activity along the shared medium (remember that the access point and the detection point corresponding to the nth input/output port are generally integrated together).

If some signal is detected along the bus, the access points deduce that other access points are transmitting packets and do not transmit up to the first useful moment in which no signal (no carrier) is present in the bus.

The intent of this procedure is to reduce the collision probability so increasing the bus throughput.

However, different from what seems as a superficial analysis, CSMA protocol does not completely avoid collisions, due to the presence of finite propagation delays along the bus. As a matter of fact, in the moment in which a certain access point senses the absence of carrier on the medium and starts to transmit it is possible that also another access point in a remote point of the bus is starting to transmit a packet whose leading edge is not yet arrived to the first access point.

Thus, a newly generated packet in a CSMA bus is vulnerable to collision during the time needed to propagate its leading edge over all the bus length and so inform all the access points of its presence. This time is called bus round trip.

Since the collision probability can be written as the probability of collision during the round trip multiplied for the ratio between the round trip and the packet duration, increasing the packet duration in a given CSMA bus causes the saturated throughput to increase monotonically up to 100% for negligible round trip with respect to the packet duration.

Moreover, if a collision occurs, any access point detects it within a round trip time and can stop any active transmission to end the collision as soon as possible and recover media availability.

This last technique is called collision detection (CD) and the protocol adopting CSMA and CD is called CSMA-CD.

As in the case of random access busses, CSMA-CD busses can also be slotted or unslotted, depending on the fact that the input packets are synchronized on a slotted bus frame or are transmitted on the bus in an asynchronous way.

From now on we will consider switch fabrics based on slotted CSMA-CD busses due to their high performances.

To evaluate analytically the throughput of a CSMA-CD bus-based switch fabric, we will assume again that the input queues are of infinite length and that high traffic conditions are verified so that a packet is always present at an input queue head [30]. Moreover, in order to evaluate the worst case, we will assume that all the access points are at one of the bus extremes and all the detection points at the other, so that any packet, independent from the origin and the destination, experiences a round trip of 2τ.

We will also assume that the time slot on the bus frame is exactly equal to 2τ. This only means that the length of a packet in the bus bit rate has to be an integer multiple of 2τ, that is, practically no restriction, considering that the packet length will be designed to be much larger than the round trip time.

At the beginning of each time slot, in a small service time $\tau_s << \tau$, where no transmission is possible, each access point senses the shared medium and, if no carrier is detected, starts a transmission with probability p.

If two or more stations attempt to access the bus in the same slot, in the successive slot a collision is detected by all the access points and any ongoing transmission is immediately stopped.

At the beginning of an available time slot (i.e., a time slot where no transmission is ongoing) the probability that one station begins a successful transmission is equal to the probability P_s that no other station starts to transmit in the same time slot. It writes, with the usual notations,

$$P_s = Np(1-p)^{(N-1)} \tag{7.18}$$

Generally, collisions can occur; thus, the number J of time slots waited by a packet at an input queue head before a successful transmission is a random variable; and since successive transmission attempts in different time slots are independent, it is a geometrically distributed random variable. This means that we can write

$$P(J = k) = P_s(1-P_s)^{(k-1)} \tag{7.19}$$

The total time $\Im$ elapsing from the completion of the transmission of a packet and the completion of the transmission of the following packet is also a random variable. The average of $\Im$ can be written as

$$\langle \Im \rangle = 2\tau \langle J \rangle + t_s = 2\tau \sum_{k=1}^{\infty} kP_s(1-P_s)^{(k-1)} + t_s \tag{7.20}$$

summing up the geometrical series and simplifying it is obtained

$$\langle \mathfrak{I} \rangle = \frac{2\tau}{P_s} + t_s = \frac{2\tau}{Np(1-p)^{(N-1)}} + t_s \tag{7.21}$$

Finally, the throughput is given by

$$\eta = \frac{t_s}{\langle \mathfrak{I} \rangle} = \frac{1}{1 + \left(1 / \left(Np(1-p)^{(N-1)}\right)\right)\left(2\tau / t_s\right)} \tag{7.22}$$

In Equation 7.22, the role of the ratio between the round trip and the packet length at the bus bit rate is clearly evidenced.

We know that the efficiency given by Equation 7.22 is maximized if the denominator is minimized. For a given ratio $2\tau/t_s$, it is verified if the term $Np(1-p)^{(N-1)}$ is maximum, thus for $p = 1/N$, obtaining

$$\eta_{max} = \frac{1}{1 + \left(1 - (1/N)\right)^{-(N-1)} \left(2\tau / t_s\right)} \tag{7.23}$$

and for large values of N the equation for the throughput further simplifies as

$$\eta_{max} = \frac{1}{1 + (e-1)\left(2\tau / t_s\right)} \tag{7.24}$$

It is to be noticed that when applying Equation 7.23 to a switch fabric implemented with today's available technology, assuming that the input queues are sufficiently long to avoid congestion for the impossibility of accepting packets at the input, the result is quite different with respect to the well-known LAN case.

Let us assume to have a 16×16 switch fabric whose bus is located in the backplane of an equipment and implemented with an optical star coupler.

Assuming that each transmitting laser transmits at 10 Gbit/s in the second window (see Chapter 3) emitting a modulated power of 1 dBm.

The overall loss of a 16×16 optical star coupler is about −15 dB for the coupler and 1 dB for propagation, so that a standard PIN detector is enough to detect the signal at 10 Gbit/s without noise problems (compare Section 5.4 from Chapter 5).

The round trip time, assuming the star coupler to be $L = 2$ m long can be written as

$$2\tau = \frac{2L}{c} + \tau_e \tag{7.25}$$

where
  c is the speed of light
  τ_e is the electronic elaboration time, that is, the time the electronic processor of the access point needs to carry out the carrier sense

Since $2L/c = 6.6$ ns, the round trip time is dominated by the electronic elaboration time. Assuming that the access point needs to observe up to 5 bits to decide if the carrier is present, we can write $2\tau \approx \tau_e = 50$ ns.

Let us also assume that the packets are 512 bytes long, that is, $t_s = 41\,\mu s$, we have, from Equation 7.22, $\eta_{max} = 0.994$.

Following the same line already considered for the random access bus–based switches, we can say that the maximum real traffic that a bus-based switch fabric can manage is half the saturated throughput divided by the number of ports.

In our case, we can assume that the line card bit rate is 622 Mbit/s while the average traffic manageable by each input port is $0.5\eta_{max}\,R/N \approx 300$ Mbit/s.

This number can be compared with the figure evaluated in the example at the end of Section 7.3.2.1, where a 16×16 switch fabric based on a random access slotted bus with an internal speed as high as 1.2 Gbit/s and line cards transmitting at 1.2 Gbit/s were able to manage only a traffic on the order of 200 Mbit/s per input port.

The comparison between these simple examples clearly shows the superiority of the CSMA-CD bus when used to build a switch fabric designed to work in a condition of high traffic.

Naturally, if the high traffic condition is not verified, having a higher throughput is less important and in these situations other parameters can influence the design of the switch fabric, like the cost. In this case, it could be possible that the simple random access bus is the preferred choice.

7.3.2.4 Switch Fabric Based on Variations of the Carrier Sense Multiple Access Bus

Up to now we have analyzed the base version of the switch fabric based on CSMA-CD bus, but several variations exist to optimize the switch fabric performances.

These variations are deeply studied in the field of LAN networks and we will simply describe some of them qualitatively, commenting on their performances without presenting an analytical derivation.

The reader interested in the mathematical models is encouraged to look at the immense repertory of technical literature on LANs, taking into account, however, that some elements are quite different in the case in which a bus protocol is used in a switch fabric, as demonstrated in the previous section [29].

In order to improve the CSMA-CD performance in the case of bus-based switch fabrics, it is possible to adopt different strategies:

- *Exploiting the fact that the propagation time along the bus is particularly short*: We have seen that this is a key element allowing the collision probability to be reduced in a standard CSMA-CD bus. In the case of a switch fabric, however, the propagation along the bus is so short that a coordination among the bus access points could be conceived.

  This coordination can be based on the transmission of an access point of a "transmitting bit" in a reserved slot of the bus frame. This allows the other access points to be informed of the intention of one of them to transmit, so as to avoid collisions.

  Naturally, when more than one access point has booked a transmission, a round-robin mechanism is needed to state the order in which they have to access the bus.

- *p Persistent bus protocol*: In this case, applicable only to slotted busses, when the CSMA-CD protocol senses no carrier on the bus, an available packet is transmitted in the available time slot only with probability p, while with probability $(1 - p)$ the access point waits a time slot to perform transmission.

- *When the inner bus structure is realized via optical transmission, WDM multiplexing can be exploited to increase the inner bus capacity*: This can be done in several ways, both with respect to the used WDM comb and with respect to the adopted protocol. The CSMA-CD protocol is almost the same that is used in single-channel solutions, but for the fact that an additional mechanism is needed to select on what wavelength the access point has to transmit. Basically two possible solutions exist: either the access point selects a random wavelength and then accesses it as in a single-channel system, or the protocol first selects the channel while sensing all the available wavelength and then it uses the standard access method on the selected channel.

  Multichannel switch fabrics based on a WDM inner bus would be very attractive due to the possibility of pushing the bus capacity to very high values. Unfortunately, the practical implementation is thwarted by the cost and complexity of tunable WDM lasers (see Chapter 5) that greatly impact on the practical value of this solution. Naturally, this situation could be completely changed by the availability of low-cost, high-speed tunable lasers, like those that are under development for tunable XFP transceivers.

In order to give an idea of the performance of switch fabric adopting different kinds of internal bus protocols, we will use the performance evaluation model reported in [29] assuming the following hypotheses:

- The switch fabric uses a slotted inner bus.
- The packet arrival process at each input port is a Poisson process.
- All the arrivals processes at different input ports are independent and have the same parameters.
- A packet arriving to a switch fabric output port can always leave the fabric.
- The switch fabric input queues are infinite so that the queue output process is also a Poisson process.
- The electronic elaboration duration dominates the round trip time, being $2\tau \approx \tau_e = 100\,\text{ns}$.
- The inner bus speed is 2.5 Gbit/s.
- The line card speed is 125 Mbit/s.
- The packets are 512 bytes long.

The effect of the persistent property of p persistent CSMA protocols for different values of the persistence probability is shown in Figure 7.21, where the normalized throughput is shown versus the average number of access points contending the transmission medium during a packet time. This parameter is naturally related both to the number of access points, that is of input/output ports of the switch fabric, and to the offered load.

As obvious, decreasing the persistence probability, the performance is quite improved in regime of high collisions up to the disappearance of the congestion regime when the persistence probability is very low.

The effect of the use of several transmission wavelengths is shown in Figure 7.22. The interesting aspect of this figure is that the saturation throughput almost remains the same increasing the number of available wavelengths; this is due to the fact that we use relative measurement units for the throughput so that the real throughput is obtained multiplying η by the offered load.

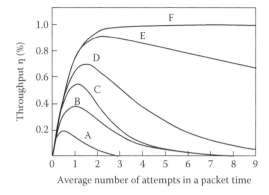

FIGURE 7.21
Comparison among the throughput dependences on the average number of conflicting access points in the case of different bus architectures. A, Unslotted random access; B, slotted random access. From C to F: p persistent CSMA. C, $p = 1$; D, $p = 0.5$; E, $p = 0.1$; F, $p = 0.01$.

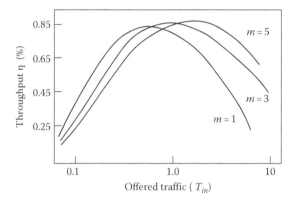

FIGURE 7.22
Relative throughput of a WDM bus-based switch fabric versus the offered traffic for different values of the number of wavelengths.

Thus, having a constant value of η that is reached at higher and higher values of the offered load by increasing the number of wavelengths indicates an absolute throughput increasing almost linearly with the number of channels.

7.3.3 Delay in Bus-Based Switch Fabrics

Analyzing bus structures in the LAN context, after the assessment of the throughput of each bus protocol, the distribution of the delay experienced by the packets while traversing the LAN is evaluated.

This is also an important issue in switch fabrics, but in the last case the overall fabric passing through delay cannot be identified with the bus transmission delay.

As a matter of fact, the overall packet delay while traversing a bus-based switch fabric is the sum of several addenda:

- The packet waiting time in the input queue
- The electronic elaboration time related to packet transmission, which is composed by the time needed to assess the transmission possibility (e.g., to sense the carrier presence on the channel) and the time needed to condition the packet for transmission (e.g., tagging it with the fabric internal header)
- The bus transmission propagation

- The average packet waiting time due to collisions in the shared medium
- The packet waiting time in the output queue (if it exists) before transmission on the line

Thus, a key role in determining the average packet delay and its standard deviation is played by the input and output queues policy and by the queues' length.

For simplicity (and also because it is a very common case) let us assume that there is no output queue in the switch fabric and the packet directly exits toward the line card that, if needed, stores it and launches it out of the switching device.

The switch fabric can then be modeled as a set of parallel queues whose servant is constituted by the fabric bus. In this model, the time elapsing for the packet preparation and physical transmission along the bus plays the role of the service time for each queue.

This model is effective in representing the switch fabric behavior, but it is difficult to manage mathematically essentially because the service processes of each input queue are statistically correlated due to the collision process on the bus.

This property causes also the queue output processes to deviate from a Poisson process, even in the case of Poisson input process and infinite queue length.

Thus, in order to analyze the distribution of the latency caused by the passage through the switch fabric, either some approximation is operated to simplify the problem or simulation is needed.

An effective approximation takes the move from the consideration that the very short length of the inner bus of the switch fabric renders the collision probability very small, especially when bus protocols designed to prevent collisions are used.

In this case, the only factor linking the processes happening in the different input queues is the time τ_w elapsing between the moment in which a packet is ready to be transmitted in the internal fabric bus and the instant in which it is really transmitted.

This time is a random variable depending on the conflict probability. Assuming to analyze a slotted switch fabric, with slot equal to t_s, the distribution of τ_w is given by

$$P(\tau_w = jt_s) = p(1-p)^{(j-1)} \tag{7.26}$$

where p is, as usual, the probability that a ready packet experiences successful transmission in a given slot and it is assumed that no policy is used to favor packets waiting from more time.

The latency induced by the switch fabric is then the sum of four terms:

1. The time T_0 needed to ready the packet to enter into the switch system (e.g., applying the internal switch header)
2. The time T_1 that the newly arrived packet occupies to wait for the current service
3. The time T_2 that the considered packet waits for the service of all the packets in its queue
4. The time T_3 elapsing from the moment in which the packet is first in its queue and ready for transmission to its effective transmission that coincides with the service time

The time T_0 is a constant, related to the electronic elaboration time, and it can be eliminated from the evaluation for the moment.

As far as the times T_1 and T_2 are concerned, they can be evaluated by considering the queue as an M/G/1 queue for which the Pollaczek–Khintchine formulas [32] hold.

In particular, the average time a packet remains in the queue, excluding the service time, is given by

$$T_1 + T_2 = \frac{(\lambda^2 \sigma^2 + \rho^2)}{2\lambda(1-\rho)} \tag{7.27}$$

where
σ^2 is the service time variance
$\rho = \lambda/\mu$ is the ratio between the average service time $1/\mu$ and the average arrival time $1/\lambda$

Naturally, it is assumed that $\rho < 1$.

The average service time and the service time variance can be deduced from Equation 7.26.

In particular the following is derived

$$\frac{1}{\mu} = t_s \langle j \rangle = \frac{t_s}{p} = T_3$$

$$\sigma^2 = t_s^2 (\langle j^2 \rangle - \langle j \rangle^2) = t_s^2 \frac{1-p}{p^2} \tag{7.28}$$

To write the final expression of the latency, it is useful to assume as independent variable the overall traffic offered to the switch fabric, coinciding with the average number of offered packets in the considered slot, whatever the incoming port, that is, $\theta = N\lambda t_s$.

Moreover, it is also useful to refer to the average delay in terms of number of bus time slots, that is, $\delta_t = T/t_s$.

Adopting these variables and substituting (7.28) in (7.27) we have

$$\delta_t = \frac{T_1 + T_2 + T_s}{t_s} = \frac{\theta(2-p)}{2p(Np - \theta)} \tag{7.29}$$

This is a simple expression that presents a vertical asymptote, as expected. As a matter of fact, when the traffic goes near Np, it means that almost every slot a packet is offered to the switch fabric from some input port.

However, the bus cannot manage more than one packet in each slot; thus, it is clear that the average delay gets higher and higher going nearer to the maximum bus load.

We have seen in Equation 7.13 that in conditions of heavy traffic $p = R_T/\xi R$ that approximate the value $1/N$ corresponding to the optimum throughput for the frame efficiency approximating to 100%. Thus, in condition of heavy traffic, Equation 7.29 can be further approximated as follows:

$$\delta_t = \frac{\theta(2N-1)}{2(1-\theta)} \tag{7.30}$$

The average normalized latency is plotted in Figure 7.23 for different values of the normalized total offered traffic and in conditions of high traffic. Different dimensions of the switch fabric are also considered.

As expected, the average delay remains quite low up to a sort of threshold value of the offered traffic, above threshold it increases rapidly up to a vertical asymptote when $\rho = 1$.

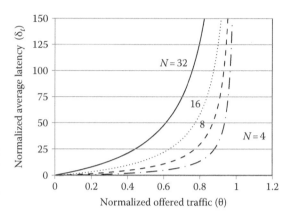

FIGURE 7.23
Average latency normalized to the packet duration (T/t_s) in a bus-based switch fabric for different values of the normalized offered traffic (λt_s) and different dimensions of the switch fabric.

7.4 Wavelength Division Switch Fabrics

Since switching can be realized both in space and in time domain, it is intuitive that switching is possible also in frequency domain [11,33].

Frequency domain switching could be useful especially for optical applications.

In principle an exact equivalent of the time domain switching based on TSI can be realized in the wavelength domain by using wavelength interchangers (WI).

A WI is a subsystem that is able to interchange two wavelengths belonging to a WDM comb without affecting the other channels. From a functional point of view, a WI is the exact equivalent of the TSI in the time domain, but this similarity does not extend to the physical implementation of the device.

As a matter of fact, nothing similar to a random access memory exists in the optical domain and all the processing on a WDM comb have to be performed using analog devices.

Thus, the WI has to be realized using an analog scheme. Different solutions have been proposed, all based on the availability of a wavelength converter, that is, a component that is able to move the carrier of a modulated WDM channel from one wavelength to another.

The most intuitive scheme for a WI is depicted in Figure 7.24. In the figure it is important to notice that the separation of the WDM channels is operated by a low-loss demux, but the reconstruction of the WDM comb by a high loss coupler.

This is due to the fact that the input ports of low-loss demuxes are frequency selective, that is, the frequency arriving at a certain demux port is fixed and cannot be changed.

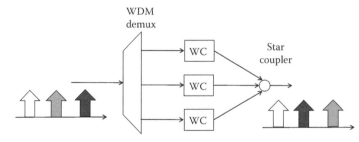

FIGURE 7.24
Functional scheme of a wavelength interchanger based on optical wavelength converters.

Due to this reason they cannot be used in the second stage of the WI, where the incoming signal frequency is not fixed.

At the present technology state, the input/output loss of a practical WI is really high as it can immediately be evaluated, for example, considering a 32λ WI. Assuming that wavelength converters are transparent and summing the loss of 5 dB due to the input AWG to the huge loss of 15 dB due to the 32 × 1 output coupler, a theoretical loss as high as 20 dB is reached, to which in a realistic case additional losses coming, for example, from connectors have to be added.

Such a loss, added to the complex structure of the WI, makes unpractical the use of such devices, even if there is a sufficiently stable technology to realize them.

A different way to realize a WI is leveraging on the integration of different optoelectronic devices onto the same InP-based chip.

In Chapter 5 we have seen that besides other components, also semiconductor optical amplifiers (SOAs) can be realized essentially with the InP laser technology.

If it is not easy to use SOAs for WDM applications, they can work much better as single-channel amplifiers, so that a WI realized with the scheme of Figure 7.25 can be realized using a small number of InP optoelectronics integrated chips (the example of a 32λ WI is considered in the figure). Moreover, several methods exist to integrate a wavelength converter into an InP circuit.

Assuming that an integrated wavelength converter will lose 4 dB, a standard SOA can easily gain 20 dB on a single channel without the need of any particular design trick, so to arrive at a global WI loss of 4 dB that is acceptable for applications.

However, this solution, although interesting, is not practical for two main reasons.

The first is that the SOA presence is paid with the introduction of ASE noise and this reduces the optical propagation performance of the signal, the second, and more important reason is that the cost of a subsystem with the structure of Figure 7.25 is huge with respect to the cost of a simple C-MOS TSI, so that up to the moment in which volumes

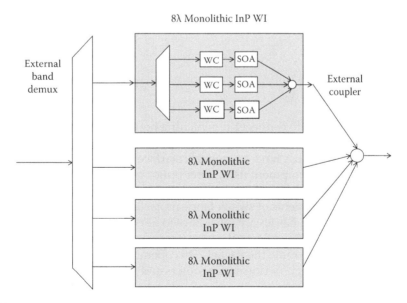

FIGURE 7.25
Functional scheme of a wavelength interchanger based on monolithically integrated subelements.

of the InP technology will be so high to justify a cost at least two orders of magnitude smaller, such a WI is feasible but not practical.

Another possible way to realize a WI is to realize it electronically. This is the most economic design and has the advantage of offering also signal regeneration. However, the final system is much more complex and expensive with respect to a similar time domain switch.

At the end, we can conclude that with present day's technology, switching in the wavelength domain using WI is theoretically interesting, but not practically applicable.

On the other hand, it is not easy to devise a wavelength domain system that is the equivalent of a TDM bus so to devise a wavelength domain dynamic switch.

Generally, the architectures proposed for wavelength domain dynamic switches are completely different and they will analyzed in more detail in Chapter 10.

7.5 Hardware Platforms for Switching Network Elements

At a first glance, the hardware platform of a switching network element is not so different from what has been described in the case of transmission systems in Chapter 6.

However, a more detailed analysis reveals a few key points that differentiate clearly a hardware platform for a transmission system from an hardware platform for a switching system.

The general architecture of a switching system platform is based on the same elements we found in the case of transmission systems: racks, sub-racks, and cards. Naturally, standards relative to racks and sub-racks are the same and, to specify with an example, an ATCA rack can be used in principle to host both a transmission and a switching system [87] (ATCA is a complete standard both for mechanical and thermal design and for network element control architecture).

The cards composing generic switching systems can be classified in four categories:

1. Interface cards
2. Switch fabric cards
3. Processing cards
4. Control and service cards

Interface cards have the role to connect the switching network element with the surrounding network both on the line side (line cards) and on the client side (client cards).

Generally, the interface cards have an optical interface so that the signal is transmitted or received from a nearby equipment using fiber optics patch cords.

The switch fabric cards host the redundant switch fabric. Even if a single card would be sufficient, due to the requirements of switch fabric protection, the two switch fabrics have to be hosted on different switch fabric cards that have to be connected through the platform.

The processing cards are the card containing hardware dedicated processors performing the operations needed to route the packet. This operation has to be performed millions of times a second; thus, in high-performance routers and switches several such cards hosting dedicated hardware are needed.

The control cards are all the cards devoted either to configure the network element (as the rack processor), or to terminate and regenerate the signal on the control and management plane, thus running either procedure forced by the central operation system or automatic protocols. Examples of the control cards are the rack controller and the service channel termination card in equipment having a few interfaces of the WDM type.

The two elements differentiating the platform of switching network elements are the power consumption and the type of backplane that is needed.

Regarding power consumption, the power needed to an electronic switch fabric card is in general much higher with respect, for example, to the power needed to a dense wavelength division multiplexing (DWDM) transponder. Talking about Clos switch fabrics, we have considered an example of a switch fabric of a global capacity of 5 THz switching 10 Gbit/s channels and we have evaluated a total power consumption of 1.4 kW. Distributing the fabric on 10 cards of an ATCA platform (the sub-rack holds 12 cards and we assume that two of them are used for control), we have a power consumption of 140 W per card. This is compatible with the ATCA standard, but gives an idea of the high power consumptions of large electronic switches.

Moreover, in packet switching applications, the line cards have also different functionalities with respect to pure transmission and detection of the signal. Generally, input queues and the first header processing are located in the line card, so that a standard line card of an IP router has a power consumption much higher with respect to a DWDM transponder, using the same transmission technology.

Regarding the backplane, the key point is that, different from DWDM systems, the signal itself have to propagate on the backplane to connect one switch fabric card with the other and to pass from the line cards to the switch cards and vice versa.

If the switch processes signals at 10 Gbit/s, this means that the backplane has to transmit signals at that speed, requiring a technology completely different from that of DWDM backplanes.

7.5.1 Fast Backplanes for Switching Equipment

Since DWDM backplanes are generally based on parallel busses, in principle, it should be possible to increase their capacity by increasing the number of parallel lines constituting the bus.

In practice the number of parallel lines that can be incorporated in a backplane is limited by several elements:

- Increasing the number of lines the interference also increases and shielding problems are harder and harder.
- The cost increases linearly with the number of parallel lines, since this is the way in which increases the number of transmitters and receivers needed to implement the bus, while using serial high-speed transmission the cost increases much less than linearly with the bit rate.
- The number of traces that can be integrated into a board is intrinsically limited by the reliability and the cost of the board creation process.

For all these reasons it is not practical to implement a backplane that has to transmit several signals at 10 Gbit/s using a huge number of parallel low-speed lines.

In order to implement a high-speed serial bus there are two possible technologies: electrical high-speed busses and optical busses.

7.5.1.1 High-Speed Electrical Backplanes

The target of a high-speed serial electrical backplane is to provide electrical busses on the backplane that are able to transmit at least at 10 Gbit/s [34].

This has to be done in a transmission line that can be represented as in Figure 7.26: the transmitter is located on one card, the receiver on another card, and the line is divided between the two connected cards and the backplane, with connectors to join the pieces.

Also, a more complex backplane architecture has been proposed, where the switch fabric is embedded at least in part into the backplane, so that the overall system architecture is simplified [35]. We will not describe this option here, but with the evolution of nanoscale integrated optics on C-MOS platform it could become a quite interesting alternative.

Returning to a backplane allowing pure connectivity among cards, the first choice to be made is the selection of the transmission format. Due to a plethora of reasons, uniformity with other transmitter/receivers applications, simplicity and so on, nonreturn to zero (NRZ) is generally chosen as the preferred modulation format.

Once the modulation format is chosen, all the components that are reported in the scheme of Figure 7.26 have to be chosen so to satisfy both the required performance criteria and the rigid cost requirements that a backplane has to fulfill.

7.5.1.1.1 Transmitter

The transmitter has to be capable to compensate jitter and dispersion introduced by the channel at such a high speed by introducing preemphasis in the transmitted pulses [36,37]. Preemphasis puts more signal energy into the high frequency components of the transmit signal to overcome the channel frequency–dependent loss characteristics. If there is a feedback channel from the receiver to the transmitter, for example, a channel used to deliver feedback from the cards to the controller on configuration operations, it is possible to adapt the preemphasis to the channel characteristics.

In order to assure that the signal is launched into the transmission line in an optimum way, without losing the shape created at the transmitter, the quality of the trace routing on the backplane PCB, near the launch point is crucial. Spurious parasitic elements due to the PCB creation processes and to other fluctuations from the design parameters must be accounted for while designing a 10 Gbit/s transmitter.

An idea of the parasitic elements impacting transmission near the signal launch point and of their origin is provided in Figure 7.27.

7.5.1.1.2 Connectors

Connectors are sources of signal reflections and crosstalk. Interference can be limited by shielding each connector pair with metal plates so to avoid coupling between different signals traveling through the backplane.

However, care has to be taken when adding metal to provide shielding because it can hurt connector performance by reducing the impedance, thus increasing reflections. The

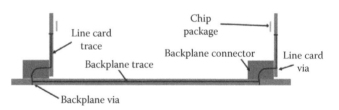

FIGURE 7.26
Elements of a high-speed backplane connection between cards of a switching network element.

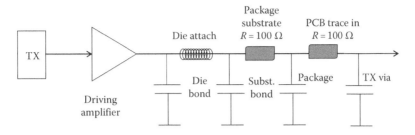

FIGURE 7.27
Example of parasitic elements in the connection between the transmitter and the backplane trace, for each parasitic element the process originating it is indicated.

use of differential signaling can help to provide low crosstalk up to 50 ps rise times that is compatible with a 10 Gbit/s transmission where the NRZ bit duration is 100 ps [38].

An undesirable effect is created in an electrical backplane when an upper layer trace is connected to another upper layer trace in the PCB-created resonant stub. These resonant stubs need to be characterized as microwave structures because they are capable of degrading system performance. As the thickness of the PCB increases, the effect of stubs gets significantly worse because of the increase in length.

The connector's termination to the PCB, generally, constrains both the maximum board thickness due to plating aspect ratios and the maximum trace width that can pass through on a given layer due to spacing. Thus, there is a trade-off between connectors' aspect ratio and fixing technique to the PCB on one side and the PCB thickness itself on the other. This trade-off has to be carefully managed by using microwave characterization of all the relevant structures not to ruin the backplane transmission characteristics.

Additionally, the structure created by the ground plane plate and via drill provides an unwanted effective capacitance that also has to be taken into account when designing the way in which traces passes through the PCB and the connectors are mechanically fixed to the cards and to the backplane.

Last, but not least, the connector reliability has to be assured over all the equipment lifetime, that can be a complex task if the material choice and the connector design are not more than accurate.

7.5.1.1.3 PCB Traces

The traces on the backplane generally account for the longest signal path and contribute the most to the overall loss budget. In order to support 10 Gbit/s signals transmission, a series of measures have to be adopted to improve traces above the standard quality [39].

- Selection of low-loss laminate materials to implement the conduction trace. As a matter of fact, material loss is the dominant effect on 10 Gbit/s channels.
- Use wide traces reduce conductor loss.
- Use a high aspect ratio for small vias.
- Crosstalk can be reduced by routing signals as differential pairs to provide differential common mode rejection.

7.5.1.1.4 Receiver

The use of receiver equalization allows a channel to operate error free, with an almost closed eye, before equalization [41]. Decision feedback equalizers (DFEs) are generally used for this application (see Chapter 6 for the description of these equalizers used in optical transmission).

The performance of the DFE improves as the number of taps increases allowing designers to trade off performance with power consumption and design complexity (that frequently is proportional to cost).

7.5.1.2 Optical Backplane

Since the key issue of switching network element backplanes is the transmission of high-speed signals, it is spontaneous to consider the opportunity to use optical transmission [41,42].

As a matter of fact, the need of using special materials with particular low losses and low parasitic capacity and the need of a more accurate process with respect to standard PCB fabrication increase the cost of very high capacity backplanes based on copper busses, creating a condition in which optical transmission can be competitive also from a price point of view.

Last, but not least, optical backplanes seem to be able to concentrate more functions with respect to the electrical one, due to the smaller transmission problems. In any case, also this option presents a number of issues.

The adoption of optical interconnects for use in standard systems environment would have to meet the following requirements:

- Optical would need to coexist with copper on a substrate in order to draw power and slow speed management interfaces [43].
- Optical connectors would have to operate under current environmental conditions in the chassis.
- Optical backplane and, in particular, optical connectors would need to be resilient, being a critical part for the whole equipment availability [44].
- The solution would have to be cost-effective.

From the aforementioned list of qualitative requirements the two critical elements for the success of optical backplane emerge: the availability of optical connectors with the same reliability of the electronic connectors used in the backplanes and ability to operate in the same conditions, and the overall cost of an optical transmission system at 10 Gbit/s.

As far as the cost is concerned, there is a strong push to reduce the cost of 10 Gbit/s transmitters and receivers also coming from other applications like the next generation gigabit passive optical network (GPON) at 10 Gbit/s (see Chapter 10).

In this field the use of vertical cavity surface emitting laser (VCSEL; see Chapter 5) seems to be very promising [45], because, differently from other applications, here there is no need to use a wavelength around 1.5 μm, but also shorter wavelengths can be used, in the second and even in the first fiber transmission window.

As far as the transmission medium is concerned, at least three options are at the designed disposal: standard single-mode glass fibers [46], multimode glass fibers [46], and plastic fibers [45].

The choice of the optical transmission medium is a critical step in designing an optical backplane, due to the fact that a lot of elements depend on it.

Naturally, standard single-mode glass fibers, used in the second transmission window, are very attractive due to the very good transmission performances (low loss, low dispersion) and also due to the experience accumulated in all the equipment designing groups in designing systems based on this kind of fibers.

This choice has however its own weaknesses: the process to incorporate in a standard PCB-based backplane a glass transmission fiber is complex, resulting in a lower yield with respect to normal PCB production and a quite high cost of the final product.

Quite complex are also the processes that are needed to interface transmitters and receivers with a so small core fiber, bringing further cost increase.

The difficulties related to single-mode fibers can be relieved by using multimode fibers in conjunction with suitable compensation of the modal dispersion at the receiver. The core of multimode fibers is much bigger with respect to the core of single-mode fibers and all the alignment processes are easier. Moreover, provided that a suitable equalization is incorporated in the receiver, lower quality sources can be used with respect to the transmission on a single-mode fiber.

The critical aspect related to the difficulties in incorporating a glass fiber into a standard PCB remains. In order to relieve this difficulty glass has to be substituted by a more resistant material. Here comes into play plastic optical fibers that seem a promising solution for this application, even if their transmission performances are not so good as glass fibers.

Research is also quite intense on the possibility to directly integrate on the backplane PCB dielectric optical waveguide, using either a deposited crystal material or a suitable polymer.

As far as connectors are concerned, optical connectors do not yet have the performances of electronics ones, but great processes have been done [47] and there is a diffused consensus that, if the present day's trend will be confirmed, there will be a moment in which optical backplane will become preferable to electrical ones, and it is not so far [48].

7.5.1.3 Optical Backplanes Based on Monolithic Optical Integration

Besides the more traditional idea of the optical backplane sketched in the previous section, a few among the larger companies of semiconductors in the world are carrying out a considerable effort in a different direction.

The idea behind these projects is that the need of structuring a communication network among cards of single equipment at very high speed is not limited to telecommunications, but is common with computer industry, manufacturing automatic equipment, and a set of other applications. This could drive large volumes of market that justify the application of monolithic integration technologies allowing low costs to be reached.

If this forecast is to be true, the optical backplane will become a real "nano-network" that is a complete high-speed network deployed on a single PCB using large scale optical integrated circuits both for transmission and detection of backplane signals on the cards and for routing and conditioning of the signal in the backplane itself.

In order to acquire more specific information, the reader is encouraged to start from [49–51].

7.5.1.4 Protocols for Very High-Speed Backplanes

Protocols for slow backplanes are not suitable to work at a high transmission speed and a suitable choice has to be done, trading off simplicity with a sufficient performance.

A sufficiently diffused consensus exists in avoiding the introduction of a brand new protocol for backplane application, but to rely on an existing solution [46].

Looking at the existing solutions, besides the continuous tendency to improve the performance of various PCI standards (see Chapter 6, the section devoted to backplane busses) the XAUI, a 10 Gbit/s bus protocol stack standardized by the 10 Gbit Ethernet task force [52] seems to have all the needed characteristics. The "AUI" portion of the name is borrowed from the Ethernet attachment unit interface. The "X" represents the roman numeral for 10 and implies 10 Gbit/s.

The XAUI is designed as an interface extender, and the interface which it extends is the XGMII, the 10 Gbit media-independent interface.

The XGMII is the standard interface which interfaces 10 GbE layer 2 and a physical layer device that supports PCS/PMA/PMD. PCS/PMA/PMD are three sublayers in the 10 GbE protocol stack.

Physical coding sublayer (PCS): This is the GMII sublayer, which provides a uniform interface to the adaptation layer for all physical media. It uses 8B/10B coding like fiber channel. In this type of coding, groups of 8 bits are represented by 10 bit "code groups." Some code groups represent 8 bit data symbols. Others are control symbols. Carrier sense and collision detect indications are generated by this sublayer. It also manages the autonegotiation process by which the NIC (network interface) communicates with the network to determine the network speed (10,100 or 1,000 Mbps) and mode of operation (half duplex or full duplex).

Physical medium attachment (PMA): This sublayer provides a medium-independent means for the PCS to support various serial bit-oriented physical media. This layer serializes code groups for transmission and deserializes bits received from the medium into code groups.

Physical medium dependent (PMD): This sublayer maps the physical medium to the PCS. This layer defines the physical layer signaling used for various media. The MDI (medium-dependent interface), which is a part of PMD, is the actual physical layer interface.

The XAUI comprises × 3.125 Gbit/s physical lanes with each using the 8B/10B [53] coding scheme and in order to handle the potential de-skew between the four lanes a new 8B/10B control word [A] was introduced. The [A] control word is periodically inserted, at exactly the same time, into each XAUI stream. In the receiving direction, the [As] are detected and used as alignment markers to resynchronize the four XAUI lanes.

The fact that this type of interface is foreseen to be used in large volumes in any type of 10 GbE application is an important element in favor of its choice.

7.5.2 Platform Volume Value

It is clear that the cost of a platform based on standard cooling and backplanes is not equal to the cost of a platform based on high-performance cooling and a very high-speed backplane.

This fact has an important impact on the cost of switching network equipment, but also has an impact on architectural choices.

In order to represent the aforementioned difference and to divide different factors contributing to it, the concept of "platform utilization cost" can be introduced. The idea is to underline the fact that when a card is plugged into an equipment to be used, not only it is necessary to pay the cost of the card, but also the cost of a certain percentage of the platform that is needed to host the card and put it in the right conditions to work.

Let us consider a standard sub-rack with its backplane and cooling system (i.e., with cooling devices mounted). Let us image to equip the sub-rack with all the control cards necessary to the system we are considering.

At the end we will have a sort of system shell, whose cost we assume to be C, ready to work from the control and thermal point of view, and with N free slots where interface and signal processing cards (comprising switching cards) have to be plugged to obtain a functional system.

Let us imagine that all the cards are equal in shape (even if the definition is easily generalized to different cards shape) and that each card will have a small thickness and is $a \times b$ cm wide.

In these conditions, the use cost U_C of one full slot of the platform is

$$U_C = \frac{C}{NA_s} \tag{7.31}$$

where A_s is the area of a card filling completely the slot.

When a card is inserted into the sub-rack occupying an empty slot or part of it (there are platforms allowing to put small cards one over the other in the same slot), it consumes a platform cost equal to

$$C_p = U_C S = \frac{C}{N} S \frac{A}{A_s} \tag{7.32}$$

where

S is the number of slots occupied in the sub-rack by the new unit
A is the card area

The presence of the areas also allows the comparison of the utilization cost of different platforms having different dimensions of the cards and, as a consequence, different backplanes and cooling systems.

In order to use Equation 7.32 it is needed to have costs of the different elements of the platform. They critically depend on the technical solutions used to implement the backplane and the cooling system, and the ratio between these costs are different among different solutions.

Thus, general results can be achieved only in a parametric way, so driving the target of the development toward the most critical elements of each solution. Thus, it is frequently useful to rewrite Equation 7.32 in the following form:

$$C_p = \frac{\lfloor C_M + C_c + C_B + C_T \rfloor S}{N} \frac{A}{A_s} \tag{7.33}$$

where the addenda composing the cost of the platform are as follows: the mechanics cost, the control cards cost, the backplane cost, and the cooling system cost.

To provide an example of the use of the platform cost of a card, let us image that a designer has to design a medium capacity backbone router, and the decision has to be made as to whether to integrate into the router platform the DWDM line cards or to integrate only gray interfaces, so to use in the network transponders to enter into the DWDM systems. The "gray" interfaces are single wavelength, short-reach interfaces, that are called "gray" to underline that the emission wavelength is generally in the second or in the third window, but there is no strict control on it.

To simplify the situation, that is in reality very complex, let us image that the relevant costs are

- C_{TX}: the cost of one transponders
- C_{GI}: the cost of one optical gray interface
- C_{int}: the cost of a DWDM interface integrated into the router line card

Moreover, let us call

- N_{int}: the number of line interfaces of the router
- N_{gr}: the number of gray optical interfaces in a line card
- N_{wdm}: the number of DWDM interfaces in a line card

It could seem that the problem is decided by the following inequality

$$C_{int} \geq C_{TX} + C_{GI} \tag{7.34}$$

whose meaning is that, if the sum of the costs of a transponder plus a gray interface is smaller than the cost of an integrated DWDM interface, the transponder solution is implemented, otherwise it is implemented with the integrated DWDM interfaces solution.

In reality, this kind of analysis is quite inaccurate, due to the great difference in the platform cost of a transponder mounted in a DWDM optimized platform and a line card interface. The correct comparison has to be carried out considering also platforms costs.

Thus, let us call P_{TX} the platform cost of a transponder, P_{int} the platform cost of a DWDM interface integrated into the router line card, and P_{GI} the platform cost of a gray interface integrated into the router line card.

Generally, it is $P_{GI} < P_{int}$ due both to the better quality electronics needed to condition a DWDM signal and to the more compact aspect ratio of gray interfaces. For example, DWDM interfaces at 10 Gbit/s would need XFPs, while gray interfaces, either extended or normal small form factor pluggables (SFPs) (compare Chapter 5).

Supposing that the difference in electronics can be neglected, calling U_R the platform cost of a router card, A_R the card area devoted to interfaces, and A_S the overall card area, we can write

$$P_{int} = \frac{U_R}{N_{int}} \frac{A_{int}}{A_s}$$

$$= \frac{U_R}{N_{GR}} \frac{A_{int}}{A_s} \tag{7.35}$$

At this point, the correct inequality that can drive the choice in the absence of other issues, is

$$C_{int} + P_{int} \geq C_{TX} + C_{GI} + P_{GR} + P_{TX} \tag{7.36}$$

At this point the solution is no more obvious. As a matter of fact, an optimized DWDM platform has in general a platform cost much smaller that a router platform, so that $P_{int} \gg P_{GR}$ and $P_{int} \gg P_{TX}$; moreover at least two gray interfaces are present on a line card in the space of one DWDM interface, that is, $N_{GR} \approx 2N_{int}$.

Just to see even better the effect of the platform cost, let us incorporate P_{TX} and C_{TX} in C'_{TX} so to rewrite Equation 7.36 like

$$C_{int} - C_{GI} - C'_{TX} \geq \frac{U_R}{2N_{GR}} \frac{A_{int}}{A_s} \tag{7.37}$$

In effect, the result is now much less obvious and with the decrease of the prices for transponders and the increase in complexity of the routers back plane, it is possible to say that the trend is that using external transponders will get more and more convenient.

7.6 On the Performances of Core Switching Machines

Core switching machines, mainly OXCs and multi service switch platform (MSSP) in the transport layer and routers in the upper layer, are the most complex systems in the network. Core routers can have routing capabilities up to 10^9 packets per second, while the larger OXCs manage a total capacity on the order of hundred Tbit/s. In order to achieve these performances the most advanced architectures have to be adopted, with a great attention to the impact that the characteristics of these big machines have on the network.

In this section we would like to carry out a brief discussion on a few key requirements that are common to core switching equipment and discuss briefly the quantitative measurement of their performance.

7.6.1 Capacity, Throughput, and Channel Utilization

Even if switching equipment has the common role of routing the information carrying signal through the network, different classes of equipment are characterized by a "capacity" that often is defined ad hoc for the specific equipment.

Since confusion can derive from this situation, in this section we want to define clearly the capacity definitions we will use in the following.

In the case of circuit switching (e.g., OXCs), the machine capacity critically depends on the switch fabric.

If the switch fabric is nonblocking, every permutation can be done among the input and output channels. In this case, it is natural to call switch capacity the maximum amount of signal in terms of bit/s that can be delivered to the system.

In the circuit layer (both OTN and SONET/SDH) all the connections are bidirectional, so that it is usual for this kind of switching machines to define the bidirectional capacity. For example, if a maximum of 100 channels at 10 Gbit/s structured in 50 bidirectional connections converge on an OXC, the capacity of the OXC will be 5 Tbit/s (and not 10 Tbit/s as it was taking into account the whole input information).

We will adopt the bidirectional capacity definition since it is almost universally used in technical literature, but also advise that it creates a few problems when comparing routers' capacity with OXC and MSSP capacity due to the fact that routers do not have the restriction of bidirectional connections.

If the inner switch fabric is blocking, the capacity of the switching element can be evaluated multiplying the earlier definition multiplied by $(1 - b)$, where b is the blocking probability.

Considering now packet systems, namely, routers and carrier class Ethernet switches, there is not a universally accepted measure of the performance and there are several possible definitions for throughput.

Among the possible definitions of throughput, we want to evidence the following:

- *Instantaneous throughput:* Let us fix a traffic condition at the input, that is, let us define a process for the packet arrivals to each router port. Let us consider a time interval T_m much longer than the maximum packet duration, but much shorter than the changing pace of the traffic statistical conditions.

  The average over the output ports of the number of packets forwarded to the output lines in the interval T_m is a random variable, whose ensemble average can be called n_{out}. Similarly, we can define n_{in} substituting arrived packets to the forwarded ones.

  The instantaneous throughput is given by $\Im_{ist} = n_{out}/n_{in}$, and it depends both on the switch performances and on the traffic at its input.

- *Instantaneous throughput in bit/s:* It is clear that, defined as in the previous section, the throughput is an adimensional variable. On the contrary, frequently the throughput is measured in bit/s. The throughput in bit/s is defined as $\Im_{bit} = n_{out}\langle L_p \rangle / T_e$ where $\langle L_p \rangle$ is the average packet length (possibly evaluated in the observation interval) and T_e the observation interval.

- *Maximum throughput*: This is the instantaneous throughput when the router is loaded with high traffic, that is in the situation in which a packet is always available at the head of every input queues to be forwarded. Thus, we will have both a relative maximum throughput and a maximum throughput in Gbit/s. The maximum throughput does not depend on the traffic characteristics, once the traffic processes are assumed uncorrelated and identical at each input port. For this reason, this is a useful definition for theoretical comparisons. Unfortunately it is difficult to measure the maximum throughput so that rarely this quantity is used in relationship to real machines.

- *Peak throughput*: This is the throughput measured on a real machine in conditions of high traffic load. Due to traffic unbalances and imperfect realization of the high traffic conditions, the peak throughput is generally smaller than the maximum throughput.

In our considerations we will refer in general to the maximum throughput, so that when we say that a router has a capacity of 300 Gbit/s, this means that its maximum throughput is 300 Gbit/s.

Naturally, the maximum throughput is generally smaller than the sum of the bit rates of the lines exiting from the switch. A router having 100 in/out ports at 10 Gbit/s (for a total transmission capacity of 1 Tbit/s) could have a maximum throughput of 600 Gbit/s (about $6 \cdot 10^9$ packets out per second).

This is due to the fact that statistical fluctuations of the traffic load create a fluctuation of the load on each output line. Thus, the transmission systems have to be dimensioned on the maximum possible load and not on the average one.

The ratio between the maximum throughput and the overall capacity of a packet machine output system is called channel utilization and it is often used besides the maximum throughput to characterize the performances.

7.6.2 Scalability

A core router or a core OXC is a long-lived equipment, corresponding to a relevant CAPEX and operational cost (OPEX) expended from the carrier to buy and operate them. Thus, they will remain in the network well beyond the time a commercial strategy lasts or a particular service is implemented and also much more than the edge machines that are more service related.

Thus, they have to provide the carrier with a real future proof equipment, characterized ideally by a small first installation cost and a cost of successive upgrades as proportional to the capacity of the upgrade as possible.

The choice of the switch fabric architecture is a key from this point of view due to the fact that the different modules of the switch fabric have to interact with each other if the nonblocking characteristic is to be maintained while expanding the router, while different line cards are almost independent, but for a loose coordination carried out by the central controller.

7.6.3 Interface Cards Density

Core OXCs and routers, especially Tbit-routers, are multi-rack machines and the reduction of the real estate needed to accommodate them is one of the important design tasks, that involve mainly line cards that occupy in almost all the configurations the major part of the platform space.

Dense line cards are not only beneficial due to the smaller occupation of platform space, but also allow evolutions of the electronics to be exploited at best. As a matter of fact, while full functionality adoption of the automatically switched optical network (ASON)/generalized multiprotocol label switching (GMPLS) control plane could push distribution of the OXC intelligence to perform correctly and on time all the needed protocols, in the routers the great part of the distributed processing is already carried in the line cards.

This trend has a very important impact on the whole telecommunication network. Core routers are characterized by completely distributed architectures. In order to exploit very high-speed queues and processors, two techniques can be used: either the transmission speed is also increased so to directly send to the elaboration part of the line card the received data streams after suitable adaptation (regeneration and physical layer overheads termination, layer 2 header stripping, and so on), or lower bit rate streams received from the transmission lines are suitably serialized taking into account the packet dimension so to feed the electronic processing unit that is faster than the transmission lines.

The second solution is better suited to optimize the transmission systems, but it is much less efficient in terms of density of the line cards. As a matter of fact, not only four 10 Gbit/s interfaces occupy much more space in the line card with respect to a single 40 Gbit/s interface, but also a certain amount of auxiliary electronics have to be deployed on the card for adaptation, serialization, and deserialization, and so on.

For this reason, router vendors have been the first to push for the adoption of 40 Gbit/s formats in transmission systems even if transmission is generally less efficient at this high bit rate.

A similar trend is now ongoing regarding 100 Gbit/s transmission. Router and switch vendors are pushing the adoption of this very high speed in transmission systems in order to simplify line cards architecture and increase their density, leaving to transmission systems vendors the problem of managing long distance transmission at this huge bit rate.

7.6.4 Power Consumption

Power consumption is a key issue in designing core switching machines. As a matter of fact, such complex machines if not suitably designed could well go completely over the limits typical of telecommunication equipment.

Two elements condition the power consumption of a core switch: continuous evolution of electronic chips and parallel increase of the switch capacity.

A composite plot reporting data from a great number of measurements on real core routers is reported in Figure 7.28 [54] showing the evolution of the power consumption of a core router versus its capacity. Bullets on the curve indicate the energy per processed bit used by the router, figure depending critically not only on the base power consumption of the CMOS, but also on the router design.

From the figure it is clear that, even in the presence of a continuous decreasing of the unitary power consumption (see also Figure 7.29 [54]) the core routers' capacity increased so rapidly that the overall power consumption not only increased continuously, but faster and faster.

Similar behavior can be also registered for core OXCs, at least as far as the switch fabric will be realized using CMOS technology.

The power consumption trend is one of the main elements on which the research on optical transparent OXCs and routers is based even if the introduction in the network of

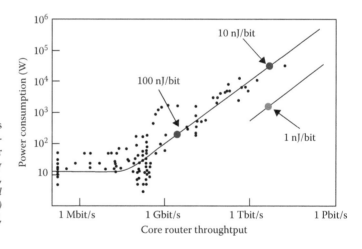

FIGURE 7.28
Historical trend of routers power consumption versus routing capacity. (After Tuker, R., Modeling energy consumption of IP networks, *European Conference on Optical Communications (ECOC 2008) Symposium*, September 21–25, 2008, Brussels, Belgium.)

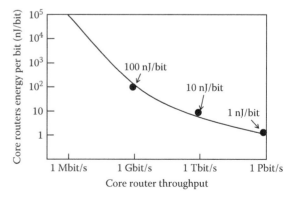

FIGURE 7.29
Historical trend of the energy needed to route a bit in a core routers versus the router capacity. (After Tuker, R., Modeling energy consumption of IP networks, *European Conference on Optical Communications (ECOC 2008) Symposium*, September 21–25, 2008, Brussels, Belgium.)

transparent OXCs and routers seem farer and farer the promise of a drastic reduction of the power consumption while the capacity has a further increase is sufficient to justify quite an intense activity in this field.

We will review the results on transparent OTN and on optical packet switching in Chapter 10.

7.6.4.1 Services

As we previously mentioned the router, besides its main forwarding function, assures a set of other services, like packet filtering and firewall. These services have to be carried out at an extremely fast speed in a core router, and all has to be designed having in mind these characteristics.

7.6.5 Availability

The amount of traffic a core switching machine manages causes a failure of the equipment to impact dramatically the network. For this reason, these machines have to be characterized by a very high availability degree that is reached by hardware duplication and software modular design.

Almost always the switch fabric is completely redundant and works either in hot standby, or (in case of packet switching) in load sharing.

In the first case, the traffic is constantly sent to both the switch fabrics and only the receiving line card selects the data coming from one of them. In the case of failure, every line card switches very rapidly on the data coming from the second fabric with so small a delay that the router working is never suspended.

In the second case, possible only for routers and Ethernet switches, the load is shared between the two switch fabrics and each line card is connected with both of them. When a failure occurs no change in configuration has to be operated in the detection side of the line cards, simply no data will arrive from the failed switch fabric and all the data will be detected from the working one. This will be obtained by switching all the traffic on the working fabric in the transmitting line card side. This switch is either driven by the receiving side that detected no signal (or a deteriorated signal) from one of the line cards or by the router central controller that detects a switch failure via alarms.

Load balancing is a bit slower in recovering from a failure with respect to hot standby, but allows both statistical fluctuations of the received traffic to be better managed and in some cases the possibility to add best effort traffic that can be lost when a failure occurs.

In large OXCs and in Tera routers, generally the two switch fabrics are also located in different racks, so to add a further protection against some types of platform failures.

The central processor is also duplicated and, generally, all the running software works in a mirroring configuration. This means that the state of the back-up processor is continuously updated to be the same to that of the main processor, comprising the fact that both processors have independent memories with replicated software and data.

In this way, a processor failure can be recovered in a very short time by activating the other processor.

Finally, line cards can be either locally protected by 1:N mechanisms, or protected in the framework of the global network protection. In this second case, especially if restoration mechanisms are used, the recovery time is longer, but there is a more efficient resources' exploitation.

Other duplicated elements are generally the power feed and the fun units, so to guarantee the equipment working in front of these types of platform failure.

7.7 Circuit Switching in the Transport Layer

The transport layer, either the OTN or the SDH/SONET layer, is characterized by long lasting connections, called circuits.

Even with the evolution of the traffic generated by the dominance of data related communications, the duration of a transport layer connection is much longer that the duration of connections at upper layers, for example, MPLS LSPs.

Thus, switching machines working at the transport layer have to switch rarely after remaining a long time in the same configuration: they are static switches.

Both in SDH/SONET and in OTN standards (see references from Chapter 3) the concept of switching at the multiplex section layer is defined as the operation carried out by a machine that accepts at the input a set of STM or OC signals in the TDM case or a set of WDM combs in the WDM case and routes them toward a defined output port without demultiplexing.

In practice, however, this type of physical layer switches are rarely used in the SDH/SONET case and never passed from labs to a product stage in the WDM case.

The physical layer switches that are deployed in the field operate on paths and optical channels (i.e., tributaries of the TDM multiplexed frame or of the WDM comb), so that the switching functionality is located at the OCh layer in the WDM case or at the path layer in the TDM case.

In this chapter we will deal only with optical channel/path layer switches.

The functionalities of a physical layer switch can be divided in three classes [26,55]:

1. Connection switching
2. Connections management
3. Connections survivability

7.7.1 Connection Switching

This is the set of functions needed to transmit within the equipment the signal incoming from an input port toward the correct output port.

Functions belonging to this class are as follows:

- Demultiplexing the incoming signal to separate the tributaries
- Separation of the service channels carrying control plane and management plane information from the incoming data plane signals
- Regeneration or amplification if required
- Switching through the equipment switching fabric
- Multiplexing of the tributaries directed toward the same output line
- Addition to each multiple section of the service channel with the relevant information

Naturally, it is possible that not all the aforementioned functions are really present. For example, if an OXC is connected to the client ports of a WDM system, it is the transport system that operates multiplexing and demultiplexing and the OXC has not to perform these functions again.

7.7.2 Connection Management

This class collects all the functions needed to run correctly the procedures and protocols needed to set up, manage, and tear down a connection.

An example of connection management is given by the connection switching control.

As a matter of fact, to prevent misrouting of entire physical layer connections, when a network manager controls the physical layer, each connection is tagged with an ID number, that is generally a field of the record containing the information about the connection. The cross-connection status is memorized into the database of the switch by a data table (at least at the logical level) relating each incoming connection ID with an input port number and an output port number.

The switch continuously monitors the ID of the connections entering the input ports and exiting the output ports to verify both that the arriving connections are correct (i.e., no switching errors occur in the previous switches) and that its own switch is transferring all the connections at the correct output port.

This operation, if the network is managed by the ASON/GMPLS control plane, is performed not with IDs and ports numbers, but with generalized labels and ports IP addresses, exploiting the potential of the GMPLS protocols. However, a completely analogous control has to be done, to prevent errors.

Another important connection management operation is related to the termination of the service channel in the OTN case and of the MS header in the SDH/SONET case and to their recreation at the output of the switch when the multiple sections are recreated.

7.7.3 Connection Survivability

All the functionalities related to protection and restoration mechanisms are collected under this class of functionalities [57].

In particular, depending on the type of survivability mechanism implemented, the protection switching can be operated by the switch or can be the responsibility of a separated card that could also be part of the transmission system.

In any case, all the physical layer procedures and protocols related to survivability are managed also by the switch.

7.7.4 Optical Cross Connect

The OXC is the physical layer switch used in mesh OTNs; the layering structure of the functionalities of an OXC was already introduced describing the OTN standard in Chapter 3 and it is repeated here in Figure 7.30 for reference.

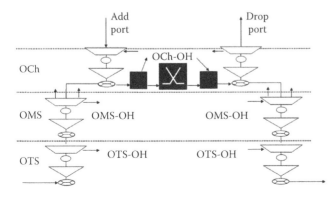

FIGURE 7.30
OTN layered architecture of an OXC.

In the figure, an OXC directly connected with the WDM line is considered so that multiplexing and demultiplexing functions are performed by the OXC.

The specific characteristic of the OXC is that the switching function is located at the OCh layer; this means that the OXC does not access the signals multiplexed into a wavelength (e.g., the LSPs, if the client layer is an MPLS network), but manage the wavelength channel as a whole.

As a consequence, the OXC terminates both the optical transport section overhead (OTS-OH) and the optical multiple section overhead (OMS-OH), both encoded into the service channel.

The OCh-OH is on the contrary transmitted transparently at the output.

There are several different types of OXCs that can be classified considering the following elements:

- How the OXC is connected with the WDM transmission system
- The technology used to realize the OXC switch fabric
- The routing properties of the OXC switch fabric

7.7.4.1 OXCs with WDM or Gray Interfaces

Since the OXC does not decompose the optical wavelength channel, both line and client interfaces have the capacity of a wavelength.

The client interfaces are generally "gray" interfaces, but for the case in which another WDM system is connected to the client lines, for example, a metro ring or a coarse wavelength division multiplexing (CWDM).

Line interfaces can be either gray, wavelength interfaces or WDM interfaces, that is, interfaces with the wavelength stability suitable for WDM transmission.

In the first case, that is schematically represented in Figure 7.31, the OXC is connected on the line side with the client input/output ports of a WDM system line terminal. Multiplexing and demultiplexing are performed by the WDM that also terminate the OTS-OH and the OMS-OH.

Since, in general, it is the OXC that hosts the control plane controller, there must be a way in which the WDM passes OTS and OMS information to the OXC controller.

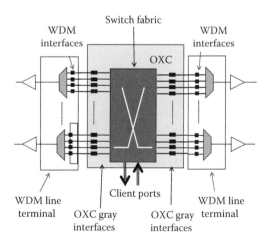

FIGURE 7.31
Block scheme of an OXC connected on the line side with the client input/output ports of a WDM system line terminal.

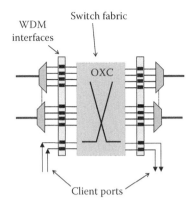

WDM interfaces

Switch fabric

OXC

Client ports

FIGURE 7.32
Block scheme of an OXC incorporating into its platform the line terminals of the input/output DWDM systems.

The other solution is schematically represented in Figure 7.32: the OXC has WDM quality line interfaces and, from a control point of view, the multiplexer and the demultiplexer, like the booster amplifier and the preamplifier, belong to the OXC.

From a pure cost point of view, the analysis of this solution is similar to the analysis presented as an example at the end of Section 7.6 and the cost aspect of the two solutions depends critically on the complexity of the OXC platform.

The main difference is that OXC line cards have not the complexity of router line cards, being essentially transponders; thus, the relation among the interface prices is a bit different from the case in the example.

Moreover, it is to be considered also the situation in which the transmission elements are located in their own sub-rack, receiving the signal on the front via a well-dimensioned fiber patch-cord routing system, so to be divided as a platform, but integrated as far as control is concerned.

Since the OXC switch fabric is huge and, in its typical configuration, occupies several sub-racks, this is many times a convenient solution from an economic point of view.

However in this case, as in other similar cases, there is another important consideration that often brings mainly incumbent carriers to choose the solution adopting gray interfaces.

Even if ITU-T has done a certain amount of work for the standardization of internal interfaces of WDM systems, the design of long-haul systems is so complex and sensible to small variations in the design parameters, that it is, in practice, impossible to generate a WDM channel with a WDM interface of a vendor and propagate it through the line (amplifiers and optical equalizers like flattening filters and dispersion compensating fiber [DCF] modules) designed by another vendor, unless a large underperformance of the system is accepted.

For this reason, if the integrated solution is selected, selecting the same vendor for the OXC and the WDM is almost mandatory, both for the first installation and for the successive upgrades.

This solution can be good for a competitive carrier, since it gives the network almost completely in the hands of a big vendor maintaining in this way a light cost structure that an independent research lab or a large network design department cannot sustain.

On the contrary, the incumbent carriers have advantage in leveraging on the market competition among the main equipment vendors and would see as a strong limitation to their autonomy installing a network architecture that right from the installation has to be sustained by single vendor.

At the end, the advantage of having a cheaper equipment architecture would be almost surely overcompensated by the impossibility to exploit competition among suppliers to have lower prices.

For this reason, it is difficult that incumbent carriers decide to adopt OXCs that are integrated with the transmission system.

7.7.4.2 OXC with an Electronic Switch Fabric

The capacity of an OXC switch fabric is impressive, both in terms of switched Gbit/s and in terms of number of channels.

A network node of order 4 must host an OXC with at least four incoming WDM and four outgoing WDM. Assuming that all the optical channels are at 10 Gbit/s and that the long-haul systems can accommodate up to 128 channels, the switch matrixes have to accommodate 512 unidirectional fluxes at 10 Gbit/s, that means about 5 Tbit/s. If the channels would be at 40 Gbit/s, the capacity would be as high as 20 Tbit/s.

An OXC designed for large core networks can happen to have even 20 in/out WDM systems with a switch capacity on the order of hundreds of Tbit/s.

In this section, we will analyze the case in which the switch fabric is realized by means of an electronic switch.

As we have seen in Chapter 6, electronic switching is probably for a long period of time more than competitive with optical switching. The 32 nm technology has allowed power consumption to be greatly reduced and the 22 nm node seems to be not so far.

The electronic solution is also rich of functionalities that can be integrated into the same chips where the switch resides, thus simplifying the design and the management of the switch fabric.

There are two main possible architectures for an electronic OXC switch fabric: scaling the signal speed or working with the same signal speed used for transmission.

In the first case, the electronic switch matrix is fed by an intermediate speed signal produced by a battery of deserializers placed in front of the switch itself. A deserializer transforms a high-speed serial signal into a set of lower speed signals that are present on space distinct lines at the output of the component. This architecture is shown in Figure 7.33.

The serial signals at the output of the deserializer are completely different from the real tributaries of the high-speed signal, since they are obtained artificially dividing the signal, without any reference to the structure of the signal itself (the deserializer is not an electronic demultiplexer for this reason).

Thus, no grooming or sub-wavelength switching can be done; the signal is simply rearranged differently to ease the switch fabric work.

Once the signal is so rearranged, the switch fabric can work on lower speed signals, but increases its dimensions correspondingly.

On the contrary, if no deserializer is used, the signal is directly sent to the switch fabric without any parallelization.

The first solution is convenient either when other kind of processing has to be done on the signal or when the transmission is at 40 Gbit/s or beyond, to reduce the required switch bandwidth.

One of the most used, architecture of the switch fabric itself is the Clos architecture that is strictly nonblocking and exploits at best the bigger crossbar chips that can be realized.

An example is reported in Figure 7.34, where it is imagined to build up a nonblocking switch fabric with a crossbar chip 200 × 200 obtaining a 10,000 × 10,000 switch. This example well represents what happens for OXCs with input channels at 40 Gbit/s. As a matter of

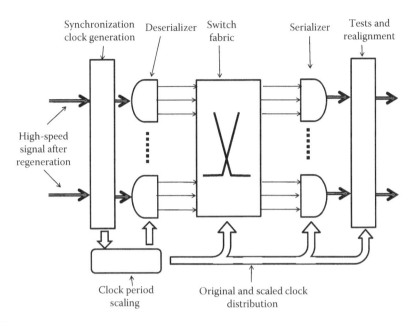

FIGURE 7.33
Architecture of an OXC based on a low internal bit rate electrical switch fabric.

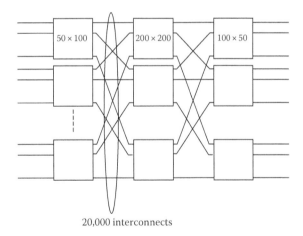

FIGURE 7.34
Scheme and interconnects number of a Clos 10,000 × 10,000 nonblocking switch fabric.

fact, designing the fabric to work at 2 Gbit/s and if four input WDM systems are foreseen with 128 wavelengths each, a switch of 10,240 × 10,240 is needed.

A key parameter of the switch matrix of an OXC is the power consumption, while the switching speed is not so important since the optical layer would operate in a circuit switching mode also under the ASON/GMPLS control plane. This means that optical connections are established for a long time and fast switching is not a requirement. This naturally assumes that protection is implemented with separate switches.

At the 22 nm technology node is reasonable to assume that the single chip consumes about 1 pJ/bit/s per switch crosspoint.

Considering an architecture with deserializers and 10 DWDM at the input, each of which carry 100 channels, we have 1000 channels at 10 Gbit/s; that is, we need a 10,000 × 10,000 Clos switch fabric with lines at 1 Gbit/s.

This is exactly the example of Figure 7.34, where there are 1.4 million crosspoints and 40,000 connections between different chips, taking into account that the fan in and the fan out of a chip also are constituted by transistors and consume power.

If we refer to an energy per transit bit of 1 pJ/bit/s, we have a global power consumption of 140 kW for the crosspoints and 40 W for the chip connections. Adding the possible power consumption of the processor implementing the manager and the arbiter, we have an overall power consumption of about 1.8 kW.

It is very interesting to verify if power consumption due to parasitic currents can be neglected. The amount of parasitic power consumption depends on the structure of the crosspoint that can be a simple transistor with its driver polarization circuit or a more complex subsystem.

As an example, we will assume that the crosspoint is realized using two flip-flops, each of which drives a single transistor used as physical switch.

The microscopic analysis of this structure arrives at a conclusion that for 22 μm it can be assumed a leakage power of 45 μW.

Assuming a simple connection structure built with three transistors, each connection has a leakage power of 20 μW.

This means that in our example, where there are 1.4 million crosspoints and 40,000 connections, the pure leakage power, that coincides with the power consumption without transiting signal, is about 200 mW, that is, it can be still neglected with respect to the 2 kW due to regular current caused by CMOS switching when bits are transmitted.

These are not high values, taking into account the footprint of the switch fabric and partially explain why still today almost all commercial OXCs have an electrical switch fabric.

A brief description of the main features of a practical OXC built using an electrical switch fabric is reported in Table 7.1.

From the table it is evident that in general the OXC is a multi-rack machine, integrating, besides the main OTN switch fabric, a series of other features like the possibility of an embedded carrier class Ethernet switch with the corresponding Ethernet interfaces or a set of line DWDM interfaces at 10 and 40 Gbit/s that are naturally compatible with the DWDM line produced by the same vendor producing the OXC.

Other kinds of OXCs could support also digital cross connect (DXC) functionalities like SDH/SONET grooming.

7.7.4.3 OXC with an Optical Switch Fabric

Since the OXC switch fabric has to manage high-speed signals and a high switching speed is generally not required, it is natural to consider the possibility to realize the switch fabric using optical switches [33,58].

Optical switches are expected to have a very low power consumption and to be able to manage high-speed channels without any need to decompose them in a larger number of slower data fluxes.

For several years this expectation has driven a large amount of research effort in the direction of large optical switches with low losses, effort that has produced several results, mainly in the field of micro electrical mechanical machine (MEMs)–based optical switches, as detailed in Chapter 5. As reported in Table 5.8, MEMs-based switches

TABLE 7.1

Summary of the Feature List of an OXC Based on an Electrical Switch Fabric

OXC for Core Transport Network			
Platform			
Mechanic	ETSI 300 119 rack	16 unit per sub-rack	3 sub-racks per rack
Backplane	Electrical bus using advanced PCI parallel protocol and hot standby protection		
Control	Duplicated network element controller in hot mirroring configuration		
	Duplicated sub-rack local controller		
	Inter-rack 10 GbE protected interconnection		
Management and Configuration			
Option 1	Centralized OTN Network manager—compatible with SONET/SDH management		
Option 2	ASON protocol suite (integrated with proprietary development when needed)		
OXC Switch Fabric			
Expandable	Strictly nonblocking	Hot standby duplication	
Switch Fabric Options			
80 Gbit/s	12 slots	1 sub-rack	1 rack working and protection
320 Gbit/s	32 slots	2 sub-racks	1 rack working, 1 rack protection
720 Gbit/s	62 slots	4 sub-racks	2 racks working, 2 racks protection
960 Gbit/s	62 slots	4 sub-racks	2 racks working, 2 racks protection
Optional Embedded Carrier Class Ethernet Switch			
Expandable	Hot standby duplication		
Switch Fabric Options			
40 Gbit/s	6 slots	1 sub-rack both working and protection	
80 Gbit/s	12 slots	1 sub-rack working 1 sub-rack protection	1 rack working, 1 rack protection
Interfaces			
STM-1	16-ports	Electrical	1:N Card protection supported
STM-1, STM-4	SFP modules	Gray optical	4, 8, or 16 ports
STM-16	SFP modules	Gray optical	16 ports
STM-64	XFP modules	Gray optical	4 ports
STM 128	Embedded TXT	Gray optical	1 port
STM-64	Embedded TXT	DWDM tunable (50 and 100 GHz)	2 ports
STM-128	Embedded TXT	DWDM tunable (200 GHz)	1 port
GbE	SFP modules	Gray optical	16 ports

as large as 56×56 are available in the market having the low power consumption of 700 mW.

Based on this kind of switches, an OXC with the block scheme reported in Figure 7.31 can be designed. Naturally, also a so-called transparent OXC could be considered, where the signal is not regenerated, but amplified at the input and output of the switch fabric. This choice would introduce a completely different network model that we will analyze, in detail, later in Chapter 11.

Let us imagine targeting a 4 input OXC terminating WDM systems with 128 channels at a bit rate of 40 Gbit/s. In order to manage this huge capacity, a 512×512 optical switch fabric has to be realized. Imagining using a Clos architecture and that the bigger available crossbar switch is a 64×64 switch, we have $r = 64$, $m = 64$, and $n = 8$ (see Section 7.2.2), so that the condition $m > 2n - 1$ is realized and the switch fabric is strictly nonblocking.

Using this design we need 64 switches 8×8 and 16 switches 64×64. Assuming that in a first approximation each crosspoint has a power consumption of 0.22 mW (700 mW for a 56×56 switch as in Table 5.8), we can estimate a total switch fabric power consumption of 15 W.

Adding the power consumption of the overall electronic control we achieve global power consumption around 20 W, much smaller that the CMOS power consumption so that, if CMOS footprint is determined by the need to dissipate the generated power, in the case of the optical switch, the footprint is determined by the physical dimensions of the switch.

Even if the promise of a clear decrease in power consumption is realized by optical switches, other elements make the optical solution difficult to implement. The first is the footprint of the resulting switch fabric.

Each 64×64 switch has an estimated dimension of $10 \times 8 \times 0.2$ cm (see Table 5.8). This means that one switch with its analog driver occupies in many platforms a whole card. Even in platforms with big cards, it is possible to estimate the need of more than 20 cards only to host the switch elements with their drivers. This is much more than the real estate needed for the equivalent electronic switch fabric.

Perhaps even more important is the cost of optical switches, which is much higher than those of electronic switches.

The use of optical switches can be made much more effective in terms of footprint and cost by observing that OXCs are characterized by the fact that they terminate DWDM systems, so that it is possible to divide the optical channels in large groups, the channel directed toward the same output DWDM system, that are directed to the same geographical location, even if on different wavelengths.

On the ground of this observation is possible to introduce in the OXC switch fabric a certain degree of blocking, maintaining the property that whatever the switch state, any input channel can access every output DWDM comb, thus being able to be directed along any OXC output directions.

The easiest way of doing this is represented in Figure 7.35 and it consists in separating inside the OXC channels having the same wavelength and assigning to each wavelength group a different switch fabric.

In our example this corresponds to deploy 128 optical switches 4×4. These kinds of switches can be realized using MEMs 2D technology in a very compact shape also integrating several switches on the same chip. Also, the drivers of 2D MEMs are quite simpler with respect to those of 3D MEMs, so contributing to the complexity and real estate sparing.

The architecture of Figure 7.35 was first proposed for optically transparent OXC where wavelength conversion is not possible, so that a general wavelength continuity has to be maintained across the switch [58].

However, in our case we are discussing the so-called opaque OXC, where transponders are located both at the input and at the output of the switch so to regenerate the signal and divide from a transmission point of view input and output WDM systems. In this case, there is no need that the inner wavelength set of the OXC coincides with the wavelength set used by transmission systems. In reality, if the exact wavelength definition of each

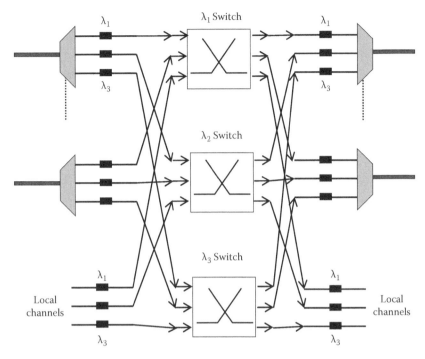

FIGURE 7.35
Architectural scheme of an opaque OXC with a wavelength planes-based switch fabric.

channel is not needed for the switches to work, there is no real need that the channels are colored inside the OXC switch fabric, since they are always spatially separated.

In general, given N input/output DWDM systems with W wavelength each, it is possible to divide the NW input channels into S groups of M channels each so that $NW = MS$ and to assign a separate optical switch to each channel group. This channel aggregation is, in general, useful when $M < W$, so that the blocking inside the OXC is less probable than in the case of $M = W$ [59, 60].

The situation $M < W$ is also favored by the fact that, in general, the output DWDM ports of an OXC are much less than the channels of each WDM comb (as seen in the example we have done in this section).

The S wavelength subsets can also be called switching planes (SP) of the OXC and the OXC switch fabric can be viewed as a set of S noninteracting switch fabrics. This abstract model of the OXC is shown in Figure 7.36.

If $S = 1$, the switch is strictly nonblocking while increasing S the blocking probability inside the switch increases. The ratio Ψ between the number of possible switch states and the number of all the permutations of the $C = WN$ input channels is given by

$$\psi = \frac{S(C/S)!}{C!} = \frac{(C\eta - 1)!}{(C-1)!} \tag{7.38}$$

where the factor $\eta = 1/S$ ($0 < \eta \le 1$) and can be called OXC inner connectivity.

Let us assume that the wavelength channels arriving at the input of the OXC have the same probability to be directed toward any output direction but have no requirements in terms of wavelength on the output DWDM system. Blocking in an OXC based on the

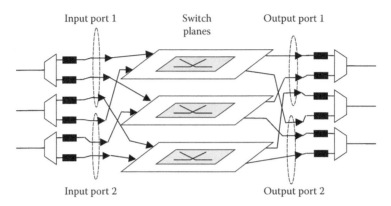

FIGURE 7.36
Architectural scheme of an opaque OXC with a switch fabric segmented in switching planes.

SP architecture occurs when a channel belonging to an SP cannot be routed toward the wanted direction, due to the fact that all the SP exits toward that direction are occupied, when free exits exist toward that direction on other SPs.

In order to evaluate the blocking probability let us assume that the OXC is in a non-blocked state before the tentative of setting up a further channel that provokes blocking.

Let us assume, as it is frequent in practice, that the network manager sets a minimum time interval between successive new connection setup in the network to allow centralized operation to be carried out.

In this condition, requests arrive at the OXC in such a way that if a sufficiently small time interval is selected the probability that two or more requests will arrive in this interval is zero.

Let us call T_{in} the probability that in the considered interval a new connection request is delivered to the OXC through a specific input direction. This probability can be called "traffic intensity" and we assume that it is independent from the specific port considered and from the state of the other ports.

If all the channels are statistically independent and all the destinations have the same probability to be selected when a new connection is required, the probability $p(j)$ that a generic channel setup request arrives in the considered time interval and is directed toward the jth direction among the possible N is $p(j) = T_{in}/N$ so that the index j can be suppressed calling it simply p.

The new channel provokes blocking, if it belongs to an SP where output is no longer free toward the channel destination while other SPs have a free output but are unreachable by the new channel.

Let us call $m(j) < M, j = 1, \ldots, SP$, the number of channels routed through the jth SP.

The probability P_s that the SP receiving the new channel cannot route it toward the correct direction coincides with the probability that c channels are already routed in that direction in the same SP.

This configuration requires that, in the considered plane (let us first imagine without the loss of generality and let us drop the useless index) $m \geq W\eta$ and its probability writes

$$P_s\left(\frac{N}{m}\right) = \sum_{k=W\eta}^{m} \binom{m}{k} p^k (1-p)^{m-k} \tag{7.39}$$

where $P_s(N/m)$ represents the probability that a blocking condition arises in the given time interval due to the arrival of a new channel conditioned to the fact that, before the new arrival, the SP processes m channels.

Saturating all the acceptable values of m it can be written

$$P_s(N) = \sum_{m=W\eta}^{M} P_s\left(\frac{N}{m}\right)$$

$$= \sum_{m=W\eta}^{M} \sum_{k=W\eta}^{m} \binom{m}{k} p^k (1-p)^{m-k} \qquad (7.40)$$

In order to evaluate the real blocking probability, it is also needed that at least one output toward the correct direction does exist in some other SP. This brings to the following equation for the overall OXC blocking probability, that we have to remember is conditioned to the fact that the OXC was not blocked at the arrival of the new connection request

$$P_B(N,S) = SP_s(N)[1 - P_s(N)^{S-1}] \qquad (7.41)$$

where N takes into account the possibility of blocking on N-independent directions.

The blocking probability is shown in Figure 7.37 for $N=4$ and different values of S. From the figure (take into account the log scale) it results that there is an abrupt transition for a threshold value of the traffic intensity between a regime of low blocking probability to a regime of high blocking probability, where the OXC functionality is greatly degraded.

In order to maintain the probability of blocking low it is needed to maintain low T_{in}, that can be done by increasing the OXC dimension.

Thus, a network level trade-off should be needed to correctly fix all the OXC parameters.

The need of this kind of trade-off and the fact that blocking always can occur has limited the diffusion of this solution with respect to the solution based on an electronic switch fabric, even if it is effective in terms of power consumption and footprint.

In order to alleviate the blocking problem without using architectures with very low values of S, several solutions have been proposed; one of them is based on the idea of adding

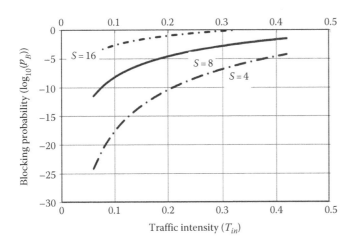

FIGURE 7.37
Blocking probability of an opaque OXC with a switch fabric segmented in switching planes versus the offered traffic intensity for different number of switch planes and different OXC cardinality (i.e., number of input/output directions).

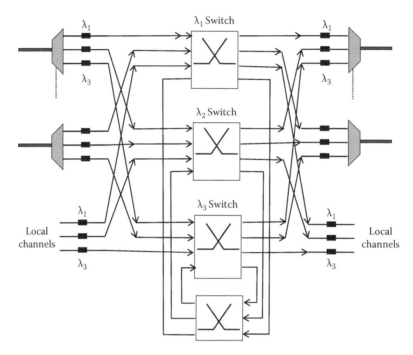

FIGURE 7.38
Alternative architectural scheme of an opaque OXC with a switch fabric segmented in switching planes (SPs) with an inter-SPs connectivity to reduce the blocking probability.

another switch plane with an $(S + 1) \times (S + 1)$ switch devoted to the communication among the different switch planes. This architecture is shown in Figure 7.38.

However, the situation qualitatively does not change and either dimension and cost of the optical switch core approach that of a nonblocking structure or blocking probability thwart the OXC performances for high traffic levels.

7.7.5 Optical Add-Drop Multiplexer

While the OXC is designed as the transport layer switch to be used in mesh networks, the OADM is used either in point-to-point links to drop channels in an intermediate location and to create branching or in optical self-healing rings [58,61].

We will mainly refer to the second application, that is the most demanding in terms of functionalities, but we will also hold all the observation for the first application unless is explicitly said that this is not true.

From a functional point of view, the OADM is mainly an OCh layer component, having network functionalities similar to those of an OXC.

In particular, in an OTN network, while the optical channels pass through the OADM without electronic regeneration, the service channel is terminated so to terminate the OM header and the OR header.

The OADM also manages the OCh protection through the self-healing ring mechanism or through analogous mechanisms in the case of point-to-point links (see Section 3.2.7).

Referring to a two fiber ring network, an OADM interfaces both the fibers and is able to add/drop optical channels from both the fibers. Thus, the principle scheme of an OADM

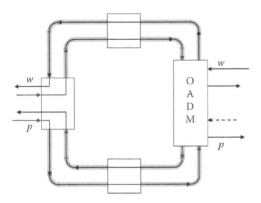

FIGURE 7.39
Position and functionality of an OADM in an OCh-DP ring. The working path of the depicted connection is called w, the protection path p.

can be depicted as in Figure 7.39 where a simple OCh-P ring is shown with four OADM nodes and the path of an OCh is represented both in the working (w) and in the protection (p) fiber.

The basic building block of an OADM is the so-called single-sided OADM (S-OADM), whose functional scheme is reported in Figure 7.40. This subsystem accepts at the input a WDM comb, extracts without electronic elaboration a set of WDM channels delivering it to a local termination, and replaces them with locally generated channels.

The ratio between the maximum number of add/drop channels and the number of channels of the line DWDM system is called add/drop degree of the OADM.

In principle, there are two possible ways of connecting two S-OADMs to make a complete OADM: the so-called vertical and horizontal architectures that are shown in Figures 7.41 and 7.42, respectively. In the figure TXT represents the interfaces of the local equipment

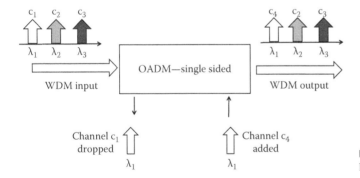

FIGURE 7.40
Basic building block of an S-OADM.

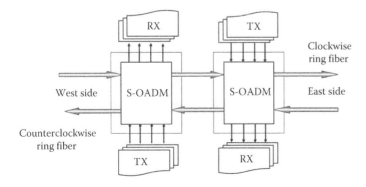

FIGURE 7.41
Vertical architecture of an OADM designed for a self-healing optical ring. Two S-OADMs are used, besides clients' transmitters (TX) and receivers (RX).

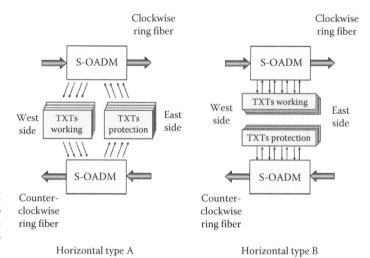

FIGURE 7.42
Two possible horizontal OADM architectures corresponding to different ways to connect transmitters (TX) and receivers (RX) to the two used S-OADMs.

that, in order to manage OCh-P (either shared or dedicated), have to be divided into working and protection interfaces. Moreover, in Figure 7.42 two possible ways to connect the local equipment interfaces to the OADM are considered.

Even if the horizontal configuration is frequently easier to realize from a technological point of view, the vertical configuration is almost the only used in practice [57] even if it requires four S-OADMs (each block of Figure 7.41 incorporate two S-OADMs).

This is due to the fact that, in case of an S-OADM failure, the vertical configuration deviates all the channels through the other S-OADM, automatically, due to the fact that they are switched on the other side of the ring with respect to the failure (see Chapter 3). This property allows the failed S-OADM to be replaced without disrupting the protected channel routes.

This is not true for horizontal configurations, where the failure of an S-OADM completely eliminates the access of the node to one of the ring fibers so that standard OCh protection mechanisms result not to protect the S-OADM failure.

It is not impossible to devise different types of routing to implement channel protection in the case where horizontal OADMs are used, but besides the fact that they should be standardized before their introduction in a product, they are all somehow underperforming with respect to the standard one.

This can be understood even without entering into the detailed explanation of the channel routing along the ring, simply looking at Figure 7.42 where the two possible ways to connect a horizontal OADM with local equipment interfaces are shown.

As a matter of fact, if the solution called A in the figure is adopted, each TXT is connected to booth the S-OADM of the node; thus, S-OADM failure cannot be recovered by any possible routing of the channel along the ring, since both working and protection TXTs have a termination on the failed S-OADM.

In the case of the configuration called B in the figure, each transponder card is connected to only one S-OADM module, but the RX and the TX paths go through different arcs of the ring, since both are routed on the same fiber. This provokes different delays in the propagation of the two directions of the same bidirectional optical channel, thing that is generally to be avoided, especially, when packet services are present in the overlay layers.

Depending on the flexibility in dropping and adding channels, S-OADMs can be classified as follows [62,63]:

- *Fixed OADM*: In this case the set of add/dropped channels is fixed and cannot be changed.
- *Switchable OADMs*: In this case the OADM has a certain number of add/drop ports, each of which is associated to one wavelength channel; the OADM can be configured so to add/drop or to pass through any of these wavelength channels while the channels that are not associated to one add/drop port can be only passed through.
- *Tunable OADM*: In this case the OADM has a certain number of add/drop ports and any wavelength channel can be added/dropped from any of the add/drop ports if the OADM is suitably configured.

Fixed OADMs are mainly used in point-to-point systems to drop channels in an intermediate location or to create a branching point or in access to CWDM rings. Switchable and tunable OADMs are instead used mainly in optical DWDM rings.

As a matter of fact, using fixed OADMs in optical rings induces several blocking situations that can be solved only via the installation of redundant client equipment and that have to be avoided in a functional network. An example is reported in Figure 7.43 where for the sake of simplicity an optical ring with only two wavelengths (the dark and the clear wavelength) is represented. The ring has four nodes where fixed OADMs are located. This fixed OADM has drop ratio equal to one; thus, they drop both the dark and the clear wavelength together.

In the bottom part of the figure the traffic matrix is represented, that is the table of the connection to implement in the ring. When implementing the connection A and B on the dark wavelength two fixed OADMs are located in A and B. Thus, when implementing the connection between C and D the clear wavelength is automatically dropped in the transit nodes A and B and the connection can be realized only regenerating the optical

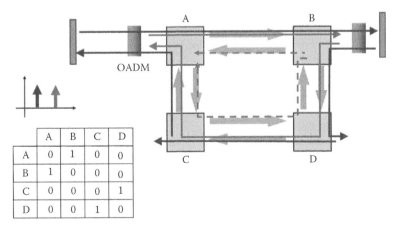

	A	B	C	D
A	0	1	0	0
B	1	0	0	0
C	0	0	0	1
D	0	0	1	0

FIGURE 7.43
Example of blocking in an OCh-DP ring due to the use of fixed OADMs. In the figure the blocking situation is removed by adding supplementary regenerators in the transit mode A.

channel in A and B and injecting again it into the ring. This is a waste of resources with respect to the transparent solution that has to be avoided.

Naturally, this could be avoided if single wavelength OADMs would be available. In real rings however, there are 90 or 120 wavelengths and having such a high number of different OADMs would complicate farther the spare part management for the carriers and the product catalogue management for the vendor.

Both switchable and tunable OADMs allow similar situations to be avoided. The difference between tunable and switchable OADM is essential in the process of setup of new connection.

In Chapter 2 we have noted as the evolution of the service bundles proposed by carriers renders the traffic in the geographic and metro networks more and more unpredictable.

Thus, when a request for setting up a new connection in the optical layer arises we cannot assume that it was foreseen on time.

Depending on the traffic already routed in the ring and on the near and far ends of the new connection a suitable wavelength is first individuated to support the connection. In order to operate the setup if switchable OADMs are used two conditions are needed: the OADM covering the selected wavelength has to be present in the two end nodes and a spare client interface has to be present exactly on the port corresponding to the selected wavelength.

The coincidence of both the condition is quite improbable, mainly because in the impossibility of following a preplanned sequence of new connections setup, it is too expensive to place spare client interfaces on all the available OADM ports in the network.

Thus, if switchable OADMs are used, the setup of new connections almost surely requires a field intervention to install new cards in the present equipment.

If the OADM is tunable, the second condition is removed: it is sufficient that an OADM with the selected wavelength in the add/drop bandwidth is present with at least one spare tunable client interface connected to whatever port. As a matter of fact, tunability allows the network control to send the desired wavelength to the port where the tunable client interface is connected.

By deploying a limited number of tunable client interfaces, the presence of tunable OADMs allows setup of new connections to be carried out frequently without the need of an immediate field intervention.

If OADMs that are not tunable are used, but the add/drop ratio is one, a careful deployment of spare tunable interfacing the network can be remotely upgraded adding several new connections without changing the equipment configuration.

The different OADM types can be all realized using multiplexer, demultiplexers, and a progressively more complex switch fabric.

Fixed OADMs are generally realized by demultiplexing the incoming DWDM comb and managing the channels one by one. In order not to impose a too high attenuation to pass through channels, frequently a subband demultiplexer is used for a first stage channel separation as shown in Figure 7.44.

In the case of switchable OADM, a set of 2×2 optical switches are placed along the add/drop lines as shown in Figure 7.45 so that the channels of the add/drop bandwidth can be either added/dropped or passed through.

In the case of tunable OADM with a reduced add/drop degree, the only variation to the architecture of a switchable device is that the set of 2×2 switches is substituted by a nonblocking $2n \times 2n$ switch, where n is the number of accessible wavelength channels as shown in Figure 7.46.

Finally, tunable OADM with unitary add/drop degree can be realized by two standard DWDM multiplexer and demultiplexer with a $2W \times 2W$ nonblocking switch fabric, where W is the number of wavelengths of the line DWDM system, as shown in Figure 7.47.

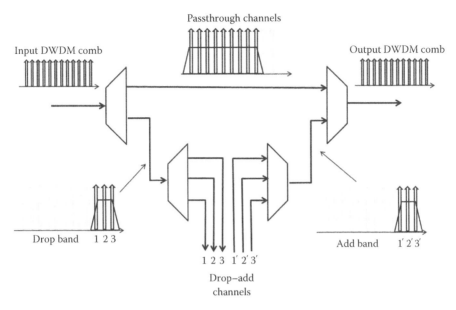

FIGURE 7.44
Scheme of a fixed OADM realized by demultiplexing the incoming DWDM comb and managing the channels one by one.

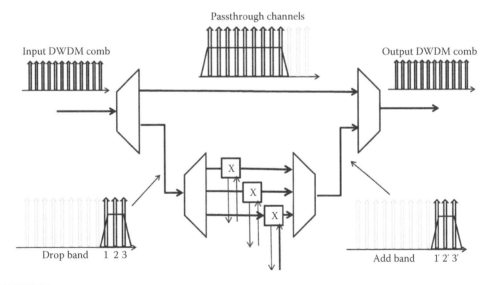

FIGURE 7.45
Scheme of a switchable OADM realized by demultiplexing the incoming DWDM comb and managing the droppable channels using optical 2 × 2 switches.

Since multiplexer and demultiplexer based on the AWG principle can be integrated with thermo-optic switches on a silica on silicon platform, the schemes of figures from Figures 7.42 through 7.45 are not only principle schemes, but real products are built with these architectures.

Nevertheless, there are a plethora of different optical technologies competing with the AWG-based design suitable for the realization of S-OADMs. Almost all these

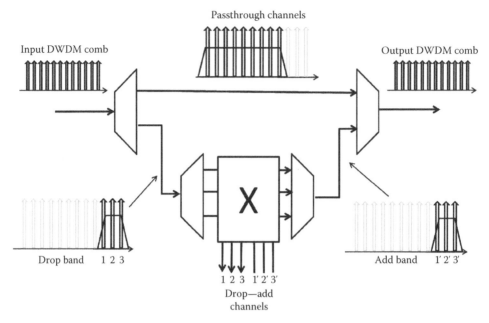

FIGURE 7.46
Scheme of a tunable OADM realized by demultiplexing the incoming DWDM comb and managing the droppable channels using an $n \times n$ optical switch, where $n < W$ is the number of droppable channels and W the number of DWDM channels used in the ring.

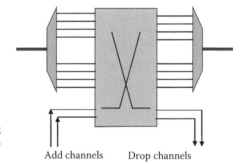

FIGURE 7.47
Scheme of a tunable OADM realized by demultiplexing the incoming DWDM comb and managing all the channels using a $W \times W$ optical switch.

technologies are based on the observation that the switch fabric that is needed into an OADM is not as big as the switch fabric inserted into an OXC. Thus, both the losses are limited allowing optical transparency of the component and the cost results much more affordable.

A technology that is quite diffused besides the plain use of multiplexers and demultiplexers is the use of optical wavelength selective switches (WSS, see Chapter 5) [63,64].

The only functionality at data layer that lacks to the WSS to be a complete S-OADM is the channel Add, that in general is realized via a coupler and an external Add multiplexer to obtain the full S-OADM as shown in Figure 7.48.

A brief summary of the performances of practical OADMs for different applications is reported in Table 7.2. All the considered OADMs are a set of cards to be integrated into a DWDM system; thus, all control and configuration is performed by the WDM controller.

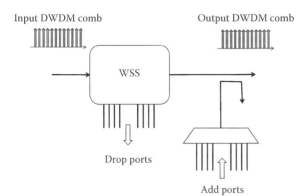

Drop ports

Add ports

FIGURE 7.48
Block scheme of a possible implementation of a tunable OADM using a WSS.

TABLE 7.2

Comparison among the Main Features of OADMs for Different Applications

Application	DWDM Metro Core Rings	Core DWDM Branching	Access CWDM Rings
Deployment	Integrated into the DWDM system	Integrated into the DWDM system	Integrated into the CWDM system
Type	Tunable	Fixed	Fixed
Implemented protection	OCh-DPRing	Dedicate 1+1	OCh-DPRing
Platform	19′ rack card	19′ rack card	ETSI N3 cabinet
Slot	One for OADM One for each EDFA	1	1
Embedded controller	No	No	No
Technology	WSS using MEMs	Silica on silicon mux/demux	Thin film filters
Number of ports	9	8	4
Wavelength range	Extended C band	Extended C band	CWDM wavelengths
Channel spacing	50 GHz	50 GHz	20 nm
Integrated amplification	Two EDFAs	No	No
Pass through loss	—	9 dB	2.8 dB
Input power range	−24.6 to −0.6 dBm	−2.6 to 9 dBm	−20 to 2 dBm
Maximum output power	17 dBm	0 dBm	—
Preamplifier gain	17–22 dB	—	—
Postamplifier gain	Fixed gain of 20.5 dB	—	—
Power requirements (W)	50	8	2

7.7.6 Add-Drop Multiplexer

Conceptually, the ADM is the TDM equivalent of the OADM, where the standard used for TDM multiplexing is SONET/SDH. In practice this is not true.

Due to the possibility of realizing big switch fabric rich of functionalities using electronics with the scope of switching TDM signals, the ADM is much richer of functionalities with respect to the OADM.

The switch functionality of the ADM, in its simpler version, is located at the SONET/SDH path layer. Thus, the ADM terminates the MSH and the RSH and demultiplexes the tributaries of the aggregated.

From a general point of view, the ADM is designed to be applied in self-healing SONET/SDH rings and, depending on the ring type, can terminate either two counter-propagating fibers or four fibers, a couple for each ring direction.

The conceptual scheme of a four fibers SONET ADM is presented in Figure 7.48. On each line side of the ADM (east and west) a couple of fibers brings working OC48 signals and another couple of fibers brings protection OC48 signals. The ADM can manage both dedicated and shared protection depending on the functionalities that are selected at the moment of the installation.

The incoming OC48 signals are detected by line cards, the MSH and RSH are terminated and the frame is demultiplexed up to the level of STS3. All the STS3 incoming in the OADM enters in a strictly nonblocking switch fabric, that is able not only of directing the STS3 paths to be locally dropped in order to be substituted with locally generated channels, but also performs a series of other functionalities.

Finally, locally directed channels are terminated on the drop cards where they are further demultiplexed and their slower tributaries are delivered to the client equipment.

Among the common additional functionalities of an ADM switch fabric, we can mention, referring to the example of Figure 7.49:

- Manage protection switching.

- Interchange upon request the position of the tributaries (STS3 in the figure example) in the OC 48 frame; this operation is useful to optimize the exploitation of the protection bandwidth in shared protection rings.

- Replay the same tributary in different positions inside the OC48 frame, so operating broadcast along the ring.

- In the case of four fibers ring, the switch fabric performs grooming optimization between paths with the same level traveling on different fibers in the same direction.

- In any case performs grooming optimization among tributaries of a level higher than the minimum (that in this case is STS3).

- Implements drop and continue switching, that is required to protect a set of interconnected rings against interconnection failures.

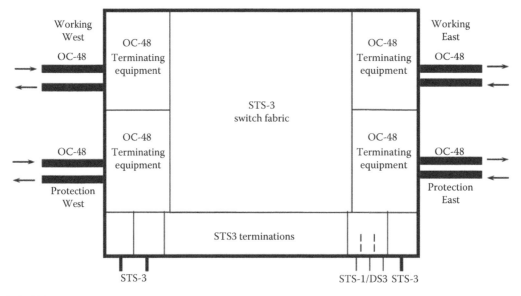

FIGURE 7.49
Functional scheme of an OC-48 four fiber SDH/SONET ADM.

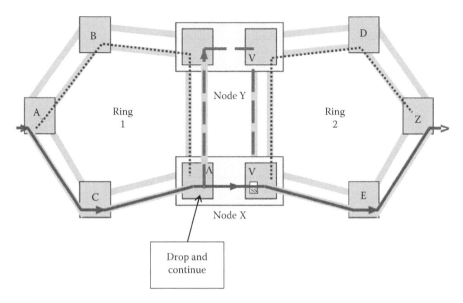

FIGURE 7.50
Double gateways protection in a double interconnected ring network using SDH/SONET. The role of drop and continue in node X is evidenced.

The last functionality deserves an explanation. The drop and continue feature of an ADM is the capability to broadcast a tributary channel both on a drop port and on the output line port. These features are used in situations like that depicted in Figure 7.50, where a network is built connecting several SONET/SDH rings.

If a failure occurs in a ring, normal ring protection starts so that the whole network can cope with one failure per ring. In order to face the failure of one of the interconnection nodes or of one of the interconnection fibers, the signal exchange between each couple of rings is performed in two distinct points.

To render this technique efficient it is needed that the signal that passes from one ring to another in one connection node is also sent to the other connection node to reply the same situation. In order to do that, the ADM in the first connection node has both to drop the signal (to pass it to the other ring) and to send to the line to forward it to the following connection node. Thus, drop and continue is needed.

In Figure 7.50 this situation is clear imagining to set up a path from node A in ring 1 and node Z in ring 2. Working path in the two rings are represented with a full line and protection paths with a dotted line (dedicated path protection is assumed and only one direction is represented, that from A to Z is shown).

The additional signal path created by the drop and continue capability of the ADM belonging to ring 1 located in node X is represented with a dashed line. In the working mode, the signal passes in node X from ring 1 to ring 2, but if the connection between the two ADM in node X fails, the signal is also sent to node Y, where it is passed from ring 1 and ring 2 to be sent again to node X along the ring 2 fiber.

Up to now we have described the features of a standard SDH/SONET ADM.

This kind of machines is more and more substituted, especially in north American networks, by the so-called Next Generation SONET equipment (also called Multi Service Platforms MSPP) [56,66].

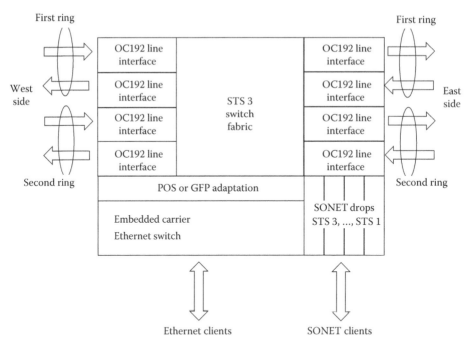

FIGURE 7.51
Functional scheme of an OC-192 four fiber MSPP.

An MSPP ADM has all the features of a standard ADM, plus a set of completely new functionalities that put it apart as a machine performing both circuit and packet switching on the same platform.

There is quite a variety of design for MSPP ADM on the market, just for a reference an example is reported in Figure 7.51.

Two main differences are immediately evident between the ADMs in Figures 7.49 and 7.51:

1. The MSPP ADM has a switch fabric with all the typical DXC characteristics that allows more than one ring to be managed both independently and, if it is beneficial, consolidating and grooming the traffic among different rings.

2. The MSPP ADM incorporate besides traditional SONET drop interfaces also a packet over sonet (POS) or a generic framing procedure (GFP) adaptation interface and a small capacity Ethernet switch that allows it to be directly interfaced with carrier class Ethernet machines; in this way the ADM performs switching operations both at layer 1 and 2 and realizes through its embedded capability the adaptation function between the two layers.

The evolution of SONET machines, however, has not stopped to MSPP. As a matter of fact, below the next generation SONET rings, almost every time the network is built on DWDM transmission. On the ground of this fact, a new generation of equipment was born integrating together DWDM and MSPP that is generally called multi service transport platform (MSTP).

It is not yet clear if MSTP systems will have success on the market, since there is quite a discussion both among carriers and among equipment vendors on their real effectiveness.

On one side, MSTP ADMs promise a sizeable reduction of the CAPEX needed to deploy the network due to the synergies present in the integration of many functionalities on the same platform.

Just to make an example, building the equipment of a network node by using a DWDM terminal with a local OADM to drop a few channels, a separated SONET ADM for each dropped channel to perform grooming and path routing at low speed, and an Ethernet switch requires three different controllers, three different cooling systems, and so on, while an MSTP ADM has only one processor, attaining also the further advantage to be naturally controlled by one management system, while three different pieces of equipment have to be separately integrated under the carrier network manager.

On the other side, two elements seem to attenuate if not cancel the advantage of integration of different function in a single equipment.

The first was already mentioned discussing opaque OXC with colored line interfaces. Since it is in practice impossible to generate a WDM channel with a WDM interface of a vendor and propagate it through the line (amplifiers and optical equalizers like flattening filters and DCF modules) designed by another vendor, integrating the DWDM into the switching equipment (an MSSP ADM in this case) is equivalent to forcing the carrier to buy also the DWDM lines from the same vendor.

Mainly incumbent carriers have all the advantage in leveraging on the market competition among the main equipment vendors to decrease prices and they see as a strong limitation to their autonomy installing a network architecture that has to be sustained by single vendor.

At the end, the advantage of having cheaper equipment would be almost surely overcompensated by the impossibility to exploit competition among suppliers to have lower prices.

Besides the considerations related to the market price dynamic, in the case of MSTP equipment, more technical consideration seems to indicate that the supposed MSTP integration advantage could be not present in practice.

Different from OXCs, MSSP equipment is generally quite compact.

In the OXC case, the switch fabric was so huge that DWDM interfaces, even if integrated from a control point of view into the same equipment, were for sure destined to have their own sub-rack specifically designed for DWDM.

In the MSTP case if the equipment has to be compact, DWDM cards like transponders and amplifiers are generally hosted in the same sub-rack designed for the ADM switch fabric. Thus, the cost of platform related to the DWDM cards is much higher than those correspondent to the installation into a platform designed for DWDM.

In some sense, it can be asserted that "precious" space from a switching platform is wasted to accommodate cards that do not need it.

This factor critically depends on the design of the platform, on the difference between a proprietary and a standard platform, and so on.

Last, but not least, another type on next generation ADM has emerged in recent years the MSSP to denote large MSPP systems, which have switching and grooming capacity of at least 300 Gbit/s. One noticeable difference is that most MSSPs do not provide granularity down to the DS1 level, and many of them are actually pure optical systems in terms of available interface units.

Even if they leverage the same technology and have similar functionalities, MSSPs are clearly directed toward a different application with respect to MSPP.

MSPP are compact machines, designed for use mainly in metro networks.

On the contrary, an MSSP is essentially a core machine, directed to carriers that are building a transport infrastructure.

Due to the different application, also requirements for the two classes of machines are different. Just to make an example, the MSPPs generally have low bit rate interfaces to connect clients from the access network and frequently have also the possibility to incorporate Plesiochronous Digital Hierarchy (PDH) interfaces to connect to the network low bit rate legacy equipment using the PDH [27,30] that is the transmission standard that was in use before SDH/SONET.

On the other hand MSSP, being oriented toward the core network, does not incorporate low bit rate interfaces, but frequently has 40 Gbit/s interfaces.

This difference can be noted from Tables 7.3 and 7.4, where the features of an MSPP/MSTP dedicated to the metro core network and of an MSSP are reported.

Finally, another particularly compact type of ADM that is gaining momentum in long-haul networks is the so-called ADM on a blade.

This is a small capacity, particularly compact MSPP (or even simple ADM) that is completely concentrated into a single blade of a DWDM system.

The use of this device is sometimes alternative to the use of the OADM, when there is the need of dropping a small part of the traffic in an intermediate location, sometimes complementary, when it is used to perform grooming and GFP adaptation of data signals before their insertions into the DWDM stream via an OADM.

In this case naturally, since the switching device is a single blade, no particular platform has to be used and its accommodation inside the DWDM system implies no problems.

TABLE 7.3

Summary of the Feature List of a Practical MSSP

MSSP			
Platform			
Mechanic	Sub-racks compatible with 19′ rack; 18 unit maximum, two sub-racks per rack		
Control	Duplicated network element controller in hot mirroring configuration		
Management	ITU ASON I-NNI		
	OIF UNI 1.0, OIF E-NNI 1.0		
	ASON compatible network manager		
Switch Fabric			
SDH fabric strictly nonblocking		Hot standby duplication	Mixed time and space domain
Switch Fabric Options			
SDH (STM 1 capacity)	160 Gbit/s	320 Gbit/s	640 Gbit/s
Ethernet switch	20 Gbit/s	30 Gbit/s	40 Gbit/s
Interfaces			
STM-1, STM-4	SFP modules	Gray optical	4 ports
STM-16	SFP modules	Gray optical	4 ports
STM-64	XFP modules	Gray optical	1 port
STM 128	Embedded TXT	Gray optics	1 port
GbE	SFP modules	Gray optical	16 ports
10 GbE	XFP modules	Gray optical	1 port

TABLE 7.4

Summary of the Feature List of a Practical MSPP/MSTP

	MSPP/MSTP		
Platform			
Mechanic	Option 1	One sub-rack compatible with ETSI 300 119 rack; 16 unit maximum	
	Option 2	Compact rack proprietary: compatible with ETSI ETSI 300 119 mounting system	
Control	Duplicated network element controller in hot mirroring configuration		
Management	Q interface		
	Local craft and remote management		
	OSI IS-IS and TCP-IP OSPF		
Switch Fabric			
Strictly nonblocking VC12 based		Hot standby duplication	Mixed time and space domain
Switch Fabric Options			
SDH (STM 1 capacity)	64 × 64	128 × 128	384 × 384
Ethernet switch	5 Gbit/s	7 Gbit/s	20 Gbit/s
Interfaces			
STM-1	16 ports	Electrical	1:N card protection supported
STM-1, STM-4	SFP modules	Gray optical	4 ports
STM-16	SFP modules	Gray optical	4 ports
STM-64	XFP modules	Gray optical	1 port
STM-64	Embedded TXT	DWDM optical (50 and 100 GHz)	2 ports
STM 16	SFP modules	CWDM—eight channels available	2 ports
Fast Ethernet	SFP module	Gray optical	22 ports
Fast Ethernet	SFP electrical	Electrical	22 ports
GbE	SFP modules	Gray optical	16 ports

7.8 Packet Switching at MPLS and IP Layers: Routers

This section is devoted to the architecture of the routers. Routers were born as Layer 3 switches, that is, switches that implement switching at the IP layer using a datagram like architecture [1].

The evolution of the data network has brought to the incorporation into the router of several functionalities that are not typical of a datagram-based network.

The main evolution from the point of view of the architecture of the router itself is the introduction into the equipment of MPLS first and then GMPLS.

MPLS and GMPLS switching are more similar to layer 2 functionalities, so configuring an IP/LSR as a machine having switching capabilities at different layers.

The separation, however, is not so sharp as it would be in a potential machine incorporating a layer 2 Ethernet switch and on top of it a layer 3 IP router. An example is constituted by QoS, that are positioned frequently in both layer 2 and layer 3, somehow making exceptions to the rule of layer functionalities separations.

This situation is historically a consequence of the quite informal environment that gave rise to Internet whose aggregation community is the IETF. The fact is, however, that this approach had a so huge success that it is now the reference point for the new network architecture (see ASON/GMPLS architecture in Chapter 8).

7.8.1 Generalities on IP/MPLS Routers and Routers Classification

The process that an IP packet undergoes while traversing a router and being directed by it into the correct propagation direction is shown schematically in Figure 7.52 and it can be synthetized as follows [1]:

- The IP packet arrives at the adaptation layer between layer 2 and layer 3 and the layer 2 header is terminated.
- The IP checksum is calculated to verify packet integrity; if the checksum is not correct the packet is canceled.
- The packet is stored into the input queue of the input interface card.
- An interrupt is sent to the routing processor to inform it that a packet is waiting to be processed.
- When the routing processor answers the interrupt, it receives the IP header from the interface card where the packet is stored.

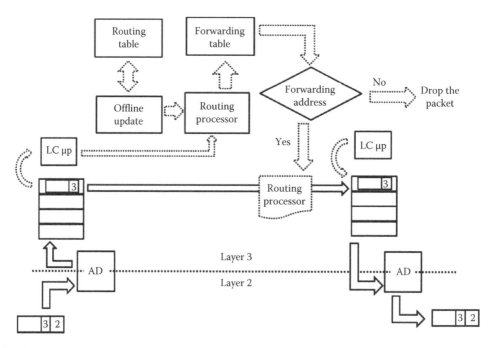

FIGURE 7.52
Schematic representation of the process a router carries out for each IP datagram. Continuous lines indicate the datagram flow, dotted lines the control and address flow. AD represents the adaptation function between layer 2 and layer 3, LC µp represents the line card microprocessor and the packet is stored in a memory (queue) located in the line card while the routing process is carried out.

- The routing processor starts a process of access to the routing and forwarding tables.
- From the access to the routing and forwarding tables, the routing processor determines where is the output card and the corresponding output queue where the packet has to be transferred.
- If no correspondence is found it means that the packet is affected by an uncorrected error so that its destination cannot be reconstructed; in this condition the packet is deleted.
- The processor looks at the destination output queue to determine if there is a free queue slot.
- If a free queue slot is present, the packet is transferred to the destination output queue.
- In the absence of a free output queue slot, the processor repeats the lookup periodically, up to the moment in which the slot is freed.
- The processor transfers the packet from the input to the output queue; three different types of transfers exist, depending on the IP address of the packet:
  - A local delivery (i.e., the destination address is one of the router's local addresses and the packet is locally delivered)
  - A unicast delivery to a single output port
  - A multicast delivery to a set of output ports
- After forwarding the processor performs two other operations:
  - *Packet lifetime control*: It adjusts the time-to-live (TTL) field in the packet used to prevent packets from circulating endlessly throughout the network.
  - *Checksum calculation*: The IP header checksum must be recalculated due to the change in the TTL field.
- The output line card confirms the correct transfer to the processor.
- The processor also regenerates the layer 2 envelop; after that the packet is transferred again at layer 2 to be sent to the output transmission system.

In case MPLS is used in the router, besides the IP forwarding table (i.e., present if the router maintains also IP routing functionalities), there is a label forwarding table that is used to build the MPLS tunnels.

The procedure to build the different table and to use them to route packets can be briefly described as follows.

The router on the ground of the network knowledge and on the working of the selected routing protocols builds the routing table. If MPLS traffic engineering is used via a centralized control, it forces some elements of the routing table to represent traffic engineered tunnels.

The routing table is continuously updated based on the evolution of the network state from a traffic and resource availability point of view. For example, failures or addition of new equipment force a recalculation of the routing table. This is why automatic discovery is so important to guarantee efficient routing.

Once the routing table is calculated, depending on the need of performing either IP or MPLS operation, an IP forwarding table or an MPLS correspondence table is constructed.

The IP forwarding table contains not only a correspondence between input and output ports so that once the IP address of a packet is known also the destination output port can be obtained, but also a set of information that are needed to forward correctly the packet into the layer 2 in the next hop. As a matter of fact, when a packet is forwarded by the routing processor and it arrives at the destination queue, before the transmission to next network hop it has to be enveloped again into the layer 2 envelop.

The MPLS label correspondence table contains the correspondence between input and output labels that allows the packet to be directed toward the correct port. The difference between the IP forwarding table and the label correspondence table is that the first is temporary, there is no connection concept in IP. Thus, changing the network condition, the change in the routing table can provoke changes in the IP forwarding table and packets directed toward the same destination (also packets belonging to the same TCP connection) are routed toward different output ports.

On the contrary, the MPLS label correspondence is a permanent thing, once a correspondence has been set up it cannot be canceled simply because the situation of the network is changed and the routing table is changed, but a complete process of teardown is needed to cancel a label correspondence. Thus, MPLS tunnels behave like virtual connections and packets having at the LSR input the same MPLS label are routed toward the same output port whatever the situation of the network is, up to the moment in which the MPLS tunnel is tear down.

An exception is of course the failure of the output link or of the output interface when MPLS restoration mechanisms automatically set up an alternative tunnel via suitable protocols.

Naturally, in case of MPLS routers MPLS label correspondences have to be distributed along the network so that a tunnel can be recognized by the various LSRs. As explained in Chapters 3 and 8 it is done via suitable labels distribution protocols.

Originally MPLS was designed to simplify the router's operation: as a matter of fact, it relies on the idea that all the datagrams belonging to the same connection and thus directed to the same destination are tagged with a label and are switched on the same route. Thus, the orgin of MPLS relies on the idea that the labels correspondence table is simpler than the IP forwarding table.

This original intention, however, is no more the most important reason pushing the adoption of MPLS especially in backbone IP networks. Even because frequently an LSR setup a so high number of tunnels that the label correspondence table is almost so complex as the IP forwarding table.

The presence of fixed tunnels conveying datagrams belonging to the same communication is instead precious to perform network monitoring and quality of service management.

As a matter of fact, the transit of datagrams along a tunnel can be easily monitored by the router that can recognize them on the ground of the MPLS labels producing statistics of errors in transmission, passing through time, loss of packets due to the transmission toward a full queue, and so on.

Moreover, a service related tag can be applied to each MPLS label that can be used to run priority algorithms in the input queues of the routers and so manage real time services.

Beyond core routing via IP forwarding or MPLS tunnels, a router carries out a set of other services that have gained great importance in the process that has transformed the Internet from a data network to a global network platform for telecom carriers.

Among these services there are

- Packet translation
- Packet encapsulation

- Packet fragmentation if needed
- Traffic prioritization and generally QoS related procedures [67]
- Users authentication (generally carried out by particular routers called broadband remote access server [BRAS])
- Packet filtering for security/firewall purposes
- Network address translation (NAT), that is needed when a private IP domain whose addresses are not universally unique have to be interfaced with the public Internet (see Section 3.3)
- Link monitoring/statistics gathering
- Proxy, that is the functionality of a router connecting different IP domains in the public network (see Section 3.3)
- Packet encryption/decryption for virtual private networks (VPNs) that is a key functionality for services directed to business

It is evident that the routing procedure is quite elaborated, but it could be completely run by a suitable software on a general purpose computer with suited line interfaces.

In practice there exist several types of routers, whose performances and capacities are completely different, and some of them are really computers with a suitable SW and line interfaces.

Just to provide a tentative list of different router classes we can identify:

- *Home routers*: They are the routers residing in the customer's house as the delimitation point between the home network and the public network of the service provider; they generally are constituted by a layer 2 switch that interface the home Ethernet and on top of it a routing engine that interfaces the public Internet. Their maximum throughput comprises in general between 1 and 100 Mbit/s.

- Home routers are connected at layer 3 with BRAS. A BRAS is the first router of the public network that a packet traverses after leaving the customer appliance or a private network. Besides routing and forwarding IP packets, they also are responsible for access authentication and customer profiling; at the beginning of the introduction of broadband services, BRAS where also big computers with suitable programs, but this approach was rapidly found insufficient to follow the needed increase in BRAS throughput; thus, dedicated hardware-based BRAS was introduced, with an evolution path in term of internal architecture similar to that of core routers.

- BRAS are connected by core routers, that are the routers that execute routing and forwarding in geographically extended networks. They are complex and high throughput machines and the need of specialized hardware to reach the required performances was clear from the beginning of Internet. Nowadays, core routers are effectively divided into two categories: Giga-routers with a routing throughput on the order of hundreds of Gbit/s and Tera-routers with a routing throughput on the order of Tbit/s. These two routers categories leverage generally different technologies, depending on the different required performances.

7.8.2 IP Routers Architecture

We have seen that the key operation carried out by a router is packet forwarding. Even if this operation could be done via software, only the low-capacity routers can use this technique and generally a dedicated hardware is needed [68–70].

The first architecture of core routers, and an architecture that is still in use for low-capacity machine, is very simple and it is reported in Figure 7.53. This architecture is not so different from that of a general purpose computer.

The incoming packets are stored in the central RAM memory after transiting from the input line card. When the CPU is ready for it, the packet is sent to a registry inside the CPU to allow all the verifications on the packets to be done (e.g., checksum calculation) and from it to a location in the RAM memory again.

While the packet is in the RAM, the CPU calculates the routing and then the packet is transmitted from the RAM to the correct exit line interface.

Node management software, routing protocol operation, routing table maintenance, routing table lookups, and other control and management protocols such as SNMP are also implemented in the CPU.

The architecture depicted in Figure 7.53 is very simple and it is still used for home routers, for example, but is limited to small capacity machines.

The reasons for which this architecture cannot achieve high performances, even if high-performance components are used, can be summarized essentially as follows.

The CPU is in charge of all the processes needed inside the router, comprising the elaboration of all the transiting packets. This makes the CPU a potential bottleneck of the system.

Even more important is the fact that packets transit two times through the bus to be written on the memory.

If we analyze the times needed to carry out the different phases of the packet processing we have

- *Packet input storage*: Neglecting the time needed to enter the input line card we have that the time elapsed for this operation is $\tau + T_{RAM}$, where τ is the bus latency due both to the waiting time and to the real propagation time, and T_{RAM} is the RAM writing time
- *Packet verification*: With the earlier definition we have $\tau + T_{CPU} + t_{el}$ where T_{CPU} is the writing time of the packet in the CPU register and t_{el} is the time needed for the packet verification

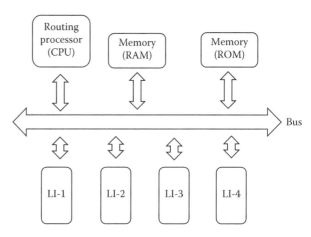

FIGURE 7.53
Router architecture based on a single centralized memory and a time division bus-based switch fabric. Line interfaces are indicated with LI.

- *Routing calculation*: Since routing calculation by access to the routing and forwarding tables is carried out contemporary to the new transfer of the packet to the RAM memory we have that the time needed for this phase is $\text{Max}(t_{ro}, \tau + T_{RAM})$, where t_{ro} is the time for the routing calculation that is dominated in this context by the RAM access time, since also the routing and forwarding tables are hosted into the RAM

- *Transfer of the packet to the output line card*: The needed time is $\tau + T_{RAM}$

If we call T the time needed to access to the RAM and write a bit, and t the time needed to elaborate the operation corresponding to a clock cycle in the CPU, we can say that T_{RAM} is proportional to T and t_{el} is proportional to t. Regarding the term $\text{Max}(t_{ro}, \tau + T_{RAM})$, faster the processor, more the term $\tau + T_{RAM}$ is dominant.

At the end, the time T_{IP} needed for the processing of a packet can be written as

$$T_{IP} = 3\tau + 3T_{RAM} + t_{el} = 3\tau + 3aT + bt \tag{7.42}$$

where a and b are constant depending on the process implementation. Equation 7.42 means that increasing the speed of the processor more and more beyond the value $(3\tau + 3aT)/b$ is useless, since the packet processing time is dominated by the bus speed and by the access time to the memory.

Not only the speed of the processors increases much faster with passing from a CMOS technology node to the other with respect to the memory writing time, but the bus access and propagation time can be improved only by increasing the bus speed, again much slower compared with the processors' speed evolution.

Thus, in order to fully exploit the CMOS continuous evolution to burst routers speed, a most efficient architecture has to be devised.

An important step ahead is done using the architecture depicted in Figure 7.54. Distributing fast processors and route caches, in addition to receive and transmit buffers, over the interface cards reduces the load on the system bus and of the CPU. Packets are therefore transmitted only once over the shared bus.

In this architecture, a router keeps a central master routing table and the satellite processors in the network interfaces each keep only a modest cache of recently used routes. If a

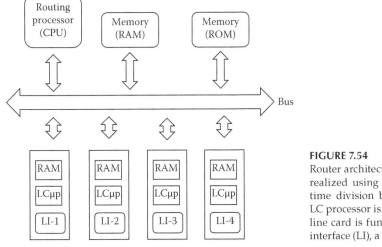

FIGURE 7.54

Router architecture based on an input queue realized using line card (LC) RAMs and a time division bus-based switch fabric. The LC processor is indicated as LCμP, so that the line card is functionally composed by a line interface (LI), a memory and a processor.

route is not in a network interface processor's cache, it would request the relevant route from the central table.

The route cache entries are traffic driven in that the first packet to a new destination is routed by the main CPU (the central route processor) via the central routing table information and as part of that forwarding operation, a route cache entry for that destination is then added in the network interface.

This allows subsequent packet flows to the same destination network to be switched based on an efficient route cache match. These entries are periodically aged out to keep the route cache current and can be immediately invalidated if the network topology changes.

With the new architecture it is easy do demonstrate that, neglecting the time needed to store a packet in the memory resident in the input line card upon the packet reception, the average time needed to process a packet is given by

$$T_{IP} = p(3\tau + 3aT + bt) + (1-p)(\tau + dt)$$

$$= (1 + 2p)\tau + 3apT + (3bp + d - dp)t \approx \tau + 3apT + dt \qquad (7.43)$$

where $p \ll 1$ indicates the probability that a packet needs to be routed by the CPU and a, b, d are constants depending on the implementation.

From Equation 7.43 it is now clear how the importance of the memory access time is quite small and the improvement of the CPUs elaboration time is key to improve the router performances. However, this condition is realized only when $p \ll 1$, that is a condition that is not always realized.

By continuously improving processors speed t, the bus becomes the bottleneck of the router, as evident from Equation 7.43.

Considering the architecture of Figure 7.54 from a different point of view, it is clear that the bus appears to be a time domain switch fabric with input queues and distributed processing.

If the switch fabric becomes the bottleneck it is needed to substitute it with a faster switch fabric.

This can be done by passing from time domain switching to space domain switching; since historically this architecture allowed to introduce routers with multigigabit capacity we will call this type of machines multigigabit routers (MGR).

The block scheme of an MGR is reported in Figure 7.55. The line cards are connected via a space division strictly nonblocking switch fabric with a set of forwarding engines realizing a distributed processing space switch architecture.

Also a bus connects all the units of the router, but it is used for commands and configuration and for exchanging information like the routing table update. Datagrams never are transmitted through the bus.

Every forwarding engine can process a different packet since each has a completely updated replica of the routing and the forwarding tables. The packets are assigned from the line cards to the forwarding engine by transferring to the engine the corresponding header for the forwarding calculation in a round-robin fashion.

Since in some applications QoS requires that packet order is maintained, the output control circuitry also goes round-robin, guaranteeing that packets will then be sent out in the same order as they were received.

The forwarding engine returns a new header (or multiple headers, if the packet is to be fragmented), along with routing information (i.e., the immediate destination of the packet). Then the input line card integrates the new header with the rest of the packet and sends the entire packet to the output line card for transmission.

Also the central processor is connected to the switch fabric since, besides caring of management and configuration of the equipment, it is the central processor that updates the routing table and distributes it to the different forwarding engine that from it derives the forwarding table.

Since the packet is never completely passed to the routing processor, the IP datagram elaboration made by a router with the architecture of Figure 7.55 has to charge the local line card processor of the checksum calculation at the packet input and other similar operations, while the forwarding engine only verifies the header to control that it really comes from an IP packet.

From this point of view, in the MGR there is a substantial step toward a really distributed architecture.

Possible limitations of the architecture of Figure 7.55 are the processing power and the amount of memory usage.

Both these limitations can be alleviated by pushing further the distribution of the router tasks among a set of processors. These are the processors on the line card, a set of specialized forwarding engines for packet forwarding evaluation [71,72], and a central equipment processor.

The principle diagram of a completely distributed processing router is presented in Figure 7.56 [73,74]. In this scheme each network interface provides the processing power

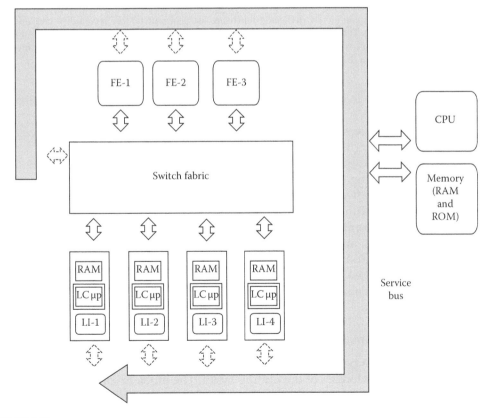

FIGURE 7.55
Distributed router architecture based on the presence of routing caches in the distributor forwarding engines (FEs). Connection between input and output ports is realized by a single, space division, switch fabric.

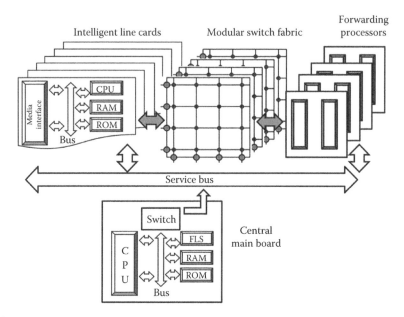

FIGURE 7.56
Complete distributor router architecture. The line cards (LC) have the architecture of a computer, with dedicated hardware for packet processing. Packet forwarding is calculated by a battery of parallel forwarding engines (FEs), realized with a completely dedicated hardware, each of which store a complete and updated replica of the forwarding table. The connection between input and output ports is realized via a modular space diversity switch fabric, while a service bus connects all the units with the central board having a computer like architecture with a processor, a ROM and RAM memory and generally a flash (FLS) memory.

and the buffer space needed for packet processing tasks related to all the packets flowing through it.

They perform the processing of all protocol functions (in addition to QoS processing functions) that lie in the critical path of data flow.

In order to provide QoS guarantees, a port may need to classify packets into predefined service classes. Depending on router implementation, a port may also need to run data-link level protocols or network-level protocols. The exact features of the processing components depend on the functional partitioning and implementation details. Concurrent operation among these components can be provided.

The network interfaces are interconnected via a high-performance switch that enables them to exchange data and control messages. The central CPU is used only to perform some really centralized tasks. As a result, the overall processing and buffering capacity is distributed over the available interfaces and the CPU.

7.8.3 Routing Tables Lookup

For a long time the main performance limitation of both core and edge routers has been the time required for the lookup of the routing tables. The problem can be defined as that of searching through a database (routing table) of destination prefixes, the longest prefix that matches the destination address of a given packet.

Longest prefix matching was introduced as a consequence of the requirement for increasing the number of networks addressed through classless inter–domain routing (CIDR).

The CIDR technique is used to summarize a block of addresses into a single routing table entry. This consolidation results in a reduction in the number of separate routing table entries.

This advantage is counterbalanced by the fact that searching for the longest prefix matching the address is much harder than searching for a perfect match between the address and the routing table entry.

As a matter of fact, the incoming packet does not bring the information on how long is the longest prefix matching its address; thus, the research has to span all the table up to the end or up to the discovery of a perfect matching, that is necessary also the longest possible.

In order to have an idea of the speed at which this operation has to be performed, let us assume that our router has five 40 Gbit/s interfaces and that the average length of an incoming IP packet is 100 bit (the minimum value is 40 bit). Let us also to assume that there are eight parallel forwarding engines and that perfect contemporary operation can be attained in the forwarding table lookup.

In these conditions, and in the high traffic approximation, each forwarding engine has to perform 250 millions accesses to the forwarding table in a second. This is a huge number also for a specialized processor, showing the critical nature of the forwarding table access algorithm.

Besides lookup speed, it is also important to look at what kind of database organization the algorithm requires.

As a matter of fact, not only the memory occupancy has to be reduced as much as possible, but it has also to be possible to update the structure fast and without interfering with the search process.

It has to be considered that many core routers have about one thousand border gateway protocol (BGP) updates per second and, due to the growth of the address space, forwarding tables growth is forecasted to be about 40 kB/year.

7.8.3.1 Binary Trie–Based Algorithms

A wide class of algorithms are based on the organization of the forwarding table like a data structure called binary trie.

In general a trie defined on the finite alphabet $A(p)$ is a tree where every tree element has a number of pointers to the attached leafs comprised from 0 to $p - 1$. The hth level of the tree corresponds to the hth symbol in a word composed of elements of $A(p)$.

The generic element of the trie represents all the words that have the prefix indicated by the route from the trie root to the considered element. If the node does not have p leafs and represents a valid prefix it is a so-called terminating node, a node coding a prefix that, for some acceptable words is the longest matching. In this case, a forwarding port addresses is stored in the data field of the record.

When the longest matching elements contained in the set of words represented in the trie with a given word in $A(p)$ have to be searched, it can be done spanning the trie levels in ascending order and moving from level hth to the next following the pointer that corresponds to the given word symbol in the hth position.

In the router case, $p = 2$ so that we have to consider a binary trie. A very simple example of binary trie is shown in Figure 7.57, where both the tabular form and the trie form of a very small forwarding table are shown.

To understand the meaning of the figure, let us image that an incoming IP packet has an address equal to 000. By visual inspection of the table on the left of the figure it is

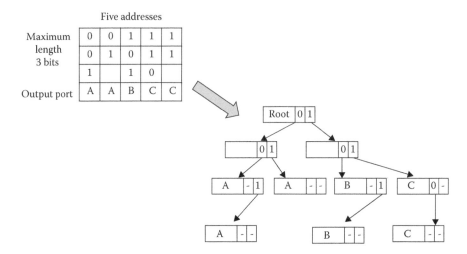

FIGURE 7.57
Example of binary trie, on the left the corresponding prefix table is reported.

immediately determined that the packet has to be sent toward port A since the longest matching is provided by 001 in the table.

Using the trie, the first bit is "0," thus from the root we move left, the second bit is zero again, thus we move left again, the third bit is a third "0," that it is not present as a pointer in the present record. Thus, the search is ended and the output port is determined by reading the data in the last accessed record, that is A.

This is surely a fast algorithm with respect to a tabular search; moreover it is quite easy to update a trie, especially if the update consists in adding new addresses.

The algorithm complexity is on the order of b, the number of bits constituting the address (i.e., 32 for IPv4 and 128 for IPv6); also the complexity implied by the addition of a new prefix is on the order of b (exactly equal to b in the worst case).

The memory space required is Mb in the worst case, where M is the number of stored prefixes.

A huge amount of work has been done to simplify from some point of view the trie algorithm [74–76].

Many attempts have been done to simplify the search in the trie starting from the observation that the forwarding table is almost every time a sparse structure, with a big number of elements with an empty data field.

A first method, particularly effective for very sparse tries is the so-called path compression that consists in eliminating from the trie all the nodes having no associated forwarding information and only one child. These nodes do exist due to the fact that not all the prefixes are possible as the first part of an IP address and they represent a forbidden prefix.

Naturally, since the trie has to be updated, the canceled nodes have to be recorded, for example, in the data field of the father nodes.

Another possibility is to shorten the trie depth by using values of p greater than 2. For example, the bits can be joined three by three and each three bit group can be considered an element of an alphabet $A(8)$ with eight elements.

Naturally, this representation of the routing table causes a decrease in the number of operations and an increase in the complexity of the single operation.

7.8.3.2 Hardware-Based Algorithms

Whatever search algorithm is used, it can be implemented a specialized hardware that runs it much faster than a general purpose processor.

However, if a specialized hardware has to be used, it is also possible to study algorithms that optimize the hardware speed.

This has been extensively done in the case of forwarding table lookup.

The most used specialized hardware is the so-called content addressable memory (CAM) (see Chapter 5). A CAM is designed such that the user supplies a data word and the CAM searches its entire memory to see if that data word is stored anywhere in it. If the data word is found, the CAM returns a list of one or more storage addresses where the word was found.

In the case of forwarding table search, a simple CAM is not sufficient, due to the fact that what is needed is not a complete identification of the incoming address with the memory content, but the longest prefix has to be found.

For this operation, the so-called ternary CAM (TCAM) is needed. TCAM allows a third matching state of "X" or "Don't Care" for one or more bits in the stored data word, thus adding flexibility to the search. For example, a TCAM might have a stored word of "10XX0" which will match any of the four search words "10000," "10010," "10100," or "10110."

The forwarding calculation using a TCAM is done with a circuit whose scheme is reported in Figure 7.58. The prefixes are loaded into a TCAM in ascending order, so that the longer between two prefixes having the same initial bits has the higher TCAM address. Moreover, the correspondence between prefixes and addresses of the forwarding port is loaded into a RAM.

When the IP address is sent to the forwarding engine, it is loaded into the TCAM register to have, at the output, all the prefixes compatible with the incoming IP address.

The TCAM is set so to give priority to the prefix that has the higher address, that due to the prefix ordering inside the memory is also the longer [77,78].

The longer prefix is then sent to the RAM where it is associated to an output port address that is selected as the forwarding port address.

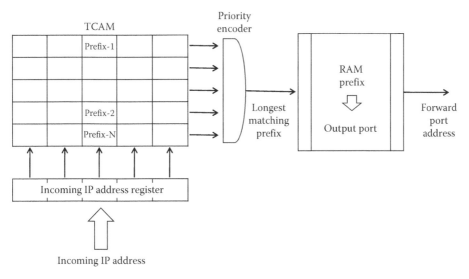

FIGURE 7.58
Block scheme of an IP forwarding engine based on a T-CAM.

The use of a TCAM to implement forwarding engines is fast and simple, but it has the drawback to require great power dissipation due to the TCAM characteristics. Since the power dissipation of a TCAM is approximately proportional to the number of cells (see Chapter 5), it is important to devise schemes to reduce the number of entries in the TCAM so to adopt lower power circuits.

A first algorithm of this type, very simple, but nevertheless quite effective, is based on the elimination of redundant prefixes.

Let us imagine that 1000110 has only one child 10001100 and both are associated to the same output port address. It is clear that, imagining the prefixes trie, the second prefix can be eliminated, attaching all its children directly under the first. This situation is quite common in a great number of forwarding tables of core and edge routers; thus, even simply eliminating redundant nodes in the prefixes trie helps a lot in reducing the number of prefixes to be loaded into the TCAM.

The second technique exploits TCAM hardware's flexibility. The mask for a routing prefix stored in TCAM consists of ones (the same number of ones as the prefix length) followed by all zeros. However, TCAM allows the use of an arbitrary mask, so that the bits of ones or zeros needn't be continuous. This technique is called mask extension because it extends the mask to be any arbitrary combination of ones and zeros. A simple example helps to describe the mask extension technique. Let us assume that the two prefixes 10010011 and 10000011 both correspond to the same route, port A. It's possible to combine the two prefixes into a single entry, with the prefix set to 100X0011, associated to the port A.

Using both the aforementioned techniques a typical forwarding table can be reduced eliminating 35% or even 45% of the prefixes so to proportionally gain in dissipated power.

Last, but not least, TCAMs have to be updated hundreds of times per second. Most routing updates are route flaps—the same prefix is added and then removed repeatedly in quick succession. It is straightforward to reduce the number of actual updates by keeping a buffer of recent route update announcements and updating only the end result. This eliminates most route flaps. Still, tens of updates per second may be required in backbone routers, so a fast update algorithm is necessary.

Several proposals have been done and the problem can be considered almost solved. It is not the scope of this brief section to review the relevant algorithms in this sector, the reader is encouraged to read the specific literature, starting from the following references [87–89].

7.8.3.3 Comparison between Forwarding Table Lookup Algorithms

In Table 7.5 the various forwarding table lookup algorithms are compared under different points of view.

The characteristics of trie search algorithms to scale complexity with the number of bits of the address render them suitable, practically, only for small capacity routers having bus-based architecture and software implemented table lookup.

In this case, algorithms based on multi-bit trie are often used for their reduced complexity.

Almost all public network routers, first of all core routers where the forwarding speed is a key, use TCAM–based forwarding engines with several design particularities intended to reduce either the TCAM required dimension or the number of different TCAM used in each line card.

TABLE 7.5

Complexity of the Different Algorithms for the
Forwarding of IP Packets

Algorithm	Worst Case Lookup	Storage	Update
1 bit trie	$O(b)$	$O(Mb)$	$O(b)$
PC-trie	$O(b)$	$O(M)$	$O(b)$
k bit-trie	$O(b/k)$	$O(2^k Mb/k)$	$O(b/k + 2^k)$
LC-trie	$O(b/k)$	$O(2^k Mb/k)$	$O(b/k + 2^k)$
TCAM	$O(1)$	$O(M)$	

The symbol $O(x)$ indicates that the number of required operations
tends to infinity increasing the dimension of the forwarding table
as x, $O(1)$ indicates that the number of required operations is
independent from the dimensions of the forwarding table.

7.8.4 Broadband Remote Access Servers and Edge Routers

Up to now we have described router functionalities and hardware architecture with a particular attention to core routers. Core routers are the highest capacity routers and for this reason, the routers where new technologies are generally applied first.

Considering the functional role in the network however, BRAS and edge routers are not less important and their evolution has been a key enabler for the diffusion of broadband services.

BRAS [79,80] serves as an aggregation point for subscriber traffic (IP, PPP, Ethernet, and ATM) and provides session termination and subscriber management functions such as authentication, authorization, accounting (AAA), and IP address assignment.

Pushed by the evolution of the service bundles offered by carriers, modern BRAS provide also advanced services—such as subscriber policy management (e.g., Web login), IP and layer-2 QoS, security and VPN capabilities, as well as full IP routing and MPLS support. From this point of view, BRAS are fully functional routers both at MPLS and at IP layer.

Edge routers [81,82] are the routers that assure non-IP traffic adaptation to the IP network and create MPLS labels and tunnels as boundary routers of an MPLS network.

The present trend is to incorporate edge router and BRAS functionalities in a single machine, so creating a high capacity and functional rich new generation BRAS.

For this reason, we will concentrate our attention on the new BRAS generation that incorporates also edge routers functionalities.

As for core routers, also BRAS have traversed an evolution where three BRAS generations can be identified, while capacity and functionalities were growing in pace with the broadband services penetration.

The first BRAS generation was composed by machines where the routing functionalities were implemented by software on a bus-based platform with a single general purpose memory and a single general purpose processor.

This kind of BRAS was used to terminate both dial-up and broadband Internet access, where broadband was practically ADSL only.

In pace with broadband network evolution, a second generation of BRAS was born implemented in hardware and having much higher performances. At the same time, functional scalability was limited as BRAS devices were still designed around centralized, processor-based architectures and were optimized for single-service–Internet access. Attempts to add advanced features such as filtering or QoS still led to performance degradation.

Around year 2004 xDSL reached a critical mass deployment and several fiber-based architectures started to be proposed mainly in Asia and in the United States.

As a response to this trend, BRAS further evolved generating third generation machines. Centralized hardware gave way to distributed architecture that allowed service providers to deliver the capacity and throughput required to support advanced broadband service delivery.

At this stage the BRAS functionality has also begun to be integrated into service edge routers so to provide, through a single machine, a rich set of functionalities, for example,

- ATM and Ethernet aggregation
- Session termination and adaptation of ATM, PVC, PPP signals to the backbone MPLS/IP network
- Authentication, authorization, accounting
- Comprehensive IP routing (with the relevant protocols like BGP, open shortest path first [OSPF], RIP)
- IP address management—DHCP server, proxy services
- Integrated layer 2 switching, both on the client side using, for example, ATM or Ethernet, and on the backbone side using carrier Ethernet or MPLS
- Policy management and dynamic per session QoS
- IP multicast routing (with the relevant protocols like PIM, MBGP, IGMP)
- Security via firewall and intrusion detection

Moreover, such complex machines started to implement reliability methods typical of core machines like 1:N interface redundancy, processor hot standby protection, and switch fabric redundancy.

Besides the trend we have described up to now, pushing BRAS toward an architecture and a capacity typical to small core routers, also an alternative trend has been recently pushed by a few BRAS vendors: the move of the BRAS toward the access, up to collocating it or, in some cases, to integrating it with the digital subscriber line access multiplexer (DSLAM) or the GPON optical line terminal (OLT).

The resulting machine is much more compact with respect to classical third generation BRAS, but has redundant and distributed architecture so to assure a high reliability degree.

It is not so clear if this trend will have success with carriers for sure; if more machines are collocated in the access central office, this could bring to an increase of OPEXs in the most critical part of the network from this point of view, while in case of integration of the BRAS functionalities into the DSLAM or the OLT, this will bring IP network much nearer to the end customer without an evident cost increase, providing a potentially attractive solution.

7.8.5 Practical Routers Implementations

In this section we will present a review of the performances and features that can be achieved with present days' technology from several router types.

As always in this book, such tables are inspired to true products, but do not reflect the characteristics and performances of a specific product. They are instead oriented to give to the reader an idea of what can be achieved with the technology we have described up to now.

Naturally, data sheets of real products are readily available contacting equipment vendors and only updated and original material from the routers building companies can be used to understand the features of a specific router implementation.

The first case we present is that of a terabit router to be used as core routers in the backbone of an incumbent carrier. The characteristics of such a machine are reported in Table 7.6.

TABLE 7.6

Summary of the Feature List of a Practical Terabit Core Router

Core Terabit Router			
Platform			
Mechanic	ETSI 300 119 rack	16 unit per sub-rack	2 sub-racks per rack
	To grow up to maximum capacity	1 rack for the switch fabric	
		1 rack for routing processors	
		1–2 racks for line cards	
Platform redundancy	Power-feed redundancy 1:1		
	Fan-tray redundancy 1:1		
	Fan-controller redundancy 1:1		
	Sub-rack control card redundancy 1:1		
	Fabric-card redundancy 1:4		
	Sub-rack communication management card redundancy 1:1		
Backplane	10 GbE electrical bus using XAUI protocol and hot standby protection		
Control	Duplicated network element controller in hot mirroring configuration		
	Inter-rack 10 GbE protected interconnection		
Management and Configuration			
SMNP support	SNMPv1		
	SNMPv2c		
	SNMPv3		
IP protocol suite	Full support for IPv4 and IPv6		
MPLS capability	Full MPLS protocol suite		
GMPLS capability	GMPLS protocol suite with UNI interface toward transport layer network elements		
Router Switch Fabric			
Properties	Strictly nonblocking space diversity switch fabric		
	Modular architecture built with 4 Tbit/s modules		
	Modules 1:4 redundancy		
	In the maximum capacity configuration a full rack is devoted to the fabric		
Scalability	From 4 to 32 Tbit/s with steps of 4 Tbit/s		
Route Processors			
Properties	Completely hardware based		
	Distributed processing up to 24 working route processors (20 working, 4 protection)		
	Route processors 1:5 redundancy		
Scalability	From 8 to 24 route processors in pace with the capacity growth		
Interfaces			
STM-16	SFP modules	Gray optical	24 ports
STM-64	XFP modules	Gray optical	6 ports
STM 128	Embedded TXT	Gray optical	2 ports
STM-64	Embedded TXT	DWDM tunable (50 and 100 GHz)	4 ports
STM-128	Embedded TXT	DWDM tunable (200 GHz)	2 ports
GbE	SFP modules	Gray optical	32 ports
10 GbE	XFP modules	Gray optical	2 ports

This is for sure the most high throughput machine that we are presenting as an example in this book, arriving to switch up to 32 Tbit/s throughput. Considering the average length of a packet of 100 bits, this means 320 billion packets a second.

This router class is the answer to the request for a global and convergent IP network, requiring all the traffic to be managed by IP/GMPLS machines.

From a hardware point of view, such a big machine has to be modular as much as possible, and this aspect is accurately considered. All the important parts of the router can grow in pace with the traffic capacity that the router has to manage. Moreover, a large amount of effort is devoted in providing reliability while avoiding a useless proliferation of redundant units.

From this point of view, 1:N redundancy is used when possible, leveraging on the modular architecture of the router elements.

From a software point of view, both traditional IP over MPLS or pure IP and GMPLS working are foreseen. If the GMPLS network architecture is chosen by the carrier, the user network interface (UNI) is implemented to realize interworking with the transport layer.

SNMP is used to manage the network element as in almost all present technology routers, but the machine is ready to be updated to more complex network management as far as ASON will be completely implemented.

A similar overview of the features of a core gigabit router is reported in Table 7.7. Under several points of view, being a core machine devoted to the same application of the terabit router, where such a huge capacity is not needed, the gigabit router is similar from the architecture point of view, but with reduced capacity.

From a platform point of view the design of the gigabit router is based on particularly wide cards (quite out of normal standards). This has allowed compacting the machine in a single rack, so avoiding the complexity of inter-rack communication.

Looking at the available interfaces, the lower capacity of the gigabit router implies the presence of lower capacity, high density interfaces that allows lower capacity client to be connected to the router.

Moving from the core to the edge, the features of a practical BRAS/edge router are summarized in Table 7.8. From a hardware point of view, this is a simpler machine with respect to core routers.

The distributed architecture is constituted by a single redundant forwarding processor working in team with the line cards processors and the central processor located in the central control card. The switch fabric is contained in a single double slot unit and has a capacity of 40 Gbit/s.

The software features, however, are quite articulated both on the client interface and on the line interface.

On the client side, since the BRAS has to interface with DSLAMs, it can accept and process at layer 2 both native Ethernet and ATM. This is so, because several DSLAMs are still ATM based and this situation will last for a long time.

Moreover, several business customers still use ATM and frame relay [30], both to guarantee a greater security (in case of customers like banks) and simply to use existing equipment. Thus, the BRAS has to manage also frame relay.

Last, but not least, several customers are simply connected to the network at layer 2 via PPP; thus, also PPP has to be supported in all its flavors, depending on the protocol stack: as a matter of fact, we have IP/PPP, ATM/PPP, Frame Relay/PPP [30].

In the view of a convergent network, also signals coming from the mobile network can be consolidated via the BRAS to be transported through the network infrastructure. This

TABLE 7.7

Summary of the Feature List of a Practical Gigabit Core Router

Core Gigabit Router			
Platform			
Mechanic	ETSI 300 119 rack	12 unit per sub-rack	2 sub-racks per rack
		1 sub-rack for the switch fabric and the processors	
Platform redundancy	Power-feed redundancy 1:1		
	Fan redundancy 1:4 (1 redundant fan out of 4 working)		
	Fan-controller redundancy 1:1		
	Sub-rack control card redundancy 1:1		
	Fabric-card redundancy 1:5 (1 card redundant out of the 4 working)		
Backplane	1 GbE electrical bus using advanced PCI protocol		
Control	Duplicated network element controller in hot mirroring configuration		
Management and Configuration			
SMNP support	SNMPv1		
	SNMPv2c		
	SNMPv3		
IP protocol suite	Full support for IPv4 and IPv6		
MPLS capability	Full MPLS protocol suite		
GMPLS capability	GMPLS protocol suite with UNI interface toward transport layer network elements		
Router Switch Fabric			
Properties	Strictly nonblocking space diversity switch fabric		
	Modular architecture built with four 120 Gbit/s modules		
	Modules 1:4 redundancy		
Scalability	From 120 to 480 Gbit/s with steps of 120 Gbit/s		
Route Processors			
Properties	Completely hardware based		
	Distributed processing 4 working route processors		
	Route processors 1:4 redundancy		
Interfaces			
STM 1	SFP electrical	Electrical	32 ports
STM1, STM 4	SFP modules	Gray optics	32 ports
STM-16	SFP modules	Gray optical	32 ports
STM-64	XFP modules	Gray optical	9 ports
STM 128	Embedded TXT	Gray optical	3 ports
STM-64	Embedded TXT	DWDM tunable (50 and 100 GHz)	6 ports
STM-128	Embedded TXT	DWDM tunable (200 GHz)	3 ports
GbE	SFP modules	Gray optical	32 ports
10 GbE	XFP modules	Gray optical	4 ports
10 GbE	XFP modules	DWDM fixed wavelength (100 GHz)	4 ports

TABLE 7.8

Summary of the Feature List of a Practical Integrated Edge Router—BRAS

BRAS/Edge Router			
Platform			
Mechanic	48.26 × 57.78 × 40.64 cm sub-rack		
	16 slots, 12 single slot units, 2 double slots units (redundant switch fabric)		
Platform redundancy	Power-feed redundancy 1:1		
	Fan-controller redundancy 1:1		
Control	Duplicated network element controller in hot mirroring configuration		
Management and Configuration			
SMNP support	SNMPv1		
	SNMPv2c		
	SNMPv3		
IP protocol suite	BGP, IS-IS, OSPF, RIP, MPLS, Virtual Routers; IPv4 and IPv6		
Mobile aggregation	Mobile IP home agent		
Encapsulation methods	IP/PPP, IP/Frame relay, IP/ATM, IP/PPP/POS, IP/Carrier Ethernet		
Layer 2 protocols	Pure PPP, ATM, Ethernet, frame relay		
Subscribers management	Section termination, AAA functionalities, security via firewall, subscriber VPN management		
Multicast	Full multicast functionality supported		
MPLS	MPLS complete protocol suite		
Router Switch Fabric			
Properties	Strictly nonblocking space diversity switch fabric		
	40 Gbit/s capacity in one double slot card		
	1:1 Switch fabric redundancy		
Route Processors			
Properties	Completely hardware based, dedicated route processor		
	Route processor 1:1 redundancy		
Interfaces			
T3	Embedded	Electrical	64 ports
STM 1 native	SFP elect.	Electrical	32 ports
STM 1 ATM	SFP elect.	Electrical	32 ports
STM 1 POS	SFP elect.	Electrical	32 ports
STM1, STM 4 native	SFP	Gray optics	32 ports
STM1, STM 4 ATM	SFP	Gray optics	32 ports
STM1, STM 4 POS	SFP	Gray optics	32 ports
STM-16 native	SFP	Gray optical	32 ports
STM-16 ATM	SFP	Gray optical	32 ports
STM-16 POS	SFP	Gray optical	32 ports
STM-64 native	XFP	Gray optical	9 ports
STM-64 POS	XFP	Gray optical	9 ports
Fast Ethernet	SFP elect.	Electrical	32 ports
Fast Ethernet	SFP	Gray optical	32 ports
GbE	SFP	Gray optical	32 ports
10 GbE	XFP	Gray optical	4 ports

justifies the presence of a software feature allowing the consolidation at layer 3 signals coming from the mobile network.

These layer 2 capabilities also reflect in the availability of a wide variety of interfaces supporting all the aforementioned formats.

On the core side, full MPLS functionality is supported to interface with MPLS networks, in this case the BRAS can also work as label edge router, being the origin and sink of the LSPs.

7.9 Packet Switching at Ethernet Layer: Carrier Class Ethernet Switches

In Chapter 3 we have seen that the main role of switching machines in the metro core area is to aggregate the signals coming from the access and deliver them to the point of presence where the AAA functions are performed. Such aggregation is performed either with TDM machines like MSPP or using carrier Ethernet switches, a solution that is attracting more and more the attention of the main carriers, especially in Europe [83].

In Section 3.2.3 we have described the principle on which the carrier class Ethernet is based. In particular we have seen that the metro Ethernet forum standards base the working of carrier class Ethernet on the concept of backbone VLAN, to which a particular VLAN identifier is associated to distinguish it from the private VLAN that a real LAN network can define.

In particular, when an Ethernet frame is received from an edge public switch or from an edge switch/router, it is enveloped within a specific public Ethernet envelope containing the B-VLAN ID and the other backbone related fields.

The carrier class Ethernet network is thus divided into virtual networks where the routing of Ethernet frames is carried out through the spanning tree algorithm. Since there are up to 4096 B-VLAN identifier, also a carrier having 5 million users has an average VLAN constituted by about 1220 users so that the spanning tree can manage it.

Routing in the backbone Ethernet network is carried out by carrier class Ethernet switches.

7.9.1 Generalities on Carrier Class Ethernet Switches

The carrier class Ethernet switch procedure when receiving a packet from a line interface can be simplified as follows:

- The Ethernet frame is obtained by terminating level 1 envelops.
- The frame is stored in the input queue, located in the line card.
- The B-VLAN tag is accessed and the frame is assigned to the correct backbone VLAN.
- The correct B-VLAN routing table is accessed using the incoming frame media access control (MAC) to find the address of the output interface where the frame has to be forwarded. This operation requires, in standard switches, the accessibility to the network spanning tree to determine allowed and forbidden routes through the network.
- The forwarding engine performs output interface lookup to detect its availability, QoS criteria are used to manage conflicts.

- When the output interface is available, the frame is forwarded to the correct output interface via the switch fabric.
- The new layer 1 envelop is created while the frame is stored in the output interface.
- The QoS is managed taking into account the QoS class of the frame and, when the output line is available, the frame is transmitted through the transmission system.

Besides Ethernet frame switching, a switch also manages the spanning tree protocol as detailed in Appendix B.

The steps listed assume that the frame belongs to a unicast service. Multicast is supported by carrier class Ethernet and several multicast services can be delivered using this technology. If the frame belongs to a multicast service, more than one output port is calculated as frame destination. In this case, the input interface creates replicas of the frame and repeats the sending operation toward all the selected output ports [84].

The first important observation is that a switch with multiple B-VLANs cannot pass Ethernet frames from a B-VLAN to the other. If this is needed, either the switch incorporates also a layer 3 engine, that naturally provides full connectivity between input and output ports, or an external router has to be connected to the switch.

From a switching point of view, a carrier class Ethernet switch is a configurable set of independent Ethernet forwarding planes, one for each B-VLAN, that are defined at a certain moment.

This structure is configurable since B-VLAN can be set up and tiered down, but once a B-VLAN is defined it is a completely independent MAC addressing plane.

Another important observation is that, differently from IP forwarding, MAC addresses do not grow in number so much while the network grows, due to the segmentation in B-VLANs. MAC address–based routing can thus be performed searching a MAC address database, without the complication of storing only prefixes that is present in IP forwarding.

Finally, a carrier class Ethernet switch has to manage all LAN Ethernet features. For example, it has to be able to recognize a standard VLAN and treat it as a LAN Ethernet switch, isolating it from the other traffic belonging to the same B-VLAN. In this sense, it has to manage a structure with two stacked VLAN layers.

If the Ethernet switch is deployed on top of an SDH/SONET layer, for example inside an MSSP, the physical layer provided by SDH/SONET is sufficient to solve all the physical layer issues, comprising integration into the new network of legacy SDH/SONET equipment.

A problem arises when, leveraging on the fact that the great majority of the traffic in the metro network is originating from Ethernet equipment, carrier class Ethernet switches are deployed directly on an optical layer.

In this case, in the quite frequent situation in which the former network architecture was based on SDH/SONET, some solution is required to manage legacy TDM equipment.

The solution is the so-called pseudowiring. In general terms a pseudowire is an emulation of a native service over a packet switched network (PSN). The native service may be low-rate TDM, SDH/SONET, ATM, frame relay, or Ethernet while the PSN may be Ethernet, MPLS, or IP (either IPv4 or IPv6). A pseudowire emulates the operation of a "transparent wire" carrying the native service.

There are now many pseudowire standards, the most important of which are IETF RFCs as well as ITU-T recommendations [85].

TDM pseudowire design to carry a real-time bit stream like SONET/SDH has, however, many specific characteristics. As a matter of fact, conventional TDM networks have numerous special features, in particular, those required to carry voice-grade telephony channels. These features imply signaling systems that support a wide range of telephony features, a rich standardization literature, and well-developed operations and management (OAM) mechanisms. All of these factors must be taken into account when emulating SDH/SONET over Ethernet.

Another critical issue in implementing SDH/SONET pseudowires is clock recovery. In native TDM networks the physical layer carries highly accurate timing information along with the TDM data, but when emulating TDM over Ethernet this physical layer clock is absent. TDM timing standards dictate precise and demanding requirements, and conformance with these requires careful design of Ethernet equipment.

A final issue that must be also addressed is SDH/SONET pseudowire packet loss concealment. Since TDM data is delivered at a constant rate over a dedicated channel, the native service may have bit errors, but data is never lost in transit. As all packet networks, also Ethernet suffers to some degree from packet loss, and this must be compensated when delivering TDM over Ethernet.

All these issues are addressed in the pseudowire standards applicable to carrier class Ethernet, and present days' carrier Ethernet switch can transport SDH/SONET signals so to allow legacy equipment to be integrated into the new metro network.

7.9.2 Architecture of a Carrier Class Ethernet Switch

From an architectural point of view, many issues relative to router design are replicated if carrier Ethernet switches design is considered.

From a functional point of view, the architecture of an Ethernet switch is based on distributed intelligence exactly like that of a backbone router.

Received frames are stored in a memory resident in the line card, after that the forwarding is performed in two steps: first the B-VLAN ID and the MAC address are read and sent to the forwarding processor. This processor first looks at the B-VLAN ID to identify the forwarding plane to which the considered frame belongs, then access the MAC addressing table using the MAC of the incoming frame to determine the address of the output port.

Generally, the B-VLAN assignment can be done via software, since the maximum number of possible B-VLAN is limited to 4096.

On the contrary, the MAC address table is generally implemented via a CAM to speed the frames forwarding.

References

1. Chao, H. J., Liu, B., *High Performance Switches and Routers*, Wiley-IEEE Press, Hoboken, NJ, s.l., 2007, ISBN-13: 978-0470053676.
2. Law, K. L. E., Leon-Garcia, A., A large scalable ATM multicast switch, *IEEE Journal on Selected Areas in Communications*, vol 5, num 3, 1997, pgs. 844–854.
3. Gebali, F., *Analysis of Computer and Communication Networks*, Springer, New York, s.l., 2008, ISBN: 9780387744377.
4. Kabaciński, W., *Nonblocking Electronic and Photonic Switching Fabrics*, Springer, New York, s.l., 2005, ISBN: 038725435.

5. Hsu, L.-H., Lin, C.-K., *Graph Theory and Interconnection Networks*, CRC Press, Boca Raton, FL, s.l., 2008, ISBN-13: 978-1420044812.

6. Katare, R. K., Chaudari, N. S., Study of topological properties of interconnects and it mapping to sparse matrix model, *International Journal of Computer Science and Applications*, 6(1), 26–39 (2009).

7. Iniewski, K., McCrosky, C., Minoli, D., *Network Infrastructure and Architecture*, Wiley, Hoboken, NJ, s.l., 2008, ISBN: 9780471749066.

8. Dimitrakopoulos, G., Chrysos, N., Galanopoulos, K., Fast arbiters for on-chip network switches, *International Conference on Computer Design, ICCD 2008*, Lake Tahoe, CA, IEEE, s.l., October 12–15, 2008.

9. Olesinski, W., Eberle, H., Gura, N., PWWFA: The parallel wrapped wave front arbiter for large switches, *Workshop on High Performance Switching and Routing*, Brooklin, NY, IEEE, New York, May 30–June 1, 2007.

10. Lee, T., A modular architecture for very large packet switches, *IEEE Transactions on Communications*, 28(7), 1097–1106 (1990).

11. Iannone, E., Sabella, R., Modular optical path cross-connect, *Electronics Letters*, 32(2), 125–126 (1996).

12. Fan, C. C., Bruck, J., Tolerating multiple faults in multistage interconnection networks with minimal extra stages, *IEEE Transactions on Computers*, 49(9), 998–1004 (2000).

13. Clos, C., A study of non-blocking switching networks, *Bell System Technical Journal*, 32(2), 406–424 (1953).

14. Hui, J. Y., *Switching and Traffic Theory for Integrated Broadband Networks*, Kluwer Academic Publishers, Norwell, MA, s.l., 1990, ISBN-13: 978-0792390619.

15. Chao, H. J., Liew, S. Y., Jing, Z., A dual-level matching algorithm for 3-stage Clos-network packet switches, *Proceedings of the 11th Symposium on High Performance Interconnects*, Stanford, CA, IEEE, s.l., August 20–22, 2003.

16. Huynh, D. T., Nguyen, H. N., A rearrangement algorithm for switching networks composed of digital symmetrical matrices, *Information Sciences*, 125(1–4), 83–98 (2000).

17. Keum, Y. W., Kim, S. C., Design and analysis of the high-performance interconnected Banyan switching fabric (IBSF), *Proceedings of the Second International Symposium on Parallel Architectures, Algorithms, and Networks*, Beijing, China, pp. 249–255, June 12–14, 1996.

18. Diaconisb, P., Grahamd, R.L., Kantor, W.M., The mathematics of perfect shuffles, *Advances in Applied Mathematics*, 4(2), 175–196 (1983).

19. Ellis, J., Fan, H., Shallit, J., The cycles of the multiway perfect shuffle, *Discrete Mathematics and Theoretical Computer Science*, 5, 169–180 (2002).

20. Kim, H. S., Leon-Garcia, A., A self-routing multistage switching network for broadband ISDN, *IEEE Journal of Selected Areas in Communications (JSAC)*, 8(3), 459–466 (1990).

21. Yu, C., Jiang, X., Horiguchi, S., Quo, M., Overall blocking behavior analysis of general Banyan-based optical switching networks, *IEEE Transactions on Parallel and Distributed Systems*, 17(9), 1037–1047 (2006).

22. Lea, C.-T.A., The load sharing Banyan network, *IEEE Transactions on Computers*, C-25(12), 1025–1034 (1986).

23. Narasimha, M. J., The Batcher-banyan self-routing network: Universality and simplification, *IEEE Transactions on Communications*, 36(10), 1175–1178 (1988).

24. Park, B.-S., Kim, S.-C., Analysis of queueing schemes for Batcher-Banyan network with speedup under various traffic loading, 1997, *International Conference on Communications, ICC 97*, Montreal, Canada, IEEE, s.l., June 8–12, 1997.

25. Khankder, Md. M. R., Jiang, X., Horiguchi, S., Blocking behavior of crosstalk-free pruned optical banyan networks, *Proceedings of the 13th International Conference on Networks*, Boston, MA, IEEE, Washington, DC, s.l., Vol. 1, p. 5, October, 2005.

26. Freeman, R. L., *Fundamentals of Telecommunications*, IEEE Press-Wiley Interscience, Hoboken, NJ, s.l., 2005, ISBN: 0471710458.

27. Obara, H., Efficient parallel time-slot interchanger for high-performance SDH/SONET digital crossconnect systems, *Electronics Letters*, 37(2), 81–83 (2001).

28. Ellanti, M. N., Gorshe, S. S., Raman, L. G., Grover, V. D., *Next Generation Transport Networks: Data, Management, and Control Planes*, Springer, New York, s.l., 2005, ISBN: 0387240675.

29. Acampora, A. S., *An Introduction to Broadband Networks*, Plenum Press, New York, s.l., 1994, ISBN: 0306445581.

30. Tasaka, S., *Performance Analysis of Multiple Access Protocol*, MIT Press, Cambridge, MA, s.l., 1986, ISBN-13: 978-0262200585.

31. IEEE /ISO, CSMA/CD Standard ISO/IEEE 802/3 (this integrates also old IEC TC83).

32. Kleinrock, L., *Queueing Systems. Volume 1: Theory*, Wiley, New York, s.l., 1975, ISBN-13: 978-0471491101.

33. Papadimitriou, G. I., Papazoglou, C., Pomportsis, A. S., Optical switching: Switch fabrics, techniques, and architectures, *IEEE Journal of Lightwave Technology*, 21(2), 384–405 (2003).

34. Zicin, S., Tunner, Z. P., Design advances in PCB/backplane interconnects for the propagation of high speed Gb/s digital signals, *Microwave Review*, 9(2), 11–18 (2003).

35. McKeown, N., A fast switched backplane for a gigabit switched router, *Business Communication Review*, 27(12), (1997).

36. Bo, D. C., Bangli, W., Kwasniewski, L., A simulator for high-speed backplane transceivers, *Proceedings of 11th International Conference on Computer MOdeling and Simulation, UKSIM09*, Cambridge, England, U.K., IEEE, s.l., pp. 589–593, March 25–27, 2009.

37. Lin, L., Noel, L. L., Kwasniewski, P. T., Implementing a digitally synthesized adaptive pre-emphasis algorithm for use in a high-speed backplane interconnection, *Proceedings of Canadian Conference on Electrical and Computing Engineering*, Ontario, Canada, Vol. 3, pp. 1221–1224, May 2–5, 2004.

38. Cartier, M., Cohen, T., Patel, G., Smith, J., Signal integrity considerations for 10.0 Gbps transmission over backplane systems, *DesignCon 2001*, Santa Clara, CA, s.n., January 29–February 1, 2001.

39. Pajovic, M. M., Jinghan Yu Potocnik, Z., Bhobe, A., Gigahertz-range analysis of impedance profile and cavity resonances in multilayered PCBs, *IEEE Transactions on Electromagnetic Compatibility*, 52(1), 179–188 (2010).

40. Chung, H., Wei, G.-Y., Design-space exploration of backplane receivers with high-speed ADCs and digital equalization, *Custom Integrated Circuits Conference, CICC '09*, San Jose, CA, IEEE, s.l., pp. 555–558, September 13–16, 2009.

41. Michalzik, R., Optical backplanes, board and chip interconnects, *Handbook of Fiber Optic Data Communication: A Practical Guide to Optical Networking*, Elsevier, Amsterdam, the Netherlands, s.l., 2008.

42. Grivas, E., Kyriakis-Bitzaros, E. D., Halkias, G., Katsafouros, S., Morthier, G., Dumon, P., Baets, R., Farell, T., Ryan, N., McKenzie, I., Armadillo, E., Wavelength division multiplexing based optical backplane with arrayed waveguide grating passive router, *Optical Engineering*, SPIE, 47(2), (2008).

43. Goossen, K. W., Fitting optical interconnects to an electrical world-packaging and reliability issues of arrayed optoelectronic modules, *Proceedings of the 17th Annual Meeting of the IEEE Lasers and Electro-Optics Society, LEOS 2004*, Rio Grande, TX, Vol. 2, pp. 653–654, November 7–11, 2004.

44. Guidotti, D. et al., Reliability considerations in parallel optical interconnects, *Proceedings of 59th Electronic Components and Technology Conference ECTC 2009*, San Diego, CA, IEEE, s.l., pp. 2081–2085, May 26–29, 2009.

45. Gwyer, D., Misselbrook, P., Bailey, C., Conway, P. P., Williams, K., Polymer waveguide and VCSEL array multi-physics modelling for OECB based optical backplanes [opto-electrical circuit boards], *Proceedings of the 5th International Conference on Thermal and Mechanical Simulation and Experiments in Microelectronics and Microsystems, 2004, EuroSimE 2004*, Brussels, Belgium, pp. 399–406, May 10–12, 2004.

46. IEEE, IEEE Standard for a High-Performance Serial Bus—2006 Amendment to IEEE 1394b/2000.

47. Bierhoff, T., Schrage, J., Halter, M., Betschon, F., Duis, J., Rietveld, W., All optical pluggable board-backplane interconnection system based on an MPXTM-FlexTail connector solution, *Photonics Society Winter Topicals Meeting Series (WTM)*, Majorca, Spain, IEEE, s.l., pp. 91–92, January 11–13, 2010.

48. Pitwon, R., Hopkins, K., Milward, D., Selviah, D. R., Papakonstantinou, I., Wang, K., Fernández, F. A. (Eds.), High speed optical backplane connectors technology, *International Symposium on Photonic Packaging, Fraunhofer IZM & VDI/VDE-IT*, s.n., Munich, Germany, 2006.

49. Williams, K. A., Glick, M., Lin, T., Roberts, G. F., Penty, R. V., White, I. H., Monolithic integration of semiconductor optical switches for optical interconnects, INTEL Research Public Report, 2004, www.intel.com (accessed: August 20, 2010).

50. Denic, S. Z., Vasic, B., Charalambous, C. D., Chen, J., Wang, J. M., Information theoretic modeling and analysis for global interconnects with process variations, *IEEE Transactions on Very Large Scale Integration (VLSI) Systems*, 99, 1–14 (2009).

51. Case, C., Interconnect Working Group activity review and targets, *ITRS Interconnect Working Group Meeting*, 2007.

52. IEEE Standards. 802.3—Section 4—Chapter 47 XGMII Extender Sublayer (XGXS) and 10 Gigabit Attachment Unit Interface (XAUI).

53. LoCicero, J. L., Patel, B. P., Line coding [book auth.]. In Suthan S. Suthersan (Ed.) *Mobile Communication Handbook*, CRC Press, Boca Raton, FL, 1999.

54. Tuker, R., Modeling energy consumption of IP networks, *European Conference on Optical Communications (ECOC 2008) Symposium*, Brussels, Belgium, September 21–25, 2008.

55. Helvoort, H. van., *The ComSoc Guide to Next Generation Optical Transport: SDH/SONET/OTN (ComSoc Guides to Communications Technologies)*, Wiley, Hoboken, NJ, s.l., 2009, ISBN-13: 978-0470226100.

56. Askarian, A., Zhai, Y., Subramaniam, S., Pointurier, Y., Brandt-Pearce, M., Cross-layer approach to survivable DWDM network design, *Journal of Optical Communications and Networking*, 2(6), 319–331 (2010).

57. Ramaswami, R., Sivarajan, K. N., Sasaki, G. H., *Optical Networks*, 3rd revised edn., Morgan Kaufmann-Elsevier, San Francisco, CA, s.l., 2009, ISBN: 9780123740922.

58. Van Caenegem, B. et al., Internal connectivity of optical crossconnects in opaque networks, *Optical Fiber Communication Conference, OFC 1999*, San Diego, CA, IEEE, s.l., Vol. 1, pp. 159–161, February 21–26, 1999.

59. Iannone, E., Bentivoglio, F., The opaque optical network, *Optical Networks Magazine*, 1(4), 24–31 (2000).

60. Arijs, P. et al., Architecture and design of optical channel protected ring networks, *IEEE Journal of Lightwave Technology*, 19(1), 11–22 (2001).

61. Indumathi, T. S. et al., Routing analysis and flexibility of ROADMs in all optical networks, *Proceedings of 24th International Conference on Advanced Information Networking and Applications Workshops (WAINA)*, Perth, Australia, IEEE, s.l., pp. 75–78, April 20–23, 2010.

62. Ennser, K. et al., Reconfigurable add/drop multiplexer design to implement flexibility in optical networks, *Proceedings of International Conference on Transparent Optical Networks ICTON 2006*, Nottingham, U.K., Vol. 3, pp. 74–77, June 18–22, 2006.

63. Strasser, T. A., Wagener, J. L., Wavelength-selective switches for ROADM applications, *IEEE Journal of Selected Topics in Quantum Electronics*, 16(5), 1150–1157 (2010).

64. Collings, B. C., Wavelength selectable switches and future photonic network applications, *International Conference on Photonics in Switching, PS '09*, Pisa, Italy, IEEE, s.l., pp. 1–4, September 15–19, 2009.

65. Hernandez-Valencia, E., Rosenfeld, G., The building blocks of a data-aware transport network: Deploying viable Ethernet and virtual wire services via multiservice ADMs, *IEEE Communication Magazine*, 42(3), 104–111 (2004).

66. Vegesna, S., *IP Quality of Service*, Cisco Press, s.l., 2001, ISBN: 1578701163.

67. Aweya, J., IP router architectures: An overview, *International Journal of Communication Systems*, 14(5), 447–475 (2001).

68. Lawton, G., Routing faces dramatic changes, *Computer*, 42(9), 15–17 (2009).

69. Csaszar, A. et al., Converging the evolution of router architectures and IP networks, *IEEE Networks*, 21(4), 8–14 (2007).

70. Chang, Y.-K., Liu, Y.-C., Kuo, F.-C., A pipelined IP forwarding engine with fast update, *International Conference on Advanced Information Networking and Applications, AINA '09*, Bradford, U.K., IEEE, s.l., pp. 263–269, May 26–29, 2009.

71. Zhu, Y., Jiang, W., Load-aware bidirectional pipeline construction for terabit IP forwarding, *Proceedings of 43rd Annual Conference on Information Sciences and Systems, CISS 2009*, Baltimore, MD, IEEE, s.l., pp. 701–706, March 18–20, 2009.

72. Shukerski, T., Lazarov, V., Kanev, I., A distributed architecture of IP routers, *International Conference on Computer Systems and Technologies—CompSysTech' 2005*, Varna, Bulgaria, Vol. 1, pp. 4–1, 4–6, June 16–17, 2005.

73. Bang, Y.-C., Lee, W.B., Choo, H., Rao, N. S. V., Architecture for internal communication in multigigabit IP routers, *International Conference on Computer Systems ICCS 2003*, Melbourne, Australia, Springer, s.l., pp. 495–503, June 2–4, 2003.

74. Hsieh, S., Huang, Y., Yang, Y., Multi-prefix trie: A new data structure for designing dynamic router-tables, *IEEE Transactions on Computers*, 60(5), 693–706 (2011).

75. Bando, M., Chao, H. J., FlashTrie: Hash-based prefix-compressed trie for IP route lookup beyond 100 Gbps, *Proceedings of INFOCOM 2010*, San Diego, CA, IEEE, s.l., pp. 1–9, March 14–19, 2010.

76. Lim, H., Yim, C., Swartzlander, E. E., Priority tries for IP address lookup, *IEEE Transactions on Computers*, 59(6), 784–794 (2010).

77. Maurya, S. K., Clark, L. T., A dynamic longest prefix matching content addressable memory for IP routing, *IEEE Transactions on Very Large Scale Integration (VLSI) Systems*, 19(6), 963–972 (2011).

78. CAMs, IP Prefix Matching with Binary and Ternary, IP prefix matching with binary and ternary CAMs, *Proceedings of 7th Consumer Communications and Networking Conference (CCNC)*, Las Vegas, NV, IEEE, s.l., pp. 1–2, January 9–12, 2010.

79. Golden, P., Dedieu, H., Jacobsen, K. S., *Implementation and Applications of DSL Technology*, Auerbach, Boca Raton, FL, s.l., 2007, ISBN-13: 978-0849334238.

80. Kazovsky, L. G., *Broadband Access Networks*, Wiley, New York, s.l., 2010, ISBN-13: 978-0470182352.

81. Yoo, S. J. B., Fei Xue Bansal, Y., Taylor, J., Zhong Pan Jing Cao Minyong Jeon Nady, T., Goncher, G., Boyer, K., Okamoto, K., Kamei, S., Akella, V., High-performance optical-label switching packet routers and smart edge routers for the next-generation Internet, *IEEE Journal of Selected Areason Telecommunications*, 21(7), 1041–1051 (2003).

82. Louati, W., Jouaber, B., Zeghlache, D., Configurable software-based edge router architecture, *Computer Communications*, 28(14), 1692–1699 (2005).

83. Gumaste, A. et al., Omnipresent Ethernet—Technology choices for future end-to-end networking, *IEEE Journal of Lightwave Technology*, 28(8), 1261–1277 (2010).

84. Xiao, S. et al., Design and implementation of aggregation platform for extended Ethernet transport and services, *Proceedings of Global Telecommunications Conference, GLOBECOM 2009*, Honolulu, HI, IEEE, s.l., pp. 1–7, November 30–December 4, 2009.

85. Support, IETF and ITU-T pseudo-wiring standards for different native signals and packet.

86. All relevant standards of the ATCA mechanics and control platform are reported in the official ATCA site: http://www.picmg.org/v2internal/resourcepage2.cfm?id=2

87. Shaha, D., Gupta, P., *Fast updating algorithms for TCAM*, Micro IEEE, vol. 21, n° 1, pp 36-47, Jan–Feb 2001.

88. Chad R. Meiners, Alex X. Liu and Eric Torng, *Hardware Based Packet Classification for High Speed Internet Routers*, Springer, june 23, 2010, ISBN-13: 978-1441966995.

89. Reddy, P. M., *Fast updating algorithm for TCAMs using prefix distribution prediction*, International Conference On Electronics and Information Engineering (ICEIE), 2010, proceedings, vol 1, pp. V1-400–V1-404.
    IETF-RFC 3985 Pseudo Wire Emulation Edge-to-Edge PWE3 Architecture
    IETF-RFC 4385 Pseudowire Emulation Edge-to-Edge PWE3 Control Word for Use over an MPLS PSN
    IETF-RFC 4448 Encapsulation Methods for Transport of Ethernet over MPLS Networks

IETF-RFC 4447 Pseudowire Setup and Maintenance-Using the Label Distribution Protocol LDP

IETF-RFC 4553 Structure-Agnostic Time Division Multiplexing TDM over Packet SAToP

IETF-RFC 4623 Pseudowire Emulation Edge-to-Edge PWE3 Fragmentation and Reassembly

IETF-RFC 4618 Encapsulation Methods for Transport of PPP/High-Level Data Link Control HDLC over MPLS Networks

IETF-RFC 4619 Encapsulation Methods for Transport of Frame Relay over Multiprotocol Label Switching MPLS Networks

IETF-RFC 4720 Pseudowire Emulation Edge-to-Edge PWE3 Frame Check Sequence Retention

IETF-RFC 4717 Encapsulation Methods for Transport of Asynchronous Transfer Mode ATM over MPLS Networks

IETF-RFC 4816 Pseudowire Emulation Edge-to-Edge PWE3 Asynchronous Transfer Mode ATM Transparent Cell Transport Service

IETF-RFC 4842 Synchronous Optical Network/Synchronous Digital Hierarchy SONET/SDH Circuit Emulation over Packet CEP

IETF-RFC 5087 Time Division Multiplexing over IP TDMoIP

IETF-RFC 5086 Structure-Aware Time Division Multiplexed TDM Circuit Emulation Service over Packet Switched Network CESoPSN

IETF-RFC 5085 Pseudowire Virtual Circuit Connectivity Verification VCCV: A Control Channel for Pseudowires

IETF-RFC 5287 Control Protocol Extensions for the Setup of Time-Division Multiplexing TDM Pseudowires in MPLS Networks

ITU-T-Y.1411 ATM pseudowires

ITU-T-Y.1412 AAL5 pseudowires

ITU-T-Y.1413 TDM pseudowires

ITU-T-Y.1414 Voice Services pseudowires

ITU-T-Y.1415 Ethernet pseudowires

ITU-T-Y.1418 Pseudowire Layer Networks

ITU-T-Y.1452 Voice Services over IP

ITU-T-Y.1453 TDM over IP

ITU-T-X.84 Frame Relay pseudowires

8

Convergent Network Management and Control Plane

8.1 Introduction

In Chapter 2 we have seen that the main target of major telecommunication carriers is to decrease the operational expenditure (OPEX) necessary to maintain and operate the network. In Chapter 3 we have seen that there are different implementations of IP over optical architecture, but it is a fact that the old architecture made by two independent networks, one for data and one for telephone, has almost disappeared everywhere.

With the convergence of transport over one architecture, independently on the service bundles offered by the carrier to different categories of customers, also the division of the network management in two completely different architectures becomes inefficient: central management for the physical transport (SDH/SONET and optical transport network [OTN]) and automatic control plane for the packet layer (IP and Ethernet) have to merge.

As a matter of fact, with the network technology evolution and the introduction of a great number of new services, static centralized management is demonstrated to be not suitable for the carriers needs.

It is true that TMN-based management had the merit of introducing a never before seen ability of controlling the network and of assuring performances and reliability, but with a traffic more and more dynamic, the great number of manual operations required by centralized network managers are a source of both costs and errors.

On the other hand, the packet layer is managed through an automatic control plane that, even if is sometimes poor in terms of reporting with respect to the carriers expectations, is very effective in decreasing costs and eliminating errors.

Thus, it is quite natural to envision a new control plane that is capable of automatically managing all the network layers in an integrated manner and is also sufficiently rich in functionalities so as to be at least comparable with traditional TMN.

Different initiatives are ongoing to create new control plane standards. In particular, ITU-T, the world standardization institute that is traditionally carrier-driven, has developed the automatic switched optical network (ASON) standard with the aim of providing a framework for the development of the convergent network control plane [1].

Also IETF, that is the facto the Internet standardization body and is traditionally vendor-driven, has created the generalized-multiprotocol label switch (GMPLS) protocol suite, a complete control plane whose standardization is in some parts still ongoing, whose target is to generalize the ideas at the base of MPLS [32, 33, 34] to be able to manage not only packet-oriented networks, but also circuit-oriented networks like the WDM layer or the next generation SONET layer.

At a first glance, it could seem that ASON and GMPLS are two competing standards and that only one of them will prevail with time, but this is completely false. As a matter of fact, ASON is not a protocol suite, even if it gives a lot of recommendations about protocols, but it is more an architecture and a framework in which the protocols of the new control plane have to fit. Thus, it is well possible to fit GMPLS protocols in the ASON architecture, even if this is not completely free of work since the two standards were born independently.

The work of fitting GMPLS into the ASON architecture is ongoing in another industry standardization group: the Optical Internetworking Forum (OIF). The OIF is issuing a series of documents called implementation agreements (AI) [2] whose role is exactly to fit GMPLS into the ASON architecture pointing out where there are difficulties, corrections to be applied to existing GMPLS protocols or different interpretations of the ASON architecture.

Moreover, OIF is promoting a wide experimentation activity whose goal is to set up test beds for the new control plane that verify the functionalities of its parts. This is a very important activity since, if there is a sort of progressive experimental verification of the functions of the new control plane, they can be introduced gradually into the carrier networks, thereby distributing the costs over a long time and allowing the new procedures to be introduced progressively.

This chapter is divided into three parts: in the first we will analyze the ASON architecture, in the second we will review the GMPLS protocol suite, and in the third we will discuss how a multilayer network based on an integrated control plane can be designed and optimized.

8.2 ASON Architecture

8.2.1 ASON Network Model

Historically the control plane accompanies the packet-switched networks, which are not supposed to have a rich management plane as the transport networks. The first point to solve if a control plane is to be developed so to manage a whole network how will the control plane protocols and the network manager interact and how will they divide the functions.

As a matter of fact, several control plane functions are performed in traditional transport networks by the central network manager so that both control plane and network manager have to be redefined in a functional sense.

The ASON architecture states that the control plane is responsible for the actual resource and connection management within an automatically switched network (ASN). The control plane resides in a distributed network intelligence constituted by the optical connection controllers (OCCs); the OCC can run in separate workstations as the element managers but generally directly reside into the control unit of the equipments dedicated to switching at each network layer. The OCCs are interconnected via the interface called network to network interfaces (NNI) and run the control plane protocol suite having the following functions:

- Network topology discovery (that is, resource discovery)
- Address assignment to an equipment port when it is discovered and address advertisement to all the networks
- Signaling (connections setup, management, and tear down)
- Connections routing

- Connection protection/restoration
- Traffic engineering
- Resource assignment (e.g., wavelength assignment in the WDM layer, time slot assignment in the SONET layer etc.)

The management plane is responsible for managing the control plane. Its responsibilities include the following:

- Configuration management of the control plane resources
- Definition of the administrative areas
- Control plane policies definition

Few of the traditional network manager functionalities should also be maintained on the data plane entities:

- Fault management
- Performance management
- Accounting and security management

The management plane is constituted by the central management unit, which runs the network manager, and the local management units, which run the element managers, and by the management network that connects all the entities of the management plane.

The central network management entity is connected to an OCC in control plane via the network management interface for ASON control plane (NMI-A) and to one of the switches via network management interface for the transport network (NMI-T).

Using these interfaces, the management plane can access to all the control plane entities via the control plane connectivity to perform control plane management and to all the data plane entities to perform reporting of data plane status and those functionalities that are responsibility of the management plane. The management structure of the ASON network is illustrated in Figure 8.1.

The network model on which the ASON standard is based is a multilayer model in which a single control plane coordinates the activity of all the layers. This vertical structure is reproduced in Figure 8.2, where there is evidence to show that all the layers define a specific physical or virtual connection, so that in its whole we can talk about a virtual connection–oriented network.

Connectionless layer can be clients of an ASON network, but cannot be managed by an ASON control plane. The ASON network is intrinsically a multilayer network and different network topologies can be defined at each layer. Let us refer just for an example to Figure 8.3 where a three layer network is represented.

At the base physical layer, there is the set of fiber cables represented by gray bold lines and the set of buildings that are the carrier points of presence in the area.

In all the buildings there is an optical cross connect (OXC), but not all the cables are equipped with WDM systems, quite a realistic situation due to the fact that cables are generally deployed with long-term plans that are completely independent of the network that will be realized on the cable infrastructure.

The WDM layer has probably different clients, one of that is an MPLS layer composed by a set of label-switched routers (LSRs). Not in all the nodes there is a LSR, but only in

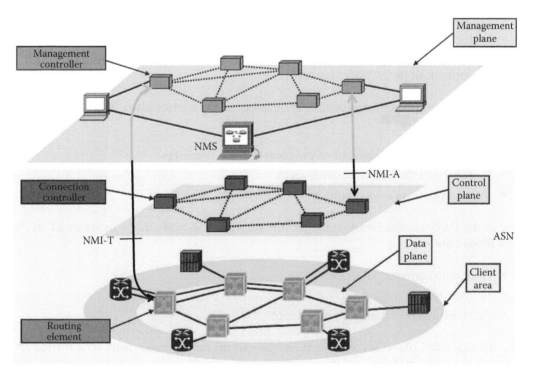

FIGURE 8.1
ASON standard layered model for the Automatic Switched Network management; standard interfaces between management and control plane (NMI-A) and between management and data plane (NMI-T) are shown.

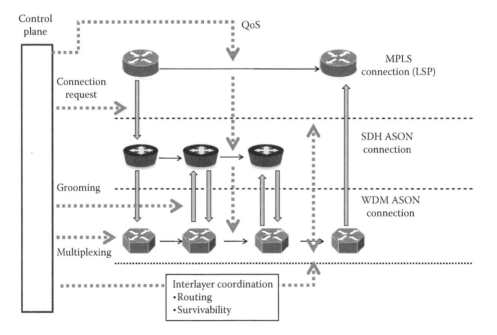

FIGURE 8.2
ASON Network model: a multilayer network in which a single control plane coordinates the activity of all the layers.

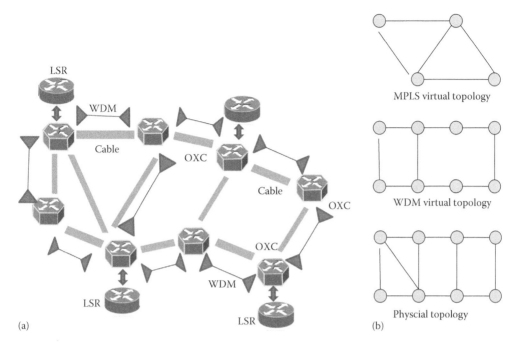

FIGURE 8.3
Scheme of a three layer network where the deployed equipment is shown in the physical topology plot (left side) and the virtual topologies at each layer are evidenced as graphs (right side).

the nodes where such a machine is needed. As a consequence of this complex network architecture, three network topologies can be proposed: a physical network topology composed by cable ducts and buildings, a WDM virtual network topology composed by WDM systems and OXCs, and a MPLS virtual network topology composed by LSRs with the connectivity that is offered to them by the underlying transport layer.

The different topologies are shown as graphs in Figure 8.3.

The situation of this example can generally be replayed in all multilayer network architecture, identifying a specific topology (physical of virtual) at each layer of the network. Identifying the correct topology is very important since in a multilayer network managed by a single control plane, every routing element refers for its routing protocols to the topology that is specific to that layer.

Under a horizontal point of view, ASON is organized in administrative domains. The administrative domain is a part of the network that has its own autonomy, for example, the network belonging to a carrier in the framework on a national multicarrier network.

The concept of administrative domain is needed to allow the network to hide its own internal data to the overall network and to encapsulate all the internal addressing into a global network address so that they cannot be seen by other administrative domains. This hiding and encapsulating strategy has a twofold function: it is needed in order to allow carriers to hide data relative to their network to competitors even if roaming among different carriers have to be provided by the control plane, but it is also needed to reduce the dimension of the routing tables in the network routing elements. As a matter of fact, when the network is divided into administrative domains, every switching element stores the complete topology of its own domain and a metatopology relative to the connections with different domains. The partition of the ASON network is shown in Figure 8.4.

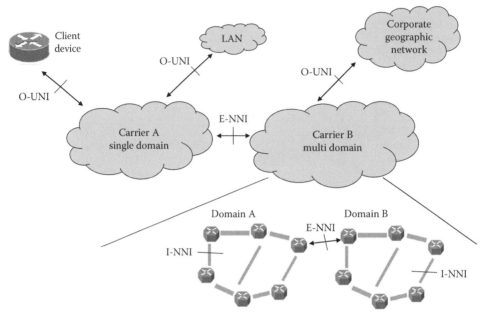

FIGURE 8.4
Domain partition of the ASON network model.

When a connection has to be set-up with a node in a different domain, the precise node address is encapsulated into the global domain address. The switch that has to set up the connection simply send a connection set-up protocol to the point of contact with the destination domain and it will be this node that is inside the domain of the connection end point to complete the connection setup establishing the path into the destination domain. This procedure is shown in Figure 8.5, where the logical working of encapsulation is shown.

Let us image that the address of every node is a couple of letters, so that node A of domain A has an address (A, A). Let us image that a connection has to be established from (A, A) to (B, B). The node (A, A) receives the instruction to open a connection to (*, B). The exact destination node address is hidden since it belongs to a different administrative domain whose topology is not known to node (A, A).

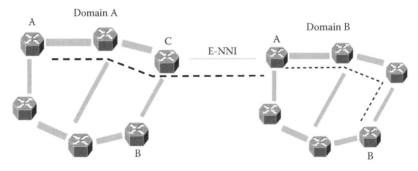

FIGURE 8.5
Example of inter-domain routing in an ASON network.

Node (A, A) has a routing table knowing that the external network to network (E-NNI) interface toward administrative domain B is between node (C, A) and (A, B) so node (A, A) sets up a link to node (C, A) that forwards the link through the interdomain interface to node (A, B).

Now Node (A, B) knows that a connection between node (*, A) is to set up with node (B, B) starting from the connection that was set up through the interdomain interface with domain A.

Node (A, B) knows exactly from the routing table the position within domain B of the node (B, B) and set up the second part of the connection, represented in the figure by the dotted line.

Besides the separation of information belonging to competitor carriers, in our example the administrative domain separation cause every node to store a topology made of six nodes: five in the same domain and one is a virtual node that is the connection with the other domain.

If all the information was shared among all the network nodes, every node should store a topology of 11 nodes. The concept of administrative domain is also a powerful instrument for the gradual introduction of the ASON architecture in carrier networks. As a matter of fact, an abrupt introduction of a new architecture in a big network like that of an incumbent carrier is not feasible. The new technology is in general introduced gradually in pace with CAPEX expenditure and market request.

Thus during the introduction of the ASON control plane, for a long time there will be parts of the carrier network that will include subsets of ASON features, possibly different subsets depending on the circumstances and on the kind of equipment the carrier has installed in that part of the network, and parts of the network that are not yet migrated.

The network has to work in this hybrid situation without any service disruption: this can be achieved by defining every area of the network where there is a different ASON-compatible technology as a different administrative domain, and interface these administrative domains with those parts of the network where ASON is not yet implemented by suitable interfaces.

Besides the administrative domains, the ASON standard defines another horizontal entity: the routing area. The routing area is generally a subpart of an administrative domain that has the characteristic to be represented by other routing areas like a single virtual node.

Thus, exactly like for the administrative domain, when a node external to the routing area wants to set up a connection with a node inside the routing area it sends the connection setup protocol to the virtual node representing all the routing area that physically corresponds to the point of contact of the destination routing area with the origination one.

Only when the connection setup protocol arrives at the gateway of the routing area it completes the connection setup inside the area. Differently from the administrative domain, routing areas are not insulated under an administrative point of view and the internal information is not completely hidden.

This is due to the different destination of the routing areas with respect to the administrative domains: routing areas are used inside a big administrative domain to reduce the complexity of the address tables of the routing entities by introducing a routing hierarchy while the administrative domain serves mainly to divide parts of the network that have to be administrated independently either because they are owned by different carriers or because they are built on different technologies.

8.2.2 ASON Standard Interfaces

In correspondence to the structure of the network in administrative domains, the ASON standard defines a set of interfaces that are conceived for compatibility between different equipments and different particular implementations of the control plane.

As a matter of fact, it should be remembered that ASON is not a complete protocol suite, but more a reference architecture. Thus it is well possible that different vendors use different specific sets of protocols, all ASON-compatible.

In this case, naturally every different protocol suite will be confined to a different administrative domain and suitable interfaces will guarantee the cross-compatibility.

The first standard interface that is the optical user network interface (O-UNI). This is the interface connecting into the data plane the client equipment with a network equipment through an optical signal.

Somehow the O-UNI is a limited functionality interface, since it has also to shield the network from any improper client activity that could either interfere with the network working or try to detect information about the network that the carrier wants to hide to the customer.

The activity of the O-UNI is thus limited to resource discovery, signaling and, naturally, data exchange. In particular, the following protocols flow through the O-UNI:

- Connection request and creation
- Connection parameters change
- Connection delete
- Client identity change
- Data transit

Just to do an example, it is through the O-UNI that the Ethernet switch belonging to a corporate network asks to the public network the creation of a virtual connection for a data center back-up operation.

In particular, the ASON-enabled Ethernet switch sends through the O-UNI a suitable signal to the multilayer node to which it is connected for specifying the request of a mission-critical data connection between two points of the network. The desired format is Gbit Ethernet and the network back-up service foresees a specific service level agreement.

The ASN uses all the control planes to provision the connection and, when the connection is correctly provisioned, a return message is returned through the O-UNI to the switch that starts the data transfer.

At the end of the back-up, the private network manager decides not to destroy the connection, but to reduce its capacity to 100 Mbit/s and to use it for another scope. This step is signaled through an O-UNI-compatible protocol to the network that makes all the steps that are needed to change not only the capacity of the connection but also the related SLA. All the public network planes are involved in the operation: the data plane has to be suitably reconfigured by the control plane protocols so as to find an efficient way to satisfy the customer request (in case the network could also decide to destroy the old connection and to set up a new one) while the management plane is engaged in guaranteeing, among the other things, that the SLAs with other customers are not violated, that the billing is done in the correct way and that all is reported to the central management place.

After a certain time also, the new connections become useless, at that point the switch of the private network uses the O-UNI-compatible protocol to delete the connection, an operation that forces the network to a connection tear down.

A second interface completely included into the data plane is the E-NNI interface, that is, the interface that is defined to connect different administrative domains.

Due to its definition, the E-NNI can connect areas of the network that are managed by different carriers, but also areas that implement different ASON-compatible protocol suites.

Due to the fact that internal data relative to an administrative domain have to be hidden to entities belonging to a different administrative domains; also the E-NNI interface has the role to allow a limited number of interactions between the different administrative domains and to avoid that one of them makes undue influences on the working of the other.

All the functionalities carried out by the O-UNI are also carried out by the E-NNI; in addition, the E-NNI has to manage cross-domain survivability.

An example is given in Figure 8.6, where end-to-end dedicated optical channel protection is carried out to protect an optical layer connection between two OXC's belonging to different administrative domains. The working path (the dotted line in the figure) is used in the normal network status to convey low-priority traffic.

Let us image that a failure happens on the working path (the continuous line path) between nodes (D, A) and (E, A). The connection originator node (A, A) performs low-priority traffic switch-off and commutes the traffic to be protected on the working path.

However, this operation has to be advertised also to the nodes of the domain B to drive resources reservation in the transit nodes and correct switching of the connection destination node.

This is a specific role of the E-NNI-compatible protocols that have to do that without allowing one domain to know hidden information from the other domain.

In conclusion, the functions carried out by the E-NNI are as follows:

- Connection request and creation
- Connection parameters change
- Connection delete
- Client identity change
- Data transit
- Protection/restoration management

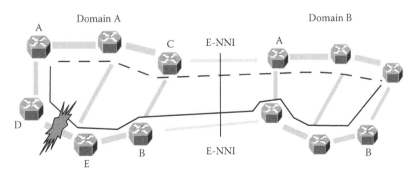

FIGURE 8.6
Inter-domain path protection via the E-NNI interface.

The third standard interface is the I-NNI (internal Network to Network Interface) that is designed to connect network equipments belonging to the same administrative domain.

All control plane protocols transiting through the I-NNI interface have the role to exchange information. While the O-UNI and the E-NNI are objects of a great standard work within ITU-T and OIF, since they assure the cross-compatibility of any specific control plane implementation and its ability to interwork with its clients, the I-NNI is specific to any control plane implementation and has not been standardized in detail.

The functionalities of the I-NNI are strictly related to the functionalities of the control plane and it can be said that the best I-NNI definition is that it has to be able to convey all the protocols constituting the control plane.

8.2.3 ASON Control Plane Functionalities

The ASON architecture, besides the general structure of the ASN network and the interfaces between different network entities, define the functionalities the protocol suite composing the control plane has to carry out.

In general, ASON does not assume that vertical interoperability will exist between different implementations of the control plane, since the internetworking is guaranteed by the possibility of assigning to any technology its own administrative domain and to connect them via the O-UNI, but if different implementations of the protocols are ASON-compatible, they have to provide the same basic set of functionalities.

8.2.3.1 Discovery

The discovery function can be divided into three different subfunctions:

1. Neighbor discovery
2. Resource discovery
3. Service discovery

The neighbor discovery is responsible for determining the state of local links connecting to all neighbors. This kind of discovery is used to detect and maintain node adjacencies; without it, it would be necessary to manually configure the interconnection information in management systems or network elements.

The neighbor discovery usually requires some manual initial configuration and automated procedures running between adjacent nodes when the nodes are in operation.

Three instances of neighbor discovery are defined in ASON, that is

1. Physical media adjacency discovery
2. Data plane layer entities adjacency discovery
3. Control entities logical adjacency discovery

Physical media discovery has to be done first to verify the physical connectivity between two ports. Generally this verification is carried out through an exchange of messages between adjacent network entities that verify both the network entities and the connecting link functionality.

The layer adjacency discovery is used for building the layer specific network topology (see Figure 8.3) that has to be stored into the local CC to support routing.

As a matter of fact, logical adjacencies between both data and control entities are created by layer adjacency discovery. Moreover, layer adjacency discovery, fixing the topology of the layer, also identifies link connection endpoints that are needed to describe connections through the cascade of links composing them, to manage such connections in correspondence to client request changes and to assure proper network working when a failure happens.

Discovery processes involve exchange of messages containing identity attributes. Relevant protocols may operate in either an acknowledged or unacknowledged mode. In the first case, the discovery messages can contain the near-end attributes and the acknowledgment can contain the far-end identity attributes. The service capability information can be also contained in the acknowledgment message.

In the unacknowledged mode both ends send their identity attributes. Recommendation G.7714 discusses the following two discovery methods:

1. Trace identifier method
2. Test signal method

In the trace identifier method, the discovery is carried out by using the trail identifiers that are included in all the overheads of the transport ITU-T standards (both in SONET/SDH and in OTN, see Chapter 2).

In particular, the trail termination points are first identified and then links connections are inferred. Just to give an example, let us imagine to have an ASON-managed next generation SONET network and to consider path layer discovery. A trace identifier discovery protocol will reside into the CC and will detect in all the network entities if the J0 field of the SONET path layer header is terminated and regenerated (the J0 field is the trail identifier in this case).

Once all the points in which the J0 is terminated and recreated are detected, the path layer logical topology is inferred by connecting the point where a value of J0 is created with the point where such a value is terminated with a virtual link (that is, links at the path layer).

It is to be observed that, under a layering point of view, the trail reports an information that is typical of the service layer, thus the trace identifier method performs discovery in layer one by exploiting also information relative to the client layer.

In the test signal method, test signals are used to directly find associations between subnetwork termination points (see network layering, Chapter 3) without discovering any service layer trails. While the use of overhead elements identifying a trail allows discovery to be carried out during normal network working, the use of data field to send test messages implies that the client data transmission is suspended. Thus the first discovery method is a type of in-service discovery, while the second is a type of out-of-service discovery.

The resource discovery has a wider scope than the neighbor discovery. It allows every node to discover network topology and resources. This kind of discovery determines what resources are available, what are the capabilities of various network elements, and how the resources are protected. It improves inventory management as well as detects configuration mismatches.

It has to be considered that, if an OXC has five DWDM output ports, this means that it switches probably more than 600 optical channels and that more than 1200 fiber patchcords (every channel comes in and goes out) physically comes out from some OXC card to be plugged on some other card or on an external equipment. Since physical connections of fiber patch-cords have to be done manually, the probability that some error occurs when a new equipment is installed or when an upgrade is done is not negligible.

Many errors can be detected via the equipment self-test, but not all, and there is a probability that an error manifests its presence much later with respect to the equipment installation, since without any specific test, it will become evident only when a certain card is really used (let us think, e.g., about redundancy cards). Thus, having an automatic procedure that through the discovery of network links at each network layer points out errors in connecting different equipments is very important to simplify the network operation.

The service discovery is responsible for verifying and exchanging information about service capabilities of the network, for example, services supported over a trail or link. Such capabilities may include, as an example, the quality of service (QoS) a certain network link is capable of providing in terms for example of error probability, connectivity restoration time in case of failure or packets average delay.

8.2.3.2 Routing

Routing is used to select paths for establishment of connections through the network. Although some of the well-known routing protocols developed for the IP networks can be adopted, it has to be noted that optical technology is essentially an analog rather than digital technology, thus routing is strongly influenced by the transparency degree of the network. To be specific, several types of transparency can be defined at different network layers. Examples are as follows:

- Service transparency: the layer ability to work ignoring the services managed by the client layer
- Protocol transparency: the ability to transport any client layer protocol
- Bit Rate transparency: the ability to work at any bit rate within a given maximum
- Optical transparency: the absence of network elements where the signal is electronically regenerated

On the ground of the aforementioned definitions, any OTN network is service and protocol transparent, while it is not bit rate transparent. The transparency that is relevant discussing about routing is the optical transparency. The optical layer of a network can be divided into optically transparent isles, which are network areas where the optical signal is never electronically regenerated.

The optical transparency areas can be so small as a single DWDM link connecting two electrical core OXCs or so big as an entire routing zone where optical interconnected rings are deployed. While the optical signal traverse a single optical transparent area, transmission impairments accumulated along the optical paths and this has to be taken into account while calculating the route (see Chapter 10 for details). Under the point of view of the routing strategies, ASON supports hierarchical, source-based, and step-by-step routing.

In the first case, CCs are related to one another in a hierarchical manner as we have seen defining administrative domains and routing areas. Source routing is based on a federation of distributed connection and routing controllers. The path is selected by the first connection controller in the routing area. This component is supported by a routing controller that provides routes within the domain of its responsibility. Step-by-step routing requires less routing information in the nodes than the previous methods. In such a case path selection is invoked at each node to obtain the next link on a path to a destination.

8.2.3.3 Signaling

Signaling protocols are used to create, maintain, restore, and release connections. Such protocols are essential to enable fast provisioning or fast recovery after failures. Signaling network in ASTN should be based on common channel signaling, which involves separation of the signaling network from the transport network. Such a solution supports scalability, a high degree of resilience, efficiency in using signaling links, as well as flexibility in extending message sets.

A variety of different protocols can interoperate within a multidomain network and the interdomain signaling protocols shall be agnostic to their intradomain counterparts. As a matter of fact, interdomain signaling is managed by the domain interface protocols related to E-NNI interface, which are completely independent of the internal protocols of the domain.

8.2.3.4 Call and Connection Control

Call and connection control are separated in the ASON architecture. A call is an association between endpoints that supports an instance of service, while a connection is a concatenation of link connections and subnetwork connections (connections crossing the border between nearby administrative domains or routing areas) that allows transport of user information.

A call may embody any number of underlying connections, including zero. The call and connection control separation makes also sense for restoration after faults. In such a case, the call can be maintained (i.e., it is not released) while restoration procedures are underway. The call control must support coordination of connections in a multiconnection call and the coordination of parties in a multiparty call. It is responsible for negotiation of end-to-end sessions, call admission control, and maintenance of the call state. The connection control is responsible for the overall control of individual connections, including setup and release procedures and the maintenance of the state of the connections.

8.2.3.5 Survivability

As detailed in Chapter 2, survivability can be attained by either protection or restoration mechanisms, or both in some cases, where protection is based on the replacement of a failed resource with a preassigned standby resource, while restoration, is based on rerouting using available but not preprovisioned spare capacity.

Since the ASON architecture is intrinsically a multilayer architecture, protection or restoration may be applied at different layers allowing very high availability levels to be achieved, but also requiring appropriate layers coordination. In the ASON architecture, protection management is in general terms a responsibility of the management plane.

The management plane performs protection configuration and failure identification and reporting after that the protection mechanism has assured service continuity. However, the control plane is not out from the protection management activity. First, the management plane should inform the control plane about all failures of transport resources as well as their additions or removals. Unsuccessful transport plane protection actions may trigger restoration supported by the control plane.

Moreover, the control plane supports the protection switching by suitable control plane protocols that are used in any case to advertise all the network entities in the data plane of the performed protection and, in some cases, necessary to coordinate protection switching itself.

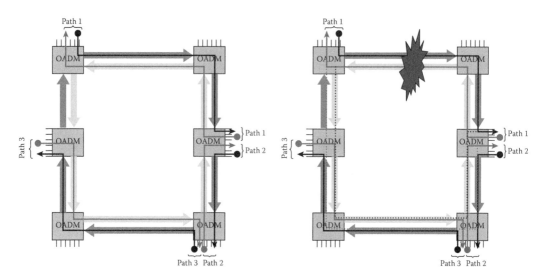

FIGURE 8.7
Protection mechanism in an OCh-SPRing.

Just to do an example, let us image to have an optical channel shared protection ring (OCh-SPRing detailed in Chapter 3). The OCH-SPRing mechanism is shown in Figure 8.7; the wavelengths on each fiber is divided into two groups: a working group and a protection group, where the working wavelengths on the outer fiber are devoted to protection in the inner fiber and vice versa. Working channels, as shown in the figure, are routed bidirectionally using both the fibers on working wavelengths.

When a failure occurs, switching happens at the OCh end and the channels affected by the failures are rerouted in the opposite ring direction by using the other fiber and the set of protection wavelengths. As it is also evident from Figure 8.7, if several optical paths are routed in the ring on the same wavelengths, the corresponding protection wavelengths have to be shared among all those paths (from where comes the name shared protection).

Moreover, also where a unidirectional failure occurs, that is a failure affecting only one direction, both the optical channels ends have to switch to change the ring configuration from working to protection. This means that, when a node detects a failure via a signal power off or a signal quality major alarm (e.g., an estimated BER greater than 10^{-10}), it has to inform the other node at the end of the affected channel of the failure. Moreover also all the other nodes in the ring have to be advertised that protection switching occurred, since the protection wavelengths are no more available.

This means that a control plane protocol has to be started from the CC supervisioning the node that detected the failure with the primary role to make handshaking with the note at the other end of the affected link. The messages are first sent on the other direction of the path: if the failure is unidirectional, they will be detected and the other node will answer on the protection path, if the failure is bidirectional also the other node has detected the failure and after some time the handshake will start on the protection path in both directions.

Before engaging the protection path for the first protection handshake, another protocol can be used to advertise all the nodes of the ring that the protection wavelengths has been reserved by the nodes involved in the failure so that no other node will try to use them. After the successful handshake of the nodes at the two ends of the affected path, which also has the scope of verifying that the protection path works correctly, the affected optical channel is switched on the protection wavelengths in both the directions.

It is clear that there is a complex activity to perform if the protection mechanism involves shared resources to coordinate protection switching in a network. Under a functional point of view, this operation of advertising and handshaking could also be forced by the management plane that can naturally reach all the nodes of the data plane. However, protection has to be carried out generally within 50 ms, or a similar short time, thus an intervention of a SW manager running on a centralized workstation is not feasible and real-time protocols are needed.

However, all the protection mechanisms are controlled and managed by the management plane. Just for an example, if protection switching does not work, it is the management plane to detect and to manage the result, invoking if is the case protection on an upper layer or resorting to restoration coordinated by the control plane.

A particular circumstance is created in the case of control plane protection in consequence of a failure either of a CC or of a control plane connection. In this case, only the source and destination connection controllers are involved in managing protection and it is suggested to avoid shared protection techniques not to be obliged to create protocols of a sort of meta-control plane.

Differently from protection, the restoration is completely managed by the control plane. During restoration, the control plane uses specific protocols to build new routes for the connections that have been disrupted from a failure starting from the spare capacity available in the network. In the ASON standard, such a rerouting service is performed on a per-routing domain basis, i.e., the rerouting operation takes place between the edges of the routing domain and is entirely contained within it. This assumption does not exclude requests for an end-to-end rerouting service.

Naturally, if restoration mechanism is foreseen to increase the network availability, the spare capacity has to be suitably distributed in the network so to permit restoration after the desired class of failures. Hard and soft rerouting services can be distinguished. The first one is a failure recovery mechanism and is always triggered by a failure event. Soft rerouting is associated with such operations as path optimization, network maintenance, or planned engineering works, and is usually activated by the management plane.

In soft rerouting the original connection is removed after creation of the rerouting connection, while in hard rerouting, the original connection segment is released prior to creation of a new alternative segment.

8.3 GMPLS Architecture

If ASON is a reference architecture that is intended to assure functionalities standardization and interoperability to different control and management plane implementations, GMPLS is the IETF-driven effort to standardize a specific control plane architecture that extends the concepts of MPLS to a multilayer network including TDM and WDM layers implementing physical circuit switching.

Within the GMPLS architecture, different GMPLS models are defined to comply with different levels of GMPLS implementation and with gradual control plane introduction in the carrier networks. The first model, intended mainly for the first implementation of the control plane, is the so-called overlay model. In this model the packet layer, implementing

MPLS as control plane and constituted by LSRs, is considered as a client of the underlying transport network that can be either a TDM network (like SDH or SONET) or a WDM network.

GMPLS is used in the overlay model as a control plane of the transport network and it communicates with the overlay MPLS control plane via the UNI interface. In the peer model, the IP/MPLS layer operates as a full peer of the transport layer. Specifically, the IP routers are able to determine the entire path of the connection, including through the optical devices.

A sort of GMPLS overlay model is obtained using ASON terminology by connecting a standard MPLS layer with an ASON transport network via O-UNI interfaces so that conceptually there is some correspondence between GMPLS and ASON configurations. As in the case of the ASON architecture, the GMPLS control plane is extended to support multiple switching layers. Moreover, also in GMPLS, there is a clear separation between the multilabel control plane (MLCP) and the data plane at each layer of the network so that at a specific layer they could have a different topology or be supported by different transmission media as shown in Figure 8.8.

Unlike control signaling in MPLS that following the IP network paradigm is always in-band, the MLCP can manage the GMPLS network out-of-band; hence control signaling need not follow the forwarded data. This implies that the MLCP can continue to function although there is a disruption in the data plane and vice versa and allows for separate control channels to be used for the MLCP. By deploying the MLCP on separate control channels, the other channels are completely dedicated to forwarding data.

In GMPLS, the MLCP utilizes routing and signaling protocols to manage all the network functionalities. These protocols are extensions to well-known protocols of the TCP/IP protocol suite. As such, the MLCP implements IPv4 or IPv6 addressing. This also applies to the data plane, but in cases where addressing is not feasible, or convenient, unnumbered links (i.e., links without network addresses) are supported. MLCP addresses are not required to be globally unique (however global uniqueness is required to allow for remote management). However, addressing in the MLCP is separated from that in the data plane.

In the GMPLS original version, the network manager has only the role of defining the network when it is initialized and of reporting network events to the central network control room. Moreover, the network manager also implements traffic engineering tools that can be run off line on the network manager workstation to verify if the network is optimized.

If the connection routing and resource assignment is not optimized, as surely will happen after a certain period of network working, the network manager is able to force the network to exit from the present configuration to assume a better one, for example, rerouting some paths. This original role of the network manager in GMPLS networks is completely different from what is prescribed from ASON and probably at the end of the

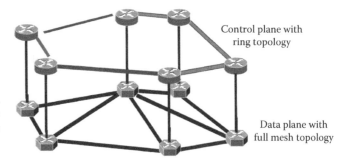

FIGURE 8.8
Communication structure dedicated to the control plane and its independence from the network on the data plane.

Control plane with ring topology

Data plane with full mesh topology

story the ASON version will prevail if not for other reasons since it is pushed from the network owners.

GMPLS uses generalized labels to perform routing of generalized label switched path (GLSP). In contrast to MPLS where labels merely represent network traffic, generalized labels represent network resources. For example, a generalized label on an optical link could identify a wavelength or a fiber, while on a packet-switched link, it would simply identify network traffic, just as in MPLS. In the lower layers, generalized labels are "virtual," meaning that they are not inserted into the network traffic, but instead implied by the network resource being used (e.g., wavelength or fiber). This is necessary since neither packets nor frames are recognized at the lowest layers that GMPLS supports.

Generalizing the label format, the conventional MPLS label has been extended to 32 bits. The labeling of ports of routing equipment in GMPLS follows a particular convention that allows the control plane to identify the functionality of an equipment from the labeling of its ports.

In particular, the ports of routing equipments are classified in several classes that are as follows:

- Fiber switching capable ports (FSC): These are the ports of equipment that can switch the entire fiber signal as a whole.

- Wavelength switching capable ports (WSC): These are ports of equipment that can switch the single wavelength signal as a whole; since generally the port is occupied by a WDM system, a demultiplexer is expected so to divide at least a certain number of wavelengths (see the OADM structure in Chapter 6) so to be able to deal with each wavelength independently.

- Time division multiplexing switching capable ports (TSC): These are the ports of equipment that can switch the single TDM frame; since generally the port is occupied by a complete TDM frame, a TDM demultiplexer is expected after the TSC.

- Layer 2 frames switching capable ports (L2SC): These are essentially the ports of Ethernet equipments.

- Packet switching capable ports (PSC): These are the ports of LSR and IP routers.

Every type of port has a different kind of generalized label so that the control plane can identify the type of port and so understand the equipment capability. An example of different types of ports is reported in Figure 8.9, where a core node hosting the overlay of an OXC and an LSP router is schematically shown. A few of the wavelength GLSPs transiting through the OXC are also sent to the LSR both for local dropping and for grooming, while other wavelength GLSP directly transit through the OXC.

This is a typical overlay node configuration, where the OXC ports are wavelength capable ports, while the LSR ports are packet capable ports. Finally ports with multiple switching capability can be defined, for example both PSC and WSC. When a GLSP requires the use of the port it also requires the functionality he needs among those available.

8.3.1 GMPLS Data Paths and Generalized Labels Hierarchy

A data path between two GLSRs is a different concept with respect to the GLSP. A GLSP consists of a sequence of consecutive labels, which, when swapped in a specific order,

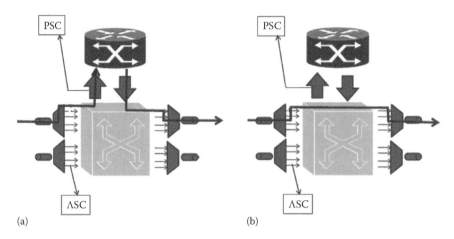

(a)   (b)

FIGURE 8.9
Different types of ports in the overlay of an OXC and an LSP router. (a) Shows a locally dropped GLSP, and
(b) a pass-through GLSP. ΛSC indicates a wavelength switch capable port.

carries data from one point in a label switched network to another. In short, a GLSP is represented by the distributed state needed to send data along a specific route.

Because GLSPs might differ in link composition, an entity requesting labels for a GLSP needs to specify three major parameters: switching type, encoding type, and generalized payload-ID (G-PID). Switching type defines how an interface switches data. Because this is always expected to be known, a switching type needs only be specified for an interface with multiple switching capabilities.

Encoding type is needed to specify the specific encoding of the data associated with the GLSP. For example, data associated with an L2SC interface might be encoded as Ethernet.

The G-PID finally defines the client layer of the GLSP. In GMPLS, bidirectional GLSPs are considered the default. Bidirectional GLSPs are established through simultaneous label distribution in both directions. Since generalized labels are nonhierarchical, they do not stack. This is because some supported switching media cannot stack. Given an optical link, for example, it is not possible to encapsulate a wavelength in another and then get it back again.

In GMPLS, tunneling data through different layers is therefore based on GLSP nesting (i.e., encapsulating GLSPs within GLSPs). Nesting of GLSPs can be realized either within a single network layer or nesting GLSPs of a layer through a GLSP of another layer. Examples of the nesting system are reported in Figure 8.10.

L Fiber = (Lλ_1; Lλ_2; Lλ_3; Lλ_4)   Lλ_4=(L1; L2; L3; L4)

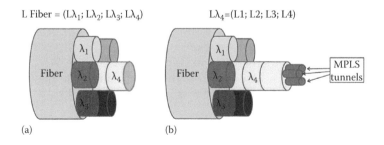

(a)   (b)

FIGURE 8.10
Examples of Generalized Labels nesting: on the left (a) labels relative to different wavelength channels are nested into the fiber label, on the right (b) labels relative to different MPLS LSPs are nested into the wavelength path label.

In particular in Figure 8.10a, a set of wavelengths are nested into a fiber link by nesting the labels associated with wavelength channels resources into the label associated to the fiber channel.

This kind of nesting allows the wavelength channels to be considered as a single entity, to be routed always together, switched together and protected together. This kind of nesting is useful, for example, if a 100 Gbit/s virtual channel is transmitted as a group of four 20 Gbit/s channels. In this case, the association made through GLSPs nesting within the same layer allows the 100 Gbit/s channel to be recognized as a single entity by the network even if it is constituted by parallel wavelength channel.

Another case in which nesting wavelength into a fiber channel is useful when a client layer requires a SLA prescribing the same network traversing time for different wavelength channels. This can be achieved by nesting them into the same GLSP.

In Figure 8.10b the nesting of several MPLS LSPs into a lower layer wavelength channel is shown. Physically the GLSPs are multiplexed into a wavelength channel using OTN frame and GFP adaptation, under a logical point of view, in order to recognize that the GLSPs will travel the network in the same route, a GLSP is built nesting the individual layer two labels into a layer one label attached to the wavelength channel resources.

In general, ordering GLSPs hierarchically within a network layer requires that GLSP are encoded in a hierarchical way. A natural GLSP hierarchy can be established based on interface types and it is also the hierarchy used to build the examples of Figure 8.10. At the top of this hierarchy are FSC interfaces followed, in decreasing order, by LSC, TDM, L2SC, and PSC interfaces. This order is because wavelengths can be encapsulated within a fiber, time slots in wavelengths, data link layer frames in time slots, and finally network layer packets in data link layer frames. As such, an GLSP starting and ending on PSC interfaces can be nested within higher ordered GLSPs.

8.3.2 GMPLS Protocol Suite

Protocols of the GMPLS protocol suite can be divided into three distinct sets based on their functionality:

1. *Routing protocols* must be implemented by the MLCP to disseminate the network topology and its traffic engineered (TE) attributes. For this purpose, open shortest path first (OSPF) with TE extensions (OSPF-TE) or intermediate system to intermediate system (IS–IS) are currently defined. To account for multiple layers, however, GMPLS needs to add some minor extensions to these existing protocols. The GMPLS routing process using OSPF-TE is explained in the following sections.

2. *Signaling protocols* are concerned with establishing, maintaining, and removing network state (i.e., setting up and tearing down GLSPs). For signaling, GMPLS can use either Resource ReSerVation Protocol (RSVP) with TE extensions (RSVP-TE) or constraint-based routing-label distribution protocol (CR-LDP). Again, supporting multiple layers requires some extensions to existing protocols.

3. *Link management* is performed by a new protocol called the link management protocol (LMP), which has been defined within the GMPLS standard. LMP can be used by GMPLS network elements to discover and monitor their network link connectivity. To enable link discovery between an optical switch and an optical line system, LMP has been further extended, creating LMP-WDM.

In the following subsections, the working of the most used among the protocols of the GMPLS protocol suite will be described in some detail so as to give an idea of how GMPLS implements its functionalities.

8.3.2.1 Open Shortest Path First with Traffic Engineering

To enable automated configuration of the controlled network, GMPLS defines an intradomain routing process. Via this routing process, the network topology and its TE attributes are disseminated within the TE domain. This routing process is implemented using routing protocols specified for the MLCP. However, this routing process is not used for routing user traffic, but only for distributing information in the MLCP. To understand this point, we have to remember that, taking as an example the optical domain, user signals are organized in optical channels that are routed like circuits in that layer. Extensions to existing protocols necessary for this routing process were therefore created in the relevant IETF and OIF standards.

Essentially, with OSPF-TE, participating generalized label switched routers (GLSRs) first establish routing adjacencies by exchanging hello messages. After routing adjacencies have been established, the GLSRs then synchronize their link state databases. This is done by exchanging database description packets.

The database description packets contain at least one database structure referred to as a link state advertisement (LSA). Different LSA types exist but all share a common 20-byte header whose structure is shown in Figure 8.11.

The primary OSPF-TE operation is to flood LSAs throughout the MLCP domain by appending them to "link state update" messages periodically sent between adjacent GLSRs. To avoid interference with any ordinary routing processes, a TE LSA is made opaque. Such an opaque LSA is a special type of LSA only processed by specific applications (e.g., the GMPLS routing process). By extending the link state database with TE information, a traffic engineering database (TED) is produced. From this TED, a network graph with traffic engineering content can be computed. Constructing a TE network graph is necessary to provide input for the constraint-based algorithms subsequently used to compute network paths.

When flooding LSAs, each OSPF-TE routing message contains a common 24-byte header, which is used to forward it. This routing message header includes information about message type, addressing, and integrity. Within this header, LSAs are then encapsulated and specific payloads appended to each LSA. In order to enable advertisement of TE attributes in opaque LSAs, the LSA payload consists of type-length-value (TLV) records. These are information records composed by three fields: two 2-byte fields, the "Type" and "Length" fields, and a variable length field, the "Value" field. Using TLVs, router addresses and TE links can be expressed.

In GMPLS, if an advertising router is reachable, a "router address"-TLV can be used to describe a network address at which this router (i.e., GLSR) can always be reached. In turn,

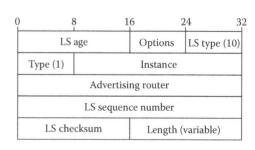

FIGURE 8.11
Common 20-byte header of OSPF-TE LSA messages.

TABLE 8.1

Record Containing the Information Needed to the MLCP to Perform Routing Operations

Sub-TLV Name	Type	Length	Value
Link local/remote identifiers	11	8 bytes	2×4 bytes local/remote link identifiers
Link protection type	14	4 bytes	1 byte for link protection (3 bytes reserved)
Interface switching capability descriptor	15	Variable	Minimum 36 bytes for ISCD information
Shared risk link group	16	Variable	$N \times 4$ bytes for link SRLG identification

the "link"-TLV can be used to abstract advertised TE links. Multiple TE attributes can be represented on each link using fields that are related to the "link"-TLV address, which are called sub-TLV. Thus, in the case of a link address, the TLV appears as the first field and pointer of a bigger record containing all the information that are needed to the MLCP about the link in order to perform routing operations.

The structure of this bigger record is reported in Table 8.1. Two links attributes of Table 8.1 are worth a comment. The first is the last attribute of the record, advertising that the link is part of a shared risk link group (SRLG). An SRLG is a group of links that will fail contemporary when a certain type of failure happens.

As an example, we can look at Figure 8.12, where typical situations are represented. Physical fiber connections are routed through multifiber cables and links having different destinations are built over a cable structure that generally has been deployed without any knowledge of the overlay network.

Thus, also links that have completely different source and destinations can share in a part of their path the same fiber cable. If the cable is cut, for example, due to civil works, all those links goes down, thus they have to be put together in the same SRLG.

The definition of SRLGs is fundamental when defining protection or restoration strategies. As a matter of fact, when the protection capacity is preprovisioned, it is fundamental to avoid that a link is protected by another link in the same SRLG, otherwise a single failure could hit both working and protection link, making the protection useless. In the case of restoration, when routing of the restoration path is performed, also links in the same SRLG has to be avoided. This is why the SRLG is indicated among the information related to the routing protocol.

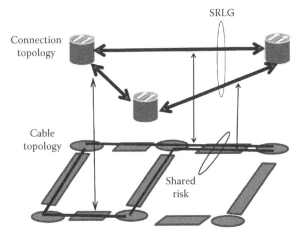

FIGURE 8.12
Concept of SRLG.

The second field that is worth a comment is the interface switching capable description. It is clear that a link will terminate on the same type of interfaces and this field let the routing protocol know what kind of interfaces there are at the link extremes. During GLSPs, this information can be very useful to balance sparing of higher layer interface use achieved by transparent pass-through of the traffic not destined to the node with transmission resources sparing achieved by grooming (see Section 8.4: Design and Optimization of ASON/GMPLS Networks).

Once the OSPF-TE has flooded the network and created all the nodes capable of routing GLSPs, a database of the routing area topology, the OSPF-TE routing engine can start every time there is a GLSP to route either if it is a new path or it is needed to carry out restoration of a failed path.

The OSPF-TE routing engine is based on the so called Dijkstra algorithm (from the name of the inventor), which is an algorithm that is able to find in a graph with tagged links the shortest path between two given nodes. In GMPLS, what should be the tagging of the links of the network is not prescribed in a mandatory way; generally every link is associated to its length and the GLSP are routed along the shortest path in term of kilometers. This is a reasonable solution since the cost of a DWDM system is roughly proportional to its length due to the presence of EDFA sites. Thus, since if I occupy preferably the shortest links I will have to upgrade them sooner, this strategy tends to minimize the cost of network upgrades.

However, in specific situations, it is possible to tag the network links with other numbers. For example, a competitive carrier leasing fibers from another company could tag the link with the leasing price so to use always the cheaper links. This direct routing through the topology of the network with the Dijkstra algorithm becomes increasingly complex while the network dimension increases, so that the routing process has to be simplified in some way if OSPF-TE has to scale up to the dimension of an incumbent carrier network. The solution to this problem is the division of the network in routing areas and the execution of the routing in a hierarchical manner.

In particular, OSPF-TE individuates in the network a set of routing entities (GLSR) that are suitably distributed to constitute the so-called network backbone. All the other GLSRs are divided in routing areas constituted by nearby nodes. Each distinct routing area has to be in contact with the backbone at least in two points to assure effective network survivability. An example of this hierarchical organization of the network is reported in Figure 8.13.

Staring from this structure, a new higher level topology (the area topology) is constructed substituting all the routing areas with virtual nodes; this new topology has for sure at least interconnection degree equal to 2 due to the presence of the backbone and to the way in which the areas have been constructed.

The routing between nodes in different areas is performed in two steps: first the route is individuated in the area topology, then, knowing for each traversed areas the incoming and the outcoming nodes, the path route is specified also within areas.

8.3.2.2 IS–IS Routing Protocol

IS–IS is designed and optimized to provide routing within a single network domain, while passage through a domain interface for multidomain GLSPs have to be realized with a different protocol (e.g., the BGP, see Chapter 3, Table 3.1 IP protocol suite). IS–IS is based on the concept of routing domains. An IS–IS routing domain is a network in which all the equipments that perform GLSP routing run the integrated IS–IS routing protocol to support intradomain exchange of routing information.

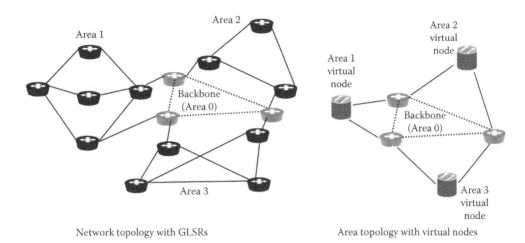

Network topology with GLSRs   Area topology with virtual nodes

FIGURE 8.13
GMPLS partition of the network in routing areas and backbone. On the right of the figure the virtual topology where each routing area is represented by a virtual node is represented.

The underlying goal is to achieve consistent routing information within a domain by having identical link-state databases on all routing entities in that area. Hence, all routing entities in an area are required to be configured in the same way.

IS–IS protocol supports a two-level hierarchy for managing and scaling routing in large networks. A network domain can be divided into small segments known as areas. The topology resulting from the substitution of each area with a virtual node and each connection between nodes in different areas with a connection between the corresponding virtual nodes is called area topology. This construction allows hierarchical routing to be carried out within the domain.

This means that the routing between nodes in different areas is performed in two steps: first the route is individuated in the area topology, then, knowing for each traversed areas the incoming and the outcoming nodes, the path route is specified also within areas. Following the same network representation, IS–IS routing entities are hierarchically ordered depending on their ability to route GLSPs only within their area or also toward other routing areas. In particular routing entities that are only able to route paths within their area are called Level 1 routers (not to confound with Layer 1 entities; this is a specific IS–IS hierarchy completely different from the network layering).

Following the same criterion, routing entities that can route GLSPs only between virtual nodes that represents routing areas are called Level 2 routers (or inter area routers) and finally routing entities that are able to route GLSPs both inside the area and between different areas are considered logically as a superposition of communicating Level 1 and Level 2 routers and are called Level 1–2 routers.

Level 2 routers are interarea routers that can only form relationships with other Level 2 routers; in other words, they can route through the area topology constituted by virtual nodes representing the routing areas. Similarly Level 1 routers can exchange information only with other Level 1 routers. Level 1–2 routers exchange information with both levels and are used to connect the interarea routers with the intraarea routers.

Integrated IS–IS uses the legacy CLNP node-based addresses [35–37] to identify routers even in pure IP environments. The CLNP addresses, which are known as network service access points (NSAPs), are made up of three components: an area identifier (area ID) prefix, followed by a system identifier (SysID), and an N-selector. The N-selector refers to the network service user, such as a transport protocol or the routing layer.

A group of routing entities belongs to the same area if they share a common area ID. Note that all routing entities in an IS–IS domain must be associated with a single physical area, which is determined by the area ID in the router address. IS–IS packets can be divided into three categories:

1. Hello packets are used to establish and maintain adjacencies between IS–IS neighbors.
2. Link-state packets are used to distribute routing information between IS–IS nodes.
3. Sequence number packets are used to control distribution of link-state packets, essentially providing mechanisms for synchronization of the distributed link-state databases on the routers in an IS-IS routing area.

Each type of IS–IS packet is made up of a header and a number of optional variable-length fields organized in records and containing specific routing-related information. The variable length fields are called TLV as in the case of the OSPF-TE from the form of the record that is organized into three fields: type, length, and value.

Each variable-length field has a 1-byte type label (the Type field) that describes the information it contains. The second field is a length information that contains the length of the third field whose content is the specific information that is carried by the TLV (see Table 8.1 for examples). The different types of IS–IS packets have a slightly different composition of the header, but the first 8 bytes are repeated in all packets. Each type of packet then has its own set of additional header fields, which are followed by TLVs.

Table 8.2 lists TLVs specified in ISO 10589 while Table 8.3 lists the TLVs introduced by IETF in RFC 1195.

Enhancements to the original IS–IS protocol are normally achieved through the introduction of new TLV fields. Note that a key strength of the IS–IS protocol design lies in the ease of extension through the introduction of new TLVs rather than new packet types.

The routing layer functions provided by the IS–IS protocol can be grouped into two main categories: subnetwork-dependent functions and subnetwork-independent functions.

TABLE 8.2

List of TLVs Specified in ISO 10589

TLV	Type	Description
Area address	1	Area address(es) of source node
Intermediate system neighbors	2	Neighboring routers and pseudonodes (appended to the link-state packet)
End system neighbors	3	Connected workstations
Partition designated level 2 intermediate system	4	Level 2 neighbors that interconnect pieces of a partitioned area with a virtual link over the Level 2 area
Prefix neighbors	5	Reachable NSAP prefixes (not including local area prefixes)
Intermediate system neighbors	6	LAN-connected routers from which IS–IS hello packets have been received (appended to LAN hello)
Not specified	7	
Padding	8	Padding for hello packets to maximum transmission unit (MTU)
LSP entries	9	Link-state information
Authentication information	10	Information for IS–IS packet authentication

TABLE 8.3

List of the TLVs Introduced by IETF in RFC 1195

TLV	Type	Description
IP internal reachability information	128	Intradomain IS–IS routes
Protocols supported	129	Protocol identifiers of supported network layer protocols (e.g., IP and CLNP)
IP external reachability information	130	Routes external to the IS–IS domain, such as those imported from other sources via redistribution
Interdomain routing protocol information	131	For transparent distribution of interdomain routes
IP interface address	132	IP address of the outgoing interface
Authentication information	133	IS–IS packet authentication (similar to Type 10, but doesn't define authentication Type 255)

The subnetwork-dependent functions involve operations for detecting, forming, and maintaining routing adjacencies with neighboring routing entities over various types of interconnecting network links. The subnetwork-independent functions provide the capabilities for exchange and processing of routing information and related control information between adjacent routers as validated by the subnetwork-dependent functions.

The routing information base is composed of two databases that are central to the operation of IS–IS: the link-state database and the forwarding database. The link-state database is fed with routing information by the update process. The update process generates local link-state information, based on the adjacency database built by the subnetwork-dependent functions, which the router advertises to all its neighbors in link-state packets.

A routing entity also receive similar link-state information from every adjacent neighbor, keeps copies of GLSPs received, and readvertises them to other neighbors. Routing entities in an area maintain identical link-state databases, which are synchronized using SNPs. This means that routing entities in an area will have identical views of the area topology, which is necessary for routing consistency within the area. The decision process creates the forwarding database by running the Dijkstra algorithm on the link-state database so selecting for each GLSP and for the forward of each control packet the shortest path in terms of whatever weight is associated to each network link in the topology database.

The IS–IS forwarding database, which is made up of only best IS–IS routes, is fed into the routing information base.

8.3.2.3 Brief Comparison between OSPF-TE and IS–IS

IS–IS and OSPF-TE are link-state protocols, that is routing protocols that perform routing on the ground of information on the state of the links of the network that is derived from a map of the network topology that each routing network element stores and updates using the protocol flooding functionality. Moreover, OSPF-TE and IS–IS have several other common characteristics:

- They use the Dijkstra algorithm for computing the best path through the network.
- They allow any link tag as weight for the Dijkstra calculation.
- They exchange via protocol packets records composed by sets of TVL fields that describe the characteristics of the link they are advertising.
- They use the concept of subnetwork and virtual node to leverage on hierarchical routing to simplify routing tables and routing computation.

- They can use multicast to discover neighboring routing equipment using hello packets.
- They can support authentication of routing updates.

As a result, OSPF-TE and IS–IS are conceptually similar.

While OSPF-TE is natively built to route IP and is itself a Layer 3 protocol that runs on top of IP, so that OSPF-TE packets are a particular type of IP packets, IS–IS is natively an OSI network layer protocol (it is at the same layer as CLNS [35–37]), a fact that may have allowed OSPF to be more widely used. IS–IS does not use IP to carry routing information messages and its packets are not IP packets.

In line of principle IS–IS does not need IP addressing, even if, since some form of addressing is needed to route GLSPs through the network, in practical applications IP addressing is almost always used in conjunction with IS–IS.

Routing entities of a network implementing IS–IS build a topological representation of the network. This map indicates the subnetworks that each IS–IS enabled equipment can reach, and the lowest cost (shortest) path to any subnetwork is used to forward IS–IS packets and to route GLSPs.

Due to its structure, IS–IS is also more scalable that OSPF-TE: given the same set of resources, IS–IS can support more routing entities in an area with respect to OSPF-TE. IS–IS differs from OSPF-TE also in the way that "routing areas" are defined and interarea routing is performed. No hierarchy similar to that present in IS–IS is present in OSPF-TE, whose routing entities are all on the same ground, capable of interarea or intraarea routing when it is needed.

In OSPF, areas are delineated on the interface such that an area border router (ABR) is actually in two or more areas at once, effectively creating the borders between areas inside the ABR, whereas in IS–IS area borders are in between routers, designated as Level 2 or Level 1–2. The result is that an IS–IS router is only ever a part of a single area. IS–IS also does not require the backbone (or Area Zero as it is sometimes called) through which all interarea traffic must pass, while OSPF-TE needs this definition to manage multiarea routing domains.

The logical view is that OSPF creates something of a spider web or star topology of many areas all attached directly to area zero and IS–IS by contrast creates a logical topology of a backbone of Level 2 routers with branches of Level 1–2 and Level 1 routers forming the individual areas.

8.3.2.4 Resource Reservation Protocol with Traffic Engineering Extensions

Because RSVP-TE inherits its design from the RSVP protocol, it is based on distributing various signaling objects. These signaling objects have been grouped. These groups contain mandatory and optional signaling objects; encapsulating the groups with a common header, distinct signaling messages are created.

When a GLSR receives a signaling message, the resident objects are examined and interpreted based upon the message type indicated by the common header. Extending the RSVP-TE protocol for GMPLS was thus a matter of generalizing existing signaling objects, including some new objects that are reported in Table 8.4, and adding some minor signaling enhancements.

Considering the RSVP-TE protocol with GMPLS extensions for signaling, each signaling message contains a common 8-byte header. The common header defines the message type followed by the encapsulated objects. Encapsulated objects are of variable length and contain a 4-byte header defining the object length, class, and type within class.

TABLE 8.4

New Objects Introduced in the Adaptation of RSVP-TE for the GMPLS

Object Name	Length	Message Description
Generalized label request	4 bytes	Path describes the requested LSP
IF_ID RSVP_HOP	Variable	Path/Resv defines what interface to label
Generalized label variable	4 bytes	Resv/ResvErr downstream label
Upstream label variable	4 bytes	Path/PathErr upstream label
Label ERO	2 bits + label	Path/Resv explicit label control
Suggested label variable	4 bytes	Path/PathErr label suggestion
Label set	Variable	Path label selection restriction

The first signaling activity of RSVP-TE is to distribute the labels that are used in the GLSPs. The mechanism used by RSVP-TE is downstream-on-demand label distribution: this means that upstream GLSRs request downstream GLSRs to select labels for the links connecting them. In this way, each GLSR acknowledges a request to install an GLSP, forward the request to the next downstream hop, and awaits the response.

As a response is returned upstream, the GLSR can install a cross-connection (i.e., state describing ingoing and outgoing label-to-interface mappings and associated network resources) for this GLSP. Here, downstream is defined as the direction from GLSP ingress to GLSP egress for a unidirectional GLSP. In the case a bidirectional GLSP is to be set up, this is done by contemporary setup of two unidirectional GLSPs.

Once all the labels are assigned, the GLSP can be set up. As a matter of fact, GLSR can route the signal incoming from an input port toward the output port with the same label and the GLSP is identified in the path topology through the vectors whose elements are the labels of the links traversed by the path in the correct order.

As for all signaling protocols, RSVP-TE operation is composed of three main functions:

1. GLSP setup that is carried out via labels distribution.
2. GLSP management, a function composed by various elements among which error management is probably the most important.
3. GLSP tear down that is carried out by labels removal.

8.3.2.4.1 GLSP Setup

Establishing bidirectional GLSPs employing RSVP-TE for signaling requires full sets of Path and Reservation (Resv) messages to be exchanged between two GLSRs. This procedure is sketched in Figure 8.14.

Initially, a sender GLSR (GLSP ingress) requests a GLSP to be set up by sending a Path message downstream to the next hop. This Path message contains an UPSTREAM_LABEL object defining the label to use in the upstream direction, objects describing the data flow, and a GENERALIZED_LABEL_REQUEST object for requesting the GLSP.

If the Path message is successfully received, the next hop then reserves path state to enable correct signaling of returning Resv messages and saves the upstream label. The next hop then selects its own upstream label, creates state for the upstream direction, replaces the upstream label in the Path message, and passes it on downstream to the next hop. This procedure is repeated until the next hop is the receiver GLSR (GLSP egress). The GLSP has now been established in the upstream direction, but no state has been saved in the downstream direction. Consequently, the receiver GLSR selects a downstream label

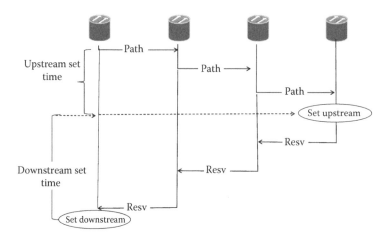

FIGURE 8.14
Establishment of a bidirectional GLSPs employing RSVP-TE for signaling; full sets of Path and Reservation (Resv) messages are exchanged between two GLSRs.

and returns a Resv message upstream. This Resv message mimics the Path message, but inserts a GENERALIZED_LABEL object defining the selected downstream label. If the Resv message is received successfully, the previous hop (the signaling direction has changed) then sets state for the downstream direction, replaces the downstream label with its own selected label, and passes the Resv message further upstream.

This procedure is repeated until the sender GLSR successfully receives the Resv message corresponding to a dispatched Path message. Now, the request has been fully established and is ready to tunnel data in both directions. This procedure is shown in Figure 8.14 while the final aspect of the GLSP as a cascade of labels is sketched in Figure 8.15.

It is possible using RSVP-TE to route a certain GLSP on a preassigner route. This is needed in order to traffic engineer the network. As a matter of fact, if a generic network is considered, the OSPF routing criterion used both by OSPF-TE and IS–IS is a suboptimum routing strategy. It is not difficult to find examples where routing a few GLSPs along a longer path allows a more efficient exploitation of the network resources.

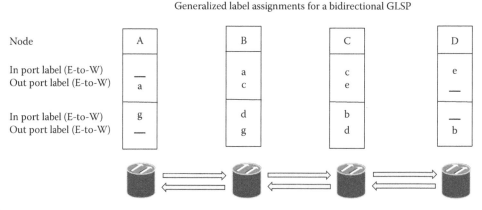

FIGURE 8.15
Generalized Label Assignment for a bidirectional GLSP.

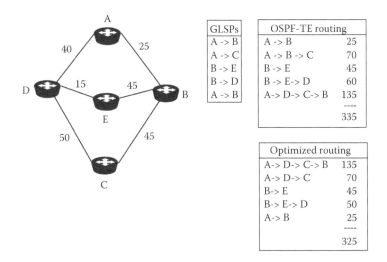

GLSPs
A -> B
A -> C
B -> E
B -> D
A -> B

OSPF-TE routing	
A -> B	25
A -> B -> C	70
B -> E	45
B -> E-> D	60
A-> D-> C-> B	135

	335

Optimized routing	
A-> D-> C-> B	135
A-> D-> C	70
B-> E	45
B-> E-> D	50
A-> B	25

	325

FIGURE 8.16
An example of sub-optimum routing performed with OSPF-TE. Each link between a couple of GLSR is assumed to have enough capacity to support only two GLSPs.

Such an example is provided in Figure 8.16. Here a network of five nodes with an incomplete mesh connection is represented. For simplicity let us image that each link is physically realized with a WDM system with two wavelengths and that all the wavelengths have the same capacity, which we assume equal to 1.

Let us also imagine to tag every link with a weight that is the "cost" of using a wavelength along that link. We could also think that the weight is the link length and that it is assumed as a cost due to the fact that WDM systems have a cost roughly proportional to length.

Such GLSPs, each of which for simplicity is imaging to have unitary capacity, so that it completely uses a single wavelength, correspond to traffic requests arriving at the network control plane one after the other so that they are routed in successive times and when one of them is routed the network control plane does not know what set of traffic requests will arrive in the future.

In this dynamic situation, typical of GMPLS networks, the OSPF-TE protocol produces the routing that is listed in the first routing table in Figure 8.16. Naturally, also IS–IS would have produced the same routing due to the use of the same routing algorithm and to the fact that this simple network is a single routing area. The overall cost of the routing is obtained by adding together all the costs of the used wavelengths and it is reported in the figure as 335.

However, it is not difficult to see that at least one alternative routing exists having a lesser cost, reported as 325 in Figure 8.16. This routing is based on the knowledge of the whole traffic request, that in reality the control plane does not have, and achieve a lesser overall cost by using two times the link DC even if it is the most expensive of the network, and obtaining in return the possibility to better distribute the weights of all the GLSPs.

This example shows that if the traffic is completely known, routing solutions better, and sometimes much better, then the routing produced by protocols like OSPF-TE or IS–IS can be produced. As a consequence, generally a traffic engineering tool is incorporated into the network manager. It has the scope to route the traffic that the network is managing in an optimized way with respect to OSPF.

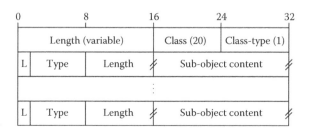

FIGURE 8.17
RSVP-TE message structure.

From time to time, the deviation of the real routing achieved by OSPF-TE or IS–IS from the solution found through the traffic engineering tool becomes important and it is mandatory to reroute a certain number of GLSPs to optimize the network and avoid waste of resources.

In order do that is must be possible to force the routing of a GLSP along a predetermined route. Using RSVP-TE this can be done introducing in the Path and Resv messages exchanged during GLSP state installation the EXPLICIT_ROUTE object (ERO, see Figure 8.17).

When used in Path messages, the ERO describes the next and previous hop for any GLSR along the explicit route. When signaling paths explicitly use an ERO, path state is not needed to indicate a reverse route, since returning Resv messages can instead be routed based upon the ERO. In practice, when dealing with large networks composed by many layers and a huge number of paths, the use of automatic routing protocols to route GLSPs is so inefficient and explicit routing would be so frequent that there is a proposal to use routing protocols to route packets in the control plane and adopting explicit routing through ERO as the default to route GLSPs.

While the earlier description of the signaling procedures presumed that no errors occurred during signaling, this is unlikely to always be true. Thus, a need for error handling messages arises. When errors occur, PathErr or ResvErr messages can therefore be signaled.

A PathErr message indicates an error in processing a Path message and is sent upstream toward the GLSP ingress. Similarly, a ResvErr message indicates an error in processing a Resv message and is sent downstream toward the GLSP egress. A GLSR receiving an error message may try to correct the error itself, if minor, or pass it further on.

8.3.2.4.2 GLSP Tear Down

RSVP-TE is a soft-state protocol. This means that it continuously sends messages refreshing timers associated with installed state. Originally designed for MPLS, the "softness" is somewhat reduced in GMPLS, however timers are still implemented.

GLSP state removal can be triggered in two ways: when a timer expires in a GLSR or by some external mechanism (e.g., manual operator or management system). To remove GLSP state using RSVP-TE, PathTear or PathErr messages are dispatched. A PathTear message is dispatched downstream following the path of a Path message, while a PathErr message is sent upstream following the path of a Resv message.

As these messages are processed by GLSRs they immediately clear, or partially clear, the GLSP state. This enhancement is specific to GMPLS and enables the GLSP egress and intermediate GLSRs to initiate GLSP state removal. Using the PathErr message for clearing state, a flag introduced by GMPLS is set to indicate that path state is no longer valid. This means that GMPLS can tear down GLSP state in both directions (both upstream

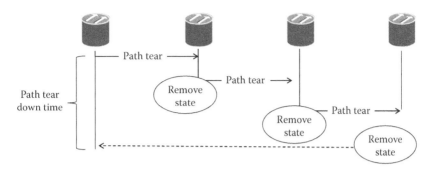

FIGURE 8.18
GLSP Tear down procedure with RSVP-TE.

and downstream). Additionally, GMPLS provides rapid error notification via the Notify message (Figure 8.18).

The Notify message can be used to inform an GLSP ingress or egress of errors, enabling them to initiate state removal in the place of an intermediate GLSR. Although the PathErr message is, strictly speaking, not needed, it can increase signaling efficiency by eliminating the need for notification.

8.3.2.5 Constrained Routing Label Distribution Protocol

In the GMPLS network, GLSR must agree on the meaning of the labels used to forward traffic between and through them. This signaling operation can be performed by RSVP-TE, but a group of equipment vendors has pushed another protocol. The CR-LDP is an extension of the protocol used for the same scope in MPLS: the Label Distribution Protocol (LDP).

CR-LDP is a new protocol that defines a set of procedures and messages by which one GLSR informs another of the label bindings it has made. This implicitly advertises the GLSP setup since it is performed by assigning label bindings across GLSRs. The idea below the introduction in the GMPLS protocol suite of this extended protocol is to incorporate traffic engineering capabilities into a well-known protocol widely used in MPLS networks. The traffic engineering requirements are met by extending LDP for support of constraint-based routed label switched paths (also sometimes called CR-GLSPs).

Like any other GLSP, CR-GLSP is a path through a GMPLS network. The difference is that while other paths are setup based on information in routing tables or from a direct intervention of the management system, the CR-LDP constraint-based route is calculated at one point at the edge of network based on criteria including but not limited to explicit route constraints or QoS constraints.

The GLSR uses CR-LDP to establish LSPs through a network by mapping network layer routing information directly to data-link layer switched paths. A Forwarding Equivalence Class (FEC) is associated with each GLSP created. This FEC specifies which type of information is mapped to that GLSP and can be used for assuring the needed QoS during the GLSP setup.

Two GLSR which use CRT-LDP to exchange label mapping information are known as CR-LDP peers and they have a CR-LDP session between them. In a single session, each peer is able to learn about the other label mappings, that is, the protocol is bi-directional.

There are four types of CR-LDP messages:

1. Session messages
2. Discovery messages
3. Advertisement messages
4. Notification messages

Session messages are used to manage the CR-LDP section through the standard section management steps:

- Section setup through handshaking between the GLSR at the section ends
- Section monitoring
- Section teardown through another specific handshaking

Using discovery messages, the GLSRs announce their presence in the network by sending Hello messages periodically. This Hello message is transmitted as a UDP packet allocated into a GLSP that always exists when a physical capacity is deployed along a link in order to convey at least control plane messages.

When a new session must be established, the Hello message is sent over TCP to have a better control of the session opening procedure. Apart from the Discovery messages all other messages are sent over TCP. Advertisement messages are used to disseminate across the network the discovery of new network entities or other modifications in the network topology that influences the labels distribution. For example, if a link fails, this has to be eliminated from the topology map that each GLSR has in the routing engine. This is discovered by processing the absence of response to all the Hello messages that are directed toward that link and is advertised to all the network routing entities via advertisement messages. The notification messages signal errors and other events of interest. There are two kinds of notification messages:

1. Error notifications: these signal fatal errors and cause termination of the session
2. Advisory notifications: these are used to pass on GLSR information about the CR-LDP session or the status of some previous message received from the peer

All CR-LDP messages have a common structure that uses a TLV encoding scheme. This TLV encoding is used to encode much of the information carried in CR-LDP messages. The Value part of a TLV-encoded object may itself contain one or more TLVs. Messages are sent as CR-LDP PDUs. Each PDU can contain more than one CR-LDP message. Each CR-LDP PDU is an CR-LDP header followed by one or more CR-LDP message.

8.3.2.6 *Comparison between RSVP-TE and CR-LDP*

From the brief description we have done, it is quite evident that RSVP-TE and CD-LDP have almost all the same functionalities, even if they are constructed using different criteria. Thus a comparison between the two protocols, which are obviously alternative in a particular implementation of GMPLS, can be useful.

The summary of this comparison is mainly derived from Ref. [3].

- *Transport Protocol*: The first difference between CR-LDP and RSVP is the choice of transport protocol used to distribute the label requests. RSVP uses connectionless raw IP or UDP encapsulation for message exchange, whereas CR-LDP uses UDP only to discover GMPLS peers and uses connection-oriented TCP sessions to distribute label requests.

- *Security*: Once the path has been established and the data is being forwarded, the frame is no longer visible to the management or control plane software at the upper layers of the network. There is minimal chance that unauthorized individuals will be able to sniff or ruin the data or even redirect them toward an unintended destination.

Data is only allowed to enter and exit the GLSP at locations authorized and configured by the control plane. Both CR-LDP and RSVP-TE have the support of MD5 signature password and authentication. CR-LDP uses TCP/IP services, which is vulnerable to denial of service attacks; therefore, any in-between errors are reported immediately.

RSVP-TE uses UDP services and the IETF standard to specify an authentication and policy control. This allows the originator of the messages to be verified (e.g., using MD5) and makes it possible to police unauthorized or malicious reservation of resources. Similar features could be defined for CR-LDP but the connection-oriented nature of the TCP session makes this less important.

8.3.2.6.1 Protocol Type

RSVP-TE is a soft state protocol. After initial GLSP setup process, refresh messages must be periodically exchanged to notify the peers that the connection is still desired. State refreshes help to make sure that GLSP state is properly synchronized between adjacent nodes, procedure that is required by the fact that RSVP-TE uses IP datagrams, so that control messages may be lost without notification.

Periodic refresh messages imply the refresh overhead: this is a sort of weakness in the protocol that makes a bit more difficult to scale it to large networks. A proposed solution to reduce the refresh messages number and the bandwidth they require from the control plane communication structure is bundling many refresh messages together. This reduces the traffic volume but the processing time is still the same.

CR-LDP is referred to as a hard state protocol. This means that all the information is exchanged at the initial setup time, and no additional information is exchanged between GLSR until the GLSP is torn down. No refresh message is needed due to the use of TCP. When a GLSP is no longer needed, decision that can be taken by the client that has delivered the request to the control plane or by the central management system, messages must be exchanged notifying all routers that the GLSP is down and the corresponding allocated resources are now free.

8.3.2.6.2 Surviving Capability in Front of Control Plane Communication Failures

RSVP-TE is based on datagram transmission of service packets using UDP; thus it is intrinsically robust versus failures in the communication capacity available for the control plane. On the other hand, CR-LDP is in principle more vulnerable versus such failure. However, this vulnerability is not always a real weakness of the protocol, since survivability strategies are generally adopted in the control plane communication network.

8.3.2.7 Line Management Protocol

The last key protocol of the GMPLS suite is the line management protocol; it is a point-to-point protocol that has the role of managing the link between adjacent nodes. As a matter of fact, traditional IP and MPLS protocol suite leverages on the physical layer to manage physical links. On the contrary, GMPLS cannot do that, since it is in charge of the entire network layering, thus a new protocol has been invented for this role. In particular, referring to an optical link, for generality with several WDM channels, LMP is aimed at the following tasks:

- Control-channel management
- Link connectivity verification
- Link property correlation
- Fault management

8.3.2.7.1 *Control Channel Management*

In order to set up a channel, the control plane must know the IP address on the far end the channel either due to a manual configuration or for automatic discovery. This role, which is a typical point-to-point function relative to a specific link, is performed by the LMP via Config/ ConfigAck messages exchange. This procedure discovers peer's IP address and establishes LMP adjacency relationship automatically. Adjacency relationship requires at least one bidirectional control channel (multiple can be used simultaneously) between peers.

In order to run the control plane through the link and establish an adjacency relationship, a control channel traversing the link is needed. This is generally defined at the control plane level by a set of parameters that are negotiated by LMP during the adjacency discovery. Once set up, the control channel maintenance is done by LMP by periodic Hello messages.

8.3.2.7.2 *Link Connectivity Verification*

A GLSR knows the local address (ID) for data links that connect its ports to the ports of a peer GLSR. In the absence of a suitable mechanism, the GLSR does not know how a peer identifies the same data links and has no a priori information about the state of the data links. Link IDs mapping between peers should be made in order to have a common unambiguous association between a link and a link ID and in order to understand signaling messages received from a peer where the link ID is that assigned from the peer.

Links ID association is performed by the LMP using test message contains two key parameters: *sender's local data_link_ID* and *Verify_ID*; the first is the ID that the GLSR assigns to a link, the second is the request done to the GLSR on the other side of the link: verify the link ID correspondence. *Verify_ID* also contains the GLSR ID.

The GLSR on the other side of the link looks at Verify_ID to understand from which adjacent GLSR the message comes from. The peer answers with *TestStatusSuccess* that contains peer's local data link ID and the data link ID from the Test message.

8.3.2.7.3 *Link Property Correlation*

This is a process of synchronizing properties of data channels (e.g., supported switching and encoding types, maximum bandwidth that can be reserved, etc.) between peers and aggregation of data channels to form one TE link done by exchange of messages containing the parameters *LinkSummary* and *LinkSummaryAck* or *LinkSummaryNack* (if the peer does not agree on some parameters). In this case a new *LinkSummary* message is sent after *LinkSummaryNack*.

8.3.2.7.4 *Fault Management*

Since LMP is the tool that GMPLS control plane uses to manage the single point-to-point link, the procedure to localize a failure and to issue a notification message to the GLSR on the other side of the failed link starts, if it is the case, the restoration mechanism is located within the functionalities of LMP. To locate a fault up to a particular link between adjacent nodes, a downstream node (in terms of data flow) that detects data link failures will send a *ChannelStatus* message to notify its upstream peer.

8.4 Design and Optimization of ASON/GMPLS Networks

The introduction of the ASON/GMPLS control plane is a real revolution in the carriers network, a new paradigm that substitutes the idea of centrally controlled network that was the strong point of the introduction of SDH/SONET. This evolution is strictly related to the gradual change in the service bundles that carriers offer to customers and consequently in the characteristics of the traffic routed through the network [4,5].

In particular the traffic is getting more dynamic and the operation of setting up a connection or reroute a connection is no more a rare event. Moreover, and probably more important, it is much more difficult to foresee the evolution of future traffic. In the telephone era, when incumbent carriers were monopolist almost everywhere in the world, the traffic was strictly related to the population and to the well-known statistical characteristics of the telephone call.

With the dominant role of data services and the mobility of the customers due to the competition of different carriers and to the variety of service offering, there is no more a known statistical distribution that helps carrier planners to forecast the traffic behavior. As a consequence, the network design has to be adapted to the new paradigm, where unpredictable traffic is routed on an automatically managed network.

8.4.1 Detailed Example: Design Target and Issues

Since the design problem for an ASON/GMPLS network is really complex comprising a great number of elements, in this section, we will carry out in detail a simple example, in order to clarify some of the main issues in designing such a network.

We will not be so formal in defining performance evaluation parameters and design procedures, and a more precise discussion of these points will be carried out in Sections 8.5 and 8.6. The main task of this section is to give a feeling of the main aspects and of the complexity of the problem working on a very simple example.

The network we will consider in this example is a multilayer network, like the ASON/GMPLS networks. We will assume for simplicity that a DWDM transport layer based on electrical core OXCs has a single client layer, an IP network. We will also assume that the IP network is composed of LSRs and managed by an MPLS control plane, and that the overlay model defined by the GMPLS standard is implemented for the overall control plane. This means that the general network node is equipped as in Figure 8.19, with an LSR connected to an electrical core OXC via a UNI interface.

The starting point is the physical topology constituted by buildings where it is possible to place network nodes and fiber cables as represented in Figure 8.20.

At the beginning, we will assume that the traffic to be routed through the network is known and that the design consists of deploying equipment on the given topology in an optimized way to support the traffic. In a second step, we will see what happens if the traffic is not known, but it is known only as traffic statistics.

To simplify the approach, we will assume that all the MPLS layer LSPs to be set-up have the same capacity and that we will set equal to one to normalize transmission capacities.

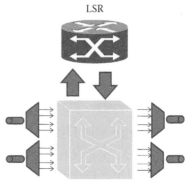

FIGURE 8.19
Scheme of the two layer node considered in the example of network dimensioning.

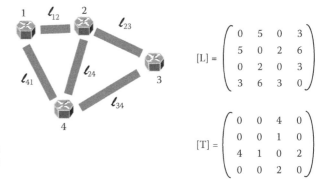

FIGURE 8.20
Physical topology, traffic matrix, and link lengths of the network considered in the example of dimensioning.

In this normalized measurement unit, a wavelength, structured with the OTN frame, has capacity $w \geq 1$.

The last element that is needed to characterize the topology is the distance matrix **[L]**, which is the matrix whose elements are the length of the physical links connecting the nodes. We are considering obviously physical length, which takes into account the real structure of the cable system and thus that can be used to design transmission systems. The value we will assume in our example for the normalized lengths of the links are reported in Figure 8.20.

The traffic also, if it is a priori known, can be represented with the matrix **[T]** of the MPLS layer LSPs routing requests, whose value for our exercise is also reported near the network topology in Figure 8.20. To deploy equipment in an optimized way means to deploy the minimum cost network that can manage with the required reliability and QoS the given traffic. For the moment we do not consider QoS, even if this is in general an important point of the design.

In order to minimize the cost, a cost model has to be done for all the equipment that is present in the network.

We will choose a simple model, in line with the nature of this example:

- The cost of LSR will be proportional to the number of ports (λ_j will indicate the number of ports of the LSR in the node j, if $\lambda_j = 0$ no router is present). The proportionality constant (unitary cost of one LSR port) will be indicated as χ_L.

- The cost of OXC will be proportional to the number of ports (κ_j will indicate the number of ports of the OXC in the node j, if $\kappa_j = 0$ no equipment is present since in the overlay model no LSR can be deployed without the transport layer switch). The proportionality constant (unitary cost of one OXC port) will be indicated with χ_x.
- The cost of the WDM will be composed by two addenda:

  1. The first term represents transponders and is proportional to the number of wavelength (ω_{jn} and will indicate the number of wavelength of the WDM on the link jn, if $\omega_{jn} = 0$ no WDM is present). The proportionality constant (unitary cost of one WDM wavelength) will be indicated with χ_w.

  2. The second is proportional to the length l_{jn} and represents the infrastructure (amplifiers and so on). The proportionality constant (cost of a unit length WDM infrastructure) will be indicated as χ_0.

With this simple model we can try to optimize the network.

In particular, we will adopt two different optimization criteria that correspond to different balances between the cost of the transport layer and the cost of the packet layer: we will try to minimize first the cost of the nodes and then the cost of the links. In both cases we will do the realistic hypothesis that $\chi_x < \chi_L$, that is, the port of an OXC is cheaper than a port of an LSR.

8.4.1.1 Basic Examples of Network Design

In this section, we will start to introduce a few basic elements of ASON/GMPLS network design neglecting two key elements: The uncertainty on the traffic forecast and the need of designing the network so as to allow some survivability strategy.

Even if these elements are key points in a real network design, at the first moment, they would mask more fundamental issues like transparent pass-through versus grooming-based design, mesh versus hub design, and finally the importance of linear cost models. In the next section, we will reconsider the design example of this section adding design for survivability and design in uncertain traffic conditions. Section 8.4.1 is intended as a preparation to the more formal discussion about ASON/GMPLS network design that we will carry out from Sections 8.4.2 through 8.4.4.

8.4.1.1.1 Node Cost Minimization and Shortest Possible WDM Links

Since we know the traffic to be routed, but we do not know the order in which the requests will arrive, we can imagine to route the GLSPs in an optimized way and then to use the constrained routing to reproduce such optimized configuration whatever be the order of the requests.

In order to minimize the node cost and, in situations in which the node cost does not change, to have the shortest possible WDM links, we will proceed as follows:

- GLSPs and OXC have to be installed in every node.
- WDM links are installed in links ℓ_{23}, ℓ_{12}, ℓ_{34}, ℓ_{14} so to form a sort of ring. Link between nodes 2 and 4 is not equipped.
- The GLSP between nodes 2 and 3 is routed through the direct route that is also the shortest route of the graph.
- The GLSPs from nodes 1 to 3 are routed through node 4 in order to have a WDM link of length 6 instead of 7 that is the length of the connection $1 \rightarrow 2 \rightarrow 3$.
- The GLSPs from 3 to 4 are also routed through the direct route.

The LSR ports are used only to generate and terminate local GLSPs and the transport is completely performed at the WDM layer. When an GLSP has to pass-through a node, it is switched by the OXC.

This kind of design produces the following numbers:

- The number of ports of LSRs is $\Sigma_j \lambda_j = 2\,T$, where T is the overall number of routed GLSPs. This is the minimum number of LSR ports: two for each GLSP in the network. The property that the lower bound of the number of higher layer GLSRs ports is equal to twice the sum of the elements of the traffic matrix holds, of course, in general and it is a useful way for a fast evaluation of the order of magnitude of the number of ports.

- The number of ports of OXCs is $\sum_j \kappa_j = 2 \sum\sum_{h,k} (1+\tau_{h,k})\lceil t_{h,k}/w \rceil = 16$ where $t_{h,k}$ is the number of GLSPs from node h to node k and it has been assumed, as reasonable, $w \gg \mathrm{Max}\,(t_{h,k})$. In the aforementioned equation $\lceil x \rceil$ is the smallest integer greater than x and $\tau_{h,k}$ is the so-called crossing factor, which is the number of intermediate nodes the GLSPs from node h to node k traverse. Naturally this number can be defined only if all the GLSPs with the same end nodes are routed through the same route, and in complex networks, this is not obvious. However, in our simple case, this definition allows us to obtain a simple expression of the number of OXC ports.

- The number of wavelengths has to be evaluated link by link, since we have assumed electrical core OXCs, thus the wavelength is physically discontinued passing through an OXC and could also change. With the assumption $w \gg \mathrm{Max}(t_{h,k})$, we have one wavelength for each direction on all the equipped link.

Thus the final expression of the network cost C is

$$C = 2\chi_L \sum\sum_{h,k} t_{h,k} + 2\chi_x \sum\sum_{h,k} (1+\tau_{h,k})\left\lceil \frac{t_{h,k}}{w} \right\rceil + \sum\sum_{h,k} (\ell_{h,k}\,\chi_0 \varepsilon_{h,k} + \chi_W w_{h,k}) \qquad (8.1)$$

where
   $\varepsilon_{h,k}$ is the link functional tag that is equal to one if a WDM system is equipped on that link and equal to 0 if no transmission system is equipped on the link;
   $w_{h,k}$ is the number of wavelengths routed along the link h-k;

Equation 8.1 is specific of the case we are studying and of the simple cost models we have chosen. However, if the integer part is neglected with the idea of taking the greatest integer at the end, it presents itself as a linear combination of the elements representing the chosen routing, $\tau_{h,k}$, $w_{h,k}$ and $\varepsilon_{h,k}$, whose coefficient depends on the cost coefficients (the χ_j parameters) and on the traffic matrix and on the link matrix elements (that are the number describing the physical topology).

This is a quite general property and with a suitable definition of the elements representing routing, topology, traffic, and cost, a similar linear relationship can be obtained also with more complex networks and more realistic cost models. This property, to which we will return in the next section, is the basic of a network design method that is applicable when a traffic forecast is available.

8.4.1.1.2 Link Cost Minimization

In order to minimize the link cost, we have to minimize the link length and the number of wavelength. This can be done by modifying the routing used in the previous example as follows. The paths from node 1 to node 3 are always routed through node 4 to minimize the route length, but arriving at node 4, they are sent to the LSR that is resident in node 4 that sent it back together with the other GLSPs, which have to travel from node 4 to node 3.

This operation is called grooming and consists essentially of using the switching capability at a higher layer (in this case the MPLS layer) in order to exploit at best the transmission capacity installed in the field. In this case, exploiting the grooming capability of the LSR in node 4 allows a wavelength to be spared on the link from 4 to 3, at the cost of eight LSR ports in node 4.

Naturally, more complex network algorithms can be devised to decide if grooming has to be performed in a certain node or not so to achieve a global network optimization.

The equation providing the network cost if this second design criterion used is

$$C = 2 \sum_{h,k} \sum (1 + \tau_{h,k}) \left[\chi_L t_{h,k} \cdot + \chi_x \left\lceil \frac{t_{h,k}}{w} \right\rceil \right] + \sum_{h,k} \sum (\ell_{h,k} \chi_0 \varepsilon_{h,k} + \chi_W w_{h,k}) \tag{8.2}$$

As anticipated, this is again a linear expression of the design parameters $\tau_{h,k}$, w_h, k, and $\varepsilon_{h,k}$ if the integer part is neglected.

8.4.1.1.3 Design with a Routing Hub

Looking at the topology, it is possible to note that two nodes, namely nodes 2 and 4, are connected with all the other nodes. This characteristic can be exploited for a particular type of design where all the nodes originating an GLSP create a connection with the hub and then the hub distributes the GLSPs to the destinations. This design neither achieves minimum cost of nodes nor minimum cost of links, but it can be useful in cases in which the overall cost is almost equally shared between links and nodes.

Moreover, several times, there are also other design objectives besides the cost minimization. For example, in the case of the presence of a hub, since all the GLSPs pass there, authentication and QoS are easier to manage.

In our example, let us set the hub in node 4. Thus all the GLSPs are sent to node 4 with a single hop connection and node 4 distributes the GLSPs to the destinations. Grooming generally is required at the hub to save wavelength at the hub exit. In our case, MPLS LSPs that from nodes 1, 4, and 2 are directed toward node 3 can be statistically multiplexed together in a single wavelength between nodes 4 and 3.

The overall number of LSR ports is in our case is $\Sigma_j \lambda_j = 48$, the number of OXC ports is $\Sigma_j \kappa_j = 12$, and the number of wavelength is 4, one for each link of the network (it is always assumed w sufficiently large). The formal expression providing the cost of the network in this third case, indicating with H the node number of the hub, is

$$C = 4 \sum_{\substack{h \neq H \\ k \neq H}} \sum \left[\chi_L t_{h,k} \cdot + \chi_x \left\lceil \frac{t_{h,k}}{w} \right\rceil \right] + 2 \sum_{h \neq H} \left[\chi_L t_{h,H} \cdot + \chi_x \left\lceil \frac{t_{h,H}}{w} \right\rceil \right]$$

$$+ 2 \sum_{h \neq H} (\ell_{h,H} \chi_0 \varepsilon_{h,H} + \chi_W w_{h,H}) \tag{8.3}$$

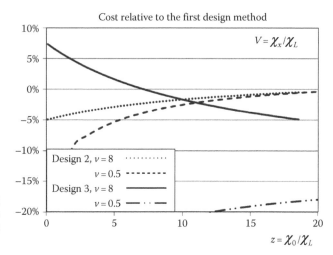

Cost relative to the first design method

$V = \chi_x / \chi_L$

Design 2, $v = 8$ ··········
$v = 0.5$ --------
Design 3, $v = 8$ ———
$v = 0.5$ — ·· —

$z = \chi_0 / \chi_L$

FIGURE 8.21
Dimensioning results: cost for second and third design methods normalized to the cost of the first method for different values of the cost parameters.

Also in this case, neglecting the integer part to return at the greater integer after achieving the noninteger result, we arrive at a linear expression of the design variables; moreover, as also expected, the hub has a completely particular role in Equation 8.3.

8.4.1.1.4 Comparison between the Design Methods

Since the design methods we have introduced aim at different kinds of network optimization they have to be compared when the cost variables of the network changes. Just to have a feeling of what happens, let us put $\chi_x = v\,\chi_L$ and $\chi_0 = 3\chi_w = z\,\chi_L$ so as to have only two cost parameters: z and v. Increasing z, the cost of the WDM is more and more prevalent, while the contrary happens decreasing z. On the other hand, the ratio between the cost of the node at layer 1 and at layer 2 is regulated by v. Figure 8.21 reports the network cost relative to the base cost obtained with the first method when the second or the third methods are used.

As intuitive, increasing z, the cost of the WDM links is more important and the design with optimized WDM links becomes preferable. This trend is almost independent of v, since the number of ports of LSRs and OXCs have almost the same trend with z.

With high values of v, an LSR port costs much more than an OXC port, and the hub design that reduces the OXC dimensions at the expense of a double passage of each GLSP through an LSR is not convenient. On the contrary, for low values of v sparing in OXC ports is meaningful, since one OXC port costs much more than an LSR port, thus the hub design is convenient.

Since we have done a very simple example, it is not so relevant the fact that in practice, considering long-haul DWDM systems based on EDFA amplifiers; z is slightly greater than one and v is small, perhaps between 0.5 and 0.2. More relevant is the observation that network optimization depends on cost balance and that frequently in the time interval between the planning and optimization of a new network and the end of its deployment, the cost framework is changed. Thus, in real design, it is important to capture qualitative trends that are constant with respect to market evolution to be able to keep the network design updated in all the circumstances.

8.4.1.2 Design for Survivability

In the previous section, we have carried out a basic example of network design without taking into account any survivability strategy. In reality, survivability has to be taken into account. In the case of ASON/GMPLS networks, both protection and restoration can be implemented and each of them can be carried out both at layers 1 and 2. Depending on what solution is chosen for survivability, the impact on the network design changes and network availability has a different value.

Since we are dealing with a simple example to underline the main points of network design, in considering survivability, we will assume to face only fiber cables cuts or DWDM line sites failure.

We will assume that the probability of a cable cut in a given time interval Δt is constant (does not depends on the absolute time period but only on the interval duration) and for a sufficiently small Δt can be written as

$$P(\Delta t) = p_f \Delta t \tag{8.4}$$

In our case, Δt is sufficiently small as far as the failure probability will remain much smaller than one and in this case it can be as long as a few months, since the probability that a single link will be broken in a transport network is generally low.

Naturally, if the number of links in the network is very high, the overall probability that one link breaks in the period can be not negligible. In particular, we can assume as a measure of the network reliability the probability that no service disruption occurs in a fixed time if the service was working at the start of the period. We will call this measure network survivability degree D. In Section 8.5, we will give a more precise definition of the standard parameter used for this kind of evaluations: the network availability.

The survivability degree of a network with N_L links without any protection and/or restoration mechanism is

$$D = 1 - P(failure, t_0) = 1 - \sum_{j=1}^{N_L} \binom{N_L}{j} p_f^j (1 - p_f)^{(N_f - j)} \tag{8.5}$$

where $P(failure, t_0)$ is the probability of a failure in the considered time interval when the network was working at the interval start (instant t_0) and it is assumed that the single link failure events are independent and their probabilities p_f equal and much smaller than 1.

It is important to note, just for an example, that if the network has 100 links and the probability p_f in the network repair time is 2×10^{-4}, the result of (8.5) is $D \approx 0.98$, that is clearly unacceptable in front of a normal request of $1 - D = 10^{-5}$.

This clarifies the need for a survivability strategy.

8.4.1.2.1 Link 1 + 1 Dedicated Protection

This is the most effective but also the most resource-hungry survivability strategy. Every single link is doubled with another identical link running physically on a different path. This is done by deploying, for example, two identical cables on different sides of a highway or of a railways so that if one is broken, the other is sufficiently far not to be involved in the accident. Every node has a protection switch in front of the receiver and the transmitted signal is doubled and sent both on the working and on the protection path (see Figure 8.22).

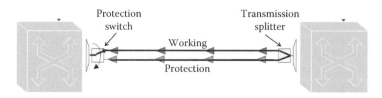

FIGURE 8.22
Block scheme of 1 + 1 link dedicated protection.

If this protection strategy is chosen, the base cost of all the fiber infrastructure is more than doubled (since for each link we have twice the number of amplifiers, DCFs, and so on and moreover we have to add the protection switch card with all the related monitoring), but the network design remains the same. The failure probability of such a protection mechanism adopted is such that within the repair time both the working and the protection links fail.

The probability of two failures on two specific links is, in a first approximation, $p_{f2} \approx p_f^2(1-p_f)^{(2N_L-2)}$.

Thus the survivability degree can be approximately expressed as

$$D = 1 - P(fail) \tag{8.6}$$

where $P(fail)$ is given by

$$P(fail) \approx N_L p_{f2}(1-p_{f2})^{(N_L-1)} \tag{8.7}$$

If the network has 100 links and p_f in the network, repair time is 2×10^{-4}; the result of is $D \approx 0.999996$, a value compliant with values required for D in real networks.

8.4.1.2.2 Optical Channel: 1 + 1 Dedicated Protection

In this case, every GLSP is routed two times in the transport layer on two optical paths such that the links of the working path are never in the same SRLG with any link of the protection path. In our example, we have to connect the following node couples: (2,3), (1,3), (3,4). For each couple of node, two completely disjoined paths are to be found following the desired design criterion.

From the topology, it is quite clear that a design with the hub is impossible in this case due to lack of sufficient topology connectivity. Considering the minimum node cost design, suitable couples of paths are listed in Table 8.5.

TABLE 8.5

List of Working and Protection Paths for the Minimum Node Cost Design

GLSP Ends	Working Route	Protection Route
(2,3)	2→3	2→4→3
(1,3)	1→4→3	1→2→3
(3,4)	3→4	3→2→4

If dedicated protection is to be implemented, the signal has to be doubled by a splitter after the OXC and sent simultaneously on the working and protection paths. Thus, looking at Table 8.5, two links have to be equipped for protection only (links 1–2 and 2–4) and six more wavelengths are needed.

In a certain number of real situations, the protection paths can be identified as follows:

- The working path is routed following the chosen routing criterion.
- A new topology is constructed by eliminating all the links traversed by the working path.
- The path is routed again using the chosen criterion in this new topology to obtain the protection path.

In the hypothesis of using the OXC switching capacity to double the signals and to feed the working and the protection paths, the cost of the network is given by Equation 8.1 wherein the counts of OXC ports and of links and wavelengths both protection and working paths have to be considered.

In a very first approximation a service disruption occurs if a working path and its protection path both fail during a repair time. Thus probability of failure of a path can be calculated starting from the number of links composing the path. In particular, let us assume to have N_p working paths (three in the example) and the same number of protection paths, each composed of H_j hops $j = 1 \ldots 2 N_p$), so that in our example, always considering the first design method, $[\mathbf{H}] = (1, 2, 1)$.

The path failure for the j-th path (that can be working if $j \leq N_p$, otherwise protection) is then

$$P(j) = \sum_{v=1}^{H_j} \binom{H_j}{v} p_f^v (1 - p_f)^{(H_j - v)} \tag{8.8}$$

so that a failure that cannot be faced occurs when during a repair time both working path and its protection path fails. The survivability degree, in a first approximation that is valid if the path failure probability is much smaller than one for all the paths, is thus given by

$$D = 1 - \sum_{j=1}^{N_p} P(j)P(j + N_p) \prod_{\substack{s \neq j \\ s \neq j + N_p}} [1 - P(s)] \tag{8.9}$$

In the example network with 100 links and p_f 2×10^{-4}, we will assume a number of routed optical paths equal to 40 with an average length of 5 hops each. With this features $D \approx 0.99996$.

8.4.1.2.3 Optical Channel: 1 + 1 Shared Protection

From the routing of protection paths in the example network we are considering, it is evident that some sharing of protection resources can be carried out, paying it with a lower efficiency of the protection mechanism because of losing capacity of facing simultaneous single failures on different paths.

In particular, in our example, in the dedicated protection scheme, two protection wavelengths are routed along the link 2–3. If facing a simultaneous failure of working connections $1 \rightarrow 3$ and $2 \rightarrow 4$ is not a requirement, a single protection wavelength can be equipped on the link 2–3 so that it is used both to protect connection 1–3 and to protect connection 2–4. The OXCs will suitably switch so to create the correct routing of the protection path.

Naturally, if dedicated protection needs only advertising through the control plane, but switches on without any protocol operation, this is not true for shared protection. As a matter of fact, the intermediate OXCs that have to switch to assign the shared protection resources to the right path have to be informed by a control plane protocol of the configuration to implement.

In this case, the network cost is that explained in the previous section, without the wavelengths and OXC ports that now are not needed due to protection resource sharing. A general expression of the survivability degree D is quite difficult to provide for the shared protection, as D depends on the effective sharing of protection resources and then on the topology.

However, a lower bound for D is immediately obtained by assuming as failure probability the probability that two working paths will fail in the repair time plus the probability that a working path and its associated protection both fail during the same repair time.

Thus the lower bound can be approximately expressed as

$$D = 1 - P(fail, 1)\,[1 - P(fail, 2)] - P(fail, 1)[1 - P(fail, 2)] - P(fail, 1)P(fail, 2) \qquad (8.10)$$

where $P(fail,1)$ and $P(fail,2)$ are the failure probabilities for the two failures cases and are given by

$$P(fail, 1) \approx \sum_{\substack{j=2}}^{N_p} \sum_{\substack{v=2 \\ v \neq j}}^{N_p} P(j)P(v) \prod_{\substack{s \neq j \\ s \neq v}} [1 - P(s)] \qquad (8.11)$$

$$P(fail, 2) \approx \sum_{j=1}^{N_p} P(j)P(N_p + j) \prod_{s \neq j} [1 - P(j)][1 - P(N_p + j)] \qquad (8.12)$$

In the example network with 100 links and p_f 2×10^{-4}, a number of routed optical paths equal to 40 with an average length of 5 hops each. With these features $D \approx 0.9996$. Shared protection can arrive in this example to lose one "9" in D, in front of a relevant sparing in protection resources when the sharing is relevant and thus D is near its lower bound.

8.4.1.2.4 GMPLS Restoration

The extreme form of sparing in survivability resources is GMPLS restoration. When restoration is set up there is no preassignment of survivability resources to specific paths, but a set of resources is assigned to restoration and left unused up to the moment a failure occurs. When a failure occurs, the control plane eliminates the failed link from the topology and reroutes the failed GLSPs using restoration resources.

In general, a complex algorithm is needed to reserve the minimum amount of restoration resources to face the desired set of failures (e.g., all the single link failures), but in the simple example we are dealing with, it is quite clear that if a spare wavelength is installed in links 1–2, 2–3, 1–4, and 3–4, whatever optical channel is involved in a single failure can be rerouted on the spare wavelengths. Thus restoration, in this specific case, to face all single-channel failure, needs 4 spare wavelengths (in each direction) and consequently 16 spare OXC ports.

When a channel is restored using restoration resources, these are occupied; thus, in our example, no other channel can be restored. As in the case of shared protection, in complex situations it is quite difficult to evaluate the exact network survivability degree, but a lower bound can be evaluated.

Such lower bound is obtained by assuming as failure probability the probability that two working paths will fail in the repair time plus the probability that a working path fails besides a single restoration reserved link.

Thus we can write, as in the previous section,

$$D = 1 - P(\mathit{fail}, 1)[1 - P(\mathit{fail}, 2)] - P(\mathit{fail}, 1)[1 - P(\mathit{fail}, 2)] - P(\mathit{fail}, 1)P(\mathit{fail}, 2) \qquad (8.10)$$

Where $P(\mathit{fail}, 1)$ is given by Equation 8.11 and $P(\mathit{fail}, 2)$

$$P(\mathit{fail}, 2) \approx \left[\sum_{v=1}^{N_R} \binom{N_R}{v} p_f^v (1 - p_f)^{(N_R - v)} \right] \left[\sum_{v=1}^{N_L} \binom{N_L}{v} p_f^v (1 - p_f)^{(N_L - v)} \right] \qquad (8.13)$$

where N_R is the number of links dedicated to restoration.

In the example network with 100 links for working paths and 30 links reserved for restoration only and $p_f\ 2 \times 10^{-4}$ the lower bound is $D \approx 0.99948$.

8.4.1.2.5 Comparison of Various Designs for Survivability

We have seen four possible single-layer survivability strategies starting from the more effective and also more expensive in terms of network resources and arriving at the most effective in terms of resources allocation but also less effective in terms of survivability degree. In any case naturally the resulting availability is at least two "9" better than the value of D achieved without any spare resource in the network.

Naturally multilayer survivability strategies can also be implemented, but it is not the scope of this simple example to go so farther and we will return to these more complex solution in a more general analysis later. Now we want to carry out on our simple network an economic comparison between the different strategies taking as a reference the minimum node cost design method.

The survivability degree for the network in Figure 8.20 if the different survivability strategies are used is reported in Table 8.6 if $p_f = 10^{-4}$.

From a network cost point of view, we can take as cost reference the cost of the network designed with the same criterion but without any survivability strategy. The relative cost

TABLE 8.6

Survivability Degree for the Network in Figure 8.20 if the Different Survivability Strategies Are Used

Survivability Strategy	Survivability Degree
Link dedicated 1 + 1 protection	0.99999996 (seven "9"s)
Optical channel dedicated 1 + 1 protection	0.9999998 (six "9"s)
Optical channel shared protection	0.9999997 (six "9"s)
Optical channel restoration	0.9999996 (six "9"s)

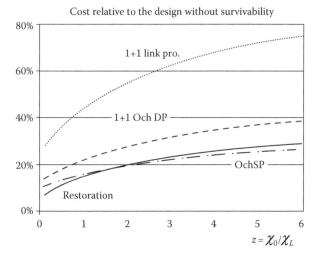

FIGURE 8.23
Dimensioning results: cost of survivable networks normalized to the cost of the first method of design without survivability for different values of the cost parameters.

of the various survivability strategies are reported in Figure 8.23 versus the normalized cost parameters v and z.

It is evident that the difference in the cost of survivability is great: the most expensive case is $1 + 1$ link protection and the dedicated $1 + 1$ channel protection. The relationship between the cost of shared protection and the cost of restoration is completely dependent on the particular topology and traffic, thus can be quite different from network to network. In any case, a qualitative idea is given by the network sharing degree, that is, the average number of working path that share a certain protection wavelength on a protection link in the shared channel protection.

It is intuitive that this parameter impacts also the quantity of spare capacity needed for restoration and in general higher the sharing degree, higher the cost advantage that restoration has on shared protection (even if counterexample can be found with a suitable choice of the topology and the traffic). In out example, the sharing degree is 1.3, quite low, thus shared protection and restoration are not so different in cost and in survivability degree. It exist even a value of z above it shared protection is cheaper than restoration. Both are however quite cheaper with respect to dedicated protection.

8.4.1.2.6 Design in Condition of Uncertain Traffic

A last step in the design of our example network before generalizing our considerations to provide design tools that are valid for every network: what happens when the traffic is not known?

If the traffic is completely unknown, no design can be done, different from placing a tentative capacity on each link and see what happen. However generally traffic estimation exists, but it is not sufficiently accurate to allow an optimum design to be realized.

In this case the design has to leverage the statistical description of the traffic fixing a maximum acceptable block probability (that is, the probability that an GLSP setup request has to be rejected) and designing the network to assure a blocking probability at most equal to the maximum acceptable.

We will detail much better this problem dealing with general design methods, for now the discussion about our small example can be considered ended. It allowed us to introduce a set of fundamental concepts for the network design:

- The description of multilayer networks in terms of virtual topologies
- The balance between grooming and transparent pass-through in multilayer nodes
- The possibility to express the network cost as a linear function of the design variables
- The necessity of deciding a survivability strategy
- A way in which network survivability can be quantified
- The different impact on costs of different survivability strategies
- Last but not least an idea of how the design can be carried out in conditions of uncertain traffic

8.4.2 Design Based on Optimization Algorithms

The most direct way to generalize the design methods we have used in our example is to set up an optimization algorithm working on a completely general multilayer topology and aimed at optimizing the cost of the network evaluated with a more accurate and general model.

Leveraging on the observation that in our example, neglecting the integer part appearing in the platform cost, the cost function was always linear in the design variables we will target a problem formulation compatible with integer linear programming (ILP), which is a well-known method to model a class of NP-hard problems whose optimization variables are integer [6–9].

8.4.2.1 Optimized Design Hypotheses

The first step to do to develop a general design procedure that can be applied to every topology and traffic request is to define the design problem data. We will assume a few very general hypotheses to restrict the problem and make it solvable.

- The infrastructure is completely given; that means that geographical position of Points of Presence of the carrier and of fiber cables are given. Equipment can be installed only in the points of presence, but for in-line amplifier sites of DWDM systems.
- We will assume that no OADM branching point exists in DWDM systems.
- We will work in terms of nodes topology: the topology in which the physical network nodes are represented by the nodes of the topology graph. A completely equivalent model can be carried out in the so called links topology, where the topology graph has nodes representing the physical network links. An example of corresponding links and node topologies are given in Figure 8.24.

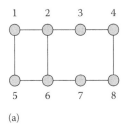

(a)

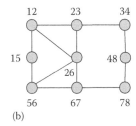

(b)

FIGURE 8.24
Example of (a) node topology and (b) corresponding link topology.

- We will image that all the fiber cables are owned by the carrier that is designing the network so that the bare use of a free fiber does not entail a cost and a fiber is always available on each route where a cable is present. However, the model can be generalized without changing the base definitions if the fibers are leased and using a fiber is a cost for itself. Also a possible limitation of fiber availability can be easily introduced without substantially changing the model.

- The topology will be meshed at every network layer; design of multiring networks is an important subject, based on different algorithms, and we encourage the interested reader to start the study on multiring networks design from Refs. [10–12].

- The layering is also fixed: physical layer is composed by a WDM transport network based on OXCs. We will assume that the OXCs are opaque and that every link between adjacent OXCs has been correctly designed under a transmission point of view so that we need not deal with transmission systems design when designing the network.

- The client layer is assumed to be an MPLS layer that interfaces the optical layer via the GMPLS overlay model interface. This means that a node comprising an LSR only cannot exist in the network since in the overlay model, the LRS cannot interface DWDM systems and OXCs directly in other nodes.

The design of optically transparent networks is an important subject that we will consider in Chapter 10. The opaque hypothesis also implies that wavelength continuity is not required in crossing an OXC, since the opaque switching machine also performs wavelength conversion.

- In principle, not always an LSR is superimposed on an OXC, and only when it is true, it is possible to communicate between layer 2 and layer 1. Since in practical networks it almost never happens that a node has no local traffic, we will assume that an LSR is always superimposed on an OXC at every node.

- Equipment type is already selected when doing network dimensioning at every network layer, thus all the cost parameters are given. In particular, the DWDM layer is composed only of one type of wavelength channel with a given capacity (let us say 10 Gbit/s).

- As far as QoS is concerned, we will simplify the situation with respect to a realistic case and we will assume that all the GLSPs are on the same level as far as the survivability is concerned. Also in this case introducing high priority (that is, protected or restored) and best effort GLSPs simply complicate a bit the model, but does not introduce any new element. Introducing realistic QoS classes on the contrary has quite an impact on the optimization model. Several studies have been done in this direction, see, for example, Refs. [13,14].

- We will also assume that a maximum number of GLSPs requests can be done between the same couple of nodes. This is a reasonable limitation, aimed at limiting the number of ports of the network nodes. All the MPLS LSPs will have the same capacity (e.g. 1.2 Gbit/s) that is assumed the unit of capacity of the network. Also this condition can be removed with a slight modification of the optimization algorithm if it is needed.

- In principle, an ideal optimization should be carried out by contemporarily assigning both working and survivability resources. We will see that the ILP problem deriving from the multilayer optimization is very complex (precisely NP-hard). Thus, to simplify the solution under a computational point of view, we will design the network in two steps: first assigning the working resources and in a second phase the survivability ones.

The decision of performing a two-step dimensioning, first working capacity and then protection or restoration, deserves some comments. Several studies in literature demonstrate that final network cost is quite near the optimum when proceeding in this way [6], while the computational complexity is sensibly reduced. Besides in terms of computation complexity, this approach has the practical advantage to allow simulation of different survivability mechanisms by designing the working part of the network only once.

We will exploit this feature in the last part of the chapter, when we will compare different survivability strategies maintaining the same traffic and network. These hypotheses are a sufficient base to lay the foundation for a general network design method based on ILP. In order to do that we have to built a general description of the network that is suitable for the scope.

8.4.2.2 Network Model for ILP

In a multilayer network model, a set of topology and traffic variables can be defined at the same way at infrastructure layer (layer 0 in our model), at the optical layer (layer 1) and at the MPLS layer (layer 2). We will call them with the same symbol with a suitable prefix. The layer prefix will be written on the left of the symbol, so that it is not confused with indices representing elements of a matrix.

For example, the number of nodes will be indicated with ^0N if we are considering the infrastructure layer, with ^1N if we are considering the optical layer, and ^2N if we are dealing with the MPLS layer. Naturally variables at different layers are related, in this case $^0N \geq {}^1N = {}^2N$.

The topology will be described by two matrixes: the adjacency matrix [^{s}A], whose dimensions are $^sN \times {}^sN$ ($s = 0, 1, 2$), and the link matrix [L], which exists only at the physical layer. The element $^s\alpha_{ij}$ of the adjacency matrix is 1 if at the sth layer the nodes i and j are connected via a layer direct connection, otherwise it is zero. The element ℓ_{ij} of the link matrix contains the length of the physical cable connecting nodes i and j at layer 0 if this cable exists, otherwise it is zero.

The traffic is defined through a traffic matrix [T] of dimension $^2N \times {}^2N$ that contains the request of traffic in terms of LSPs to route in the MPLS layer, its element τ_{ij} represents in particular the number of LSPs to route between node i and node j of the MPLS layer (layer 2).

If the traffic forecast is not sufficiently reliable, the matrix [T] has to be derived from the traffic statistics with a statistical model. For the moment we will assume that [T] is given, we will analyze in the next section a draft method to manage traffic uncertainly.

We will call $T_i^M = \mathrm{Max}(\tau_{ij})$, ($i = 1,\ldots, {}^2N$), ($j = 1,\ldots, {}^2N$) the maximum number of LSPs requests starting from node i towards another node. Generally T_i^M is a small number since we imagine to design a transport network where the LSPs are high speed (e.g., 1 Gbit/s). In any case $T_i^M \leq M$, which is the maximum number of LSPs required between the same nodes.

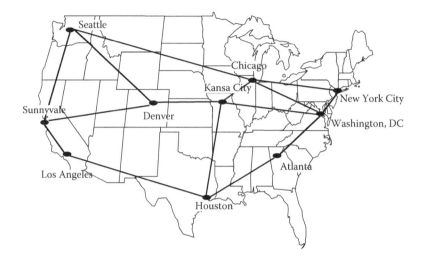

FIGURE 8.25
Reduced version of the North American network used for the examples of this chapter. The link length is assumed equal to the highway distance between the corresponding cities.

The maximum number T_M of LSPs to be routed between a couple of nodes is given by $T_M = \text{Max}\left(T_i^M\right)$. The routing is represented by the routing matrixes [^sR], [Λ]. The matrixes [^sR] ($s = 1, 2$) have dimensions $^2N \times {}^2N \times {}^sN \times {}^sN \times T_M$ and it is defined as follows: the element $^sr_{ijkdv}$ is one if the vth LSP request between the nodes i and j transit through the link k-d at the layer s. In principle this is a huge matrix, but in reality it is very sparse. For example, in a network like the pan American network of Figure 8.25 with 10 nodes at all the layers, supposing $T_M = 3$, it seems a matrix with about 30,000 elements.

However, in the network we have chosen as an example and in general in any network, it is possible to set a maximum number of hops per LSP and it is quite small (generally much smaller than N). For example, optimizing with random traffic the network of Figure 8.25 the maximum number of hops results to be 3 at optical layer and 2 at the MPLS layer. As a matter of fact, more hops generally means longer transmission and more costly DWDM.

Let us assume that H is the maximum number of elements different from zero in one slice of [^sR] obtained by $^sr_{ijkdv}$ varying k and d and maintaining constant the other indices. Let us also assume that it is the same for all the layers. Due to the meaning of the elements of [^sR], H results to be the maximum number of hops of a GLSP on layer s.

The matrixes [^sR] have a maximum number of $^sN^2H^2$ elements different from zero, which in the previous example is 900 for $s = 1$ and 400 for $s = 2$.

Moreover, as we will see in detail, the property that all GLSP are bidirectional can be used to further reduce the number of variables, arriving to 450 for $s = 1$ and 200 for $s = 2$ out of a total of 30,000. Thus, if it is needed to compact the representation of the matrixes [^sR], it is possible to adopt a sparse matrix representation. This can be also beneficial to speed up searches in the matrix and thus calculation.

One way to do that is to represent [^sR] through a $^2N \times {}^2N \times M$ pointers matrix and to append to each pointer a list of the only nonzero elements with the first two and the last indexes equal to the indexes of the pointer. The records of the list will be H^2 and each of them will contain the third and the fourth indexes of the nonzero element of [^sR].

This representation has also an important physical meaning that helps in implementing the algorithm. The first two indices are the two ends of an LSP while the last index is the order number of that LSP among all those with the same ends. Thus every pointer represents an LSP and it points to records representing the links traversed at layer s by that LSP.

The computation complexity of the final algorithm is a different issue, which is also triggered by the great number of design variables and we will discuss it at the end of the section. Returning to the definition of the routing matrixes, it is relevant to underline that if an LSP of the MPLS layer is routed through the hth node of the optical layer while it is merged in an optical channel, but it is not sent to the local LSR, we will have $^1r_{ijkhv} = 1$, but $^2r_{ijkhv} = 0$ since it does not travel through the node in the MPLS layer, but only in the optical layer. This shows as in the multilayer representation of routing the grooming is automatically comprised.

The matrix $[\Lambda]$ is a $^1N \times ^1N$ matrix and λ_{ij} represents the number of wavelength connecting node i with node j to route the LSPs. The wavelength channels connecting OXCs and transporting LSPs that travel only at the optical layer without being extracted by the wavelength up to their destination can be considered GLSPs in the optical layer.

The elements of the matrixes $[^sR], [\Lambda]$ are the design parameters and the solution of any optimized design has to provide values for these elements. However, this determines also the number of variables of our ILP problem and from this point of view, this is clearly quite a big number. Thus frequently simplifications are introduced to decrease the number of variables. We will discuss a few ways to simplify the problem when we will have a complete algorithm formulation.

Considering the cost model, we will use a generalization of what we have assumed in our example in the last section. Every equipment has a cost constituted by two terms: a platform and an interfaces term.

- DWDM systems cost is given by $\left(\chi_{w0}l\left(1 + \lceil w/M_w \rceil\right) + \chi_{w1}w \right)$ where l is the system length and w is the number of wavelengths, while χ_{w0} and χ_{w1} are constants typical of the system. M_w is the number of wavelengths supported by a single system and $\lceil x \rceil$ indicates the smaller integer greater than x. We will also call W the normalized capacity of one wavelength in term of LSPs (maintaining for definition equal to one the capacity of each LSP).

- The cost of OXC is given by $\left[\chi_{x0}\left(1 + \left\lceil \dfrac{q}{M_q} \right\rceil\right) + \chi_{x1}q \right]$ where q is the number of bidirectional line ports of the OXC, M_q is the number of bidirectional line ports supported by a single platform unit (e.g., a single rack), and $\lceil x \rceil$ indicates the smaller integer greater than x. The parameters χ_{x0} and χ_{x1} are normalized cost constants typical of the OXC.

- The cost of LSR is given by $\left[\chi_{L0}\left(1 + \lceil u/M_u \rceil\right) + \chi_{L1}u \right]$ where u is the number of bidirectional line ports of the LSR and M_u is the number of bidirectional line ports supported by a single platform unit (e.g., a single rack). The parameters χ_{L0} and χ_{L1} are constants typical of the LSR.

The number u_j of bidirectional ports of the jth LSR can be calculated as the sum of the locally generated LSPs plus the number of transit LSPs that are sent to the LSR to perform grooming optimization.
Thus

$$u_j = \sum_{i=1}^{2N} \sum_{k=1}^{2N} \sum_{h=1}^{2N} \sum_{v=1}^{M} \left({}^2r_{jikhv} + {}^2r_{khjiv} \right) \tag{8.14}$$

Similarly, the number q_j of bidirectional ports of the jth OXC is given by

$$q_j = \sum_{k=1}^{^1N} \lambda_{jk} \tag{8.15}$$

Finally, neglecting the $\lceil x \rceil$ operator to maintain the linearity of the cost function, the ILP target function writes

$$C = \sum_{j=1}^{^2N} \left[\chi_{L0} + \left(\chi_{L1} + \frac{\chi_{Lo}}{M_u} \right) \sum_{i=1}^{^2N} \sum_{k=1}^{^2N} \sum_{h=1}^{^2N} \sum_{v=1}^{M} \left({}^2r_{jikhv} + {}^2r_{khjiv} \right) \right]$$

$$+ \sum_{j=1}^{^1N} \left[\chi_{x0} + \left(\chi_{x1} + \frac{\chi_{x0}}{M_q} \right) \sum_{k=1}^{^1N} \lambda_{jk} \right]$$

$$+ \sum_{j=1}^{^1N} \sum_{k=1}^{^1N} \left[\chi_{w0} \, l_{jk} + \left(\chi_{w1} + \frac{\chi_{w0} l_{jk}}{M_w} \right) \lambda_{jk} \right] \tag{8.16}$$

The target function is linear in the design variables as desired. In this condition the resulting problem is an ILP problem if suitable constraints are set. Arriving at an ILP requires elimination from the cost function the lower integer part to the addenda representing the platform cost of the equipment. This is not a small approximation and it is worth a comment.

Generally, deploying a new DWDM system or a new OXC is an expensive operation with respect to adding a new wavelength to an existing DWDM or a new card to an existing OXC. This is exactly due to the integer part operation that we have neglected in the cost function: that operator represents the fact that the whole platform cost is paid when the first equipment card is installed.

Thus, while in practice the deployment of a new system is always avoided if possible, in our problem, this would not be true, since the first card will have the same cost of all the others. This approximation is important mainly in the case of DWDM systems whose platform has a high cost due to amplifiers and other line equipment.

In specific cases, this approximation can cause the solution to be far from optimum. This situation is detected when there are a lot of DWDM systems or OXCs and LSRs with a very low number of wavelengths or interfaces, only one in the limit case.

In this case, it can be beneficial to create a postprocessing routine that tries to eliminate these systems by rerouting the wavelengths on already existing equipment and verifies if the cost of the network can be lowered in this way. We do not detail how such postprocessing can be implemented, technical literature on such heuristics is very rich, and the interested reader can start to approach the problem from [15]. To obtain the results presented in this chapter, no postprocessing is used since looking at the obtained solutions shows that the platform approximation is never so important to change the results significantly.

Considering the constraints, we have to maintain them linear if we want to formulate an ILP problem. The first set of conditions is the flow continuity across the nodes at every layer. This means that at every layer the traffic entering the node from client and line ports has to balance the traffic exiting from client and line port. Looking at the LSR, the continuity, coinciding with the bidirectional nature of the LSR, can be written as

$$\left(^2r_{ijkhv} - {}^2r_{jikhv}\right) = 0 \quad (\forall i, j, h, k \in [1, {}^2N], \forall v \in [1, M]) \tag{8.17}$$

$$\left(^2r_{khijv} - {}^2r_{khjiv}\right) = 0 \quad (\forall i, j, h, k \in [1, {}^2N], \forall v \in [1, M]) \tag{8.18}$$

It is important to note that even if (8.17) and (8.18) are constraints, they can be imposed directly on the routing matrix with the effect of reducing the independent variables and decreasing the overall complexity. This is particularly easy if the sparse representation is used for the routing matrix, where LSPs appear explicitly. In this case, only the variables representing an LSP in one direction have to be considered, so halving number of elements of the pointers matrix.

While looking at the OXC we have two continuities, one in terms of wavelengths and one in terms of LSPs. In terms of wavelengths we can write

$$(\lambda_{jk} - \lambda_{kj}) = 0 \quad (\forall j, k \in [1, {}^1N]) \tag{8.19}$$

which means that the wavelength matrix is symmetric.

In terms of LSPs, the bidirectional condition is written as

$$\left(^1r_{ijkhv} - {}^1r_{jikhv}\right) = 0 \quad (\forall i, j, h, k \in [1, {}^1N], \forall v \in [1, M]) \tag{8.20}$$

$$\left(^1r_{khijv} - {}^1r_{khjiv}\right) = 0 \quad (\forall i, j, h, k \in [1, {}^1N], \forall v \in [1, M]) \tag{8.21}$$

Last but not least the continuity at the layers' boundary has to be set. Remembering that an LSR is always present where there is an OXC, we can say that the LSPs traversing the LSR (that is, passed from layer 1 to layer 2 in the nodes) will be in general a subset of the LSR traversing the OXC. This subset has to coincide with the set of LSPs carried by a certain number of wavelengths.

Mathematically this condition is expressed as

$$^1r_{ijkhv} \geq {}^2r_{ijkhv} \quad (\forall i, j, h, k \in [1, {}^1N], \forall v \in [1, M]) \tag{8.22}$$

Moreover, the number of LSPs terminated in a node has to coincide with the corresponding entry of the traffic matrix, thus

$$\sum_{h=1}^{^2N} \sum_{k=1}^{^2N} \sum_{v=1}^{M} {}^2r_{ijkhv} = \tau_{ij} \tag{8.23}$$

$$\sum_{h=1}^{^1N} \sum_{k=1}^{^1N} \sum_{v=1}^{M} {}^1r_{ijkhv} = \tau_{ij} \tag{8.24}$$

After the bidirectionality, it is needed to assure that the number of wavelengths deployed on each link is sufficient to support the traffic routed along that link and to assign the LSPs to the wavelengths so that passing a set of LSPs between the layers at a certain node coincides with the termination of a certain number of wavelengths in that node. This constraint requires to introduce the number of wavelengths terminated in a node and the number of wavelengths that bypass the node without termination.

We will call $[\Omega] \equiv \{\omega_{kh}\}$ the $^1N \times {}^1N$ matrix containing the wavelengths bypassing node k when arriving from link h-k. The requirement of a certain wavelength number is written as

$$W\lambda_{kh} \geq \sum_{i=1}^{{}^1N1} \sum_{j=1}^{{}^1N} \sum_{v=1}^{M} {}^1r_{ijkhv} \quad (h \text{ and } k = 1,\ldots, {}^1N) \tag{8.25}$$

while the condition on the wavelength termination, utilizing the fact that the locally dropped LSPs are the same both on layer 1 and on layer 2, can be written as

$$W\sum_{h=1}^{{}^2N1} \omega_{kh} \geq \sum_{j=1}^{{}^2N1} \tau_{kj} + \sum_{i=1}^{{}^2N1} \sum_{h=1}^{{}^2N} \sum_{j=1}^{{}^2N} \sum_{v=1}^{M} {}^2r_{ijkhv} \tag{8.26}$$

and finally

$$\omega_{kh} \leq \lambda_{kh} \tag{8.27}$$

where W is the maximum number of MPLS LSPs that can be multiplexed into a wavelength.

After the constraints on bidirectionality and capacity, we have to impose the GLSPs continuity through the network. This means two things:

1. If the routing matrix tells that a certain GLSPs travels at layer s from node h to node k with one hop, these nodes have to be contiguous in their own layer, which means that $^s\alpha_{hk}$ have to be equal to 1.

2. Successive hops have also to be contiguous, that is, the GLSP can go from node h to node k and then from node k to node j, but not from node j to node v, since after the previous step it was in node k.

First, let us introduce the maximum number of hops ^sD allowed to the GLSP in the network at layer s (it can be ^sN in the extreme case).

The first condition involves obviously the adjacency matrix at the given level, and is written as

$$^sr_{ijkhv} = {}^sr_{ijkhv}{}^s\alpha_{hk} \quad (j \text{ and } i = 1,\ldots, {}^1N, v = 1,\ldots, M) \tag{8.28}$$

which says that if $^1r_{ijkhv} = 1$, implying that the vth path routed through the nodes j and i passes through the link h-k, thus the link h-k has to exist.

In order to express the second condition we can notice that a matrix $[^sr_{ijkhv}]$ where k and h are the running indexes while the other are fixed represents the path of a GLSP that, by definition, has no loops. Thus only one element per line and one element per column can be different from zero, which is the same as saying that the path enters into the hth node of the topology only one time and exits from it only one time.

The previous conditions can be written in term of constrain compatible with ILP as

$$\sum_{h=1}^{^sN} {}^sr_{ijkhv} = \sum_{k=1}^{^sN} {}^sr_{ijkhv} = 1$$

(8.29)

If the condition (8.29) is a constrain of the problem, we are sure that only one element is different from zero among the elements of a column and of a row.

The last step is to express the fact that the path of an GLSP has to be made by adjacent links, that is, the set of nonzero elements ${}^sr_{ijkhv}$ with i, j, v fixed has to be constituted by concatenated elements like $({}^sr_{ijihv}, {}^sr_{ijhkv}, {}^sr_{ijkjv})$.

This condition can be expressed by the use of the adjacency matrix. From the nodes adjacency matrix, a links adjacency matrix $[{}^s\mathbf{B}]$ can be easily derived, whose elements ${}^s\beta_{ijk}$ are equal to one only if the two adjacent links i, j, and j, k both exist. In particular, it is immediately seen that ${}^s\beta_{ijk} = {}^s\alpha_{i,j} {}^s\alpha_{j,k}$.

Thus the condition we are searching is written as

$$^s\beta_{h,k,x} \left({}^sr_{ijhkv} + {}^sr_{ijkxv} \right) = \left({}^sr_{ijhkv} + {}^sr_{ijkxv} \right)$$

(8.30)

The meaning of this constraint is the following. Starting from ${}^sr_{ijikv}$, which are the unique nonzero elements with the first and third index equal to i, the second equal to j and the fifth equal to v, a chain of elements of $[{}^s\mathbf{R}]$ can be constructed by adding to each element the following element having the same first, second, and fifth index and the fourth index moved at the third place.

These last conditions also define a unique nonzero element, so that, if a valid path has been found between i and j, it can be constructed in this way. This also implies that both ${}^sr_{ijhkv}$ and ${}^sr_{ijkxv}$, representing adjacent links of the path, are equal to 1. From this property derives the fact that Equation 8.26 is satisfied.

A similar continuity constraint has to be written across a node for the wavelengths that are not passed to the upper layer by the OXC. In particular, the number of passing through wavelengths entering the OXC has to be equal to the number of passing through wavelengths exiting the OXC.

This condition however is automatically satisfied by the coexistence of the constrain (8.18) and by the property $\omega_{kh} = \omega_{hk}$ deriving from the bidirectional nature of the GMPLS GLSPs. The equations we have derived up to now are sufficient to dimension the network with optimized resources to sustain the traffic, but without any survivability strategy.

In order to add survivability resources a further dimensioning step has to be done. This step is sensibly different if protection or restoration is used. We will limit ourselves to describe the algorithms that can be used in this second step, leaving to the reader the construction of the optimization equations.

8.4.2.2.1 Optical 1 + 1 or 1:1 Link Protection

This is the simplest case; every optical link is protected through a similar link connecting the same nodes, but running on a different path (e.g., on the opposite sides of an highway or of a railway). The 1:1 protection differs from the 1 + 1 only because in the first case low-priority traffic can be routed on the protection capacity when it is not used while in the second case the working signal is sent also on the protection path to render recovery against failure the fastest possible.

In this case the transmission capacity of the links and the switching capacity of the nodes at the optical layer increments by 100%. The only attention to be paid is not to duplicate uselessly the OXC platforms using when possible for protection platform-free space that is present in the platforms already installed.

8.4.2.2.2 Optical 1 + 1 or 1:1 GLSP Protection

At the optical layer, the GLSPs are the wavelength paths. A wavelength path starts when a set of LSPs with the same destination are aggregated by an LSR in a wavelength and sent to the collocated OXC and terminates when the wavelength is extracted by the optical layer to be delivered to the destination LSR.

There is no data structure in our network model that represents directly the optical GLSPs, thus they have to be derived from the information we have. In particular, the following criterion can be used.

Every LSP is routed through the network in a known way so that it is possible to divide the LSPs into groups so that each group is formed by LSPs starting for node k, traveling through the same set of nodes on the same links and ends with the node h. They can be either LSPs that really start for a layer three request in k and end in h or LSPs that are injected in the optical layer in k or in h or both after grooming at the MPLS layer.

The extreme case is that of an optical GLSP that starts in node h from the aggregation of LSPs collected by other optical GLSPs via grooming at layer two and ends in node k where the LSPs are distributed among other optical GLSPs to be transported to destination.

Once the groups are constructed, every group corresponds to one or more optical GLSPs, depending upon the group cardinality and on the number of LSPs a wavelength can accommodate. Once a map of the optical GLSPs is created, a protection path is completely disjoined from the working path has to be preprovisioned for each of them.

To do that three matrixes have to be defined:

1. The optical traffic matrix [Q] whose elements q_{hk} represents all the layer 1 GLSPs (the wavelength paths previously calculated) connecting node h with node k.
2. The working routes matrix [ψ] whose generic element ψ_{ijhkv} is equal to one if the vth working wavelength path starting from node i and terminating in node j transits through the link h-k.
3. The protection routes matrix [Φ] whose generic element ϕ_{ijhkv} is equal to one if the vth protection wavelength path starting from node i and terminating in node j transits through the link h-k.

The first two matrixes are easily derived from the solution of the working optimization problem, the third matrix elements are the parameters of a new single layer ILP problem where the traffic [Q] is routed in the optical topology and the condition that working and protection paths have to be disjoined is guaranteed imposing the condition $\psi_{ijhkv} \phi_{ijhkv}=0$.

8.4.2.2.3 Optical Shared Protection

In this case, before going on with the design of the protection, a data structure has to be constructed where the working optical GLSPs are divided into sets so that the GLSPs of one set are routed on paths completely disjoined by the GLSPs of the other set.

After that, when performing the optimization, for every link assigned for protection, the sets of the working GLSPs protected through that link are determined and tagged on the link. To assign the protection capacity for a certain working GLSPs, let us say the GLSP with ID equal to 123 belonging to the fourth group, all the protection links that are not tagged with the fourth group are considered available capacity to route the protection GLSP.

8.4.2.2.4 MPLS 1 + 1 or 1:1 Protection

In this case, for each MPLS LSP, a protection path has to be preprovisioned that is completely disjoined from the working path. The optimization is carried out using the same approach already used for 1+1 optical protection with a further step: the possibility that new wavelengths have to be introduced due to the need of routing the protection LSPs. Naturally before adding resources to the network the unused resources already present have to be exploited. A new wavelength is introduced on a certain optical link only if no wavelength exists where the working LSP is not multiplexed and with enough free capacity to host the protection LSP.

8.4.2.2.5 MPLS Shared Protection

In this case, in analogy to what was done for the shared optical protection, before going on with the design of the protection, a data structure has to be constructed where the working LSPs are divided into sets so that the LSPs of one set are routed on paths completely disjoined by the LSPs of the other set.

After that, when performing the optimization for every link assigned for protection, the sets of the working LSPs protecting through that link are determined and tagged on the link. To assign the protection capacity for certain working LSPs, not only the free capacity left by the routing of the working traffic is considered available, but also all that link that has been introduced to route protection paths and that are not tagged with the group to which the link under analysis belongs.

8.4.2.2.6 Restoration at Optical Layer

In case of use of restoration at the optical layer, the OXC manages the restoration procedure in such a way that the MPLS layer does not detect the failure if possible. Different types of restoration exists, depending on when the restoration resources are assigned to the LSPs affected by a certain failure event. Among them we will consider three cases:

1. *Type 1.* Restoration resources are reserved and the connection is set up after the failure. In other words, the optical GLSPs affected by the failure are rerouted in the network as if they were new requests of connection delivered directly at the optical layer. If new connection requests arrive during the restoring phase, they go in competition with the GLSPs to be restored.

2. *Type 2.* Reservation is made before the failure (but without a links with a specific failure event) and the path setup is performed after the failure. If new connection requests arrive during the restoring phase, no competition with the GLSPs to be restored happens due to the assignment of the restoration resources to survivability only.

3. *Type 3.* Resources are reserved to survivability and a certain number of restoration paths are set-up. When a failure occurs, the affected paths are rerouted using the prepared restoration paths. This is the most rapid restoration, based on the setup of a certain number of prereserved paths. Naturally it is also the most resource-hungry restoration technique.

When restoration is used, an effective way to operate is similar to that of shared protection, but for the fact that in Type 1 restoration even the parts of the working GLSP that are not affected by failure are considered available as restoration paths.

In order to take into account this strategy, the following algorithm can be used to design the restoration capacity: for each link in the topology graph, the link is eliminated and the GLSPs involved are determined. These GLSPs constitute a capacity to be routed through the network without the failed link. This can be done with an ILP algorithm to optimize the resources use. Generally the cost of the restoration part of the network has to be recalculated since the final solution will not coincide with one of the partial solutions.

From a complexity point of view, it could seem that restoration design is much more complex than working capacity assignment, due to the fact that one ILP optimization is required for each network link. This is generally not true due to two facts: the first is that only the optical layer is involved in the routing, grooming is not permitted during this type of restoration to avoid multilevel signaling. Moreover, only a few elements of the traffic matrix are different from zero, contributing to lower computation complexity.

8.4.2.2.7 *Restoration at GMPLS Layer*

The procedure is quite similar to that carried out for restoration in the optical layer, with the same three restoration types. The difference is that now there are the MPLS LSPs to be restored, so that we have to face a two-layer optimization. This allows grooming to be used also when designing restoration resources generally with a better result with respect to the same restoration strategy adopted at the optical layer.

On the other hand, the complexity of the whole design algorithm is quite higher.

8.4.2.2.8 *Final Comment on Survivability Resources Design*

We have presented a set of methods to design survivability resources independently from working resources. This is for sure a suboptimum procedure, but as we have seen has a certain number of advantages. In any case, several optimization methods have been proposed in literature that does not divide the planning of working and survivability capacity. They are strictly related to a specific survivability method, but in principle, provide an optimum solution to the planning problem.

Naturally, the complexity of such a design method is higher with respect to what we have used here and generally some approximation is in any case used to make the problem solvable. For a discussion on different joined optimization methods, see Refs. [6,8,9].

8.4.2.3 *ILP Design Complexity*

The ILP optimization model we have presented in the last section is very explicit in representing the capacity routed on every link, which is useful in those cases in which DWDM links have to be designed after the network general optimization.

Other model exists that are a bit less complex from a computational point of view [16–18], however, whatever model is defined, the optimization of a multilayer mesh network is an NP hard problem, which means that the problem nature implies an exponential complexity increase with the key scale variable, which in our case is the number of nodes. Thus it is important to understand how the computation time can be shortened and what does it mean in terms of accuracy of the found solution. A very good analysis on this point is reported in Ref. [6] and we encourage the reader interested to detailed results on specific examples to start from this reference. Here we will cite and justify intuitively a few techniques for ILP complexity reduction without reporting a detailed quantitative analysis.

The first observation is that ILP goes uniformly toward the optimum solution, thus there is a relevant part of the time used by the algorithm to refine an already good solution. There is a measure of how much the configuration found in an intermediate state is far from the final optimization, which is the so-called solution optimality gap to be evaluated comparing the fully optimized solution with the intermediate one and measuring the cost difference in percentage with respect to the optimum cost.

It has been demonstrated in Ref. [6] a quite general result regarding algorithms like this: if an optimality gap of the order of 3% is accepted the computation time can be significantly shortened and even quite large networks can be analyzed.

In our case, trying to find a full optimal solution in the case of the network with 15 nodes and 25 links with a dense traffic matrix requires a vector computer or a very long time. However, in few hours of working on a well-equipped workstation a solution with a cost within 3% from the optimum can be obtained.

In practice, of course, the optimum solution is not known and the optimality gap cannot be evaluated exactly. However, observing the solution evolution for different running times it is not difficult to estimate the order of magnitude of the solution refinement versus the running time.

Even accepting a certain optimality gap, networks with a number of nodes in the order of 11–13 depending on the traffic matrix and the number of links are the limit for a reasonable execution time. If greater networks have to be analyzed, some more important simplification has to be done.

We can divide the possible simplifications in two classes: physical simplifications related to the particular nature of the problem at hand and mathematical simplifications, which is the mathematical approximation of the original ILP problem that can be applied to all the networks.

8.4.2.3.1 *Physical Approximations of the ILP problem*

The most important class of simplifications are those aimed at reducing the number of design variables. Generally such simplifications work very well on certain types of network and traffic while they give a poor performance in some specific cases. Now we report a list of possible simplifications that has been successfully used to design large networks, but only direct experience can achieve complete confidence when used on a generic topology.

- A first possible simplification is to decouple the design of the MPLS layer and of the optical layer. In this case, two different ILP problems result that are both quite simpler with respect to the joined design problem.

  This approach is very effective when grooming has not to be used intensively to increase the effective utilization of the optical channels. In the extreme case in which no grooming is used and MPLS layer and optical layer have separated survivability mechanisms, it is exact. Unfortunately, this is not the generic situation and in real networks excluding grooming brings to a suboptimal design. Naturally smaller the *W*, more difficult an effective grooming, more performing the method of disjoined design.

- Another important simplification is possible when the network is divided into zones and only a few LSRs are routed through different zones. In this case, the problem can be divided into a different problem for each zone and a global problem only for the interzone LSRs.

The single-zone problem can be solved exactly being relative to a network with much less nodes, while if the interzone GLSPs are really a small quantity of the traffic, they can routed through a suboptimum algorithm without affecting greatly the final network cost.

- If W is small (that is, the LSR are high speed and a small number of that fit into a wavelength) and the network is poorly connected, further simplification can be achieved assuming that all the paths having the same extremes are routed through the same route. In this case, the routing matrix loses the last dimension, since all the LSRs are routed together. This decreases the number of elements of the routing matrix by a factor M, so simplifying the resulting computation.

- If W is great and a large number of GLSPs are routed toward each couple of nodes (that is, the traffic matrix elements are large), the problem can be simplified by creating sets of GLSPs and routing them together.

8.4.2.3.2 *Mathematical Approximations of the ILP problem*

Since it is well known that an ILP problem can be very hard to solve, several mathematical studies have been developed to analyze possible approximations of the problem that reduces its complexity [19,20].

A simple method to obtain an upper-bound and a lower-bound to the solution of the ILP problem is to use the so-called relaxation method. This method simply consists in eliminating the condition that the variables have to be integer, substituting them with real variables. In the particular case in which there is a variable that can be zero or one, this is substituted with a real variable in between zero and one.

The first case happens in our model, for example, for the number of wavelengths in a link and the second for the elements of the matrix $[^s\mathbf{R}]$.

With these substitutions, the problem becomes a linear programming problem, whose complexity is polynomial, thus solvable also in high-dimensional cases. At the end, the optimum variables will not be integer in general, but the optimum of the LP problem will be a lower bound of the real problem solution; this lower bound is called LP-relaxation bound [20].

In order to derive an upper bound of the solution, the following algorithm can be used:

- Construct a problem-acceptable solution by rounding the LP optimum variables to the upper or lower integer as suited.

- Substitute these variables into the cost function to obtain the cost relative to the constructed network design.

- The obtained cost is an upper bound of the solution of the ILP problem.

The first step of the algorithm is generally not unique and, if several possible networks can be designed rounding to integer value the LP solution, the tighter upper bound is represented by the smaller cost corresponding to one of such networks. This upper bound is called rounding bound [20].

If the upper-bound and the lower bound are sufficiently near, the rounded variables constitute a good approximation of the effective ILP optimum solution. These simplification methods can make solvable a problem that is not solvable searching an exact solution, but the general nature of NP hard problem remains. This is why, in order to analyze large networks or to analyze a large number of alternatives, completely different design methods have been introduced based on heuristic approaches. These methods provide suboptimum solutions, but are much less complex than ILP optimization.

8.4.2.4 Design in Unknown Traffic Conditions

The design method we have detailed up to now relies on the fact that a reliable prediction of the traffic to be routed in a considered period of time (e.g., between two major network upgrades) is well known and that this traffic is constituted by connections that will last up to the end of the considered period.

In this case, the network design does not depend on the order in which the connections are opened, since the traffic matrix is available from the beginning. This condition is called deterministic static traffic.

The traffic model can be altered essentially in two ways:

1. Substituting the absolute knowledge of the traffic to be routed in the network with a partial knowledge of the traffic matrix; for example, the element of the matrix can be known only with a great approximation (like $5 < \tau_{ij} < 8$) or they can be random variables with a known statistical description. In this case the design is performed in unknown traffic conditions.
2. Instead to be semipermanent, the connections are setup and tear-down several times in the considered observation period on the ground of some statistics. In this case the network is designed under dynamic traffic conditions.

GMPLS sets up semipermanent connections in the transport and MPLS layers, thus dimensioning in conditions of dynamic traffic does not occurs in the bottom layers of the network. On the contrary, if the dimensioning was performed for a TCP/IP network, since TCP set-up and tear down continuously connections design in dynamic conditions would be needed.

On the other hand, the consolidation of competition among carriers and in general the uncertainty on the penetration of new services renders the knowledge of the traffic matrix less and less precise so that dimensioning is often performed in conditions of unknown traffic.

In order to use a dimensioning algorithm in these conditions, it is needed to set up a method to manage the unknown traffic condition. The problem is widely studied in literature and in general we would like to evidence three possible methods [21]:

1. A priori adjustment
2. A posteriori adjustment
3. Probabilistic approach

In the first two methods, the idea is simply to use a safety margin to face possible fluctuations of the traffic matrix. This safety margin reflects at the end in an overdeployment of network resources with respect to the optimum solution dictated by the average traffic matrix. The difference between the two methods is that in the first the average traffic matrix is augmented by the safety margin and then the optimization is carried out as usual.

In the second method all the optimization is carried out on the ground of the average traffic and at the end more resources are deployed with respect to the optimization output. The key issue in both these methods is how to choose the safety margin to be sure to reduce suitably the network rejection probability (that is, the probability a connection request is rejected by the network) while containing the deployment cost.

A discussion based on statistical approach is reported in Ref. [21] where it is demonstrated that it is possible to select an optimum safety margin starting from hypothesis on the statistical distribution of the traffic both in a priori and in a posteriori adjustment. However, even when the margin is chosen in this way, in particular cases, its use leads to excessive overdeployment in some areas of the network where in other areas resource shortage could still happen for modest fluctuations of the traffic variables. Sometimes the problem is solved considering different safety margins in different zones of the network, but other times this gives no advantage.

In any case, after the design of the network using the safety margin–driven overdeployment, simulations have to be generally performed to verify the effectiveness of the design in reducing the rejection probability.

Better results are achieved via the probabilistic method, which uses directly the knowledge of the traffic statistic distribution.

The easier case is that of uniform traffic variation.

In this case the random traffic matrix $[\mathbf{T}]$ can be written as follows:

$$[T] = x[T_0] \tag{8.31}$$

where

$[T_0]$ is the average traffic matrix.

x is a random variable with average equal to one and known probability distribution

In this case, the problem can be formulated as follows: the network has to be designed so that the probability that a connection request is rejected is smaller than a certain limit, called P_R.

The solution is straightforward once the design tool is available. Calling $p(x)$ the probability density of the variable x, it is possible to find a value x_0 so that

$$\int_{x_0}^{\infty} p(x)\,dx = P_R \tag{8.32}$$

If the network is dimensioned for the traffic matrix $[T] = x_0[T_0]$, the rejection probability is surely smaller or equal to P_R.

The situation gets more complex if the traffic matrix depends on more than one random variable, but in line of principle the solution method is the same. Let us assume that $[\mathbf{X}]$ is a random matrix, with $2N \leq V \leq 2N^2$ nonzero elements, and whose average is the identity matrix $[\mathbf{I}]$. Let also assume to know the joined probability density functions of the V random variables that are elements of $[\mathbf{X}]$, called $p(x_1, x_2, \ldots, x_V)$.

If the traffic matrix can be written as $[T] = [X][T_0]$, the method used for the single random variable can be directly generalized. Naturally it is not easy to derive the model of the traffic matrix and its distribution so that, in the absence of other information, a Gaussian model is often adopted. It is possible to imagine more complex models for the random nature of the traffic, for example, based on Markov chains [21]. In this case, the design gets more complex and we direct the reader to the relevant bibliography for a study of the problem.

8.4.3 Routing Policies–Based Design

As we have seen, ILP-based optimization algorithms are quite computationally complex, representing in reality, a particular formulation of an NP hard problem. If they are largely used for dimensioning of networks, it is not possible to adopt them when network dimensioning is only a step of a complex procedure, for example, of a comparison between different network architectures or the study of the network sensitivity to a set of design variables.

In this case, it could be necessary to repeat the dimensioning hundreds of times, which is quite hard using an ILP algorithm if the network is not particularly simple. Other times it is needed to have a first impression of the network deployment with a fast calculation, without resorting to the complexity of an ILP program. A plethora of heuristic network design algorithms have been proposed to be used in these cases, much less computationally hard than the ILP optimization.

A class of design algorithms based on suboptimum design methods is those of the routing-based design. These algorithms are inspired by the automatic configuration of the network and work as follows:

- A traffic matrix is individuated either through forecast of through statistical elaboration.
- A routing algorithm is also fixed for working capacity at all layers.
- Starting from the upper layer, connections are routed assuming the availability of infinite capacity in the network.
- At the end of the routing, the nodes and the links are inspected and the equipment required to sustain the performed routing is deployed.

Every time that the dimensioning depends on the order in which the paths are routed, the dimensioning is performed in several cases and the cheaper solution is selected. Moreover the algorithm can be easily generalized to include design of survivability capacity both in cases of protection and restoration.

For example, in case of 1 + 1 optical layer protection, the protection capacity is routed after the working one. In particular, the optical path protecting a link (i,j) is routed in the optical layer virtual topology where the link (i,j) has been eliminated and this procedure is repeated for all the links. Once the protection capacity has been provisioned it is considered occupied exactly like the working one in the successive routing operations, to reflect the 1 + 1 protection.

On the other hand, if shared protection has to be planned, the algorithm is similar, but the protection capacity is considered free for use in successive routing of protection optical paths with a procedure similar to that adopted for shared protection in the section dealing with ILP.

A great number of routing policies has been proposed as basis of the network design. We will introduce and comment few of them [15] limiting the comment to the design of the working capacity and of the optical layer only. Extension to a multilayer network is immediate.

8.4.3.1 OSPF Protocol

This is the simpler choice, suggested by the fact that is one of the most diffused routing protocols. Moreover, it is possible to shape the working of the protocol by tagging the network links with a suitable tag and, in case, tagging also the crossing through a node.

The algorithm corresponding to pure OSPF for the routing, as it is generally used in dimensioning tools, remembering that in this case infinite resources are assumed, can be described as follows:

- Evaluate the shortest path between source and destination with the given graph metrics.
- If on every link there is an incomplete DWDM system with available wavelengths, route the GLSP and go to END.
- If there is no wavelength available on one or more links, route the GLSP on the shortest path by opening new DWDM systems where needed.
- END

The operation of opening new DWDM systems is generally an expensive operation since it requires the deployment of new equipment in the central office and the deployment in the field of optical amplifier places.

Thus sometimes effort is paid to avoid the aperture of a new DWDM system by exploring the possibility of routing the new GLSP on a second shortest path. In this case the algorithm is the following (we will call it M-OSPF: Modified OSPF):

- Evaluate the shortest path between source and destination with the given graph metrics.
- If on every link there is an incomplete DWDM system with available wavelengths, route the GLSP and go to END.
- If there is no wavelength with sufficiently capacity available on one or more links, eliminate those links from the topology.
- Find the shortest path in the new topology.
- If on every link there is an incomplete DWDM system with available wavelengths, route the GLSP and go to END.
- Return to the original topology and to the original shortest path.
- Add a new DWDM system when needed and route the GLSP.
- END

Naturally this is only one of the possible flavors that OSPF like algorithms can have when used in network design tools. For example, we have listed a version that opens a new DWDM if the second shortest path is also unavailable, but this is not needed. It would be possible to look for the third shortest path and so on. Finally, in this paragraph we have considered explicitly the cost of the links, but the cost of the nodes can be added using the same criterion.

8.4.3.2 Constrained Routing

A different type of routing can be applied when designing a network, more oriented to limit the overall network cost. This is the cost constrained routing whose algorithm can be described as follows:

- Set an order for the GLSP setup requests.
- For each GLSP, find all the paths connecting source and destination.
- With an opportune cost model compute the incremental cost related the possibility of routing the new GLSP on every possible route.
- Among the considered paths chose that with the minimum incremental cost.
- END

Several routing algorithms can be devised introducing different kind of constrains, for example, on the QoS, on the capacity and so on. A class of interesting constrains are the so called dynamic once. A routing constraint is called dynamic when the tags that are put on the links and drive the constrained routing are not constant but change while the situation of the network evolves.

Naturally also in this case an order in which the setup of the GLSP is request has to be fixed. A classical example is constituted by the CAPEX related to the routing of a new wavelength on a certain link. This is generally constant, but when the routing of the wavelength implies the deployment of a new DWDM system, then it becomes very high so limiting as much as possible the opening of new DWDM systems.

Another typical application of dynamic constrains is possible when the number of fibers connecting the network nodes are not unlimited, for example, since the carrier is leasing them, and the bare use of a fiber has its own cost. In this case the cost of using a link for a new wavelength depends on the link length, on the number of fibers available along the link and on the leasing price if applicable. In this condition a dynamic link cost tag is very effective in performing a correct network optimization.

The design methods based on routing strategies are only one of the classes of heuristic algorithms that has been devised to design GMPLS networks. A more complete analysis can be done starting from the following biographies [22–24].

8.4.3.3 Comparison among the Considered Algorithms

In comparing the different heuristics we have introduced, we will take as a reference the results obtained using the ILP optimization.

The comparison is based on sets of 10 random networks designed to cover an approximately square area of the dimension of Italy for all the considered values of the number of nodes. All the considered networks have a connectivity (average number of disjoined path between couples of nodes) around 2.2 and the design does not include any survivability mechanism.

The traffic matrix is uniform and composed of three GLSPs between each couple of OXC. When the order of the connection request is required the dimensioning is repeated in 10 random orders and the lower cost is accepted. The comparison result is reported in Figure 8.26. All costs are normalized with respect to the optimized cost obtained with the ILP.

The first observation is that there is a good performance of the routing-based algorithms, the cost does not change more that 20% from the value obtained from ILP. The reason essentially has to be found in two characteristics of the considered networks:

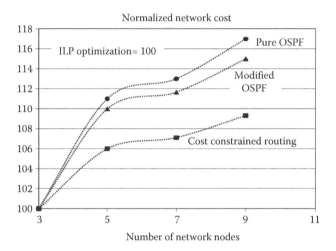

FIGURE 8.26
Comparison among routing based design based on a set of 10 random network for all the considered values of the number of nodes. All the considered networks have a connectivity around 2.2 and the design do not include any survivability mechanism. All costs are normalized with respect to the optimized cost obtained with the ILP.

connectivity and geographic extension. The connectivity is low, generally two paths between any couple of node with a few exceptions, thus a smart optimization algorithm cannot do much better than a simple OSPF having a little number of alternatives. Moreover the number of nodes is also low, so that any advantage achieved in node dimensioning has little impact.

The transmission systems are not so long, so that the platform cost (proportional to the system length) has less impact with respect to a continental network like the North American one. This also explains why the performance of the M-OSPF is so near to the simple OSPF.

It is interesting to note that the situation depicted in the figure is common to a significant part of the national networks in Europe and Asia, but for connectivity that is generally slightly higher, while a completely different problem is the design of a U.S. or a Chinese network due to the greater extension and number of nodes.

8.5 GMPLS Network Design for Survivability

Since core network equipment processes a huge amount of traffic, survivability against failures is a key characteristic of carrier networks. Transport networks based on SDH/SONET or on OTN assures a high degree of reliability by using protection mechanisms, like self-healing rings, to recover the traffic affected by a cable cut or by a node failure in a so short time not to influence the packet network.

On the other hand, extensive use of protection means devoting a main part of the CAPEX needed to build the network to built a back-up capacity. Survivability strategies are also implemented at higher layer with respect to transport, at the MPLS or Ethernet layer, for example. Due to the poor interworking between transport and upper layers, potential coordination between survivability strategies at different layers is generally not fully exploited at present days, creating sometimes quite inefficient situations.

An example is provided by the configuration in which dedicated optical channel protection is applied at the OTN transport layer and restoration with prereserved back-up is applied at the MPLS layer. To implement restoration, for each LSP the MPLS asks the OTN layer to provide a back-up capacity, which is prereserved so as to be associated to one or more LSPs once a failure happens.

On the OTN side, this request is not different from any request of MPLS layer and the provided capacity will be protected at the OTN layer. Thus, in case of failure, for each failed LSP, the OTN will devote three times the back-up capacity: one time for OTN protection, one time to provide capacity for MPLS restoration, and one time to protect the capacity provided for MPLS restoration.

The introduction of a convergent control plane is expected to improve the network survivability exploiting possible interlayer coordination while decreasing the cost of the back-up equipment. In this section, we will use the methods we have introduced in the previous section to analyze the design and performances of different multilayer and single layer survivability strategies applicable in GMPLS networks.

In particular, we will analyze the overlay GMPLS model, so that no pure LSR node can exist, since nodes of the Layer 2 can exchange information with layer one OXCs only through the UNI interface and not through the NNI. The OXC will be considered opaque so that electronic wavelength conversion and regeneration are performed in the nodes. We will also assume that partial equipment failures are all managed with redundancy at the equipment level; the only equipment-related failure we will consider will be complete node failure.

Realistic core networks are generally constituted by a mesh layer 0 topology with a connection degree between 2 and 2.5. A connection degree greater than 2 is needed to implement effective survivability strategies, but a too high connection degree is avoided due to the high cost of installing in the field long cable ducts to create connections between far nodes.

We will assume as test network a reduced version of the network of Figure 3.2, which is shown in Figure 8.25. Some nodes are eliminated from the real network both to rise the connection degree (in the real network there are also apparently single connections that are in reality doubled by parallel cable ducts deployed along different routes) and to lower the complexity of ILP design reducing to 10 the number of nodes.

We will carry out the survivability performance evaluation in two steps:

1. The network is designed using the ILP procedure with a target traffic matrix. The design is performed in two steps: first the design of a network without survivability, then in a second phase equipments and transport capacity is added to allow the chosen survivability strategy to be implemented. Normalized cost parameters for the network design are summarized in Table 8.7.

2. Once the network is dimensioned and loaded, a simulation of different failure scenarios is performed so to assess the capability of the chosen survivability strategy to react. This simulation is a discrete time slot simulation, where the evolution in time of the network behavior is simulated starting from the instants in which a failure event happens. These instants are randomly generated. In this way it is possible also to evaluate correctly parameters like the availability that imply the knowledge of the failure and repair process time dynamic. In the simulation, the network is loaded as foreseen by the ILP design and no further connection request arrives during the simulation period.

TABLE 8.7

Normalized Cost Parameters for the Network
Design and Performance Evaluation Examples
Presented in This Section (per Span of 75 km)

χ_{x1}	2
χ_{x0}	30
χ_{L1}	3
χ_{L0}	40
χ_{w0}	4.5
χ_{w1}	1.5

The process of failure events is considered a Poisson process with average interarrival time equal to 1000 days. In every failure instant, the program extracts one of three possible failure scenarios, each one with its own probability:

1. Single link failure (88%)
2. Double link failure (11%)
3. Triple link failure (1%)

Once the failure scenario is fixed, a group of failed links is randomly individuated and the failure and successive recovery simulated in the time domain. Besides the solution we have chosen in this section, it is also possible to develop analytical models for the survivability strategy performance evaluation. A detailed comparison among the results of analytical models and time domain simulations is carried out in Ref. [25] and we recommend the interested reader to start from this quite complete work to study this subject.

8.5.1 Survivability Techniques Performance Evaluation

The first parameter related to the effectiveness of a survivability strategy is the resulting network availability. In Section 8.4.1, we have introduced a concept somehow similar to network availability in an operative manner as the probability that no service disruption happens during a given time and we have called it survivability degree. It was useful at that time not to complicate further an example, but now we have to use more complete standard definitions.

ITU-T Rec. E.800 defines the so-called instantaneous availability $A(t)$ of a system as the ratio between the up time T_{up} (that is, the time in which the system worked correctly) and the total system working time. Since the complement of the up time is the down time (that is, the time T_{dw} in which the system was out of work) it can be written as

$$A(t) = \frac{T_{up}}{T_{up} + T_{dw}} \tag{8.33}$$

The ratio between the up time and the total operation time of the system can also be interpreted as the probability of founding the system working looking at it in a random instant. This is another possible definition of the availability of a system.

Besides the instantaneous availability, depending generally from time due to the fact that old systems tend to go out of work more frequently than new systems, the average availability is widely used, which is the average probability that in a fixed a time interval the system works in an instant of this time interval.

Mathematically it can be written as

$$\langle A(t, t + \Delta t)\rangle = \frac{1}{\Delta t}\int_{t}^{t + \Delta t} A(\tau)d\tau = 1 - \frac{1}{\Delta t^2}\int_{t}^{t + \Delta t} T_{dw}(\tau)d\tau = 1 - \frac{\langle T_{dw}(\tau)\rangle}{\Delta t} = \frac{\langle T_{up}(\tau)\rangle}{\Delta t} \quad (8.34)$$

where the total observation time has been identified with Δt, the up time is written as $\Delta t - T_{dw}$, which is the difference between the total observation time and the downtime and the function $T_{dw}(\tau)$ is a function defined only in the observation interval and equal to one when the system is out of work and to zero when it is correctly working.

Equation 8.34 can be read as the normalized average up time divided by the observation time, which is also the average probability to find the system working, observing it in a random instant during the observation time.

In the analysis of failures affecting systems that does not deteriorate sensibly with time, it is often assumed that the system working stabilizes up to a steady state in which the system characteristics does not depends on time. In this case, that reflects sufficiently well what happens in communication systems, a steady-state availability is defined as

$$A = \lim_{t \to \infty} A(t) \quad (8.35)$$

In order to compare different survivability methods we will evaluate the corresponding value of the average steady state availability of the routed GLSPs using a simulation program. In order to evaluate the average GLSP availability, let us number the GLSPs (optical or MPLS depending on the case) that are routed through the network in an arbitrary order. Let us imagine that the simulation simulates N_F failure events by extracting the failure starting instants t_j with $j = 1$ to N_F.

The simulation, in correspondence of each failure event, evaluates the duration of the different phases of protection or restoration and recovery for each individual GLSP and for each individual failure event. Let us consider the k-th GLSP; in correspondence of the generic h-th simulated failure it can be in one of three states:

1. Not involved in the failure
2. Restored after the failure via the network survivability mechanism
3. Impossible to restore via the survivability mechanism, restored when the network return in the working state that is when the failure is fully repaired

Let us assume a multiple survivability mechanism: for example, shared protection at the optical layer and Type 1 restoration at the MPLS layer coordinated through a hold-off time so as to avoid interlayer problems.

Simulating N_F failure scenarios, let us assume that $N_{0,k}$ simulated failure events do not involve the k-th GLSP, in $N_{1,k}$ events, corresponding to the initial instants t_{h1}, it is recovered in a time $T_{k,h}^{R1}$ via the first survivability mechanism (in our example also the faster) while in $N_{2,k}$ events, corresponding to the initial instants t_{h2}, it is recovered in a time $T_{k,h}^{R2}$ via the second survivability mechanism (in our example also the slower).

Finally, in $N_{3,k} = N_F - N_{0,k} - N_{1,k} - N_{2,k}$ failure events, the k-th GLSP is not recovered so that it turns up again only when the failure is completely repaired after a time $T_{k,h}^{C}$.

Thus the steady-state availability of the k-th GLSP can be evaluated as

$$A_k = 1 - \frac{1}{T_T}\left(\sum_{h=1}^{N_{1,k}} T_{k,h}^{R1} + \sum_{h=1}^{N_{2,k}} T_{k,h}^{R2} + \sum_{h=1}^{N_{s,k}} T_{k,h}^{C}\right) \tag{8.36}$$

where T_T is the total simulated time that is the time elapsing from the simulation MPLS layer n start $t_1 = 0$ to the instant in which, after the last failure event, the network is again in the working state.

Considering the average availability with respect to the N_P GLSPs that are routed at the considered layer through the network it is finally obtained:

$$\langle A \rangle = \frac{1}{N_P}\sum_{k=1}^{N_P} A_k = 1 - \frac{1}{N_P}\frac{1}{T_T}\sum_{k=1}^{N_P}\left(\sum_{h=1}^{N_{1,k}} T_{k,h}^{R1} + \sum_{h=1}^{N_{2,k}} T_{k,h}^{R2} + \sum_{h=1}^{N_{s,k}} T_{k,h}^{C}\right) \tag{8.37}$$

In cases in which both the recovery time and the failure repair time do not depend on the particular GLSP under examination and on the specific nature of the failure (that is, do not depends on the instant t_{hj}), Equation 8.37 becomes very simple.

Let us assume that we are observing a so long time that, for example, $N_{1,k}$ can be approximated as $N_{1,k} = P_1(F)N_F$, where the probability $P_1(F)$ is the probability that, assuming to have a failure event, the considered GLSP is involved in the failure and can restored by the first survivability mechanism and analogously $N_{2,k} = P_2(F)\,N_F$ and $N_{3,k} = [1 - P_1(F) - P_2(F)]\,N_F$.

With these definitions Equation 8.38 becomes

$$\langle A \rangle = 1 - \frac{N_F}{T_T}\left\{P_1(F)T^{R1} + P_2(F)T^{R2} + [1 - P_1(F)]T^{C}\right\} \approx 1 - \frac{1}{\tau_F}\langle T \rangle \tag{8.38}$$

where

τ_F is the interarrival time of the failure events.

$\langle T \rangle$ is the average unavailability time of the generic GLSP in correspondence to a failure.

Besides network availability, when complex survivability strategies are implemented so that multiple failure can be faced, the so-called resilience factor (RF) is useful to evaluate the effectiveness of a survivability strategy. Let us imagine that a specific survivability strategy is able to face all single link failures (property always true for effective strategies) and a subset of multiple failure events.

Calling $S_k \equiv \{s_1, s_2, \ldots\}$ the set of all possible $\binom{m}{k}$ unordered sequences of k links selected among the m cable ducts at layer 0, the kth order resilience factor is defined as [26]

$$RF_K = \frac{1}{\binom{m}{k}}\sum_{j=k}^{\binom{m}{k}}\frac{lsp_r(s_j)}{lsp(s_j)} \tag{8.39}$$

where

$lsp(s_j)$ is the cardinality of the set of working GLSPs that are interrupted by the contemporary failure of all the links in the set S_j

$lsp_r(s_j)$ is cardinality of the subset of those GLSPs that are restored without causing service disruption by the network protection mechanism

The normalization factor is derived from the number of links m.

The resilience factor of kth order can be also interpreted, if the failure events have the same probability, as the probability that an GLSP is restored in the presence of a random kth order failure event, and complete protection against k contemporary failures imply $RF_K = 1$.

Since a network that is able to face k contemporary failures is in general also capable to manage a lower number of failures, the overall resilience factor RF is sometimes defined as

$$RF = \sum_{k=1}^{F} RF_k \tag{8.40}$$

where F is the maximum number of contemporary failures that can be partially recovered and for our assumptions $R_1 = 1$.

From the earlier discussion, it results that a key point in evaluating the survivability of a network is the analysis of the time that various elements of the recovery process, either protocols of hardware mechanisms, take to do their work.

Thus a detailed analysis of the recovery process is needed. When a failure occurs the network goes into the recovery state. The network in this state performs a series of operations aimed to perform recovery of the lost connections and return of the network to the normal state after the failure has been repaired.

These phases can be described as follows:

- *Phase 1*: Fault detection
- *Phase 2*: Fault localization and isolation
- *Phase 3*: Fault notification
- *Phase 4*: Recovery (protection/restoration)
- *Phase 5*: Reversion (normalization)

Fault detection, localization, and notification phases are often joined together and called fault management. Reversion, also known as normalization process, is the mechanism bringing back the network from the recovery state, where the failed connections are reestablished using spare capacity, to the normal state. This process can be performed only after the physical repair of the failure and, among other elements, comprise the switch back of the connections affected by the failure to a working path.

It is to note that not always the working path after reversion is the same of those before reversion: as a matter of fact some restoration mechanisms reroute the connections in the reversion phase without any knowledge of their route before the failure occurs.

In order to perform failure recovery, alternative backup paths are required. These backup paths can be either set-up prior the failure or a posteriori. Given such possibilities, a range of different subphases in an a priori or a posteriori manner exists within the recovery process. The subphases and the associated elapsed times are illustrated in Table 8.8.

TABLE 8.8

Definition of the Failure Management Phases and the Corresponding Times

T_{RRj}	Overall restoration recovery time	Restoration (any type)	Total recovery time for restoration scheme type j ($j = 1, 2, 3$)
T_{PROT}	Overall protection recovery time	Protection (any type)	Total recovery time for protection schemes
T_{PL}	Time with packet losses	Any technique	Time in which there are packet losses
T_{SW}	Recovery time before switchover	Any technique	Time required before starting the switchover of traffic
T_D	Delay associated to the recovery time	Any technique	Delay between the last packet from the working path and the last packet of the back-up path
T_{DET}	Fault detection time	Any technique	Time to detect the fault
T_{HOF}	Hold-off time	Multilayer or multi-strategy survivability	Time to allow the lower layers to recover the failure
T_{NOT}	Fault notification time	Any technique	Time to inform to the node responsible of the switchover that a failure has occurred
T_{BR}	Backup routing time	Restoration type 1	Time for new backup creation and routing
T_{BS}	Backup signaling time	Restoration type 1 and 2	Time required to activate the backup path
T_{BRA}	Backup assigning time	Restorations types 2 and 3	Time for pre-provisioned backup assignment to the failed path
T_{BSA}	Backup signaling time	Restoration type 3	Time for advertisement of the pre-provisioned backup assignment to the failed path
T_{BA}	Backup activation before the switchover	Any technique	
T_{SW}	Switch over time	Any technique	Time to switch the traffic from a working path to the backup path
T_{CR}	Recovery completion time	Any technique	Time to complete the fault recovery, i.e., the time it takes the first packet to arrive from the backup path to the merging node
T_{RDET}	Initial path recovery detection time	Any technique	Time to detect that the working path has been repaired
T_{RNOT}	Initial path recovery notification time	Any technique	Time to notify about the working path recovery
T_{SWB}	Switchback time	Any technique	Time taken to switch the traffic from the backup path back to the working path

The restoration recovery time T_{RR} depends on the type of restoration is used (compare Section 8.4.2.2). In particular we have

$$\text{Type 1}: \quad T_{RR} = T_{DET} + T_{HOF} + T_{NOT} + T_{BR} + T_{BS} + T_{BA} + T_{SW} + T_{CR}$$

$$\text{Type 2}: \quad T_{RR} = T_{DET} + T_{HOF} + T_{NOT} + T_{BRA} + T_{BS} + T_{BA} + T_{SW} + T_{CR} \quad (8.41)$$

$$\text{Type 3}: \quad T_{RR} = T_{DET} + T_{HOF} + T_{NOT} + T_{BRA} + T_{BSA} + T_{BA} + T_{SW} + T_{CR}$$

On the contrary, the protection recovery time T_{PR} does not include the backup routing and backup signaling since the backup path was preestablished:

$$T_{PR} = T_{DET} + T_{HOF} + T_{NOT} + T_{BA} + T_{SW} + T_{CR} \quad (8.42)$$

When a failure occurs, connections are lost until the traffic is switched over to the backup path. The amount of lost information due to a connection loss is a function of T_{PR} or T_{RR} and the number of LSRs directly corrupted by the failure. Generally, losses cannot be totally avoided by most of the protection mechanisms exceptions being $1 + 1$ protection.

In the case of protection techniques, the time required to start the switchover is expressed as:

$$T_{Rec\,sw} = T_{DET} + T_{HOF} + T_{NOT} + T_{BA} \qquad (8.43)$$

In case of restoration the switchover can be avoided: after restoration the new routes becomes working routes and when the failed capacity is restored it is included among the back-up resources.

8.5.2 Protection versus Restoration

In this section, we will consider the case of a single layer (the optical layer). One survivability mechanism is implemented in the network, either a type of protection or a type of restoration. Since link-dedicated protection is quite trivial within a GMPLS context, we will consider the following forms of optical path protection:

- End-to-end 1 + 1 Dedicated Protection
- End-to-end 1 + 1 Shared Protection
- End-to-end 1:1 Shared Protection

As far as restoration is concerned, as we have already anticipated in the previous section, three methods of restoration are considered:

1. End-to-end GLSP restoration with complete re-provisioning (*type 1 restoration*):
   A recovery path is not established before a failure. After the fault notification, the ingress layer 2 node starts the procedure to establish the whole failed GLSP to the egress node once again. This is a working involving both layer 2 and layer 1. However, it is not guaranteed that the process will succeed because the bandwidth is not prereserved and could be unavailable (it will not be possible to establish a recovery path). Two resource reuse procedures are associated to this kind of restoration technique: *stub reuse* and *stub release*. The stub reuse procedure consists in the fact that the recovery path may use parts of the working path not affected by the failure. The stub release procedure consists in the possibility that resources related to faulty connections are freed to be used by different connections for their recovery.

2. End-to-end GLSP restoration with presignaled recovery bandwidth reservation and no label preselection (*type 2 restoration*):
   The bandwidth on the links that will constitute the recovery path is reserved by using signaling protocols, although the recovery path is not established. After a failure, the labels are chosen and the path is established by using a signaling protocol and, if necessary, the OXCs are rearranged. Even new routing through layer 2 topology could happen to exploit LSR grooming capability. The reserved resources may be shared among different recovery paths, but only when they are not affected by the same failures.

3. End-to-end GLSP restoration with presignaled recovery bandwidth reservation and label preselection (*type 3 restoration*):
Analogous to the method described earlier only faster, since the labels are assigned before the failure. Therefore, the necessity of OXC switching is signaled immediately after the fault notification.

   These restoration methods can be applied at each layer of the network (but at layer 0 naturally), in this section we will consider them applied at layer 1 and we will compare their performances with those of the different protection mechanisms first under a qualitative point of view and then with a quantitative comparison.

8.5.2.1 Bandwidth Usage

Dedicated protection uses exactly the same bandwidth for protection that for working GLSPs. Shared protection is a bit more efficient. In 1:1 protection, the spare bandwidth is used for low-priority traffic, generally best effort services. In this way also the protection bandwidth is used during the normal network working.

Restoration with presignaled GLSPs and reserved bandwidth is generally slight better than shared protection. On the contrary, restoration with complete reprovisioning exploits the bandwidth at best.

8.5.2.2 Recovery Time

All protection mechanisms, due to the preprovision of the protection path, are very fast. The SDH/SONET standard of 50 ms between the failure and the complete service restoration is the reference for these methods. Restoration with presignaling and bandwidth reservation has a speed similar to protection, while the other two forms of restoration in general are quite slow.

8.5.2.3 Specific Protocols

Optical path 1 + 1 protection requires no real-time protocol, even if obviously the protection setup has to be advertised at the network elements and at the network manager. However, these communications does not condition the speed of the protection mechanism that is completely hardware.

In case of shared protection, a dedicated specific protocol has to be realized in order to manage the protection resources sharing and coordinate the protection switch on and switch off. The SDH/SONET automatic protection switching protocol (APS) is an example of such a protection dedicated protocol. Restoration processes are completely managed by the control plane and uses control plane standard protocols.

8.5.2.4 QoS Issues

In the cases of shared protection and, mainly, of restoration with complete reprovisioning of the path, it is difficult to guarantee some elements of the QoS, i.e., the network traversing time. If services with a high-level QoS are concerned, other survivability methods should be used.

8.5.2.5 Quantitative Comparison

To perform a quantitative comparison, for each survivability strategy we will consider three merit parameters. First is the survivability cost efficiency C_E, which is the percentage of the overall cost expended for working equipment.

In other words,

$$C_E = \frac{\text{total network capex} - \text{capex for survivability dedicated equipment}}{\text{total network capex}} \qquad (8.44)$$

This parameter represents somehow the results of the design. Besides C_E we will consider the average GLSP availability, or better the so-called unavailability ($U = 1 - A$), and finally the parameter RF_2 to characterize the behavior in front of double failure events.

The simulation was carried out using the parameters summarized in Tables 8.7 and 8.9 and the main results, which are weakly dependent on traffic as far as the network is correctly dimensioned, are summarized in Table 8.10.

The first thing to be noticed is that dedicated protection absorbs about 64% of the network cost to deploy protection resources. This is due to the fact that the network connectivity is quite low so that protection paths are quite longer with respect to working paths. This causes DWDM systems used for protection to be much more costly than those used for working traffic. Moreover the protection routes also have an higher number of hops, forcing the nodes to devote a large amount of resources to process transit protection paths.

The situation is only a bit better for shared protection. Again, the poor network connectivity, which is frequently a characteristic of real geographical networks, forces protection

TABLE 8.9

Design Parameters for the Design for Survivability of the North American Network

Input/Output LSP between Two Nodes-Distribution	Random Uniform Distribution
Input/Output LSP between two nodes-min	10
Input/Output LSP between two nodes-max	30
Wavelength channel bit-rate R (Gbit/s)	10
LSPs capacity (Gbit/s)	1,25
Number of LSP per wavelength	8
Number of wavelength per DWDM system	32
Links average repair time (days)	8
Links average repair time—mission critical traffic	4
Average time between link failures (days)	365
Dedicated protection switching time (ms)	40
Shared protection switching time (ms)	50
Average type 1 restoration recovery time (ms)	2000
Standard deviation of type 1 restoration recovery time (ms)	100
Average type 2 restoration recovery time (ms)	500
Standard deviation of type 2 restoration recovery time (ms)	15
Average type 3 restoration recovery time (ms)	100
Standard deviation of type 3 restoration recovery time (ms)	12
Hold off time (ms)	60
Maximum OXC wavelength ports	512
Maximum LSR optical ports	15
Maximum LSR output LSP	120

TABLE 8.10

Main Results of the Performance Evaluation
of Different Survivability Strategies

	C_E	$1 - A$ ($\times 10^{-5}$)	R_2
1 + 1 Protection	36%	0.14	82%
1:1 Protection	36%	0.14	82%
Shared protection	49%	0.26	20%
Type 3 restoration	65%	0.45	15%
Type 2 restoration	65%	1	15%
Type 1 restoration	68%	2.5	11%

paths to be frequently quite longer with respect of working path and sharing the protection capacity helps only partially. The result is that even when using shared protection a bit more that 50% of the network CAPEX is used to set up protection capacity.

The situation is completely different considering restoration at the MPLS layer. As a matter of fact, while optical channel protection is constrained to maintain the optical channel in its wholeness protecting it as a single entity, MPLS restoration does not have this constraint.

What happens is that when restoration is in place and a link fails, layers 1 and 2 communicate through the common control plane and the GLSPs are individually restored not only by opening other optical paths exploiting free resources in layer 1, but also filling free spaces in already created optical paths and exploiting grooming to better distribute the traffic to be restored.

This causes a negligible increase of the cost of layer 2 (less than 2%) but the effect is strong and the final efficiency of restoration is near 70% if only the optical layer is considered or about 65% if all the network is taken into account.

The first type of restoration, without any preassignment of resources is slightly better due to the stub reuse and stub release procedures, while in the case of the other two restoration techniques, under the point of view of efficiency they have the same performances.

The value of the asymptotic availability, which we will use as a parameter, depends not only on the protection or restoration scheme, but also on the characteristics of the link failure process, on the repair time and on average time passing from the failure discovery to a complete traffic restoration.

In case of protection the time needed to restore a connection is assumed deterministic, while in case of restoration the recovery time is assumed a Gaussian random variable due to the need of running signaling protocols for the last two models of restoration and both routing and signaling for the first case. The time this protocol takes to provision and reserve restoration resources depends on the number of nodes to be traversed by the restored path and on the number of GLSPs to be restored. As a matter of fact, in our model, restoration manages each GLSP as an independent connection to be routed using available layer 1 capacity. As expected, the best availability is reached using dedicated protection, networks using bare rerouting restoration have the lower availability. Last, but not least, Table 8.10 shows the value of the reliability coefficient related to double failures.

The only scheme that has a reasonable probability to survive to a double failure is the dedicated optical channel protection. As a matter of fact, if two links will fail within a link repair time so that a double failure situation occurs, the only case in which traffic is lost is the case in which one failure affects a working path and the other the corresponding protection.

The values of RF_2 are much smaller in the case of shared protection, due to the fact that in this case the traffic is lost either if both working and protection path fails and if two working paths sharing the same protection resources fails. In the considered case, the sharing degree (number of protection links shared by at least two working paths normalized to the number of protection links) is quite high.

A similar comment can be done regards restoration types 2 and 3, since in these cases restoration resources are reserved for restoration, even if not coupled with a particular working path. In this case the sharing is high so that RF_2 is small.

8.5.3 Multilayer Survivability Strategies

The step beyond the use of a single survivability strategy is to use more than one contemporary method, in case at different layers. In the last section, the restoration was effectively based on cooperation between layers 1 and 2, due to the fact that all the MPLS LSPs hit by a failure were rerouted starting from their MPLS origin in layer 2. However, much better can be done if a coordinated multilayer strategy is adopted to manage a recovery mechanism at each layer.

8.5.3.1 Multilayer Survivability

Let us image that shared protection is adopted at Layer 1 with a network design suitable to face all single link failures. In case a multiple failure occurs and some GLSPs cannot be restored by layer 1 protection, layer 2 LSRs manage a restoration process to exploit free transport resources to reroute the affected LSPs.

Naturally the two processes have to be coordinated in order to avoid those problems that can occur in a bad designed multilayer restore mechanism. The most important phenomenon to be avoided is protection instability: if a path is protected at layer 1 and then also restored at layer 2, restoration can cause switching off optical power on the layer 1 protection path. Layer 1 protection mechanism could interpret the fact as a new failure and try to protect the considered and other paths against it, with service disruption and the possibility to start an oscillatory phenomenon between the two layers.

The simpler way to coordinate two reliability mechanism located at different layers is to introduce a sufficiently long hold-off time at the upper layer so to give to the lower layer sufficient time to stabilize. More sophisticated methods however exists to avoid the increment of recovery time due to the fact that the hold-off time is the same for any failure event. These methods are based on the use of signaling protocols to advertise all the relevant network entities of the started protection switch.

It is also interesting to compare, at last in a qualitative way, multilayer reliability with a multistrategy single layer reliability, where, for example, both dedicated protection and wavelength level restoration are carried out by OXCs directly at the optical layer.

The basic multilayer approach advantage is that, in case the optical reliability mechanism does not work or is compromised by a multiple failure, the successive step is to route MPLS LSPs individually, exploiting the small granularity to perform more efficient restoration.

This is a principle similar to that of grooming, where complexity at layer 2 is traded with more effective exploitation of transmission capacity at layer 1. We will analyze multilayer survivability with the same method used in the previous section: we will design first the network for the required traffic, so deriving information on the network cost.

TABLE 8.11

Efficiency and Availability of Different
Multilayer Survivability Strategies

Only Optical Layer Resources	C_E	$1 - A$ ($\times 10^{-6}$)
1 + 1 protection	36%	1.4
SP-R multilayer survivability	25%	0.8
R-R multilayer survivability	45%	1.3
Restoration type 2	65%	10

We will use a three stage optimization, first optimizing the network for working traffic, then dimensioning the bottom layer survivability and at the end the upper layer survivability. The average availability is evaluated by simulation as in previous section. Two multilayer methods will be considered, assuming simple coordination through an hold-off time. The first method, which for simplicity we will call SP-R, is based on 1 + 1 optical GLSP at layer 1 and type 1 restoration at layer 2. This means that after a failure event, the MPLS LSR at layer 2 wait a certain time to let the optical layer to end its recovery attempt. After this time, all the LSR that has not been recovered by the optical layer has to be restored using already the available network resources to route them.

The second method, which briefly we will call R-R, comprises wavelength level type 3 restoration at layer 1 and LSR type 3 restoration at level 2. The survivability method efficiency and the average GLSP availability are reported in Table 8.11 in the conditions detailed in the previous section.

The two multilayer methods are both better with respect to single-layer methods, even if under different points of view. The R-R method attains almost the same availability of 1 + 1 optical protection, but at a much lower cost in terms of redundant equipment in the network. The SP-R method is quite expensive, but has the highest availability among the methods analyzed up to now.

Not only double failures can be efficiently faced, but also a few of triple and quadruple failure events are recovered and a good recovery capability also exists in front of node failures. This property is shown in Table 8.12, where resilience factors of different degree are evaluated.

Besides the factor relative to multiple link failures, the resilience factor RF_n relative to node failure is also reported. This new factor is defined in line with the definition of RF_k as

$$RF_n = \frac{1}{N} \sum_{j=1}^{N} \frac{lsp_n^r(j) + lsp_n^0(j)}{lsp_n(j)} \tag{8.45}$$

TABLE 8.12

Resilience Degrees of Various Order for Two
Multilayer Survivability Strategies

RF	SP-R	R-R
RF_2	1	1
RF_3	0.85	0.82
RF_4	0.77	0.71
RF_n	0.9	0.88

RF_n indicates the factor relative to full node failure.

where

N is the node number

$lsp_n(j)$ is the total number of GLSPs processed in the absence of any failure by node j

$lsp_n^r(j)$ is the subset of $lsp_n(j)$ composed by the GLSPs that are restored by the network survivability procedure after the failure of node j

$lsp_n^0(j)$ is the subset of $lsp_n(j)$ that are started and terminated locally in the failed node

The term $lsp_n^0(j)$ is added to the numerator of Equation 8.45 so to obtain a definition that gives one in case of full recovery, since local GLSPs cannot be recovered with any mean if the node fails.

8.5.3.2 QoS-Driven Multilayer Survivability

From a survivability point of view, multilayer restoration R-R and multilayer protection-restoration SP-R are completely satisfactory, assuring excellent resilience against a large number of failure events and a very high GLSP survivability. Against this good performance, the cost to pay in terms of additional equipment to be deployed to provide the required spare capacity is very high. The cost of the network equipment is almost multiplied by two in the case of R-R and by more than two in the case of SP-R.

Sometimes it could be the right choice, due to the fact that the customer is willing to pay a high price to have a nine "9" survivability, it is the case of banks, other business customers like insurance, companies providing network based back-up of strategic data and so on.

In other cases however, this is clearly an excessive target bringing an excessive cost. This consideration, besides the very flexible structure of GMPLS, allowing to characterize GLSPs with a certain number of different tags, suggest that it should be the customer requirements, summarized in the service level agreement to drive the type of survivability to implement and consequently the service availability and cost.

In our model for QoS-driven multilayer survivability we will start mainly from Ref. [27], where a similar model was introduced and detailed. The main differences are that we will exploit multilayer structure of the network to benefit, when needed, to the possibility of layer 2 to manage reduced granularity LSPs with respect to the high capacity optical GLSPs and that we will assume that the network is designed and optimized for the specific level of traffic.

Even if in real networks, every LSPs has a specific bandwidth request, so that different classes of LSPs exist with respect to the capacity, we will continue to model the network as having a single type of LSP having fixed capacity. This assumption simplifies greatly the model and does not impact essentially results on survivability.

We will imagine, on the contrary, that when the layer 3 makes the request of a new LSP to layer 2 through the UNI interface, this request is characterized not only by the indication of the two end nodes, but also by the request of a certain QoS that, for what is our interest in this section, consists in the request of one of the available survivability classes.

In order to maintain a better control on the use of network resources, the network manager is imagined to force a constraint on the control plane, forcing not to allow the traffic allocated into a certain service class to go over a fixed percentage of the maximum traffic that the network can accept. Only the class one traffic, which does not require spare capacity, is allowed to grow as much as possible.

We will image that the network will offer to the client layer five possible classes of survivability, whose description is given in the following.

1. *Service Class 0: Low Priority Traffic*
   - This is constituted by unprotected LSPs that are routed using exclusively spare capacity that is planned to be used for shared protection or for restoration. When a failure occurs, the low-priority traffic that is routed onto the spare capacity, which has to be exploited for recovery is blocked, and the connection is lost.
   - This is the lower service class in terms of availability.

2. *Service Class 1: Best Effort Traffic.* This service class collects all the services, like web browsing and many other consumer services, that run onto a specifically provisioned capacity, but are not protected. If a failure occurs, on a working link where a best effort LSP is routed, it is lost up to the repair of the link.

3. *Service Class 2: Standard Traffic.* This is a class composed by LSPs that are protected via a type 1 restoration at the MPLS layer. It is important to notice that services of the classes 1 and 2 can run on the same physical fiber and even in the same optical GLSP. As a matter of fact, class 2 traffic restoration is managed by MPLS on layer 2. Thus if a fiber containing both class 1 and class 2 services is cut, the LSR that are at the ends of the involved GLSPs will start the rerouting of the failed class 2 LSPs, while will start signaling over the GMPLS network and through the UNI interface toward layer 3 to advertise that the involved best effort LSPs have been lost. In this case RF_1 is very near to one, but not exactly equal.

4. *Service Class 3: High Availability Traffic.* This traffic is protected via a multilayer protection with the R-R method. This class of LSPs has a very high resilience toward many type of failure events, as we have evidenced in the last section, besides a complete recovery ability toward single and double link cut. Class 3 services cannot run on the same fibers where classes 1 and 2 services run, since they are protected also at the optical layer by a restoration mechanism. We will see that they can run in the same fibers of class 4 services, but not in the same wavelength.

5. *Service Class 4: Mission Critical Traffic.* This is the traffic that cannot be lost, but in cases of cataclysmic disasters like extended earthquakes or strategic nuclear attacks. The protection of these LSRs is with the SP-R method but with type 2 instead that type 1 restoration at MPLS layer. Moreover the SLA relative to this traffic type foresees a repair time of only 4 days against any type of link failure. It is to be noticed that LSPs belonging to class 4 can run on the same fibers where class 3 services run, but not on the same wavelengths. As a matter of fact, wavelengths accommodating class 4 services have to be protected at the optical layer as a whole, but this is not true for wavelength containing LSPs of class 3.

From the earlier discussion, it results that, as a consequence of the differentiation of the classes of service of the GLSPs, the wavelengths are also differentiated.

In particular, there are the following types of wavelengths:

- Type 1 wavelengths: Containing working LSPs of classes 1 and 2 and spare capacity to be used by Layer 2 restoration processes. This last capacity can be occupied by low-priority traffic (service level 0).

- Type 2 wavelengths containing service level 3 LSPs and spare capacity to be used by Layer 2 restoration processes. This last capacity can be occupied by low-priority traffic (service level 0).

- Type 3 wavelengths containing service level 4 LSPs. Both protection and working paths are considered type 3 wavelengths, since they have no difference under a traffic load point of view but for the fact that at the receiver the working path is selected whenever possible.

- Type 4 wavelengths that convey spare capacity to be used by restoration at layer 1 and can support low-priority traffic.

To dimension a network with a so complex traffic demand and survivability technique with IPL optimization in only two steps, one for capacity managing working traffic the other for capacity for use as a back-up is naturally possible under a theoretical point of view, but difficult to do with a realistic computation machine. It is to be noted however that the introduction of different classes of wavelength, practically segments the optical layer in different optical plans that have OXCs in common.

Thus it can be devised an algorithm based on the following steps:

- Step 1: The first virtual network is dimensioned starting from the given layer 0 topology and assuming as traffic-only mission critical traffic. The fibers that will be equipped in this network will all be Type 3 fibers, all devoted to carry either working or protection Service Class 4 LSPs.

- Step 2: A second virtual network is dimensioned starting from the same layer 0 topology and assuming as traffic only high-availability LSPs. The fibers that will be equipped in this network will all be Type 2 and Type 4 fibers and are tagged so to be possible to recognize their nature.

- Step 3: A third virtual network is dimensioned starting from the same layer 0 topology and assuming as traffic only standard traffic and best effort GLSPs. The fibers that will be equipped in this network will all be Type 1 and Type 2 fibers and are tagged so to be possible to recognize their nature. This is the only step mixing two types of traffic, but it is managed easily due to the fact that best-effort traffic does not require back-up capacity.

  In particular, the substeps are organized as follows:

  - A whole traffic matrix for the working traffic is created by merging standard and best-effort LSPs.
  - The overall traffic matrix is used to dimension the virtual network for the working part.
  - The traffic matrix relative to standard traffic only is retrieved and used to deploy protection capacity.

- Step 4: OXC synthesis. The OXC occupying the same topological position in each virtual network are analyzed by merging them so to minimize their cost. The only exception is the OXC managing mission critical traffic that is not merged with other. The cost of these OXCs is on the other end recalculated doubling the cost of the interfaces to simulate 1 + 1 local interface protection. As far as the switch matrix is concerned, it is assumed to be doubled for protection in all the deployed OXCs and that this is already included in the cost of the OXC platform.

TABLE 8.13

Maximum Percentage of Traffic That
Can Belong to a Specified QoS Class in
the Example Discussed in This Section

Mission critical	5%
High reliability	20%
Standard	35%
Best effort	40%
Low priority (in addition to 100%)	30%

To show an example of OXC merging, let us image that in a certain node we have the following OXCs:

- Virtual network 2: Number of out/in wavelength ports 22, platform 1
- Virtual network 3: Number of out/in wavelength ports 12, platform 1
- Final OXC design: Total number of ports 32, 1 platform (a second platform has to be eliminated)

- Step 5: LSR synthesis. The LSR occupying the same topological position in each virtual network are analyzed by merging them so to minimize traffic. The operation is similar to that performed for OXC, but with no exceptions.

- Step 6: Network synthesis. The fibers cannot be merged due to the limitations we have underlined previously, so this step only consists of putting all the network information into a single data base.

- Step 7: Low-priority traffic is routed. It is to be noted that no further capacity is deployed and if the present back-up capacity is not sufficient to route a low priority traffic request, it is rejected by the network.

As in the other cases, after dimensioning a simulation is performed, where all the classes of service are present as foreseen by the network design. In the example whose results are reported in this section the mix of different LSR is reported in Table 8.13.

The result of the design and simulation is summarized in Figure 8.27 where the efficiency (indicating the CAPEX cost of the survivability) is related to the average availability of the selected types of LSR.

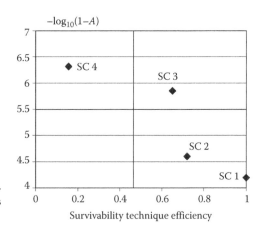

FIGURE 8.27
The efficiency (related to the CAPEX cost of the survivability) is related to the availability of the selected types of LSR for different types of Service Class (SC).

Looking at this figure a further efficiency reduction due to the OXC platform duplication is taken into account in the CE of the SP-R method.

8.6 Impact of ASON/GMPLS on Carriers OPEX

It is completely evident that the introduction of the ASON/GMPLS model is a sort of revolution in the way in which the carrier network is conceived and managed. The rationale of this revolution is in the great need of big carriers of reducing the OPEX related to the network so to prepare themselves to further network evolutions related to new services.

Intuition tells us that if all the network is managed through an automated and unique control plane, OPEX should decrease with respect to a situation in which the network layers are managed separately and in managing the transport layer several manual operations are needed.

However, we also know that intuition is not always reliable, thus carriers have carried out several studies to quantify the OPEX advantages related to the introduction of the new network paradigm. This quantification is also needed since this technology step implies the investment of relevant capital needed to buy new equipment or to update the deployed equipment so to be compatible with the new control plane. Besides equipment capitals, training of network personnel to the new procedures, change of carriers organization to reflect the new network structure are all costs that have to be paid to implement the new control plane.

A so big project cannot be executed without a detailed business plan pointing out costs and benefits and describing the impact of the project on the carrier business. Among the studies that has been done to try to determine the benefits of ASON/GMPLS we will follow the work presented in Refs. [28,29] both in the method and in the results.

In order to evaluate the OPEX expensed a model of the process implemented by the carrier to manage the network has to be done. This process is composed not only of technical activities, like network maintenance and technical personnel training to use and maintain new equipment, but also by sales and marketing activity to promote and sell new services, administration activity to control the company general working and other nontechnical processes that are fundamental for a good carrier working.

In order to model this complex set of activities, the model developed *i* [28,29] is based on an activity graph, which correlate various activities in terms of cause and effect. In particular, each activity is connected with other activities with oriented branches, tagged with a probability. This is the probability that if the carrier is carrying on the activity at the beginning of the branch, the activity at the end of the branch also will be needed.

A very simple example is given in Figure 8.28 where the graph relative to the process of repairing a failed card in an equipment is shown in detail. In Figure 8.29 a small part of the graph of Figure 8.28 is isolated to show how every element of the graph is tagged with a record containing the main element that are needed to evaluate the impact of the considered step on the cost of the overall activity represented by the graph.

The considered example regards a spot process of the reactive type: this means that the process happens at random instants as a reaction to a causing event, in this case the individuation of a failed card in an equipment. All spot events have a trigger; if the event is of the reaction type the trigger is an external event, if the event is a direct action (e.g., the planned substitution of a working but obsolete equipment) the trigger is the arrival of the time at which the action was planned.

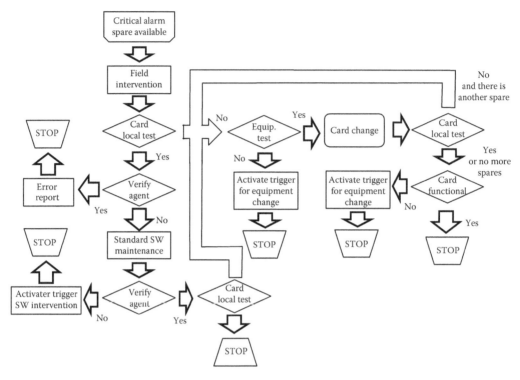

FIGURE 8.28
Activity graph relative to the process of repairing a failed card in an equipment.

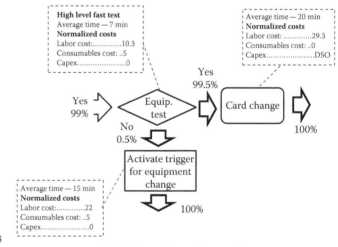

FIGURE 8.29
A small part of the graph of Figure 8.28
showing associated tags.

DSO = Depending on the specific occurence

Besides spot processes, carriers implement a great number of recurring processes that periodically happen. In this case the graph does not have a trigger, but it is a closed loop containing a delay to regulate the periodicity of the event. A third type of processes are the continuous processes; these are the activities that never stop and constitute the base of the carrier work.

In particular circumstances a process can activate other processes starting their trigger. This causes a strong dependence of the operational processes one from the other. Some dependencies however can be neglected since they are so rare that do not impact the OPEX in a significant way.

The analysis proceeds by creating a complete description of the business activity of a big carrier via process diagrams and analyzing how the processes and the corresponding diagrams are changed by the introduction of ASON/GMPLS.

Thus operational models of the two carriers are defined before and after ASON/GMPLS. Since we tag every graph branch with a probability and every graph node with a double tag, we can calculate the cost and time needed to carry out the activity. For can every activity we can define an activity cost as the weighted average of the cost of every possible route through the activity graph connecting the activity start with the activity end. The weight of a route is naturally the overall probability that the route will take place.

Beside the activity cost, an average activity time is also evaluated, which is sometimes needed to take into account the fact that some activities cannot start again if they are still ongoing. Once the models are set up, we can evaluate the overall OPEX expenditure of the carrier. This is estimated by adding all the costs of parallel recurring or continuous activities.

As far as the spot activities are concerned, their costs are weighted with the probability of the trigger, so to take into account the fact that they are not continuous. Here we will not report the real activity graphs of the most complex activities, and much more details are reported in Refs. [28,29].

We will limit our self to list the main activities and the impact of the introduction of ASON/GMPLS on them. The main activities are all periodical or continuous, thus we will not repeat this every time.

- *Cost of failure free network operation*: This is the base cost to take the network operational in a failure-free situation: it includes the cost of electrical power, building leasing for central offices, fiber leasing if it is the case, permissions for traversing private properties or to bring cables in private buildings, taxes on the ownership of the infrastructure and so on. This cost is heavily impacted by ASON/GMPLS due to the change of several equipments and to the substitution of preprovisioned protection with restoration in several areas of the network with a consequent reduction of the number of network equipments.

- *Routine operations:* This is the cost related to all actions aimed to maintain the network in an efficient working status, maintaining the ability to face a possible failure. The main actions performed here aim at monitoring the network and its services. Therefore, the actions involved include the following:
  - Periodic verification of the equipment performances and services quality versus existing SLAs.
  - Management of the spare parts warehouses and activity to assure spare parts availability when a failure occurs.
  - Software management in terms of patches and new releases installation and bugs tracking.
  - Alarms management with all the activities aimed to failure identification, verification, and repair.
  - Network configuration management in terms of keeping track of changes in the network derived from failures or from equipment substitution due to preventive replacement of old or obsolete equipment.

Routine operations are influenced by ASON/GMPLS in two different ways. On one side, faster and more effective network reconfiguration allows the reduction of failure-related service disruptions and consequently a more effective and less expensive process of intervention when a failure occurs.

On the other hand, software management will become more complex and expensive due to the introduction of a much more complex control plane.

- *Operational network planning:* This is the set of actions that aim to maintain the network efficient and optimized. In a traditional transport network they are mainly manual operations carried out with the support of the TMN network manager and of off-line network optimization tools, often included into the TMN software. If ASON/GMPLS is used, these actions are mainly constituted by traffic engineering that, as we have seen in Section 8.3, are needed not to allow the network to slide very far from an optimized GLSPs routing. Since the ASON/GMPLS network is more complex, it is probable that this activity will be more expensive, even if it will be supported by more powerful tools.

- *Marketing:* This is the traditional activity of customer base enlargement and customer maintenance through service promotions, new billing schemes introduction and advertisement and so on. In correspondence with the introduction of ASON/GMPLS, it is reasonable to assume that new service bundles will be introduced so increasing the marketing expenses. On the other hand, even if this does not appear in our model, this will also bring new revenues.

- *Service management*: These are all the activities related to offering, delivery, and managing end user services. Since this is a category quite important both for its absolute impact on carrier OPEX and for the impact of ASON/MPLS on it, it is worth doing a detailed analysis dividing this activity into subactivities.

  - *Service offering:* This is the process related to the preparation of the offering for the end customer in terms of service SLA and price. The sales department negotiates the terms and conditions of the offer with the customer, and does an inquiry as to whether the connection request can be handled by the standard mechanisms and infrastructure.

    For nonstandard connection inquiries, a separate project is triggered. Then the offer is sent to the customer.

  - *Service provisioning*: This is the process related to the effective service setup after a contract has been subscribed with the end customer. After contract registration, a service provisioning project starts involving various parts of the carrier organization. At the end of the project, after a final test of the service by the carrier test department, the final customer is enabled to use the new service. A detailed action graph for service delivery is reported in Ref. [29].

  - *Service cease*: This is the process related to the discontinuation of a service after the end of the contract with the end customer.

  - *Service move or change*: Often moving or changing a connection is one of the most complex operations. It involves all three previous processes: contract update, new connection setup, and release of the previous connection. The customer's request for change is handled by the sales department as a service offer process, checking again for the availability of resources. The sales department then generates orders for the service provisioning and cease process that are implemented through coordination from the Project Management department. At the same time, the client is receiving updates on the new installation.

Technologies automating some of the network operation allow the cost for service provisioning to be significantly reduced, because the signaling can be done via standardized interfaces (UNI and NNI), without requiring manual intervention. This means that the cost of setting up a new connection decreases greatly. In this case, the service offer and provisioning processes will change fundamentally. Since service delivery will now be automated and executed on the pure machine level, correct agreements and regulations must be negotiated by the sales department and implemented well before in the form of SLAs. The use of GMPLS technologies and the possibility to offer dynamic services are strongly interconnected issues.

The strongest impact of dynamic services is on the pricing and billing process. Fixed price services (e.g., leased lines) will definitely be cheaper in pricing and billing than dynamic services. For dynamic services, it is much more difficult (and thus more expensive) to correctly assign costs to customer accounts. Calculating a new price for a new service is more expensive than just applying a traditional pricing scheme.

As a conclusion, provisioning of new services, especially when the setup of a new connection is needed, is quite cheaper both due to the easier process and to the absence of errors due to the elimination of manual operations. On the other hand, the process carried out by the sales department is more complicated and thus more expensive.

In analyzing the activity graphs and the related costs, it becomes evident that the greater OPEX cost element is the labor cost and it is also the most impacted by the transition to ASON/GMPLS.

Thus we will start our analysis from labor cost. This will be evaluated by detailing exactly the level of personnel that is needed and estimating the labor cost on the average over European nations and over United States. This average is done both using official statistics and doing a set of interviews with carriers to confirm them [29]. Naturally there is a difference between labor cost in Europe and in United States, but the percentage impact of ASON/GMPLS is not relevantly different [30,31].

It is natural to divide the carriers in two categories: incumbent carriers and competitive carriers. Incumbent carriers are characterized by a large, well-structured organization by a set of coded processes (probably relater to global quality certifications) and by the ownership of almost all the network they use to deliver services.

On the contrary, competitive carriers have a lighter organization and the general cost of a process is smaller both for the simpler organization and for the lighter structure. On the other hand, competitive carriers have higher costs in terms of consultancy services from the vendors since generally a set of activities that an incumbent operated through its own organizations are delegated to vendors by competitive carriers.

In Figure 8.30 the labor cost related to service management for a European incumbent carrier is shown considering a traditional network architecture and an architecture implementing ASON/GMPLS control plane.

The costs are divided into the processes we have considered in the previous analysis and the cost related to each process is further divided considering the carrier professional group carrying on the work. Costs are reported in normalized units.

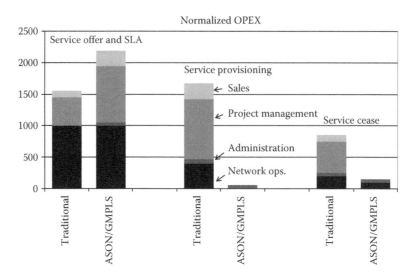

FIGURE 8.30
Service management labor cost for an incumbent carrier considering a traditional network and an architecture implementing ASON/GMPLS control plane. (After Pasqualini, S. et al., *IEEE Commun. Mag.*, 6, 28, 2005.)

The first observation is that network automation reduces drastically the network management costs needed in a traditional network, for example, to provision a new connection or to reroute a connection. These operations that require a relevant amount of manual work in a traditional network are completely automated via the control plane.

The counterpart is that the carrier front end, especially the sales group, has to face a more difficult work, absorbing more costs than in the traditional case. This is evident in Figure 8.31 where labor cost for service management processes is reported per professional group.

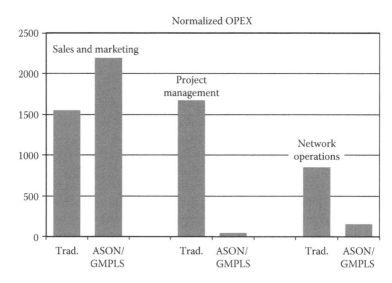

FIGURE 8.31
Service management labor cost for processes per professional group. (After Pasqualini, S. et al., *IEEE Commun. Mag.*, 6, 28, 2005.)

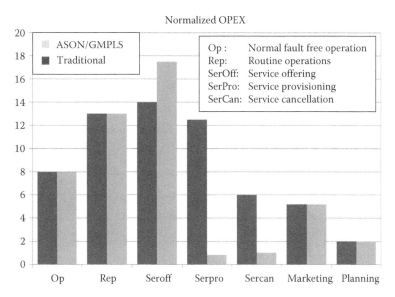

FIGURE 8.32
Global result for the overall OPEX of an incumbent carrier. (After Pasqualini, S. et al., *IEEE Commun. Mag.*, 6, 28, 2005.)

Taking into account all the changes, labor cost related to service management decreases about 52% with the introduction of ASON/GMPLS. A similar analysis can be carried out with North American costs deriving a reduction of the labor cost related to service management of about 42%. This is due to the fact that the difference between high-level and low-level employees in terms of compensation is much more evident in North America. The spared personnel in the network operation area is mainly constituted by technicians, warehouse workers, and similar professional figures that are located in the middle or lower part of the organization while the number of managers is less affected.

Thus the percentage OPEX spared is smaller. However, the result is in any case impressive, taking into account that we have considered the most critical part of the carrier OPEX.

The global result, obtained considering the overall OPEX of an European carrier, is shown in Figure 8.32. The overall sparing related to the introduction of ASON/MPLS is 24% (20% with U.S. numbers).

This is an important result if we consider that the annual OPEX of an average incumbent carrier is around 70% of the overall carrier expenditure, a figure of the order of €3 billion for a medium European incumbent carrier. The results in the case of a competitive carrier are only slightly different due to the different weight of internal process structure.

We will give only, as an example, the result relative to the service management for a competitive European carrier. The plot, analogous to that of Figure 8.30, is reported in Figure 8.33. The percentage OPEX spare is of the same order of magnitude of that forecasted for the incumbent, demonstrating that, if ASON/GMPLS will maintain its promises it is a good step in the right direction.

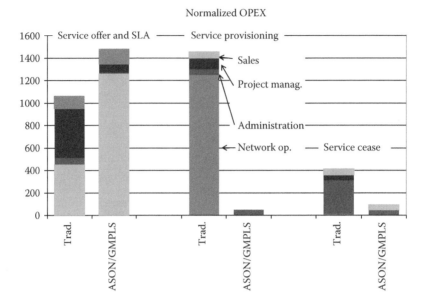

FIGURE 8.33
Global result for the overall OPEX of a competitive carrier. (After Pasqualini, S. et al., *IEEE Commun. Mag.*, 6, 28, 2005.)

References

1. ITU-T Recommendations regarding ASON standard architecture

   G.8080/Y.1304: Architecture for the automatically switched optical network (ASON)

   G.807/Y.1302: Requirements for automatic switched transport networks (ASTN) Call and Connection Management

   G.7713/Y.1704: Distributed call and connection management (DCM)

   G.7713.1/Y.1704.1: DCM signalling mechanism using PNNI/Q.2931

   G.7713.2/Y.1704.2: DCM signalling mechanism using GMPLS RSVP-TE

   G.7713.3/Y.1704.3: DCM signalling mechanism using GMPLS CR-LDP Discovery and Link Management

   G.7714/Y.1705: Generalized automatic discovery techniques

   G.7715/Y.1706: Architecture and requirements of routing for automatic switched transport network

   G.7716/Y.1707: Architecture and requirements of link resource management for automatically switched transport networks

   G.7717/Y.1708: ASTN connection admission control. Other Related Recommendations

   G.872: Architecture of optical transport networks

   G.709/Y.1331: Interface for the optical transport network (OTN)

   G.959.1: Optical transport network physical layer interfaces

   G.874: Management aspects of the optical transport network element

   G.874.1: Optical transport network (OTN) protocol-neutral management information model for the network element view

   G.875: Optical transport network (OTN) management information model for the network element view

G.7041/Y.1303: Generic framing procedure (GFP)

G.7042/Y.1305: Link capacity adjustment scheme (LCAS) for virtual concatenated signals

G.65x: Series on optical fibre cables and test methods

G.693: Optical interfaces for intra-office systems

G.7710/Y.1701: Common equipment management function requirements

G.7712/Y.1703: Architecture and specification of data communication network

G.806: Characteristics of transport equipment. Description methodology and generic functionality

2. OIF Implementation Agreements for ASON/GMPLS coordination

OIF-UNI-01.0: User Network Interface (UNI) 1.0 Signaling Specification (October 2001)

OIF-UNI-01.0-R2-Common: User Network Interface (UNI) 1.0 Signaling Specification, Release 2: Common Part

OIF-UNI-01.0-R2-RSVP: RSVP Extensions for User Network Interface (UNI) 1.0 Signaling, Release 2 (February 2004)

OIF-UNI-02.0-Common: User Network Interface (UNI) 2.0 Signaling Specification: Common Part

OIF-UNI-02.0-RSVP: User Network Interface (UNI) 2.0 Signaling Specification: RSVP Extensions for User Network Interface (UNI) 2.0 (February 2008)

OIF-CDR-01.0: Call Detail Records for OIF UNI 1.0 Billing (April 2002)

OIF-SEP-01.0: Security Extension for UNI and NNI (May 2003)

OIF-SEP-02.1: Addendum to the Security Extension for UNI and NNI (March 2006)

OIF-SLG-01.0: OIF Control Plane Logging and Auditing with Syslog (November 2007)

OIF-SMI-01.0: Security for Management Interfaces to Network Elements (September 2003)

OIF-SMI-02.1: Addendum to the Security for Management Interfaces to Network Elements (March 2006)

OIF-E-NNI-Sig-01.0: Intra-Carrier E-NNI Signaling Specification (February 2004)

OIF-E-NNI-Sig-02.0: E-NNI Signaling Specification (April 2009)

OIF-ENNI-OSPF-01.0: External Network-Network Interface (E-NNI) OSPF-based Routing - 1.0 (Intra-Carrier) Implementation Agreement (January 2007)

OIF-G-Sig-IW-01.0: OIF Guideline Document: Signaling Protocol Interworking of ASON/GMPLS Network Domains (June 2008)

3. Iqbal, A. A., Mahmood, W., Ahmed, E., Samad, K., Evaluation of distributed control signaling protocols in GMPLS, *Proceedings of the Fourth International Conference on Optical Internet (COIN05)*, Chongqing University of Posts and Telecommunications, Chongqing, China, May 29-June 2, 2005.

4. Duarte, S., Martos, B., Brestavos, A., Almeida, V., Almeida, J., Traffic characteristics and communication patterns in blogosphere, *Proceedings of the International Conference of Weblogs and Social Media*, Boulder, CO, s.n., March 26–28, 2007.

5. Sheluhin, O., Smolskiy, S., Osin, A., *Self Similar Processes in Telecommunications*, Wiley, Chichester, England, s.l., 2007. ISBN-13: 978-0470014868.

6. Bigos, W., Cousin, B., Gosselin, S., Le Foll, M., Nakajima, H., Survivable MPLS over optical transport networks: Cost and resource usage analysis, *IEEE Journal on Selected Areas in Communications*, 25(5), 949–961 (2007).

7. Nabil, N., Mouftah, H. T. Optimum planning of GMPLS transport networks. *International Conference on Transparent Optical Networks - ICTON 2006*, Nothingham, s.n., pp. 70–73, June 18–22, 2006.

8. Sabella, R., Settembre, M., Oriolo, G., Razza, F., Ferlito, F., Conte, G., A multilayer solution for path provisioning in new-generation optical/MPLS networks, *Journal of Lightwave Technology*, 21(5), 1141–1155 (2003).

9. Tornatore, M., Maier, G., Pattavina, A., WDM network optimization by ILP based on source formulation, *Proceedings of IEEE INFOCOM*, New York, USA, pp. 1813—1821, June 23–27, 2002.

10. Proestaki, A., Sinclair, M. C., Design and dimensioning of dual-homing hierarchical multi-ring networks, *IEE Proceedings on Communications*, 147(2), 96–104 (2000).

11. Song, Y., Wool, A., Yener, B., Combinatorial design of multi-ring networks with combined routing and flow control, *Computer Networks: The International Journal of Computer and Telecommunications Networking*, 41(2), 247–267 (2003).

12. Binetti, S., Bragheri, A., Iannone, E., Bentivoglio, F., Mesh and multi-ring optical networks for long-haul applications, *IEEE Journal of Lightwave Technology*, 18(12), 1677 (2000).

13. Ash, G. R., *Traffic Engineering and QoS Optimization of Integrated Voice & Data Networks*, Morgan Kaufmann, San Francisco, CA, s.l., 2006. ISBN-13: 978-0123706256.

14. Anjalia, T., Scoglio, C., A novel method for QoS provisioning with protection in GMPLS networks, *Computer Communications*, 29, 757–764 (2006).

15. Minoux, M., Networks synthesis and optimum network design problems: Models, solution methods and applications, *Networks*, 19(3), 313–360 (2006).

16. Tornatore, M., Maier, G., Pattavina, A., WDM network design by ILP models based on flow aggregation, *IEEE Transactions on Networking*, 15(3), 709–720 (2007).

17. Scheffel, M., Kiese, M., Stidsen, T., *A Clustering Approach for Scalable Network Design*, Munich University of Technology, Munich, Germany, 2007. Deliverable of German Ministry of Education and Research Project ID 01BP551.

18. Nguyen, H. N., Habibi, D., Phung, V. Q., Efficient optimization of network protection design with p-cycles, *Journal of Photonic Network Communications*, 19(1), 22–31 (2010).

19. Lucentini, M., Spaccamela, A. M., Approximate solutions of an integer linear programming problem with resource variations, *Lecture Notes in Control and Information Sciences*, 18, 207–219 (1979).

20. Schrijver, A., *Theory of Linear and Integer Programming*, Wiley, New York, 1998. ISBN-13: 978-0471982326.

21. Verbrugge, S., Colle, D., Pickavet, M., Demeester, P., Common planning practices for network dimensioning under traffic uncertainty, *Design of Reliable Communication Networks (DRCN) 2003*, Banff, Alberta, Canada, October 19–22, 2003.

22. Sugiyama, R., Takeda, T., Oki, E., Shiomoto, K., Network design method based on adaptive selection of facility-adding and path-routing policies under traffic growth, *Global Telecommunications Conference, GLOBECOM 2008*, New Orleans, LA, IEEE, s.l., USA, November 30–December 3, 2008.

23. Naas, N., Mouftah, H. T., Efficient heuristics for planning GMPLS transport networks, *Canadian Conference on Electrical and Computer Engineering, CCECE 2007*, Vancouver, British Columbia, Canada, s.n., April 22–26, 2007.

24. Fukumoto, T., Komoda, N., Optimal paths design for a GMPLS network using the Lagrangian relaxation method, *Symposium on Computational Intelligence in Scheduling, SCIS '07*, Honolulu, HI, IEEE, 2007.

25. Lacković, M., Mikac, B., Analytical vs. simulation approach to availability calculation of circuit switched optical transmission network, *Proceedings of the 7th International Conference on Telecommunications, ConTEL 2003*, Zagreb, Croatia, IEEE, s.l., Vol. 2, pp. 743–750, 2003.

26. Sone, Y., Nagatsu, N., Imajuku, W., Jinno, M., Takigawa, Y., Optical path restoration scheme escalation achieving enhanced operation and high survivability in multiple failure scenarios, Network Innovation Laboratories, Public Presentation, NTT, 2007.

27. Ricciato, F., Listanti, M., Salsano, S., An architecture for differentiated protection against single and double faults in GMPLS, *Photonics Networks Communications*, 8(1), 119–132 (2004).

28. Kirstadter, A., Iselt, A., Winkler, C., Pasqualini, S., A quantitative study on the influence of ASON/GMPLS on OPEX, *International Journal of Electronics and Communications*, 60(1), 1–4 (2006).

29. Pasqualini, S., Kirstädter, A., Iselt, A., Chahine, R., Verbrugge, S., Colle, D., Pickavet, M., Demeester, P., Influence of GMPLS on network providers' operational expenditures: A quantitative study, *IEEE Communications Magazine*, 6, 28–34 (2005).

30. USA Bureau of Labor, National Compensation Statistical data, 2010, Public Report, http://www.bls.gov/data/ (accessed: July 2, 2010).

31. Organization for Economic Cooperation and Development, Unit Labor Cost working paper including detailed zone aggregation methodology, 2010, Public Report, http://stats.oecd.org/mei/default.asp?lang=e&subject=19&country=OEU (accessed: July 2, 2010).

32. IETF RFCs related to the MPLS Architecture
IETF-RFC2205: Braden, R. (Ed.), Resource Reservation Protocol (RSVP)—Version 1 Functional Specification, September 1997, http://www.ietf.org/rfc/rfc2205.txt
IETF-RFC2370: Coltun, R. The OSPF Opaque LSA Option, July 1998, http://www.ietf.org/rfc/rfc2370.txt
IETF-RFC2748: Durham, D. (Ed.), The COPS (Common Open Policy Service) Protocol, January 2000, http://www.ietf.org/rfc/rfc2748.txt
IETF-RFC2903: Laat de, C., Gross, G., Gommans, L., Vollbrecht, J., Spence, D., Generic AAA Architecture, Experimental, August 2000, http://www.ietf.org/rfc/rfc2903.txt
IETF-RFC2904: Vollbrecht, J., Calhoun, P., Farrell, S., Gommans, L., Gross, G., de Bruijn, B., de Laat, C., Holdrege, M., Spence, D., AAA Authorization Framework, August 2000, http://www.ietf.org/rfc/rfc2904.txt
IETF-RFC3630: Katz, K. (Ed.), Traffic Engineering (TE) Extensions to OSPF Version 2, September 2003, http://www.ietf.org/rfc/rfc3630.txt
IETF-RFC4203: Kompella, K. (Ed.), OSPF Extensions in Support of Generalized Multi-Protocol Label Switching, October 2005
IETF-RFC4204: Lang, J. (Ed.), Link Management Protocol (LMP), October 2005, http://www.ietf.org/rfc/rfc4204.txt
IETF-RFC4940: Kompella, K., Fenner, B., IANA Considerations for OSPF, July 2007

33. Griffith, D., Rouil, R., Klink, S., Sriram, K., An analysis of path recovery schemes in GMPLS optical networks with various levels of pre-provisioning [aut. libro]. In Z. Zhang, A. K. Somani (Eds.) *Proceedings of SPIE Optical Networking and Communications*, Orlando, USA, September 7–11, 2003.

34. Resea, B., Multiple hierarchical protection schemes for differentiated services in GMPLS networks, *Proceedings International Conference on Information Technology: Research and Education, ITRE 2003*, Newark, NJ, USA, August 10–13, 2003.

35. Aweya, J. IP router architectures: An overview. *International Journal of Communication Systems*. Whiley, 14 (5), 447–475 (2001).

36. Lawton, G. Routing laces dramatic changes. *Computer. IEEE*, 42 (9), 15-17 (2009).

37. Csaszar, A., et al. Converging the evolution of router architectures and IP networks. *Networks. IEEE*, 21 (4), 8–14 (2007).

9

Next Generation Transmission Systems Enabling Technologies, Architectures, and Performances

9.1 Introduction

In Chapter 6 we have discussed Dense Wavelength Division Multiplexing (DWDM) technologies for implementing high capacity very long reach optical transmission systems.

In that chapter, it has been shown as transmission systems with an overall capacity in excess of 2 Tbit/s and a reach longer than 2000 km can be designed using direct detection and in-line optical amplification.

From the point of view of the product capacity-reach it seems that these figures are sufficient to satisfy the needs of telecom carriers for the time being.

As a matter of fact, the need of a new generation of transmission systems does not emerge from the need of increasing the transmission capacity or the system reach, but comes from other reasons.

The fast increase of the traffic in American and European networks is due neither to the increase of the network nodes, whose number remains almost constant, nor to the increase of the network subscribers, that is very slow. What is driving the traffic increase is the increase in the bandwidth required by each subscriber.

This means that the traffic pattern remains statistically constant, while the service bundle offered by carriers differentiates, requiring to the transport layer to increase more and more the number of wavelengths routed along the same network routes.

If the number of wavelengths along a network route is huge, they are demultiplexed and multiplexed uselessly several times, since they should not to be separated. Moreover, every time they traverse a wavelength switch, several management controls have to be done to monitor the correct routing of every wavelength, even if they are directed toward the same direction. Last, but not least, the switching error probability increases with increasing the number of wavelengths.

The request to bundle together in the same network entity (a high capacity transport connection) a great amount of traffic directed toward the same end node is thus natural, and the most direct way to do that is to adopt a higher bit rate. The introduction of 40 Gbit/s in the network was done to alleviate this problem, but it is not enough. Thus a higher speed of the order of 100 Gbit/s is required.

Besides network design needs, hardware design requirements also call for a bit rate much higher than 10 Gbit/s.

Even if the increasing quality of transceivers generated signals will probably bring to a reduction of transponders footprint, the DWDM systems real estate is in any case huge. A 200-channel system using 10 Gbit/s transmission and C + L band amplification needs 100 working and 100 protection transponders even if XFPs are used to compact two channels per transponder.

On a standard Advanced Telecommunications Computing Architecture (ATCA) platform, this means 13 subracks and thus 5 racks only for transponders. This is a huge real estate, greater that that needed for a 30 Tbit/s router.

The only way that seems possible to achieve a strong footprint reduction is to increase greatly the bit rate. For example, assuming that a 100 Gbit/s transponder will occupy a two rack-units card (four times the footprint assumed for 10 Gbit/s), a system with a capacity of 2 Tbit/s using 100 Gbit/s channels will occupy 80 rack units for transponders, that is 5 ATCA subracks, much less than 13 needed in the 10 Gbit/s case.

Moreover, if the power consumption per unit capacity will continue to decrease with increasing the bit rate as that occurred when moving from 2.5 to 10 Gbit/s and from 10 to 40 Gbit/s, a strong decrease of power consumption is also probable if the bit rate is increased up to 100 Gbit/s.

Finally, the convergence on IP networking renders requirements related to routers architecture very important. As detailed in Chapter 7, in order to fully exploit the advances in electronics, routers line cards need to increase the speed of the memories implementing the input queues and of the associated processors.

Also this trend calls for a line bit-rate increase so that useless multiplexing and demultiplexing stages in the line cards need not be forcefully implemented.

If all the above reasons push the evolution of transmission systems toward an increase of the bit-rate, designing systems at 100 Gbit/s is a formidable challenge.

As a matter of fact, the presence of a great number of deployed transmission lines designed for 10 and 40 Gbit/s imposes a set of quite stringent requirements to practical 100 Gbit/s transmission systems.

In particular, 100 Gbit/s channels should be transmitted for distances of the order of magnitude of 1000 km on existing lines and should coexist with 10 and 40 Gbit/s channels without requiring disruption of present services.

Meeting these requirements is surely not possible with the same transmission technologies adopted at lower rates, due to the way in which various transmission impairments affects their performances; thus a complete redesign of transmission systems is needed.

In this chapter we will analyze the various alternatives for 100 Gbit/s transmission, taking into account different applications in the telecommunication network.

9.2 100 Gbit/s Transmission Issues

The first step to do in order to approach the problem of 100 Gbit/s long haul transmission is to understand impairments thwarting conventional transmission at a such a high speed to individuate potential solutions.

In general, solutions to transmission impairments can be attained either via the use of suitable devices to eliminate or compensate the impairing effect or shaping the transmission format and the detection strategy to be less sensible to the considered effect.

We will consider compensation technique in this section while advanced modulation formats and detection techniques will be reviewed in the next two sections.

9.2.1 Optical Signal to Noise Ratio Reduction

The first effect of the bit rate increase is the SN_o reduction due to the increase of the required optical filter bandwidth.

The SN_o is inversely proportional to the optical bandwidth, thus it can seem that the impact of increasing the bit-rate by a factor 10 with respect to 10 Gbit/s reflects in the need of an SN_0 10 dB greater.

Even if this is a good first approximation, two effects could reduce the needed SN_0 increase. First, the signal statistical distribution is not Gaussian. Second, and potentially more important, while in practice it is very difficult in a 10 Gbit/s system to maintain the optical bandwidth equal to the bit rate, it is much easier at 100 Gbit/s. Thus the extra optical noise power is greater at 10 Gbit/s.

In order to verify the impact of the bit-rate increase on the needed SN_o, let us use the correct error probability model introduced for amplified systems in Chapter 6.

Assuming that no pattern effect is present and that thermal noise at the receiver is negligible with respect to the Amplified Spontaneous Emission (ASE) noise, the error probability expression (6.65) relative to an intensity modulated system with in-line amplifiers simplifies as

$$BER = \frac{1}{2} \frac{e^{-(\gamma SN_0 - 2M_{oe})}}{\left(2M_{oe}/\gamma SN_o\right)^{2M_{oe}}} + \frac{1}{2} \frac{e^{-SN_0 M_{oe}[X(SN_0)/(1+X(SN_0))-\gamma X(SN_0)/M_{oe}]}}{[1+X(SN_0)]^{2M_{oe}}} \tag{9.1a}$$

$$X(SN_o) = -\gamma SN_o + M_{oe} + \sqrt{M_{oe}}\sqrt{(\gamma SN_o + M_{oe})} \tag{9.1b}$$

(compare Equations 6.65a and b)

where

  $C_{th} = \gamma SN_o/M_{oe}$ is the threshold value in terms of a fraction of the signal to noise ratio at
    the optical level
  M_{oe} is the ratio between the unilateral optical bandwidth and the electrical bandwidth
    (often approximated with the bit rate R)

Moreover, the saddle point approximation has been further simplified assuming $SN_o \gg M_{oe}$.

Naturally, for each value of SN_o the decision threshold (thus γ) has to be optimized.

To reflect the practical difficulty in implementing stable optical filters with a very small bandwidth, we will assume that at 10 Gbit/s, a filter with 60 GHz bandwidth ($M_{oe} = 6$) is used, corresponding in wavelength to a bandwidth of about 0.5 nm at 1.55 μm. The filter bandwidth is increased at 120 GHz (1 nm) for the 40 ($M_{oe} = 3$) and to 200 GHz at 100 Gbit/s ($M_{oe} = 2$).

Passing from $M_{oe} = 6$ to $M_{oe} = 3$ to $M_{oe} = 2$ implies that the SN_o corresponding to a bit error rate (BER) of 10^{-12} passes from 15.4 to 18.3 to 20 dB.

This is a correction not completely insignificant to the value $\Delta SN_o = 10$ dB that was estimated at first glance if a fixed error probability has to be maintained passing from 10 to 100 Gbit/s.

Naturally, this does not mean that the performances get better by increasing M_{oe}.

As a matter of fact, if we take into account the increase of the ASE bandwidth related to the increase of M_{oe}, we discover that the power needed to achieve an error probability of 10^{-12} slightly increases by 6% passing from $M_{oe} = 2$ to $M_{oe} = 3$ and increases again by another 3% passing to $M_{oe} = 6$.

In any case, considering the SN_o, any change in M_{oe} has to be considered.

In Chapter 6, we have seen that in an optically amplified line, the ASE spectral density is proportional to the number of spans, thus to the link length.

Let us imagine having a link 2500 km long designed for 10 Gbit/s, constituted by 50 spans of 50 km each. If the Optical Signal to Noise ratio has to increase by 10 dB passing to

100 Gbit/s, this means that, once the transmitted power is fixed, whose value is determined by the nonlinear effects, the link has to be shortened by a factor 10. In other words, the combination of nonlinear effects limiting the transmitted power and the increase of ASE noise causes a signal at 100 Gbit/s to propagate for a distance 10 times smaller with respect to a signal at 10 Gbit/s.

Besides improving the amplifiers design to reduce the noise factor, there is no other way of compensating the ASE effect. On the other hand, we know from Chapter 4, that optical amplifiers are quite near the ideal quantum performances, thus there is no great improvement to attain from amplifiers design, at least as far as quantum coherent states are propagated into the fiber.

It is possible to design parametric amplifiers, where the ASE noise spectral density is not uniform along the different polarizations. If a parametric amplifier is used to reduce the ASE along a polarization at the expense of that along the orthogonal one, the signal could be transmitted along the low ASE polarization thus improving the SN_o.

Unfortunately, the resulting quantum state of the optical field at the output of such an amplifier is not a coherent state and such nonclassical field states tend to asymptotically transform into a coherent state during a lossy propagation due to the combination with the void state (see Chapter 4, Basic Theory of Optical Amplifiers section for the quantum model of the power loss). Thus this solution, which could be very attractive for other applications, is not practical for telecommunications where long distance and thus high propagation loss is required.

The only possible solution is to try to shrink the transmitted signal bandwidth by using suitable transmission formats. As a matter of fact, if the transmitted symbols are M instead that two, that is, if multilevel transmission is used, the bandwidth of the transmitted signal with a given bit rate results smaller with respect to that of a binary signal.

With a good approximation, if Nonreturn to Zero (NRZ) transmission is used and the transmitted symbols are statistically independent, M-levels transmission has a unilateral optical bandwidth of the order of $R/\log_2(M)$ almost independently from the modulation format (excluding a few cases of frequency modulation). In order to gain 10 dB with multilevel transmission, $2^{10} = 1024$ symbols are needed, a thing quite difficult with current technology.

If the target is to gain 4 dB, to perform from an ASE point of view like a 40 Gbit/s transmission, it is sufficient to transmit $2^3 = 8$ symbols, which is an approachable task, even if not easy.

Thus, multilevel modulation is one of the main design strategies to arrive to a practical 100 Gbit/s long distance Wavelength Division Multiplexing (WDM) system whose performances would be comparable with a 40 Gbit/s or even with a 10 Gbit/s.

Another possible strategy is using coherent detection. In a standard Intensity Modulation–Direct Detection (IM-DD) system, the main noise term contributing to the electrical current fluctuations is the beat term between the incoming signal and the ASE noise.

In an ideal coherent detection, the current is the frequency downshifted replica both in amplitude and phase of the incoming optical field: as a matter of fact, an ideal coherent detector has to track the variations of the incoming polarization to always optimize the signal detection.

Since no beat term is present, the main electrical noise term is proportional, in module and phase, to the ASE noise.

In this situation, the detection problem is practically moved in the electrical domain, where Shannon sampling can be used to reduce the noise bandwidth from $2R$ to R thus gaining 3 dB. This is almost what is needed to achieve the performances of a 40 Gbit/s transmission, at least from a noise point of view.

Even if an ideal homodyne coherent detection would need an optical phase-lock loop (PLL), a very complex circuit even for a laboratory demonstration, we will see in Section 9.4 that it is possible to avoid the use of such a complex optical system, at the cost of more complex electronics to be used at the receiver.

9.2.2 Fiber Chromatic Dispersion

9.2.2.1 Impact of Chromatic Dispersion on 100 Gbit/s Transmission

Chromatic dispersion is so a bad problem even at 10 Gbit/s that uncompensated systems are dispersion limited at about 36 km. However, due to the deterministic nature of this phenomenon, a carefully designed compensation map can almost completely cancel the problem at 10 Gbit/s, but for the limitation posed to the span length from the interplay between dispersion and nonlinear effects.

At 100 Gbit/s the situation is completely different. As a matter of fact, not only the second order dispersion has a big impact on transmission, but also the third order dispersion (the so-called dispersion slope) cannot be neglected.

The effect of the dispersion slope is evident from Figures 9.1 and 9.2. The result of the simulation of a perfectly linear transmission of a WDM 100 Gbit/s NRZ signal through a

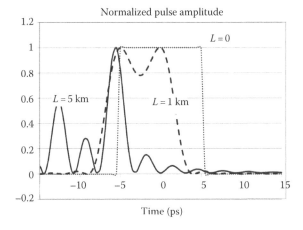

FIGURE 9.1
NRZ pulse at 100 Gbit/s after linear propagation without any PMD. The effect of dispersion slope is evident in the asymmetric form of the pulse after propagation.

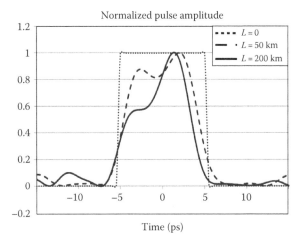

FIGURE 9.2
NRZ pulse at 100 Gbit/s after linear propagation without any PMD. The spectrum is centered on the dispersion zero of the used DS fiber to show the effect of dispersion slope.

link of DSF fiber is reported in Figure 9.1 without polarization mode dispersion (PMD). The WDM comb is composed of 16 channels at 200 GHz spacing and the dispersion zero is exactly in correspondence with the carrier of the eighth channel.

In particular, the pulse shape of the 16th channel is reported for a dispersion slope of $0.05\,\mathrm{ps^3/}$ km and different link length. All the pulses are normalized to their maximum amplitude.

The pulse distortion due to the dispersion slope is quite evident, and after 5 km propagation, the greatest part of the transmitted energy is out of the bit interval while the pulse shape is completely destroyed.

This means that dispersion compensation has to be performed channel by channel, taking into account the exact dispersion value in the channel spectral position.

But this is not enough, as shown in Figure 9.2. In this figure, similar to Figure 9.1 for its construction, the center channel of the WDM comb is considered, having the dispersion zero in its center wavelength.

Nevertheless, the fact that the channel spectrum is so wide causes a pulse distortion due to the third order dispersion even if the average dispersion in the channel bandwidth is zero. The intersymbol interference (ISI) caused by third order dispersion is already important after 50 km propagation and after 200 km propagation a relevant quantity of the pulse energy is out of the bit interval.

The need of per channel compensation depends from the width of the NRZ spectrum that is about 200 GHz. Since multilevel modulation shrinks the spectrum of the transmitted signal, it helps to attenuate the penalty dependence on third order dispersion.

As an alternative, or even to integrate the effect of spectrum width reduction, per channel compensation can be applied.

The cheaper method to operate per channel dispersion compensation is to use an electronic equalization circuit. The exact compensation of a precise wavelength behavior of the dispersion coefficient is more difficult if electronic dispersion compensation is used in conjunction with Direct Detection since the nonlinear detector characteristic has to be taken into account. This difficulty is removed if coherent detection is used, due to the complete proportionality of the photocurrent to the optical field, both in amplitude and phase. In this condition, if linear fiber propagation can be assumed, the propagation channel is linear even considering the detector characteristic, thus allowing far more efficient dispersion compensation to be applied.

9.2.2.2 Tunable Optical Dispersion Compensator

A great amount of research has been devoted also to Tunable Optical Dispersion Compensators (TODC) for application at 40 and 100 Gbit/s. A TODC is an optical transparent component that is able to compensate a selectable amount of dispersion (generally with a selected slope) on one or more high speed channels.

TODC are divided into single channel and multiple channels devices, where the first are suitable to compensate a single channel, while a whole WDM comb can be simultaneously compensated by the latter.

Single channel TODC [1] are mainly studied for intermediate reach, very high speed systems, where electronic dispersion compensation is not enough due to the nonlinear characteristics of the photodiode, but cost constraints or other considerations (related for example to real estate or power consumption) render unsuitable coherent detection.

On the other hand, the advantage expected by multichannel TODC is the ability to compensate the dispersion of a whole WDM comb with a single optical device, without the need of demultiplexing the optical channels.

Thus, multichannel TODCs are complementary to the electronic compensation and the most probable solution to deal with chromatic dispersion of large bandwidth optical signals is a contemporary use of TODC for in-line compensation plus electronic fine equalization at the receiver.

This solution, based on periodic compensation along the optical transmission line, has the advantage of avoiding combination between dispersion and nonlinear distortion that unavoidably occurs if all equalization is placed at the receiver and that renders electronic equalization much more difficult and expensive.

In order to be used in this way, the multichannel TODC have to be simple from a design point of view and must be able to compensate the typical amount of dispersion occurring in a long haul system span.

One of the more direct ways to realize a multiple channel TODC is shown in Figure 9.3 and consists in dividing the incoming optical channel between the two branches of an interferometer where two different dispersion elements are placed. The amount of optical power sent in each branch is regulated by a selectable directional coupler [2].

The transfer function of the system is given by

$$H(\omega) = \frac{\alpha}{\sqrt{2}} e^{i(k_1\omega^2 + \tau_1\omega)} + \frac{1}{\sqrt{2}} \sqrt{1-\alpha^2} e^{i(k_2\omega^2 + \tau_2\omega)} \tag{9.2}$$

where

k_i indicates the coefficient of dispersion of the dispersive elements

τ_i is the overall delay along the two interferometer branches

α is the tunable splitting ratio, while the factor $1/\sqrt{2}$ takes into account the unavoidable loss of the beam splitter

The module of the transfer function writes

$$|H(\omega)|^2 = \frac{1}{2} + \frac{\alpha}{2} \sqrt{1-\alpha^2} \cos(\Delta k\omega^2 + \Delta\tau\omega) \tag{9.3}$$

where $\Delta k = k_1 - k_2$ and $\Delta\tau = \tau_1 - \tau_2$.

The module of the spectral response of the interferometer is a periodic function, as obvious, whose period is frequency dependent due to the quadratic behavior of the object of the cosine function. The optical bandwidth depends on the frequency dependent term and it is maximum where this term is zero, that is, when $\alpha = 0$ or $\alpha = 1$.

On the other hand, the optical bandwidth is minimized when the third term in Equation 9.3 is maximum, that is, $\alpha = 1/\sqrt{2}$.

In order to determine the group delay of the interferometer and then the overall dispersion characteristic, the phase contribution $\Phi(\omega)$ of the response in Equation 9.2 has to be calculated and derived with respect to the angular frequency.

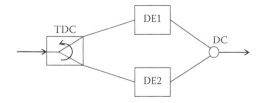

FIGURE 9.3
Block scheme of an interferometric TODC: DE1 and DE2 represents two different dispersion elements.

The result depends again on the factor $A(\omega) = (\Delta k \omega^2 + \Delta \tau \omega)$ that appears under a cosine function.

It is possible to select Δk and $\Delta \tau$ so that in a determined optical bandwidth, the approximation is verified: $A(\omega) = (\Delta k \omega^2 + \Delta \tau \omega) = 0$.

Assuming to be in this condition, the differential delay expression of the interferometer can be quite simplified, up to arriving at the following expression

$$\frac{d\Phi}{d\omega} = \frac{(\alpha^2 + \alpha\sqrt{1-\alpha^2})(k_1\omega + \tau_2) + (1 - \alpha^2 + \alpha\sqrt{1-\alpha^2})(k_2\omega + \tau_2)}{2\pi(1 + 2\alpha\sqrt{1-\alpha^2})} \tag{9.4}$$

Equation 9.4 has the form expected by a TODC, since it linear in ω and depends on an adjustable parameter, which is α.

Just to make an example, let us consider a fiber interferometer where the dispersion is achieved by fiber brag gratings.

The device parameters are summarized in Table 9.1. From the table it results that the condition of $A(\omega) = 0$ is attained introducing a length difference of about 1.1 μm among the gratings at a wavelength of 1.55 μm. The behavior of $\cos[A(\omega)]$ is represented in Figure 9.4 versus the wavelength deviation with respect to 1.55 μm.

TABLE 9.1

Physical Parameters of the Fiber Components Used in the Example of Wideband Tunable Dispersion Compensator

λ_0 (μm)	1.3	
λ (μm)	1.55	
$\Delta\omega$ (cy/s)	-3.7×10^{13}	
ω_0 (cy/s) (null dispersion)	2.3×10^{14}	
ω (cy/s)	1.9×10^{14}	
	Grating 1	Grating 2
L (mm)	80.0000	79.94
τ (ps)	266.66	266.48
Dispersion (ps²/mm)	5	0

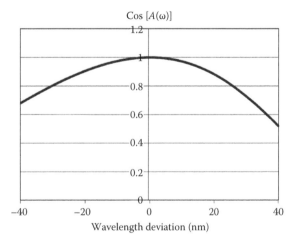

Cos $[A(\omega)]$

FIGURE 9.4
Bandwidth of the response of an interferometric TODC.

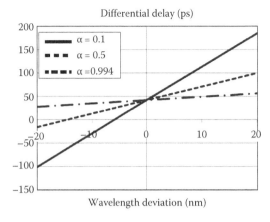

FIGURE 9.5
Differential delay versus the wavelength for an interferometric TODC and for different values of the setting parameter α.

The TODC in this example has a bandwidth of about 40 nm around the reference wavelength (corresponding to 1.55 μm) that is a very good bandwidth to operate as multi-wavelength TODC in DWDM systems.

The group delay is shown in Figure 9.5 versus the wavelength deviation for different values of α. The group delay behavior is, in this ideal case, linear with the wavelength with a very good approximation, thus confirming the good performance of this architecture in conjunction with fiber Brag gratings.

Naturally, in practical implementations, several nonidealities affect the behavior of the TODC, like reflectivity of the gratings, ripple in the phase characteristics of the dispersion elements, losses of the whole device, and so on. A detailed analysis of such impairments besides a practical implementation of a compensator and a transmission experiment are presented in [2].

Moreover in [2] it is also shown that a compensator with the architecture of Figure 9.3 can also be used as a single channel compensator coupling the signal directly in the electrical domain. In this case, the compensator is an alternative to electronic compensation.

Several other architectures have been proposed for the TODC, having different characteristics.

A class of compact TODC are based on the use of Array Waveguides (AWGs) [3,4] since these devices provide contemporary demultiplexing and different dispersion contribution channel per channel. Moreover, if suitably designed, the same AWG can be used both for multiplexing and for demultiplexing the DWDM comb increasing the dispersion contribution and gaining in compactness.

An example of AWG architecture is presented in [3] and is reported in Figure 9.6. The component working is quite simple. The incoming optical field is decomposed by the AWG, which divides its spectrum slices, its spectral phase around the center frequency is adjusted by the lens located near the spectral plane, reflected by the mirror, and recombined in the AWG to regenerate the field. Depending on the AWG design, this TODC can be used either to compensate interchannel dispersion due to the dispersion slope in particularly wide spectrum transmissions or multichannel dispersion in WDM systems.

In the first case however, the distortion due to the spectral analysis of the channel and successive composition of the spectral components has to be taken into account. The compensator tuning is performed by using a cooler and exploiting the thermal dependence of the optical characteristics of the silica composing the AWG and the lens structure.

An improvement of the performance of the TODC can be attained by realizing several lenses monolithically integrated into the structure of the chip so that the whole device is

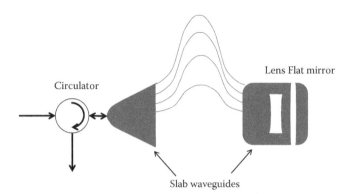

FIGURE 9.6
TODC based on a cyclic AWG.

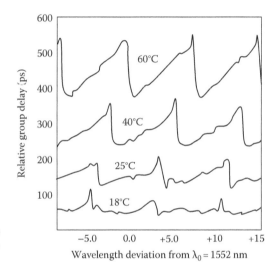

FIGURE 9.7
Differential Delay of the AWG based TODC versus
the temperature. (After Ikuma, Y. and Tsuda, H.,
IEEE J. Lightwave Technol., 27(22), 5202, 2009.)

a planar silicon on silica-integrated circuit. The characteristics of an experimental realization of the device are reported in Figure 9.7 [3].

Another technology used to implement experimental TODCs is based on discrete optics and on technologies similar that used in 3D Micro Electrical Mechanical machines (MEMs), Liquid Crystal on Silicon (LCOS) switches and in Wavelength Selective Switch (WSS) components (see Chapter 5 for MEMs switches and Chapter 6 for WSS).

An example of TODC using this technology is reported in [5] and its block scheme is shown in Figure 9.8. A circulator deviates the whole DWDM comb into an AWG where the channels are demultiplexed. At the AWG output, a free space optics directs the demultiplexed channels on a bulk grating and, at the grating output, a lens focalizes the channels on an LCOS switch. The LCOS reflects back the channels that traverse the system again in the opposite direction to be reinjected by the circulator into the transmitting fiber after multiplexing by the AWG.

The AWG, the LCOS deflector, and the Brag grating, all provide dispersion and, if the dispersion axis of the grating is orthogonal to that of the AWG and the two dispersion axes of the LCOS are parallel and orthogonal to the AWG axis, it is possible to combine the dispersion contributions to have different dispersions for the different channels.

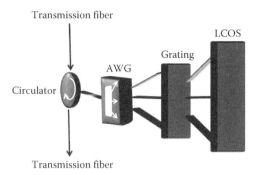

FIGURE 9.8
Discrete optics TODC based on LCOS technology.

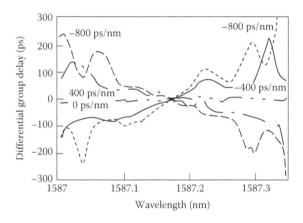

FIGURE 9.9
Differential delay of the LCOS based TODC. (After Seno, K. et al., 50-Wavelength channel-by-channel tunable optical dispersion compensator using combination of arrayed-waveguide and bulk gratings, *Conference on Fiber Communication (OFC), Collocated National Fiber Optic Engineers Conference, 2010 (OFC/NFOEC)*, San Diego, CA, IEEE/OSA, s.l., 2010, pp. 1–3.)

In the paper where this architecture is proposed [5] the TODC design is performed for application in the L band at 40 Gbit/s.

The measured dispersion performances for a single channel among the 50 that can be simultaneously processed by the device are reported in Figure 9.9 for different driver voltages of the LCOS. The ripple that can be observed in the experimental characteristics is attributed to the imperfect alignment of the LCOS device, a hypothesis that is reinforced by the presence of an analogous ripple in the amplitude transmittance [5]; in any case, the ability of the device to attain a range of dispersion between −800 and 800 ps/nm results clearly from the figure.

9.2.3 Fiber Polarization Mode Dispersion

9.2.3.1 Impact of Polarization Mode Dispersion on 100 Gbit/s Transmission

A first evaluation of the impact of PMD on transmission of high speed channels has been carried out in Chapter 6 and an example of performance evaluation is reported in Figure 6.12 for channels at 40 and at 100 Gbit/s.

Using the parameters of Figure 6.12 (see Table 6.1) it is possible to evaluate the transmission length at which the PMD-induced penalty is 2 dB, that we can call maximum PMD transmission distance. The value of the maximum PMD transmission distance is reported in Figure 9.10 versus the bit rate for different values of the PMD parameter. Dots are values calculated with the model of Chapter 6 and assuming linear propagation. The

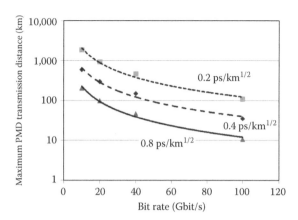

FIGURE 9.10
Maximum PMD transmission distance versus the bit-rate for different values of the PMD parameter.

curves are interpolated with a function inversely proportional to the bit rate through a constant.

From the figure it is quite evident that the fit is very good, almost in all the interesting range of values for the PMD parameter.

Thus we can conclude without a big error that the maximum PMD transmission distance is inversely proportional to the bit rate.

From Figure 9.10 it is also clear that the impact of PMD is very important at 100 Gbit/s: the maximum PMD transmission distance for a PMD parameter of 0.2 ps/km$^{1/2}$ is about 100 km.

Also in this case, since the dependence of the PMD penalty on the bit rate is due to the relationship between the bit rate and the optical bandwidth, both adoption of multilevel modulation and coherent detection are beneficial for the same reasons discussed in the above section.

9.2.3.2 Polarization Mode Dispersion Compensation

PMD can be naturally compensated both electronically and optically. The main issue in PMD compensation is the fact that it is a random phenomenon, thus requiring adaptive compensators to follow its fluctuations.

Electronic compensation circuits generally have such a feature, thus being suitable for compensating PMD, besides other ISI-related impairments like chromatic dispersion. For example, in Figure 5.53, it is shown as both feedforward equalizer/ decision feedback equalizer (FFE/DFE) and maximum-likelihood estimation (MLE)-based compensators can reduce the PMD-induced penalty for 10 and 20 Gbit/s signals.

Scaling the bit rate at 100 Gbit/s poses however a great challenge to electronic compensation, due to the great speed at which the compensator has to work.

Even if traveling wave FFE/DFE compensators have been realized and demonstrated as effective in compensating PMD at 40 Gbit/s [6], no electronic equalization has been realized on a serial signal at 100 Gbit/s.

For this reason, optical PMD compensation has a particular importance at very high speed.

As a matter of fact, due to the random nature of the PMD, the instantaneous performance of the system can be also completely ruined by PMD in the occasion of unfortunate fluctuations of the Jones matrix even if the average penalty is quite low.

To understand this phenomenon, let us consider the PMD-related penalty as defined in Figure 6.12: it was evaluated by starting from the error probability for a specific value of the PMD broadening $\Delta\tau$ and averaging with respect to $\Delta\tau$. It is clear that, while $\Delta\tau$ fluctuates, the system can experience instantaneous values of the penalty either smaller or larger than the average value.

To evaluate the impact of these fluctuations, generally, the so-called outage probability P_O is used, that is, the probability that the instantaneous value of the PMD-induced penalty is higher than the value used in the system design, or equivalently (assuming all processes ergodic), the percentage of time in which the PMD-induced penalty is higher than the design value.

In order to approach the problem of reducing the outage probability, it is important to notice that when launching a pulse into a fiber link with PMD, the received pulse width depends on the launched state of polarization (SOP), and in particular on its position with respect to the input Principal States of Polarization (PSPs), following Equation 4.29. The effect of the launched polarization state on the pulse broadening is represented in Figure 9.11 [7] where the pulse spreading ratio β is shown versus the normalized average pulse width α.

The parameters β and α are defined as

$$\beta = \sqrt{1 + \frac{\sigma_{\Delta\tau}^2}{\tau_{in}^2}} \tag{9.5}$$

$$\alpha = \frac{\langle\tau\rangle}{\tau_{in}}$$

where

τ_{in} is the half width at half maximum of the input pulse

$\langle\tau\rangle$ is the average half width at half maximum of the output pulse

$\sigma_{\Delta\tau}^2$ is the variance of the half width at half maximum of the output pulse as defined in Equation 4.29

Looking at Figure 9.11, it results that there is quite a difference between launching the input pulse along the worst and along the best polarization state. Thus, a way to attenuate

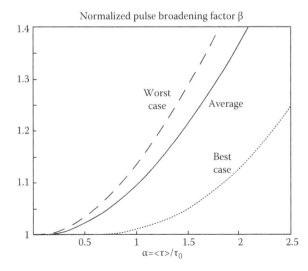

FIGURE 9.11
Dependence of the variance of the pulse width at the output of a PMD affected fiber on the launch polarization state: best, average and worst cases are shown. (After Sunnerud, H. et al., *IEEE Photon. Technol. Lett.*, 12(1), 50, 2000.)

the penalty due to PMD via the elimination of the worst cases is simply to apply a fast random polarization scrambling.

If the polarization scans almost all the possible states during a bit period, the pulse spreading will be, in a very first approximation, always the average one and the worst cases will be avoided.

In order to evaluate the effectiveness of this method, let us image that the outage probability is $P_O = 10^{-3}$ in the absence of fast polarization scrambling and that, in order to avoid mixing of the PMD effect with self-phase modulation (SPM), we will introduce periodic scrambling along a 300 km long line at 100 Gbit/s with a PMD of 0.1 ps/km$^{1/2}$ (see Figure 9.11).

The effect on the outage probability of periodic polarization scrambling via equally spaced scramblers placed along the line is shown in Figure 9.12 versus the number of scramblers [8].

Polarization scrambling technique is quite effective in reducing penalty fluctuations due to the random nature of the PMD, with the additional advantage of being able to process a whole WDM comb without demodulation, but while the bit rate increases, it is more and more difficult to perform a perfect scrambling, that is, scramble at a so high speed that almost all the polarizations states are traversed during the bit time. Let us imagine that the scrambling speed is not sufficient, in particular, let us assume that it is 20% smaller than the speed needed to span all the possible states of polarization.

The resulting outage probability is also reported in Figure 9.13 [8] and the result is that, even if the scrambling is much less effective, the outage probability is in any case much smaller than the case without scrambling if a sufficient number of scramblers is used.

The fact that PMD fluctuations induce outage also suggests the fact that errors due to these fluctuations form bursts. In case of transmission channels causing error bursts, the use of specific burst correcting forward error correcting codes (FECs) is quite effective.

Systems using specific FECs and fast polarization scrambling can be optimized to greatly reduce PMD penalty fluctuations up to very low values of the outage probability [9,10].

If the average penalty due to PMD is acceptable from a system design point of view, a reduction of the outage probability below a required lower limit is enough to assure a correct system working.

FIGURE 9.12
PMD related outage probability versus the number of in line scramblers, both ideal and slow scramblers (see text) are considered. (After Lid, X. et al., Multichannel PMD mitigation through forward-error correction with distributed fast PMD scrambling, *Optical Fiber Communication Conference (OFC 2004)*, Los Angeles, CA, IEEE, s.l., 2004, Vol. 1, pp. WE2 1–3.)

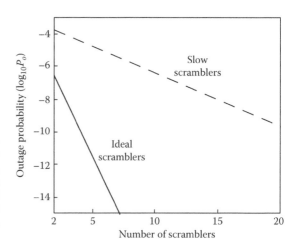

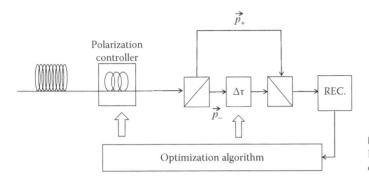

FIGURE 9.13
Block scheme of a first order PMD compensator.

If the average PMD penalty is too high, an effective PMD compensation is needed that is able to reduce both the outage probability and the average penalty. The scheme of a simple PMD compensator, a so-called first order compensator, is reported in Figure 9.13. At the transmission fiber output a Polarization Controller is set to split the signal between fast ($\vec{p}_-$) and slow ($\vec{p}_+$) PSP components. The components along the PSPs are mutually delayed by the differential group delay (DGD) $\Delta\tau$ and then recombined. The delay is generally obtained by rotating the PSPs to match the axes of a tunable birefringence element.

Both the PSP decomposition and the DGD have to be adjusted via an adaptive algorithm to match the slowly varying transmission link characteristic [11–14].

Several algorithms have been devised to simplify the apparent complexity of the PMD compensator and several designs have been presented that are quite effective in perfectly compensating the PMD in the central frequency of the signal spectrum.

Obviously, to compensate a whole WDM comb, it has to be demultiplexed and compensated channel by channel. Thus, a first order PMD compensator like that shown in Figure 9.13 is generally designed to be deployed immediately before detection.

A first order PMD compensator has two main limitations. The first is related to the nonlinear evolution of the signal along very long fiber links. When the linear and nonlinear polarization evolutions are mixed, it is not yet possible to distinguish them and a linear compensator is not yet able to correctly recover the signal [11]. This means that in order to use effectively such compensators, the propagation has to be maintained as linear as possible (but for the soliton case that is considered in detail in [12]).

More important, by increasing the bit rate, the bandwidth of the PSPs starts to be of the order or even smaller than the optical bandwidth of the signal. As a matter of fact, from Chapter 4, we derive that the PSP bandwidth in a Standard Single Mode Fiber (SSMF) if of the order of 100 GHz, while the bilateral optical bandwidth of a 40 Gbit/s NRZ signal is 80 GHz and that of a 100 Gbit/s NRZ signal is 200 GHz.

In such a condition, intrachannel variations of the PSP causes a relevant signal distortion even if a first order PMD compensator is used.

A more accurate compensation can be attained by approximating the PMD with a second order expression in ω instead of a first order one [15] and designing a second order PMD compensator. A possible scheme of a second order PMD compensator is reported in Figure 9.14 [16,17].

The effectiveness of second order compensation is quite improved for high bit rate signals with respect to first order compensation as shown in [18]. A comparison between different first order and second order PMD compensators based on various control algorithms and compensating elements is reported in Figure 9.15 [18] where the outage probability P_O

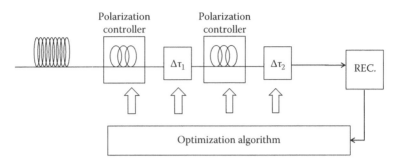

FIGURE 9.14
Block scheme of a second order PMD compensator.

is reported as a function of the design average PMD penalty ΔSN_e (dB) for a bit rate of 100 Gbit/s, NRZ transmission, and an average PMD-induced DGD of 3.8 ps (deriving, e.g., from a 100 km long link with $D_{PMD} = 0.38$ ps/km$^{1/2}$).

Comparing Figure 9.15 with literature, it is to take into account that the penalty is frequently expressed as instantaneous Eye Opening Penalty (EOP). Since evolution of the PMD is much slower than the bit-time, the PMD DGD can be considered almost constant for a great number of consecutive bits. In this situation, there is a direct relationship between the instantaneous EOP and the SN_e penalty once the link characteristics are fixed. As a matter of fact, indicating with I_0 and I_1 the average values of the photocurrent decision samples corresponding to the two transmitted bits, and with I_{00} and I_{10}, the same values in the absence of PMD, we can write (see Chapter 6)

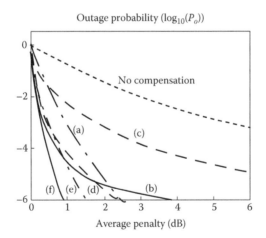

FIGURE 9.15
Comparison between different first order and second order PMD compensators based on various control algorithms and compensating elements. (After Heismann, F., *IEEE Photon. Technol. Lett.*, 17(5), 1016, 2005.) (a) Conventional FO-PMDC with variable differential phase delay. (b) PMDC based on the Kogelnik–Bruyère model described in [19]. (From Kogelnik, H. et al., *Opt. Lett.*, 25(1), 19, 2000.) (c) PMDC based on the truncated EEF described in [20]. (From Eyal, A. et al., *Electron. Lett.*, 35(17), 1658, 1999.) (d) PMDC described by [18]. (From Heismann, F., *IEEE Photon. Technol. Lett.*, 17(5), 1016, 2005.) (e) PMDC formed by two concatenated First Order-PMDCs, wherein the first introduces a fixed DPD equal to the mean DGD in the fiber and the second a continuously variable DPD [21]. (From Yu, Q. et al., *IEEE Photon. Technol. Lett.*, 13(8), 863, 2001.) (f) PMDC described by [19,21]. (From Yu, Q. et al., *IEEE Photon. Technol. Lett.*, 13(8), 863, 2001; Kogelnik, H. et al., *Opt. Lett.*, 25(1), 19, 2000.)

$$EOP = \frac{I_1 - I_0}{I_{10} - I_{00}} \tag{9.6}$$

$$SN_e = \frac{I_1 - I_0}{\sigma_1 - \sigma_0}$$

where σ_1 and σ_0 are the overall noise variance corresponding to the two transmitted bits. Assuming that the noise does not change, the following equation is immediately derived:

$$\Delta SN_e(\text{dB}) = 10\log_{10}\left(\frac{SN_e}{SN_{eo}}\right) = 10\log_{10}(EOP) \tag{9.7}$$

The PMD compensators considered in Figure 9.15 are

- Conventional first order-polarization mode dispersion compensation (FO-PMDC) with variable differential phase delay
- PMDC based on the Kogelnik–Bruyère model described in [19]
- PMDC based on the truncated exponential expansion form (EEF) described in [20]
- PMDC described by [18]
- PMDC formed by two concatenated First Order-PMDCs, wherein the first introduces a fixed differential phase detector (DPD) equal to the mean DGD in the fiber and the second a continuously variable DPD [21]
- PMDC described by [19,21]

The performance of second order compensators depends in an important measure on the control algorithm and on the effectiveness of the compensation device, but it is in any case quite better than the performance of a single order compensator.

The number of stages of the PMD compensator can be increased to three or more in order to achieve high order compensation and target better performances. More compensation stages however also mean a more complex control algorithm and a more expensive device (keeping in mind that PMDCs are per channel devices).

The performance improvement and the control complexity of high order compensation devices have been studied in several papers [22]. An estimation of the possible performance of a third order PMD compensator is reported in Figure 9.16 [22]. When compared to a relevant gain passing from a first to a second order compensator, the gain achieved when passing from a second to a third order component is not so pronounced while it is paid with a sensible increase in the complexity. Thus it is possible to conclude that system design should try as much as possible to use second order compensator (if optical compensation is needed) and passing to third order compensators only if it is really needed.

9.2.4 Other Limiting Factors

9.2.4.1 Fiber Nonlinear Propagation

Increasing the signal bandwidth has essentially no impact both on Brillouin (at least if the carrier is maintained in the spectrum) and on Raman effects.

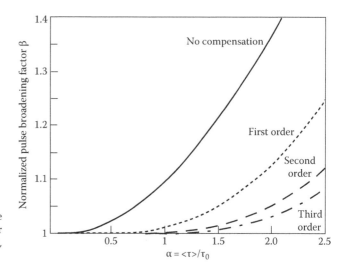

FIGURE 9.16
Estimation of the possible performance of a third order PMD compensator. (After Kim, S., *IEEE J. Lightwave Technol.*, 20(7), 1118, 2002.)

While the impact of Cross Phase Modulation (XPM) and Four Wave Mixing (FWM) can be managed by a suitable channel spacing and residual dispersion, SPM depends on the individual channel and can become the dominant nonlinearity and the system limiting factor besides PMD.

In this case also, the reduction of the signal bandwidth occupancy seems to be the only way to go to reduce SPM impact besides the possibility of an effective electronic compensation via some kind of coherent detection or of signal predistortion (compare Chapter 5).

9.2.4.2 Timing Jitter

Due to the very short bit time, timing jitter assumes a particular importance in 100 Gbit/s serial systems. We can decompose the jitter into three factors:

1. Transmission jitter
2. Detection clock jitter
3. Propagation jitter

The transmission jitter depends on the fluctuations of the transmitter clock that causes the center instant of the pulses sent to the modulator to fluctuate randomly around the average value. This problem can be faced with a combination of a local improvement of the clock electronics and a more performing network timing distribution in case synchronous time-division multiplexing (TDM) is used.

Analogously, the receiver clock jitter depends on the imperfection of the receiver clock electronics that causes the sampling instant before the decision device to fluctuate randomly around the average value. Also in this case, the problem can be faced with a combination of a local improvement of the clock electronics and a more performing network timing distribution.

The propagation jitter is caused by a different phenomenon and affects systems where SPM is partially compensated by a residue chromatics dispersion [23]. This jitter contribution becomes evident for pulses with a full width at half maximum (FWHM) of the order of few ns, that could be used in return-to-zero (RZ) 100 Gbit/s systems.

The propagation jitter is a combination of the jitter due to pulse interaction and to the jitter due to ASE to pulse nonlinear combination.

The first effect is a reduced form of the Gordon-Haus jitter that appears in soliton propagation. This jitter caused by attraction between nearby pulses becomes more and more effective by increasing the Chromatic Dispersion–SPM compensation degree [24,25], and comes into play especially when strong under-compensation of chromatic dispersion is used to counterbalance SPM. The other cause of propagation jitter is the phase noise produced by the nonlinear ASE to signal coupling and it is known as Gordon-Mollenauer effect [26].

9.2.4.3 Electrical Front End Adaptation

A serial 100 Gbit/s signal has a huge bandwidth and it is not easy to device receivers with a low-pass bandwidth of 100 GHz. Distortions in the photocurrent directly affect the receiver performances via ISI and loss of power in the bit interval. Great progress have been made in the last years, but this is yet an important challenge [27].

9.3 Multilevel Optical Transmission

In the previous section we have seen that one of the main problems in NRZ serial transmission of 100 Gbit/s signals is the huge optical spectral width.

The NRZ 100 Gbit/s spectrum is 200 GHz wide (unilateral bandwidth 100 GHz), causing a great amount of ASE noise to enter the receiver and a great sensitivity both to chromatic dispersion slope and to PMD.

Moreover, nonlinear Kerr effect depends also on the signal bandwidth, worsening the problem of transmitting NRZ 100 Gbit/s signals.

Several alternative modulation formats can be used to reduce the 100 Gbit/s spectral width, such as single side band modulation or duobinary modulation (see description and References in Chapter 6), but the most effective and flexible way to reduce signal bandwidth is multilevel modulation [28,29].

In general, multilevel modulation consists in coding the binary signal to be transmitted in a signal that consists of a series of symbols extracted by an alphabet A of M symbols so that the two signals are equivalent under an information point of view.

This means that an invertible code must exist that establishes a correspondence between the input bit stream and the stream of elements from the alphabet A. In this condition, if the stream of transmitted symbols is correctly received, the original bit stream can be univocally reconstructed.

If this is true, it is also possible to evaluate, starting from the correlation function of the bit stream, the resulting correlation function of the symbol stream to be transmitted [30].

In the case of 100 Gbit/s application, all electronic coding and decoding operations have to be carried out by very high speed electronics; thus, it is realistic to restrict the analysis of multilevel modulation to a particular case of modulation formats, allowing simple coding/decoding and modulation/demodulation operations to be performed.

This is the class of the so-called "instantaneous" modulation formats, characterized by the property that, if the incoming bits are statistically independent, also the symbol stream derived from coding is composed of statistically independent symbols.

In general, the multilevel coding is performed by dividing the incoming bit stream in words of n bits and coding each word with a symbol from an alphabet of $M = 2^n$ symbols.

Even if this is a particular class of multilevel modulations, it is still sufficiently rich to contain remarkably different modulation formats.

9.3.1 Optical Instantaneous Multilevel Modulation

In Chapter 6, we have introduced the so-called quadrature representation of the single mode electromagnetic field that propagates in an optical fiber.

Let us call $\vec{p}_+$ and $\vec{p}_-$ two orthogonal unitary polarization vectors (e.g., the output polarization principal states or two orthogonal linear polarizations). In general, once the slowly varying component of the optical field is neglected, the transversal shape that is constant during propagation (see Chapter 4), can be written as

$$\vec{E}(t) = [e_1(t) + ie_2(t)]\vec{p}_+ + [e_3(t) + ie_4(t)]\vec{p}_- \tag{9.8}$$

The four real functions $e_j(t)$ ($j = 1, 2, 3, 4$) are the so-called four quadratures of the optical field in the base $(\vec{p}_+, \vec{p}_-)$ and they are the independent degrees of freedom that can be exploited to modulate the field.

The optical field power $P(t)$ can be expressed as the sum of the square of the quadratures, that is,

$$P(t) = e_1^2(t) + e_2^2(t) + e_3^2(t) + e_4^2(t) \tag{9.9}$$

Exploiting (9.9), it is possible to introduce the so-called quadrature space, that is, the space having the quadratures as coordinates. This is a Euclidean four dimensional space, where the instantaneous value of the field is represented by a four-vector whose norm is the field power. In this space, the set of field states having the same power are thus spherical surfaces whose radius is the square root of the power itself.

A general instantaneous multilevel modulation of the field can be represented as following

$$e_j(t) = \sum_{k=-\infty}^{\infty} s_{j,k} g_j(t - kT) \quad (j = 1,2,3,4) \tag{9.10}$$

where
  $g_j(t)$ is the pulse used to modulate the jth quadrature
  $s_{j,k}$ is one of the set of amplitudes $(s_1, \ldots, s_{Nj})$ adopted for the modulation of $e_j(t)$, where we can assume that $N_j = 2^{nj}$, so that

$$M = N_1 N_2 N_3 N_4 = 2^n \quad \text{with} \quad n = n_1 + n_2 + n_3 + n_4 \tag{9.11}$$

In many practical cases, for example, M-ary Quadrature Amplitude Modulation (M-QAM) and M-ary Phase Shift Keying (M-PSK), the same pulse is used to modulate all the quadratures, thus we will assume this condition and we will neglect the index for $g_j(t)$, calling the modulation pulse simply $g(t)$. In any case, it is straightforward to generalize the derivations that follow to the case of different modulating pulses for the different quadratures.

Once a general description of this class of modulation formats is set up, we have first to demonstrate that a practical modulator for such signals can be built and then evaluate the transmitted signal power spectral density to verify the expected shrink with respect to a standard NRZ signal at the same bit rate.

A possible scheme of a modulator for a general modulation format of this class is reported in Figure 9.17.

This is a cascade of an amplitude, a phase, and a polarization modulator [31]. It is clear that in practice, when particular modulation formats are adopted, this will not be the optimum modulator and other schemes will be used. Just to consider only one aspect, the modulators cascade causes a great loss to be introduced on the field to be transmitted.

However, the scheme of Figure 9.17 demonstrates that whatever modulation format of this class is considered, a possible modulation scheme exists.

As far as the spectrum occupation is considered, we can define the so-called optical field spectrum matrix $\mathbf{S}$ in the base $(\vec{p}_+, \vec{p}_-)$ through its elements as

$$S_{h,j}(\omega) = \Im\left\{\left\langle s_{j,k}g(t-kT)s_{h,k'}g(t-k'T)\right\rangle\right\} \tag{9.12}$$

where
$\Im()$ indicates the Fourier transform
$<x>$ indicates the ensemble average

Since the pulse $g(t)$, in order not to provoke ISI, is completely contained in the symbol interval and due to the independence among symbols transmitted in different intervals, (9.12) rewrites

$$S_{h,j}(\omega) = \langle s_{j,k}s_{h,k}\rangle\Im\{\,|\,g(t)\,|^2\} = G(\omega,T)\langle s_{j,k}^2\rangle\delta_{j,h} \tag{9.13}$$

where
$G(\omega, T)$ is the power spectral density of the transmission pulse
$\langle s_{j,k}^2 \rangle$ is the average of the square of the transmitted symbols
$\delta_{j,h}$ is one if $j = h$, otherwise it is zero

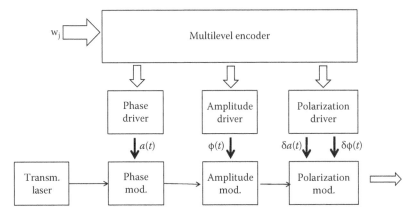

FIGURE 9.17
Block scheme of the generic modulator for instantaneous multilevel modulation formats.

The power spectral density $S_+(\omega)$ and $S_-(\omega)$ of the two polarization components of the transmitted signal (9.8) can be evaluated starting from the elements of the matrix $\mathbf{S}$. In the hypothesis of instantaneous modulation, from Equation 9.13 it is deduced that $\mathbf{S}$ is always a diagonal matrix so that

$$S_+(\omega) = S_{1,1}(\omega) + S_{2,2}(\omega) = S_{0+}G(\omega, T)$$
$$S_-(\omega) = S_{3,3}(\omega) + S_{4,4}(\omega) = S_{0-}G(\omega, T)$$

$$(9.14)$$

where
$$S_{0+} = \left\langle s_{1,k}^2 \right\rangle + \left\langle s_{2,k}^2 \right\rangle$$
$$S_{0-} = \left\langle s_{3,k}^2 \right\rangle + \left\langle s_{4,k}^2 \right\rangle$$

$T = n/R$, where R is the bit rate and n is defined by Equation 9.11

In the particularly important case of NRZ modulation, $g(t)$ is a square pulse and (9.14) becomes

$$S_u(\omega) = S_{0u}\frac{n}{2\pi R}\frac{\sin^2(n\omega/R)}{(n\omega/R)^2} \quad u \in (+,-)$$

$$(9.15)$$

This is quite a common form of an NRZ signal, whose spectrum width is equal to $B_o = 2R/n$ thus confirming the rule anticipated in Section 9.1 for the spectral width of an instantaneous multilevel modulation.

9.3.2 Practical Multilevel Transmitters

Almost in any practical case, the transmitter depicted in Figure 9.18 has a too high insertion loss and practical modulation formats are also selected on the ground of the possibility of designing a low loss–high efficiency modulator.

In this section, we will present a few particular modulation formats that are important in practice.

9.3.2.1 Multilevel Differential Phase Modulation (M-DPSK)

This is probably one of the simpler cases that generalize to multilevel transmission binary differential phase shift keying (DPSK) that we have introduced in Chapter 6: the phase of the transmitted signal is modulated using multilevel differential modulation. Let us call w_j the jth world of n transmitted bits and $\Delta\theta = 2\pi/2^n$.

The M-DPSK signal transmitted in the kth symbol interval and modulated with w_j writes

$$\vec{E}(t, k, w_j) = \sqrt{P_0}e^{i\theta(k)}\eta_T(t, k)\vec{p}_+$$

$$(9.16)$$

where $\eta_T(t,k)$ is the Heaviside function that is equal to one for $t \in ((k-1)T, kT)$ and zero elsewhere and $\theta(k) - \theta(k-1) = j\Delta\theta$.

As in the binary case, the name differential comes from the fact that the transmitted symbol is encoded into the phase difference between signals in two consecutive symbol intervals.

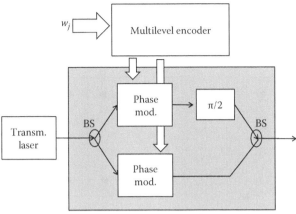

Dual driver mach-zehnder phase modulator

FIGURE 9.18
Dual driver Mach–Zehnder modulator used to obtain a multilevel DPSK signal.

In terms of quadrature components, (9.16) writes

$$e_1(t) = \sqrt{P_0} \cos[\theta(k-1) + j\Delta\theta]\eta_T(t,k)$$

$$e_2(t) = \sqrt{P_0} \sin[\theta(k-1) + j\Delta\theta]\eta_T(t,k) \qquad (9.17)$$

$$e_3(t) = e_4(t) = 0$$

Equation 9.17 shows that the points representing the transmitted symbols in the quadrature space are all distributed on a circle on the e_3, e_4 plane whose center is in the origin and whose square radius is the transmitted power, which is constant due to the pure phase modulation.

A typical characteristic of a multilevel constellation is the minimum distance d in the quadrature space between a couple of symbols. It is quite intuitive that this distance will be related to the noise robustness of the modulation format. Considering for example binary IM, the constellation is done by two points and the distance is $d = \sqrt{P}$, so that we can write $SN_o = d^2/P_{ASE}$, where P_{ASE} is the ASE power.

In this case, we have simply

$$d = \sqrt{P} \sin\left(\frac{2\pi}{M}\right) \qquad (9.18)$$

A practical transmitter for the M-DPSK signal is presented in Figure 9.18, conceptually, this is simply a phase modulator, in practice almost only dual drive integrated modulators are used to reduce distortions (see Chapter 5).

9.3.2.2 Multilevel Quadrature Amplitude Modulation (M-QAM)

This modulation format is widely used in radio applications and consists in the amplitude modulation of the two complex quadratures of one of the field polarizations.

The points representing the states of the transmitted field in the quadratures space are comprised in one coordinate plane and form a square lattice: a few examples are provided in Figure 9.19.

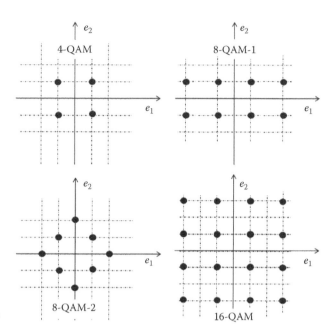

FIGURE 9.19
Examples of squared QAM constellations.

In terms of quadrature, the components of the transmitted field can be written as

$$e_1(t) = \sqrt{P_0}\, s_{1,k}\, \eta_T(t,k)$$

$$e_2(t) = \sqrt{P_0}\, s_{2,k}\, \eta_T(t,k) \tag{9.19}$$

$$e_3(t) = e_4(t) = 0$$

where $s_{j,k}$ is the amplitude of the symbol transmitted along the jth quadrature in the kth symbol interval.

The maximum distance of QAM modulation depends on the adopted pattern and it can be easily evaluated starting from the fact that the power of an optical field is the square of its distance from origin. The minimum distance for a few common QAM patterns is reported in Table 9.2.

A general scheme for an optical M-QAM modulator is reported in Figure 9.20: a laser emits a linearly polarized optical field that is divided into two fields with a phase difference equal to π by a directional coupler. Each field is independently modulated by an amplitude modulator and then one of them is delayed to introduce a further phase difference of $\pi/2$. At that point, the two fields are combined again by a directional coupler that introduces a further phase difference of π, so that the total phase difference if $2\pi + \pi/2$ and the two combined fields are exactly the quadratures of the M-QAM signal.

Such a scheme is suitable to be integrated in a single optical chip, for example, using lithium-niobate or indium phosphide (InP) technology.

Several different modulators have been proposed for particular M-QAM constellations. An interesting scheme is that of the hybrid modulator for 64-QAM proposed in [32]. The principal scheme of the modulator is shown in Figure 9.21 besides the way in which the 64-QAM constellation is formed in the different stages of the modulator. The fact that the modulator is composed of a parallel of three dual drivers phase modulators

TABLE 9.2

Normalized Constellation Minimum Distance for
M-QAM

Modulation Format	Minimum Distance (Normalized to $\sqrt{P}$)
4-QAM	$\sqrt{2} \approx 1.14$
8-QAM-1	$\sqrt{\dfrac{2}{5}} \approx 0.632$
8-QAM-2	$\dfrac{1}{\sqrt{2}} \approx 0.707$
16-QAM	$\dfrac{\sqrt{2}}{3} \approx 0.471$
32-QAM	$\sqrt{\dfrac{2}{29}} \approx 0.263$
64-QAM	$\dfrac{\sqrt{2}}{7} = 0.202$
128-QAM	$\dfrac{2}{\sqrt{335}} = 0.111$
256-QAM	$\dfrac{\sqrt{2}}{15} = 0.094$
1024-QAM	$\dfrac{\sqrt{2}}{31} = 0.046$

that has to be driven to produce a four-level phase modulated signal renders the modulator control easy and the loss relatively small.

The most interesting property of this modulator is, however, that it can be realized by using hybrid integration with two Silica on Silicon chips and a modulator array composed of three phase modulators (let us remember that the modulator of Figure 9.21 is only apparently simpler, since each amplitude modulator is in reality a couple of phase modulators on the

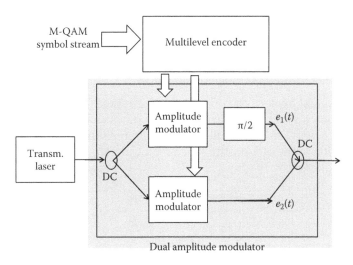

FIGURE 9.20
Block scheme of an M-QAM modulator.

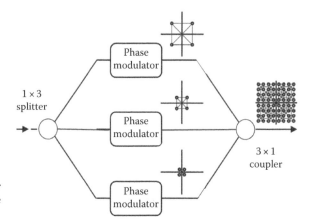

FIGURE 9.21
Architecture of a hybrid modulator for 64-QAM realized with a parallel of three phase modulators.

1 × 3 splitter

Phase modulator

Phase modulator

Phase modulator

3 × 1 coupler

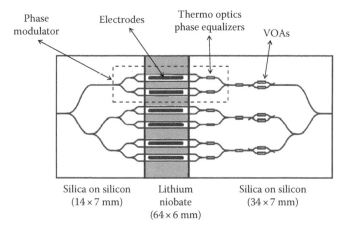

FIGURE 9.22
Scheme of the layout of a hybrid Si-SiO$_2$ and Lithium Niobate-integrated optical circuit for the implementation of the modulator represented in Figure 9.20. (From Yamazaki, H. et al., *IEEE Photon. Technol. Lett.*, 22(5), 344, 2010.)

Phase modulator   Electrodes   Thermo optics phase equalizers   VOAs

Silica on silicon (14 × 7 mm)   Lithium niobate (64 × 6 mm)   Silica on silicon (34 × 7 mm)

branches of an interferometer). The scheme of the hybrid integrated chip is reported in Figure 9.22 where the Silica on Silicon chips are in white and the Lithium–Niobate chip is in gray.

With the same technology, also other kinds of multilevel modulators has been designed and prototyped, like a 16-QAM modulator [33] and a 4-DPSK modulator [34] demonstrating its flexibility.

9.3.2.3 Multilevel Polarization Modulation (M-PolSK)

This modulation format exploits the property of the optical field to be polarized to switch the field polarization among a certain number of states that represents the transmitted symbols. In order to represent optical field states with the same power and phase, but with different polarization in the quadrature space it is needed to represent the surface defined by the following equations:

$$e_1(t) = \sqrt{P} \cos(\alpha)\cos(\theta)$$

$$e_2(t) = \sqrt{P} \cos(\alpha)\sin(\theta)$$

$$e_3(t) = \sqrt{P} \sin(\alpha)\cos(\theta + \varphi) \qquad (9.20)$$

$$e_4(t) = \sqrt{P} \sin(\alpha)\sin(\theta + \varphi)$$

where the three angles φ and θ define the field polarization. Naturally, the surface is a part of a three-sphere due to the characteristic of the fields described by Equation 9.20 to have the same power. In particular, it is that part corresponding to all the points with the same absolute phase.

Due to the characteristics of this surface, it is not so easy to use quadrature representation to describe polarization modulation.

In optics however, when there is the need for describing the polarization evolution, the Stokes parameters representation is frequently used.

The Stokes parameters are defined as

$$S_1 = e_1^2(t) + e_2^2(t) - e_3^2(t) - e_4^2(t)$$
$$S_2 = 2e_1(t)e_3(t) + 2e_2(t)e_4(t) \tag{9.21}$$
$$S_3 = 2e_1(t)e_4(t) - 2e_3(t)e_2(t)$$

It is easy to verify that, for a monochromatic perfectly polarized beam it is

$$P^2(t) = S_1^2(t) + S_2^2(t) + S_3^2(t) \tag{9.22}$$

so that the SOP of a field can be represented as a point in the so-called Stokes space, which is an Euclidean three dimensional space having the three Stokes parameters as coordinates. In this space, the equal power surface is again a sphere, as in the quadrature space, but the sphere radius in this case is equal to the square of the field power. The sphere of normalized radius (assumed to be equal to one) in the Stokes space is called the Poincare sphere.

The position on the Poincare sphere of the main SOPs are shown in Figure 9.23.

These positions can be derived easily by rewriting the Stokes parameters expressions using spherical coordinates in the Stokes space as follows

$$S_1 = P\cos(2\phi)\cos(2\theta)$$
$$S_2 = P\sin(2\phi)\cos(2\theta) \tag{9.23}$$
$$S_3 = P\sin(2\theta)$$

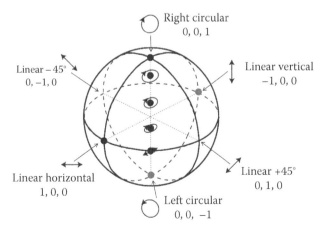

FIGURE 9.23
The Poincare sphere and the position of the main states of polarization.

The angular coordinates in the Stokes space, if defined as in Equation 9.23 and in Figure 9.23 are related to the angles determining the form of the polarization ellipse as shown in Figure 9.24 as it is possible to demonstrate starting from the definition (9.23) of the Stokes parameters.

From (9.23) and the angles definition of Figure 9.24, the position on the Poincare sphere of any SOP can be individuated.

Up to now we have supposed the field completely polarized, as an M-PolSK signal at the output of the modulator.

During propagation however, ASE is added to the field. The noise is randomly distributed among the two polarizations, thus introducing a field depolarization. In this case, the polarization degree can be defined as

$$\sigma_p^2 = \frac{s_1^2(t) + s_2^2(t) + s_3^2(t)}{P^2} < 1 \tag{9.24}$$

Where P is the overall field power and the definition of the Stokes parameters becomes

$$S_1 = P\sigma_p \cos(2\phi)\cos(2\theta)$$
$$S_2 = P\sigma_p \sin(2\phi)\cos(2\theta) \tag{9.25}$$
$$S_3 = P\sigma_p \sin(2\theta)$$

If the light is composed of a fraction of perfectly polarized light and a fraction of perfectly depolarized light (like a PolSK signal plus ASE noise), the power of the polarized component P_p and the power of the unpolarized component P_u are given by

$$P_p(t) = P(t)\sigma_p = \sqrt{S_1^2(t) + S_2^2(t) + S_3^2(t)} \tag{9.26}$$

$$P_u(t) = \left[P(t) - P_p(t)\right] = P(t)(1 - \sigma_p) \tag{9.27}$$

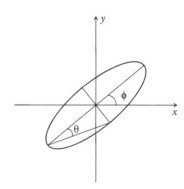

FIGURE 9.24
Relations between the angular coordinates in the Stokes space and the angles defining the polarization ellipse in the physical space.

so that, if the Optical SN coincides with the ratio between the polarized and the unpolarized part of the field, it can be written as

$$SN_o = \frac{\sigma_p}{1 - \sigma_p} \tag{9.28}$$

The general instantaneous multilevel Polarization modulation format (M-PolSK: Multilevel Polarization Shift Keying) is obtained by choosing a constellation on the Poincare sphere and associating to each of the constellation points a transmitted symbol.

We will introduce in the following a Stokes parameters receiver for the M-PolSK [35] that can be demonstrated to be the optimum receiver for this type of modulation format [36]. We will see that, also in this case, what is relevant to evaluate the effectiveness of the constellation is the minimum distance in the Stokes space.

Limiting our analysis to constant power modulation, the problem of determining the best possible constellation can be formulated as the problem of distributing M points on a sphere so that the minimum distance between two of them is maximized.

This problem is not easy to be solved analytically and, as far as I know, there is no closed form solution for a generic number of points. A numerical solution for $M = 4$, 8, and 16 is presented in [37] where both the angular coordinates indicating the position of the points on the Poincarè sphere and the normalized distance matrixes are reported.

The optimum constellations are in a certain sense regular, but do not reproduce the vertices of polyhedrons or other known tridimensional forms. On the other hand, in [36] regular constellations are proposed for M-PolSK.

Considering the vertices of regular polyhedron in the space, the possible M-PolSK formats the minimum distance in the Stokes space and the minimum distance for the optimum configuration with the same number of points are listed in Table 9.3.

TABLE 9.3

Minimum Normalized Distance in the Stokes Space for the Configurations Derived from Regular Polyhedron in the Stokes Space and the Minimum Distance for the Optimum Configuration with the Same Number of Points

PolSK Symbols #	Regular Solid	Minimum Distance, d	Optimum, d_{opt}	Deviation $(d - d_{opt})$%
4	Tetrahedron	$4/\sqrt{6} \approx 1.633$	1.633	0%
6	Octahedron	$\sqrt{2} \approx 1.414$	—	—
8	Cube	$2/\sqrt{3} \approx 1.155$	1.21	4.5%
12	Icosahedron	$\dfrac{4}{\sqrt{10 + 2\sqrt{5}}} \approx 1.051$	—	—
16	—	—	0.86	—
20	Dodecahedron	$\dfrac{4}{\sqrt{3}(1 + \sqrt{5})} \approx 0.716$	—	—
60	Snub dodecahedron[a]	≈ 0.464	—	—
120	Icosidodecahedron	$\dfrac{2}{\sqrt{31 + 12\sqrt{5}}} \approx 0.263$	—	—

[a] There are four Archimedean solids with 60 vertices and the snub Dodecahedron is the best under a minimum distance point of view.

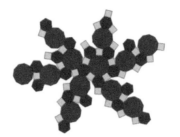

Icosidodecahedron

FIGURE 9.25
Two of the Archimedean solids used to build high order M-PolSK constellations. In the figure, both the plane development and a perspective three dimensional drawing are represented.

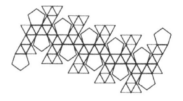

Snub dodecahedron

To construct regular constellations, Pythagorean and Archimedean solids are used as appropriates to achieve a given number of symbols. When several such solids exist with the same number of vertices, the best from the minimum distance point of view is selected [38,39].

Two of the less known regular solids cited in Table 9.3 are represented both in plane development and in tridimensional representation in Figure 9.25.

For the only three cases in which the optimum constellation is approximately known, it is evident that the gain in terms of minimum distance is relatively small. This will bring us to consider from now on the regular configurations listed in Table 9.3.

For each regular constellation, the power spectral density of the transmitted signal along two reference orthogonal polarizations can be evaluated with standard methods and the complete expression is reported in [37].

The way in which the transmitted power is divided among the two polarization components depends obviously on the choice of the polarization reference frame and, if the field at the output of a transmission fiber is considered, on the fiber instantaneous Jones matrix.

However, the power spectral density in any case exhibits a central zone comprised between the first two spectrum zeros, which are always symmetric with respect to the origin, where almost 92% of the power is comprised. The two considered zeros are at $\omega = \pm 2\pi R/M$ thus confirming also for M-PolSK the general rule that the unilateral optical bandwidth can be assumed equal to $B_o = R/M$ so that the spectrum width is $2R/M$.

Under the point of view of modulation feasibility, a simple polarization modulator is needed to produce a PolSK signal. A polarization modulator can be realized using electro-optics materials exactly the same way a phase modulator is realized (see Chapter 5) [36,40].

Polarization modulation can be also implemented in conjunction with differential encoding. In this case, the information is coded into the change of the polarization vector in the Stokes space between two adjacent symbol intervals [41,42].

This further coding does not alter the structure of the transmitter, but for the addition of the differential encoder, and in a few cases is quite useful to simplify the receiver.

9.3.2.4 Multilevel Four Quadrature Amplitude Modulation (M-4QAM)

This modulation format exploits the possibility of modulating all the optical field degrees of freedom. The generic M-4QAM signal can be written as

$$e_1(t) = \sqrt{P_0}\, s_{1,k} \eta_T(t,k)$$
$$e_2(t) = \sqrt{P_0}\, s_{2,k} \eta_T(t,k)$$
$$e_3(t) = \sqrt{P_0}\, s_{3,k} \eta_T(t,k) \tag{9.29}$$
$$e_4(t) = \sqrt{P_0}\, s_{4,k} \eta_T(t,k)$$

where $s_{j,k}$ is the amplitude of the symbol transmitted along the jth quadrature in the kth symbol interval.

A set of constellations for M-4QAM can be obtained simply replying on the two coordinates couples corresponding to the two polarizations the reference constellations of M-QAM.

It is not difficult to demonstrate that this procedure does not produce the optimum constellations from a minimum distance point of view, but is useful to simplify the modulator, that in this case is really a key issue.

As a matter of fact, if these configurations are chosen, the M-4QAM signal can be obtained by a modulator with the block scheme reported in Figure 9.26, which is constituted by two parallel M–QAM modulators.

Since M-QAM modulators can be built as in Figure 9.22, the same technology can be used for M–4QAM modulators.

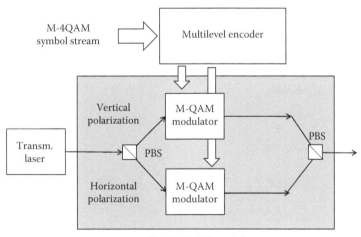

Dual M-QAM modulator

FIGURE 9.26
Block scheme of a M-4QAM modulator composed of two parallel M-QAM modulators. This modulator scheme is mainly suitable for modulation formats where the constellation is the combination of two squared QAM constellations.

If such regular constellations are chosen, the normalized amplitudes $s_{j,k}$ of the four quadratures are extracted from four sets X_j whose elements can be expressed as a function of four integers numbers L_j and four indices h_j, as follows:

$$s_{j,k} \in X_j \equiv \{(2h_j - L_j - 1) : h_j = 1, 2, \ldots, L_j\} \quad j = 1, 2, 3, 4 \tag{9.30}$$

It is clear that, from (9.30) it derives

$$M = \prod_{j=1}^{4} L_j \tag{9.31}$$

and that the ratio between the minimum distance d and the transmitted power P of a M-4QAM constellation is given by

$$\sqrt{P} = d \sqrt{\left(\frac{L_1 - 1}{2}\right)^2 + \left(\frac{L_2 - 1}{2}\right)^2 + \left(\frac{L_3 - 1}{2}\right)^2 + \left(\frac{L_4 - 1}{2}\right)^2} \tag{9.32}$$

A few key parameters of some M-4QAM formats of this class are summarized in Table 9.4. For comparison, in the table data relative to M-QAM are also reported. It results that using M-4QAM is always advantageous, but the gain gets smaller and smaller while M increases if squared configurations are used. Just as an example, for $M = 16$ the gain of squared 16-4QAM versus 16-QAM is 3.2 dB while it is reduced to 0.7 dB if $M = 256$.

If it is important to gain as much as possible from the constellation points distribution a better constellation has to be selected.

A great amount of work has been done by mathematicians to study forms and properties of four dimensional constellations [43,44].

TABLE 9.4

Minimum Normalized Distance in the Quadrature Space for Squared M-4QAM Constellations

(L_1, L_2, L_3, L_4)	Modulation Format	Minimum Distance (Normalized to $\sqrt{P}$)	M-QAM
$(2, 2, 1, 1)$	4-4QAM	$\sqrt{2} \approx 1.14$	$\sqrt{2} \approx 1.14$
$(2, 2, 2, 1)$	8-4QAM	$\dfrac{2}{\sqrt{3}} = 1.155$	$\dfrac{1}{\sqrt{2}} \approx 0.707$
$(2, 2, 2, 2)$	16-4QAM	1	$\dfrac{\sqrt{2}}{3} \approx 0.471$
$(4, 2, 2, 2)$	32-4QAM	$\dfrac{1}{\sqrt{13}} \approx 0.277$	$\sqrt{\dfrac{2}{29}} \approx 0.263$
$(4, 2, 4, 2)$	64-QAM	$\dfrac{1}{\sqrt{25}} \approx 0.2$	$\dfrac{\sqrt{2}}{7} = 0.202$
$(4, 4, 4, 2)$	128-4QAM	$\dfrac{1}{\sqrt{62}} = 0.127$	$\dfrac{2}{\sqrt{335}} = 0.111$
$(4, 4, 4, 4)$	256-4QAM	$\dfrac{1}{9} = 0.111$	$\dfrac{\sqrt{2}}{15} = 0.094$

M-QAM distances are reported for comparison.

In the particular case of constant power modulation we can use to define the M-4QAM constellations the vertices of convex, regular polyhedrons in four dimensions, as defined in [45]. Table 9.5 reproduces the structure of Table 9.4, but using the new M-4QAM constellations.

Even better results can be achieved with a careful optimization of more complex constellations [46,47].

It is out of the scope of this book to discuss in detail all the implications of the optimum choice of a four dimensional constellation, here we simply limit ourselves to a few examples based on a simple principle: creating a dense lattice of points via the Cartesian product of bidimensional QAM constellations and then select a subset of points for the final four dimensional constellation.

We will use a simple rule to construct our constellation as the Cartesian product of partitioned QAM constellations in which points are alternatively taken and discarded. An example of this kind of constellation is reported in Figure 9.27, where the generic M-QAM constellation is divided into two partitions: gray points that are taken as components in the (e_1, e_2) plane and black points that are taken as components in the (e_3, e_4) plane.

TABLE 9.5

Minimum Normalized Distance of Constant Power M-4QAM Constellations Obtained from the Vertices of Four Dimensional Regular Polyhedrons

Polyhedron	Modulation Format	Minimum Distance (Normalized to $\sqrt{P}$)	M-QAM
16 Tetrahedral	8-4QAM	$\sqrt{2} \approx 1.414$	$\dfrac{1}{\sqrt{2}} \approx 0.707$
8 Hexahedral	16-4QAM	1	$\dfrac{\sqrt{2}}{3} \approx 0.471$
24 Octahedral	24-4QAM	1	
600 Tetrahedral	120-4QAM	$\dfrac{2}{\sqrt{5}+1} \approx 0.618$	
120 Dodecahedral	600-4QAM	$\dfrac{\sqrt{2}}{\sqrt{5}+3} \approx 0.270$	

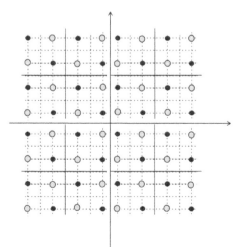

FIGURE 9.27

Partition of a square M-QAM constellation to form by Cartesian product an efficient M-4QAM constellation. Gray and black points individuates the two partitions used to build the final constellation.

TABLE 9.6

Minimum Normalized Distance in the Quadrature Space for
M-4QAM Constellations Obtained by the Cartesian Product of
Regular Partitions of Squared QAM Constellations

QAM Sub-Set Number of Points	Modulation Format	Minimum Distance (Normalized to $\sqrt{P}$)	M-QAM
8	64-4QAM	0.4	0.202
18	324-4QAM	0.329	
32	1024-4QAM	0.286	0.046

The minimum distances relative to the constellations that are built with this algorithm are reported in Table 9.6.

The results reported in Tables 9.4 through 9.6, besides the outcome of Equation 9.18 for M-DPSK are compared in Figure 9.28. An interesting observation emerges from this figure, which is useful to guide the decision about the number of levels to select when designing a multilevel system based on quadrature modulation. Increasing the number of available degrees of freedom passing from M-QAM to constant power M-4QAM to full M-4QAM is effective in maintaining an high distance among constellation points at high number of levels, while if the number of levels is low, a constellation with a lesser number of degrees of freedom can perform sufficiently well.

9.3.3 Multilevel Modulation Receivers

Since all possible optical field modulation formats can be seen as modulations of one or more of the field degrees of freedom, a receiver that is able to detect all the field quadratures is a universal receiver for optical modulated fields.

Since the quadratures have to be detected with their own sign, to distinguish points in similar positions, but in different quadrants in the quadrature space, coherent detection is needed. This is performed by beating the incoming signals with a local laser (called local oscillator [LO]) at a slightly different wavelength to obtain an intermediate frequency signal that can be processed with standard intermediate frequency (IF) electronics.

This has also the advantage of allowing very effective electronic compensation of fiber propagation effects, first of all polarization fluctuations.

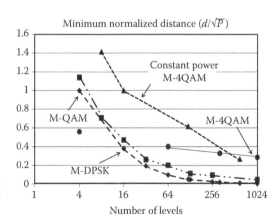

FIGURE 9.28
Comparison on different modulation formats based on quadrature modulation on the ground of the minimum distance between couples of points.

Naturally, this will not be the optimal receiver for all the cases, and sometimes, better performances could be achieved with receivers tailored specifically for the considered modulation format.

This is the case of both M-DPSK and M-PolSK, where better receivers can be designed with respect to the universal receiver.

On the other hand, the quadrature receiver can be demonstrated to be the optimum receiver for both M-4QAM and M-QAM modulation formats, independently from the chosen constellation [28].

Thus, in this section, we will first describe the universal quadrature receiver and then specific optimal receivers for M-DPSK and for M-PolSK.

9.3.3.1 Four Quadrature Receiver

The block scheme of the four quadrature receiver is reported in Figure 9.29. The scope of the receiver is to detect the transmitted quadrature vector for each symbol interval and compare it with the constellation in the symbol space to produce an estimate of the transmitted symbol.

The receiver is divided into two sections: an optical front end (depicted in Figure 9.29a), whose scope is to detect the received field quadratures, and an electronic processing stage (whose architecture is reported in Figure 9.29b), whose scope is to invert the effect of fiber propagation to reconstruct the transmitted field.

A key component in the optical front end is the $\pi/2$ optical hybrid that is used, after the separation of the incoming field linear polarization components, both to insulate the quadratures of each polarization and to combine the LO with the incoming field before balanced detection.

Optical hybrids has been produced with several technologies [48–50] and their performances are quite good and stable.

The four fields $\vec{E}_j$ ($j = 1, 2, 3, 4$) obtained after polarization splitting and branching with $\pi/2$ hybrids can be written as

$$\vec{E}_j(t) = [\xi_j(t) + \eta_j(t)\vec{\kappa}(t) \cdot \vec{\varsigma}_j]\vec{\varsigma}_j \tag{9.33}$$

where the information bearing quadratures are indicated with $\xi_j(t)$ and the ASE optical noise quadratures with $\eta_j(t)$. The linear polarization unitary polarization vectors directed along the quadrature space reference axes are indicated with $\vec{\varsigma}_j$ ($j = 1,2,3,4$) while the ASE polarization vector, that is a fast function of time due to ASE complete depolarization, is indicated with $\vec{\kappa}(t)$. Finally "." indicates the scalar product.

The field emitted by the LO at $\pi/4$ with respect to the local receiver axis is divided into its polarization components by a polarization beam splitter and coupled with the received field components by a set of four $\pi/2$ optical hybrids.

In a practical receiver, the set of hybrids can be substituted by a single monolithic hybrid realized in silica on silicon technology. A possible architecture for such a hybrid is shown in Figure 9.30.

Also the set of eight PIN Photodiodes that are needed to realize four balanced receivers can be realized with a diode bar in a single package so that the cost is not eight times the cost of a PIN diode but, for high volumes, could be less than two times.

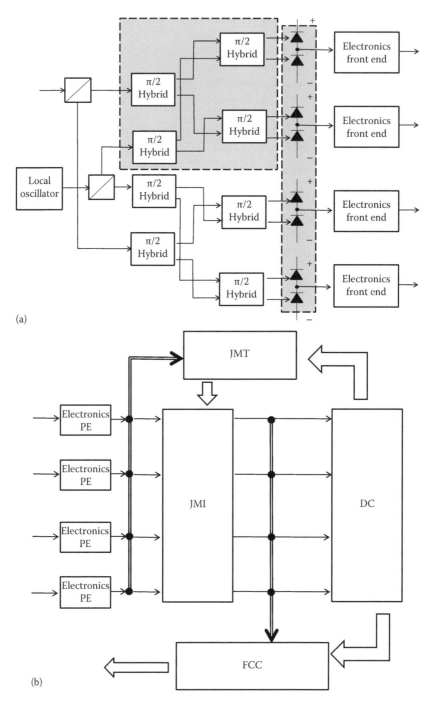

(a)

(b)

FIGURE 9.29

Block scheme of the quadrature coherent receiver: (a) optical front end; (b) electronic processing. In the optical front end, the gray part can be integrated in an integrated optics circuit: a monolithic Si-SiO$_2$ chip for the hybrids and an InP chip the photodiodes array.

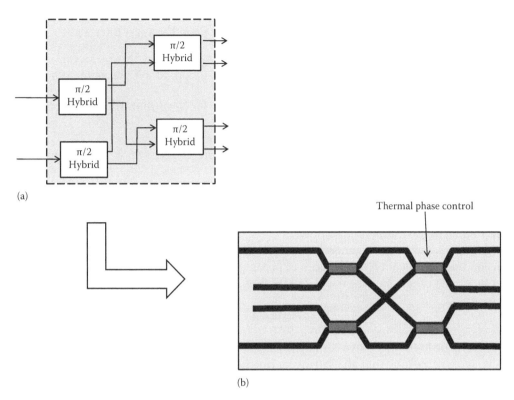

(a)

Thermal phase control

(b)

FIGURE 9.30
Schematic layout of the hybrids Si-SiO$_2$ chip. (a) Represents the functional scheme of the chip and (b) a simplified layout.

At the exit of the balanced receivers, the four photocurrents are modulated at the intermediate frequency $\Delta\omega$ and further electronic demodulation allows the four field quadratures to be derived.

In order to allow this operation to be performed with high performances, the intermediate frequency $\Delta\omega$ has to be stabilized. This can be achieved by a frequency lock loop driven by the wavelength of the incoming signal [31,51].

If the LO frequency is exactly locked to the frequency of the incoming signal the IF demodulation is not needed and the photocurrents are directly in baseband. In this case, the coherent receiver is called homodyne [31].

Under a processing point of view, heterodyne receiver is more complex, but it has been used in radio frequency systems for a long time to alleviate the difficulty of designing wideband components with a very low low-pass frequency (ideally working up to continuous wave [CW]). In optical receivers, heterodyne detection was proposed for the same reason when coherent receivers was first proposed.

With the progress of electronics and with the possibility of digital design also for analog functions, this difficulty is no more so important and the simple homodyne structure seems preferable.

For this reason, up to now we will assume $\Delta\omega = 0$. In this case, the electronics front end reduces to a baseband filter fixing the electrical bandwidth $B_e \approx R_S$ of the receiver, R_S being the symbol rate, not differently from the case of a direct detection receiver.

The photocurrents in the homodyne receiver have the following expressions:

$$c_j(t) = 2RP_L[\xi_j(t) + \eta_j(t)] + RP_L^2 + R[\xi_j(t) + \eta_j(t)]^2 + n_j(t) + \rho_j(t) \qquad (9.34)$$

where
  R is the photodiodes responsivity, which is assumed to be the same for all the photodiodes
  P_L is the LO power, which can be written as $P_L(t) = P_{L0}[1 + r(t)]$ if the LO RIN $r(t)$ has to be considered (see Chapter 5)
  $n_j(t)$ is the shot noise component coming from the jth receiver
  $\rho_j(t)$ is the thermal noise component coming from the jth receiver

If the coherent receiver has to work with the best performances, the LO power has to be much higher both of the incoming signal power and of the so-called "thermal" power. The thermal power is defined as the optical power that would cause a photocurrent with a power equal to the thermal noise power.

The second condition is generally easily realized, while the first could impose limitations on the amplification of the receiver preamplifier.

In these conditions, Equation 9.34 can be simplified as

$$c_j(t) = 2RP_L[\xi_j(t) + \eta_j(t)] + n_j(t) \qquad (9.35)$$

Moreover, due to the fact that a complex electronic processing has to be performed after filtering, we will assume that all electronics functions will be performed via digital circuits. This requires a sampling of the baseband signal before filtering at a sufficiently high sampling rate. On the ground of the experience in electronic compensation, we will assume that five samples per pulse are sufficient to allow effective electronic processing and we will assume a sampling rate $R_e \approx 5R_S$.

This means that in order to work at 100 Gbit/s with a 64-4QAM, the sampling rate has to be about 83 GHz, a very challenging speed, but probably within the reach of very high speed complementary metal–oxide–semiconductor (C-MOS) 22 nm application-specific integrated circuits (ASICS).

Increasing the number of levels, in order to transmit 100 Gbit/s using 512-4QAM with constant power (obtained by selecting 512 levels from the 600-4QAM reported in Table 9.4), the required sampling rate goes down to about 55 GHz, a speed of the order of that used in experimental systems with electronic nonlinear precompensation (experiments with sampling rates around 50 GHz are reported as an example in Chapter 6).

After baseband filtering, the four photocurrents are proportional to the quadratures of the field arriving at the receiver, but for the shot noise contribution.

A part the ASE noise, the quadratures of the received field, is related to the transmitted field quadratures by the fiber propagation characteristics causing chromatic dispersion, polarization rotation and PMD, and nonlinear distortions.

Let us assume in a first moment that fiber propagation is perfectly linear.

The role of the electronic postdetection processing is to compensate as much as possible linear propagation effects to recover the transmitted signal quadratures, to drive the frequency lock of the LO, and to estimate the transmitted symbol.

These functions can be recognized in the scheme of Figure 9.29b, where fiber-induced polarization rotations are compensated by multiplying the received quadratures by the inverse of the fiber polarization rotation matrix (the Jones matrix) in the Jones Matrix Inversion (JMI) block, and pulse distortion by residual chromatic dispersion and PMD are compensated via the Pulse Equalizer (PE).

The Jones matrix evolution at the first order (i.e., without the ω dependence that causes PMD) is tracked by the Jones Matrix Tracking (JMT) block, while the frequency lock loop driving signal is generated by the Frequency Control Circuit (FCC); finally the transmitted symbol is estimated by the Decision Circuit (DC).

The key elements of the electronic section of the receiver are the JMT and the FCC; electronic techniques to compensate linear pulse distortions are already reviewed in Chapter 6, and nothing more has to be added here for this particular case, while Decision Strategies in the general case of a multilevel signal will be discussed in Section 9.4.

9.3.3.1.1 *Jones Matrix Tracking Circuit*

Since propagation is assumed linear, Equation 4.8 describes the fiber propagation so that the transmitted and the received quadratures are related by the following equation:

$$\begin{bmatrix} \zeta_1(\omega) + i\zeta_2(\omega) \\ \zeta_3(\omega) + i\zeta_4(\omega) \end{bmatrix} = e^{i\delta(\omega)} \begin{bmatrix} m_{xx}(L,\omega) & m_{xy}(L,\omega) \\ m_{yx}(L,\omega) & m_{yy}(L,\omega) \end{bmatrix} \begin{bmatrix} e_1(\omega) + ie_2(\omega) \\ e_3(\omega) + ie_4(\omega) \end{bmatrix} + \begin{bmatrix} \eta_1 + i\eta_2 \\ \eta_3 + i\eta_4 \end{bmatrix} \tag{9.36}$$

where the fiber link length has been indicated with L and the residual dispersion after the online dispersion compensation is indicated with $\delta(\omega)$.

The elements $m_{i,j}(L, \omega)$ of the Jones matrix are random complex numbers that fluctuates very slowly with respect to the modulation rate and that are related one with the other by the property of the Jones matrix to be unitary. Exploiting the property that the square norm of an optical field is invariant after a transformation with the matrix **M**, that is, the optical energy is invariant to polarization rotations, it is easy to demonstrate that **M** depends only on three independent parameters and that it can be rewritten as follows (35):

$$\mathbf{M} = \begin{vmatrix} m_{xx}(L,\omega) & m_{xy}(L,\omega) \\ m_{yx}(L,\omega) & m_{yy}(L,\omega) \end{vmatrix} = \begin{bmatrix} \varepsilon e^{i\vartheta} & \sqrt{1-\varepsilon^2}\,e^{i\varphi} \\ -\sqrt{1-\varepsilon^2}\,e^{-i\varphi} & \varepsilon e^{-i\vartheta} \end{bmatrix} \tag{9.37}$$

$$= \begin{vmatrix} \cos(\alpha)e^{i\vartheta} & \sin(\alpha)e^{i\varphi} \\ -\sin(\alpha)e^{-i\varphi} & \cos(\alpha)e^{-i\vartheta} \end{vmatrix}$$

The third expression of the Jones matrix, depending on three independent angles $(\alpha, \vartheta, \varphi)$ clearly express the fact that linear polarization evolution can be represented as a rotation in the quadrature space.

The polarization rotation compensation algorithm implemented in the block JMI simply multiplies the input filed quadrature vector by the inverse of **M** (i.e., equal to $\mathbf{M}^+$ since it is a unitary matrix). The first estimate of the matrix **M** is performed during the system initialization phase by transmitting a field periodically switching among two orthogonal polarizations and observing the field at the output of the receiver.

Once the Jones matrix estimate is initialized, it is updated every T_U seconds by the algorithm implemented in the block JMT.

The update interval is also determined in the initialization phase and it has to be at least 10 times shorter than the typical polarization fluctuations time T_p and at least $10^6\,M$ times longer than the symbol interval.

The interval T_p is defined as the minimum time interval needed to change at least one of the output quadratures of 1% when the field at the system input is constant.

The requirement $T_U > 10^6 M$ assures that, since the symbols are assumed statistically independent and equally probable, averaging over this interval the quadratures of the received field, the obtained value has a Gaussian distribution with a standard deviation about three orders of magnitude smaller that the average [52].

The requirement $T_p > 10\,T_U$ assures that the fluctuations of $\mathbf{M}$ are tracked by the algorithm with a small tracking error.

In real 100 Gbit/s transmission systems, the symbol interval is generally smaller than 100 ps, while T_p ranges from hundreds of milliseconds to minutes [53–55] depending on the type and collocation of the fiber cable, on the season, and similar factors.

An update time T_U of the order of 100 μs is thus suitable to assure independency from the modulation and very precise JMT.

The update algorithm works as follows. The constellation points are divided in three categories C_1, C_2, and C_3 constituted by almost the same number of points.

At every point of the category C_j (j = 1, 2, 3) let it call $P_{j,i}$ (i = 1, …, n_1) a rotation matrix is associated, let it call $\mathbf{R}_{j,i}$, which rotates the reference frame in the quadrature space to transform the coordinates of $P_{j,i}$ in the reference point Π_j.

The coordinates of the reference points are

$$\Pi_1 = (1,0,0,0)$$

$$\Pi_2 = (0,1,0,0) \tag{9.38}$$

$$\Pi_3 = (0,0,1,0)$$

These matrixes depend only on the chosen constellation and are stored in a look-up table.

Every time a symbol is received, a quadrature vector $[\zeta_1, \zeta_2, \zeta_3, \zeta_4]$ is present at the input of JMT. In correspondence to these quadratures, the DC produces an estimate of the transmitted symbol and passes it to the JMT. The JMT uses this estimate to associate to the received quadrature vector a matrix $\mathbf{R}_{k,i}$.

At this point the JMT circuit multiplies the vector $\mathbf{X} = [\zeta_1, \zeta_2, \zeta_3, \zeta_4]$ by $\mathbf{R}_{k,i}$ obtaining an estimate of the position of the reference point Π_k.

These estimates are averaged in the time T_U to eliminate almost completely the effect of the ASE noise and of the modulation. An average estimate of the positions of the reference vectors Π_k is thus obtained. This estimate, that we call $\langle \mathbf{RX} \rangle_k$, depends on the differential rotation happened in the preceding updated interval during fiber propagation.

At this point, the inverse Jones matrix update, which coincides with the differential rotation of the quadrature space, can be obtained solving the following linear system:

$$\begin{pmatrix} \Pi_1 \\ \Pi_2 \\ \Pi_3 \end{pmatrix} = \begin{pmatrix} \Delta\mathbf{M}^+ & \mathbf{0} & \mathbf{0} \\ \mathbf{0} & \Delta\mathbf{M}^+ & \mathbf{0} \\ \mathbf{0} & \mathbf{0} & \Delta\mathbf{M}^+ \end{pmatrix} \begin{pmatrix} \langle \mathbf{RX} \rangle_1 \\ \langle \mathbf{RX} \rangle_2 \\ \langle \mathbf{RX} \rangle_3 \end{pmatrix} \tag{9.39}$$

where
$\Delta\mathbf{M}^+$ is a unitary 4 × 4 real matrix
$\mathbf{0}$ is the 4 × 4 null matrix
$\langle \mathbf{RX} \rangle_k$ and Π_k are four components vector in the quadrature space.

System (9.39) is generally largely redundant, since only three out of the 16 parameters of the matrix $\Delta\mathbf{M}^+$ are independent variables, but it is necessary to consider three reference points (and not only two as indicated in [56]) to manage the degenerate case of a four dimensional rotation that leaves the plane containing two reference points unchanged in the quadrature space.

This particular case arises from the fact that differential rotations in a four dimensional space cannot be described as rotations around an axis, as it happens in the three dimensional case. Rotation around an axis is an idea fostered by our experience, but it is only a coincidence that any rotation in three-space can be determined by an axis in three-space.

For example, let us consider the idea of rotation in a two dimensional space. The "rotation axis" seems perpendicular to the space where the rotation happens: it is not even contained in it, and we can think about it only because we continuously image our plane immerged in a three dimensional space.

Rotations, whatever the space dimensionality is, are more properly thought as parallel to a plane. This way of thinking is consistent with two dimensions (where there is only one such plane), three dimensions (where each rotation "axis" defines the rotation plane by coinciding with the normal vector to that plane), and with four dimensions.

Defining rotations in this way evidences also the fact that, since in a four dimensional space there is a plane completely orthogonal to the rotation plane, the points on this plane are not affected by the rotation, so that, coming back to our case, if both our reference points belongs to that plane they are not able to individuate the rotation matrix.

Thus, it is not possible to eliminate a priori some equations from the system (9.39); it has to be simplified every time isolating three independent equations and solving them to obtain the parameters of the matrix ΔM^+. Once evaluated, this matrix is added to the running estimate of the inverse Jones matrix without any disruption of the detection process.

9.3.3.1.2 Local Oscillator Frequency Lock

The scope of the LO control loop is to stabilize the LO frequency to maintain the photocurrent in baseband while the wavelength of the transmitting laser fluctuates.

Different systems have been devised to perform this operation, but the greater number of them can be divided into two categories: electronically driven and injection locking based.

In the first case, the LO is tuned to track the incoming signal via a decision driven electronic feedback circuit that modulates the laser current, in the second case, part of the incoming optical signal, in case suitably filtered, is directly injected into the LO.

The scheme of an electronic LO circuit is reported in Figure 9.31, where it is evidenced that the loop has to be driven by a signal somehow related to the frequency different from the LO and the signal.

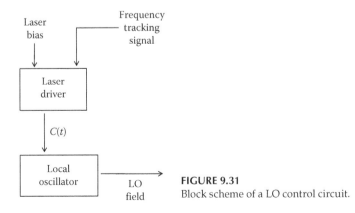

FIGURE 9.31
Block scheme of a LO control circuit.

This signal is derived in the block FCC of Figure 9.29b. As a matter of fact, modulation can be interpreted as a rotation of the quadrature space that associates to each symbol represented by the point P_k a rotation matrix $\mathbf{Y}_k$ that transforms the reference point $(1, 0, 0, 0)$ in P_k. The matrixes $\mathbf{Y}_k$ depends only on the used constellations and can be stored in a look-up table.

In the presence of a small frequency difference between the signal and the LO, the signals after JMI can be written as

$$c_1(t_j) = R[e_1(t_j) + \eta_1(t_j)]\cos(\Delta\omega t_j) + R[e_2(t_j) + \eta_2(t_j)]\sin(\Delta\omega t_j) + n_1(t_j)$$

$$c_2(t_j) = R[e_2(t_j) + \eta_2(t_j)]\cos(\Delta\omega t_j) - R[e_1(t_j) + \eta_1(t_j)]\sin(\Delta\omega t_j) + n_2(t_j)$$

$$c_3(t_j) = R[e_3(t_j) + \eta_3(t_j)]\cos(\Delta\omega t_j) + R[e_4(t_j) + \eta_4(t_j)]\sin(\Delta\omega t_j) + n_3(t_j)$$

$$c_4(t_j) = R[e_4(t_j) + \eta_4(t_j)]\cos(\Delta\omega t_j) - R[e_3(t_j) + \eta_3(t_j)]\sin(\Delta\omega t_j) + n_4(t_j)$$

$$(9.40)$$

where t_j indicates the discrete time after sampling in the electronics front-end.

The four dimensional vector whose coordinates are $c_k(t_j)$ is multiplied in each time interval by the $\mathbf{Y}_k^+$ thus obtaining only noise on the third and fourth components, while the first and second components contains the frequency deviation. Adding $c_1'(t_j)$ and $c_2'(t_j)$,

$$c_D(t_j) = \sqrt{2}R\cos\left(\Delta\omega t_j - \frac{\pi}{4}\right) + n_D(t_j) \tag{9.41}$$

is obtained, where all the noise terms are collected into $n_D(t_j)$ that results, with a very good approximation, a white Gaussian noise.

This function is now independent from the transmitted signal and it can be used to drive the frequency tracking circuit.

9.3.3.2 M-DPSK Optimum Receiver

Although the quadrature receiver can work with M-DPSK signals once a suitable decision rule is implemented, it is not the optimum receiver, which is the receiver assuring the minimum error probability in the presence of optical noise.

Moreover, the quadrature receiver is very complicated in architecture and requires high speed electronics, with all the consequences in terms of costs and power consumption.

Thus, it is also practically very important to design a simpler receiver for M-DPSK signals.

The receiver scheme is shown in Figure 9.32 and is the analogous of the binary DPSK receiver shown in Chapter 6, but for the fact that to better use the received power a balanced configuration has been adopted.

The idea is to use the signal as a sort of LO by delaying it of a symbol interval and leveraging on the differential encoding at the transmitter.

If the transmitted signal has the form (9.16), at the receiver the signal writes as

$$\vec{E}(t, k) = \sqrt{P}e^{i\theta(k)}\eta_T \vec{\epsilon} + \eta_T \vec{\eta}_{ASE}(t) \tag{9.42}$$

where
  k indicates the considered symbol interval
  $\theta(k)$ depends on the symbol transmitted in the kth symbol interval
  $\vec{\epsilon}$ is the normalized field polarization vector

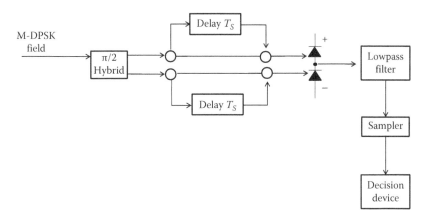

FIGURE 9.32
Block scheme of a balanced M-DPSK receiver. (From Yamazaki, H. et al., *Photon. Technol. Lett.*, 22(5), 2010.)

the ASE noise contribution is represented with $\vec{\eta}_{ASE}(t)$ (while it is to remember that η_T represents the function equal to one in the symbol interval and zero elsewhere)

The field on the upper branch of the receiver, after splitting, delay, and recombination, writes as

$$\vec{E}(t,k) = \sqrt{P}\alpha\left\{e^{i\theta(k)} + e^{-i\theta(k-1)}\right\}\eta_T\,\vec{e} + \alpha\eta_T\vec{\eta}'_{ASE}(t) \tag{9.43}$$

where α is the optical chain loss, and an analogous expression holds for the field in the lower branch and

$$\vec{\eta}'_{ASE}(t) = \eta_{ASE}(t) - \vec{\eta}_{ASE}(t - T_S) \tag{9.44}$$

Detecting the fields with a balanced receiver gives the following photocurrent

$$c(t) = RP\alpha^2\cos\left(2\pi\frac{j}{M}\right)\eta_T + 2R\sqrt{P}\alpha^2\eta_T Re\left\{\left[e^{i\theta(k)} + e^{-i\theta(k-1)}\right]\vec{\eta}'_{ASE}\right\} + \alpha^2\left|\vec{\eta}'_{ASE}(t)\right|^2\eta_T + n_s(t) \tag{9.45}$$

where $n_s(t)$ is the shot noise term.

In equation (9.45) the symbol transmitted in the k-th symbol interval is represented by j, that determines the phase of the cosine function. The photocurrent is then sampled after low pass filtering and the samples are sent to the decision device.

The receiver we have described is much simpler than the whole quadrature receiver: no LO and no polarization tracking are needed due to the use of signal self-beating.

This is one of the reasons that qualify M-DPSK modulation as a good candidate for long haul 100 Gbit/s systems, where limiting the system cost is frequently much more important to push performances up to the limit.

9.3.3.3 M-PolSK Receivers

As in the case of M-DPSK, in the case of polarization modulation also, the quadrature receiver is not optimum and a direct detection receiver offering better performances can be designed [37,36].

The possibility to perform direct detection also contributes to simplify the receiver structure eliminating the need of the LO, but polarization tracking is needed since the transmitted information is coded in the polarization itself.

The idea on which optimum receivers, both coherent and direct detection, are designed is to detect the Stokes parameters to be able to operate processing and decision in the Stokes space.

The block scheme of the optimum M-PolSK receiver is presented in Figure 9.33, and it is composed of a front end producing four photocurrents that are proportional to the Stokes parameters and by an electronic processing section that is in charge to compensate polarization fluctuations, in the coherent case to drive the LO frequency locking and to perform symbol decision.

The block scheme of the coherent front end is reported in Figure 9.34; it is a combination of an heterodyne optical receiver and of square law type electronic demodulation that allows the Stokes parameters of the incoming field to be detected.

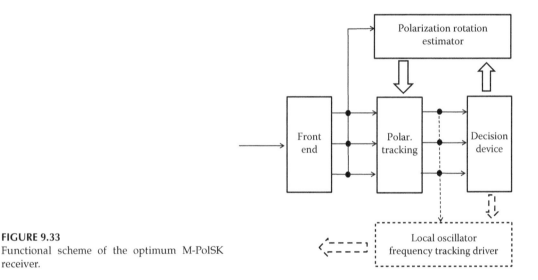

FIGURE 9.33
Functional scheme of the optimum M-PolSK receiver.

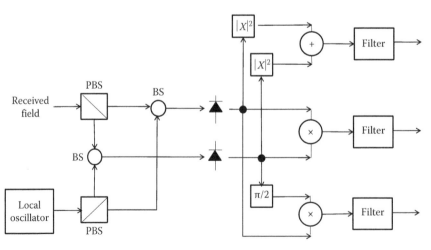

FIGURE 9.34
Coherent front end of a PolSK receiver; BS, Beam Splitters; PBS, Polarization Beam Splitters.

As far as Direct Detection is concerned, implementing an optical front end that is able to detect the Stokes parameters is equivalent to search the structure of a direct detection polarimeter. Several such polarimeters are used in practice, based on different principles, and the optical front end scheme of choice depends essentially on practical considerations.

An example of direct detection front end is reported in Figure 9.35: this structure is based on the analysis of the incoming field by means of quarter and half wave plates and polarization beam splitters. The input electrical field is divided by means of a polarization beam splitter into its linear polarization components and then these are combined by means of a couple of standard beam splitters of splitting ratio ε_1, a couple of standard beam splitters of splitting ratio ε_2, and a couple of polarization beam splitters as shown in the figure. If the optical path is accurately controlled, the two polarization components are combined in phase (technically means with the same phase) before detection on the second photodiode and in quadrature on the third one.

This structure is difficult to be realized via discrete components, due to the need of a very accurate control of the length of the light paths. On the other hand, it is quite suitable to be implemented by hybrid integration, where all the splitting structure shown in gray in the figure is integrated in a single silicon on silica chip. In this case, the phase control can be implemented by thermal phase controller and the structure results quite stable.

At the output of the interferometric polarimeter, the photocurrents results to be a linear combination of the Stokes parameters, thus inverting this relation, an estimate of the Stokes parameters is obtained.

Another interesting example of direct detection polarimeter is reported in [57] and it is based on the analysis of the incoming field by means of half and quarter wave plates. This structure does not require optical path length control and it is promising for discrete components implementation.

Finally, a polarimeter can be designed based on the fact that partially reflecting mirrors reflects the orthogonal linear polarization in different ways. Its schematic block diagram is shown in Figure 9.36.

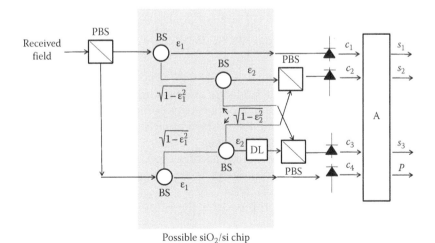

Possible siO$_2$/si chip

FIGURE 9.35
Direct Detection front end of a PolSK receiver based on half and quarter wave plates: BS indicates unbalanced Beam Splitters; PBS, Polarization Beam Splitters; DL, Delay Line; A is the matrix relating the photocurrents to the Stokes parameters.

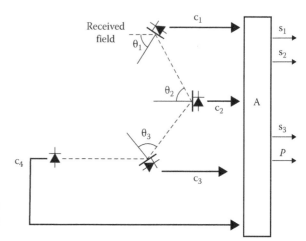

FIGURE 9.36
Direct Detection front end of a PolSK receiver based on partially reflective coating of four photodiodes.

Three photodetectors are coated with a reflection coating characterized by reflection parameters ψ_k, ρ_k, and κ_k ($k = 1, 2, 3$) while the fourth photodetector is antireflection coated so that all the optical power incident on it is detected. In particular, ρ_k indicates the reflectance of the surface for circular or unpolarized light, $\kappa_k\,exp(i\psi_k)$ is the ratio between the complex reflection coefficient for linearly polarized light along parallel and perpendicular directions to the local incidence plane.

In all the cases we have considered up to now, a direct detection polarimeter produces four photocurrents that changes while the polarization of the received field changes. It is possible to demonstrate that in all the cases we have considered, there is a linear relation between the Stokes parameter and the currents at the output of the polarimeter that does not depend on the received polarization, thus it is possible to characterize the polarimeter with a matrix A so that, if the four currents at the polarimeter output are collected in a vector C, we can write [58,59]

$$\begin{pmatrix} S_1 \\ S_2 \\ S_3 \end{pmatrix} = AC \qquad (9.46)$$

Since the current at the output of the electrical demodulation in the coherent receiver shown in Figure 9.34 is also proportional to the Stokes parameters and the proportionality is a particular type of linear dependence, we will assume Equation 9.46 as valid for any type of optical front end in an M-PolSK receiver.

As far as the electrical processing section is concerned, it is in charge to perform both fiber polarization tracking and, in the coherent case, of generating the driving signal for the frequency lock of the LO.

Different algorithms have been proposed for these functions and we will not review them in detail.

Some algorithms use the same ideas we explained dealing with the quadrature receiver. In this case, the modulation is eliminated by using a decision driven algorithm that inverts the modulator operation starting from the estimate of the transmitted symbol [35].

Other algorithms use the idea that the constellation can be designed so that the average value of the Stokes vector over the entire constellation is not the null vector.

In this case, the modulation can be eliminated by simply averaging the modulated field over a time long with respect to the symbol time: the result is an average field whose polarization fluctuations are only due to the fiber propagation.

In this case, simply observing the variations of such an average vector, the fiber-induced polarization rotations can be compensated [35].

The last class of algorithms has the very important advantage not to require electronics speed comparable to the symbol rate, but it is not applicable to all the constellations, requiring an average Stokes vector different from zero.

9.3.4 Ideal Performances of Multilevel Systems

In this paragraph, we will compare different multilevel modulation formats considering only the limitation deriving from ASE. This approach is justified by the fact that in optically amplified systems, generally an optical preamplifier is placed in front of the receiver, so that the received power is sufficiently high to neglect both receiver thermal noise and shot noise.

Naturally, in real systems, several other elements thwart the system performances; however, a first approach based only on noise consideration is quite useful to have a first comparison among different systems.

As in the case of binary systems, also the performances of multilevel systems are evaluated considering the bit error probability. As a matter of fact, in real applications, the original message is almost always binary and multilevel coding is used only to ease the transmission.

Let us assume, as usual, that errors are independent from the transmitted symbol and from the pattern transmitted in the previous symbol intervals. The bit error probability P_e and the symbol error probability P_S are related by the equation

$$P_e = \frac{\langle e \rangle}{\log_2 M} P_S \tag{9.47}$$

The average number of wrong bits <e> consequent to a symbol error depends on the coding, that is, on the association between bit patterns and points in the multilevel constellation. This association is generally called tagging of the constellation.

In order to have the minimum error probability, it is useful to tag a constellation in such a way that nearby points corresponds to nearby bit words. In the case of squared constellations, tagging is easy due to the fact that the constellation can be divided into the Cartesian product of one dimensional constellations.

If the number of one dimensional constellations is n (e.g., two for M-QAM, or four for M-4QAM), also the bit word has to be divided into n parts and each part is the tag of one one dimensional constellation.

This tagging technique is shown in Figure 9.37 in the case of 16-QAM. In the case of more complex configurations, tagging can be much more complicated, and it is also possible that in a few cases the adjacency of symbols corresponding to very different bit words cannot be avoided.

An upper bound for the number of erroneous bits in a wrong symbol can be obtained assuming that, when a symbol is wrong, the receiver produces a random symbol among the $M - 1$ different from the transmitted one.

In this case, if $n = \log_2 M$,

$$\langle e \rangle = \frac{1}{M-1} \sum_{k=1}^{n} \binom{n}{k} = \frac{1}{M-1} \sum_{k=0}^{n-1} \frac{n(n-1)!}{k!(n-1-k)!} = \frac{M \log_2 M}{2(M-1)} \tag{9.48}$$

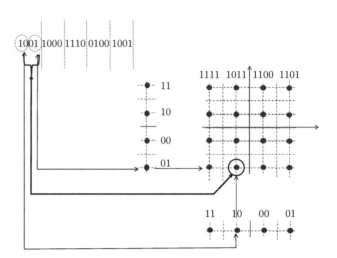

FIGURE 9.37
Tagging of a squared constellation.

where besides the definition of n the following combinatorial property has been used:

$$\sum_{k=0}^{r} \binom{r}{k} = 2^r \qquad (9.49)$$

Thus, the following upper bound can be used for the error probability, if the symbol error probability is known:

$$P_e = \frac{M}{2(M-1)} P_S \qquad (9.50)$$

As far as the symbol error probability is concerned, it depends on the decision criterion adopted to produce an estimate of the transmitted symbol.

We will consider here only the so-called "hard decision" criteria, where the estimate of the transmitted symbol is produced by the decision device without taking into account the presence of a FEC. After hard decision, the produced symbol stream (or often the produced bit stream after multilevel decoding) is passed to the FEC for error correction [43,60].

Traditionally, soft decision criteria, where decision and FEC decoding are not divided, but the decision device uses the FEC redundancy also to take the better possible estimate of the transmitted message, has been considered too complex to be used at optical transmission speed. This situation is changing with the increase of the speed of C-MOS electronics and it is possible that soft decision decoders will also become available at optical speeds quite soon [60].

Due to the better exploitation of the information contained in the received message, generally, soft decision performs better than hard decision, with a gain of the order of a couple of dBs at the typical optical SN_o level.

A hard decision-based decision device for a multilevel modulation format is based on the division of the space in which the multilevel constellation is created (e.g., the quadrature space or the Stokes space) in decision zones, let us call them $\mathfrak{D}_j$ with $j = 1, 2, \ldots, M$, each of which contains only one constellation point.

The decision device simply associates the received signal to the constellation point inside the same decision zone. Naturally, to achieve good performances, the decision zones have to be optimized in order to minimize the overall error probability.

Assuming that the noise distribution is isotropic around any constellation point, the optimum decision zones coincides with the so-called Voronoi zones of the multilevel constellation and they can be individuated with the so-called elementary cell method [46,61].

The method works as follows. In order to determine the Voronoi zone of the jth point of the constellation, as a first step the considered point is connected with $M − 1$ oriented segments with all the other constellation points. Let us call Π_j the jth point and $\overline{\Pi_j\Pi_k}$ the segment connecting the jth point with the kth point.

From now on, as in the notation $\overline{\Pi_j\Pi_k}$, we will indicate with upper-line capital letters like $\overline{AB}$ vectors in an abstract space individuated by the end points (A and B in the example), coherently with the notation that we have adopted, where $\bar{X}$ indicates a vector in an abstract space, while $\vec{E}$ a physical vector (for example the electrical field).

Calling $\bar{Z}$ the vector representing the received field, let us construct the projections of $\bar{Z}$ over $\overline{\Pi_j\Pi_k}$ and call it $z_{j,k}$. It is a random variable, due to the presence of the noise, and it can be defined as the following probability

$$P\left(Z_{j,k} > C_{j,k}/\Pi_j\right) = \int_{C_{j,k}}^{\infty} p\left(z_{j,k}/\Pi_j\right) d\sigma_{j,k} \tag{9.51}$$

where
$c_{j,k}$ is a threshold fixed on $\overline{\Pi_j\Pi_k}$
$z_{j,k}$ is the curve coordinate along $\overline{\Pi_j\Pi_k}$
$p(z_{j,k}/\Pi_j)$ is the probability density function of the projection $z_{j,k}$

Let us optimize the threshold $c_{j,k}$ to minimize the probability $P(j,k)$ that is given by

$$P(j,k) = \int_{C_{j,k}}^{\infty} p\left(z_{j,k}/\Pi_j\right) d\sigma_{j,k} + \int_{C_{j,k}}^{\infty} p\left(z_{k,j}/\Pi_k\right) d\sigma_{j,k} \tag{9.52}$$

Once optimized the threshold, let us consider, for any $k = 1, 2, \ldots, M$ and different from j, the plane orthogonal to the segment $\overline{\Pi_j\Pi_k}$ and passing by $c_{j,k}$. In this way, $M − 1$ planes are individuated. The optimum decision zone is the minimum volume individuated by such planes.

As an example, the optimum decision zones for 4-QAM are the four quadrature plane quadrants, shown in Figure 9.38.

Once the decision zones have been optimized, the decision procedure leads to the following equation for the symbol error probability:

$$P_S = \frac{1}{M} \sum_{K=1}^{M} \int_{\cup_{j \neq k} \mathcal{D}_j} p\left(\bar{Z}/\bar{C}_k\right) d\bar{Z} \tag{9.53}$$

where
$\bar{Z}$ indicates the points representing in the constellation space the received field
$\bar{C}_k$ indicates the constellation kth point
$p(\bar{Z}/\bar{C}_k)$ is the probability density function that $\bar{Z}$ is received when $\bar{C}_k$ is transmitted
Integral in $d\bar{Z}$ is a volume integral into the constellation space

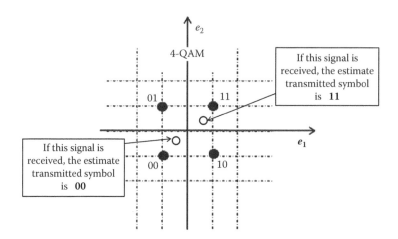

FIGURE 9.38
Hard decision for 4-QAM.

Several techniques have been devised to manage integrals like that in Equation 9.53, comprising an n-dimensional extension of the Saddle point approximation [61]. However, since we will deal in any case with low symbol error probabilities, the easier way to go is to evaluate the integral (9.53) using the so-called union bound.

The union bound is an asymptotically exact approximation of an integral in the form (9.53) that exploits the definition of the optimum decision zones to approximate the integral as follows

$$P_S = \frac{1}{M} \sum_{k=1}^{M} \int_{\bigcup_{j \neq k} \mathfrak{D}_j} p\left(\overline{Z}/\overline{C}_k\right) d\overline{Z} \approx \frac{1}{M} \sum_{k=1}^{M} \sum_{\substack{j=1 \\ j \neq 1}}^{M} \int_{C_{j,k}}^{\infty} p\left(z_{j,k}/\Pi_j\right) d\sigma_{j,k} \qquad (9.54)$$

Equation 9.54 shows as the union bound reduces the decision process to a set of binary decisions along the direction connecting the points of the constellation.

Generally, the noise distribution has an exponential tail, thus the main contribution to the error probability is given by the first neighbors of the considered point. If it is possible to limit the evaluation only to this contribution, Equation 9.54 further simplifies yielding

$$P_S = \frac{1}{M} \sum_{k=1}^{M} \int_{\bigcup_{j \neq k} \mathfrak{D}_j} p\left(\overline{Z}/\overline{C}_k\right) d\overline{Z} \approx \frac{1}{M} \sum_{k=1}^{M} \gamma_k \int_{C_k}^{\infty} p\left(z_k/\Pi_j\right) d\sigma_k \qquad (9.55)$$

where
γ_k is the number of first neighbors
z_k is the projection of the received vector along the direction connecting the considered
point with one of the first neighbors
σ_k is the curve coordinate along this direction

Finally, if the constellation is a regular lattice, where all the points have the same number of first neighbors located at a fixed distance and if the noise has the same distribution whatever symbol is transmitted (situation very common for regular lattices), a further simplification of (9.55) is possible since the optimum threshold, for symmetry reasons, is located at half the distance between first neighbors.

Thus in this case it is possible to write

$$P_S = \gamma \int\limits_{d/z}^{\infty} p(z/\Pi)\, d\sigma \qquad (9.56)$$

where d is the distance between first neighbors and the other symbol has the same meaning as in (9.56) with useless indices eliminated.

In the remaining of this paragraph we will use the above equations to analyze the ideal performances of 100 Gbit/s amplified multilevel systems using different types of modulation and detection.

The general system scheme used for this evaluation is reported in Figure 9.39, which is similar to Figure 6.18, but for the presence of multilevel modulation. We will assume the spans to be of the same length and, due to the presence of the preamplifier, we will neglect shot and thermal noise.

As seen in Chapter 6, the ASE noise spectral distribution is given by

$$S_{ASE} = \left\{ \frac{\alpha L_s}{\ln(G_{il})}(G_{il}-1)NF_{il} + (G_p - 1)NF_p + (G_b - 1)NF_b \right\}\hbar\omega \qquad (9.57)$$

where
L_s is the span length
G_{il} is the in-line amplifier gain
G_b is the booster amplifier gain
G_p is the preamplifier gain
NF_{il} is the in-line amplifier noise factor
NF_b is the booster amplifier noise factor
NF_p is the preamplifier noise factor

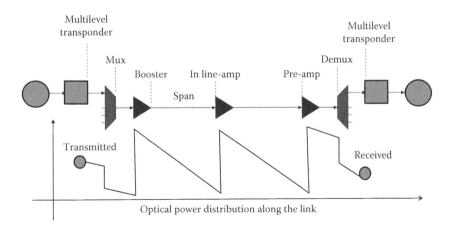

FIGURE 9.39
Architecture of the amplified system used to evaluate theoretical performances of the considered modulation formats at 100 Gbit/s.

We will not follow in detail all the calculations that are needed to arrive at the error probability for each format/receiver. The methods are all explained in Chapter 6 so that for the interested reader it should not be difficult to reconstruct the steps to arrive to the results.

9.3.4.1 M-QAM and M-4QAM with Quadrature Receiver

First we will analyze the coherent quadrature homodyne receiver.

In this case, as evident from Equation 9.35, the dominant noise term is the ASE component of the photocurrent that results amplified by the local oscillator.

This term is linear in the ASE amplitude, thus resulting in a Gaussian white noise at the electrical level. Since attenuations due to the receiver are experienced both by the signal and by the ASE noise, these do not appear in the error probability, at least as far as the ASE term is the dominant noise.

Thus, the electrical signal to noise ratio SN_E is given by

$$SN_E = \frac{P}{S_{ASE}R_S} \tag{9.58}$$

where the baseband has been assumed equal to the symbol rate. Using the approximation (9.56) with γ equal to the average number of first neighbors for the symbol rate and the bound (9.50),

$$P_e = \frac{\gamma M}{2(M-1)} Q\left(d\sqrt{\frac{SN_E}{2}}\right) \tag{9.59}$$

is obtained where d is the normalized minimum distance of the constellation. From Equation 9.59 the importance of maximizing the constellation minimum distance is clearly seen.

In Figure 9.40, the error probability is plotted for a multilevel M-QAM system using a quadrature receiver. Due to the rapid decrease of the minimum distance with the number of constellation points, the performances get rapidly worse increasing the number of levels. On the other hand, for a low number of levels, this is an effective modulation format.

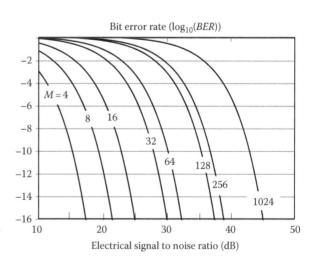

FIGURE 9.40
Plot of the ideal performance of M-QAM for various values of the number of levels.

Better performances at high number of levels are attained by the constant power M-4QAM, whose constellations are derived from regular convex solids in the four dimensional space.

In Figure 9.41, the error probability for this kind of modulation is shown, deriving the constellation with 64 points from a random selection of points among the vertices of the 600 Tetrahedral (120 vertices) and the constellation with 512 points from a random selection of 512 among the 600 vertices of the 120 Dodecahedral. No attempt has been made to optimize the choice, although there is a large literature about segmenting of complex constellations and the interested reader is encouraged to start from [62–64] if is interested in methods for an optimum subconstellation selection.

Increasing the number of levels, the performance is far better with respect to M-QAM, due to the better exploitation of the available degrees of freedom.

Finally, in Figure 9.42, variable power M-4QAM performances are represented in the case of the constellations of Table 9.5. The high number of degrees of freedom renders these modulation formats extremely efficient at very high number of levels, even if at low number of levels the performances are not superior to that of other quadrature based formats.

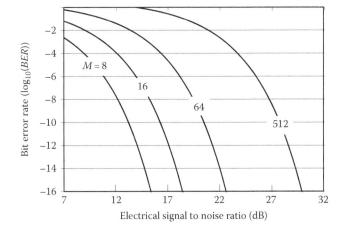

FIGURE 9.41

Plot of the ideal performance of constant power M-4QAM for various values of the number of levels

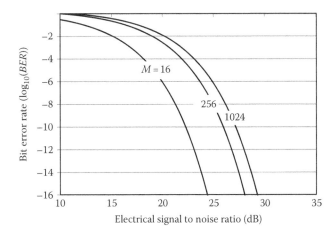

FIGURE 9.42

Plot of the ideal performance of variable power M-4QAM for various values of the number of levels in the case of the constellations of Table 9.5

9.3.4.2 M-PolSK with Stokes Parameters Receiver

Following the receiver structure to determine the expression of the decision variable and its statistics, it is possible to demonstrate the following properties:

- The decision variable is a quadratic form of the input field quadratures
- The decision variable is independent from the polarimeter matrix A if the ASE noise is dominant
- The decision variable statistic is independent from the fiber polarization rotation; this property derives from the fact that the ASE power is uniformly distributed on the Poincarè sphere among all the polarization states

The last two properties, which are a key to evaluate the performance of various PolSK systems, deserve a comment.

Let us consider the evolution of the M-PolSK signal under the polarization point of view representing it in the Stokes space. Let us indicate with $\bar{S}$ the Stokes vector having for components the three Stokes parameters that are coordinates in the Stokes space.

If the Launched signal is represented with $\bar{S}_i$, after fiber propagation, the Stokes vector $\bar{S}_o$ in front of the receiver can be written as

$$\bar{S}_o = M\bar{S}_i + \bar{n} \tag{9.60}$$

where

M is the Muller matrix (the Stokes space version of the Jones matrix)

$\bar{n}$ is the ASE noise vector in the Stokes space

Both coherent and direct detection receivers have an optical front end constituted by a polarimeter generating a photocurrents vector $\bar{C}_0$ whose elements are proportional to the Stokes parameters. This means that a nonsingular matrix A can be found so that, neglecting distortions and the photodiode sensitivity,

$$\bar{C}_0 = A\bar{S}_o \tag{9.61}$$

The polarimeter is followed by an electronic processing section that has the role of inverting the matrix to obtain a set of photocurrent $\bar{C}_1$ proportional to the received field Stokes parameters.

This means that, whatever be the polarimeter, coherent or incoherent, the currents in front of the electronic polarization compensation section have the following expression:

$$\bar{C}_1 = A^{-1}\bar{C}_0 = \bar{S}_o \tag{9.62}$$

Remembering that $\bar{S}_o$ does not depend on the receiver structure, we can conclude that the decision variable is exactly the same in every ideal Stokes parameters receiver and thus all these receivers have the same performances.

At this point, it is useful to choose the receiver structure that makes easiest the evaluation of the BER, and it is for sure the coherent receiver.

Due to the square law detection operated at the electrical level, the decision variable is still a quadratic form of the field quadratures.

In these conditions, the saddle point approximation (see Chapter 6) can be used to evaluate the error probability if the union bound is used. As a matter of fact, this approximation allows us to decompose the decision process in a set of independent binary decisions.

Moreover, among all the addenda entering in the union bound expression of the symbol error probability we can maintain only those relative to first neighbors.

The characteristic function of the decision variable relative to the decision on the segment among a couple of first neighbors is given by

$$F_c(s) = \frac{\exp\left[-SN_o(d^2 M_{oe}s/4)((1-4s^2)/(1-d^2s^2))\right]}{(1-d^2s^2)^{2M_{oe}}}$$ (9.63)

where

M_{oe} is the ratio between the optical and the electrical bandwidth
SN_o is the optical signal to noise ratio
d is the distance between first neighbors in the Stokes space

A simple closed form expression of the bit error probability it is not easy to achieve, due to the complexity of the calculation involved by the saddle point approximation.

Since a simple formula is important for quick evaluation, we have derived an approximated form of the error probability assuming a high signal to noise ratio and the condition $SN_o \gg M_{oe}^2$ that is always realized in practice.

Moreover, we have decided to put $M_{oe} = 2$, so that the electrical and the optical SN coincides (the difference is smaller than 2%).

Expressing the error probability as a function of SN_e allows us to consider the dependence of the BER on M_{oe} as slow and, for $0.5 < M_{oe} < 4$ negligible in a first approximation (see Chapter 6).

Under these hypotheses, the bound optimization parameters s_0 can be approximated with a simple expression

$$s_0(SN_e, d) = \left(\frac{1}{SN_e}\right)^{(a_2d^2 + a_1d + a_0)} + \left(b_5d^5 + b_4d^4 + b_3d^3 + b_2d^2 + b_1d + b_0\right)$$ (9.64)

where the numerical value of the approximation parameters are reported in Table 9.7. The ratio between the approximation standard deviation and the average is smaller than 0.6% in all the parameter domain suitable for the applications.

TABLE 9.7

Values of the Approximation Parameters That Are Present in the Expression of the Error Probability of Direct Detection M-PolSK

b_5	−35.241		
b_4	15.202		
b_3	−26.489		
b_2	23.732	a_2	−0.1117
b_1	−11.320	a_1	0.4341
b_0	2.809	a_0	0.7513

Once $s_0(SN_e, d)$ is evaluated, in the approximation of high SN the optimum threshold is always 0.5 SN_e and the saddle point approximation can be further simplified yielding

$$P_e = \frac{\gamma M}{2(M-1)} \frac{e^{-SN_e s_0}\left[(4-d^2)/2\left(d^2 s_0^2 -1\right)+23/10\right]}{\left(1-s_0^2 d^2\right)^4} \tag{9.65}$$

The accuracy of the approximation can be verified comparing the results with the exact saddle point expression evaluated by numerical solution of the bound parameter equation and of the resulting expression of the error probability reported in [37]. The error is never above 0.3 dB as far as the error probability is lower than 10^{-3} in all the cases reported in [37].

The performances of coherent detection M-PolSK are shown in Figure 9.43 using the configurations whose parameters are reported in Table 9.3, the 32 level system constellation is derived by randomly selected 32 points out of the 60 vertices of the snub-dodecahedron and the 64 points constellation by randomly selecting 64 out of the 120 vertices of the Icosidodecahedron. In the figure, the performances of a coherent detection M-PolSK system using the optimum constellations reported in [35,37] are also plotted in dashed lines. It is evident that the sensitivity gain related to the use of the optimum constellation with respect to a more regular suboptimum one is small and probably in practice it is cancelled by the difficulty to reproduce accurately the optimum configurations due to their more complex structure.

9.3.4.3 M-DPSK with Direct Detection Receiver

Last of the modulation formats we have considered is multilevel DPSK with direct detection receiver.

Due to its structure based on signal self beating, the receiver belongs to the class of quadratic receivers. In this case, the problem of determining the symbol error probability from the characteristic function can be exactly solved as already underlined in Chapter 6 for the binary case. Passing from the symbol error probability to the bit error probability, we obtained

$$P_e = \frac{M}{2(M-1)} e^{-SN_e d^2/2} \tag{9.66}$$

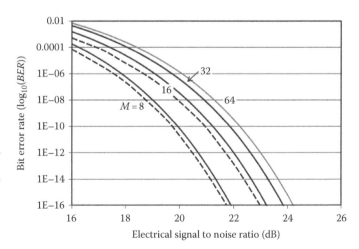

FIGURE 9.43
Plot of the ideal performance of PolSK with Stokes Parameters Receiver and for various values of the number of levels and $M_{oe} = 2$. Results obtained with optimum constellations reported in [35,37] are plotted too in dashed lines.

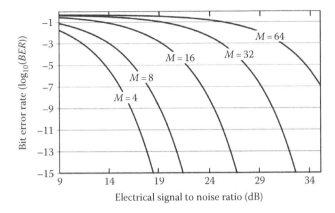

FIGURE 9.44
Ideal performances of M-DPSK for different values of the number of levels M.

Equation 9.65 is used to evaluate the performance of an M-DPSK system that are shown in Figure 9.44. As expected, this modulation format gets rapidly worse by increasing the number of levels, due to the fact that it is based on the modulation of a single degree of freedom. For a low number of levels, however, the penalty with respect to other formats is smaller and in practice, especially at very high speed as 100 Gbit/s, it could be more than counterbalanced by the simplicity of the receiver with respect, for example, to the quadrature receiver.

9.3.4.4 Comparison among Different Modulation Formats

In Figure 9.45, the sensitivity of the various systems we have analyzed are represented for a BER of 10^{-4}, typical of applications with FEC. Collecting the results, they can be divided into two sets, with different and well individuated behaviors with respect to the number of levels.

This is essentially due to the fact that the BER trend with the number of levels is essentially a metric property of the constellation, thus depending on the characteristics of the space to which the constellation belongs.

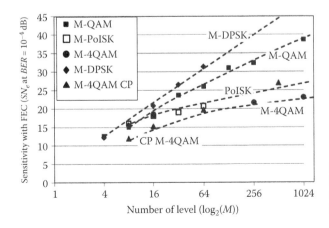

FIGURE 9.45
Sensitivity of different multilevel modulation formats in ASE-limited conditions at $BER = 10^{-4}$.

In order to have better trends, it seems that either the number of degrees of freedom has to be increased, or the space metrics has to be changed (passing, e.g., from a plane to a sphere), the first step being more effective than the second.

9.3.5 Coherent Receivers Sensitivity to Phase and Polarization Fluctuations

At the end of this section devoted mainly to the performances of multilevel transmission systems, it is necessary to consider an important element regarding coherent detection that comes into play any time the transmission system cannot be considered ideal.

We will take as a reference the receiver presented in Section 9.3.3.1, that is, the universal quadrature receiver, but what we will present in this section can be adapted at almost any coherent receiver with marginal modifications to take into account the receiver structure.

The main advantage of coherent detection is that the set of electrical currents at the output of the receiver are proportional to the field quadratures, but for the noise. This property allows information to be coded into any degree of freedom of the optical field thus fully exploiting its transmission potential.

On the other hand, this also means that any form of random fluctuation of the field is detected by the receiver and constrains its performances.

Here we will consider the two main noises that can limit the performance of a coherent receiver while in a direct detection receiver these noises can be generally neglected: the phase noise and the polarization noise.

9.3.5.1 Phase Noise Penalty for Coherent Quadrature Receiver

Several causes can generate phase noise on the optical signal before detection: generally, the most important are the frequency fluctuations of the transmitting laser and of the LO and the nonlinear phase noise coming during propagation from Kerr effect. We will consider here the phase noise derived from the frequency fluctuations of the lasers.

The phase noise has a complex degrading effect: it reduces the power available for symbol estimation on each quadrature, introduces a rotation of the (e_1, e_2) and (e_3, e_4) planes in the quadrature space creating crosstalk between the corresponding quadratures, and finally it creates imperfect frequency lock between the LO and the received field carrier.

As a matter of fact, Equation 9.34 is rewritten in this case as

$$c_1(t_j) = R[e_1(t_j) + \eta_1(t_j)]\cos(\varphi_T - \varphi_L) + R[e_2(t_j) + \eta_2(t_j)]\sin(\varphi_T - \varphi_L) + n_1(t_j)$$

$$c_2(t_j) = R[e_2(t_j) + \eta_2(t_j)]\cos(\varphi_T - \varphi_L) - R[e_1(t_j) + \eta_1(t_j)]\sin(\varphi_T - \varphi_L) + n_2(t_j)$$

$$c_3(t_j) = R[e_3(t_j) + \eta_3(t_j)]\cos(\varphi_T - \varphi_L) + R[e_4(t_j) + \eta_4(t_j)]\sin(\varphi_T - \varphi_L) + n_3(t_j) \tag{9.67}$$

$$c_4(t_j) = R[e_4(t_j) + \eta_4(t_j)]\cos(\varphi_T - \varphi_L) - R[e_3(t_j) + \eta_3(t_j)]\sin(\varphi_T - \varphi_L) + n_4(t_j)$$

where

φ_T represents the transmitter phase noise

φ_L is the phase noise of the LO

The problem of an accurate evaluation of the impact of phase noise on this class of receivers has been studied extensively; several models have been devised with different levels of accuracy and complexity for all the relevant modulation formats. It is not our intention here to either develop an accurate model or to compare different models in different situations; the interested reader is encouraged to personally analyze the rich technical literature starting from the literature reported in [31].

In practical high performance systems, very high quality tunable lasers are used to reduce cost of spare parts, to ease the LO operation, and to minimize the effect of phase noise via a very narrow linewidth. In general, the linewidth is smaller than a few MHz and in the external cavity, the linewidth is of the order of 500 kHz.

In this condition, we have to expect that in general the phase noise impact will be small, which is a necessary condition for our draft design. Thus we will introduce a very simplified model for the analysis of the phase noise impact, which holds in the low penalty regime. This model will allow us to evidence all the elements that are important and also to obtain a closed form expression of the sensitivity penalty.

In the low penalty condition, Equation 9.67 can be linearized with respect to the phase noise obtaining

$$
\begin{aligned}
c_1(t_j) &= R[e_1(t_j) + \eta_1(t_j)] + R[e_2(t_j) + \eta_2(t_j)](\varphi_T - \varphi_L) + n_1(t_j) \\
c_2(t_j) &= R[e_2(t_j) + \eta_2(t_j)] - R[e_1(t_j) + \eta_1(t_j)](\varphi_T - \varphi_L) + n_2(t_j) \\
c_3(t_j) &= R[e_3(t_j) + \eta_3(t_j)] + R[e_4(t_j) + \eta_4(t_j)](\varphi_T - \varphi_L) + n_3(t_j) \\
c_4(t_j) &= R[e_4(t_j) + \eta_4(t_j)] - R[e_3(t_j) + \eta_3(t_j)](\varphi_T - \varphi_L) + n_4(t_j)
\end{aligned}
\tag{9.68}
$$

From (9.68), it is clear that the phase noise presence in the regime of small ratio between the linewidth and the symbol rate has a first effect of creating an additional noise composed of two terms: one proportional to the signal power and the other to the ASE.

The noise term proportional to the signal power is particularly dangerous, since its effect cannot be compensated by increasing the transmitted power.

As far as the effect on the frequency lock is concerned, Equation 9.41 is substituted in the presence of phase noise with

$$
c_D(t_j) = \sqrt{2} R \cos\left[\Delta\omega t_j - \frac{\pi}{4} + \varphi_T(t_j) - \varphi_L(t_j) \right] + n_D(t_j)
\tag{9.69}
$$

so that, assuming an ideal frequency lock, a simple first order digital filter to limit the lock loop bandwidth and neglecting in conditions of low penalty the unlock probability, the estimated frequency deviation $\Delta\omega_e$ can be expressed as

$$
\Delta\omega_e = \Delta\omega + \delta\omega = \Delta\omega + \frac{1}{N_I} \sum_{j=1}^{N_I} \frac{\varphi_T(t_j) - \varphi_L(t_j) - \varphi_T(t_{j-1}) + \varphi_L(t_{j-1})}{\Delta t}
\tag{9.70}
$$

where
Δt is the sampling rate
N_I is the lock loop bandwidth in terms of number of integrated samples so that the loop bandwidth is $B_I = 1/N_I \Delta t$

This error in the estimated angular frequency causes a further element to be added to the photocurrent. In small phase noise regime, Equation 9.69 becomes

$$c_1(t_j) = R[e_1(t_j) + \eta_1(t_j)] + R[e_2(t_j) + \eta_2(t_j)](\varphi_T - \varphi_L + \delta\omega t_j) + n_1(t_j)$$

$$c_2(t_j) = R[e_2(t_j) + \eta_2(t_j)] - R[e_1(t_j) + \eta_1(t_j)](\varphi_T - \varphi_L + \delta\omega t_j) + n_2(t_j)$$

$$c_3(t_j) = R[e_3(t_j) + \eta_3(t_j)] + R[e_4(t_j) + \eta_4(t_j)](\varphi_T - \varphi_L + \delta\omega t_j) + n_3(t_j)$$

$$c_4(t_j) = R[e_4(t_j) + \eta_4(t_j)] - R[e_3(t_j) + \eta_3(t_j)](\varphi_T - \varphi_L + \delta\omega t_j) + n_4(t_j)$$

(9.71)

Since we have assumed to use a small linewidth laser, we can assume that its linewidth is Lorentian with a very good approximation.

For a laser with a Lorentian linewidth $\Delta\nu$, the following equation holds [31]:

$$\Delta\upsilon = \frac{1}{2\pi} S_\varphi(0) \qquad (9.72)$$

where $S_\varphi(0)$ is the frequency noise spectrum evaluated for zero frequency (see Chapter 5).

Moreover, the following relation exists between the phase noise variance and the frequency noise spectrum:

$$\sigma_\varphi^2 = \frac{2}{\pi}\int_0^\infty S_\varphi(\omega)\frac{1-\cos(\omega T_S)}{\omega^2}d\omega \approx \frac{2}{\pi}S_\varphi(0)\int_0^{2\pi B_\varphi}\frac{1-\cos(\omega T_S)}{\omega^2}d\omega = \frac{T_S}{\pi^2}S_\varphi(0)F(T_S B_\varphi) \qquad (9.73)$$

where T_S is the symbol interval, B_φ is the laser $S_\varphi(\omega)$ spectral width (of the order of 2–3 GHz; see Chapter 5) and

$$F(x) = \int_0^x \frac{1-\cos(2\pi\upsilon)}{\upsilon^2}d\upsilon \qquad (9.74)$$

Assuming that both transmitter and LO have the same linewidth, the variance of the phase fluctuations in a bit interval can be expressed as a function of the laser linewidth as follows:

$$\sigma_\varphi^2 = \frac{2}{\pi}\frac{\Delta\upsilon}{R_s}F\left(\frac{B_\varphi}{R_s}\right) \qquad (9.75)$$

As far as the angular frequency lock error is concerned it is a Gaussian zero mean process. Its variance can be calculated easily taking into account that the summation can be simplified so that only the terms in the first and the last instants of the interval remain.

$$\sigma_{\delta\omega}^2 = \frac{4}{\pi}\frac{\Delta\upsilon}{R_s}B_l^2 F\left(\frac{B_\varphi}{R_s}\right) \qquad (9.76)$$

where
B_l is the lock loop bandwidth

It derives a general result: better performances of the LO lock loop are achieved if it is possible to find a loop bandwidth sufficiently large to efficiently follow the lasers parameters fluctuations but much smaller than the symbol rate.

In the low penalty limit, we will assume that the error probability expression versus the SN_e does not change, thus the penalty can be evaluated evaluating the SN_e.

We can always rotate the reference system in the quadrature space so that the e_1 axis is along the segment joining the two first neighbors we are considering. In this way, it is $e_2 = e_4 = e_3 = 0$ and the decision variable C, at the first order in the phase noise, has the following form:

$$C = \sum_{j=1}^{4} c_1(t_d)$$

$$= R[e_1(t_j) + \eta_1(t_j)] - R[e_1(t_j) + \eta_1(t_j)](\varphi_T - \varphi_L + \delta\omega t_d) + \sum_{j=1}^{4} n_{j1}(t_d) \qquad (9.77)$$

To evaluate the electrical SN_e we will neglect the additive noise with respect to the ASE contribution and all the terms of second order in the noise.

Moreover we note that

$$\left\langle (\varphi_T - \varphi_L + \delta\omega t_d)^2 \right\rangle = \frac{4}{\pi} \frac{\Delta v}{R_s} F\left(\frac{B_\varphi}{R_s}\right) \left[1 + \frac{B_I}{R_s} + \left(\frac{B_I}{R_s}\right)^2\right] \qquad (9.78)$$

due to the fact that $\delta\omega$ and $\varphi_T - \varphi_L$ are not uncorrelated and their product cannot be set to zero.

Finally

$$SN_e = \frac{Pd^2}{S_{ASE}R_s + (4/\pi)Pd^2(\Delta v/R_s)F\left(B_\varphi/R_s\right)\left[1 + \left(B_I/R_s\right) + \left(B_I/R_s\right)^2\right]} \qquad (9.79)$$

The behavior of the SN_e in the presence of Phase noise for M-4QAM is plotted in Figure 9.46.

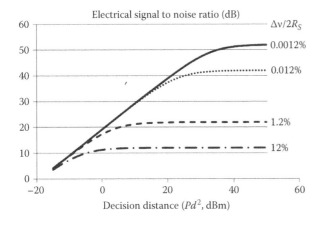

FIGURE 9.46
SN_e penalty due to phase noise for M-4QAM in the approximation of small impairment.

The most striking feature of the curves representing the SN_e behavior is the presence of a clear plateau due to phase noise. When the plateau is reached, increasing the power does not change the SN_e due to the fact that the phase noise term also depends on the power.

9.3.5.2 Depolarization Penalty for Coherent Quadrature Receiver

The working of the coherent quadrature system is based on the assumption that the evolution of the signal polarization is linear so that a linear estimate of the channel can completely compensate the polarization rotation.

If a nonlinear polarization evolution is stimulated by the high transmitted power, this is no more true and a further sensitivity penalty is induced by this phenomenon.

In a first approximation, we can divide polarization fluctuations between fast nonlinear and slow linear. Due to the fact that the linear polarization tracker has a small bandwidth, it passes through the fast part of polarization fluctuations.

Kerr-induced polarization fluctuations, depending on the transmitted message, can be seen as a random process inducing random time depolarization of the received field [65].

As far as the induced depolarization degree is small, we can imagine that it is passed through by the polarization tracker so that practically only the linear fluctuations are tracked and the nonlinear depolarized field appears to fluctuate randomly after multiplication by the inverse of the Jones matrix and have to be dealt as signal dependent noise.

Calling δ_{POL} the nonlinear depolarization degree, which is the reciprocal of the polarization degree defined in Equation 9.34, the received optical field can be written as

$$\vec{E}(t) = E(t)\left[\sqrt{1+\delta_{POL}}\left(ae^{i\psi}\vec{p}_+ + \sqrt{1-a^2}\,\vec{p}_-\right) + \sqrt{\delta_{POL}}\,\vec{\varsigma}(t)\right] \tag{9.80}$$

where

   $\vec{\varsigma}(t)$ is a randomly varying unitary vector determined by nonlinear polarization evolution that has a bandwidth larger than R_S

   a and ψ are the Jones matrix parameters (along with an absolute phase here set to zero) that the polarization tracker has to estimate

Even if the correlation between the depolarized and the polarized part of the field is not zero, a simple way to evaluate the impact of nonlinear polarization evolution is to consider the depolarized terms if it was a noise with a colored spectrum.

In particular, the expressions of the photocurrents after polarization recovery becomes

$$c_1(t_j) = R\sqrt{1-\delta_{POL}}\,[e_1(t_j) + \eta_1(t_j)] + \sqrt{\delta_{POL}}\sum_{h=1}^{4} A_{1h}[e_h(t_j) + \eta_h(t_j)] + n_1(t_j)$$

$$c_2(t_j) = R\sqrt{1-\delta_{POL}}\,[e_2(t_j) + \eta_2(t_j)] + \sqrt{\delta_{POL}}\sum_{h=1}^{4} A_{2h}[e_h(t_j) + \eta_h(t_j)] + n_2(t_j)$$

$$\tag{9.81}$$

$$c_3(t_j) = R\sqrt{1-\delta_{POL}}\,[e_3(t_j) + \eta_3(t_j)] + \sqrt{\delta_{POL}}\sum_{h=1}^{4} A_{3h}[e_h(t_j) + \eta_h(t_j)] + n_3(t_j)$$

$$c_4(t_j) = R\sqrt{1-\delta_{POL}}\,[e_4(t_j) + \eta_4(t_j)] + \sqrt{\delta_{POL}}\sum_{h=1}^{4} A_{4h}[e_h(t_j) + \eta_h(t_j)] + n_4(t_j)$$

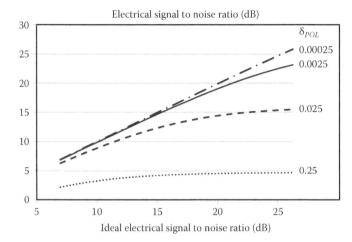

Electrical signal to noise ratio (dB)

Ideal electrical signal to noise ratio (dB)

FIGURE 9.47
SN_e penalty due to depolarization in the approximation of small impairment for M-4QAM.

The 4 × 4 matrix A is defined stating from the quadratures of $M^+\bar{\varsigma}(t)$.

Due to its definition, A is a unitary matrix like M and it can also be interpreted as the random rotation induced by the combination between linear and nonlinear effects.

The expression of the photocurrent is formally similar to that derived for the case of phase noise, thus following the same derivation and assuming perfect linear polarization tracking yields

$$SN_e = \frac{Pd^2(1-\delta_{POL})}{S_{ASE}R_s + Pd^2\delta_{POL}} = \frac{SN_{eo}(1-\delta_{POL})}{1+SN_{eo}\delta_{POL}} \tag{9.82}$$

where SN_{eo} is the ideal value of SN_e in the same circumstances. The effect of depolarization is very strong, as shown in Figure 9.47, introducing a plateau in the SN_e.

9.4 Alternative and Complementary Transmission Techniques

We have seen that multilevel modulation formats constitutes a very promising transmission technique to design long distance very high capacity optical links.

Besides multilevel modulation, there are a set of transmission techniques that can further improve the system performances, either helping to decrease multilevel sensibility to some transmission impairment or providing the way for a further ideal performance improvement.

The most important of these techniques is Orthogonal Frequency Division Multiplexing (OFDM), a technique for low speed channels multiplexing that simultaneously greatly reduce the sensitivity to linear causes of ISI and increase sensibly the bandwidth efficiency exploitation.

Besides OFDM, PDM and polarization diversity are often used to exploit the polarized nature of the electromagnetic field without an explicit polarization modulation and to decrease the sensitivity of the system to channel interference, respectively.

Also pulse polarization diversity is sometimes used, especially in RZ systems, to prevent the jitter caused by interaction between nearby pulses.

In this section, we will briefly introduce all these techniques.

9.4.1 Orthogonal Frequency Division Multiplexing

OFDM is used extensively in broadband and radio communication because it is an effective solution to ISI caused by a dispersion during signal propagation [66].

This becomes increasingly important as data rates increase to the point that the memory of the channel is not limited to a string of three bits, but extends to longer bit series. In this case, equalization is a complex matter requiring elaborated algorithms like mean-squared error (MSE) or signal back-propagation.

In contrast, the complexity of OFDM, using serial modulation and frequency domain equalization, scale well as data rates and dispersion increase.

The basic concept of OFDM is simple: data are transmitted in parallel on a number of different frequencies. Due to the parallelization of the data flux, the symbol period is much longer than that of a serial system with the same total data rate.

Due to the longer symbol period, the channel memory gets shorter, or even disappears completely, since dispersion only affects one symbol. In most OFDM implementations, any residual ISI is removed by using a form of guard interval between adjacent bit intervals called a cyclic prefix.

It could seem that this is simply frequency multiplexing, similar to what happens in WDM systems, but there are key differences between standard WDM and OFDM.

In standard WDM, the signals transmitted on different wavelengths are completely uncorrelated and guard intervals are used in between adjacent channels in order to reduce crosstalk. WDM multiplexing is performed using an analog optical device, and the same for demultiplexing.

OFDM on the other hand, is based on the digital generation of the multiplexed signal by densely packing channels that are generated to be orthogonal from a mathematical point of view.

Since the channels can be separated using their orthogonality, there is no need of guard intervals among them.

The generation of the multiplexed signal and the demultiplexing is operated via Fast Fourier Transform (FFT) by a dedicated processor; thus it is a computationally efficient and fast operation that can incorporate, if needed, equalization in the frequency domain.

It is not our scope to review in detail OFDM techniques, but it is quite clear that OFDM could become a useful technique to design very high speed long haul optical systems [67,69].

Naturally, OFDM also presents critical aspects, which have to be faced to design effective systems.

Among the technical criticalities of OFDM, there are the sensitivity to phase noise [69] and to unbalance between the quadratures at the receiver, for example, caused by differences between receiver hybrids [66].

As a matter of fact, it is easy to imagine that if the different carriers of an OFDM signal are constructed selecting amplitude and phase so that nearby carriers results to be orthogonal, any phenomenon introducing an uncontrolled phase element can alter this balance causing strong interference between nearby channels at the receiver, when they are demultiplexed in the frequency domain.

Since SPM introduces a nonlinear form of phase noise, the SPM sensitivity is another critical aspect of OFDM modulation.

The fact that the OFDM system has to be designed with a great care for the linearity both of transmitter and receiver and of fiber propagation is also due to the characteristics of the OFDM signal to have a great ratio between peaks and average value of the amplitude. These big fluctuations immediately create distortion if nonlinearity is encountered [70].

Distortion of this kind can be classified as Out Of Band Power and In Band Distortion. Out of band power not only creates a power penalty due to the loss of a part of the available power, but also could provoke interference if this power is reflected again in the signal bandwidth during the nonlinear propagation through the system [66].

Band distortion creates interference between the carriers.

Another potential critic aspect of OFDM is the cost of the transmitter and the receiver compared with a traditional transmitter or receiver of an IM-DD system.

However, this element tends to disappear more and more while traditional WDM systems are forced to use complex maximum likelihood sequence estimation (MLSE) or back propagation systems to compensate for ISI. As a matter of fact, these complex electronic equalization methods rely on high speed electronics, frequently on specific ASICS, so that the cost of OFDM is no more so high with respect to this kind of advanced IM-DD systems.

Since orthogonality among nearby channels is assured by the carrier, any modulation format can be used in principle on the subcarriers of an OFDM signal. Since OFDM is also used to increase the efficiency in the use of the available bandwidth, it is often used in conjunction with multilevel modulation on the subcarriers. A frequently used modulation format is M-QAM OFDM.

9.4.2 Polarization Division Multiplexing

Considering multilevel modulation, we have seen that several multilevel formats exploit polarization to convey information.

This not only allows a high efficiency to be reached by the multilevel constellations, but also requires a certain degree of complexity at the transmitter and at the receiver. Polarization division multiplexing (PDM) is a way to exploit polarization with a smaller complexity: it consists in transmitting two channels at the same frequency but on orthogonal polarizations.

The transmitter does not need, at least in principle, a polarization modulator and the receiver is simpler than an M-PolSK or an M-4QAM receiver.

Intuitively, the greater weakness of PDM is the need of tracking the incoming polarization: nonlinear depolarization, polarization differential losses, and similar phenomena affect the channel recovery at the receiver in the same measure they affect M-PolSK or M-4QAM systems.

9.4.3 Channel and Pulse Polarization Diversity

If there is no need of using polarization to convey information, polarization diversity can be used to decrease interference between nearby channels in the WDM comb.

If this technique is used, nearby channels are transmitted on orthogonal polarizations so that, as far as polarization evolution during fiber propagation can be considered linear, they do not interfere at the receiver. As a matter of fact, since linear polarization evolution during fiber propagation can be described as a random rotation in the Stokes space, the property of two polarization states to be orthogonal is maintained.

Moreover, also nonlinear crosstalk caused by FWM and XPM does not take place with nearby channels since these effects both require that the beating between the interfering channels is different from zero.

This technique is less sensitive to the presence of a nonlinear contribution to the evolution of polarization since, as far as we are in a regime of small nonlinearity, it implies only the resurgence of a small interference percentage.

Polarization diversity can be used in a different manner also. If NRZ with a small duty cycle is used, nonlinear interaction between adjacent pulses in the pulse stream at a certain frequency can manifest, with a stronger and stronger efficiency while SPM and chromatic dispersion compensation is more and more effective (see Section 9.2.4 for more details and the references).

This phenomenon causes a jitter in the transmitted pulse stream that is to be controlled. One of the possibilities to control this effect is to transmit adjacent pulses on orthogonal polarizations. If this technique is used generally adjacent WDM channels are synchronized so that pulses on adjacent channels are orthogonal in polarization in order to achieve a global interference minimization.

9.5 Design Rules for 100 Gbit/s Long Haul Transmission Systems

Up to now we have analyzed "ideal" systems, where the ASE noise is the only limit to system performances.

In Chapter 6, we have seen that this is not a realistic model and many other impairments have to be considered in designing a transmission system.

In this section, we will face the problem of designing realistic systems in a variety of situations. In this way, we will assess the real potentialities of the different system architectures when practical requirements are set.

We know that a large number of long haul systems are installed all around the world, thus the requirement of accommodating 100 Gbit/s systems on transmission lines designed for lower rates is natural.

The great majority of the system we will consider will be designed to work on a transmission line originally conceived for 10 or 40 Gbit/s channels. For the parameters of such lines, we will consider the examples we have developed in Chapter 6.

At the end of the section we will also present some elements on the design of 100 Gbit/s systems with a newly optimized transmission line.

9.5.1 Practical Multilevel Systems: Transmitting 100 Gbit/s on a 40 Gbit/s Line

We will assume to have a line designed for 40 Gbit/s and to design a 100 Gbit/s transmission system that can be substituted to the 40 Gbit/s channels by only changing the transponder (in case accommodating the new transponders in a separated ad hoc subrack).

We will require that the system can be scaled substituting one by one the 40 Gbit/s channels, up to a complete substitution that increases the overall system capacity of a factor 2.5.

In order to have available all the parameters of the original 40 Gbit/s system, we will refer to the system described in Section 6.3.2.6: this is a WDM 40 Gbit/s system 2000 km long with a maximum of 22 channels in the extended C band and a channel spacing of 200 GHz.

The first point is that, in order to be easily multiplexed with the 40 Gbit/s channels, the 100 Gbit/s signal must have an optical bandwidth equal, or preferably smaller, to that of a 40 Gbit/s.

Thus, to avoid useless system complications, we will select a 8-DPSK NRZ modulation whose unilateral optical spectrum is 33 GHz wide, well compatible with the 40 GHz of the original channels. Direct detection by signal self-beating is also chosen.

The overall ratio between the transmission line optical bandwidth at 40 Gbit/s and the electrical bandwidth is $M_{oe} = 1.5$. Thus the available unilateral optical bandwidth is 60 GHz at 1/e. Since we have assumed to have Gaussian AWG, the cascade of multiplexer and demultiplexer is Gaussian again.

This means that for the 8-DPSK channel, $M_{oe} = 1.81$. The ideal SN_e needed to reach a BER of 10^{-12} with this value of M_{oe} is 20.5 dB.

Using the relation between optical and electrical signal to noise ratio

$$SN_e = \frac{SN_o^2 M_{oe}}{2SN_o + (2 - (1/2M_{oe}))} \tag{9.83}$$

we obtain the ideal optical signal to noise ratio required for $BER = 10^{-12}$ to be about 18.5 dB.

Carrying out the evaluation of the ASE limited performances, using parameters of Tables 9.8 and 9.9, an ASE limited signal to noise ratio of $SN_o = 22.3$ dB is obtained.

At this point, the first design step, that is, the power budget, can be carried out.

9.5.1.1 Power Budget

It is not difficult to see that the power budget is verified both at the transmitter and at the receiver since the power at the booster input is sufficient to saturate it but below its maximum input power and the power at the preamplifier out is well above the limit power that matches after the receiver chain the sensitivity of the photodiode, but it is also below the power needed to saturate the considered avalanche photo-detector (APD).

In this condition, the total power exiting from the amplifier at every span input is 18 dBm overall, that means 4.57 dBm per channel.

TABLE 9.8

Parameters of the Transmitter and the Receiver of the 8-DPSK System Dimensioned Transmit 100 Gbit/s on a 40 Gbit/s Line

Receiver Parameters	
Photodiode sensitivity	−15 dBm
AWG loss	7 dB
Si-SiO2 chip	8 dB
Patchecords and other losses	1 dB
Transmitter	
Laser emitter power	10 dBm
Modulator loss	5 dB
AWG loss	7 dB
Other losses	1 dB

TABLE 9.9

Characteristics of the Preexisting 40 Gbit/s
Line Where the 100 Gbit/s Signal Has to Be
Accommodated

Span length	50,000 m
Spans number	40
Attenuation	0.25 dB/km
Power at the input of a span	18 dBm
Power per channel at the span input	4.6
Power at the span end	−2 dBm
Power at the span end per channel	−15.4 dBm
Noise Figure (all amplifiers)	5 dB

9.5.1.2 Penalty Analysis

9.5.1.2.1 Residual Chromatic Dispersion

The residual chromatic dispersion at the link end is 500 ps/nm, which is 275 ps in the bandwidth of 0.55 nm of the 8-DPSK signal.

The chromatic dispersion penalty induced on 8-DPSK transmission with self-beating detection has been analyzed in [71]. The procedure is the same outlined in Chapter 6. The ISI has to be taken into account by considering the memory effect introduced by chromatic dispersion.

The ISI reduces the dynamic range of the modulation in a deterministic way, thus the penalty can be evaluated as average EOP.

This penalty is plotted in Figure 9.48. From the figure it is derived that the line residual dispersion implies a penalty larger than 3 dB, which for sure cannot be considered small. Thus per channel dispersion compensation is needed.

This compensation is realized in two steps: a fiber-based per channel dispersion compensator is placed in front of the transponder (physically in the ad hoc subrack where are allocated the 100 Gbit/s transponders) to compensate 200 ps of chromatic dispersion with the correct slope, the remaining correction is realized via an electronic dispersion compensator. Using a high dispersion dispersion-compensation fiber (DCF), 200 ps are attained for a 33 GHz wide signal with about 4 km of fiber; thus both cost and occupied space are not very large even if this is a per channel compensator.

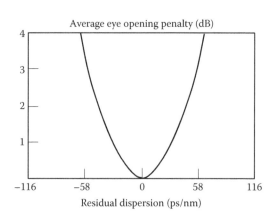

FIGURE 9.48
Chromatic dispersion-induced penalty in a 100 Gbit/s 8-DPSK with self beating detection. (After Seimetz, M. et al., *IEEE J. Lightwave Technol.*, 25(6), 1515, 2007.)

This design is due to the fact that the electronic equalizer will be useful to compensate other ISI causes also, like PMD or some forms of nonlinearities, thus it has not to be overloaded by the need of compensating a great residual dispersion.

A study on design of electronic compensation in M-DPSK systems is reported in [72] where an optimized algorithm is proposed for small state number MLSE to be applied to M-DPSK systems.

To adapt to our case the results in [60] requires passing from a bandwidth of 20 GHz for a 4-DPSK at 40 Gbit/s to a bandwidth of 33 GHz from an 8-DPSK at 100 Gbit/s; this can be done by scaling the penalty taking into account the different distances on the phase circle between nearby constellation point and scaling the dispersion corresponding to a certain penalty of 1.65^2, which is about a factor 2.7.

The resulting penalty curve is shown in Figure 9.49, and for the residual dispersion of 75 ps the penalty is 1.25 dB.

9.5.1.2.2 Polarization Mode Dispersion

Under a qualitative point of view, the impact of PMD is the same suffered by other modulation format. A detailed analysis has been carried out in [74,75], taking into account both the average penalty and the outage probability due to instantaneous bad relation among the SPS and the transmitted polarization.

Also PMD of a standard SSMF has to be compensated and since an electronic dispersion compensator is already foreseen for the residual dispersion, the easier way is to design the compensator also for PMD compensation.

This is done in [72], adopting the MSLE already described in the previous section, the PMD related penalty results as in Figure 9.50. Thus, for $\Delta\tau = 0.1\sqrt{1500} \approx 4$ ps, the associated penalty is 0.25 dB.

9.5.1.2.3 Kerr-Induced Degradation (SPM, XPM, and FWM)

The effect of SPM is considered in [71] in great detail. Naturally, the overall penalty depends both on the span input power (that is assumed the same for each span) and on the number of spans.

The SPM penalty plot deduced in the regime of small penalty per span is reported in Figure 9.51 taking into account also the mitigating effect of residual dispersion. The

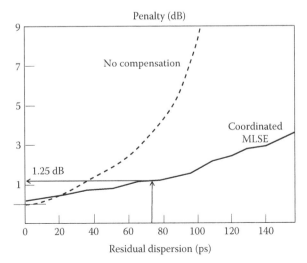

FIGURE 9.49
Penalty curve after optical and electronic residual dispersion compensation in the case of a 100 Gbit/s 8-DPSK system transmitted on a line designed for 40 Gbit/s. (After Salehi, J.G. and Proakis, M., *Digital Communications*, 6th edn., McGraw Hill, s.l., 2008.)

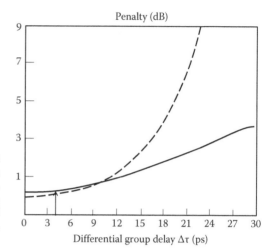

FIGURE 9.50
Penalty curve after electronic PMD compensation in the case of a 100 Gbit/s 8-DPSK system transmitted on a line designed for 40 Gbit/s. (After Freckmann, T. et al., Joint electronic dispersion compensation for DQPSK, *Proceedings of Optical Fiber Conference/International Optical and Optoelectronics Conference OFC/IOOC 2008,* San Diego, CA, IEEE/OSA, s.l., 2008, p. OTuO.6.).

sensitivity to the nonlinear phase distortion due to SPM is greater for M-DPSK with respect to IM (see Chapter 6) due to the number of levels and to the fact that the information is coded in the phase; thus it is directly affected by the Kerr-induced phase fluctuations.

Nonlinear induced crosstalk can be classified as in case of IM systems in XPM-induced and FWM-induced crosstalk. In both cases, the decrease of the spectrum width with respect to the 80 GHz for which the system is designed helps alleviate the effect.

XPM-induced crosstalk is studied in [75] where it is demonstrated that, in our conditions, SPM is dominant. Nevertheless, XPM is particularly dangerous for this modulation format, due to the fact that the power is constant.

This can be intuitively understood looking at Equation 4.41 that expresses the nonlinear phase in the case of a WDM comb. While in the case that the intensity of the channels fluctuates, the walk off due to dispersion generates a mixed situation in which some channels do not contribute due to the absence of power; this does not happen in the case of a constant power modulation format.

Naturally, polarization fluctuations and phase differences due to walk off still work to reduce XPM efficiency, but in the same general situation, XPM is more efficient for constant power modulation formats.

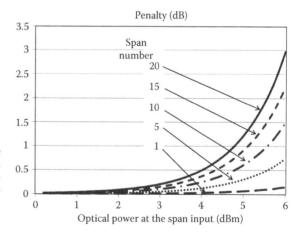

FIGURE 9.51
SPM-induced degradation in the case of the 100 Gbit/s 8-DPSK transported on a 40 Gbit/s line. In the curves, the interplay of SPM with the residual dispersion is taken into account. (After Seimetz, M. et al., *IEEE J. Lightwave Technol.*, 25(6), 1515, 2007.)

Starting from the model reported in [75] and imposing the condition of small penalty, it is possible to determine the penalty in our conditions.

For high power channels, XPM causes a plateau, not differently from other power dependent effects, but in the range of the possible power in this class of systems, the penalty is so low that the plateau can be neglected.

As far as FWM is concerned, the same model used in Chapter 6 can be reapplied here: FWM-induced crosstalk can be considered as an additional noise, due to its variability that is almost random for the effect of the walk-off. Reapplying the evaluation of Chapter 6, the FWM-induced penalty results to be about 0.5 dB.

9.5.1.2.4 Linear Crosstalk

The analysis of the impact of the linear crosstalk can be done in complete analogy to what was done in Chapter 6 for IM systems. However it has to be taken into account that the worst case for linear crosstalk is constituted by a 100 Gbit/s channel with only 40 Gbit/s channels around, as a matter of fact, the binary 40 Gbit/s channels have a wider spectrum so causing higher linear interference.

The evaluation has been done with the method of [74] both for the case of all DPSK channels and for the case of a single 100 Gbit/s channel with 40 Gbit/s channels around. The penalty curve is plotted in Figure 9.52.

In our case, it results a penalty of about 0.15 dB for the case of complete substitution of the channels with 100 Gbit/s 8-DPSK (when $\Delta\omega/2\pi R_S = 6$) and 0.3 in the case of an isolated 8-DPSK channel among a 40 Gbit/s comb (where $\Delta\omega/2\pi R_S = 5$).

9.5.1.2.5 Penalty Analysis Summary

The summary of the penalty analysis is reported in Table 9.10. The estimated SN_o required to reach the desired performances is 21 dB, smaller than the about 22.3 dB obtained from the draft design.

Second step is the simulation, which here is carried out considering the evolution of the deterministic part of the signal and adding the effect of noise analytically.

This confirms the effectiveness of the draft design and the fact that the 8-DPSK 100 Gbit/s signal performs quite better with respect to the original binary 40 Gbit/s signal for which the link was designed.

This is mainly due to the smaller value of the signal optical bandwidth, which is 33 GHz instead of 40 with a remarkable difference of 19%, but also the low value of the required SN_e has its role.

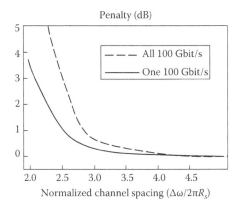

FIGURE 9.52

Linear crosstalk-induced penalty both in case of a single 100 Gbit/s channel and in case of a complete substitution of all the channels with 100 Gbit/s.

TABLE 9.10

Summary of the Penalty Analysis for the 8-DPSK 100 Gbit/s Designed to Work with a Line Optimized for 40 Gbit/s NRZ

100 Gbit/s 8-DPSK NRZ on a 40 Gbit/s link		
System length	2000 km	
Channel speed	100 Gbit/s	8-DPSK—Direct Detection, span length 50 km, Standard FEC
Channel spacing	200 GHz	Extended C band, 38 nm in the third window, that is 4560 GHz, 22 channels at 100 Gbit/s
Ideal SN_e (dB) ($B_e = 4.8R_s$)	20.0	
Optical to electrical bandwidth ratio	1.8	
Ideal SN_o (dB)	18.5	
	Margin in dB	
Transmitter dynamic range	1.0	DPSK Transponder with phase TiNbLi modulator
Linear crosstalk	0.5	Gaussian selection filter (single stage AWG), $B_o = 4.8R_S$ Channel spacing $\Delta f = 6R_S$ (see Figure 5.2)
Nonlinear crosstalk	0.5	Use of SSMF to thwart FWM, post-compensation with DCF
SPM and chromatic dispersion interplay	1.75	40 Gbit/s optimized dispersion map (see Table 6.5)
		Per channel optical and electronic dispersion compensation at the receiver. Used an MLSE equalizer (see Figure 9.49)
PMD	0.25	MLSE equalizer at the receiver
Aging	1	—
System margin	3	—
Polarization effects	1	Polarization depending losses, connectors misalignments, etc.
Total SN post code	27 dB	—
Code gain (dB)	6	
Total SN pre-code	21	To verify the transmitted power for SPM assessment To confirm with simulation

This example shown as 8-DPSK is quite promising for transmission of 100 Gbit/s signals on links designed for 40 Gbit/s.

9.5.2 Practical Multilevel Systems: Transmitting 100 Gbit/s on a 10 Gbit/s Line by 4QAM

This is a much more difficult problem with respect to that faced in the last section. In order to match the signal bandwidth for which the system is designed, the optical bandwidth should be reduced of a factor 10, and this means using 1024 levels at least.

However, similar to almost all the systems designed for 10 Gbit/s, our reference system also has an optical pass-band (essentially determined by the multiplexer and the demultiplexer) greater than the bandwidth of a 10 Gbit/s signal: as a matter of fact $M_{oe} = 2$.

Since the available unilateral optical bandwidth is 20 GHz (the channel shape is approximately Gaussian) we can also accommodate a signal with a wider spectrum. As a trade-off between spectrum width and required signal to noise, we will choose a 64 levels 4QAM with a unilateral optical bandwidth of 16.6 GHz thus reducing M_{oe} to 1.2.

The line data are summarized in Table 6.2.

9.5.2.1 Ideal Signal to Noise Ratio Requirements

The ideal SN_e required to achieve an error probability of 10^{-12} can be deduced from Figure 9.42 for constant power 64-4QAM obtaining 21.8 dB.

The symbol rate is 16.6 Gsymbol/s, thus using NRZ transmission, the optical unilateral signal bandwidth is 16.6 GHz.

With the resulting value of $M_{oe} = 1.2$, the required SN_o is equal to 24 dB.

9.5.2.1.1 Power Budget

The parameters of the components that are used at the transmitter and at the receiver are reported in Table 9.11, where it is imagined that the receiver is built using Si-SiO$_2$ chips as shown in Section 9.3.2.4.

At the transmitter, in order to overcome 15 dB of loss of modulator and transmitter passives, a high power external cavity tunable laser is selected, to assure 13 dBm of emitted power and a very small linewidth, which in line with the best commercial lasers we will assume equal to 500 kHz.

In this way, the power on the booster amplifier is −7 dBm per channel, that is 7.4 dBm total. This power needs an attenuation not to damage the booster, but it is good to face tolerances and aging.

At the receiver, the total receiver loss is 21 dB, which requires the use of an APD with a sensitivity of −25 dBm over an optical bandwidth of 20 GHz.

Fixing the power on the APD equal to −19.5 dBm to be able to neglect thermal and multiplication noise in the power budget, the power per channel at the preamplifier output is 0.5 dBm, for a total power at the preamplifier output of 20 dBm, the same power exiting from the preamplifier when the system works at 10 Gbit/s.

9.5.2.1.2 Draft Design SN Estimation

The estimation of the ASE limited optical signal to noise ratio can be carried out as in all the other cases: evaluating the ASE power in the line after the design.

In our case, the ASE power is −16.9 dBm, which provides an SN_o of 17.5 dB at the preamplifier output in the assumed hypothesis for the channels power.

TABLE 9.11

Transmitter and Receiver Data for 1024-QAM Designed to Be Used on a 10 Gbit/s Line

Receiver Parameters	
Photodiode Sensitivity	−28 dBm
Polarization beam splitter loss	2 dB
AWG loss	9 dB
Integrated optics chiploss	8 dB
Patchecords and other losses	1 dB
TOTAL RECEIVER OPTICAL LOSS	20 dB
Transmitter	
Laser emitter power	13 dBm
modulator loss	5 dB
AWG loss	9 dB
other losses	1 dB

Adding an advanced FEC gain of 11 dB an SN_o of about 28.5 dB is obtained, which is greater than the required value of 24 dB.

However, the difference between the power budget value, comprising the FEC gain, and the theoretical value is only 3.5 dB. So small a value does not allow a normal amount of penalties to be allocated, thus even if theoretically the system can work, in practice, it is not feasible due to nonideal propagation effects and to the margins necessary to take into account components characteristics fluctuations and aging.

In order to increase the SN_o to allocate the penalties, it is needed to increase the power per channel along the line without increasing the noise. Moreover, the impact of nonlinear effects has to be carefully controlled.

This can be done in two ways: either regulating the line amplifiers and the preamplifier to increase the output power or decreasing the number of channels so that the available power can be distributed among fewer channels.

In both cases naturally, great attention has to be paid not to increase the power per channel too much to generate a high penalty due to nonlinear effects.

Here we imagine, as almost always in practical systems, that amplifiers configuration can be controlled within certain limits via the system element manager.

As already done in the case of 40 Gbit/s transmission (see Chapter 6), we reconfigure the in line amplifiers so to decrease the output power up to 18.5 dBm. Moreover we divide by half the number of channels fixing to 100 GHz the channel spacing. This contemporary increases the power per channel at the output of in line amplifiers up to 2 dBm and takes under control linear and nonlinear crosstalk that otherwise will increase due to the passage of the optical signal bandwidth from 10 to 16.6 GHz.

The resulting configuration conveys in any case the huge capacity of 4.5 Tbit/s, with an upgrade of a factor 5 with respect to the 900 Gbit/s of the original system.

The SN_o calculated from the power budget increases to 30 dB, taking into account 11 dB of super FEC gain thus providing a margin of 5 dB to allocate penalties.

9.5.2.2 Penalties Analysis

As usual in this section all the relevant penalties coming from the different system impairments are evaluated and added to verify that the overall SN_o is sufficiently high to guarantee a correct system working.

9.5.2.2.1 Transmitter Dynamic Range

Due to the small margin available, the transmitter is particularly accurate, reducing the dynamic range penalty from 1 dB, which is considered a standard transmitter performance, to 0.3 dB. This is done with an accurate balance of dual arms modulators used in the transmitter [32].

9.5.2.2.2 Linear Crosstalk

We have an optical bandwidth of the transmitted signal of 16.6 GHz with a spacing of 100 GHz and a Gaussian filter of 20 GHz. Thus the ratio between the optical bandwidth and the symbol rate is 1.2 and the normalized spacing is about 6. In this condition, repeating the procedure carried out in Chapter 6 where the linear crosstalk is considered like an additional noise source, a penalty of 0.2 dB is obtained.

9.5.2.2.3 Phase Noise

Both the used optical sources, the transmitting laser and the LO, are external cavity lasers, with a linewidth of 500 kHz.

The global linewidth to symbol rate ratio is then 3×10^{-5}. In this condition, the phase noise is completely negligible.

9.5.2.2.4 Nonlinear Depolarization Effect

The depolarization occurring during fiber propagation due to Kerr effect is analyzed in [65,77]. Applying the method reported in [65] to our case, the propagation of a 16.6 GHz signal along 3000 km with a span length of 70 km and a power at the in-line amplifier output of 2 dBm, the depolarization factor δ_{pol} is about 0,0025.

Equation 9.81 can be rewritten as a function of the SN_o, obtaining for our system working with a pre-FEC SN_o equal to 19 dB a plateau due to nonlinear depolarization equal to 26 dB and a penalty due to depolarization equal to 0.9 dB.

The plateau is sufficiently far from the system working point, but the penalty is not so small, considering the small margin we have.

9.5.2.2.5 Nonlinear Crosstalk

Even if the power propagating along the line is quite high, the nonlinear crosstalk is limited by the increased channel spacing.

The impact of FWM can be estimated by the model detailed in Chapter 6 obtaining a penalty of 0.15 dB, while XPM results to be negligible

9.5.2.2.6 Residual Chromatic Dispersion, Self-Phase Modulation and PMD

Residual chromatic dispersion, SPM and PMD are the relevant ISI sources in this system.

As far as PMD is concerned, using the model of Chapter 6, it is easy to determine that the penalty due to PMD is completely negligible in this case (below 0.1 dB).

The situation is quite different regarding the interplay between SPM and DGD.

The system is designed optimizing the dispersion map for a transmission at 10 Gbit/s. Thus postcompensation is used, with an under-compensation of 1020 ps/nm that in our case means an uncompensated dispersion of 141 ps. Taking into account that the transmitted pulses are 60 ps long residual dispersion requires per channel compensation.

Coherent detection renders available both amplitude and phase of the received signal so that very smart electronic processing is possible, simultaneously targeting the equalization of several impairments [51,78].

Here we will use a suboptimal design: that is, we will assume that the ISI equalizer (caring for chromatic dispersion, PMD, and any form of nonlinear ISI), the LO frequency locker, and the decision device are separated circuits that do not exchange information, but for the link between decision and frequency lock.

As a matter of fact, decision driven frequency lock is operated so that the decision device provides a feedback to the frequency lock control signal generator. A wide review of all the available alternatives that is very useful to analyze other design possibilities is reported in [77].

As far as the ISI equalizer is concerned, we will assume to use a small number of states MLSE with the objective to compensate all the ISI deriving from the interplay between residual dispersion and SPM.

Such an MLSE can be based on a small number of sample per bit; in literature it is clearly demonstrated that a number of samples greater than 4–5 per bit interval is of small utility if the number of states of the MLSE is not huge [78] and three samples seems to be the optimum value for small states MLSE [79].

Thus a sampling rate of 50 Gsamples/s is assumed, which is within the reach of very fast 32 nm CMOS electronics, as demonstrated by literature on predistortion compensation (see Chapter 5).

A very general method for the evaluation of the performances of an MLSE in the case of multilevel transmission is reported in [79]. This can be applied in our case with suitable modification that takes into account both the different receiver and the modulation format.

The results presented here can be compared with those in [79] taking into account that, in a very first approximation, the ISI-induced penalty depends only on the ratio between the pulse broadening and the pulse width and on the distance between first neighbors in the signal constellation. This is evident also from the data reported in [79].

Thus, the penalty due to ISI for quadrature coherent detection can be plotted introducing the ISI index already used in the previous section as the ratio between the pulse broadening $\Delta\tau$ and the pulse width $T_S = 1/R_S$.

Considering a 16 states MLSE and a 4 states MLSE, as described in [79], the ISI-induced penalty is represented in our case in Figure 9.53 for different numbers of MLSE states.

In order to use data in Figure 9.53, the ISI index has to be related to the physical parameters of the link, taking into account the interplay between SPM and chromatic dispersion.

In our case, we are propagating an NRZ signal. The pulse broadening due to the interplay between SPM and dispersion can be simulated and since this effect depends on the pulse energy, the broadening will be different from pulse to pulse since different energies are transmitted in correspondence to different symbols.

This effect could be taken correctly into account by evaluating the broadening for each pulse of the constellation and defining a different error probability for different symbols. For sake of simplicity, and also since a very detailed analysis of this effect is not so useful due to the great number of other approximations, we will simply consider the worst case broadening. The worst case could be not relative to the most powerful pulse, depending on the partial compensation between SPM and dispersion at the different powers; all the constellations have to be considered in any case (compare Chapter 6).

The ISI is plotted versus the power at the input of a span in Figure 9.54 that is obtained by simulating the propagation of a single pulse through the line. From the figure, it is evident that an optimum system working point exists, as anticipated in Chapter 4, in correspondence of which the pulse broadening is minimum due to the interplay between SPM and DGD. In this case, the optimum transmitted power is about 0 dBm.

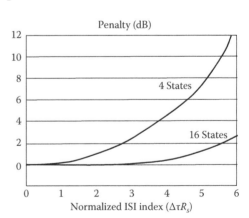

FIGURE 9.53
Penalty due to ISI versus the ISI index for different types of low number of states MLSE. (After Alfiad, M.S. et al., *IEEE J. Lightwave Technol.*, 27(20), 4583, 2009.)

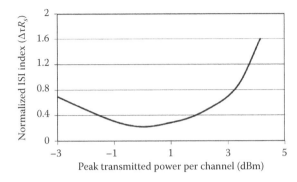

FIGURE 9.54
ISI index versus the power per channel at the input of a span.

The system however cannot work with this low transmitted power, which does not assure a sufficient SN to be attained. Thus a greater ISI has to be accepted. At the power at the span input equal to 2 dBm, the ISI factor is about 0.45, that is, the pulse width increases almost 50%.

From Figure 9.53, if we select an MLSE with 16 states, this ISI factor implies a penalty of 0.1 dB.

9.5.2.2.7 Penalties Analysis Summary

The summary of the penalty analysis relative to this case is reported in Table 9.12, the resulting pre-FEC SN_0 is 18.65 dB, slightly smaller than the 19 dB deriving from the draft design, thus suggesting that the system is feasible.

In this case, simulation is particularly important due to the small margin between the draft design and the result of the penalties analysis.

As usual we have performed a simulation of all deterministic phenomena adding the noise through a theoretical model. The simulation result completely confirms the analysis, providing a situation slightly better than that depicted by the penalty analysis, confirming that the bulk of the adopted approximations are worst cases.

9.5.3 Practical Multilevel Systems: Transmitting 100 Gbit/s on a 10 Gbit/s Line by PolSK

In the above section, we have designed the 100 Gbit/s system so that the spectrum fits into the available optical bandwidth. This has brought us to use a great number of levels, with the consequent complexity at the system terminals and the need of maintaining a high line power to achieve the required high SN_0.

In the following section, we will select another option: we will accept a strong distortion of the 100 Gbit/s signal due to system filters that we will then try to recover through an ISI equalizer.

This will allow us to maintain the number of levels lower thus simplifying the system terminals.

9.5.3.1 Draft Design and Power Budget

Reducing the number of levels quite below 64 means to increase the bandwidth above the value for which the system has been designed thus causing pulse distortion due to narrow optical filtering.

Thus pulse spreading due to tight optical filtering increases the ISI parameter and, with a suitable design of an equalizer, it could be at least partially corrected at the receiver.

TABLE 9.12

Transmitter and Receiver Data for 8-DPSK Designed to Use a 10 Gbit/s Line

100 Gbit/s 64-4QAM NRZ on a 10 Gbit/s Link		
System length	3000 km	
Channel speed	100 Gbit/s	64-DPSK—Coherent Detection, span length 70 km, Super FEC with 11 dB gain
Channel spacing	100 GHz	Extended C band, 38 nm in the third window, that is 4560 GHz, 45 channels at 100 Gbit/s
Ideal SN_e (dB) ($B_e = 4.8R_s$)	21.0	
Optical to electrical bandwidth ratio	1.2	
Ideal SN_o (dB)	24	
	Margin in dB	
Transmitter dynamic range	0.3	DPSK transponder with phase TiNbLi modulator
Linear crosstalk	0.2	Gaussian selection filter (single stage AWG), $B_o = 1.2R_s$ Channel spacing $\Delta f = 3R_s$ (see Figure 5.2)
Nonlinear crosstalk	0.15	Use of SSMF to thwart FWM, post-compensation with DCF
SPM and chromatic dispersion interplay	0.1	10 Gbit/s optimized dispersion map (see Table 6.5)
		Per channel optical and electronic dispersion compensation at the receiver. Used an MLSE equalizer (see Figure 9.49)
PMD	0.0	MLSE equalizer at the receiver
Aging	1	—
System Margin	3	—
Nonlinear depolarization	0.9	
Polarization unbalance	0.2	Polarization depending losses, connectors misalignments, etc.
Total SN post code	29.65 dB	—
Code gain (dB)	11	
Total SN pre-code	18.65	To verify the transmitted power for SPM assessment To confirm with simulation

The unilateral bandwidth of the signal generated after 8-PolSK modulation is around 33 GHz, against the 20 GHz available on the line (the system is designed with $M_{oe} = 2$ for a 10 Gbit/s channel).

9.5.3.1.1 Ideal SN_e and SN_o

In this case, the required SN_e to reach an error probability of 10^{-12} can be deduced from Figure 9.43 resulting to be 20.7 dB.

Also in this case, the corresponding SN_o can be obtained in the usual way since M_{oe} is smaller than one, but with greater than 0.5 (in a very first approximation the ratio between the optical bandwidth and the electrical one is 0.61). As a matter of fact, beyond 0.5 the equation we are using to relate the optical and the electrical SN is no more valid.

The requirement in terms of ASE-limited SN_o results to be 25.8 dB.

9.5.3.1.2 Power Budget and Polarization Dynamics

In order to try a design using Direct Detection, we will select interferometric receiver. The relevant characteristics of receiver components are summarized in Table 9.13; all the power budget equations are fulfilled as can be easily derived.

TABLE 9.13

Transmitter and Receiver Data for 8-PolSK Designed to
Use a 10 Gbit/s Line

Transmitting Chain	
Source laser maximum emitted power	13 dBm
Laser linewidth (external cavity)	500 kHz
Mach–Zehnder Lithium Niobate modulator loss	7 dB
Multiplexer (multistage AWG × 90) loss	9 dB
Patchcords, connectors, monitoring taps loss	1 dB
Detection Chain	
High efficiency APD sensitivity at 33 Gsymbols/s	−19.5 dBm
Demultiplexer (multistage AWG ×90) loss	9 dB
90° Hybrid loss	3 dB
Splitters loss	3 dB
Delay line loss (low loss component)	1 dB
Patchcords, connectors, monitoring taps loss	1 dB

The situation would be very similar if the coated diodes receiver would be chosen.

The ASE power in the signal bandwidth is −15.4 dBm so that SN_o = 17.44 dB.

This value, even after the addition of 11 dB gain of a super-FEC, is only a bit higher than the required 25.8 dB leaving no sufficient margin for the penalties allocation. As a matter of fact we have to expect at least 6 dB of penalties and margins.

In order to increase the power budget we can follow the same line used for the 4QAM and cut by half the number of channels.

Changing two 10 Gbit/s channels with one 100 Gbit/s, a comb of 100 GHz spaced channels at 100 Gbit/s is obtained formed by 45 channels as in the case of 64-4QAM.

This allows, with an amplifier output power slightly reconfigured from 20 to 20.5 dBm, to bring the power per channel at the preamplifier output to 4 dBm, bringing the SN_o from 17.44 to 21 dB.

So high a power per channel, however, creates a very detrimental effect for a polarization modulated system.

Applying the method reported in [65] to our case, the propagation of a 33 GHz signal along 3000 km with a span length of 70 km and a power at the in-line amplifier output of 4 dBm, the depolarization factor δ_{pol} is about 0,01.

Equation 9.28 tells us that, every time there is a part of the information coded in the field polarization, that is, every time there is the need of polarization tracking, high speed polarization fluctuations introduces a plateau in the SN_o and thus in the BER. This plateau is expressed by Equation 9.28 itself.

The only way to avoid this plateau would be to increase the speed of the polarization tracker to follow these fast fluctuations also, but in the case of Kerr-induced depolarization this is not possible due to the speed of the effect.

In our case, the plateau is 20 dB; thus nonlinear polarization fluctuations set an upper limit to the system SN_o that is too low to allow transmission at 3000 km.

The only way to go is to cut the link length.

In order to do that, let us imagine placing a double terminal in an intermediate site to regenerate all the signals at half the length.

In this situation, maintaining 90 channels at 100 Gbit/s, we have two identical systems 1500 km long with a capacity of 9 Tbit/s.

Maintaining the amplifiers output power at 21 dBm, the power per channel at the input of a span is 1.45 dBm and the depolarization index is δ_{pol} = 0.0012 for a limit SN_o = 29.2 dB, a quite high number.

The effective value of the SN_o estimated through the power budget and the noise power evaluation is 21.6 dBm that becomes 32.6 after the super-FEC gain.

This leaves a margin of 5.5 dB for the allocation of the penalties. This margin is not very high, but should be sufficient in our case.

Also in this shorter configuration, it is easy to demonstrate that with the figures reported in Table 9.13, the power budget equations are satisfied both at the receiver and at the transmitter.

9.5.3.2 Penalties Analysis

9.5.3.2.1 Intersymbol Interference (Residual DGD, SPM interplay, PMD, Optical Filtering)

ISI is mainly caused by three factors:

1. SPM and residual dispersion interplay
2. PMD
3. Tight optical filtering

In a regime of small pulse distortion we will assume that the overall ISI index can be obtained from the individual contributions.

Dividing SPM and dispersion from PMD we can write

$$p_{out}(t) = \hat{A}_{re}\hat{P}\hat{K}\hat{A}_{tr}\,p_{in}(t) \tag{9.84}$$

where the four operators acting on the envelop of the input pulse $p_{in}(t)$ are: the transmitter AWG ($\ddot{A}_{tr}$), the receiver AWG ($\ddot{A}_{re}$), the interplay between dispersion and SPM taking into account in-line compensation ($\hat{K}$), and the PMD ($\hat{P}$).

To each process is associated a pulse spreading τ so that, for example,

$$\tau_{re}^2 = \int t^2\hat{A}_{re}p_{in}(t)dt - T^2 \tag{9.85}$$

where T is the width of the original pulse. Since NRZ pulses are symmetric around their axis and the first filter of the chain is Gaussian (thus symmetric too) the overall pulse broadening can be found in a first approximation using the formula

$$\tau_{out} = \sqrt{\tau_{re}^2 + \tau_p^2 + \tau_k^2 + \tau_{tr}^2} \tag{9.86}$$

as graphically show in Figure 9.55. This rule is rigorously true for a cascade of linear processes with one nonlinear instantaneous process (see Appendix 3). Here we approximate our model assuming a concentrated effect for the interplay between SPM and dispersion, while they are distributed phenomena.

This will allow us to proceed with the penalty analysis considering the different causes of ISI separately.

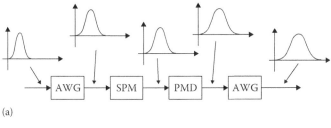

(a)

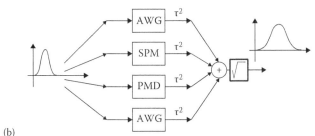

(b)

FIGURE 9.55
Approximation used to evaluate the overall ISI index of the system. (a) Physically, all the ISI-related phenomena works in series; (b) in the approximation they work in parallel and at the end, the pulse widening are added.

The analysis can be done with the same methods used for IM systems (see Chapter 6), taking into account the fact that the rapid polarization fluctuations decrease the SPM efficiency. The ISI index in our case for 8-PolSK is reported in Figure 9.56.

The influence of PMD results to be negligible, as it can be demonstrated evaluating the ISI index by multiplication of the PMD parameter by the system length and comparing with the symbol interval.

As far as the ISI index is relative to optical filtering it can be determined by simulation by simply reconstructing the relevant signal and filtering it with the filter transfer function simulating the cascade of optical filters along the transmission line (essentially the two AWGs).

The ISI index deriving from the too narrow filtering is reported in Figure 9.57.

At this point, even from a simple look at the ISI indexes, it is quite clear that transmission cannot happen without an ISI equalization.

The first choice for our design would be an MLSE. Unfortunately, the balanced MLSE requires at least three samples per symbol; thus a minimum sampling rate of 99 Gsamples/s would be required, that currently is beyond the possibility of so complex a circuit.

This consideration not only excludes the MLSE, but also all the digital compensation strategies at the receiver, which needs oversampling of the received signal.

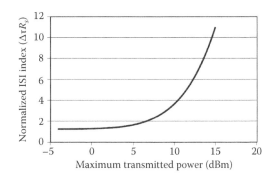

FIGURE 9.56
Normalized ISI index due to the interplay between SPM and chromatic dispersion in our 8-PolSK transmission example.

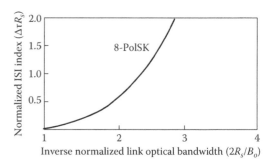

FIGURE 9.57
Normalized ISI index due to tight optical filtering in 8-PolSK assuming Gaussian filtering. (After Jiang, Z. et al., *Nat. Photon.*, 1, 463, 2009.)

The only possible solution is an analog per channel compensation. A very interesting method that can solve this kind of problems has been presented in [80]. The idea is to use an optical pulse shaper to predistort the optical signal at the transmitter in order to obtain the wanted pulse shape.

An AWG is used as spectrum analyzed so that, once the spectral components of the transmitted signal are separated, they can be filtered with an amplitude and phase mask to create an approximation of the Fourier transform of the wanted pulse. After that, the spectral components are recombined with another AWG.

If only small changes in the channel characteristics has to be compensated, the intensity and phase masks can be tuned, for example, by temperature controller, if fast channel estimation and tracking is needed, they can be substituted with modulators controlled by a DSP.

The scheme of this optical per channel compensator is presented in Figure 9.58 [81]. The entire pulse shaper has been also realized in a single InP-based monolithic chip, demonstrating the potentiality of this technique [82].

In principle, the chosen solution can compensate whatever amount of ISI can be generated; in practice the amount of ISI that can be compensated is limited by both component and system limitations.

The main limitations to the compensation method efficiency are the imperfect AWG working that causes the produced set of fields to deviate from the components of the Fourier transform of the input field, the needed approximation intrinsic in working with a discrete number of spectral components, and the presence of nonlinear propagation.

Let us call $F_0(\omega)$ the ideal Fourier transform of the impulse to be transmitted, so that at the receiver it is compressed to completely eliminate the ISI.

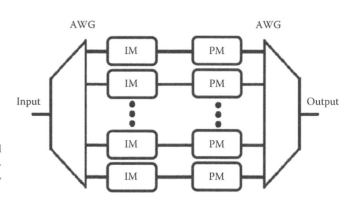

FIGURE 9.58
Block scheme of per channel optical pulse shaper used to compensate ISI. (After Jiang, Z. et al., *Nat. Photon.*, 1, 463, 2009.)

The waveform generator approximates $F_0(\omega)$ with the following function

$$F(\omega) = rect_\Omega(\omega) \sum_{j=-v/2}^{v/2} [F_0(j\Delta\omega) + \phi_j] G_{\Delta\omega}\left(\omega + \frac{j\Delta\omega}{2}\right) \tag{9.87}$$

where
Ω is the global bandwidth
$G_{\Delta\omega}(\omega)$ represents the AWG response
$\Delta\omega$ is the step between two consecutive frequency samples, while the number of samples is equal to $v+1$.

Imagining to realize the pulse shaper with an InP integrated circuit, ϕ_j is the process-caused intrinsic error in reproducing the complex numbers $F_0(j\Delta\omega)$.

The potential performance of the precompensation method depends on the bandwidth and the number of samples at the AWG output, besides the precision with which the complex values $F_0(j\Delta\omega)$ are reproduced by the intensity and phase masks/modulators.

In Figure 9.59, an example of the ability of a pulse shaper based on Gaussian AWGs is shown. In Figure 9.59a, the spectrum to be reproduced is shown, while in Figure 9.59b it is superimposed to the spectrum at the output of the shaper of Figure 9.58 that is assumed to be realized with 100 frequency slices and Gaussian AWGs. The simulation parameters are an optical bilateral bandwidth of 200 GHz and typical tolerances for InP Mach–Zehnder Modulators that uses similar technology steps.

Since the configuration above is quite near to an ideal shaper even with practical parameters, it results that the main limitation to the compensation capability of this method is due to nonlinear effects.

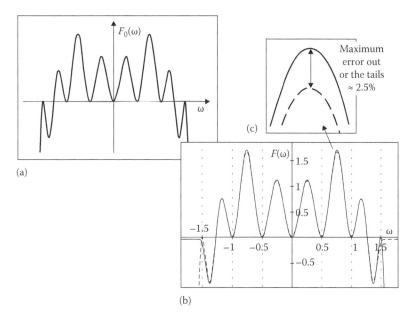

FIGURE 9.59
Example of the shaper accuracy in reproducing a complex spectrum: (a) test spectrum to be reproduced; (b) comparison of the spectrum created by the shaper with the original spectrum; (c) particular of one of the spectrum peaks where the maximum error is located if the tails far from the main spectrum band are not considered.

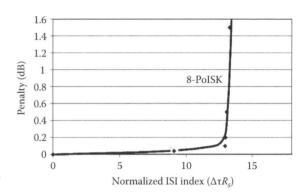

FIGURE 9.60
Residual penalty after pulse optical precompensation versus the ISI normalized index.

Passing through all the states from the predistorted form and the final form, the pulse presents in correspondence to some link sections sharp peaks that creates a great amount of nonlinear effect. These are not simulated when the final shape of a normal transmitted pulse is inverted to find the correction function; thus the shape of the predistorted pulse is not perfect.

This is also the cause of an expected marked difference between the behavior of phase modulation and polarization modulation. Polarization modulation is practically a form of double amplitude modulation and the impact of the nonlinear phase variations are less evident when the predistorted pulse propagates along the fiber link.

Passing to the compensation capacity, it has been assessed by simulation of the deterministic part of the single channel fiber propagation through the link that we are considering with a split step algorithm to determine the output pulse distortion.

This has been precompensated by a suitable transfer function and the output SN_o penalty has been derived and plotted in Figure 9.60.

Since the technology errors ϕ_j are random variables, the penalty curves plotted in the figure are worst case curves, where all the errors are in the same direction and at the 3σ point of their distribution.

The compensation capability of the method is impressive: to arrive to have a penalty of 3 dB the pulse has to broaden in noncompensated propagation up to 13 times its initial width: no practical postcompensation would be capable of doing that.

At this point, all the elements are available to evaluate the ISI penalty.

The ISI index due to the interplay between chromatic dispersion and SPM is about 0.9 while, as far as the tight filtering is concerned, the relative bandwidth is 0.6, implying a relative ISI index of 0.3.

Applying the ISI indexes summation rule, which holds with so small indexes, the total ISI is 0.95.

From Figure 9.60, we see that at such ISI, the optical compensator works very well and the residual penalty is lower than 0.1 dB.

9.5.3.2.2 Linear Crosstalk

Linear crosstalk is very important, due to the large spectra of the chosen modulation formats with respect to the channel spacing that is determined by the already deployed system.

A penalty of 1.0 dB is evaluated using the same model introduced in Chapter 6.

9.5.3.2.3 Nonlinear Crosstalk (FWM and XPM)

Chromatic dispersion under compensation makes XPM to be less efficient. However, the effectiveness of this design tool is not as good as in the case of 10 Gbit/s due to the fact that the dispersion map results are not optimized for the new modulation formats.

Applying the same model we have applied in Section 9.3.5, we derive in any case a negligible XPM penalty.

As far as FWM is concerned, the same model used in Chapter 6 can be reapplied here and the FWM-induced penalty is evaluated equal to 0.1 dB.

9.5.3.2.4 Nonlinear Polarization Fluctuations

If nonlinear polarization fluctuations are present, they have a smaller effect with respect to the case of coherent detection. As a matter of fact the fast fluctuations of the Jones matrix that cannot be tracked by the tracking circuit are due to its tight bandwidth.

Thus the overall effect is to smooth the Stokes parameters of a factor $(1 - \rho_{pol})$ and to transfer a part of the signal power proportional to the depolarization ratio to the noise.

Besides the plateau we have already introduced, the penalty relative to the depolarization can be written as in Equation 9.81 resulting to be 0.6 dB.

9.5.3.2.5 Summary of the Penalty Analysis

The results of the penalty analysis are summarized in Table 9.14.

At the end of the penalty analysis, the required SN_o is exactly identical to that evaluated by the draft design.

TABLE 9.14

Summary of the Penalty Analysis for the 8-DPSK and 8-PolSK 100 Gbit/s Designed to Work with a Line Optimized for 10 Gbit/s NRZ

	8-PolSK	
Phenomenon	(dB)	Notes
Link length (km)	1500	
Max number of channels	90	9 Tbit/s capacity
Ideal SN_e (dB)	20.7	
Electrical to optical bandwidth ratio	0.6	NRZ transmission
Ideal SN_o	26	
Signal dynamic range	0.3	Higher modulators screening and accurate balance in the design.
Linear crosstalk	1.0	Channel spacing 100 GHz
Nonlinear crosstalk	0.1	SPM dominant regime, XPM can be neglected
Intersymbol interference (SPM, DGD, PMD)	0.1	
Nonlinear polarization	0.6	Due to the modulation format
Polarization dependent losses and other effects	0.5	A few of polarization effects are considered in detail
Aging	1	
System margin	3	
Total w/o FEC	32.6	
FEC gain (Turbo Code)	11	
Required SN_o	21.6	

This lack of margin is not really important since we have to remember that a margin of 3 dB is already inserted in the penalty analysis. On the other hand, the simulation is a bit better than the penalty analysis, confirming the system feasibility.

9.5.4 Practical Multilevel Systems: Native 100 Gbit/s Ultra-Long Haul Systems

The last dimensioning exercise we will do is the design of a native ultra long haul 100 Gbit/s system.

9.5.4.1 Draft Design

We will use Raman-assisted transmission and the scheme of the generic system span will coincide with that of Figure 9.61, a familiar scheme from Chapter 6, but without in-line Erbium Doped Fiber Amplifiers (EDFAs). The only EDFA is present as booster at the transmitter.

The intention is to limit the ASE as much as possible and to maintain a low power along the line to control nonlinear effects.

We will also adopt a span length of 30 km in order to decrease the ASE power as much as possible.

The line results to be composed of 257 spans, each of which is formed by 30 km of SSMF and 5 km of DCF pumped to obtain Raman gain. The Raman pumps are extracted at the DCF end so that propagation in the SSMF is completely passive.

The whole system is thus 9000 km, long. Since the design is inspired to a submarine application, no OADM is foreseen in the nodes.

As far as modulation is concerned, we select constant power 16-4QAM with coherent homodyne receiver. The value of the electrical signal to noise ratio allowing to reach a BER of 10^{-12} is in this case $SN_e = 17.5$ dB, really a low value for a 16 levels modulation format. The unilateral optical bandwidth of the signal is 25 GHz while the overall channel bandwidth as seen by the single channel is 38 GHz so that $M_{oe} = 1.5$ and the required optical signal to noise ratio is $SN_o = 18.7$.

Due to the fact that the signal has a 25 GHz large spectrum on each side of the carrier, the channel spacing is selected to be 100 GHz, so that 45 channels are transmitted in the extended C band.

Precompensation is adopted, placing the DCF at the transmitter, and a residual dispersion of 1000 ps/nm is left to compensate channel by channel to partially counterbalance nonlinear effects during propagation.

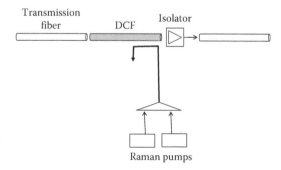

FIGURE 9.61
Block scheme of the line site of the Ultra Long Haul 100 Gbit/s system presented in this chapter.

TABLE 9.15

Transmitter and Receiver Data for an Ultra Long Haul System at
100 Gbit/s Adopting Constant Power 16-4QAM with Coherent
Detection

Transmitting Chain	
Source laser maximum emitted power	16 dBm
Laser linewidth (external cavity)	250 kHz
Mach–Zehnder Lithium Niobate phase modulator loss	7 dB
Mach–Zehnder Lithium Niobate Polarization modulator loss	6 dB
Multiplexer (multistage AWG × 90) loss	7 dB
Patchcords, connectors, monitoring taps loss	1 dB
Detection Chain	
High efficiency APD sensitivity at 25 Gsymbols/s	−20.5 dBm
Demultiplexer (multistage AWG × 90) loss	7 dB
90° Hybrids + splitters + insertion of an integrated chip	10 dB
Patchcords, connectors, monitoring taps loss	1 dB

9.5.4.1.1 Power Budget and SN_e

The main parameters of the components at the transmitter and at the receiver are reported
in Table 9.15. With 16 dBs emitted power and a transmitter loss of 20 dB, the power at the
booster input is 11.5 dB that is sufficient to saturate the booster without damaging it.

The booster gain results to be 6 dB and at the booster output and at the input of every
span after the precompensation and preamplification, the signal power is 2 dBm per channel, that is, 18.5 dBm total power.

At the receiver, the EDFA preamplifier is not needed since the Raman stage provides all
the needed gain of 17.5 dB, being dimensioned the same way of the in-line Raman amplifiers.

Moreover, a good APD has the required sensitivity at 25 Gsymbols/s to satisfy the power
budget.

The pumps of the Raman amplifiers can be optimized both for flatness and for gain;
however, also gain flattening filters with a very low loss (1.5 dB) are foreseen to improve
the gain flatness.

The design SN_e results to be 22.4 dB, with a great margin with respect to the ASE limited
value of 17.5 dB.

This is mainly due to the small value of the span bringing the system nearer to the optimum design that would be a continuous distributed amplification (see Chapter 6). Just to
verify this fact, if the design is done again with a span of 70 km, the SN_e goes down to 15 dB,
rendering in practice the transmission impossible.

It is to be observed that, under an SN point of view, the system could be much longer, up
to 10,000 km and more. Moreover, a quite higher SN is achieved by slightly increasing the
power at the span input. For example, bringing the power per channel at the span input at
3 dBm and the system at 10,000 km, the SN_e is 25.78, better than that we have now.

Both these improvements are impossible due to nonlinear effects, mainly SPM and nonlinear polarization rotation, which are the main limitations to the system performances.

9.5.4.2 Penalties Analysis

The penalties analysis in this case presents several new elements with respect to what is
done in Chapter 6 and in the previous sections.

9.5.4.2.1 Linear Crosstalk

The analysis of the impact of the linear crosstalk can be done in complete analogy to what was done in Chapter 6 for IM systems. In our case, the channels are quite near (100 GHz with a symbol rate of 25 Gsymbols/), thus even if the Gaussian filters are quite tight, the linear crosstalk penalty is quite high, of the order of 1.0 dB.

9.5.4.2.2 Residual Chromatic Dispersion, Self-Phase Modulation, and PMD

The link as it is designed uses dispersion under-compensation to balance at least in part the SPM phase shift. The residual dispersion of 1500 ps/nm requires per channel dispersion compensation.

We repeat here the assumption already done for the 64-4QAM to maintain decision and ISI compensation independent and to adopt decision driven LO phase lock.

Also in this case, a small number of states MLSE is suitable to correct the ISI coming from different dispersion sources and from SPM.

A sampling rate of 75 Gsamples/s is assumed, which should be within the reach of very fast 22 nm CMOS electronics.

Following the same procedure already used for 64-4QAM, the relation between the ISI index and the penalty after equalization is represented in Figure 9.53.

Moreover, the ISI index due to PMD and to the interplay between DGD and SPM are reported in Figures 9.62 and 9.63.

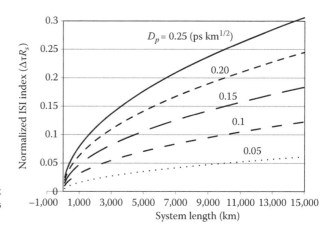

FIGURE 9.62
Pulse broadening due to PMD: ISI index versus the link length for different values of the PMD parameter.

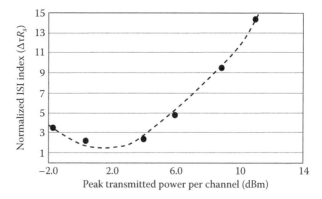

FIGURE 9.63
Inter-symbol interference due to interplay between SPM and DGD in the Ultra Long Haul 100 Gbit/s system considered in the text: ISI index versus the transmitted peak power per channel.

Figure 9.63, obtained by the simulation of nonlinear pulse propagation, and considering that it takes into account the interplay between dispersion and SPM, indicates an optimum power for which the ISI index is minimum.

In our case, the residual dispersion is quite high (1000 ps/nm) and the effective link length defined regarding nonlinear effects is of the order of 3598 km (257 spans and 14 km per span).

In these conditions, with a bilateral signal bandwidth of 50 GHz, both SPM and dispersion have a strong broadening effect inducing chirps with opposite sign so that the compensating effect is evident.

The best working point is around 2 dBm, which is exactly the chosen working point for our system.

However, even at the best point, the link is so long that the pulse in free propagation completely loses its shape and widens more than 50%.

Summarizing, at 2 dBm per channel, the ISI index is 1.5 due to SPM and chromatic dispersion and 0.25 due to PMD, for a global ISI index of 1.55.

From Figure 9.53, we conclude that adopting a 16 states MLSE, the corresponding penalty is about 0.2 dB.

9.5.4.2.3 Nonlinear Crosstalk: XPM and FWM

Also in this case, following [75], the system is in the SPM-dominated regime due to the adopted chromatic dispersion map.

As far as FWM is concerned, the same model used in Chapter 6 can be reapplied here: FWM-induced crosstalk can be considered as an additional noise, due to its variability that is almost random for the effect of the walk-off.

The effect of both XPM and FWM are however enhanced by the channel proximity, by the great bandwidth, and by the long distance. Globally, the FWM penalty is 1.3 dB and the XPM penalty is 0.4 dB.

9.5.4.2.4 Transmitting Laser and Local Oscillator Phase Noise

We will imagine using external cavity tunable lasers and, as a reference, we will assume the lasers parameters values listed in Table 5.5. The linewidth listed in the table is 550 KHz, really small with respect to the symbol rate of 25 Gsymbols/s. In our case, $\Delta\nu/R_s = 0.0002\%$. Looking at Equation 9.46 we see that the phase noise plateau is well beyond 55 dB, meaning, that at the required values of SN_e, the phase noise induces only a small power penalty that from Equation 9.78 can be evaluated in 0.1 dBm.

9.5.4.2.5 Depolarization due to PMD and Nonlinear Propagation

The analysis of the behavior of δ_{POL} where a WDM signal propagates through the fiber has been carried out in [65,77].

In particular, applying the methods developed in [65] to the present case, the plot reported in Figure 9.64 is obtained for the penalty due to the depolarization.

In the case under analysis where the power per channel at the span input is 2 dBm, the depolarization due to nonlinear effects is 0.0065.

The plateau due to depolarization is located for this value of δ_{POL} at $SN_o = 21.8$ dB, thus it is very near to the system working point, even if both the dispersion map and the transmitted power are optimized.

However, taking into account the gain of a super FEC of 11 dB and the value of 17.5 dB of ASE-limited required SN_e, a careful design should manage to make the system work.

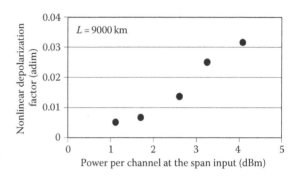

FIGURE 9.64
Nonlinear depolarization factor in the 9000 km
16-4QAM system under analysis versus the
power at the input of the span.

The penalty derived from the nonlinear polarization evolution is 3 dB and it is by far the largest penalty. In the presence of so large a penalty, the penalty addition rule in principle does not hold.

However in this case, this is the only penalty that is so high; so we will go on assuming that the penalties can be added, using the simulation of signal propagation as validation test.

9.5.4.2.6 *Polarization-Dependent Losses*

A slight amount of polarization dependent loss is always present in transmission systems due to both slightly misaligned connections and to components characteristics [83].

Due to the critical system design, this point has to be considered in detail and not included into the margins.

As a matter of fact, polarization dependent losses have a strong impact on all the very high performance systems where part of the information is coded into the field polarization [85,86].

The penalty due to the polarization dependent loss can be analyzed as in [84] as a function of the unbalance between the average amplitudes of the received field components along the reference polarization unit vectors.

The analysis shows that in order to maintain the penalty within 0.5 dB it is needed to maintain the unbalance within 2.5% with typical components parameters and system assembly accuracy, that is, to have a dichroism ratio (ratio between the difference of the two polarization components power and their sum) within −19 dB.

9.5.4.2.7 *Penalties Analysis Summary*

The summary of the penalties analysis is reported in Table 9.16.The required SN_o results to be 18.8 dB.

The draft design value is 22.4 dB, thus the system seems feasible.

Nevertheless, it is a key point to verify the assumptions by simulation. From simulation, the required SN_o results to be about 2 dB lower than that generated by the penalty addition rule, showing the fact that we are at the limits of validity of this method.

Analyzing simulation data, we can see that several points emerge where different phenomena interact. The main interactions are

- All forms of crosstalk are a bit less effective due polarization fluctuations, channel asynchronous clock, and nonlinear polarization fluctuation
- The worst case model for the linear crosstalk results to be too pessimistic
- Also the FWM penalty model results to be pessimistic mainly due to the walk-off caused by polarization fluctuations

TABLE 9.16

Summary of the Penalty Analysis for the 100 Gbit/s Ultra Long Haul System Adopting Constant Power 16-4QAM with Coherent Detection

	Native 100 Gbit/s Ultra-Long Haul System	
	16-4QAM	Constellation constituted by the vertices of an 8H hexahedral
	Constant Power	
System length	9000 km	Span length 30 km SSMF + 5 km DCF – 257 spans
Channel speed	100 Gbit/s	Bilateral optical bandwidth 50 GHz, symbol rate 25 Gbsymbols/s
Channel spacing	100 GHz	45 channels in extended C band – 4.5 Tbit/s capacity
Ideal SN_e (dB)	17.5	
M_{oe}	1.2	Overall optical bilateral bandwidth 76 GHz
Ideal SN_o (dB)	18.8	
	Margin in dB	
Transmitter dynamic range	0.3	External cavity tunable laser for transmission
Linear crosstalk	1	Overall Gaussian channel filtering function, $B_o = 2.2R_s$
Nonlinear crosstalk	1.7	FWM dominates
CD-SPM-PMD	0.2	16 Stages MLSE, total ISI factor <2
Phase noise	0.1	High output power external cavity laser, linewidth 250 kHz
Nonlinear polarization	3	SN_o and BER plateau at SN_o = 21.8 dB
Aging	1	
System margin	3	
Polarization effects	0.5	Taken into account in detail. Required an unbalance between the quadrature amplitudes not larger than 2.5%
Total SN post code	29.6	
Code gain (dB)	11	
Total SN pre-code	18.6	

9.6 Summary of Experimental 100 Gbit/s Systems Characteristics

Since currently there is no diffused offering of products for long distance transmission at 100 Gbit/s we cannot provide the list of realistic characteristics generated from a synthesis of the characteristics of real products.

Thus, we will do a sort of summary of a few experimental implementations of high speed optical transmission.

In Table 9.17, a few experiments using multilevel modulation are summarized. In all the cases, but the first, PDM is used, that is, two channels are sent on orthogonal polarizations and on the same bandwidth.

It is easy to observe that the signal produced, just for an example, using PDM 16-QAM is not so different from a 256-4QAM obtained by the Cartesian product of two 16-QAM constellations, but for the fact that the signals on the coordinate plans (e_1, e_2) and (e_3, e_4) are synchronous in time in the 256-4QAM transmission, while they are asynchronous in the case of PDM 16-QAM.

For 256-4QAM obtained with full QAM constellations, we see that the normalized minimum distance is 0,111 while it is 0,471 for one of the QAM channels.

As a matter of fact, the absence of correlation between the transmitted constellations destroy the capability of 256-4QAM to compress the signal bandwidth of a factor eight

TABLE 9.17

Review of the Main Features of a Few 100 Gbit/s Experiments Using Instantaneous Multilevel Modulation

				Serial Multilevel Transmission		
Modulation Format	Channel Capacity	WDM Parameters	Distance	Link Characteristics	Notes	References
NRZ PMD QPSK	100 Gbit/s	16 (C band)	2417 km	Recirculation over installed SSMF link 63 km long, all EDFA amplification	Coherent homodyne detection, manual polarization control, phase diversity No in line DCF	[1]
NRZ PMD QPSK	100 Gbit/s	164 (C + L bands) 50 GHz spacing	2550 km	Recirculation over SSMF link 65 km long, lumped Raman amplification in DCF before and after the span	Coherent homodyne detection, manual polarization control, phase diversity Off-line electronic equalization and post-processing	[2]
NRZ PMD QPSK	100 Gbit/s	155 (C+L bands) 50 GHz spacing	7200 km	Ultra low loss fiber ($\alpha = 1.66$ dB/km), mixed Raman and EDFA amplification, Raman back-propagating pumps, span 90 km	Coherent homodyne detection, phase and polarization diversity	[2]
NRZ PMD QPSK	100 Gbit/s	36 (C band)	401 km	Unrepeated link Remotely pumped Raman amplifier. Raman pump 5.5W at the far end at 1276 and transferred at 1430 through second order Raman. 30dBm output EDFA booster	Dispersion pre-compensation at the transmitter with DCF. Polarization and phase diversity coherent receiver. Electronic ISI equalization	[5]
NRZ PMD 16-QAM	171 Gbit/s	432 (C + Exte. L bands)	240 km	80 km long spans, mixed EDFA and Raman amplification with Raman back-propagating pumps. Extended L band	Digital coherent detection, homodyne with phase and polarization diversity	[6]

at the expense of a reduction in the receiver sensitivity, while PDM simply transmits two channels in parallel, each of which reduces the bandwidth only of a factor four but has also the sensitivity of a format with much less levels.

A common factor for all the experiments presented in the table is the coherent detection. As a matter of fact, the additional sensitivity provided by coherent detection is an important plus in 100 Gbit/s systems.

Coherent receiver is generally realized by digital signal processing, so that an Analog to Digital converter at a speed of the order of 50 GHz is generally employed immediately after the detection.

In the table, there is one Ultra Long Haul system, conveying an impressive capacity by the use of both C and L band.

The line site of this system is exactly like the scheme of Figure 6.28, but for the presence of three Raman pumps, which are needed to have a flat gain over all the huge exploited bandwidth, and the fact that two EDFA in parallel provides gain in C and L band. The long span (90 km) is compensated by the very low attenuation fiber (0.166 dB/km).

The first two systems presented in the table are for long haul application. The first experiment is interesting since its scope was not to transmit a huge capacity but to verify what dispersion compensating scheme is optimum for such a system. The result, achieving the performances reported in the table, was that the optimum performance is attained by electronic equalization via nonlinear equalizer without any DCF in line.

It is possible that this result depends on the particular system scheme, a low number of channels, recirculating loop, and so on; in any case it is an interesting suggestion.

The third system is a sort of prototype of a potential long haul native 100 Gbit/s system, showing all the potentialities of the employed technologies.

The fourth system is a long single span system. It uses the remote Raman pumping technique with an impressive Raman pump injecting 5.5 W power into the fiber to create gain of about 140 km far from the end terminal.

The fourth system is a demonstration of the possibility to go beyond 100 Gbit/s, with a comb of 175 Gbit/s channel transmitted on more than 400 km of fiber.

References

1. Shabtay, G., Mendlovic, D., Itzhar, Y., Optical single channel dispersion compensation devices and their application, *Proceedings of European Conference on Optical Communications ECOC 2005*, Glasgow, U.K., Vol. 3, pp. 321–323, 2005.
2. Drummond, M. V. et al., Tunable optical dispersion compensator based on power splitting between two dispersive media, *IEEE Journal of Lightwave Technology*, 28(8), 1164–1175 (2010).
3. Ikuma, Y., Tsuda, H., AWG-based tunable optical dispersion compensator with multiple lens structure, *IEEE Journal of Lightwave Technology*, 27(22), 5202–5207 (2009).
4. Seno, K. et al., Tunable optical dispersion compensator consisting of simple optics with arrayed waveguide grating and flat mirror, *IEEE Photonics Technology Letters*, 21(22), 1701–1703 (2009).
5. Seno, K. et al., 50-Wavelength channel-by-channel tunable optical dispersion compensator using combination of arrayed-waveguide and bulk gratings, *Conference on Fiber Communication (OFC), Collocated National Fiber Optic Engineers Conference, 2010 (OFC/NFOEC)*, San Diego, CA, IEEE/OSA, s.l., pp. 1–3, 2010.
6. Yang, K., Liu, J., Zeng, X., Electronic dispersion compensation for PMD in 40-GB/s optical links, *Communications and Photonics Conference and Exhibition (ACP)*, Shanghai, China, IEEE, s.l., pp. 1–2, 2009.

7. Sunnerud, H., Karlsson, M., Andrekson, P. A., Analytical theory for PMD-compensation, *IEEE Photonics Technology Letters*, 12(1), 50–52 (2000).

8. Lid, X., Xie, C., van Wijngaarden, A. J., Multichannel PMD mitigation through forward-error correction with distributed fast PMD scrambling, *Optical Fiber Communication Conference (OFC 2004)*, Los Angeles, CA, IEEE, s.l., Vol. 1, pp. WE2 1–3, 2004.

9. Liu, X., All-channel PMD mitigation using distributed fast polarization scrambling in WDM systems with FEC, *Conference on Optical Fiber Communication and the National Fiber Optic Engineers Conference, 2007, OFC/NFOEC 2007*, Anaheim, CA, IEEE/OSA, s.l., pp. ONH4 1–3, 2007.

10. Klekamp, A., Werner, D., Bülow, H., Study of different 40Gbit/s FECs regarding PMD mitigation efficiency by fast polarization scrambling, *Conference on Optical Fiber Communication and the National Fiber Optic Engineers Conference, 2008, OFC/NFOEC 2008*, San Diego, CA, IEEE/OSA, s.l., p. JWA54, 2008.

11. Bulow, H., Limitation of optical first-order PMD compensation, *Optical Fiber Communication Conference, 1999, and the International Conference on Integrated Optics and Optical Fiber Communication, OFC/IOOC '99*, San Diego, CA, Technical Digest, IEEE/OSA, s.l., Vol. 2, pp. 74–76, 1999.

12. Midrio, M., First-order PMD compensation in nonlinearly dispersive optical communication systems via transmission over the principal states of polarization, *IEEE Journal of Lightwave Technology*, 17(2), 2512–2515 (1999).

13. Linares, L. C. B., von der Weid, J. P., Comparison of first order PMD compensation techniques, *Proceedings of the Microwave and Optoelectronics Conference, 2003, IMOC 2003*, Parana, Brazil, IEEE, s.l., Vol. 2, pp. 1019–1022, 2003.

14. Noe, R., Sandel, D., and Mirvoda, V., PMD in high-bit-rate transmission and means for its mitigation, *IEEE Journal of Selected Topics in Quantum Electronics*, 10(2), 342–355 (2004).

15. Shieh, W., On the second-order approximation of PMD, *IEEE Photonics Technology Letters*, 12(3), 290–292 (2000).

16. Neukirch, U. et al., Time-resolved performance analysis of a second-order PMD compensator, *IEEE Journal of Lightwave Technology*, 22(4), 1189–1200 (2004).

17. Merker, T., Schwarzbeck, A., Meissner, P., Analytical calculation for PMD compensation up to second order, *European Conference on Optical Communications (ECOC 2001)*, Amsterdam, the Netherlands, Vol. 3, pp. 352–353, 2001.

18. Heismann, F., Improved optical compensator for first- and second-order polarization-mode dispersion, *IEEE Photonics Technology Letters*, 17(5), 1016–1018 (2005).

19. Kogelnik, H., Nelson, L. E., Gordon, J. P., Jopson, R. M., Jones matrix for second-order polarization mode dispersion, *Optics Letters*, 25(1), 19–21 (2000).

20. Eyal, A., Marshall, W. K., Tur, M., Yariv, A., A new representation of second order polarization mode dispersion, *Electronics Letters*, 35(17), 1658–1659 (1999).

21. Yu, Q., Yan, L.-S., Xie, Y., Hauer, M., Willner, A. E., Higher order polarization mode dispersion compensation using a fixed time delay followed by a variable time delay, *IEEE Photonics Technology Letters*, 13(8), 863–865 (2001).

22. Kim, S., Analytical calculation of pulse broadening in optical higher order PMD compensation, *IEEE Journal of Lightwave Technology*, 20(7), 1118–1123 (2002).

23. Draca, D. et al., Timing jitter as a performance limiting factor when signal propagated along a nonlinear and dispersive fiber, *6th International Conference on Telecommunications in Modern Satellite, Cable and Broadcasting Service, TELSIKS 2003*, Nis, Serbia and Montenegro, Vol. 1, pp. 83–86, 2003.

24. Marcuse, D., An alternative derivation of the Gordon-Haus effect, *IEEE Journal of Lightwave Technology*, 10(2), 273–278 (1992).

25. Mei, Z., Joint time and frequency analysis for unchirped Gaussian pulses in nonlinear optical fibers, *International Conference on Communications, Circuits and Systems Proceedings*, Guilin, China, IEEE, s.l., Vol. 1, pp. 1978–1982, 2006.

26. Demir, A., Nonlinear phase noise in optical-fiber-communication systems, *IEEE Journal of Lightwave Technology*, 25(8), 2002–2032 (2007).

27. Schubert, C. et al., Integrated 100-Gb/s ETDM receiver, *IEEE Journal of Lightwave Technology*, 12(1), 122–130 (2007).

28. Ibrahim, S. K., *High Speed Optical Communication Systems: Study of Multilevel Modulation Formats*, VDM Verlag, Germany, s.l., 2008, ISBN-13: 978-3639041279.

29. Ohm, M., *Multilevel Optical Modulation Formats with Direct Detection*, Shaker Verlag GmbH, Germany, s.l., 2006, ISBN-13: 978-3832253578.

30. Djordjevic, I., Ryan, W., Vasic, B., *Coding for Optical Channels*, Springer Verlag, New York, s.l., 2010, ISBN-13: 978-1441955685.

31. Betti, S., De Marchis, G., Iannone, E., *Coherent Optical Communications Systems*, Wiley, New York, s.l., 1995, ISBN-13: 978-0471575122.

32. Yamazaki, H. et al., 64QAM modulator with a hybrid configuration of silica PLCs and LiNbO3 phase modulators for 100-Gb/s applications, *IEEE Photonics Technology Letters*, 22(5), 344–346 (2010).

33. Sakamoto, T., Chiba, A., Kawanishi, T., 50-Gb/s 16 QAM by a quad-parallel Mach–Zehnder modulator, *European Conference on Optical Communications ECOC 2007*, Berlin, Germany, p. PD2.8, 2007.

34. Yamada, T., Sakamaki, Y., Shibata, T., Kaneko, A., 86-Gbit/s differential quadrature phase-shift-keying modulator using hybrid assembly technique with planar lightwave circuit and LiNbO devices, *LEOS Annual Meeting*, Montreal, Canada, IEEE, s.l., p. ThDD4, 2006.

35. Betti, S., De Marchis, G., Iannone, E., Multilevel coherent optical systems based on Stokes parameters modulation, *IEEE Journal of Lightwave Technology*, 8(6), 1127–1136 (1990).

36. Benedetto, S., Poggiolini, P. T., Multilevel polarization shift keying: Optimum receiver structure and performance evaluation, *IEEE Transactions on Communications*, 42(2/3/4), 1174–1186 (1994).

37. Betti, S., De Marchis, G., Iannone, E., Polarization modulated direct detection optical communication systems, *IEEE Journal of Lightwave Technology*, 10(12), 1985–1997 (1992).

38. Watts, E. F., *Descriptive Geometry*, 2009 (Re-edition), ISBN: 9781443730006.

39. Sutton, D., *Platonic and Archimedean Solids*, Walker & Company, New York, s.l., 2002, ISBN-13: 978-0802713865.

40. Bull, J. D. et al., Ultrahigh-speed polarization modulator, *Conference on Lasers and Electrooptics (CLEO/QUELS 2005)*, Baltimore, MD, IEEE, s.l., Vol. 2, pp. 939–941, 2005.

41. Benedetto, S., Poggiolini, P., Theory of polarization shift keying modulation, *IEEE Transaction on Communications*, 40(4), 708–721 (1992).

42. Nazaraty, M., Simony, E., Stokes space optimal detection of multidifferential phase and polarization shift keying modulation, *IEEE Journal of Lightwave Technology*, 24(5), 1978–1988 (2006).

43. Gitlin, R. D., Hayes, J. F., Weinstein, S. B., *Data Communications Principles*, Plenum Press, New York, s.l., 1992, ISBN 0306437775.

44. Forney, G. D., Jr., Wei, L.-F., Multidimensional constellations. I. Introduction, figures of merit, and generalized cross constellations, *IEEE Journal of Selected Areas in Communications*, 7(6), 877–892 (1989).

45. Coxeter, H. S. M., *Regular Polytopes*, Dover Publications Inc., New York, s.l., 1973, ISBN 0486614808.

46. Forney, G. D., Multidimensional constellations. II. Voronoi constellations, *IEEE Journal of Selected Areas in Communications*, 7(6), 941–958 (1989).

47. Laroja, R., Farvardin, N., Tretter, S., On SQV shaping of multidimensional constellations—High rate large dimensional constellations, Technical research report, Systems Research Center, University of Maryland at Harvard, College park, MD, s.l. 1992, http://drum.lib.umd.edu/bitstream/1903/5187/1/TR_92-5.pdf (accessed: September 8, 2010).

48. Yanagisawa, M. et al., A new planar lightwave circuit platform for cost-effective optical hybrid modules, *Laser and Electrooptical Society Annual Meeting—LEOS 1996*, Boston, MA, Vol. 1, pp. 77–78, 1996.

49. Jeong, S.-H., Morito, K., Compact InP-based 90° hybrid using a tapered 2 × 4 MMI and a 2 × 2 MMI coupler, *International Conference on Indium Phosphide and Related Materials (IPRM)*, Takamatsu, Japan, pp. 1–4, 2010.

50. Liu, X. et al., Multi-carrier coherent receiver based on a shared optical hybrid and a cyclic AWG array for terabit/s optical transmission, *IEEE Photonics Journal*, 2(3), 330–337 (2010).

51. Zhang, S., Kam, P. Y., Yu, C., Chen, J., Laser linewidth tolerance of decision-aided maximum likelihood phase estimation in coherent optical M-ary PSK and QAM systems, *IEEE Photonics Technology Letters*, 21(15), 1075–1077 (2009).

52. Mendenhall, W., Beaver, R. J., Beaver, B. M., *Introduction to Probability and Statistics*, Duxbury Press, Belmont, CA, s.l., 2008, ISBN-13: 978-0495389538

53. Namihira, Y. et al., Dynamic polarization fluctuation characteristics of optical fiber submarine cables under various environmental conditions, *IEEE Journal of Lightwave Technology*, 6(5), 728–738 (1988).

54. Wuttke, J., Krummrich, P. M., Rosch, J., Polarization oscillations in aerial fiber caused by wind and power-line current, *IEEE Photonics Technology Letters*, 15(6), 882–884 (2003).

55. Bulow, H. et al., Measurement of the maximum speed of PMD fluctuation in installed field fiber, *Optical Fiber Communication Conference, and the International Conference on Integrated Optics and Optical Fiber Communication, OFC/IOOC '99*, San Diego, CA, IEEE/OSA, s.l., Vol. 2, pp. 83–85, 1999.

56. Cusani, R., Iannone, E., Salonico, A. M., Todaro, M., An efficient multilevel coherent optical system: M-4Q-QAM, *IEEE Journal of Lightwave Technology*, 10(6), 777–785 (1992).

57. Pikaar, T., van Bochove, A. C., van Deventer, M. O., Fraukene, H. J., Groen, F. H., Fast complete polarimeter for optical fibers, *European Fiber Optic Conference and LAN Conference EFOC/LAN 1989*, Amsterdam, the Netherlands, pp. 206–209, 1989.

58. Hecht, E., *Optics*, 4th edn., Addison Wesley, Reading, MA, s.l., 2001, ISBN-13: 978-0805385663.

59. Azzam, R. M. A., Arrangement of four photodetectors for measuring the state of polarization of an optical field, *Optics Letters*, 10(4), 309–311 (1985).

60. Salehi, J. G., Proakis. M., *Digital Communications*, 6th edn., McGraw Hill, New York, s.l., 2008, ISBN-13: 978-0071263788.

61. Stokes, J. W., Ritcey, J. A., *Evaluation of Error Probabilities for General Signal Constellation*, Electrical Engineering Department, Washington University, Saint Louis, MO, s.l., 1999.

62. Huber, K., Efficient utilisation of multilevel signal constellations, *Proceedings of the International Symposium on Information Theory, 1998*, Cambridge, MA, IEEE, s.l., 1998.

63. Wu, H.-C., Chang, S.Y., Constellation subset selection: Theories and algorithms, *IEEE Transaction on Wireless Communications*, 9(7), 2248–2257 (2010).

64. Huang, S.-H., Wu, H.-C., Chang, S. Y., Novel efficient algorithms for symmetric constellation subset selection, *Proceedings of the International Communication Conference—ICC 2010*, Cape Town, South Africa, IEEE, s.l., pp. 1–5, 2010.

65. Iannone, E. et al., Performance evaluation of very long span direct detection intensity and polarization modulated systems, *IEEE Journal of Lightwave Technology*, 14(3), 261–272 (1996).

66. Armstrong, J., OFDM for optical communications, *IEEE Journal of Lightwave Technology*, 27(3), 189–204 (2009).

67. Yi, X. et al., Tb/s coherent optical OFDM systems enabled by optical frequency combs, *IEEE Journal of Lightwave Technology*, 28(14), 2054–2061 (2010).

68. Sano, A. et al., No-guard-interval coherent optical OFDM for 100-Gb/s long-haul WDM transmission, *IEEE Journal of Lightwave Technology*, 27(6), 3705–3713 (2009).

69. Peng, W.-R., Analysis of laser phase noise effect in direct-detection optical OFDM transmission, *IEEE Journal of Lightwave Technology*, 28(17), 2526–2536 (2010).

70. Qiu, W. et al., The nonlinear impairments due to the data correlation among sub-carriers in coherent optical OFDM systems, *IEEE Journal of Lightwave Technology*, 27(23), 5321–5326 (2009).

71. Seimetz, M., Noelle, M., Patzak, E., Optical systems with high-order DPSK and star QAM modulation based on interferometric direct detection, *IEEE Journal of Lightwave Technology*, 25(6), 1515–1530 (2007).

72. Freckmann, T., González, C. V., Ruiz-Cabello Crespo, J. M., Joint electronic dispersion compensation for DQPSK, *Proceedings of Optical Fiber Conference/International Optical and Optoelectronics Conference OFC/IOOC 2008*, San Diego, CA, IEEE/OSA, s.l., p. OTuO.6, 2008.

73. Wang, J., Kahn, J. M., Impact of chromatic and polarization-mode dispersions on DPSK systems using interferometric demodulation and direct detection, *IEEE Journal of Lightwave Technology*, 22(2), 362–371 (2004).

74. Park, S.-G. et al., On the WDM transmissions using multilevel (M>4) DPSK modulation format, *IEEE Photonics Technology Letters*, 17(7), 1546–1548 (2005).

75. Kim, H., Cross-phase-modulation-induced nonlinear phase noise in WDM direct-detection DPSK systems, *IEEE Journal of Lightwave Technology*, 21(8), 1770–1774 (2003).

76. Matera, F., Mecozzi, A., Settembre, M., Light depolarization in long fiber links, *Electronics Letters*, 31, 473–475 (1995).

77. Savory, S. J., Digital coherent optical receivers: Algorithms and subsystems, *IEEE Journal of Selected Topics in Quantum Electronics*, 16(5), 1164–1179 (2010).

78. Rozen, O., Sadoz, D., Katz, G., Levy, A., Mahlab, U., Dispersion compensation of self phase modulation impairment in optical channel using MLSE, *International Conference on Transparent Optical Networks, ICTON 2008*, Athens, Greece, IEEE, s.l., Vol. 1, pp. 178–181, 2008.

79. Alfiad, M. S. et al., Maximum-likelihood sequence estimation for optical phase-shift keyed modulation formats, *IEEE Journal of Lightwave Technology*, 27(20), 4583–4594 (2009).

80. Geisler, D. J. et al., Single channel, 200 Gb/s, chromatic dispersion precompensated 100 km transmission using an optical arbitrary waveform generation based optical transmitter, *Optical Fiber Communication Conference OFC 2010*, San Diego, CA, IEEE/OSA, s.l., p. OWO4, 2010.

81. Jiang, Z., Huang, C., Leaird, D. E., Weiner, A. M., Optical arbitrary waveform processing of more than 100 spectral comb lines, *Nature Photonics*, 1, 463–467 (2009).

82. Jiang, W., Soares, F. M., Seo, S.-W., Baek, J.-H., Fontaine, N. K., Broeke, R. G., Cao, J., Yan, J., Okamoto, K., Olsson, F., Lourdudoss, S., Pham, A. A monolithic InP-based photonic integrated circuit for optical arbitrary waveform generation, *Optical Fiber Communication Conference 2008*, San Diego, CA, IEEE/OSA, s.l., p. JThA39 (2008).

83. Fukada, Y., Probability density function of polarization dependent loss (PDL) in optical transmission system composed of passive devices and connecting fibers, *IEEE Journal of Lightwave Technology*, 20(6), 953–964 (2002).

84. Betti, S. et al., Dichroism effect on polarization-modulated optical systems using Stokes parameters coherent detection, *IEEE Journal of Lightwave Technology*, 8(11), 1762–1768 (1990).

85. Xie, C., Mollenauer, L. F., Performance degradation induced by polarization-dependent loss in optical fiber transmission systems with and without polarization-mode dispersion, *IEEE Journal of Lightwave Technology*, 21(9), 1953–1957 (2003).

10

Next Generation Networking: Enabling Technologies, Architectures, and Performances

10.1 Introduction

In Chapters 8 and 9, we have reviewed what is a widely agreed view of the roadmap for the introduction of new generation control and management planes and of new generation transmission systems.

In both the cases, the evolution is directly pushed by telecommunication market requiring to increase the margin of the carriers via a decrease of the huge network operational costs (OPEX).

When the discussion moves toward network equipment and the data plane, the situation is different.

Not only is there no generally agreed view of the evolution of the data plane, but are also different proposals on the table, partly complementary and partly alternative.

Trying to make a synthesis, three different views emerge of the networking evolution, all representing a revolution more than an evolution.

Naturally, each of these alternative views of the future networks has several different interpretations and depends in a different way on the technology evolution.

10.1.1 Digital Optical Network

From the point of view of network management, the SDH/SONET remains the reference for all carriers. The strength of this standard is that the signal passes continuously through network elements that are capable of performing digital signal processing like regenerators, line terminals, ADM, and so on. In this way, the network control is complete and detailed.

In a wavelength division multiplexing (WDM) network, all-optical elements are on the contrary analog, as the amplifiers and the optical add drop multiplexers (OADMs). It is much more difficult to implement monitoring and management functions through an analog equipment, and so nodes containing analog optical elements are somehow functionally poorer with respect to digital network elements.

On the other hand, when WDM was introduced, there was so huge a cost difference between the optical elements and the corresponding electrical one that there was no choice.

However, with the great improvement in electronics, and with the possibility, at least on a III-V platform, of integrating optics and electronics components on the same chip, the situation could change.

Carriers and equipment vendors pushing the digital optical network solution leverage on this evolution to propose a new physical layer architecture based not on analog optics but on digital electronics in order to be able to enrich signal processing and network management functionalities.

In this view of the network, amplifiers are again substituted by monolithically integrated regenerators managing several wavelengths simultaneously and integrating OADM-like functionalities. This new physical layer will be able to interact in a more intelligent way with the upper IP/MPLS layer, thanks to its processing functionalities allowing effective interlayer coordination for a more optimized generalized multiprotocol label switching (GMPLS) network architecture.

Naturally, the key technology to be developed to target this kind of network is monolithical integration of electronics and optics in order to reduce the cost of the equipment.

10.1.2 Optical Transparent Network

If no revolution happens in the field of integration between electronics and optics, optical equipment will remain cheaper than the corresponding electronic equipment, due to its ability to deal simultaneously with a set of very-high-speed signals.

This is why when the WDM revolution invested the telecommunication networks, there was a diffused agreement on the fact that electronics was destined to disappear from the physical layer, substituted by an "all-optical" transparent network.

Even if a huge amount of research and industrial development has been done in this direction, the idea of a completely analog physical layer faces important problems when the implementation of control plane and management functions is considered.

Moreover in a network whose node regenerates the signal, the network design and optimization are decoupled by the design of the transmission systems (see Chapters 6 and 8). In a transparent network, wavelength routing has to be performed taking into account transmission impairments so as to avoid routes that are not feasible from the transmission point of view.

These difficulties, besides the great progresses of electronics, have quite attenuated the enthusiasm for the optical transparent solution. Nevertheless, this is still one of the possible network evolutions.

10.1.3 Optical Packet Network

The reason why optics seems so suitable to scale up the capacity of the network is its ability to deal with very-high-speed signals without the need of dividing them into lower speed components.

On the other hand, the transparent network architecture has all the problems of a completely analog layer.

This problem was clear since the beginning of the use of optics for transmission, and from that moment, a revolutionary solution was proposed: let us devise digital optics.

The idea is to completely change the network architecture by eliminating the time division multiplexing (TDM) physical layer based on circuit switching substituting it with a packet switched optical network based on packet switching optical machines.

These optical routers have to deal directly with some sort of optical burst or packet at a very high speed, performing all the operations that are needed to store and forward them.

Such a deep revolution in the network could use a suitable form of GMPLS, due to the flexibility of this control plane, and could also incorporate into the optical routers higher layer electronics functions depending on the definition of the relation between the optical bursts or packets and the traditional IP and Ethernet protocol stacks.

At the state of the art, however, the key point of this solution is not the network architecture, but the optical technology allowing digital functions like storing, header recognition and elaboration, and so on, to be performed. A huge amount of research has been devoted to this solution, both from technology and networking points of view.

Naturally, several variations have been proposed for all the alternatives we have listed, and even among the proposers of one of the three solutions, there is no complete consensus on how it should be implemented.

In this chapter, we will review the basics of the three network architectures we have introduced, both from technology and networking points of view.

It will be naturally impossible to detail all the different variations in which a particular idea has been proposed, but we will try to make clear the advantages and disadvantages of each solution and how far the technology is from what would be needed for a real deployment of such a network.

10.2 Optical Digital Network

The idea that is at the basis of the optical digital network (ODN) is to decouple the service layer (generally layers 2 and 3) from the physical aspects of the transport layer by hiding the dense wavelength division multiplexing (DWDM) structure of optical transport with the structure that the group that first introduced the idea call bandwidth virtualization [1].

The concept of bandwidth virtualization consists essentially in the ability to offer to the client layers a wide variety of possible channel capacities to maintain always a high utilization of the transmission channel.

The client layer requests are transformed in tributary channels by a suitable adaptation layer and are multiplexed onto the wavelengths that are used for transmission.

In this way, the client layer even does not know what the wavelength capacity is, but has the possibility of occupying, each time, only the needed capacity.

The channels set up as a consequence of the client requests are characterized by a quality tag, which determine among other things what kind of survivability strategy is assigned to the channel among those available at the optical layer.

Moreover, since every node, even in-line sites, is able to regenerate the signal, a very good monitoring of the signal quality can be done at each transit node, and coordination with the client layers (e.g., to implement multilayer protection) is easier.

Besides bandwidth virtualization, the fact that the signal is processed at the digital level in every node allows performing efficient grooming to maintain a high degree of utilization of the transmission system and to introduce flexibility whenever it is needed.

As a matter of fact, tunable OADMs or wavelength converters (WCs) cannot be realized via complex analog techniques, but they can be implemented via electronic processing.

From a functional point of view, the scope is the same of next generation SDH/SONET, but for the fact that the necessity to maintain legacy compatibility is not present in this case.

The huge difference is in the idea implementation, based on the integration between optics and electronics on the same integrated circuit platform, so as to drastically reduce the cost of electronic elaboration.

At the state of the art, silicon photonics is not ready for this task. The major players in the field of silicon photonics are engaged in research for other kinds of applications, and in any case, silicon phonics components do not have yet the required quality to be applied in long-haul (LH) transmission.

Besides these "optical problems," the integration of silicon high-performance components on the same chips of optical components is for sure not given, requiring substantial evolution of the fabrication processes.

For the above reasons, integration has been carried out using InP-based platform.

10.2.1 Optoelectronic Integration: ODN Enabling Technology

In order to overcome the cost reasons that pushed the telecommunication community to try to eliminate electronics from the transmission systems, the cost of opto-electronics integrated circuits has to decrease. A progressive integration of more functionalities and an increase of the volume will help recover and overcome the cost gap with optical transparent components.

With regard to functionalities, a lot of different functionalities have been integrated into a single chip built on InP platform.

Tunable sources are a reality both in the form of laser arrays and in the form of multisection distributed Bragg reflectors (DBRs) (see Chapter 5). In the first case, the chip containing the lasers also integrates a tunable combiner, generally, a microelectrical mechanical machine (MEM) switch. In the second case, a modulator and a semiconductor optical amplifier (SOA) are frequently integrated in the laser chip so as to demonstrate the ability to put several different optical components on the same chip.

The most important integrated component for the processing of the optical signal is the so-called photonics integrated circuit (PIC). This highly integrated component is devised to drastically decrease the capital expenditure (CAPEX) that is due to the repetition of costly high-speed optical interfaces every time the signal has to be converted from optical to electrical.

The functional scheme of a transceiver realized with a transmitting and a receiving PIC is represented in Figure 10.1. In the transmitting PIC, a bar integrating a certain number of lasers is coupled with a multiplexer/demultiplexer (generally an array waveguide [AWG]) that launches the multiplexed channels into the output fiber. The scheme of a receiving PIC is identical, with a demultiplexer, a bar of photodiodes, and with the WDM signal incoming from the input fiber. Both the coarse wavelength division multiplexing (CWDM) and DWDM PIC can be realized, and similarly to CWDM small form factor pluggable (SFP), it seems possible to implement CWDM PIC without thermal stabilization.

Two possible technologies exist to realize PICs: a complete monolithic integration on an InP substrate and hybrid integration. In this last case, the AWG is realized in another integrated optical technology (e.g., high index contrast Si SiO_2 obtained with a high doping of the glass with Germanium) so as to interface the fiber on one side and the laser bar on the other.

Both the technologies have advantages and disadvantages: monolithic technology has to face a relatively big area circuit compared with the small wafers used in InP foundries, so uniformity is necessary. Hybrid technology realizes the part of the component with the wider area using Si wafers, at least 8 in. large versus the 3 in. of InP wafers, so attenuating uniformity problems, but introduces the difficult operation to be performed chip by chip by interfacing two different chips in the package.

In both the cases, however, what is expected is a drastic price reduction of interfaces integrated into a PIC with respect to discrete interfaces with the same performances. An estimate of the price of CWDM PICs with respect to the price of CWDM SFP optical interfaces with the same performances is reported in Figure 10.2 [2]. Even a greater sparing is expected for DWDM transceivers.

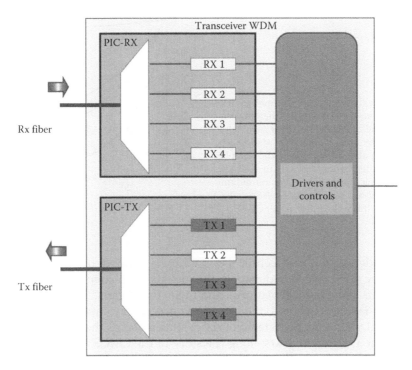

FIGURE 10.1
Functional scheme of a transceiver realized with a transmitting and a receiving PIC.

From an experimental point of view, a 10 source 10 Gbit/s PIC has been not only realized but also incorporated into a commercial system [3] demonstrating the feasibility of the technology.

Integrated circuits, putting together optics and electronics, have been realized using HEMT transistors. The cost of the individual transistor is high due to its structure and the power consumption is much higher with respect to complementary metal oxide transistors (CMOSs). However if a circuit is integrated with a large number of other optical and electronic components and produced in large volumes, it is possible that the overall cost reduces.

All these elements, besides a great number of realizations of different components using this platform, tell us that from a technical point of view, optoelectronic integration sufficiently function rich to design network elements for the ODN can be carried out.

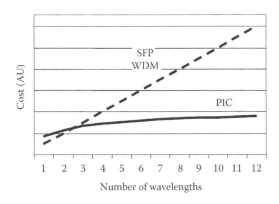

FIGURE 10.2
Expected price reduction of interfaces using PICs with respect to discrete interfaces with the same performances.

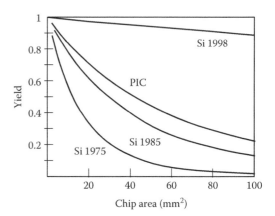

FIGURE 10.3
Typical industrial yield for the CMOS technology versus the area of the considered circuit at different years along the story of the CMOS industry. For each value of the considered die area, the value of the defects density for the integrated InP optoelectronics circuits in 2005 is shown. (After Joyner, C. et al., Current view of large scale photonic integrated circuits on indium phosphide, *Optical Fiber Communication Conference OFC 2010*, San Diego, CA, IEEE, s.l., 2010, p. paper OWD3.)

In order to understand if this technical possibility is ready to be traduced in products, there are two key points to be understood: the yield of such complicated optoelectronics integrated circuits using the InP platform, and their cost model.

As far as the yield is concerned, the issue has been deeply studied. In Figure 10.3, the typical industrial yield for CMOS technology is shown versus the area of the considered integrated circuit at different years along the story of the CMOS industry. Naturally, the wider the circuit area, the smaller the yield.

For each value of the considered die area, the value of the defects density for the integrated InP optoelectronics circuits in 2005 is shown [4].

However, even if InP optoelectronic integrated circuits start to manifest a good industrial stability, the fact that they will follow the same evolution that has characterized CMOS circuits is all but obvious.

Relevant differences exist between the two technologies, the most important being the fact that an international industry consortium leads the evolution of CMOS along a clear roadmap, while nothing similar so far exists in optoelectronics InP circuits.

A coherent roadmap allows players in different segments of the CMOS value chain to be synchronized. For example, while a new process is under preliminary research, contemporary process equipment vendors design industrial machines for it and chip designers evaluate the impact in terms of opportunities and risks on the present situation.

The players of the InP value chain are not driven by optoelectronic applications due to the high performance needed in this field and the small volume corresponding to the application in layer one core machines.

This does not mean that this solution cannot be deployed; it simply means that we cannot expect a fast diffusion and price decrease that characterized microelectronics.

A similar conclusion can be obtained by a simple cost model of this kind of integrated circuits. We do not want to correctly model the industrial cost, but only bring into evidence a few factor that heavily influences the industrial costs scaling with volumes. For this reason, beyond a simplification of the cost terms, we will neglect the fact that by increasing the volume, the yield generally increases due to a stabilization of the process.

This is an effect that exists both for InP optoelectronics integrated circuits and for CMOS circuits.

Let us assume that our integrated optoelectronic circuit is a PIC.

This is a DWDM circuit, which is to be stabilized in current by at least placing a heater inside the package. Moreover, the package has to have specific heat conduction characteristics and must accommodate high-speed transmission lines to deliver ten signals at 10 Gbit/s to the internal components.

The industrial cost of a component exiting from the line of preliminary production can be divided in three main terms that scale quite differently increasing volumes.

1. *Chip Industrial Cost.* It depends on the particular product under consideration; we assume that it is around 40% the industrial cost of the PIC. In order to justify this rough assumption, let us remember that in a standard DFB laser, the industrial value of the active chip is smaller than 10% of the cost of the whole devices. If large volumes of PICs are absorbed by the market, the industrial cost decreases abruptly, almost linear with the volume. This is due to the fact that the greater part of the industrial cost is constituted by the depreciation of the technological facility that is divided among the number of produced pieces.

2. *Bill of Material of the Final Component.* This is the cost of all the elements used for building the PIC, except the active chip. The package, the fiber V groove, the alignment optics, the thermistor (if present), and so on are all accounted for here. The value of this bill of material (BOM) out of the global industrial cost of the component can be around 40% too. The BOM decreases if volume increases, but not so rapidly as the chip cost. A decreasing law, much slower than linear is reasonable, at least up to the moment in which the volume is huge, in the order of several millions. Since we are dealing with highly specialized chips for DWDM applications, a volume of ten million pieces does not exist in the market. In case this architecture becomes successful, typical volumes for one chip producer would be on the order of 300,000 (one chip realizes the interface for 10 wavelengths).

3. *Work Cost.* Since a set of operations have to be carried out piece by piece, for example, fiber active coupling, the contribution of the work to the cost is not negligible, and so we will assume it to be about 15% of the total.

If we accept the rough assumptions given earlier, we obtain a simple cost model of a DWDM PIC that produces the behavior of Figure 10.4. In this figure, the behavior of a similar

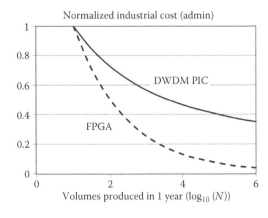

FIGURE 10.4
Estimated cost trend of DWDM PICs compared with the cost trend of high-performance FPGAs.

model applied to a high-performance field programmable gate array (FPGA) is also reported for comparison.

The result is that the FPGA cost goes down much faster than the PIC cost and reaches an asymptote beyond one million pieces as low as 0.8% of the initial preproduction pieces.

As far as the PIC is concerned, the slope is smaller and the asymptote is reached at about 50 million pieces (a volume that do not exist all over the market) and is around 20% of the preproduction costs.

At the volume of 100,000 pieces, that is reasonable as an order of magnitude for the DWDM PIC application, the cost is a bit lower, 40% of the preproduction cost, while the high-performance FPGA at the same volume has a cost as low as 1% of the preproduction cost.

This clear difference is due to the combined effect of several factors:

- Work cost importance in the case of PICs
- Importance of the BOM in the case of PICs
- Single application (telecommunication dedicated) versus the flexibility of FPGA that allows much higher volumes
- Lack of global roadmap and standardization

The first two elements have been already commented on. As far as the last two are considered, they push the BOM of the FPGA to decrease more rapidly with volume, since the circuit producer can leverage on the standardization to share similar BOM elements among different products so increasing the effective volume offered to the supplier and further lowering the price.

10.2.2 Optical Digital Network Architecture and Design

In this section, we will briefly summarize the architecture of both the data plane and the control and management planes of the ODN.

In this analysis, we will assume that the technology evolution will completely solve the issues considered in the previous section so to allow a wide ODN deployment.

10.2.2.1 ODN Control and Management Plane

As far as the management and control planes are concerned, the ODN, if widely deployed, will be based on a GMPLS protocol suite.

The processing ability of each ODN network element will easily allow the elaboration of distributed protocols and a reporting to the management.

However, the use of PICs in the network elements is not irrelevant for the control plane protocol suite. Naturally, the specific protocol elements that have to be adapted at the new hardware depend on the way in which the PIC is used.

To better understand the implications of this situation, we will focus our attention on two cases, which are probably the most important:

1. The PIC simply substitutes a number of DWDM interfaces in a 10 Gbit/s line card.
2. The PIC is used to implement 100 Gbit/s transmission by parallelizing 10 wavelengths in a 100 Gbit/s card.

In the first case, the GMPLS path at the transport layer is transported via a single 10 Gbit/s wavelength. Each wavelength represents a branch of an independent GMPLS route, even if they are generated by the same physical device.

On top of this, every wavelength multiplexed different channels as defined in Chapter 3, Section 3.2.7.

The fact that the PIC is a single physical component, implies that when one logical port experiments a failure (e.g., due to a laser failure) all the GMPLS paths traveling through the PIC have to be switched in protection, even if all but one of them are correctly working. This operation is needed to assure that the failed PIC can be substituted without affecting the traffic. In order to manage this operation at the protocol level, a correspondence has to be stated among all the paths traveling through a PIC, for example, introducing a new header field.

In the second case, a single PIC corresponds to a single logical GMPLS port, thus having a single address. The wavelengths generated by the same PIC have to be considered by the control plane as a single path. For example, they have to be routed along the same route and travel always through the same PICs. Moreover, when one of them fails, all of them have to be rerouted along the same protection route, since logically those are the same channel.

This problem can be solved also with the present version of GMPLS, where this situation is considered (see Chapter 8).

Taking into account both these elements would require quite an update of GMPLS routing and restoration protocols at the physical layer that have to be carried out by IETF and OIF before a real ODN deployment can happen.

A similar problem concerns the ITU-T automatically switched optical network (ASON) standard, where the mechanisms of wavelength aggregation needed to accommodate the use of PICs in the network are similarly not completely foreseen.

In the case of ASON, that is not a complete protocol suite, but quite a detailed network architecture, probably the needed work is easier and could be included in the ITU-T continuous work for the update of the standards.

10.2.2.2 ODN Physical Layer Sub-Layering

From a physical layer point of view, both an SDH/SONET frame and an optical transport network (OTN) frame can be adopted, so as to leverage on the widely diffused circuitry built on these standards.

From a layering point of view, even if it uses largely Optical/Electrical/Optical conversion, the ODN lacks the partitioning of the physical layer into two distinct and autonomous sub-layers typical of synchronous digital hierarchy (SDH) over WDM networks.

This simplification is useful to maintain a simple control plane and to avoid redundant signaling due to the need of synchronization between the physical sub-layers.

Due to the concept of bandwidth virtualization, the ODN would be equipped with a complex and flexible adaptation layer for the interface with the client layers.

The adaptation function has the role of receiving a bandwidth allocation request from the suitable upper layer protocol together with the indication of the quality of service (QoS) class required for the physical connection to be set up.

Once the request is received, it is matched with a channel at the physical layer equipped with the survivability mechanism most suited to assure the required QoS level among those available at physical layer.

If multilayer survivability is possible, it is the adaptation function that receives the coordination protocol from the client layer and transforms it into a physical layer protocol to be elaborated by the relevant physical layer entities. When the physical layer network elements produce a feedback to be passed to the client layer, the adaptation function would be to translate it in the client layer protocol.

In order to match the bandwidth request coming from the client layer with a physical layer channel, the adaptation function can operate in two ways, depending on the type of channels that can be transmitted through the ODN.

A first possibility assumes that the capacity of the ODN channels are prescribed, so that there exists a list of possible channel bit rates, and all the ODN channels have to travel at a bit rate comprised in the list.

It is possible to increase flexibility to include a large number of channel bit rates, but it is sure that sooner or later, a client requesting a connection at a non included bit rate will come out.

The adaptation function in this case has the role to optimize the frame efficiency by wisely using the available bit rates.

Alternatively, it is not impossible to conceive the continuous creation of channels with a bit rate exactly equal to the requested capacity plus the redundancy introduced by a frame header.

In this case, however, the adaptation function is simplified, but all the management of the tributary channels over all the ODN is quite complicated.

For example, dynamic TDM multiplexers and demultiplexers have to be considered and grooming operations are much more complex since they face a potentially unlimited set of tributary bit rates.

On the ground of the network architecture that we have introduced, a possible sub-layering of the ODN physical layer could be represented as in Figure 10.5, even if there is not yet sufficient agreement on this structure to start even a preliminary study opening the way to its standardization.

The upper layer of the ODN is the channel layer, where data coming from the client layers are organized in channels that are the elementary entity of the ODN.

As we have seen, there is a plurality of possible channel speeds and several available survivability strategies that can be applied at the channel layer.

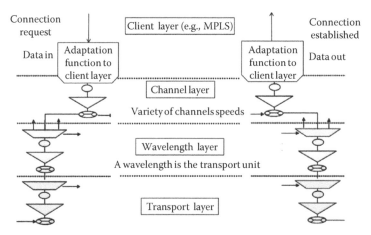

FIGURE 10.5
Scheme of a possible sub-layering of the ODN physical layer.

All the operations on the channels, grooming, for example, are carried out at the channel layer.

Below the channel layer, we have the wavelength layer where all the operations involving a whole wavelength, comprising channel multiplexing in the wavelength frame and channel demultiplexing before passing them at the upper layer, are carried out.

Finally, the transport layer manages the whole WDM comb, and is in charge of transmitting it from the near end to the far end of the transmission system.

No switching at the transport layer its foreseen in this version of the ODN architecture.

10.2.2.3 ODN Network Elements and Data Plane

The main physical layer switching elements of the ODN are the optical cross connect (OXC) and the OADM.

A possible OXC architecture that seems suited for application in the ODN exploits the interface sparing allowed by PICs and uses fast CMOS circuits to implement a lower speed digital switch fabric using deserializers to parallelize the data on the single wavelengths.

The scheme of this type of ODM OXC is reported in Figure 10.6. From the figure, it is clear that the OXC is equipped with two superimposed switch fabrics, one working on wavelength and the other on channels.

Using the upper fabric, the OXC is able to implement grooming, while the presence of the lower one significantly reduces the overall number of needed cross-points with respect to a machine working only on channels.

Even if an OXC is a powerful machine allowing efficient grooming and channel-based restoration to be implemented, it is also a complex machine to develop and an expensive equipment to install.

Thus, it will be not a surprise if the first ODN networks are built with concatenated rings. If from a technological point of view an ODN OADM is quite an innovative machine, from

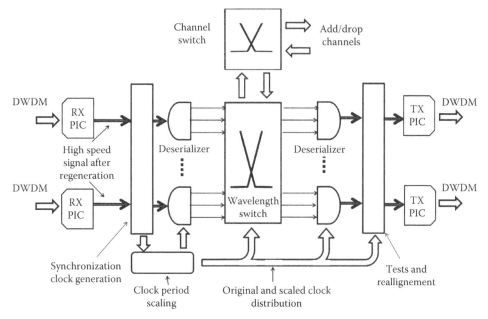

FIGURE 10.6
Possible architecture of an ODN OXC.

a functional point of view it sums up all the functionalities of a next generation SDH/SONET ADM and of a tunable OADM.

Due to the possibility of managing whole wavelengths at a wavelength layer or single channels at the channel layer, the ODN OADM, can manage two types of rings: wavelength rings (WRings) and channel rings (CRings).

As far as WRings are concerned, they are similar to OTN self-healing rings. The elementary unit transmitted and managed at the nodes is a wavelength, even if two elements make ODN WRings superior to OTN Rings.

The first property consists in the fact that since wavelengths are regenerated at each node, the logical wavelength path that is set up along the ring does not require physical wavelength continuity: wavelength conversion can be operated even at each node. The fact that the wavelength can change even at every node along the path is really beneficial in the exploitation of the backup capacity if shared protection is used. Moreover, optimum use of all the wavelengths can be done along the ring.

The second property is the ability to perform grooming at the channel layer that completely lack to OTN OADM.

Considering CRings, these are a structure completely new: under certain points of view they are similar to SDH/SONET path layer rings. Differently from SDH/SONET, however, a single machine, the ODN OADM, manages several superimposed rings together, being able to change the channel routing from one ring to the other in order to better exploit the available transmission capacity.

In a CRing, channels can be moved from one wavelength to the other at every traversed node. The content of one wavelength can be exchanged with the content of another wavelength, and channels, not wavelengths, are added and dropped.

Thus, in this new type of ring, the wavelength is only a useful way to divide the available bandwidth so as to allocate the information, but it has no logical relevance at the networking level. The role of the wavelength is the same role of a frame time slot in an SDH/SONET ring.

From a logical point of view, if all the wavelengths disappear, and a single channel at very high speed would substitute them exactly, the same ring working would be realized, with the only difference that TDM instead of WDM multiplexing would be used.

A possible architecture of an OADM for a CRing is represented in Figure 10.7.

On this ground, it is clear that wavelength protection does not exist in CRings; these rings can guarantee survivability either through WDM protection or through channel protection.

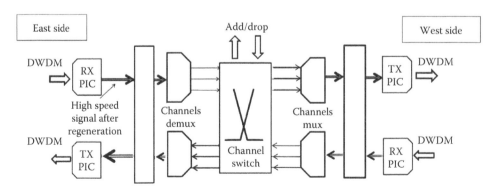

FIGURE 10.7
Possible architecture of an OADM for a CRing.

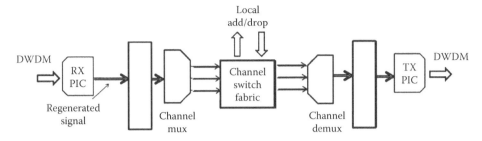

FIGURE 10.8
Possible scheme of a line terminal in an ODN.

Channel protection is completely analogous to the path protection in SDH/SONET rings, considering the WDM comb as a super-frame. Two- and four-fiber rings can be defined, and in both cases, besides dedicated protection, shared protection can be adopted

The case of WDM protection is analogous to the SDH/SONET MS ring, and also in this case, two- and four-fiber rings can be considered with dedicated and shared protection.

The block scheme of an ODN regenerator is shown in Figure 10.8: regeneration happens when the bits are regenerated via a detection and information estimate. Before the injection of the regenerated signal into the output fiber, a signal processing section assures the monitoring of the signal characteristics and, if needed, can constitute a branching point for the transmission system.

10.3 Transparent Optical Transport Network

When the revolution related to optical transmission hit the market, it seemed that optics was on the edge to completely substitute electronics in the telecommunication network.

On the wave of this forecast, the original idea of transparent optical transport network (T-OTN) was born [5]. At the very beginning, the T-OTN was conceived as a whole geographical transport network without any regeneration point, only constituted by analog transmission systems and analog OXCs.

At the edge of this network, there were the electronic edge machines that were used for signal conditioning and interface with local and access networks.

This idea is shown in Figure 10.9, and one of the first transparent OXC architectures that was proposed is shown in Figure 10.10.

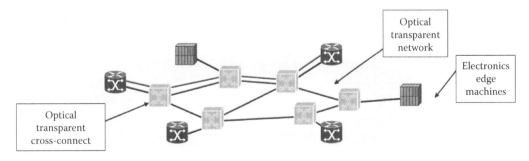

FIGURE 10.9
Scheme of an optical transparent transport network.

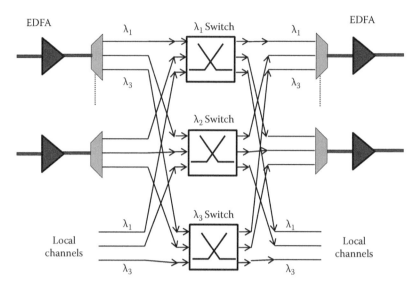

FIGURE 10.10
Architecture of one of the first proposed transparent OXCs firstly introduced in the framework of the EU project MWTN (Multi Wavelength Transport Network). (After Iannone, E. and Sabella, R., *IEEE J. Lightwave Technol.*, 14(10), 2184, 1996, where the references where this OXC was introduced for the first time can be found).

The expected advantages of the transparent optical architecture were as follows:

- *Bit rate, format, service transparence*: This means that completely different signals were conceived coexisting on different wavelengths [6].
- *Low power consumption:* On the assumption, that is still valid, that analog optics consumes less power to switch with respect to digital electronics [7].
- *Almost infinite bandwidth:* Due to the huge bandwidth of optical fibers that also assures that the network is future proof.
- *CAPEX sparing*: Based on a forecast of enormous growth of the volumes of optical components due to the growth of the telecommunication market and on the spare due to the elimination of intermediate optical–electrical–optical interfaces. This hypothesis implies that a market price of optical equipment would rapidly decrease [6,8].

In front of these expected advantages, the transparent optical network also presented important problems [9]:

- *Management*: If the signal travels transparently through all the networks, it is impossible to read its header at a switch input and output; thus, in the most radical implementation, switching errors are detected only at the network edge by the electronics edge equipment; for this reason, they are quite difficult to identify and to insulate [10].
- *Transmission characteristics*: Wavelength routing has to take into account transmission characteristics not to route a wavelength on a path that for some reason (length, number of nodes, and so on) does not allow correct transmission [11].
- *Sub-wavelength grooming* is impossible.

Moreover, as we have seen in Chapter 7, the OXC of Figure 10.10, as all the architectures suggested for transparent operation, has a certain blocking probability.

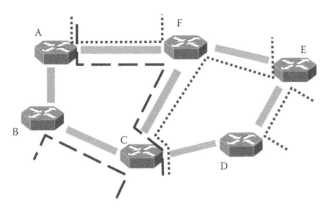

FIGURE 10.11
Example of network blocking due to the need of wavelength continuity. The figure is discussed in detail in the text.

It is a consequence of the fact that transparency implies wavelength continuity also in the passage through an OXC. This means that the transparent network appears as the superimposition of a number of independent networks equal to the number of wavelengths in the DWDM transmission systems.

In this situation, blocking occurs where a path characterized by a physical wavelength continuity connecting two designated edge ports of the network does not exist, but it would exist if the wavelength continuity were not required.

An example of this blocking situation is reported in Figure 10.11. In the figure, it is supposed that the WDM systems have two wavelengths represented by dotted and dashed lines.

Let us assume that a new wavelength is to be set up, connecting nodes A and E. Two links exit from node A, one of which (A–F) is fully exploited. Thus, the new path has to start on link A–B. Once in B, the wavelength has to reach D and finally E.

This is the only available path, since the link C–D is also completely exploited.

Unfortunately, this path does not allow wavelength continuity to be realized since wavelength 1 is used in the link D–E, while the wavelength 2 is used in the link B–C.

The impossibility of creating the required wavelength path is a network blocking, since if wavelength conversion is made available into the OXCs, the path would be exploitable.

Both routing sensitive to transmission impairments and blocking due to the wavelength plane structure could be solved with the help of two optical devices: the transparent regenerator and the wavelength converter (WC). A great amount of research has been devoted to these key technologies, but to my knowledge, no commercial product has been introduced into the market up to now.

In the following sections, we will analyze some key technological results in these two areas, and then we will discuss a few transparent OXC architectures among the great number of proposals that have been made.

Finally, we will consider the problem of design and optimization of an optical transparent network.

We will close the section with a few considerations on optical time division multiplexing (OTDM).

10.3.1 Enabling Technologies for the Transparent Optical Transport Network

Besides the improvement in the dimension of switches and in general in the loss of all the optical elements, the effectiveness of the T-OTN depends critically on the availability of two optical components: the transparent regenerator and the WC.

A standard electronic regenerator is based on the so-called three R (3R) operation: repower, reshape, retime. This means that the output signal is ideally amplified to the required power, the pulses are reshaped at the original form and the jitter is eliminated.

This operation is carried out by simply receiving the signal and retransmitting it; thus, the regenerator is bit rate and modulation format specific.

Even though 3R optical regenerators have been investigated for application in the transparent optical network [12], an optical transparent regenerator is generally imagined to accept a vast variety of optical signals, characterized by a different bit rate and modulation format, but it is not required to eliminate jitter (this is called a 2R regenerator) [13].

As we have seen, a WC is a device that changes the central wavelength of a signal ideally without altering it in other ways.

Regenerators and WCs are both strongly nonlinear devices; thus, it is natural to try to exploit the nonlinear behavior of some known device.

Several proposals have been put forward regarding the implementation of such devices, based on fibers nonlinearities [14,15], on nonlinear loop mirrors [16], and other nonlinear elements. Among these proposals, those based on SOAs have the advantage of being based on a component that can be produced in large volumes with a scalable technology exhibiting a strong nonlinearity with a characteristic time on the order of the bit time of a 10 Gbit/s signal.

We will briefly review the architecture and potential performances of SOA-based components that well represent the state of the art in this field.

10.3.1.1 Nonlinear Behavior of Semiconductor Amplifiers

SOAs, as described in Chapter 4, are essentially Fabry–Perot semiconductor lasers whose facets are antireflection coated. Amplification by stimulated emission happens in the intrinsic zone of an inversely polarized PIN junction.

Since optical transitions happen between large bands where the electrons or holes energy levels are almost continuous, the gain bandwidth is quite large, and for applications in the third window, the local gain can be assumed in a first approximation independent from the frequency.

Besides the amplitude nonlinearity due to gain saturation that is present in all the amplifiers, SOAs are also characterized by another type of nonlinearity.

The refraction index of the waveguide that constitutes the active zone of the amplifier depends, on one side, on the bulk material characteristics and the waveguide structure, and, on the other, on the density of the excited electrons that is on the injected current.

When relevant stimulated emission happens into the amplifier, the density of the electrons decreases due to the emission of photons and also the refraction index changes.

This means that SOAs show a phase nonlinearity that depends on the optical power into the active region.

As in the case of Kerr effect during fiber propagation (see Chapter 4), this nonlinearity gives rise to a complex behavior of the field at the amplifier output so that different "effects" are introduced in correspondence to particular forms of the field at the amplifier input.

Also, in this case, we can speak of Self Phase Modulation (SPM), Cross Phase Modulation (XPM) and Four Wave Mixing (FWM).

It is not the scope of this brief presentation to develop a complete model of a SOA in a regime of nonlinear behavior. The interested reader is encouraged to consult the huge technical literature dealing with this subject [17–19].

For the scope of understanding the basic mechanisms giving rise to the possibility of designing WCs and regenerators, a simple model will be enough. We will use the dynamical equations of a two-level amplifier, reflecting the wideness of the semiconductor bands by eliminating the frequency dependence of the local gain.

Thus, we will write the following equations for our amplifier (see Chapter 4).

$$\frac{\partial g}{\partial t} = \Im - \frac{g_0 - g}{\tau} - \frac{g}{\tau} \frac{P(t,z)}{P_{sat}}$$

$$\frac{dE}{dz} + \frac{1}{v_g} \frac{dE}{dt} = \left[\frac{1}{2} g(t,P) + in(z,t) \right] E \tag{10.1}$$

where
  g is the net local gain (that means that it contains the losses also)
  g_0 is the unsaturated gain
  n is the refraction index whose complete expression is

$$n(z,t) = n_0 - \alpha \, \Gamma \, g(t,P) \tag{10.2}$$

where
  α is the linewidth enhancement factor
  n_0 is the bulk index
  Γ is the superposition integral between the guided mode and the density of the
    electrons

The local saturation power is indicated with $P_{sat}(t,z)$ and the characteristic relaxation time of the electron population with τ.

Finally, the optical power is indicated with $P(t,z)$ and the optical field with $E(t,z)$.

10.3.1.2 Wavelength Converters and Regenerators Based on Cross-Gain Modulation

In amplifiers designed to boost the signal power, the saturation power is maintained not too low by designing the amplifier with a high carrier density in the inversion region and a short active waveguide.

If the amplifier is used exploiting gain saturation, this effect is enhanced by designing a long amplifier. As a matter of fact, SOAs for this application are often longer than 1 mm.

The simpler scheme of WC based on SOA gain saturation is reported in Figure 10.12 [20]. A laser, generally a tunable DBR, is integrated into the same chip with a SOA and the laser emission is injected into the SOA via a coupler.

The incoming signal is injected into the SOA too via the other coupler input at a different frequency with respect to the continuous wave generated by the local laser, and generally after amplification through another low-gain SOA, to reinforce the gain saturation effect. If the power of the local laser is suitably chosen, the SOA is at the edge of saturation without input signal. When the signal pulse arrives, the SOA gain saturates and the output is canceled.

In this way, the signal is reported on the local wavelength inverting ones with zeros.

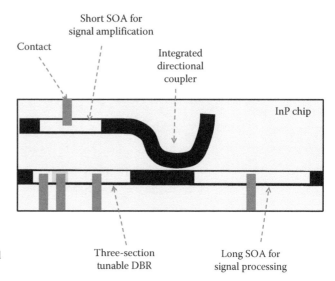

FIGURE 10.12
Scheme of a monolithically integrated WC based on SOA gain saturation.

Moreover, due to the sharp saturation threshold, if the incoming signal pulses are affected by relevant inter-symbol interference (ISI), this is partially canceled by the device that also has regeneration capabilities.

The response of a WC based on the amplifier with the parameters in Table 10.1 is reported in Figure 10.13. The input signal, at the wavelength of 1550 nm, is shown in Figure 10.13a. The incoming signal is affected by a strong additive noise (the optical SN is about 10 dB), and if detected, would cause an important ISI effect. As a matter of fact, the dynamic range is as low as 8.8 dB and the ISI index accumulated during propagation is 18%.

The incoming signal is amplified by a first SOA with a linear gain of 10 dB, and then it is inserted into the nonlinear SOA with a peak power of 14 dBm. The probe, 100 GHz apart, is produced by the integrated DBR and injected into the SOA at a power of −5 dBm.

The behavior of the amplifier gain is shown in Figure 10.13b, where gain saturation in consequence of the high modulated signal is clearly shown. The output signal at the probe wavelength is reported after the filtering with a resonant filter in Figure 10.13c. It is inverted with respect to the original signal, and besides wavelength conversion, its transmission quality is quite improved. The optical SN is about 17.5 dB, with a gain of 7.5 dB with respect to the input signal, the dynamic range is about 14.8 dB, with a gain of 6 dB with respect to the input signal, and the ISI index is reduced to 5%.

TABLE 10.1

A Few Important Parameters of a Very Long, Highly Nonlinear SOA

SOA length	1 mm
Carriers lifetime	50 ps
Superposition integral	0.5
Bulk diffraction index	3.2
Linewidth enhancement factor	10
Local unsaturated gain (enclosing losses)	$6 \times 10^4 \ m^{-1}$
Local saturation power	5 mW

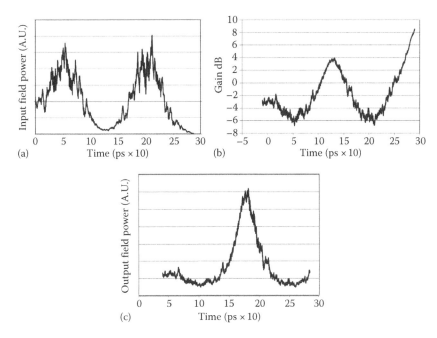

FIGURE 10.13

Time domain response of a WC based on an SOA amplifier with the parameters in Table 10.1. The input signal is plotted in (a), the gain behavior during signal amplification is reported in (b), while the output signal, inverted and regenerated, is shown in (c).

The potential of this device is interesting, but it also has a certain number of issues.

Even if in a simple experiment the signal inversion is not a big problem, in a network this feature obliges the network control plane to insert in the wavelength identifier a field that is updated every time the wavelength channel travels through a regenerator and that at the end is used to determine if the binary signal has to be inverted or not.

This would be a completely new feature that has to be inserted in the OTN or in the GMPLS protocols.

Besides this difficulty, the component performances also have some problems.

Since the modulation of the probe is obtained via gain compression, there is a minimum dynamic range and a minimum ISI index that can be attained and that are almost always present at the device output. This is due to the fact that the SOA characteristic time is on the order of magnitude of the bit interval.

In the case of Figure 10.13, the better attainable dynamic range is about 14 dB. If a signal with a dynamic range of 20 dB is injected into the SOA, the output signal is degraded instead of improved, being the dynamic range of the output signal almost equal to 14 dB.

Last, but not least, in order to work properly, the incoming signal probably has to be amplified as in the example of Figure 10.13. Every amplification adds noise, and it is to be avoided as much as possible in a transparent network.

10.3.1.3 Wavelength Converters and Regenerators Based on Cross-Phase Modulation

Instead of using the amplitude nonlinearity, the phase nonlinearity can be used in order to achieve a regenerating WC [21].

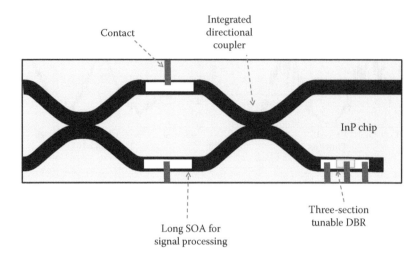

FIGURE 10.14
Scheme of a monolithically integrated WC based on SOAs in a Mach–Zehnder configuration.

A possible device scheme, with copropagating probe, is reported in Figure 10.14 [22,23]. Two SOAs are disposed along the branches of a Mach–Zehnder interferometer (MZI); the input signal is injected by one of the input ports of the interferometer, the probe by the other.

A counter-propagating configuration can also be devised, where the probe is inserted in parallel to the output signal and propagates in the opposite direction with respect to the input signal. However, it is demonstrated that in terms of bandwidth, the copropagating solution is quite superior if all the other parameters are kept constant [24].

In this case, gain saturation has to be avoided; thus, there is no need of amplifying the input signal unless it is really low at the device input.

The idea at the base of the design is that in the absence of power incoming from the signal port, the interferometer is balanced so as to cancel the optical field at the output port. If the signal is injected, the nonlinear phase changes the balance of the interferometer creating constructive interference at the output.

Not only in this way the signal is not inverted, but in principle a dynamic range as low as desired can be attained due to the interferometer structure of the device.

A demonstration of the possibilities of this device is reported in Figure 10.14, where the output of a regenerator/WC of interferometric type is compared with the input signal. The characteristics of the SOA integrated in the chip are reported in Table 10.2, besides other

TABLE 10.2

A Few Important Parameters of an SOA Designed to Enhance XPM While Taking XGM as Low as Possible

SOA length	$400\,\mu m$
Carriers lifetime	$10\,ps$
Superposition integral	0.5
Bulk diffraction index	3.2
Linewidth enhancement factor	10
Local unsaturated gain (enclosing losses)	$6 \times 10^4\,m^{-1}$
Local saturation power	$50\,mW$

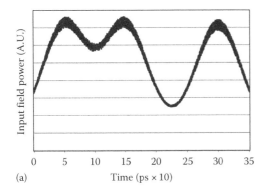
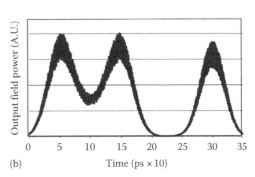

(a) Time (ps × 10) (b) Time (ps × 10)

FIGURE 10.15

Time domain response of a WC based on SOAs in a Mach–Zehnder configuration. The input signal is plotted in (a), while the output signal, with reduced ISI, is shown in (b).

component parameters. This time, the probe is quite below the local saturation in order to avoid a strong gain saturation that would interfere with the XPM.

The input signal is represented in Figure 10.15a. It is quite a noisy 10 Gbit/s signal, with an optical SN of 13 dB, but the main feature is the strong ISI, corresponding to an index of 50%.

The signal at the output of the regenerator/converter after filtering with an interferometric filter is shown in Figure 10.15b. The reduction of the ISI is evident, due to the nonlinear characteristic of the converter. As a matter of fact, the ISI factor is reduced to 7%.

As far as the noise is concerned, the optical SN ratio is decreased due to the fact that there is no relevant gain saturation, and the SOAs introduce their own amplified spontaneous emission (ASE) contribution. The output optical SN is about 11 dB with a degradation of 2 dB.

10.3.1.4 Wavelength Converters Based on Four-Wave Mixing

The components based on cross gain modulation (XGM) and XPM are effective both as partial regenerators and WCs with a few limitations.

The most important limit of such converters is that they perform "incoherent" conversion, generating intensity modulated signals.

If a different type of signal is needed at the converter output, the converter scheme has to be complicated inserting other functionalities and getting worse and worse with the fabrication yield.

The alternative is to use FWM that generates new frequency components when two or more signals at different frequencies are injected into the SOA [25,26].

While converters based on XGM or XPM can provide amplification, depending on the conditions, the FWM efficiency is generally smaller than one; thus, other means have to be found to compensate the losses.

On the other hand, the conversion operation is intrinsically coherent. Moreover, a WDM comb can be amplified in a single pass through a single amplifier [27].

The layout of a monolithically integrated WC based on FWM in a SOA is reported in Figure 10.16, and it is similar in the structure to the layout of the other SOA-based WCs.

The pump laser is a tunable DBR laser integrated into the WC chip, the amplifier is a very long SOA since in this case it can be shown that the nonlinear effects generating FWM are enhanced sensibly by increasing the interaction length [28].

A typical configuration of the spectra of signal, pump, and the main FWM product is also shown in Figure 10.16. In this plot, it can be noted that the FWM operation creates a

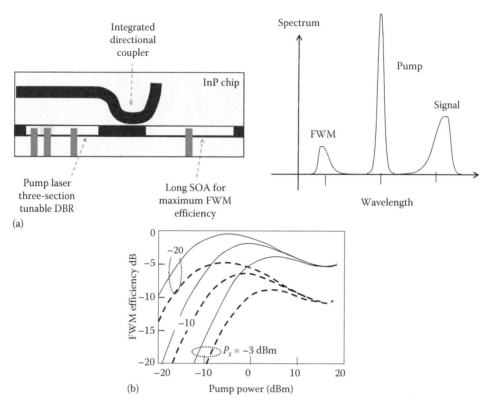

FIGURE 10.16

Scheme of a monolithically integrated WC based on FWM in an SOA. The schematic component layout and the spectrum of the output field before filtering are shown in (a) while the conversion efficiency versus the pump power or a shift of 420 GHz in an SOA amplifier, 1 mm long, is shown in (b). Continuous and dashed lines represent efficiency for a signal frequency on the right and on the left of the pump frequency (After D'Ottavi, A. et al., *IEEE Photon. Technol. Lett.*, 7(4), 357, 1995.).

mirrored replica of the signal spectrum, thus creating a phase conjugated signal at the new frequency.

The analysis of the FWM efficiency in an SOA is quite complex and cannot be carried out in any case on the ground of the simplified model of the amplifier we have used up to now.

As a matter of fact, different causes generate FWM [28]. At low detunings between the signal and the pump (10–15 GHz), the main mechanism is the carrier density pulsation induced by the pump-signal beating inside the active region. At higher values of detuning, FWM is mainly caused by nonlinear gain and index saturation due to intraband carrier dynamics. Two intraband mechanisms are prevalent in causing FWM—spectral hole burning (SHB) and carrier heating (CH). The first is dominant if the pump frequency is below the amplifier gain peak, the second is dominant above the gain peak.

An example of conversion efficiency for a shift of 420 GHz in an SOA amplifier, 1 mm long, whose parameters are reported in [28], is shown in Figure 10.16(b) [28]. From the figure it results that a conversion efficiency near 0 dBm can be attained at the practical value of the frequency shift.

As in the case of all WCs based on SOAs, a FWM WC also has a noise figure; however, due to the strong saturation conditions in which the amplifier generally works, it is much smaller that the noise figure of line amplifiers [28].

10.3.2 Transparent Optical Network Elements

Besides the structures we have briefly reviewed, a great number of other proposals have been made to design all-optical regenerators and WCs. In the following sections, we will consider these elements as functional blocks of the network that performs a certain functionality while adding noise to the signal. Naturally, when considering regenerators, we will have to consider that the 2R regeneration is imperfect.

From now on, we will move our attention to networking related subjects.

The first point to analyze is related to the network elements of the T-OTN.

The first network element is the transmission system, based on DWDM, but with an open design. This means that the input interfaces of the systems will generally receive an optical signal, and after conditioning it with an all-optical interface containing among other things a filter and perhaps a regenerator, directly launches the optical signal along the line for transmission.

All the issues we raised in Chapter 7 talking about this alternative are still valid. This means that either a new type of all-optical interfaces will be standardized so that all the DWDM systems will be produced starting from this standard, or the entire transparent part of the network will have to be produced by the same vendor.

For switching in ring networks, T-OTN uses transparent OADMs. They are available today and the only transparent optical network that is today deployed in the field is composed by two or three transparent self-healing rings.

Since we have described this solution in Chapters 4 and 7, we will not return to OADMs here.

Last, but not least, the OXCs are the main switching element in the T-OTN.

As a matter of fact, as we have seen at the end of Chapter 4, well connected (with a connectivity between 2 and 2.5) and very wide networks, like the LH networks of the incumbent carriers, have to be designed using mesh architecture if the potentiality of the network has to be completely exploited.

The first and more intuitive transparent OXC architecture is shown in Figure 10.10 [29]. We have already commented on the fact that networks entirely realized using such OXC are divided into a number of superimposed and separated networks, one for each wavelength.

The further step to avoid this separation is to install at each output port of the OXC a WC.

In this way, the space division switch matrix simply determines what output direction the wavelength has to go, while it is the output WC that sets the right physical wavelength to correctly feed the output DWDM system [30].

This OXC architecture is shown in Figure 10.17. In this figure, WCs are also used on local ports, assuming to have frequency-sensitive receiver. If this is not true, as in the great part of the state-of-the-art receivers, WCs on drop ports are not needed.

The scheme of Figure 10.17 has NM WCs, where N is the number of wavelengths for each output system, and M the number of output DWDM ports. In a typical OXC, it could be $N = 128$, $M = 5$, so 640 WCs are needed.

In order to reduce this number while maintaining accessibility to a WC at every incoming wavelength, the architecture of Figure 10.18 can be adopted [30]. In this architecture, the number of WCs is reduced by the use of the two switches that allows every incoming channel to exploit every available WC to reach every free output.

A similar solution can also be adopted to incorporate electronic 3R regeneration to be exploited when needed.

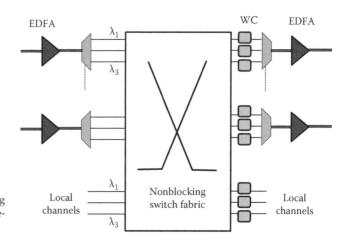

FIGURE 10.17
Block scheme of a fully non-blocking transparent OXC realized using wavelength converters.

Besides the base schemes we have shown, several other architectures have been proposed for transparent OXCs, targeting different equipment performances[30–32]. A great majority of them are based on a space division switch fabric, but a few proposals based on a wavelength division switch fabric have also been done [31].

Since an OXC is complex equipment, it is not easy to make a synthesis of its performances in few numbers.

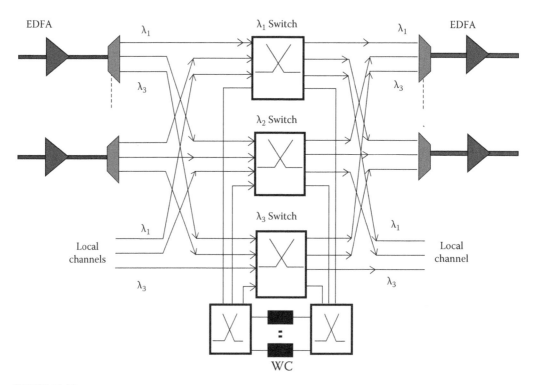

FIGURE 10.18
Transparent OXC architecture allowing the deployment of a limited number of WCs, but preserving full WC accessibility for each incoming channel.

A first group of OXC performance indicators is constituted by its transmission characteristics: its loss in case of passive equipment or its gain and ASE power in the case of an active one, the equivalent filter that represents the spectral effect of the OXC crossing on a wavelength, and possible polarization-dependent loss and dispersion.

In particular, losses or noise and filtering effect are quite important, since if not suitably controlled, they could drastically reduce the transmission reach through the network.

Another physical effect that is peculiar of transparent OXCs is the introduction of crosstalk among channels traveling in different directions [33].

The transit of a set of channel through a transparent OXC (or another transparent equipment, like an OADM) causes two types of crosstalk: in-band and out-of-band (InB, OutB) crosstalk [34].

OutB crosstalk is the same phenomenon that is present in simple DWDM systems, consisting in the fact that due to the infinite spectra tail that characterizes both the channels spectra and the optical filters, every time a DWDM comb is demodulated to divide the component wavelength channels, every channel is affected by the interference of the nearby channel power that is captured by the tails of the selection filter.

In the OXC schemes that we have introduced, this effect mainly happened at the input, where a set of AWGs demultiplexes the incoming DWDM signals.

The second crosstalk source is the so called InB crosstalk and is mainly due to the nonperfect isolation of the outputs of the switch matrixes that are present in the OXC core.

Let us consider the situation of Figure 10.19, where a part of the switch fabric is shown beside two paths at the same wavelength.

The paths of Figure 10.19 have the same wavelength and are directed to different output ports and give rise to InB crosstalk.

In this case, the interfering signal is completely transferred in the bandwidth of the useful signal after a simple attenuation. Its shape is thus that of a modulated signal. For this reason, it is useful to distinguish two kinds of InB crosstalk: the crosstalk from the same signal that arrives at the switch input following secondary paths besides the main one (see Figure 10.19) and the InB crosstalk coming from other channels. The first contribution can interfere with the useful signal if the phases are matched. The reciprocal phases of the two channels are to be considered random variables due to phase noise and to the different paths; thus, coherent ripples can be generated in the pulse experimenting interference (see Figure 10.20), and the distribution of the crosstalk penalty has the shape of Figure 10.21. The result is that in the presence of the same interfering channel suppression, interference can be occasionally twice stronger with respect to OutB crosstalk.

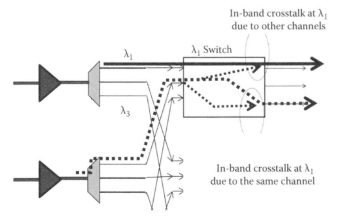

In-band crosstalk at λ_1 due to other channels

λ_1

λ_1 Switch

λ_3

In-band crosstalk at λ_1 due to the same channel

FIGURE 10.19

Example of InB interference from the same and from other channels in the switch fabric of a transparent OXC.

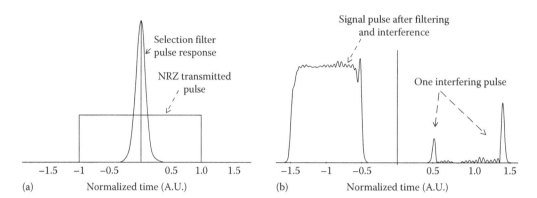

FIGURE 10.20
Time domain example of InB crosstalk; the shape of the ideal NRZ pulse and of the pulse response of a typical AWG (a Gaussian filter) are shown in (a), while the information bearing pulse shape is compared with the pulse shape of the InB interfering pulse in (b).

Remaining in the area of physical performances, the power consumption per processed wavelength and the real estate are important parameters impacting the OPEX of the carrier using the OXC.

Looking at the switch fabric, the wavelength and the fiber modularity are very useful properties (see Chapter 7) [32,35,36] besides the latching property.

Considering the inner connection architecture, the blocking probability in a given condition is another OXC important feature, like the connectivity degree (defined in Chapter 7 as the number of all the possible permutations between input and output ports divided by the number of theoretical permutations).

Another quality factor related to the inner connectivity is the percentage of the processed bandwidth that can be dropped in the OXC location.

To give a few examples, let us consider from a performance point of view, the three OXC architectures that we have introduced. For the sake of simplicity, let us call them pure wavelength planes (PWPs), pure wavelength conversion (PWC), and limited wavelength conversion (LWC).

In order to set up the comparison, let us assume that the LH DWDM line systems have W wavelengths located in the extended C band, so that every WC can move any channel to any wavelength, and that the OXC is of degree N. In the numerical examples, they will be $W = 90$, $N = 5$.

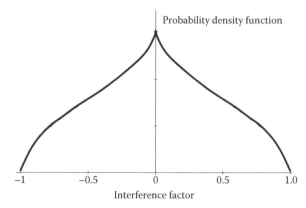

FIGURE 10.21
Probability distribution of the InB self-interference penalty versus the interference factor, the random nature of the penalty is due to the random phase difference between the useful and the interfering channels.

10.3.2.1 PWP Transparent OXC: Example of Performances

Since N is rarely greater than ten, we can assume to use a 2D MEMs switch. Looking at Table 5.8, we can assume a power consumption of 20 mW per switch with a switch crossing loss of $a_s = 4$ dB.

Thus, the traversing loss of the PWP is $A_{PWP} = 18$ dB, where the AWGs are considered to lose 5 dB each and a certain margin has been allocated for the backplane, the patchcords, and so on.

As far as the amplifiers are concerned, the situation is analogous of a two stage amplifier with a great interstage loss. It seems not opportune here to perform the optimization of the amplifier chain, not to deduce a result too dependent on amplifier parameters. It seems more reasonable to adopt a simpler solution, that is, assuming that the amplifier at the output compensates the losses of the span after the OXC, which will be L km long, while the amplifier at the input is used to compensate the OXC losses.

In this way, the OXC appears from the transmission point of view not more than an exceptionally noisy amplifier.

At the end, calling α the fiber attenuation, the gains G_1, G_2 of the pre- and post-amplifiers are

$$G_1 = 2a_{AWG} + a_s + a \quad \text{(dB)}$$
$$G_2 = \alpha L \quad \text{(dB)}$$

(10.3)

Moreover, we have to take into account that the output amplifier also has to host the DCF that has to compensate the dispersion of the fiber span following the OXC. Thus, it is reasonable to assume that the input amplifier is single stage with a noise factor $NF_1 = 5$ dB, while the output is double stage with a noise factor $NF_2 = 7$ dB. At this point, we can write

$$NF_{OXC} = NF_1 \frac{(G_1 - 1)}{(G_2 - 1)} \frac{G_2}{G_1} + NF_2 \approx NF_1 + NF_2$$

(10.4)

The switch fabric is wavelength but not fiber modular, and it is also not latching due to the chosen switching technology.

Since we have 90 switches, the power consumption is 1.8 W for the switch fabric, plus the control electronic and the drivers that we assume to burn 500 mW for a complete estimation of the switch fabric power consumption.

Passing to the inner connectivity, we will start from the hypothesis already done in Chapter 7 that the network manager fix a minimum time interval between the arrival of two connection requests to the same OXC. From Chapter 7 we also know that the blocking probability for random connection requests is

$$P_B(N, W) = W P_s(N)[1 - P_s(N)^{W-1}]$$

(10.5)

where, (p) is the probability that in the observation time a request arrives on the considered wavelength and is directed toward the considered output; it is

$$P_s(N) = \sum_{m=1}^{N-1} \sum_{k=1}^{m} \binom{m}{j} p^j (1-p)^{m-j}$$

(10.6)

10.3.2.2 PWC Transparent OXC: Example of Performances

Let us assume the same switch fabric already considered for the PWP and WCs based on SOAs in a MZI configuration. We have to remember that such WCs, in principle, cannot be traversed by the signal without changing wavelengths.

Thus, we will assume to adopt at the input the configuration of Figure 10.22, where the 2×2 switch is used to bypass the WC if the signal is not to change wavelength.

If this solution is used, the OXC becomes nonblocking as far as the inner blocking probability is concerned. Naturally, a connection request has still to be rejected if the target output link is already saturated, but no switch architecture can send more signals into a saturated transmission system.

To elaborate a power budget, we need to decide the switch fabric architecture. Let us assume realizing our $WN \times WN$ (450×450 in our case) with a strictly nonblocking Clos switch.

If we use MEMs 3D switches, we can select the three characteristic numbers of a Clos switch as $n = 10$, $r = 45$, $m = 56$. The condition $m \geq 2n + 1$ is fulfilled and the switch fabric is nonblocking.

We need to construct the switch fabric out of 56 matrixes 45×45 and 20 matrixes 10×56.

Let us assume that the power consumption of these matrixes depends on the greater between the number of input ports and the number of output ports (compare the 3D architecture). Thus we can say, by comparing with Table 5.8, that a single 10×56 matrix has a power consumption of 700 mW, while a 45×45 matrix burns approximately 600 mW.

Also adding here 500 mW of drivers, control electronics, and so on, the overall power consumption of the equipment is approximately 48 W.

Considering for every component of the switch fabric, a loss of 2.5 dB, the switch crossing implies a loss of 10 dB, comprising a margin of 2.5 dB for connections, back plane, and so on.

Adding the AWG and following the same design criterion considered for the PWP OC, we have $A_{OXC} = -22.5$ dB.

In this case, however, we can exploit the WC to have a small gain, and, on the other side, we have to take into account its noise factor.

A reasonable gain for the overall WC is 6 dB, taking into account all the subsystems of Figure 10.22, with a noise factor $NF_{WC} = 6$. Thus, the gain G_1 can be written as $G_1 = 1/G_{WC}A_{OXC}$, and we have the following equation expressing the overall OXC noise factor:

$$NF_{OXC} \approx NF_1 + NF_{WC} + NF_2 \qquad (10.7)$$

so that with our figures, $NF_{OXC} = 10.8$ dB.

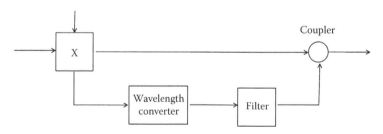

FIGURE 10.22
Adopted WC configuration needed to allow no wavelength conversion if needed.

10.3.2.3 LWC Transparent OXC: Example of Performances

This case is intermediate between the two we have analyzed up to now.

The switch fabric is composed by W switches $(N + H) \times (N + H)$ and one switch $WH \times WH$, with H the dimension parameter of the OXC.

The $WH \times WH$ switch supports WH WCs, putting them at the disposal of the incoming connection requests.

Since both N and H cannot be so large, the switch of dimension $N + H$ can be realized in 2D technology, while the switch of dimension WH has generally to be a multistage switch. In our case, we will use 3D MEMs switches of dimension $W \times W$; thus, we will need $2H^2$ such switches if we want to maintain the cross-bar structure. In all our numerical examples, we will assume $H = 2$.

Using the same approach used for the other OXC, we can calculate the power consumption of the switch fabric.

Each $N + H$ switch has a power consumption of 175 mW, while the switches supporting the WC have a power consumption of 1.1 W each.

At the end, adding as usual 500 mW for the support electronics, the switch fabric power consumption is 21.65 W.

Traversing the OXC, some wavelengths are not converted, while others are. Naturally, the two paths do not have the same attenuation so that the OXC power budget depends on the connection inner routing.

Let us consider the worst case, and let us assume that the adopted WCs are always those based on SOAs and MZI. The OXC global loss, assuming the matrixes loss equal to 2.2 and 2.5 dB, is 25 dB comprising the margin for patchcords, monitoring couplers, etc.

The overall noise factor for the worst case is

$$NF_{OXC} \approx NF_1 + NF_{WC} + NF_2 \tag{10.8}$$

That is 10.8 dB in our case.

As far as blocking is considered, blocking probability in this case is not zero, but it should be much smaller than that of the PWP. In particular, a connection request is blocked when, in the presence of a free wavelength in the designated output line, no direct connection is possible and all the WCs are already occupied.

So let us start from the probability that a connection request is sent to a WC. Excluding the case of sophisticated routing strategies, it happens when, in the absence of the WC, the request is rejected. Concentrating the attention on one particular wavelength switch, this probability is exactly $P_s(N)$ evaluated in Equation 10.6. If $H = 1$, the WCs are N, and the probability that all the WCs are occupied is constituted by all the possible combinations of N wavelengths taken among W wavelength plans that need to use a WC since they are in blocking situation.

Let us focus the attention on one of the wavelength planes, and let us consider the following events:

- The path that was send to the WCs from this wavelength plane is terminated before the arrival of other potentially blocking requests.
- The potentially blocking request is directed toward the same output where it is directed to the path that was sent to the WC from the same wavelength plane.
- The potentially blocking request is directed toward a different output with respect to the path that was send to the WC from the same wavelength plane.

In order to evaluate the probabilities of these events in a simplified manner, let us suppose that since the connection requests are semipermanent, the probability of the first event is negligible. It is easy to verify that this approximation gives an upper bound for the blocking probability.

The probability that a WC is used coincides with the probability that an incoming request cannot find a direct path and that an indirect path exists. In this case, the indirect path (passing from a different wavelength) can be reached if a WC is available.

Let us assume that the $h(j)$ wavelength out of the incoming number creates collision on the jth output.

The probability of this event is

$$P(h,j) = \binom{NW}{h} p^h (1-p)^{(NW-h)} P_S^h \left(1 - P_s^{(nW-h)}\right) \tag{10.9}$$

Let us introduce a set of N numbers $h(v)$ whose sum is comprised between HW and NW and that individually are within W. All the possible configurations of such numbers are the possible configurations of incoming channels (maximum one for each wavelength) that have the need of using the WCs and occupy all the available WCs.

The cardinality of this set is $D = N^{NW} - N^{HW}$.

Let us tag the element of the set with an integer number x comprised between one and D. This will be indicated with $h_x(v)$, so that $h_3(2)$ is the second element of the third configuration.

The probability that the HW WCs are all occupied writes with this definition as

$$P_{WC}(H) = \sum_{j=1}^{D} \prod_{v=1}^{N} P(h_j(v)) \tag{10.10}$$

where

$$p(h_j(v)) = \binom{N\,W}{h_j(v)} P_{oc}^{h_j(v)} (1-P_{oc})^{(NW-h_j(v))} \tag{10.11}$$

and

$$P_{oc} \approx p(1-p)^{(NW-1)} P_S \left(1 - P_s^{(NW-1)}\right) \tag{10.12}$$

The probability $P_{WC}(H)$ is difficult to evaluate analytically for the great number of terms composing it. However, it is possible to find an approximation considering that p has to be sufficiently small to have a functional OXC.

Thus, in a first approximation, we can neglect in the expression of $P(h_j(v))$, all the terms where $h_j(v)$ is greater than one.

In this way, only the $D' = N(N-1) \ldots (N-H)$ configuration remains and the approximated expression of $P_{WC}(H)$, considering also P_s small, becomes

$$P_{WC}(H) \approx W \sum_{j=H}^{N} p^j P_s^j (1-p)^{(NW-j)} \left(1 - P_s^{(NW-j)}\right) \sum_{i=1}^{h} \binom{j}{i} \tag{10.13}$$

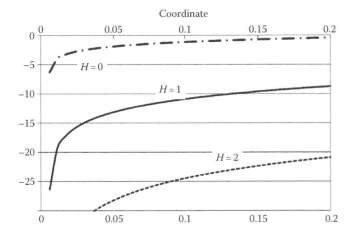

FIGURE 10.23
Blocking probability versus the offered traffic for $N = 6$ and $W = 90$. Three cases are considered, no WCs ($H = 0$), W WCs ($H = 1$), and $2W$ WCs ($H = 2$).

At this point, it is not difficult to evaluate the approximate expression of the blocking probability. The blocking probability $P_B(N,W)$ is the probability that a new connection request arrives when all the WCs are occupied finding a blocking situation. It is a joined probability that there is no direct internal connection and that all the WCs are occupied; thus,

$$P_B(N,W) = WP_s(N)[1 - P_s(N)^{W-1}]\, P_{WC}(H)$$

(10.14)

The blocking probability is shown in Figure 10.23 versus the offered traffic for $N = 6$ and $W = 90$. Three cases are considered, no WCs ($H = 0$), W WCs ($H = 1$), and $2W$ WCs ($H = 2$).

The strong reduction of the blocking probability is evident even with $H = 1$, and in many cases, more WCs are really not needed.

10.3.2.4 Final Comparison

All the performance parameters that we have evaluated analyzing the three OXC architectures are summarized in Table 10.3, where the numerical values correspond to the parameters reported in the text for each OXC.

If one of the expected values of the T-OTN is to drop dramatically the power consumption, we have a confirmation from table 10.3 that this is possible. We should not forget that a DXC power consumption is on the order of kW.

TABLE 10.3

Comparison among the Three Considered Transparent OXC Architectures

	PWP	PWC	LWC
Equivalent noise factor dB	9	10.8	10.8
Switch fabric power consumption (W)	1.8	48	21.7
Latching switch fabric	No	No	No
Fiber modular switch fabric	No	No	No
Wavelength modular switch fabric	Yes	No	No
Blocking probability	High	0	Low

This is an important property of optical equipment. On the other hand, the OXC architecture that we have analyzed has a low degree of modularity.

WCs, even if installed in a limited number, are very beneficial to reduce the OXC blocking probability.

10.3.3 Transport of Control Plane and Management Plane Messages

Besides providing the basic switching functionality, the OXCs have also to run control plane protocols and to be able to correctly transmit management plane information to the central management of the network. The main vehicle that the T-OTN uses to build the control and management plane communication infrastructure is the DWDM service channel, a spare wavelength that is out of the bandwidth of the amplifier used by DWDM, and that contains both the OMS and the OTS, both terminated after every hop in the network.

However, in a transparent network, it is not easy to transmit the information relative to the optical channel (OCh).

This is essentially due to the fact that the header of the OCh, which in the T-OTN coincides with the wavelength channel, has to be transmitted in such a way that it is impossible to divide it from the data relative to the channel.

This is necessary to monitor switching errors via the monitoring at the input and at the output of an OXC of the channel ID and comparing it with the OXC routing table. If an error occurred, for example, for a bad working of a switch matrix in a previously traversed OXC, the input port from which the channel is arriving will be different from that reported in the routing table and the error will be detected.

If the OCh header is included in the service channel, an error could send the wavelength containing the data in the wrong direction, but the service channel in the right direction.

In this condition, sooner or later, the OCh data will arrive at an edge machine where the TDM frame is decoded and the error is detected. At that point, however, the OCh arriving at the wrong edge could be switched in the wrong direction in any of the traversed OXCs and the error would be much more difficult to detect.

In principle OCh header could be coded into a TDM frame with the data over the wavelength carrier. In every node where the OCh header has to be read or regenerated, part of the signal is split and sent to a receiver to extract the OCh header from the frame.

This method is, however, not practical, since it would require a receiver at the data speed for each channel to be monitored in every monitoring point.

In order to be feasible, a solution for the OCh header coding has to require either analog narrow band components, or a low bit rate receiver.

Several techniques have been proposed to carry the OCh header with the wavelength data, including

- Pilot tones [37,38]
- Low-frequency subcarrier modulated data channels [39]
- Optical code division multiplexed data channels [40,41]

10.3.3.1 Pilot Tones

This technique is simple and targets the association to the wavelength channel only of an identifier. The association is realized by modulating the amplitude of the data carrying signal with a slow carrier whose frequency is exactly the channel ID. Thus, all the channels

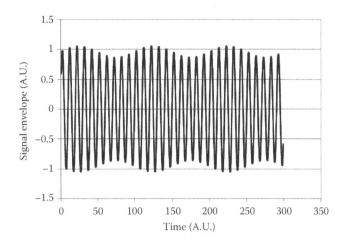

FIGURE 10.24
Time behavior of the signal after pilot tone modulation.

in the network are associated to a pilot tone frequency that allows each network element to recognize them.

After modulation, the time shape of the signal looks like the plot of Figure 10.24.

The modulation depth is generally between 1% and 10% to maintain the unavoidable SN penalty within 0.5 dB, moreover, when erbium doped fiber amplifiers (EDFAs) are present, in choosing the pilot tone frequencies, the long time response of EDFA is to be taken into account.

As a matter of fact, if a pilot tone frequency is chosen below 50 KHz, this starts to be sufficiently slow enough to modulate the excited ion population into the amplifier, thus causing intermodulation between the pilot tones of different DWDM channels traversing the amplifier.

Using pilot tones, the presence of the wavelength channel along the correct route in the network can be monitored even at an EDFA site, with the monitoring system shown in Figure 10.25.

An unbalanced tap coupler (e.g., a 1:10 coupler) can split a small amount of signal and send it to an electrical spectrum analyzer. The spectrum analyzer is set to filter out the

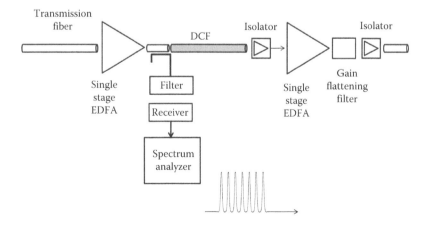

FIGURE 10.25
Scheme of the monitoring system that can be used to detect the transit of the correct WDM channel through a line site using pilot tones.

data carrying modulation and to retain the pilot tone from which the channel ID and the channel power can be deduced.

Pilot tones technique has the main advantages of being simple and inducing a little penalty on the signal. Disadvantages are the impossibility of monitoring the quality of the signal or including in the header attached to the signal other information as the QoS class, the payload, and so on.

Due to this last characteristic, when the pilot tone technique is used, an OCh header is also included into the service channel to carry all the information that is not supported by pilot tones.

10.3.3.2 Low Frequency Subcarrier Modulated Data Channel

This technique relies on the fact that the header occupies a much less bandwidth with respect to the data; thus, if it is transmitted somehow in parallel to data, the bit interval is much longer.

This fact is exploited by modulating the header data onto a low frequency carrier and then the resulting analog signal onto the wavelength channel signal.

Generally, phase or frequency modulation is selected to avoid the sensitivity penalty implied by intensity modulation.

The available bandwidth for this signaling method depends on the wavelength channel data rate and arrives at 150 kHz for transmission at 10 Gbit/s.

The header associated with this signaling technique can be accessed with a receiver similar to that used for pilot tones so that information can be retrieved at each site of the network. Moreover, the transmission is quite robust versus dispersion and other transmission impairments due to its low speed.

10.3.3.3 Optical Code Division Multiplexing

This technique relies on the use of orthogonal codes for the transmission of the useful channel and the header in the same bandwidth.

Let us consider an alphabet $\mathcal{A}\{a_j\}$ with the structure of a finite field, composed by ℓ elements. Using the alphabet letters, let us build a code that is a collection $\mathcal{C}\{(a_{1,i}, \ldots, a_{n,i}), i = 1, \ldots, c\}$, of c code words composed by n elements each.

The code $\mathcal{C}$ is called normalized orthogonal if, whatever couple of code words is selected, its product is null and the norm is one, that is,

$$\sum_{i=1}^{n} a_{i,j} \, a_{i,k} = \delta_{j,k} \tag{10.15}$$

where $\delta_{j,k}$ is the Kronecker symbol that is one for $j = k$ and zero otherwise.

A great amount of studies has been done on orthogonal codes due to their utility in code division multiple access (CDMA) communications [42].

This technique to convey the OCh header is constructed like a wideband CDMA, where the wideband signal (i.e., the data transported on the wavelength) is superimposed on the same frequency bandwidth by a short bandwidth signal (i.e., the header) via the use of an orthogonal code.

In particular, let us select a binary orthogonal code having M different words, where M is greater or equal to the number of channels to be identified in the network. It is not difficult if the length of the code is sufficient, since once given the code length N, and the code words weight W (number of ones, assumed the same for each code word), the number of different orthogonal words C_W with the given length N and weight W is

$$C_W = \frac{(N-1)!}{W!(N-W)!} \tag{10.16}$$

that is, for example, that a 32 bit code with weight 10 has about 44,000 orthogonal words.

Let us assign to each channel a specific code word.

Calling $s_j(t)$ the signal corresponding to the jth channel in the network, $T = 1/R$ the data bit duration, and $T_H \gg T$ the header duration, the following signal is built at the transmitter of the jth channel

$$s_j(t) = \sum_{i=1}^{n} [a_{i,j} h_{v(i)} \, rect_T(t - iT + \tau_j) + b_i \, rect_T(t - iT)] \quad v = \left\lfloor \frac{iT}{T_H} \right\rfloor \tag{10.17}$$

where
$h_v(i)$ are the bits of the header
b_i are the bits of the data
$rect_T(t)$ is the Heaviside rectangular function

Every header bit occupies a time interval much longer with respect to a signal bit, while a code bit occupies exactly a signal bit interval, so that there are several repetitions of the code word associated with the channel within the header duration.

As an example, the signal and the code can run at 10 Gbit/s while the header is simply 100 Mbit/s, so that a header bit contains 100 signal bit intervals. If the header is composed of 40 bits, and a 16 bit code is used, the code word associated with the channel is repeated 250 times in the header duration interval.

When the header is to be decoded, part of the signal is split via an unbalanced directional coupler, and after conditioning via optical filtering and other electronics functions up to sampling, it is divided in M replicas and convolved with M signals given by

$$d_h(t_i) = \sum_{i=1}^{n} a_{i,j} \, \delta(t - iT) \quad h = 1, \dots, M \tag{10.18}$$

The result of the convolution on the kth branch of this processing is a series of samples $x(t_i)$ given by

$$x_k(t_i) = \sum_{m=1}^{n} s_j(t_m) \, d_k(t_{i-m}) = \sum_{m=1}^{n} a_{m,j} \, a_{i-m,k} \, h_{v(m)} + \sum_{m=1}^{n} a_{i-m,k} b_m \tag{10.19}$$

The first term of Equation 10.18 can be evaluated using the orthogonal property of the code. It presents a sharp peak in correspondence to $m = j$, allowing the individuation of the channel ID (j in this case) and the reading of the other contents of the OCh header. After the convolution, this term has a bandwidth equal to $R_H = 1/T_H$.

The other signals, with $m \neq j$, and the second term involving the information bearing message result in a sort of pseudo-noise term with a bandwidth of $R + R_H$ and a power composed by two terms: the random correlation of the signal containing b_m with the code word and a term due to non ideal orthogonal nature of the physical signals representing the code words.

At this point, the header signal can be reconstructed by digital low-pass filtering. In the following paragraphs, we will neglect the impact of nonidealities, and we will assume that the terms containing convolutions of code words are rigorously zero, while the signal term gives a pseudo-noise contribution that in the worst case has a power equal to $P_s R_H/R$.

Besides interference with the signal, the header is affected by the ASE noise, whose spectral density is N_{ASE}.

The optical signal-to-noise ratio required for the header detection depends on the modulation format. Assuming using a differential phase shift keying (DPSK), it has to be $SN_H = 14\,\mathrm{dB}$.

Since the overall noise spectral density is $N_{TOT} = N_{ASE} + P_S/R$, the ratio between the required power of the header signal and of the information bearing signal in the same point where the header is read, has to be

$$\frac{P_H}{P_S} = SN_H \frac{R_H}{R} \frac{M_{he}}{M_{oe}} \left(\frac{M_{oe} N_{ASE} R}{P_S} + 1 \right) \approx \frac{SN_H}{SN_o} \frac{R_H}{R} \frac{M_{he}}{M_{oe}} \tag{10.20}$$

where

SN_o is the signal optical SN in front of the header receiver and it is assumed $SN_o \gg 1$ to neglect its inverse in the parenthesis

the terms M_{he} and M_{oe} represent the ratio between the unilateral optical bandwidth and the electrical bandwidth (supposed equal to the bit rate) for the header signal and the information bearing signal, respectively

Assuming the data used in the previous example and $SN_e = 17\,\mathrm{dB}$, since it is a pre-FEC value, we have $P_H/P_S = 20\,\mathrm{dB}$.

This number is important since, when detecting the data at the end of the wavelength path, the header signal is an unwanted InB disturbance; thus, penalty on data transmission exists.

Considering that the code spreads the header over the whole data bandwidth, when the data bandwidth is detected by a conventional receiver, the optical signal-to-noise ratio can be written in a first approximation as

$$SN_o = \frac{P_S}{M_{oe} R N_{ASE} + P_S SN_H (R_H/R)(M_{he}/M_{oe})((M_{oe} N_{ASE} R/P_S) + 1)} \tag{10.21}$$

From Equation 10.21, it is easy to evaluate in a first approximation the penalty induced by the signaling method in conditions of high SN

$$Pen = \frac{1}{1 + (SN_H/SN_{oi})(R_H/R)(M_{he}/M_{oe})} \tag{10.22}$$

where SN_{oi} is the required data SN_o is the absence of signaling.

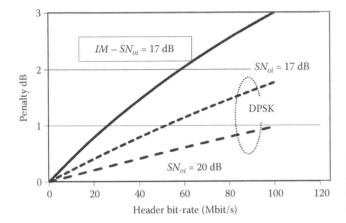

FIGURE 10.26
Penalty versus the header bit rate for both DPSK and IM modulation of the header signal and for different values of the signal SN_{oi}. For simplicity $M_{he} = M_{oe}$.

To understand the situation with a realistic example, let us assume $R = 10\,\text{Gbit/s}$ and $SN_{oi} = 17\,\text{dB}$.

The penalty versus the header bit rate is shown in Figure 10.26 for both DPSK and IM modulation of the header signal, for different values of the signal SN_{oi}, and assuming for simplicity, $M_{he} = M_{oe}$.

From the plot, it is clear that if the penalty is to be contained within 1 dB, the header bit rate should not exceed a value comprising between 25 and 100 Mbit/s.

This signaling transport method is interesting for the freedom that it gives to the designer to include information in the OCh head, but it is affected by the penalty on the data carrying signal and by the need of having a data rate decoder also, when only the header has to be read.

10.3.4 Design of a Transparent Optical Network: ILP Optimization

By its definition, the T-OTN is a static switched network; thus, a method based on integer linear programming (ILP), similar to that adopted for static networks in Chapter 8, seems more appropriate for its design.

Nevertheless, there are significant differences between the dimensioning of a network based on DWDM transmission and electrical switching and the dimensioning of a T-OTN.

We will not derive all the ILP equations for the transparent network dimensioning, as we have done in Chapter 8, for the case of an "opaque" network.

We will, instead, point out the differences between the two dimensioning problems so that deriving the equations becomes only a formal problem.

10.3.4.1 Integer Linear Programming to Dimension Transparent Optical Transport Networks

The hypotheses at the base of this static optimization approach are stated in Chapter 8 as follows:

- The infrastructure is completely given; that means that geographical position of Points of Presence of the carrier and of fiber cables are given.
- We will work in terms of nodes topology: the topology in which the physical network nodes are represented by the nodes of the topology graph.

- We will imagine that all the fiber cables are owned by the carrier that is designing the network so that the bare use of a free fiber is cost free and that a fiber is always available on each route where a cable is present.
- The topology will be meshed at every network layer.
- The layering is also given: in our examples, an MPLS layer is imagined to be superimposed on the optical layer.
- In all the nodes where an OXC is deployed at layer one, an LSR is deployed at layer 2.
- Equipment type is already selected when doing network dimensioning at every network layer; thus, all the cost parameters are given.

Once these hypotheses are set, the cost of the model for the network elements can be set on the line of what is done in Chapter 8.

Thus, the dimensioning can be divided into two steps: dimensioning of working capacity and dimensioning of protection/restoration capacity.

All the constraints of Chapter 8 still hold:

- Traffic continuity across the nodes at every layer and bidirectional GLSP
- Traffic continuity across the layer boundaries
- Number of wavelength deployed on each link sufficient to support the LSP traveling through the link
- Contiguity of the link constituting an LSP path at every layer
- Contiguity of the nodes at the extremes of each LSP hop

In the case of a transparent network, however, these constraints are not enough. As a matter of fact, whenever a WC is not present, the physical continuity of a wavelength has to be assured.

Moreover, the transmission performances of the transparent paths that are dimensioned have to be verified, since it is not at all guaranteed by a dimensioning algorithm minimizing the cost of the network.

10.3.4.2 *Problem of Wavelength Routing and the Use of Wavelength Converters*

As far as the first problem is concerned, the more rigorous way to face it is to divide the matrix $[^1R]$ (see Chapter 8) into wavelength planes and to require that up to the moment in which a WC is traversed, the path have to belong all to the same wavelength plane.

In practice, this is equivalent to adding to the matrix elements another index indicating the physical wavelength.

This is a rigorous approach, but the problem becomes quite complicated due to the increase of the number of optimization variables (the elements of $[^1R]$ that is multiplied by the number of wavelengths that in general is on the order of 100).

A less rigorous approach, less complex under a calculation point of view, is to divide the dimensioning and wavelength assignment problem.

This approach is especially useful when a few WCs can be deployed in the field.

The way to proceed in this case is to divide the problem into two parts.

In the first part, the dimensioning of the network will happen without taking into account the continuity of the physical wavelength.

In a second part, another optimization problem is set up to assign a physical wavelength to the wavelength paths.

In both cases, both if the wavelength continuity is imposed directly on the complete problem, and if the problem is split in two parts, when wavelength continuity cannot be enforced, two actions can be adopted.

The first is to add new cards to the present equipment, or when this is needed, a new network element. The second is to place a WC to allow the transfer of the channel on an available wavelength.

Generally, WCs are placed in the OXC, but if OADMs or the line terminal that does not interface a local OXC exists, these are other possible locations for WCs.

The problem of the optimum use of WCs is an important problem for transparent networks, and a great amount of research has been devoted to pointing out the general characteristics of the problem and the algorithms to solve it.

Also, in this case, a network optimization based on ILP can be set up by simply taking into account that the wavelength path can change physical wavelength passing through an OXC only if a WC is present in that OXC and if not all the available WCs are already occupied by other wavelength paths.

It is, in any case, intuitive that this ad hoc use of the WCs is not the best possible, and an optimized deployment of WCs will give better results. We will analyze this problem in detail using cyclic algorithms, that being much less complex from a computational point of view with respect to ILP, are better suited for such an investigation.

10.3.4.3 *Problem of Transmission Impairments and the Use of Regenerators*

In the case of transparent network, as we anticipated in the introduction of this section, there is no certainty that the route individuated by a routing algorithm looking to minimize the network cost is also acceptable from a transmission point of view.

This means that it is not possible to decouple the problem of the link dimensioning (comprising the placement of electronic regenerators where needed) from the routing of wavelength paths. The solution to this mixed problem is generally called Impairments Aware Routing and Wavelength Assignment (IA-RWA).

The simpler case in which transmission characteristics have to be included into the design of the transparent network is the linear case, when nonlinear propagation effects can be neglected.

This case in practice is quite common, even if the performances of very long DWDM systems are dominated by nonlinear effects.

As a matter of fact, a great majority of the links of carriers in Europe or in Asia are shorter than 1000 km, so that the transmitted power can be maintained sufficiently low to neglect nonlinear effects.

In this case, assuming a line rate of 10 Gbit/s, the dominant transmission impairments are ASE noise, linear crosstalk, and chromatic dispersion. In order to insert in the model these effects, near the matrix [**L**] introduced in Chapter 8 to take into account the link lengths, other two matrixes can be defined to take into account dispersion and noise for every link and every OXC. If all the links are realized with the same DWDM system and using the same fiber, a single parameter replaces the two matrixes: dispersion and noise equivalent factor per kilometer for the links and channel crosstalk (both InB and OutB) and the noise equivalent factor for the OXCs.

To simplify the situation, we will assume from now on in this section the last hypothesis.

Let us imagine that all the links are already dimensioned, that is, the EDFA amplifiers and the DCF are already individuated and their position stated. The only thing that the dimensioning algorithm can do is to deploy them in the preselected location or do not deploy any equipment in that location depending on the fact that the link using the considered location is loaded or not.

Moreover, let us assume that only perfect regenerators are used (implemented by optical–electronic–optical conversion) to simplify the situation with respect to the case of optical regenerators/WCs that realize the regeneration function in an imperfect way.

Regenerators can be located only at OXC locations if a suitable OXC is deployed. Moreover, there is a maximum number of regenerators that each OXC can host, depending on the number of ports and on its degree.

This is, for sure, a suboptimum approach, mainly due to the fact that a better design of the transmission systems would be realized if the dispersion compensators were deployed so as to realize an optimized map. However, in a transparent network this is practically impossible due to the fact that the real route that signals will follow is not known when the network is built.

Let us also assume that all the deployed systems and OXCs are of the same type, and that only one OCh speed is present in the network.

Under these simplifying hypotheses, we can write the condition

$$SN_o + \sum_{j=1}^{H} Pen_j < SN_{rif} \quad \text{(all in dB)} \tag{10.23}$$

where
  SN_o is the optical SN determined by the transmitted power and the ASE noise
  Pen_j are the penalties coming out from different sources
  SN_{rif} is the reference optical SN

Equation 10.22 can be generally written as a linear equation in the optimization variables of the ILP optimization problem.

In particular, it is to remember that with the definitions of Chapter 8, the number of hops of the path followed by the vth connection between nodes i and j can be written as

$$^s h_{i,j,v} = \sum_{h=1}^{^sN} \sum_{k=1}^{^sN} {}^s r_{i,j,h,k,v} \tag{10.24}$$

Since the hops are all equal (but for the last, but we will neglect this specific case), and the loss due to a traversed OXC are all of the same magnitude, the transmitted power on the link i,j depends only on these two parameters.

Following the way we have designed the OXCs, the span loss is compensated by the OXC net gain; thus, the overall ASE noise can be written as

$$N_{i,j,v} = \hbar \omega_0 G_T NF_{OXC} B_o \left(\sum_{h=1}^{^sN} \sum_{k=1}^{^sN} {}^s r_{i,j,h,k,v} - 1 \right) \tag{10.25}$$

where the net noise factor of the OXC (NF_{OXC}), the optical bandwidth (B_o), and the span gain (the opposite of the span loss, G_T) can be evaluated before the run of the optimization algorithm.

The penalties are generally caused in this case by two sources: InB and OutB crosstalk and residual dispersion besides a system margin that has to be taken into account for aging and components specifications variability.

As far as crosstalk is considered, it causes a sort of additional noise. This is almost true for the OutB crosstalk, that is a signal completely uncorrelated from the useful channel, while for the InB crosstalk, especially if generated by the channel itself, this could be far from reality. However, we know that if InB crosstalk is a critical element, a worst case approach can be adopted, where the noise approximation is not used and the InB crosstalk power is multiplied by two and subtracted by the useful power to simulate destructive interference.

This prudent approach, however, produces a nonlinear constraint and thus cannot be used in an ILP base algorithm.

In the case of the Gaussian noise approximation, once given the signal-to-interference ratio χ_i for the three causes of crosstalk in a single OXC (χ_1 for the out-of band crosstalk, χ_2 for the InB from other channels, and χ_3 for the InB from the same channel), and noticing that crosstalk happens only in the traversed OXCs whose number can be calculated from the elements of the matrix [${}^1\mathbf{R}$], a new equation for the noise is obtained

$$N_{i,j,v} = \left(\hbar\omega_0 G_T NF_{OXC} B_o + P_m \sum_{i=1}^{3} \chi_i \right)\left(\sum_{h=1}^{sN} \sum_{k=1}^{sN} {}^s r_{i,j,h,k,v} - 1 \right) \tag{10.26}$$

where P_m is the maximum power allowed by the transmitter, always in the hypothesis, to neglect nonlinear effects.

Once the structure of the receiver is known, for example, if there is an error detecting code (EDC), and the value of the DCF dispersion (that has to be known a priori), the residual dispersion penalty can be generally neglected. Calling μ the margin (in linear units), Equation 10.22, rewritten in linear units, yields to this expression

$$\left(\hbar\omega_0 G_T NF_{OXC} B_o + P_m \sum_{i=1}^{3} \chi_i \right)\left(\sum_{h=1}^{sN} \sum_{k=1}^{sN} {}^s r_{i,j,h,k,v} - 1 \right) < P_m \mu \tag{10.27}$$

Equation 10.27 is a linear constraint that can be easily inserted in an ILP optimization algorithm.

The situation changes completely if nonlinear propagation is important in the network. In this case, the nonlinear induced penalty depends on the span length and on the transmitted power.

This means that the equivalent of Equation 10.25 becomes a nonlinear relationship between the coefficients of the matrix [${}^1\mathbf{R}$], and the ILP cannot be used to solve the resulting problem.

This consideration has to be combined with the fact that the ILP design is quite more cumbersome if carried out for a transparent network with respect to an opaque one, due to the need of representing the inner structure of the OXC in order to deal with the use of WCs, the need of managing the physical wavelength continuity, and the need of adding several constraints (one for each wavelength path) in order to take into account transmission impairments.

In Chapter 8, we have optimized an opaque network with 15 nodes and more than 30 links on a good workstation simply accepting an optimality gap of about 3%. If we consider a T-OTN on the same topology taking into account linear transmission impairments, our algorithm does not converge in 24 h of work.

This is the reason [43,44] why several heuristic algorithms are born to face this problem with optimization methods less computationally complex than the ILP and capable of incorporating fiber nonlinearities. We will present only one class of these methods, and we will use it to make a study on the problem of WC placing.

10.3.5 Cyclic-Based Design Algorithms and Wavelength Converters Placement

As the ILP optimization is aimed at producing a real network optimization, cyclic-based algorithms aim at producing suboptimum acceptable results with a relatively low computation complexity [45].

Besides the adoption of heuristic procedures to determine the network design, these algorithms also perform the design of the optical layer alone, so that interlayer coordination for protection, grooming at the client layer, and all the other interlayer functionalities cannot be included in the design.

Moreover, design for survivability is performed in any case in two stages where similar algorithms are used: first the working capacity is obtained in an optimized way, then a new separated optimization allows the spare capacity to be determined.

On the other hand, these algorithms are suitable for the incorporation of complex transmission models.

The name "Cyclic" is based on the fact that this algorithm cycles more and more times on the set of paths to be routed (i.e., the traffic matrix) to perform various operations.

We will present a particular form of cyclic algorithms; several possible variations exist, some of them suited to adapt the algorithm better to a certain type of network.

All the algorithms of this class are composed of two steps, except for the design of a network completely equipped with WCs for which the first step is sufficient because wavelength continuity is not required.

10.3.5.1 Full Wavelength Conversion Cyclic Algorithm

The first algorithm of this group is that suited for networks where WCs are present on each OXC port so that OXCs are strictly nonblocking. This algorithm will also be the first part of the more complex algorithms for the cases in which WCs are either absent or present in small numbers.

The first algorithm, introduced by [46] can be described as follows:

- A flag is set to determine if regenerators can be used if needed to decrease the network cost or not.
- The initial path routes are found using the Dijkstra algorithm [35].
- Every path is analyzed from a transmission point of view; if a path is not acceptable, either the corresponding request is rejected or an electrical regenerator is added to the OXC at half the path to regain transmission quality. The choice depends on the flag set at the beginning of the algorithm.

- The *maximum link set* is determined: the links belonging to this set are called *maximum links*, and correspond to those links that support the maximum number of paths.
- The paths crossing the maximum number of maximum links are rerouted: this rerouting aims at making lower the cardinality of the set.
- For each link, the number of unused wavelengths is evaluated, and the set of the most inefficient links is found. This set consists of those links characterized by the highest waste of frequencies.
- The paths containing the maximum number of inefficient links are found: these paths are the candidates for the rerouting.
- The paths individuated in the previous step are rerouted using again a Dijstra algorithm with a link weight w_{ij} for the link from node i to node j. The weight depends on the parameter $c_{ij} = p_{ij} Mod(W)$ where p_{ij} is the number of wavelength paths accommodated in the link and W is the number of wavelengths per DWDM system. The meaning of c_{ij} is clear: it is the number of utilized wavelengths of the DWDM system that is not fully loaded, running along the considered link. The particular expression of the weight is $w_{i,j} = p_{i,j}$ if $c_{i,j} = 0$, otherwise $w_{i,j} = c_{i,j}^{-1}$.
- Every path rerouted in the previous step is analyzed from the transmission point of view. If the new path is not acceptable, either it is not rerouted (if regenerators are not selected as possible network elements) or two solutions are considered from the cost point of view: in the first the path is not rerouted, in the second a regenerator is added along the path. As always, the less costly solution is adopted.

This algorithm uses a heuristic procedure to attain a double target. With the first Dijstra routing, it tries to minimize the length of the paths, thereby minimizing the number of network elements that are needed to route and transmit them. This simple step can occasionally create inefficiencies, where in order to maintain short paths, a new DWDM system is opened

These inefficiencies are eliminated in the second part of the algorithm when the less utilized DWDM systems are eliminated rerouting the path traveling through them on longer routes where equipment is already available.

The efficiency of this algorithm is quite good where high-capacity LSPs are used so that a single wavelength carries only a small number of LSPs. In this case, grooming is not very effective, and except for the inefficient cases we have mentioned, maintaining a short path route for the wavelength paths is effective in minimizing the network cost.

Another element that determines the accuracy of the algorithm is the network connectivity. In low connectivity networks, there is little opportunity to improve the network cost by routing wavelength paths on different routes so that a shortest path routing corrected for the low utilization of DWDM systems is effective in optimizing the network. The higher the connectivity, the more inaccurate the algorithm results with respect to a rigorous optimization.

A final observation—the test on the transmission performances of the considered paths are done a posteriori with respect to path routing and no hypothesis is done on the system evaluation model that can be as simple or as complicated as suits the application.

This regarding the working paths. At this point, either protection or restoration capacity has to be deployed. As specified at the beginning, this is done with a second design step with respect to the routing of a working capacity.

As far as some restoration types are concerned, the following algorithm is used

- For each link construct the network topology without that link.
- Determine the paths affected by the link elimination.
- Reroute the affected path with the Dijstra algorithm on the new topology (where the failed link is eliminated) adding, where needed, new equipment.
- Analyze the new paths from the transmission point of view. If a path is not acceptable, either the notification is done so that the corresponding working path cannot be restored if the considered link fails or an electrical regenerator is added to the OXC at half the path to regain transmission quality. The choice depends on the flag set at the beginning of the algorithm..
- Construct a table associating the link failure with a restoration path for all the affected wavelength paths in case the restoration strategy requests preassignment.
- After the determination of the backup path for all the working paths, reconsider the whole network (working and restoration capacity).
- For the new network, repeat the procedure of creating new topologies cutting one link.
- For all affected restoration paths, try to find another route with a smaller cost. If this path exists, there is the possibility of rerouting.
- The rerouted path is analyzed from the transmission point of view. If it is not acceptable, either it is not rerouted (if regenerators are not selected as possible network elements) or two solutions are considered from a cost point of view: in the first the path is not rerouted, in the second a regenerator is added along the path. As always, the less costly solution is adopted.

Shared protection is planned in a similar manner.

10.3.5.2 No Wavelength Conversion Cyclic Algorithm

If no wavelength conversion is available, the algorithm that we have described in the previous section is not enough.

However, it is a possible starting point. As a matter of fact, the algorithm for the case in which wavelength conversion is not present works this way:

- Route working path as if wavelength conversion was present everywhere.
- The obtained path routes are divided into the minimum number of sets of routes (named layers), such that in each layer, as many links as possible are occupied, and no wavelength path share links with wavelength paths of the same layer.

    This operation is carried out by a trial and error procedure, creating sets and analyzing them to find the better set from the considered point of view. The trial and error procedure can be simplified by observing that

    - Partitions that put in the same set a lot of wavelength path crossing the same links can generally be discarded immediately, since each wavelength path in the same set will have the same wavelengths, thereby forcing the design to open several DWDM systems on the same link.

- Partitions that comprise sets with a small number of paths and sets with a large number are generally less efficient.
    Let us call B the average number of layer. In doing so, take into account that if regenerators are allowed, they can change the wavelength of the path they regenerate.
- Assign a layer number randomly to each layer so that each number appears once.
- Assign a wavelength cyclically to each layer according to the layer number.
- Evaluate the system cost according to the wavelength assignment performed in Step 4.
- Iterate from Steps 3–5 a certain number of times, finding the layer assignment that gives the minimum network cost

A similar algorithm is used for the backup capacity. We will list only the version for restoration:

- Determine restoration capacity as if WC where available everywhere.
- From the restoration table, assign to each restoration path, the same wavelength of the corresponding working path.
- When capacity sharing among paths with different wavelengths occurs, either introduce new wavelengths as needed if regenerators are not allowed, or compare the solution with the new wavelength with the solution with one regenerator that changes the wavelength of the conflicting backup path to comply with the conditions of sharing. As usual, assume the most economic solution.

It is to be observed that this algorithm is much more trial and error with respect to those presented for networks with nodes completely equipped with WCs. However, the effectiveness of this algorithm in going near the rigorous optimum network is very good. It is to take into account, that if no WC is present, the network is de facto divided into planes. A wavelength path is completely contained in a single plan.

In these conditions, the problem is in a first approximation divided into one problem for each wavelength situation in which a trial and error algorithm works well.

10.3.5.3 Partial Wavelength Conversion Cyclic Algorithm

This is the case in which the OXC accommodates only a small number of WCs, in general proportional to the number of wavelengths of a WDM system.

In this case, the so-called global arrangement algorithm is used, and which is described in [47]. This strategy makes use of the approach used in the case of the absence of WCs in a recursive manner. First of all, the strategy used without WCs is adopted, and the links with the maximum number of unused wavelength channels are individuated: these are the inefficient links.

For each inefficient link, a certain number of WCs is introduced, so as to fill, as much as possible, the wavelength channels on the used transmission fibers.

For each WC placed in the network, the new wavelength to be assigned to the converted path is determined by once again adopting the WP algorithm on a suitable subnetwork. In this way, some fibers in the network remain completely unused and can be eliminated, thereby reducing network costs. This algorithm is run several times, up to the moment in which the final results are stabilized.

A similar procedure is applied for the backup capacity.

10.3.5.4 Cost Model and Transmission Feasibility in Cyclic Algorithms

Up to now, we have not specified the target that the cyclic algorithms try to minimize. It is naturally, a cost-related target, and in particular, it coincides with the CAPEX cost of the network.

Even the model adopted in the last section talking about ILP linear optimization can be used, and we will adopt this choice.

As far as the transmission performance evaluation is concerned, we will still use Equation 10.22, but without the constraint of considering only linear phenomena.

The main problem in considering LH systems in a transparent network is that the dispersion map cannot be really optimized, since it is not a priori known what will be the routes of the channels passing from a certain link

This uncertainty also thwarts the use of predistortion based on back-propagation and of all the similar electronic compensation methods, due to the fact that the channel could change the routing through the network.

We will solve this problem simply by placing in the receiver, a strong adaptive nine taps Feed Forward Equalizer (FFE) + nine taps Decision Driven Equalizer (DFE) electronic compensator. This component introduces a back-to-back penalty, but assures a good blind compensation of all the ISI sources.

A slight dispersion under-compensation will be operated in any case in every single span. This will damage short links, but will be beneficial for longer ones that are at greater risk from a transmission point of view.

Since a further crosstalk contribution is added by the OXCs, the channel spacing will be 100 GHz with a bit rate of 10 Gbit/s.

In synthesis, the penalties we will consider in the case of NRZ modulation, that is the only case that we consider here, are summarized in Table 10.4, besides a few needed system parameters.

TABLE 10.4

Penalties in the Case of NRZ Modulation and a Few System Parameters

System			
Transmitted power	DBR tunable with integrated SOA	16 dBm	
Power injected in Fiber	AWG, modulator and passives loss	5 dBm	
Reference SN_o pre-FEC (used ITU-T standard FEC)		12 dB	
Optical bandwidth	Depends on the number of OXC	$40/\sqrt{N_{OXC}}$ GHz	Cascading effect
Penalties			
Transmitter dynamic range	Fixed	0.5 dB	IM MZ LiNbO3
Linear crosstalk	Depends on the OXC number	$0.22 + N_{OXC}$ dB	Gaussian AWG
FWM and XPM	Depends on the number of spans	$0.012 + N_{span}$ dB	70 km span
Residual DGD and SPM	Depends on the number of spans	See text	Strong blind EDC
PMD	Channels at 10 Gbit/s	0	Strong blind EDC
Polarization effects	Fixed like a margin	1 dB	
Aging	Fixed like a margin	1 dB	
System margin	Fixed	3 dB	

A few observations on Table 10.4 are useful to underline differences between opaque and transparent networks from a transmission point of view.

The first point concerns the optical bandwidth. While in a stand-alone system, this is a fixed parameter, here, due to fact that a cascading of filters generates a progressive reduction of the bandwidth, the optical bandwidth and the ratio M_{oe} depend on the number of filters traversed by the signal (in our case, this corresponds to twice the number of OXCs).

The second observation concerns the impact of interaction between residual differential group delay (DGD) and SPM. Several attempts have been made to derive a phenomenological curve that matches the dependence by the span number and the span length for a reasonable configuration of the system.

However, this depends in a non-negligible way on the general system parameters, first of all on the fiber type and the modulation chirp. Probably, the only way to face this point is to repeat this evaluation every time a parameter changes.

It is in any case evident that transmission in a transparent network cannot be planned a priori in an effective manner as in an opaque network both for the fact that the wavelength path routes are not a priori known and also for the influence of InB crosstalk that adds up along the path every time an OXC is traversed.

10.3.5.5 Example of Network Design and Role of Wavelength Converters

We would like to close the considerations about cyclic algorithms with an example of their use.

The computational complexity of a cyclic algorithm is lower with respect to an ILP optimization, both for very large networks (cyclic algorithms are not NP complete) and for a small number of nodes.

The complexity of cyclic algorithms is strongly influenced by the traffic to be routed due to the necessity of cycling on the set of paths to be routed.

In order to propose an example of the use of these algorithms, we will consider a possible topology of the Italian core network whose topology we already used in Chapter 3 and that we will repeat here for reference in Figure 10.27.

It is interesting to note that the test network has 32 nodes; thus, it would be very difficult to analyze it using an ILP algorithm, also if only working capacity is considered and the optical and MPLS layers are separately designed.

We propose results for the working capacity dimensioning without considering the second step of the algorithm that allows completing the project by adding capacity to support the selected survivability method.

Here, we are interested in analyzing what happens when the different strategies of placing of WCs are applied.

In Figure 10.28, the normalized cost for different designs of the optical transparent network on the Italian topology are compared: systems with 128 and 64 wavelengths are considered besides the three types of OXCs described in the previous section.

Electronic regenerators are not allowed, that is a rational strategy due to the geographic area occupied by the network.

The network cost is essentially composed of the following terms:

- DWDM platform (EDFAs, common parts, AWGs)
- DWDM transparent transponders (adapt the signal to the transmission with a filter and control the power with a variable attenuator)

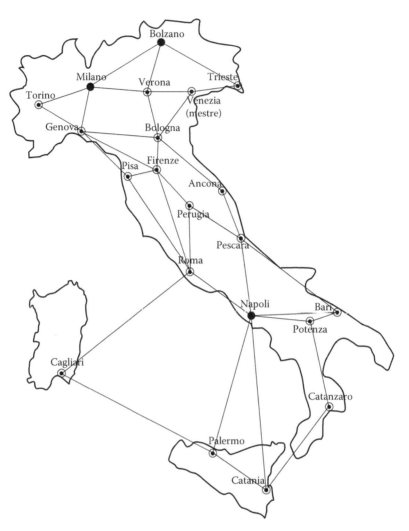

FIGURE 10.27
A possible topology of the Italian core network.

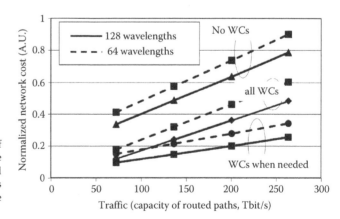

FIGURE 10.28
Normalized cost for different design of an optical transparent network on the Italian topology; systems with 128 and 64 wavelengths are considered besides the three types of OXCs described in the previous section 10.3.2.

- OXC platform
- OXC WCs
- OXC space diversity switch fabric

Comparing the cases with 128 and 64 wavelengths, the intuitive result is that the more wavelength a system allocates in the same bandwidth (in our case all channels are in the C band), the lower the cost.

As a matter of fact, all cost factors are the same, but the DWDM platform is almost cut to half.

More complex is the situation that creates the application of different WC placing strategies.

Starting with the case without WCs, the only way to avoid a traffic request refusal is to increase the dimension of the switches and, consequently, the number of DWDM systems. As a matter of fact, if we compare the overall number of ports of the OXCs of all the networks in analogous conditions, we find that using PWP OXCs there are more ports than using PWC OXCs.

Thus, in the PWP case, the average OXC machine, as well as the network, is more expensive.

Comparing the PWC and the LWC case, the algorithm used is able to design the network with the LWC OXCs in a way that the number of OXC ports is the same at every node.

In this condition, the DWDM infrastructure is the same in the two cases, and the only relevant cost difference does exist regarding the OXC switch fabric.

In the PWC case, if the fabric is assumed to be a cross bar matrix, we have $(NW)^2$ cross-points and NW WCs. In the typical LWC case, we have $W(N + 1)^2$ cross points, and generally, W WCs.

If the cost of a WC is K times the cost of a cross point (where K is on the order of 900), the ratio ρ between the cost of the network using PWC and LWC is

$$\rho \approx \frac{N^2 W^2 + KNM}{W(N+1)^2 + KW} \approx \frac{NW + K}{N + K/N} \tag{10.28}$$

This very simple analysis also shows that by increasing N, the advantage of the LWC over the PWC is about W, which in general is a large number.

This advantage is all based on the fact that after a sufficient number of cycles (not so much in a practical network), using LWC means placing only the needed WCs achieving the same OXC dimension in terms of line ports with respect to the case in which PWC is used. On the other hand, the switch matrix of an LWC is much simpler than that of a PWC and the number of deployed converters is much lower.

In order to better understand what the number of deployed WCs is in case LWCs are used, in Figure 10.29, the number of WCs deployed into the LWC is represented as a percentage of the number of line ports versus the number of wavelengths per DWDM system in different traffic conditions (total capacity of all the 10 Gbit/s deployed in the wavelength path).

Apparently from the figure it could seem that increasing the traffic, the number of WCs decrease, that is clearly false. In the figure, the number of WCs is represented in percentage with respect to the overall OXC number of ports, that is, a parameter increasing almost linearly with traffic. Thus, even if the percentage decreases increasing traffic, the absolute value increases. For example, passing from 144 to 240 Tbit/s and using DWDM systems

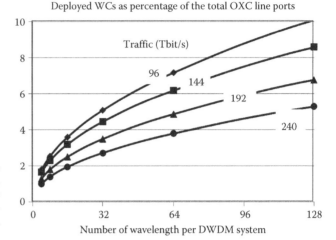

Deployed WCs as percentage of the total OXC line ports

FIGURE 10.29
Number of WCs deployed into the OXCs as a percentage of the line ports versus the number of wavelengths per DWDM system in different traffic conditions (total capacity of all the 10 Gbit/s deployed optical GLSPs).

with 128 wavelengths, the WC percentage passes from 8.5% to 5.3%. The number of ports of the OXC increases about a factor 1.8, so that the average number of WCs in a degree four OXC is about 43 in the first case, and 50 in the second case.

Considering the number of wavelength, if it is increased maintaining the traffic constant, the partitioning of the network in the wavelength subnetworks is more and more pronounced, causing a greater need of wavelength conversion.

Last, but not least, another kind of insight into the impact of wavelength conversion can be attained by considering the so-called wavelength reuse factor R_W. In our case, the wavelength reuse factor is defined as the average number of wavelength paths the node processes divided by the average number of loaded wavelengths per link.

If we call $\lambda_{i,j}$ the number of active wavelengths of the link i,j, and $d_{i,j}$ is one of (j is a second index of d) the links that exist between the nodes i and j, N the number of nodes, and N_p the number of routed paths, the wavelength reuse factor can be written as

$$R_W = \frac{N_P}{N(N-1)} \frac{\sum_{j=1}^{N} \sum_{i=j}^{N} d_{i,j}}{\sum_{j=1}^{N} \sum_{i=j}^{N} \lambda_{i,j}} \tag{10.29}$$

It is to be noted that Definition (10.37) tries to catch, in a quantitative way, the effect of WCs on the possibility of using wavelengths on links where they could not be used without WCs. But it is a different parameter with respect to the wavelength reuse that is generally defined in literature, in the case of dynamic network dimensioning [48].

The plot resulting from our definition is reported in Figure 10.30 for the example topology we are using.

As intuitive the wavelength reuse factor increases decreasing the number of wavelengths per DWDM system, since the real connectivity of the system increases (the number of wavelength planes decreases) and increases increasing the traffic up to an asymptote, depending on the considered network.

Comparing a situation where the OXC is equipped with WCs on every port and a situation where WCs are not used, we can appreciate the effectiveness of the use of WCs,

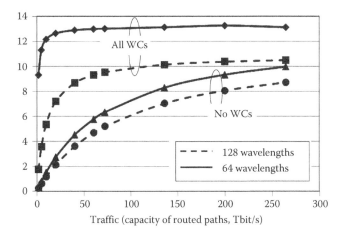

FIGURE 10.30
Wavelength reuse factor versus traffic for the same networks in Figure 10.29.

both in reaching the asymptote at small traffic volumes and in increasing the value of R_W. In particular, the smaller the number of wavelengths per DWDM system, the higher the effect of having WCs.

10.3.6 Translucent Optical Network: Design Methods and Regenerators Placing Problem

Even if a transparent network is very well designed, it remains that transparent transmission has a limited range (even if on the order of 1000 km), and if a numerous set of possible paths go beyond this "optical length," in order to set up the considered connection, regeneration is needed.

Since optical regeneration is not so efficient to be really competitive with electronic regeneration, we will assume in this chapter to talk about electronic 3R regenerators.

When 3R electronic regenerators are an important part of the transmission layer, we call the network "translucent" [49].

Three architectures have been proposed for translucent networks in the literature:

1. An architecture where electronic regenerators appear only as a support to transmission: This is essentially a transparent network, with a sparse number of regenerators to aid transmission along the worst links [50].

2. An architecture where the nodes are translucent: This means that every node has a small electrical switch sitting on top of the optical transparent switch, as in Figure 10.31. The electrical switch can be used to locally regenerate those channels that have very far destinations, but it can also be used to perform grooming if it is based on a lower speed with respect to the wavelength channel. Thus, this version of the translucent network gets nearer to a two-layer network, at least from a few points of view.

3. Last, but not least, an architecture where transparent islands are connected via opaque gateways: This is a fully developed two-layer architecture where the fact that gateways regenerate signals is only one function of this machine, and often not the more important [51].

Among the different possible architectures of the translucent network, we consider here the nearer to the full transparent network, that is, those architectures in which regenerators

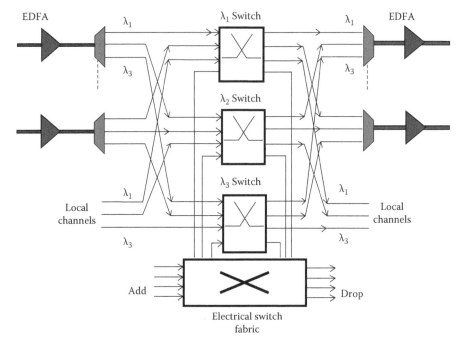

FIGURE 10.31
Translucent node architecture—every node has a small electrical switch sitting on top of the optical transparent switch.

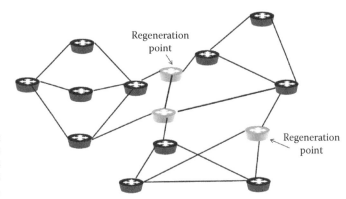

FIGURE 10.32
Translucent architecture in which regenerators are present only as a transmission aid on long links and all the network management is exactly that of a transparent network.

are present only as a transmission aid on long links and all the network management is exactly that of a transparent network.

An example of this architecture is reported in Figure 10.32.

In this representation, we will call transparent paths the paths that do not experiment regeneration (even if they could traverse regeneration points), and translucid paths those paths that are regenerated at least once.

The dimensioning of a translucent network of the kind we are considering is generally performed in two steps: placing of regenerator points and IA-RWA.

Several algorithms are presented in literature to carry out the first phase. Due to the fact that in this kind of network the link transmission performances are dominant in the process of placing regenerators and in RWA, many of these algorithms use the shortest path routing policy both in the first and in the second part of the design.

The key point is to assign to every link a correct weight to take into account all the relevant phenomena.

The strategy we want to illustrate here is based on the assumption that propagation can be considered linear all over the network, at least in a first approximation.

In the first phase, a weight is linked to each node and to each connection of the network graph that is equal to the degradation in the SN_o that a signal undergoes passing through the link or the node.

Let us assume that the wavelength-carried signal is generated at the transmitter without noise, that is, with infinite SN. Let us also image that the design of the transmission system is uniform all through the network, so that everywhere the inter-EDFA span is L km long and the EDFAs are all similar with a gain $G = \alpha L$, where α is the fiber attenuation in dB/km. Finally, let us design the links with a bit of DGD under-compensation that is completely recovered by the EDC at the receiver.

The SN_o at the output of a wavelength path will be in the ideal condition in which the gains are exactly selected to compensate losses (the gain of the preamplifier to compensate the losses at the receiver)

$$SN_o = G_{pre} + P_{in} - \Sigma_i N_i \quad \text{(all in dB)} \tag{10.30}$$

where

N_i is the noise power introduced by traversing the ith entity (link or cross-connect) along the path

G_{pre} is the preamplifier gain that is a system parameter

In order to consider the transmission impairments, it is possible to tag every link and node with the generated noise power N_i and consider it as a cost related to the traversing of the node or link.

At this point, the first step of dimensioning can be carried out as follows:

- Every couple of nodes is considered.
- The shortest path is calculated between the couple of nodes.
- The final SN_o is evaluated for the shortest path:
  - If SN_o is greater that the reference SN_{ref} (e.g., SN_{ref} =11 dB with standard FEC)
    - The path can be transmitted transparently.
  - If SN_o is smaller that the reference SN_{ref}
    - The path is a translucent path and the nodes candidate to be regeneration nodes have to be identified.
- For each translucent path
  - The path is divided into transparent subpaths in all the possible ways that employ the minimum number of regenerators.
- For each candidate node, its candidate degree is evaluated, corresponding to the number of translucent paths traveling through it and asking for regeneration.
- Regenerators are placed in the candidate nodes starting with those with a higher degree.

- For each placed regenerator, the affected translucent paths are transformed in partial paths by dividing them in two at the newly inserted regeneration point.
- When the new paths are created, they are classified transparent or translucent like for the original paths.
- When no more translucent paths are present, the regenerator placing stops and the first optimization phase ends.

After the regenerator placing also in the traffic matrix all the translucent paths are eliminated by substitution with transparent sub-path. At this point, IA-RWA can be performed with conventional methods. In our case, a cyclic algorithm is used.

As an example, we will study a network whose topology coincides with the NSFNet, a pan-American network [49] with 16 nodes represented in Figure 10.33.

This network is heavily influenced by transmission-related constraints both for the long distances and for the number of OXCs. As a matter of fact, OXC crossing adds both noise and crosstalk, so that on many paths, the OXC contribution to the degradation of the SN_o is largely dominant.

In order to carry out an example dimensioning on the network of Figure 10.33, we will assume that the OXCs are characterized by the parameters of Table 10.4 and that the links are equipped with amplifier sites 60 km spaced with amplifiers with the same characteristics of that used in the OXCs.

Considering the dimension and the scope of the network, we could imagine having DWDM systems with 128 channels distributed between C and L bands so as to work with a channel spacing of 100 GHz.

The traffic matrix has been constructed assuming that each node originates paths toward all the other nodes in the network, and that the number of paths originated by node i toward node j is proportional to the product of the degree of i and j. As an example, since the network has 18 nodes, there are $N_c = 153$ distinct couples of nodes. If it is stated that the overall number of wavelength paths routed through the network is $N_p = 24,000$, the paths routed between nodes i and j are

$$n_{i,j} = \left\lfloor \frac{c_i c_j}{A} \frac{N_p}{N_c} \right\rfloor$$

(10.31)

where
c_i is the degree of the ith node
constant A is determined by the condition that the overall number of paths is N_p, that becomes

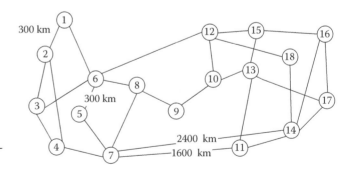

FIGURE 10.33
NSFNet network topology: A pan-American network with 16 nodes.

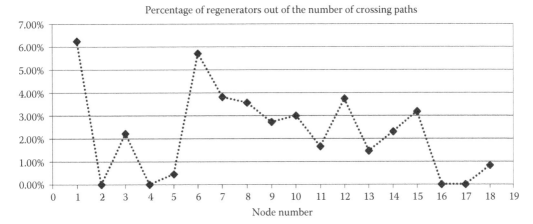

FIGURE 10.34
Regenerators deployed per node as a percentage of the number of paths traveling through the node.

$$A = \sum_{i=1}^{N} \sum_{\substack{j=1 \\ j \neq i}}^{N} \frac{c_i c_j}{N_c} \tag{10.32}$$

The result of the first dimensioning phase is reported in Figure 10.34 in terms of regenerators deployed per node as a percentage of the number of paths traveling through the node. With our assumption, this percentage remains almost constant varying the traffic load, due to the fact that all the elements of the traffic matrix increases of the same relative amount.

It is evident that almost all the nodes host regenerators confirming the difficult transmission situation of the network. Naturally, the number of regenerators is smaller, frequently much smaller, than the number of wavelength paths crossing the node.

The consequence, we can derive from Figure 10.34, is that when the network is so extended, transparency tends to fade due to the need of installing regenerators almost in all the nodes.

10.3.7 Summary: The Transparent Optical Network Status

The idea of a transparent transport network has been around for more than 20 years, a sufficient period to develop all the tools (hardware and software) needed to implement such an innovative solution.

The main advantages that were devised at the beginning, mainly the great sparing in energy, are still there.

Nevertheless, the forecasted revolution of transparent optics did not happen and while the time goes on seems less and less probable.

Several elements contribute to this apparently surprising result, and we will try to analyze them.

The first point is, for sure, the great CAPEX that would be needed to completely change the network core technology, at a time when the access network is also rapidly evolving.

From the bubble end (see Chapter 2), carriers have greatly reduced the CAPEX to focus on the exploitation of the installed equipment.

The transparent revolution would be clearly in the opposite direction.

Moreover, the new technology seems weak from a management point of view. Implementing T-OTN would require a complete change in the way the network is monitored, both as definition and as transport of monitoring variables and of management commands.

The OPEX is strongly related to the management effectiveness of the network, and there is the risk that a solution built with the intention of reducing the OPEX, in the long run, will increase them in the early operational years due both to lack of instruction about the new procedures and the lack of experience on the part of equipment vendors.

On top of these elements, there is a possibility that if transparent networking is adopted, transmission trunks cannot be designed according to the required specifics, thereby losing the optimized design that today's systems have.

In addition, 100 Gbit/s is clearly quite difficult to manage without a specifically designed link, and there is a diffused opinion that 100 Gbit/s is difficult to install in LH links in a transparent network.

As far as bit rate and format transparency is concerned, pure transparent networks, for sure, can have this property, but as soon as a few regenerators have to be deployed to cope with long distances or huge OXCs, the full transparency is immediately lost.

On the other side, standardization works very well in the core network allowing the limitation of the typology of possible routers and OXCs interfaces by defining standard transport formats.

To allow conveying into the network whatever data format through transparency would be a disaster from the monitoring point of view and would complicate a lot the network software that should be ready to deal with one among a set of different formats.

Thus, today, format and bit rate transparency appears less an advantage with respect to the early 1990s.

All the aforementioned reasons have up to now stopped the development of the optical transparent network and have contributed to convincing a large number of telecom players that this deployment will never happen.

Even if the market evolution seems against the transparent network, this technology has still a card to play and considering the strong relevance that environmental issues are also taking in the field of telecommunications, this could be the winning card.

As a matter of fact, even if electronics has taken huge strides ahead, optical components still have a much smaller power consumption. If power consumption issues become one of the main drivers of the network and if governments introduce green contribution for carriers that demonstrate less power consumption and a higher taxation for carriers consuming too much power, transparent optical networks could become again an interesting possibility.

10.4 Transparent Optical Packet Network (T-OPN)

Not long after the success of the T-OTN idea, at least at the research level, another even more fascinating idea was launched. In the mid-1990s "packet switching" was mainly ATM, though the success of IP was announced by the commencement of the penetration of internet both in business practice and in private homes.

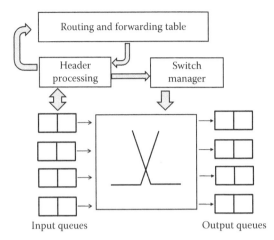

FIGURE 10.35
Optical packet switch architecture.

Input queues   Output queues

In that situation, the idea of a packet optical transparent network was launched, as if the transparent transport network was destined to last a few years: the time needed to develop the real next generation network.

It is not surprising that the first research attempt on the transparent optical packet network (T-OPN) was defeated by the lack of even the most basic components needed for packet processing.

As a matter of fact, packet processing is intrinsically a digital operation, while optical equipment is essentially analog.

An optical packet switch can be, in general, represented as in Figure 10.35, and a simplified process corresponding to the arrival of a packet, and its forwarding to the following node of its path through the network can be described as follows:

- The interface with the line detects at the correct layer the arrival of a packet.
- If the input queue is full, the packet is discarded and the event is registered.
- If there is free memory in the input queue, the packet is stored in the line card.
- The line card processor reads the packet header and verifies if it is correct.
- If the packet header is correct, the destination address is extracted and sent to the routing processor.
- The routing processor accesses the routing and forwarding tables and determines the destination output port.
- The switch is configured to allow the packet to reach the correct output port.
- The packet is transmitted to the output port via the switch and stored in the output queue.
- The packet is conditioned for transmission in the line and leaves the router.

It is clear that no optical equipment can carry out a set of such complex operations that are tailored for digital processors more than for optical systems.

On the other hand, the target of the research on the optical packet switching is not the execution of all the operations that happens in a router at the optical level, but to never have the need of transforming the packet payload from optical to electrical during the router passing through process.

In this way, on one side, it would be possible to transmit a very high bit rate (potentially far above 100 Gbit/s if OTDM is used), and on the other, the capacity of optical systems to have a much lower power consumption with respect to an analogous electronic system is exploited.

Thus, a possible scheme of an optical packet switch working can be that of Figure 10.36: the incoming optical packet is stored in an optical memory. The packet header, exploiting the fact that it has a much smaller number of bits with respect to the payload and thus it can be transmitted at the same time at a much slower bit rate, is detected and sent for processing to the line card processor.

All the forwarding processes are carried out electronically while the packet is stored in the line card. When the destination output port has been calculated, the packet is transferred from the input optical queue to the correct output optical queue via the optical switch and is sent out through a DWDM system.

It is possible that some operation has to be done on the packet payload, for example, the verification of an EDC.

These operations, however, have to be maintained as simple as possible and have to be performed by digital optics with simple algorithms.

This scenario, depicted in Figure 10.36, is naturally only one of the possible scenarios for the introduction of the optical packet systems, and at the end of this section, we will try to describe at least another couple of possibilities.

It is, however, clear that the key to enabling whatever form of packet optical network is the development of a new generation of very-high-speed optical components suitable to be integrated onto an optical packet switch.

The main components are

- Optical memories
- Dynamic optical switches

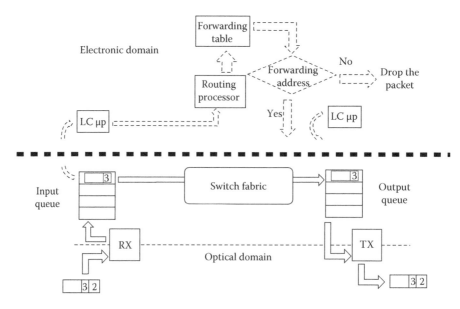

FIGURE 10.36
Possible scheme of an optical packet switch working.

- Optical logic gateways for operation on the payload
- Packets synchronizers (in the realistic case of asynchronous transmission)

Moreover, even if OTDM is not necessary for an optical packet switching, it is really complementary, creating very-high-speed TDM streams that would be very difficult to manage with electronic systems.

In the following section, we will review part of the research activity on these components, while in the sections following thereafter, we will discuss both a few optical packet network architectures and the way in which a potential optical packet layer can be integrated into the whole network.

10.4.1 Transparent Optical Packet Network Enabling Technologies

10.4.1.1 Optical Memories

This is a key component if a packet switch has to be realized. To better clarify the possible optical technologies to realize a memory, we can distinguish two types of memories:

1. Memories whose scope is to delay the signal for a given time, and in some cases, to maintain the order of the incoming signals while delaying them
2. Memories that store the data carried out by the signal and make the access possible to any part of the data at any moment

One example of memories of the first type is the First In First Out queue, which is realized so that a signal entering the first cell has to travel through all the cells to exit from the last. The most important example of the second type is the random access memory (RAM).

A simpler example of the first type of optical memory is a linear amplified ring, as shown in Figure 10.37. This is simply a large ring where the signal is inserted through a directional coupler that also has the role of point of extraction.

Along the ring, a low-gain SOA compensates part of the loss of the ring (comprises those due to the passage into the coupler) and a switch has the role of closing and opening the memory.

The scheme is simple, but it is easy to notice that if we want the single passage loss to be small, the ring design has to be carried out with particular care so to avoid lasing of the structure.

If the design of the hardware is characterized by a few critical parameters, the scheme of Figure 10.37 has the advantage of being able to store an optical packet also for a very long time, up to the moment in which the packet intensity cannot be any more amplified effectively for its very low SN.

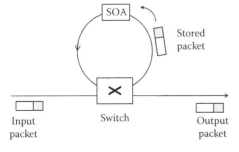

FIGURE 10.37
Example of the first type optical memory: A linear amplified ring.

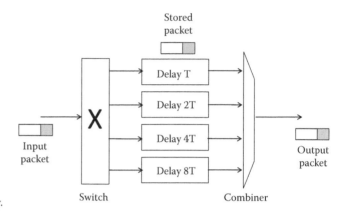

FIGURE 10.38
Delay line-based optical memory.

A completely different behavior is shown by the scheme of optical memory reported in Figure 10.38. Here, the optical packet enters into the memory through a 1×N switch that sends the packet to one out of N different memories. The memories are simply delay lines with different length so that the packet storage time is decided by the switch position.

This type of optical memory is much simpler to design, since no possibility of oscillations exists, but cannot store the packet for a time longer that the maximum delay implemented by the delay lines.

Combining the two principles, that is the recirculating memory and the delay line memory, any memory of the first type can be built.

In particular, in order to build an FIFO queue (First In First Out), first an elementary memory cell has to be built, as is shown in Figure 10.39a. The elementary cell has three states, as common to all memories: "write," "store," and "read." The schemes of the three states are shown in Figure 10.39b.

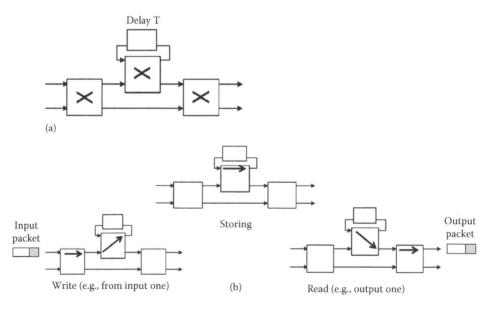

FIGURE 10.39
Three states elementary optical memory cell: (a) architecture; (b) the three cell states.

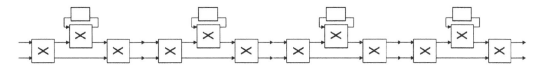

FIGURE 10.40
Scheme of an FIFO optical queue.

Cascading a certain number of elementary cells a queue is realized where, when a cell is free since the content is moved to the next cell, its content is substituted by the content of the cell placed before, but for the first cell that is enabled to receive packets from the input lines. The scheme of the queue realized with optical elements is reported in Figure 10.40.

In order to forecast possible performances of such a memory, we have to decide the technology that is used to implement them. We will consider both silicon-based 2D MEMs and InP-based SOAs, targeting optical technologies that are compatible with monolithic integration and, thus, volume production.

To evaluate the power consumption of such a queue, it is necessary to start from the elementary cell. Let us take 2D MEMs technology, so that it can be assumed that a single cross-point has a power consumption of 5 mW (see Table 5.8) and the overall power consumption of the cell is 60 mW. However, it must be taken into account that 2D MEMs are "slow" switches that are not always suitable for such applications.

Very fast switches can be built using SOAs [52], with a much higher power consumption on the order of 4 W for a 2 × 2 switch, so that all the memory blocks will consume 12 W.

It clearly appears that there is somehow a trade-off between power consumption and speed that is common to all these kinds of applications.

It is interesting to compare such figures with the power consumption of a possible very-high-speed electronics queue. If we assume that we have a queue of four elements, and that the line rate is NRZ 100 Gbit/s, it can be built with a RAM placed after a deserializer that reduces the 100 Gbit/s incoming flux in 40 fluxes at 2.5 Gbit/s. A very-high-speed RAM with a memory of 256 Mbyte working at 2.5 Gbit/s can consume less than 500 mW [53]. The deserializer cannot be built all in CMOS, due to the very high input bit rate.

The common architecture of these devices is constituted by an SiGe front end and a CMOS core. In total, a deserializer, like the one needed here, will consume up to 1.05 W in total assuming it is composed of one deserializer from 100 to 4 × 25 Gbit/s [54] and four deserializers from 25 to 5 Gbit/s. Since we can imagine that a parallel output of the queue can be used by the line card, we do not insert the serializer at the queue output, so that the overall electronic queue power consumption results to be on the order of 1.05 W.

This is only an example, but it tells us that in the field of memories, it is not so straightforward to realize optical components that have lower power consumption with respect to electronics one even if the transmission line speed is as high as 100 Gbit/s.

Another important point to verify is the switching speed, that is, the time the memory takes to pass from one state to the other. As a matter of fact, we have to take into account that a packet, one kbyte long, traveling at 100 Gbit/s, has a duration of 10 ns. Thus, the memory should commute in a time at least of the same order of magnitude of a packet duration.

Optical switches based on MEMs have a switching time on the order of 1 ms; thus, they are quite far from this application. As far as SOAs are concerned, short amplifier can arrive at a commutation time of 1 ns [52], so they are completely suited for optical packet switching.

Last, but not least, it is important to evaluate the required delay line length. The packet has to be delayed minimum a packet length not to create interference on the delay line.

If we use an optical fiber that has a refraction index of 1.5, the length required to delay a signal for 10 ns is about 2 m, that is, not a long fiber piece. If the delay line is realized by a waveguide in the InP chip where the SOAs are realized, it comes to be about 85 cm. Such a waveguide could be realized with the spiral design that is used for long waveguides in the SiO_2 platform, but due to the small dimension of the InP wafers, this will probably not be economically convenient with respect to the integration with an external waveguide or with an external fiber.

Up to now, we have discussed the possible architecture and performances of a simple optical queue. In real routers, a First In First Out queue is not enough. In a practical case, it is probably needed to control the correctness of the data in the packet payloads by means of an EDC, and in any case, QoS rules sometimes force a packet that is not in the first cell to be extracted from the memory.

In order to implement this type of operation, memories of the second type have to be realized.

For this task, a different elementary cell has to be used, that is shown in Figure 10.41. This is quite similar to the queue elementary cell, but for the fact that input and output 2×2 switches are substituted by 3×3 switches. This allows the elementary cell to have more states, in particular to have two write states, one for each possible input port, and two read states, one for each output port, as is shown in Figure 10.41b.

When the elementary cells are arranged in series, connecting the first and the second port of the Input/output switches, but not the third, they form a queue with the possibility of inserting a packet in an intermediate position (if it is free naturally) and extracting a packet from an intermediate position. This type of queue is shown in Figure 10.42.

This queue has all the required functionalities to be used in a packet switch line card.

What we have presented is a general discussion around a possible implementation of optical memories suited for use in all-optical packet switches. Several other types of memories have been proposed [55], which are also compatible with large-scale integration like

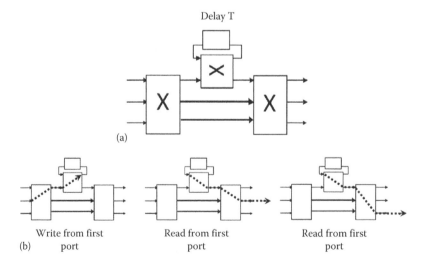

FIGURE 10.41
Five states elementary optical memory cell: (a) architecture; (b) three cell states out of the five possible.

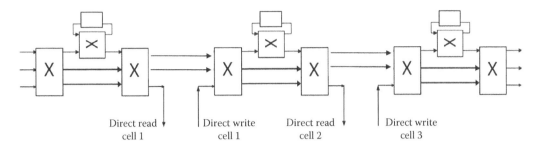

FIGURE 10.42
Scheme of an FIFO optical queue with the possibility of inserting and extracting the packets to/from intermediate positions.

those based on photonics crystals [56,57] and SOI technology [58]. It is not possible to review, here, all the proposals and the interested reader is encouraged to access the huge technical literature on this subject.

10.4.1.2 Switches: Two Examples of All-Optical Switch Fabric

Switches for packet switching are not the same as those used in circuit switching; thus, many solutions we have reviewed talking about OXCs are not usable here.

The first specific requirement is the switching speed, which has to be on the order of one nanosecond or less to be suited for the application.

The other requirement is a low dynamic power consumption, which is sometimes different from the static power consumption. As a matter of fact, in the first case, the switch is assumed to operate rarely and the power is used to maintain the switch in its state, while in the dynamic case, it switches continuously.

The consequence is that as anticipated, when talking about memories, MEMs and LCOS switches are not usable.

Using SOAs small switches with a response time on the order of 1 ns can be realized and even faster switches can be realized using nonlinear Kerr effect in optical fibers [59,60].

None of these solutions, however, is suitable for the implementation of large switches using the traditional architectures based on space division, both from the point of view of power consumption, and frequently, of the real estate. It is sufficient to realize that an SOA with suitable characteristics to be used as a switch cross-point in a dynamical switch consumes 1 W; thus, a 128 × 128 Clos matrix based on SOAs would consume about 9 kW, completely out of the target of an optical packet switch.

For these reasons, several switch fabric architectures have been studied that do not need large space switches and are suitable for implementation with small and very fast switches like those that can be built with SOAs.

An example of the architecture of an all-optical packet switch is the KEOPS switch, designed within the KEOPS ACTS project [61].

A functional scheme of the KEOPS switch fabric is reported in Figure 10.43. The switch consists of three sections:

1. Wavelength encoder block
2. Buffer and broadcast block
3. Wavelength selector block

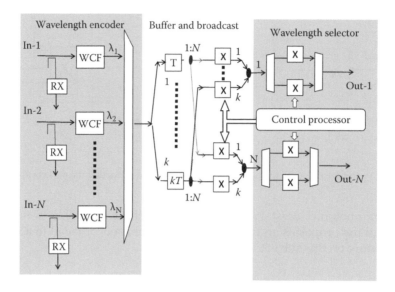

FIGURE 10.43
Functional scheme of one of the switch fabrics for optical packet switching proposed in the KEOPS European project. (After Guillemot, C. et al., *IEEE J. Lightwave Technol.*, 16(12), 2117, 1998.)

The wavelength encoder block consists of WCs, one per input. Each WC encodes its packets on a fixed wavelength. We have as many WCs as input wavelengths; thus, if we have four DWDM systems with 90 wavelengths at 10 Gbit/s, we have 360 WCs each of which has to accept any wavelengths in the C band and commute it on a fixed and unique wavelength at the input of the switch fabric.

Even if the WCs work at a wavelength spacing of 25 GHz in C and L bands (and it is quite a difficult task), there are available no more than 180 wavelengths, and this, in practice, fixes the dimension of the KEOPS all-optical switch: no more than 180 wavelengths input. Thus, as often happens, the all-optical design is more and more effective while the line rate of the single wavelength increases.

As a matter of fact, there is no problem in accepting four fully loaded systems with 45 wavelengths at a bit rate of 40 Gbit/s or four systems each of which transport 20 wavelengths carrying an NRZ 100 Gbit/s signal.

The cell buffer block comprises K fiber delay lines followed by a space switch stage realized using SOAs as fast gates. As a matter of fact, the only role of this stage is either to block or to pass-through the respective wavelength following the commands of the central electronic processor.

The last block, the wavelength selector block, consists of demultiplexers that forward the different outputs to SOA gates before the signals are recombined, thus selecting packets from the correct inputs. The broadcasting principle makes a copy of each packet at any delay, for a total of k copies, enabling the flexible management of the buffered packets flows.

A copy is also available at each output, so that this architecture easily supports multicasting.

The example of the KEOPS packet switch is only one of the ways in which the principle of broadcast and selection can be applied in conjunction to WCs to avoid the use of great space division matrixes.

This simplification is, however, paid with the great number of WCs that are needed, that makes the setting up of discrete component equipment extremely difficult, but could become less important if integration on the InP platform develops volumes and technologies for large-scale integration.

Since collisions can happen inside the KEOPS switch, an arbiter has to reside in the electronic control processor. However, this is not different from the arbiter of an electronic switch, so no difficulty is hidden under this need.

Naturally, the scheme of Figure 10.43 simply catches the KEOPS switch fabric. In the KEOPS project, all the elements of the packet switched network were developed using all-optical signal processing, and it is instructive to study this complex design to have a feeling of the type of problems that can be encountered if the whole network has to be designed.

Another simpler example of switch fabric, based on the combination of wavelength division and space division, is shown in Figure 10.44.

Here, all the incoming packets are wavelengths converted on the wavelength that they will have on the output DWDM systems and sent to one of the possible output directions via a $1 \times N$ dynamic switch, whose implementation should be feasible, since generally, the degree of such a node is on the order of 4–6. Naturally, the switches are driven by an electronic control so to avoid too many packets being sent toward the same output, at the same time. In order to manage contentions, queues are present at each input.

NW lines arrive at each output, where N is the node degree and W the number of wavelengths of each input DWDM system; however, only W among them can transmit a real information and the packets arriving from them all have different wavelengths.

Thus, every line is simply multiplexed with the others using a passive combiner after amplification by a linear SOA or a single channel EDFA. This amplification is needed due to the fact that in a normal case in which $W = 90$ and $N = 5$, we have a 450×5 splitter whose loss is about 26.5 dB.

The ASE introduced by the amplifier is partially compensated in the transparent path of the packets through the network by the adoption at the switch input of a regenerating WC, like a converter based on XGM in SOAs.

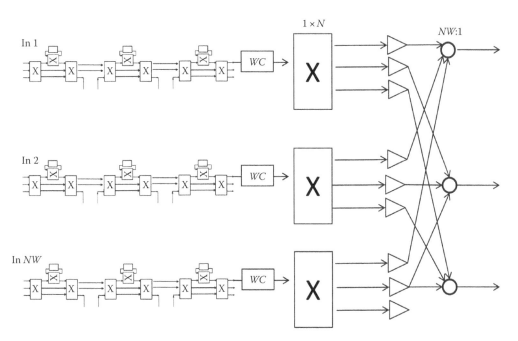

FIGURE 10.44
Example of switch fabric based on the combination of wavelength division and space division switching.

10.4.1.3 Digital Optical Processing

Even if massive amounts of digital operations are always carried out by electronic processors, the need of digital optics is unavoidable if an all-optical packet switching is considered.

Even at bit rates at which electronic processing is still possible, like 40 Gbit/s, a few operations on the packets payload are needed and they have to be carried out by a simple optical logic if the payload is not to be converted into the electronic form.

Moreover, several prototypal implementations of all-optical packet switching has envisioned very high speed for the wavelength channels, at which header processing also has to be carried out by a dedicated optical logic [62–64].

If the signal speed is within 10 and perhaps also 40 Gbit/s, SOAs are very attractive to provide the nonlinear elements for all-optical logic gates.

The simpler scheme adopting SOAs is shown in Figure 10.45: A single SOA is designed so as to have very low saturation power (e.g., short and highly doped in the p and n zones surrounding the active region) can constitute a NOT gate if a probe constituted of a periodical sequence of pulse is injected at low power next to the signal (Figure 10.46a).

If two signals that have almost the same power and are both RZ modulated in a synchronous way are injected besides the periodic probe, we obtain an NOR gate, that is, a gate generating a pulse at the output only when both the incoming signals are low (Figure 10.46b).

Many other schemes can be devised using SOAs, for example, the Mach–Zendher configuration already used for WCs can be reproduced [65], and combining devices based on

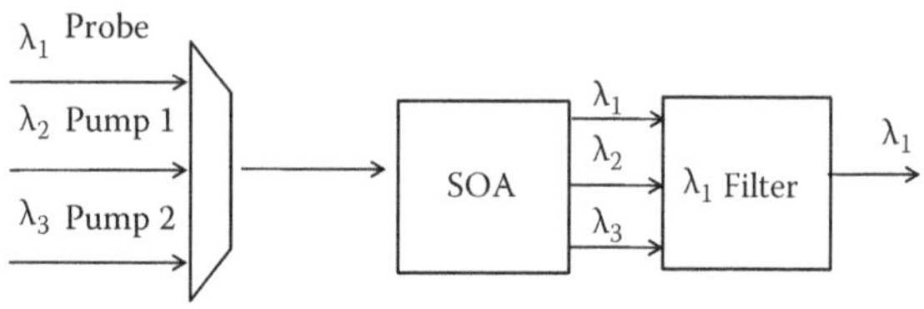

FIGURE 10.45
All-optical gate based on a single SOA. Depending on the input signal, it can work both as a NOT and as a NOR.

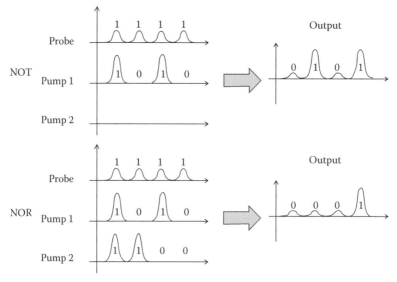

FIGURE 10.46
NOT and NOR configurations of the all-optical gate of Figure 10.45.

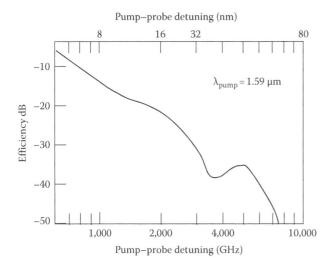

FIGURE 10.47
Example of an FWM response of a suitably designed SOA. (After Scotti, S. et al., *IEEE J. Selected Areas Quantum Electron.*, 9(6), 746, 1997.)

XPM and XGM, also complex logical circuits, for example, for the comparison of a given address with the address of an incoming packet can be implemented.

In part, the major limitation of the use of incoherent nonlinearities in SOAs can be overcome by using FWM. As a matter of fact, FWM happens only if a pulse is present contemporary on the pump and on the probe; thus, it works like an AND logic gate if both pumps and probe are pulse-modulated signals.

We know that FWM in SOAs is a coherent effect so that the output is proportional in phase and amplitude to the probe; moreover, due to very-high-speed nonlinearities, like two photons absorption, FWM happens also with probe and pump with very different carrier wavelengths (results up to 80 nm have been achieved, as shown in Figure 10.47 [66]).

On the other hand, the FWM in SOAs has an efficiency smaller than one, and that, generally, decreases increasing the distance between the probe and pump. The example of a very wide FWM efficiency curve is reported in Figure 10.48 [66] where the efficiency is represented with the ratio between the output FWM power and the input probe power.

Moreover, observing Figure 10.48, it is possible to notice that not only the FWM efficiency varies sensibly as a function of the pump wavelengths, but the shape of variation depends strongly on the detuning.

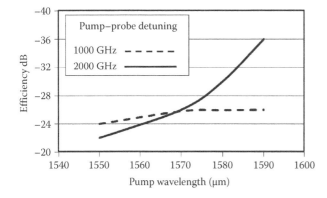

FIGURE 10.48
Two values of a very wide FWM efficiency curve versus the pump wavelength. (After Scotti, S. et al., *IEEE J. Selected Areas Quantum Electron.*, 9(6), 746, 1997.)

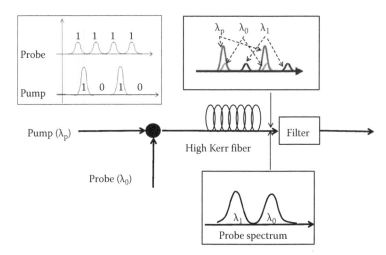

FIGURE 10.49
Possible scheme of an all-optical gate realized exploiting a high nonlinear coefficient fiber.

In order to achieve higher efficiency at very high speeds, nonlinear elements adopting the fiber need to be used, instead of the SOA. Fiber nonlinear dynamics contain Kerr effect that is practically instantaneous as far as telecommunications are concerned.

Several possible gate structures have been proposed based on special and on standard fibers. One of the simpler architectures is shown in Figure 10.49. It is based on a single high nonlinear coefficient fiber that was optimized so to minimize the required length and the spurious absorption does not require other special components [67].

Let us call $A = \{a_k\}$ and $B = \{b_k\}$ the bit streams that have to be processed. Let us also image that A and B are coded into two synchronized RZ IM signals modulated at 10 Gb/s by means of picosecond pulses.

The two signals are launched in a suitable length of small core area fiber that presents an enhanced Kerr effect with respect to SSMF.

The spectrum of the probe, that is assumed to be of sufficiently low power so as not to influence the phenomenon, experiences an XPM-induced frequency shift that is proportional to the slope of the pump signal power profile.

The probe spectrum at the fiber output is composed of two lobes. One lobe is identical to the probe spectrum before XPM and a second lobe is its spectrally shifted replica.

If a band pass filter removes the spectral components that are shifted by XPM and keeps the original frequency components, we get a signal corresponding to the operation

$$\bar{A} \text{ and } B = \{(a_k \oplus 1) \odot b_k\} \tag{10.33}$$

where $\oplus$ and $\odot$ represent the sum and multiplication between integer numbers represented in base two.

On the other hand, if the filter removes the original frequency components and keeps the XPM shifted spectral components, the output signal corresponds to the operation

$$A \text{ and } B = \{a_k \odot b_k\} \tag{10.34}$$

Using such gates, either based on SOAs or on Kerr effect in fibers, complex elaboration systems have been constructed in laboratory experiments.

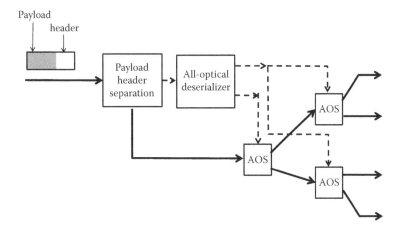

FIGURE 10.50
Block scheme of a header recognition and packet routing processor. (After Kurumida, J. et al., All-optical header recognition sub-system based on SOA-MZI switches, *Pacific Rim Conference on Lasers and Electro-Optics, 2005. CLEO/Pacific Rim 2005*, Tokyo, Japan, IEEE, s.l., 2005, pp. 1790–1791.)

A good example is constituted by the header recognition and packet routing processor presented in [68], whose functional scheme is reported in Figure 10.50.

The base hypothesis is that the switch fabric of the packet switch is a Banyan network (see Chapter 7), so that routing is performed stage by stage looking at a single bit of the packet address. In the case depicted in the figure, we see a 4 × 4 processor, but it is simply an example that can be scaled to every power of two.

In principle, the working of the system is easy—the header is separated by the payload and parallelized, so that every bit can drive the corresponding stage of the switch fabric that is assumed to be built with all-optical switches.

Much less trivial is the implementation of the whole system using all-optical components. The block scheme of the header payload divider is presented in Figure 10.51.

The core of the subsystem is constituted by a monolithic InP chip with two SOAs on the branches of a four input MZI. At the output, two tunable filters are used to insulate the payload and the header on the two interferometers output ports.

The packet is injected into the second MZI input while into the first input is injected a control signal as long as the packet header and as strong as needed to switch off the interferometer at its end.

The control signal on the last input is a pulse quite strong that is useful to force a carrier change into the SOA so to arrive at a switching time smaller than the single SOA.

Naturally, the MZI structure also works as a frequency converter; thus, the header and the payload can be recovered separately with tunable filters at the output.

The block scheme of the all-optical deserializer that has to divide the header (in our case two bits) in parallel streams is shown in Figure 10.52.

In line with the principle, it could be possible to replicate, at a different time scale, the circuit that divides the header from the payload. In this case, however, there are two bits to be deserialized and it would be difficult to introduce the pulse that in this case should elapse earlier than the bit time.

The adopted scheme is based on XGM in SOAs, exploiting the fact that SOAs that have to work as a single gate can be fabricated with a set of characteristics suitable to shorten the response time.

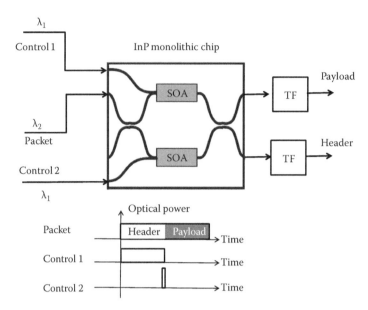

FIGURE 10.51
Block scheme of a possible header payload divider.

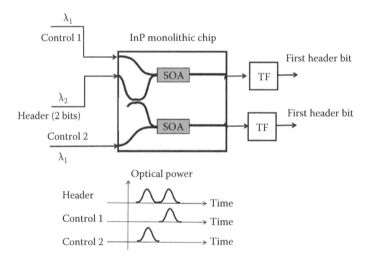

FIGURE 10.52
The block scheme of a possible all-optical deserializer that has to divide the header (in our case two bits) into parallel streams.

Thus, the scheme of the deserializer is again constituted by an integrated circuit in InP containing two couplers and two SOAs optimized to work as fast gates.

The signal is distributed among the SOAs and is here coupled with a pump depressing the gain in correspondence to the bit to cancel.

At the output, the signal is cleaned by a filter.

As far as the all-optical switch is concerned, what is needed is a 1 × 2 switch whose configuration is fixed by an optical control signal. This component also can be realized with a MZI having SOAs on the two branches, as presented in Figure 10.53. The system is

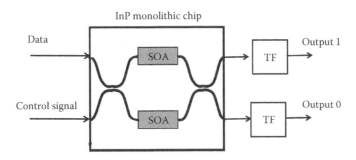

FIGURE 10.53
SOA based 1 × 2 fast all-optical switch cell.

completely identical to a WC; this is the reason why, at the output, it is possible to select the data on the desired port simply by correctly commanding the SOAs through the optical control signal without any need of tuning the output filters.

Thus, we have described how all the elements of a switch fabric with a complete optical control can be realized using SOAs.

It is at this point interesting to understand the number of SOAs and some basic performances of the resulting all-optical packet switch.

Let us imagine having four input DWDM systems bringing 45 channels with an optical bandwidth of 40 GHz (unilateral as always) and a data bit rate of 100 Gbit/s achieved using a suitable multilevel modulation format.

Let us also imagine that slotted packet transmission is adopted, so that only a random jitter thwarts the synchronism of different signals. Naturally, this condition can be realized only if a suitable synchronization network is superimposed on the data network as happens in SDH, for example.

For the moment, we will neglect jitter, even if it can become a real problem at high transmission rates.

In order to apply the architecture we have described, we will assume that our switch is realized by an input queued Banyan network and conflicts are simply managed by discarding conflicting packets and leaving upper layers to recover packet loss.

We will have 180 input channels after demultiplexing; thus, we need 180 input queues. If the queues are realized as described in this section, assuming that no QoS has to be implemented, we have 720 SOAs in the memory loops and 2160 switch 2 × 2 that, due to the required speed, we can imagine building also with SOAs, increasing the number of SOAs in the front stage of the switch to 9360.

The switch driving signal has to be prepared by the header processor that has to potentially manage 180 packet per slot.

Moreover, a Banyan network has $N \log(N)$ crosspoints if N is the number of ports; thus, we need 935 SOAs for the switches and 360 for the ports. Moreover, at each output port, a WC has to be present to adapt the channel wavelengths to the wavelengths available on the output DWDM system. If SOA MZIs are used, this means that another 360 SOAs are needed.

Globally, we need about 11,000 SOA.

With present day technology, it is impossible to build an equipment integrating 11,000 SOAs, but if InP technology evolves strong and fast, something similar could happen.

Moreover, we have to consider that the signal in a typical passage through the switch will traverse at least 20 SOAs. They are in quite different configurations with different gains. Tracing a generic state of the switch, and assuming that the system is designed to have an overall gain equal to 0 dB, the overall noise figure of the switch results to be around 40 dB, that is generally too much to allow the use in a transparent network.

This underlines how a scheme like the one we have analyzed is far from the possibility of being transformed into a product. Nevertheless, it demonstrates that all-optical packet switching is in principle possible, even if important technological steps are still to be realized.

10.4.2 Final Comment on the All-Optical Packet Network

In this section, we have presented just an example of possible implementation of optical packet switching network based on optical SOAs.

We have made no attempt to optimize the various schemes and to evaluate their performances.

A large number of different schemes for the various subsystems have been proposed in literature, based on optical fibers [69,70], or on other types of different nonlinear materials [71,72].

A large amount of research has also been devoted to switch architecture in order to fulfill the requirements of a modern packet switched network, and to adapt it as much as possible to the possibilities of all-optical components [73–75].

Even the network architecture and the methods to optimize the network design, if a particular version of packet switching technique and a particular switch architecture are used, have been the object of a great research effort [76,77].

Discussing the collocation of the T-OPN into a global network architecture, several ideas have been proposed, from a statistical multiplexing transport layer (like the so-called optical burst network [78]) to a network layer simply integrating the physical layer with layer 2.

However, up to now, all this work has had no impact on the telecommunication business, and with the passing time the fact that electronics goes ahead faster and faster decreases the probability that all-optical packet switching will become a real network technology.

Nevertheless, the results of all these researches have been numerous, and they were useful in specific applications, sometimes beyond the intentions of the research group that generated them.

Not only have several elements regarding the design and the possibilities of optical components and subsystems been discovered that are precious also in other applications, but digital optics, if not in telecommunication networks, is also founding several applications in the field of radars and in general microwave and millimeter wave applications [79,80].

Perhaps, if all the research on all-optical packet switching was not available, these applications would have much more difficulties, or they would not be born.

Today, the open research line for groups studying optical packet switching is to concentrate on what happens at very high bit rates, like 800 Gbit/s, where even multilevel transmission with a great number of levels does not reduce the bandwidth so much as to render electronics more competitive.

It is not clear if such systems (transmission and switches) will be introduced following a clear market request and when it will happen, but the history of photonics up to now has always demonstrated that increasing the bit rate over the single wavelength is beneficial from a CAPEX point of view.

It is out of the scope of this book even to review all the experiments where different methods have been proposed to multiply and demultiply, with optical means, very high bit rate channels and have also introduced all-optical node architectures suitable for this scope. The interested reader is encouraged to access the literature on this subject, starting with [81–83].

References

1. Melle, S., Building agile optical networks, *Optical Fiber Communications Conference—OFC 2008*, San Diego, CA, paper NME2, March 2008.
2. Iannone, E., Next generation wired access: Architecture and technologies. In K. Iniewski (Ed.) *Internet Networks*, Taylor & Francis, Boca Raton, FL, s.l., 2010.
3. Hurtt, S. et al., The first commercial large-scale InP photonic integrated circuits: Current status and performance, *65th Annual Device Research Conference*, South Bend, IN, IEEE, s.l., p. 183, June 18–20, 2007.
4. Joyner, C. et al., Current view of large scale photonic integrated circuits on indium phosphide, *Optical Fiber Communication Conference, OFC 2010*, San Diego, CA, IEEE, s.l., p. paper OWD3, March 21–25, 2010.
5. Cochrane, P., Heckingbottom, R., Heatley, D., The hidden benefits of optical transparency, *IEEE Communication Magazine*, 32(9), 90–97 (1994).
6. Carpenter, T. et al., Maximizing the transparency advantage in optical networks, *Optical Fiber Communication Conference, OFC 2003*, Atlanta, GA, IEEE, s.l., Vol. 2, pp. 616–618, March 2003.
7. Murakami, M., Oda, K., Power consumption analysis of optical cross-connect equipment for future large capacity optical networks, *11th International Conference on Transparent Optical Networks, ICTON '09*, Sao Miguel (Azores), Portugal, IEEE, s.l., June 2009.
8. Staessens, D. et al., Assessment of economical interest of transparent switching, *European Conference on Optical Communications, ECOC 2005*, Glasgow, U.K., Vol. 4, pp. 861–862, September 29, 2005.
9. Tkach, R. W. et al., Fundamental limits of optical transparency, *Optical Fiber Communication Conference and Exhibit, OFC '98*, San Jose, CA, IEEE, s.l., pp. 61–62, February 22–27, 1998.
10. Marciniak, M., Transparency of optical networks: How to manage it? *International Conference on Transparent Optical Networks, ICTON 1999*, Kielce, Poland, IEEE, s.l., pp. 85–88, June 2009.
11. Saradhi, C. V., Subramaniam, S., Physical layer impairment aware routing (PLIAR) in WDM optical networks: Issues and challenges, *IEEE Communication Surveys and Tutorials*, 11(4), 109–130 (2009).
12. Devaux, F., All-optical 3R regenerators: Status and challenges, *17th International Semiconductor Laser Conference, 2000*, Hyatt Monterey, Monterey, CA, IEEE, s.l., pp. 5–6, September 25–28, 2000.
13. Hainberger, R. et al., BER estimation in optical fiber transmission systems employing all-optical 2R regenerators, *IEEE Journal of Lightwave Technology*, 23(3), 746–754 (2004).
14. Matsumoto, M., Performance comparison of all-optical signal regenerators utilizing self-phase modulation in fibers, *Optical Fiber Communication Conference, OFC 2003*, Atlanta, GA, IEEE, s.l., Vol. 1, pp. 108–109, March 2003.
15. Chestnut, D. A., Taylor, J. R., Fibre optical wavelength-converter employing cross-phase modulation and low-threshold non-adiabatic Raman compression, *Electronics Letters*, 39(15), 1133–1134 (2003).
16. Rauschenbach, G. A. et al., All-optical pulse width and wavelength conversion at 10 Gb/s using a nonlinear optical loop mirror, *IEEE Photonics Technology Letters*, 6(9), 1130–1132 (1994).
17. Connelly, M. J., Wideband semiconductor optical amplifier steady-state numerical model, *IEEE Journal of Quantum Electronics*, 37(3), 439–447 (2001).
18. Connely, M. J., *Semiconductor Optical Amplifiers*, Springer, New York, s.l., 2010, ISBN-13: 978-1441949479.
19. Baghban, H., Rostam, A., *Nanostructure Semiconductor Optical Amplifiers: Building Blocks for All-Optical Processing*, Springer, s.l., New York, 2010, ISBN-13: 978-3642149245.
20. Ribeiro, N. S., Gallep, C. M., Conforti, E., Wavelength conversion and 2R-regeneration using one semiconductor optical amplifier with cross-gain modulation compression, *Conference on Lasers and Electro-Optics, 2008 and 2008 Conference on Quantum Electronics and Laser Science, CLEO/QELS 2008*, San Jose, CA, p. paper YWA28, May 2008.

21. Durhuus, T. et al., All optical wavelength conversion by SOA's in a Mach–Zehnder configuration, *IEEE Photonics Technology Letters*, 6(1), 53–55 (1994).
22. Raybon, G., Mikkelsen, B., Koren, U., Miller, B. I., Dreyer, K., Boivin, L., Chandrasekhar, S., Burrus, C. A., 20 Gbit/s all-optical regeneration and wavelength conversion using SOA based interferometers, *Optical Fiber Communication Conference, OFC 1999*, San Diego, CA, IEEE, s.l., Vol. 4, pp. 27–29, February 1999.
23. Masanovic, M. L. et al., Demonstration of monolithically-integrated InP widely-tunable laser and SOA-MZI wavelength converter, *International Conference on Indium Phosphide and Related Materials*, Santa Barbara, CA, IEEE, s.l., pp. 289–291, 2003.
24. S. C. Cao and J. C. Cartledge, Characterization of the chirp and intensity modulation properties of an SOA-MZI wavelength converter, *IEEE Journal of Lightwave Technology*, 20(4), 689–695 (2002).
25. D'Ottavi, A. et al., Wavelength conversion at 10 Gb/s by four-wave mixing over a 30-nm interval, *IEEE Photonics Technology Letters*, 10(7), 952–954 (1998).
26. D'Ottavi, A. et al., Four-wave mixing in semiconductor optical amplifiers: A practical tool for wavelength conversion, *IEEE Journal of Selected Topics in Quantum Electronics*, 3(2), 522–528 (1997).
27. Contestabile, G., Presi, M., Ciaramella, E., Multiple wavelength conversion for WDM multicasting by FWM in an SOA, *IEEE Photonics Technology Letters*, 16(7), 1775–1777 (2004).
28. D'Ottavi, A. et al., Efficiency and noise performance of wavelength converters based on FWM in semiconductor optical amplifiers, *IEEE Photonics Technology Letters*, 7(4), 357–359 (1995).
29. Johansson, L., Thylen, S., Optical switching for multiwavelength systems: Progress in Europe, *Optical Fiber Communication Conference, OFC 1994*, San Jose, CA, IEEE, s.l., p. Paper ThN3, 1994.
30. Iannone, E., Sabella, R., Optical path technologies: A comparison among different cross-connect architectures, *IEEE Journal of Lightwave Technology*, 14(10), 2184–2196 (1996).
31. Jourdan, A. et al., Design and implementation of a fully reconfigurable all-optical crossconnect for high capacity multiwavelength transport networks, *IEEE Journal of Lightwave Technology*, 14(6), 1198–1206 (1996).
32. Iannone, E., Sabella, R., Modular optical path cross-connect, *Electronics Letters*, 32(2), 125–126 (1996).
33. Zhou, J. et al., Crosstalk in multiwavelength optical cross-connect networks, *IEEE Journal of Lightwave Technology*, 14(6), 1423–1435 (1996).
34. Shen, Y., Lu, K., Gu, W., Coherent and incoherent crosstalk in WDM optical networks, *IEEE Journal of Lightwave Technology*, 17(5), 759–764 (1999).
35. Ramachandran, M., Optical switching and routing, *Optical Fiber Communication Conference, OFC 2001*, Anaheim, CA, IEEE, s.l., pp. paper TuV2.1–TuV2.3, March 2001.
36. Matrakidis, C., Politi, C., Stavdas, A., Modular broadcast-and-select optical crossconnects and their physical layer modelling, *International Conference on Photonics in Switching*, Heraklion, Greece, October 2006.
37. Ji, H. C., Park, K. J., Kim, J. K., Chung, Y. C., Optical path and crosstalk monitoring technique using pilot tones in all-optical WDM transport network, *Asia Pacific Optical Communication Conference*, Beijing, China, SPIE, s.l., Vol. 4584, November 12–16, 2001.
38. Borzycki, K., Labeling of signals in transparent optical networks, *International Conference on Transparent Optical Networks, ICTON 2003*, Warsaw, Poland, IEEE, s.l., pp. 166–169, June 29– July 03, 2003.
39. Murakami, M., Imai, T., Aoyama, M., A remote supervisory system based on subcarrier overmodulation for submarine optical amplifier systems, *IEEE Journal of Lightwave Technology*, 14(5), 671–677 (1996).
40. Chi, N., Carlsson, B., Jianfeng, Z., Holm-Nielsen, P., Peucheret, Ch., Jeppesen, P., Transmission properties for two-level optically labeled signals with amplitude-shift keying and differential phase-shift keying orthogonal modulation in IP-over-WDM networks, *Journal of Optical Networking*, 2(2), 365–367 (2003).
41. Wen, Y. G., Zhang, Y., Chen, L. K., On architecture and limitation of optical multiprotocol label switching (MPLS) networks using optical-orthogonal-code (OOC)/wavelength label, *Optical Fiber Technology*, 8, 43–70 (2002).

42. Harte, L., *Introduction to Code Division Multiple Access (CDMA): Network, Services, Technologies, and Operation*, Althos, Fuquay-varina, NC, s.l., 2004, ISBN-13: 978-1932813050.

43. Christodoulopoulos, K., Manousakis, K., Varvarigos, E. A., Cross layer optimization of static lightpath demands in transparent WDM optical networks, *Workshop on Networking and Information Theory, ITW 2009*, Volos, Greece, IEEE, s.l., pp. 115–119, June 2009.

44. Hashiguchi, T. et al., Integer programming assisted optimization of optical design for transparent WDM networks, *Optical Fiber Communication Conference, OFC 2009*, San Diego, CA, IEEE, s.l., March 22–26, 2009.

45. Sabella, R., Iannone, E., Listanti, M., Berdusco, M., Binetti, S. W., Impact of transmission performance on path routing in all-optical transport networks, *IEEE Journal of Lightwave Technology*, 16(11), 1965–1971 (1998).

46. Nagatsu, N., Okamoto, S., Sato, K., Optical path cross-connect system scale evaluation using path accommodation design for restricted wavelength multiplexing, *IEEE Journal on Selected Areas in Communications*, 14(5), 893–902 (1996).

47. Ramaswami, R., Sivarajan, K. N., Routing and wavelength assignment in all optical networks, *Transaction on Networking*, 3, 489–500 (1995).

48. Ramaswami, R., Sivarajan, K. N., Sasaki, G. H., *Optical Networks*, 3rd revised edn., Morgan Kaufmann—Elsevier, San Francisco, CA, s.l., 2009, ISBN 9780123740922.

49. Gagnaire, S., Al Zahr, M., Impairment-aware routing and wavelength assignment in translucent, *IEEE Communication Magazine*, 5, 55–61 (2009).

50. Patel, A. N. et al., Traffic grooming and regenerator placement in impairment-aware optical WDM networks, *14th Conference on Optical Network Design and Modeling, ONDM 2010*, Kyoto, Japan, IEEE, s.l., February 1–3, 2010.

51. van As, H. R., Driving optical network innovation by extensively using transparent domains, *International Conference on Transparent Networks, ICTON 2010*, Munich, Germany, IEEE, s.l., p. paper We.C1.4, June 27–July 1, 2010.

52. Alphion, Technology brief—Qligth I switches, 2010, www.alpphion.com (accessed: November 10, 2010).

53. Fujitsu, RAM power consumptions comparison, 2008, available on www.fujitsu.com, accessed on November 10, 2010.

54. Wu, K.-C., Lee, J., A 2 × 25Gb/s deserializer with 2:5 DMUX for 100Gb/s Ethernet applications, *Digest of Technical Papers of IEEE International Solid-State Circuits Conference (ISSCC), 2010*, San Francisco, CA, IEEE, s.l., pp. 374–375, February 7–11, 2010.

55. Bogoni, A. N., Prati, G., All-optical memories and flip-flops, *International Conference on Photonics in Switching*, Hokkaido, Japan, IEEE, s.l., August 4–7, 2008.

56. Yanik, M. F. et al., Sub-micron all-optical memory and large scale integration in photonic crystals, *Conference on Lasers and Electro-Optics Europe, CLEO/Europe, 2005*, Munich, Germany, IEEE, s.l., p. 588, June 12–17, 2005.

57. Shinya, A. et al., All-optical memories based on photonic crystal nanocavities, *International Conference on Photonics in Switching, PS '09*, Pisa, Italy, IEEE, s.l., September 15–19, 2009.

58. Kumar, R. et al., An ultra-small, low-power all-optical flip-flop memory on a silicon chip, *Optical Fiber Communication Conference, OFC 2010*, San Diego, CA, IEEE, s.l., March 21–25, 2010.

59. Marembert, V. et al., Investigation of fiber based gates for time division demultiplexing up to 640 Gbit/s, *Optical Fiber Communication Conference, OFC 2006*, Anaheim, CA, IEEE, s.l., March 2006.

60. Brzozowski, L., Sargent, E. H., All-optical analog-to-digital converters, hard limiters, and logic gates, *IEEE Journal of Lightwave Technology*, 19(1), 114–119 (2001).

61. Guillemot, C. et al., Transparent optical packet switching: The European ACTS KEOPS approach, *IEEE Journal of Lightwave Technology*, 16(12), 2117–2133 (1998).

62. Minh, H. L., Ghassemlooy, Z., Ng, W. P., Ultrafast header processing in all-optical packet switched-network, *7th International Conference on Transparent Optical Networks, ICTON 2006*, Nottingham, U.K., Vol. 2, pp. 50–53, June 2006.

63. Wang, J. P. et al., Demonstration of 40-Gb/s packet routing using all-optical header processing, *IEEE Photonics Technology Letters*, 18(21), 2275–2277 (2006).

64. Yuang, M. C., Lin, Y.-M., Wang, Y.-S., A novel optical-header processing and access control system for a packet-switched WDM metro ring network, *IEEE Journal of Lightwave Technology*, 27(21), 4907–4915 (2009).

65. Liu, H. et al., *Novel Scheme of Header Extraction Based on SOA-MZI with Asymmetric Control Light*, Shanghai, China, IEEE, s.l., pp. 1–7, 2009.

66. Scotti, S. et al., Effects of ultrafast processes on frequency converters based on four-wave mixing in semiconductor optical amplifiers, *IEEE Journal on Selected Areas in Quantum Electronics*, 9(6), 746–748 (1997).

67. Qiu, J. et al., Reconfigurable all-optical multilogic gate (xor, and, and or) based on cross-phase modulation in a highly nonlinear fiber, *IEEE Photonics Technology Letters*, 22(16), 1199–1201 (2010).

68. Kurumida, J. et al., All-optical header recognition sub-system based on SOA-MZI switches, *Pacific Rim Conference on Lasers and Electro-Optics, CLEO/Pacific Rim 2005*, Tokyo, Japan, IEEE, s.l., pp. 1790–1791, 2005.

69. Kikuchi, K., Taira, K., Sugimoto, N., Highly-nonlinear bismuth oxide-based glass fibers for all-optical signal processing, *Optical Fiber Communication Conference, OFC 2002*, Anaheim, CA, IEEE, s.l., pp. 567–568, March 2002.

70. Futami, F., Highly nonlinear fibers for ultrahigh-speed optical signal processing, *Optical Fiber Communication Conference, OFC 2008*, San Diego, CA, IEEE, s.l., February 24–28, 2008.

71. Luther-Davies, B., Breakthroughs in nonlinear optical materials for signal processing applications, *Joint conference of the Opto-Electronics and Communications Conference, 2008 and the 2008 Australian Conference on Optical Fibre Technology, OECC/ACOFT 2008*, Sydney, Australia, IEEE, s.l., 2008.

72. Valdueza-Felip, S. et al., Novel nitride-based materials for nonlinear optical signal processing applications at 1.5 μm, *IEEE International Symposium on Intelligent Signal Processing, WISP 2007*, Alcala de Henares, Spain, 2007.

73. Calabretta, N. et al., All-optical packet switching and label rewriting for data packets beyond 160 Gb/s, *IEEE Photonics Journal*, 2(2), 113–129 (2010).

74. Das, G. et al., Output-buffered all optical packet switch, *The Joint International Conference on Optical Internet and Next Generation Network, COIN-NGNCON 2006*, Jeju, Korea, pp. 83–85, July 2006.

75. He, T. et al., Optical arbitrary waveform generation based optical-label switching transmitter with all-optical label extraction, *Optical Fiber Communication Conference, OFC 2010*, San Diego, CA, IEEE, s.l., March 21–25, 2010.

76. Banerjee, D., Mukherjee, B., Wavelength-routed optical networks: Linear formulation, resource budgeting tradeoffs, and a reconfiguration study, *IEEE Transaction on Networking*, 8(5), 598–607 (2000).

77. Kaneda, S. et al., Network design and cost optimization for label switched multilayer photonic IP Networks, *IEEE Journal on Selected Areas in Communications*, 23 (8), 1612–1619 (2005).

78. Gauger, C. M., Mukherjee, B., Optical burst transport network (OBTN)—A novel architecture for efficient transport of optical burst data over lambda grids, *2005 Workshop on High Performance Switching and Routing*, Hong Kong, China, pp. 58–62, 2005.

79. Tonda-Goldstein, S. et al., Optical signal processing in Radar systems, *IEEE Transaction on Microwave Theory and Techniques*, 54(2), 847–853 (2006).

80. Loic, M. et al., Optical functions for microwave signal processing in radar, Communications and surveillance systems, *Radar Conference—Surveillance for a Safer World, 2009*, Bordeaux, France, IEEE, s.l., 2009.

81. Glesk, I., Sokoloff, J. P., Prucnal, P. R., Demonstration of all-optical demultiplexing of TDM data at 250 Gbit/s, *Electronics Letters*, 30(4), 339–341 (1994).

82. Morioka, T. et al., 1 Tbit/s (100 Gbit/s × 10 channel) OTDM/WDM transmission using a single supercontinuum WDM source, *Electronics Letters*, 32(10), 906–907 (1996).

83. Yao, M., Xu, L., Li, Y., Zhang, J., Chen, M., Chen, X., Han, M., Lou, C., Gao, Y., All optical clock extraction and demultiplexing in 4 × 10 GHz OTDM system, *International Conference on Communication Technology Proceedings, WCC–ICCT 2000*, Beijing, China, Vol. 1, pp. 402–405, 2000.

11

The New Access Network Systems
and Enabling Technologies

11.1 Introduction

Different factors have pushed the evolution of the access network from its copper-based architecture to a fiber-based one.

Besides bandwidth increase impelled by the introduction of new services (see Chapter 2), the penetration of ultra-broadband connections stresses the capacity of existing copper cables to support a high number of very fast xDSLs. In many cases, copper cables deployed before the diffusion of broadband access presents a non-negligible pair-to-pair interference when xDSL is used. Often, this phenomenon sets a limit to the number of connections possible in a given area, depending on the quality and length of the cables [1].

Moreover, the need of the main carriers to reduce operational cost (OPEX) also drives toward a fiber-optic-based access network. As a matter of fact, optical transmission allows a completely passive access infrastructure to be implemented; much longer spans can be realized, and even the possibility of a delayering of the peripheral part of the network appears (see Chapter 3).

Completely replacing the access infrastructure is perhaps the most CAPEX-intensive operation that an incumbent carrier can face. Such carriers have several million terminations (e.g., in a country like Italy, the network of Telecom Italia has about 20 million terminations) and their copper network reaches almost everywhere, both in cities and in the countryside.

In this situation, it is clear that the deployment cost is the main parameter characterizing a given access solution, and a huge effort is directed toward cost reduction.

In case civil work is needed to deploy new tubes where fiber cables have to be installed, this is by far the most expensive part of the deployment of the new access infrastructure. However, this situation is not so common in North America, Europe, and the most advanced Far East countries, especially in the most populated areas where network deployment is likely to start.

Over the last 20 years, fiber deployment activity has been constant in the major cities [2]. Moreover, even when fibers are not available, tubes are already in the ground, greatly facilitating fiber deployment. For these reasons, we will not deal with civil work in this chapter; readers interested in this aspect can find a starting point in the bibliography [3,4].

Beyond the capacity delivered to the end user and the cost, a variety of other merit parameters characterize an optical access network. Due to the huge investment needed for the deployment, next generation access infrastructure has to be future proof. Thus, the access infrastructure has to accommodate changes in the service bundle offered to the end user and the related evolution of the network equipment. Nevertheless, the investment needed to deploy the new network cannot be made available in a short time.

The convergence of all the telecommunication services toward the Internet paradigm opens the possibility of unifying data transport on the same infrastructure and collecting traditional wireline residential and business users, together with mobile users. Such a convergence is possible to connect the access points of the mobile network (the base stations) to the infrastructure of the wireline. This requires a further penetration to the fiber infrastructure, besides the capability of supporting a large variety of termination equipment.

From the bandwidth management point of view, the copper network assures physical separation among signals belonging to different users in the whole access segment: each user has his own twisted pair. This is a good situation from several points of view. The privacy of the user signal is completely guaranteed because it is impossible to detect the signal belonging to a user from another user terminal. Moreover, the concept of unbundling is strictly related to this characteristic of the copper network, assuring, in this case, perfect separation among signals belonging to different carriers. For these reasons, it would be desirable to reproduce this situation as much as possible in the next generation access network.

11.2 TDMA and TDM Overlay Passive Optical Network

As we have seen in Chapter 3, both due to the need of a careful use of already deployed fibers and tubes and to reflect the broadcast nature of a few key services like IP-TV, passive optical network (PON) is an almost universal choice for the deployment of the fiber access infrastructure.

Using a PON means that the fiber is deployed in a tree topology, often called the optical distribution network (ODN).

Since part of this medium is shared among the different channels, a shared medium access strategy is needed.

While almost all the possible strategies have been proposed, TDMA is by far the most used and all the PON products on the market are based on TDMA.

Recently WDMA has also been considered and a couple of products appeared on the market using this technique.

Differently from TDM PONs that are standardized in detail by ITU-T and IEEE, no consolidated standard still exists for WDM-PON, and this is an obstacle for the development of this kind of access system. In any case, ITU-T also works on WDM-PON standards and a first set of elements have been already defined, even if we are far from a rich standard like those for TDM PONs.

11.2.1 TDM PON Classification

An extended standardization activity has been carried out on TDM-PON by a dedicated group of ITU-T, called full service access network (FSAN). Different standards follow the evolution of both optical technology and switching protocols, passing from APON and BPON, based on ATM, to GPON, capable of conveying multigigabit signals using either ATM or Ethernet protocols, thanks to its adaptation layer called GEM (GPON encapsulation method).

TABLE 11.1

Summary of the Main Parameters of the ATM PONs, Ethernet PON, and GPON

	Protocol	Data Rate	Reach	Split Ratio	Standard	Channel Insertion Loss
EPON	Ethernet	1.25 Gbit/s symmetrical	1000BASE-PX10—10 km 1000BASE-PX20—20 km	16 Normal 32 Permitted	IEEE802.3ah	(Power budget) PX-10U—23 dB PX-10D—21 dB PX-20U—26 dB PX-20D—26 dB
BPON	ATM	622 or 155 Mbit/s Downstream @1490 nm 155 Mbit/s Upstream @1310 nm	Maximum 20 km	32 Maximum	ITU-T 983.3	Class A: 20 dB Class B: 25 dB Class C: 30 dB
GPON	ATM or Ethernet	2.488 or 1.44 Gbit/s Downstream 2.488 or 1.44 Gbit/s or 622 or 155 Mbit/s Upstream	Maximum physical reach 20 km	64 Maximum 32 Typical	ITU-T 984.2	Class A: 20 dB Class B: 25 dB Class B+: 28 dB Class C: 30 dB

Several North American carriers are looking to GPON to advance their FTTx rollout efforts so that GPON access is under deployment besides the older BPON infrastructure that is present in few local areas.

A different situation exists in Europe, where incumbent carriers generally have not yet committed to a clear strategy. However, a great number of FTTx field tests based on GPON are ongoing.

In the field of Ethernet-based PON, IEEE has emitted an important standard (called GEPON) that has been largely adopted mainly in the Far East. NTT, the major carrier in Japan, and one of the largest telecommunications carriers in the world, began deploying GE-PON-based FTTH access network equipment in 2003. In 2004, the IEEE ratified the IEEE 802.3ah Ethernet in the First Mile specification that fully defined the GE-PON technology deployed by NTT and later adopted by other carriers in Japan and Korea. A synthesis of the main TDM-PON standards is reported in Table 11.1, with some key optical parameters specified for each standard.

In the following discussion, we will mainly focus on GPON, due to its adoption in North America and probably in Europe also, where it is quite improbable that an ITU-T standard would be neglected in preference for an Ethernet-based GEPON.

Several considerations on transmission impairments and other elements are, however, common between GPON and GEPON.

11.2.2 GPON Architecture and Performances

The GPON standard [5], as all other standards for TDM-PONs, is based on unidirectional transmission, conveying downstream and upstream signals on different wavelengths on the same fiber. In this case, the effect of reflection is limited by the great wavelength separation between the two signals.

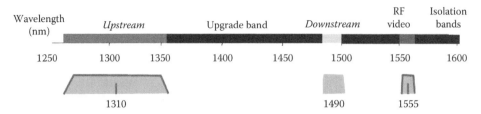

FIGURE 11.1
GPON standard wavelength plan including the case in which the video overlay is used.

In addition to the GPON version with two wavelengths, there is a GPON standard conveying two downstream wavelengths. One of them (1490 nm) is used to convey the GPON downstream signal, the second wavelength (1550 nm) is conceived for the distribution of analog CATV. This solution has been adopted in North America where CATV is practically the TV standard, and this solution allows the customers to maintain their home TV sets. The frequency plan of the GPON standard is shown in Figure 11.1.

Under a link budget point of view, GPON standard defines different link budget classes, whose parameters are reported in Table 11.1. A new class of performances will probably be standardized called C+ with a link budget higher than 33 dB.

Splitter loss depends mainly on the number of output ports on the splitter and adds about the same loss whether traveling in the downstream or upstream direction. Each splitter configuration is assigned a particular maximum split ratio loss, including connectors, defined by the ITU G.671 standard and Telcordia GR-1209.

When using the 16 or 32 splitting ratio, the standard completely defines all the needed splitter characteristics. Since this is not true for 1×64 splitters, network designers must use a single 1×2 splitter interfacing two 1×32 splitters to make up the 1×64 configuration.

11.2.2.1 GPON Transmission Performances

Using Class B optics only leaves 5.35 dB of propagation loss. Therefore, even with the best fiber access infrastructure, where the spectral attenuation can be assumed to be 0.31 dB/km, only a 17.25 km PON network is achievable without including any of the connectors within the local exchange.

A GPON reach of 20 km is specified for class B+ and C GPONs.

It is very interesting to investigate the physical effects thwarting GPON transmission, since the situation is much more complex than in the WDM-PON case.

In the case of A and B classes, the GPON span is essentially limited by the power budget. It is to be noted that in optical access networks, reducing the ONU cost is a key issue. For this reason, in A and B classes, GPON Fabry–Perot laser diodes (FP) are generally used at the ONU, while a PIN photodiode is present at the OLT.

FP lasers are multimode lasers, so that mode partition noise due to the mode statistical fluctuation of the optical source arises [6]. This effect, if not, suitably reduced by a careful design of the lasers, can greatly limit the GPON span. For these reasons, the key characteristics of FP lasers to be used in GPON are specified in the standard, so as to be able to match the required performances.

In the case of B+ and C classes, the situation is different. As a matter of fact, modal dispersion arises in the downstream link [7], constituting a first span limitation. In order to face this effect, two ways are possible. The most diffused solution is to substitute FP with distributed feedback lasers (DFBs), accepting a cost increase. Specific DFB lasers are designed for use in

GPON, whose design requirements are relaxed with respect to similar devices used in dense wavelength division multiplexing (DWDM) long haul systems in order to reduce the cost.

As a matter of fact, perfect single mode operation is not strictly needed to eliminate modal dispersion. As an alternative to DFBs, the adoption of Electrical Dispersion Compensators (EDC) in the OLT has been also proposed and demonstrated [8].

It is interesting to observe that standard forward error correction is practically ineffective to correct the effects of modal dispersion. As a matter of fact, when the fluctuation in the modal structure of the source causes errors, they are concentrated in long bursts. It is possible to design a code conceived to correct long bursts of errors [9], but these codes are generally not used in optical communications for the decreased effectiveness in correcting random errors.

As far as the downstream is concerned, in order to achieve the required power budget with the standardized classes of emitted power, the ONU detector has to be an avalanche photodiode (APD). This also impacts on the cost of B+ and C GPON classes.

Another important phenomenon appears when the analog video overlay is present. In this case, due to the relevant power of this new channel, the Brillouin effect arises [10] so that a careful design of the optics is needed to reach the specified performances.

11.2.2.2 GPON Frame and Adaptation Protocol

As in the case of all implementations of the TDM-PON, also in the GPON, different users are assigned different time-slots in a TDM frame and bidirectional transmission on the fiber infrastructure is realized using different wavelength so that only a single fiber tree is needed, as shown in Figure 11.2 [11].

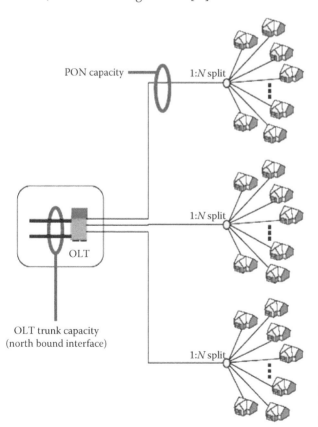

FIGURE 11.2
Topology of the set of GPON ODNs managed by a single OLT.

In this figure, a realistic OLT is shown where a set of PONs is managed by the same equipment. Hundreds of PONs can be terminated on the same OLT, optimizing on one side real estate and power consumption, and on the other, the packet multiplexing operated by the OLT switch.

Due to the topology of the ODN, a further issue arises: the different end users are at different distances from the splitter at the tree branching point. In this condition, each end user device (the ONU), has to transmit with a certain delay with respect to the reference time-frame in order to allow correct reconstruction of the signal at the branching splitter. This delay is continuously updated by a particular PON control procedure that measures the optical distance between the ONU and the branching point and drives the ONU asynchronous transmission.

This procedure is implemented into the physical layer PON processor that is generally a part of a single chip running all the physical layer protocols, comprising maintenance and critical alarms.

Another important element of the logical architecture of the GPON is the GEM: the GPON enveloping method, allowing almost every packet-oriented format to be conveyed through the PON, without the inefficiencies that were characteristics of ATM, that were deputed to do this in APON and BPON [11].

The layering of a PON network transporting Ethernet over ATM, native Ethernet and Ethernet over GFP (generic framing) is shown in Figure 11.3, where the role of the GEM is evident.

Another important physical layer procedure implemented on all the GPONs is dynamic bandwidth allocation (DBA) [12,13].

DBA allows upstream timeslots to grow and shrink based on the distribution of upstream traffic loads and operates on a timescale of milliseconds. Of course, the total of all timeslots cannot be greater than the length of a single upstream frame.

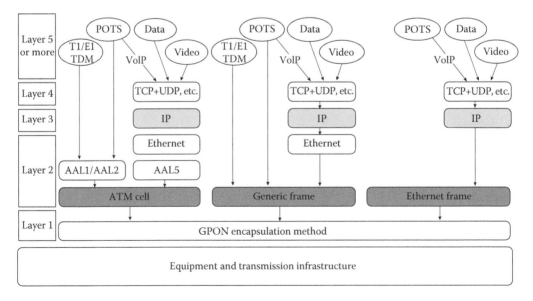

FIGURE 11.3
Layering of an access network based on GPON transport.

GPON Downstream frame is divided in transmission containers (T-CONTs), that are the upstream timeslots. There are much more T-CONTs than ONUs and DBA works exploiting this property.

An ONU must have assigned at least one T-CONT, but most have several T-CONTs whose number is managed by the DBA following the bandwidth needs of the different users.

Besides a drastic enhancement in the upstream bandwidth exploitation, DBA also allows a possible service level agreement (SLA) with a PON customer to be managed and controlled.

This is a very important characteristic, allowing a shared medium-based transmission system to be used to serve business customers with high requirements.

The ITU-T standard does not prescribe the specific DBA algorithm to be used, but defines two classes of possible algorithms.

If one of the so-called status reporting (SR) DBA is used, all ONUs report their upstream data queue occupancy, to be used by the OLT calculation process. Among the T-CONT assigned to the same ONU, each individual element may have its own traffic class, corresponding to the SLA of the conveyed service. By combining the queue occupancy information and the provisioned SLA of each T-CONT, the OLT can effectively optimize the upstream bandwidth allocation.

If, on the other hand, a non-status reporting (NSR) DBA is used, ONUs do not provide explicit queue occupancy information. Instead, the OLT estimates the ONU queue status, typically based on the actual transmission in the previous cycle.

For example, if an ONU has no traffic to send, it transmits idle frames during its allocation period. The OLT would then observe the idle frames and decrease the bandwidth allocation to that ONT in the following cycle. In the opposite case, the OLT constantly increases the allocation size until idles are detected, slowly adjusting to growing traffic.

Under a performance point of view, SR algorithms are superior in achieving high efficiency, as can be easily understood. However they are more complex to implement, also for the need of managing the return signals from all the ONUs.

11.2.2.3 GPON Capacity per User

The evaluation of the GPON capacity per user is not simple in that it depends critically on the bundle of services provided to the users. As a first step, let us assume that all the users have the same service profile composed by a 500 Mbit/s broadcast signal and a set of unicast services.

Let us start with an evaluation of the downstream capacity per user.

Due to the intrinsic broadcast capability of the GPON fiber infrastructure, the broadcast signal can be delivered in a dedicated part of the downstream TDM frame that will be received by all the end users.

The remaining part of the downstream frame (2 Gbit/s in our example) has to be divided among the users. Assuming a splitting ratio of 32, a bare physical bandwidth of 62.5 Mbit/s is assigned to each user. However, in order to evaluate the effective bandwidth per user, we have to take into account the statistical multiplexing at the end of the OLT that acts on the low activity of the sources and the statistical multiplexing at the ONU switch that grooms signals.

In the case of a realistic bundle of services (see Chapter 2) statistical multiplexing at the OLT allows a gain factor of the order of 4–5 over the bare physical bandwidth assigned to the user.

The statistical multiplexing at the ONU guarantees, for sure, that the ONU port can manage, with its queuing capacity, peaks of traffic of the order of the maximum available physical bandwidth with a very low packet rejection probability.

Thus, it can be said that thanks to the low services activity and statistical multiplexing inside the ONU and the OLT, the user has the same system experience in terms of service quality and speed that he would have with deterministic TDM multiplexing and service activity equal to one with a reserved bandwidth of about 300 Mbit/s completely dedicated to point-to-point services.

As a matter of fact, we have to remember that every broadcast service can be distributed in a separated broadcast channel exploiting the topology of the ODN.

In our case, besides the perceived bandwidth of 300 Mbit/s, a 500 Mbit/s bandwidth is also used by streaming broadcast services.

The result of these calculations could change if the downstream signal is mainly constituted by unicast video streaming (e.g., Video on Demand [VoD]).

Increasing the VoD, it could be possible that the OLT grooming is not yet enough to maintain such a highly perceived bandwidth. However, it is to take into account that DBA can be applied not only to the upstream, but also to the downstream, and it is effectively done in several GPON designs.

In the presence of several VoD channels, DBA is particularly useful, since it permits the allocation of a reserved bandwidth to the VoD service exclusively where and when it is present.

We can try to quantify the effectiveness of the DBA, in the downstream, in two ways: either by maintaining the low activity bundle we have supposed at the beginning and by trying to evaluate as the user perceived bandwidth increases, or by adding to the low activity scenario, a relevant number of VoD channels and by trying to calculate the maximum number of VoDs compatible with the overall performances required by the network.

The first evaluation can be done with the tools of the queue theory, and it results that assuming Poison processes for the start of a call on the different services and an exponential service duration, the gain related to SR DBA is something between 2 and 4.

This means that if the ONU is sufficiently equipped to manage this situation, the perceived capacity evaluated is around 600 MHz.

However, this perceived bandwidth is related to the low activity bundle, and is not useful. Generally, carriers are more interested in the possibility of accommodating the VoD without changing the GPON hardware.

Frequently, in front of a market forecasts for the fruition of new services, the carrier is interested in evaluating the maximum possible penetrations of the VoD before the need of increasing the physical access capacity arises.

Since this type of evaluation is quite instructive, we will follow it in certain detail, exposing an evaluation that a tier one carrier has shown at a conference in the framework of a discussion on the GPON capacity to face the pace of fast traffic growth [14].

The calculation refers to a typical urban area served by a central office. In the area, there are 400 users divided into three classes:

1. Thirty percent of the users (120 units) have only telephone service; this means that they have an IP phone; the average traffic produced by such users is 30 mErlang per customer.

2. Thirty percent of the customers (120 units) have the same telephone service considered above, plus a fast Internet browsing service with a peak guaranteed bandwidth of 1 Mbit/s.

3. Forty percent of the users (160 subscribers), on top of the service characteristics of the second class, have IP-TV service.

The different services are implemented so that the IP phone is coded for a peak bandwidth of 26 kbit/s while the IpTv is coded, both for the standard and the high definition channels, within a bandwidth of 16 Mbit/s. This is a conservative hypothesis, mainly assumed to simplify the calculation. However, almost nothing changes with a more accurate modeling of the IpTv bandwidth.

Finally, it is assumed that in the peak hour, 50% of the subscribers who have IpTv will use it.

It is to take into account that the IpTv is a multicast channel, thus the overall traffic for IpTv is, on an average, 0.5 Erlang per customer.

Carrying out classical traffic computations, the average traffic partition in the peak hour is obtained as shown in Figure 11.4. This figure also shows the efficiency and the weakness of current GPON downstream bandwidth management.

As a matter of fact, if low activity services should be added, there is plenty of bandwidth to support a good diffusion. The effective browsing time can be halved as far as the transmission speed is concerned, for all the users, consuming only a small part of the available bandwidth, and even a doubling of the number of IpTV customers will present no problems.

The problem comes when services with high activity, as VoD or pair-to-pair video streaming, are considered.

For example, the residual capacity allows 70 contemporary VoD channels to be distributed, and if there is an important penetration of VoD, with more than one VoD-enabled TV in high spending families, this figure would not be enough.

However, this scenario does not seem to be realistic in the near future, thus, with the service bundles that can be envisaged, the GPON seems to have quite a sufficient downstream capacity.

Let us now consider the upstream capacity. Even eliminating broadcast, presently, several services require less upstream bandwidth with respect to downstream, but in the near future, this trend could change due to the growing importance of point-to-point services.

In the case of the upstream, the user activity is even smaller that the average activity of the downstream, the ONU performs statistical multiplexing on the line and the OLT queuing guarantees the absence of lost packets.

Moreover, in the downstream direction, the DBA allows peak requests from specific ONUs to be allocated using the capacity flexible assignment.

Thus, on top of a bare average physical bandwidth per user equal to about 39 Mbit/s, we have to take into account a gain equal approximately to 4–5 from the ONU and a further gain of about 2 coming from the DBA.

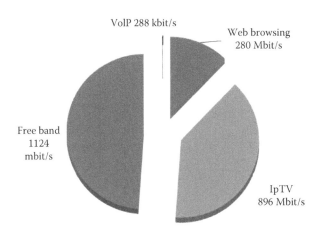

FIGURE 11.4
Occupied and free bandwidth in our example of GPON installed in a populated area.

At the end, assuming that the ONU gains a bit less than the OLT and that an SR DBA is used, the user has the same system experience in terms of service quality and speed that he would have with deterministic TDM multiplexing and service activity equal to one with a reserved bandwidth of about 320 Mbit/s available.

This calculation seems strange, but it is a fact that generally, GPON users do not perceive the transmission capacity unbalance of the network.

11.2.2.4 Functional Structure of a GPON OLT and ONU

In order to implement, with a suitable hardware and software structure, the functionalities we have seen, the GPON OLT and ONU have rich functional structures that assure both control and management functionalities and quality of service (QoS) implementation [15].

The functional structure of the OLT is shown in Figure 11.5. From a functional point of view, the OLT is divided into three areas that are often referred to as shells.

The core shell is the functional area terminating the physical layer on the side of the ODNs.

We have to remember that each OLT generally terminates several GPONs, also a few hundred in the case of big OLTs.

Thus, it is important to design this shell in a compact and functional way. Besides the physical interfaces with transceivers and all the transmission and detection electronics, the multiplex and demultiplex functions that are needed to create and terminate the GPON frame are also located in the core shell.

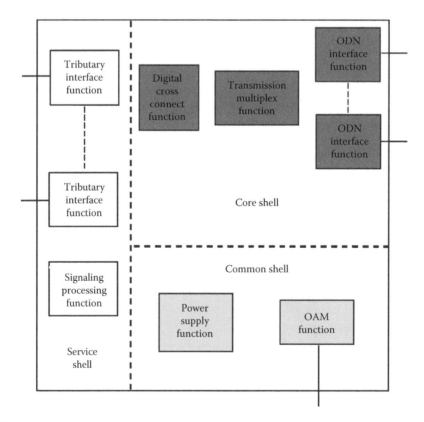

FIGURE 11.5
Functional representation of a GPON OLT.

The ranging protocol needed for the correct synchronization of all the signals sent toward the splitting center is also part of the interface functionalities.

Besides this transmission function, the core shell contains the so-called OLT Cross Connect. Originally, it was a digital cross connect since in the OLT, the SDH/SONET standard was applied.

In more modern OLTs, it is common for a carrier class switch to function as a cross connect inside the OLT. It has the role of grooming all the traffic coming from the user with lines that terminates in the OLT so as to best exploit the backhauling lines of the OLT. The DBA is managed in this part of the OLT.

The common shell includes all the management and maintenance functions, and from a hardware point of view, the power supply also belongs to this shell.

Finally, all the functionalities related to data back hauling are located in the service shell. Among them, we can note the termination of the core network signaling and the interfaces with the backhauling lines.

Also, the functional structure of the ONU, plotted in Figure 11.6, can be divided in the same shells, where now the service shell is toward the home network and the user appliances.

The partition of the main ONU functions among the three shells is quite recent and is shown in Figure 11.6.

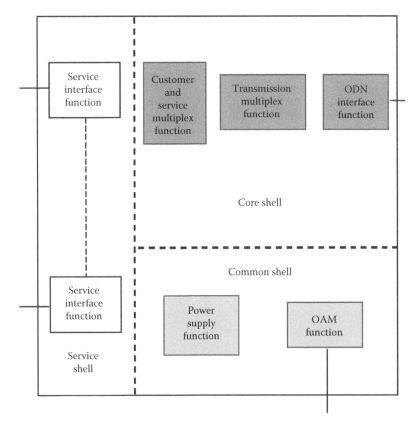

FIGURE 11.6
Functional representation of a GPON ONU.

11.2.3 NG-PON Project and the GPON WDM Overlay

One of the main targets of tier one carriers is to burst individual fruition of services with respect to broadcast fruition (that in the field of television means VoD with respect to Broadcast IpTV).

This is because broadcast can also be done, even with a smaller bandwidth, by means of satellites and traditional terrestrial television, and the customers are used to having the first for steadily decreasing prices, and the second for free.

Moreover, where a diffused cable TV exists, like in the United States, the IpTV is perceived by the customers as something they already have and it is difficult to obtain from it the marginal increase that the carriers need.

Finally, the importance that pair-to-pair customer activities are assuming, is something that puts value in the intrinsic bidirectional nature of the telecommunication network; thus, it can be a good opportunity for carriers.

This consideration also points the attention toward the weaker aspect of the GPON, as we have seen in the example of the last section.

Thus, a great amount of work is being undertaken in the standardization bodies (mainly in the ITU-T FSAN group) to standardize one or more versions of the so-called NG-PON (Next Generation PON).

Since the deployment of the GPON in vast areas is considered certain, all the work is based on the assumption that the new access transmission equipment, while providing a much greater capacity, exploits the same ODN deployed for GPON.

Besides this simple, but important requirement, a group of carriers in ITU-T have proposed that the entire standardization of the NG-PON be based on a much more severe set of requirements targeting a smooth and gradual transition from GPON to NG-PON. This set of requirements, that can be called Compatibility Conditions (CC), can be synthesized as follows [16]:

- *Coexistence*: NG-PON systems need to coexist on the same fiber with today's GPON
  - Allows existing GPON subscribers to be individually migrated to NG-PON on an as-needed basis without disrupting other users on the PON
  - Relies on the deployment of G.984.5 compliant ONTs today and NG-PON systems using G.984.5 enhancement band wavelengths
- *Digital capacity*: At a minimum, the capacity of NG-PON systems should be 10 Gbit/s downstream and 2.5 Gbit/s upstream. Alternatively, a system that overlays multiple GPON systems on the same ODN is also considered an acceptable solution.
- *Loss budget*: At a minimum, NG-PON is expected to operate over "Class C" ODNs (30 dB loss). Class C++ (>32 dB) operation with optical pre/post optical amplification is also considered a possible solution for long reach access.
- *Split ratio*: A minimum of 64 ways split should be supported, but for some applications (e.g., office consolidation) a 256-way (or higher) split may be needed. In the last case, optical amplification is suggested to overcome power budget limitations.
- *Reach*: A 20 km physical reach through a passive ODN is required and at least a 60 km reach using optical amplification is recommended.

If the CCs greatly simplify the carrier problem in evolving from one PON generation to the next, they pose a difficult problem to the equipment vendor.

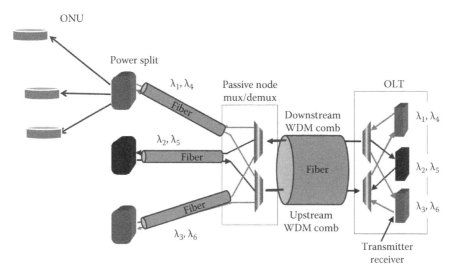

FIGURE 11.7
Architecture of the WDM overlay of a certain number of GPONs with colored interfaces. In this case, the ODN has two branching points: a WDM branching and a GPON branching.

The simpler way to respect the CCs seems to be through the exploitation of WDM technology starting from the ITU-T standard allocation of the fiber bandwidth in the access area summarized in Figure 11.1.

The idea is to use the empty bandwidths to allocate, besides existing GPONs, other GPONs that differ from the standard today only for the value of the upstream and the downstream wavelengths.

An example is the use of the 1505–1545 bandwidth (i.e., the great part of the C band) to convey a certain number of "colored" GPON on the wavelengths of the DWDM standard.

The network architecture coming out from this idea is plotted in Figure 11.7.

In order to try to smooth the traditional difficulties that are always present when designing a DWDM bidirectional system, the considered bandwidth has to be divided into two sub-bands (upper and lower wavelengths) and a guard interval is maintained between the bandwidths.

Let us assume that a 200 GHz spacing is used to simplify the optical system design, and that a guard interval of 400 GHz is maintained between the upstream and the downstream optical bandwidths.

Thus, it results that we can put 10 channels in the upstream bandwidth and 10 channels in the downstream bandwidth, using exactly the same ODN and without in any way disturbing the already existing GPON.

This solution is very good for the fidelity to (CCs), that means to guarantee an easy adoption by the carriers, and this, for sure, provides a good way of upgrading the network in pace with the penetration of broadband services.

On the other hand, functionally, the capacity delivered to a single user is the same as the GPON, unless more than one ONU is allocated at the user's premises.

Moreover, and definitely much more important, the overall cost of this solution heavily depends on the cost of the DWDM colored interfaces and on their market.

As a matter of fact, while a traditional PON has a single OLT interface for each served ODN, this version of the NG-PON has one interface for each served wavelength.

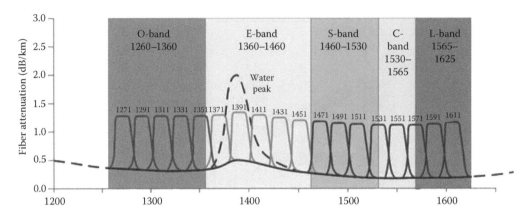

FIGURE 11.8
ITU-T standard definition of the average attenuation of an SSMF fiber with the names of the standard bandwidth and the standard CWDM channels. The water absorption peak is evidenced by the dashed line.

A way to alleviate the problem related to the interface cost is to adopt CWDM, whose standard grid is reported in Figure 11.8. As is evident from the figure, the water peak is unfortunately located around 1400 nm in the SSMF, almost the only fibers that are installed in the access area, which is just in the middle of the upgrade band (see Figure 11.1).

This renders the CWDM almost useless for use, since the number of frequencies that can be used is really small.

If the condition to interact with an already existing GPON in the same ODN is removed, the situation changes completely. In this case, the use of CWDM is quite effective, allowing eight colored GPONs (16 wavelengths) to be superimposed on the same ODN so creating a simple and powerful structure that takes advantage of the simplicity of the transmitter and receiver that are sold in great volumes for datacom and other applications.

However, the main problem, in term of cost of this architecture, is the need for colored sources both at the OLT and at the ONU. As a matter of fact, not only colored (and temperature stabilized in the DWDM case) sources have a much higher cost with respect to the simple sources used in standard GPON, but the management of the spare parts that have to be available for fast maintenance in case of failure is more complex, increasing the OPEX related to these equipments.

This problem can be removed in two ways. The first is to introduce a sufficiently cheap tunable laser suitable for this application.

In this case, the management of spare parts is the same as in the gray case. Today, tunable lasers, existing only for DWDM applications, are still too costly, but the situation could change.

As a matter of fact, on one side VCSELs seem to be almost ready for application in the third window, and tunable VCSELs have been obtained and tested in labs experiments [17].

On the other hand, the technology of lasers bars (see Chapter 5) could open the way to the realization of a CWDM tunable laser, which is an impossible task if designed with traditional tunable architectures.

The second method to solving the problem of having colored interfaces at the ONU is more on the system side.

At the OLT, 2N lasers (one for each different ODN terminating on the OLT) generate 2N wavelengths that are divided into upstream and downstream bandwidths. Downstream wavelengths are directly modulated and transmit the signal toward the end users.

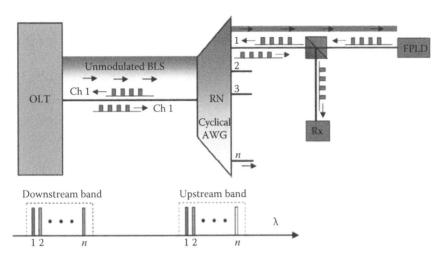

FIGURE 11.9
Implementation of a colored ONU in a PON using WDM multiplexing (either a WDM overlay of GPONs or a real WDM-PON) using the injection locking of a Fabry–Perot laser diode with a seed coming from the OLT at the correct wavelength.

Upstream wavelengths are transmitted unmodulated up to the end user. Every ONU is equipped with an identical, colorless Fabry–Perot Laser.

When an unmodulated external optical beam is injected into the cavity of an FP Laser, it will experience the internal gain from the semiconductor material. This process results in a reflected and amplified signal that contains the data modulation imparted by the pump current.

Since the injected signal can be much larger than the internally generated spontaneous noise, the majority of the optical output power will be at the wavelength of the injected signal. The normal multiwavelength spectrum of the FP is transformed into a quasi single-wavelength spectrum similar to that of a DFB laser. This narrowband output signal can then be efficiently transmitted through a WDM communication channel. The external injected wavelength is called the "locking" or "seeding" wavelength.

This mechanism is shown in the plot of Figure 11.9.

It is to be noted that since the upstream signal is amplified by the FP gain, it is possible to design the system so that it is not disturbed by small reflections of the seed arriving in the downstream direction. Last, but not least, a single AWG can be used in this system, processing the signal in two directions. This is possible with the so-called cyclical AWGs [18] that are devised to contemporary multiplex one bandwidth in one direction and demultiplex the other in the other direction.

In this way, a colorless FP is forced to emit a colored signal by the incoming wavelength containing the wanted modulation and constituting the upstream signal coming from the ONU. It can be noted that in this WDM-PON architecture also, every used wavelength is emitted by a colored source. However, all the sources are collected in the OLT, and the ONU is colorless.

Thus, the management of the spare parts and in general the system maintenance is simplified.

11.2.4 XG-PON

Another straightforward idea in preparing the network to meet a great increase in bandwidth request is simply to increase the bit rate from 2.5/2.25 Gbit/s to, for example, 10/2.5 Gbit/s.

This is the way that has been preferred for standardization by FSAN.

After finalizing the XG-PON1 basic specification, FSAN successfully carried it to ITU-T and the set of requirements was standardized as in the G.987 series.

The important choice in anticipating the standardization of the architecture relying on an higher bit-rate to achieve a greater capacity was not done arbitrarily, on the contrary, all the solutions presented by FSAN members or known in literature have been evaluated especially from the point of view of the asymmetry between the real per user bandwidth, and the XG-PON has been evaluated as the most suitable with the most common bundles of services [19].

Moreover, the design has been built on the requirements of the CCs, so that an XG-PON can be installed on an ODN where a normal GPON is running without disrupting the GPON operation.

These requirements have conditioned the choice of the spectral plan of the XG-PON, bringing us to the solution reported in Figure 11.10.

Three classes of XG-PON are foreseen even if the standardization mainly concerns the first class, which is considered the first to be deployed.

1. *XG-PON1*
    a. NRZ both upstream and downstream
    b. Downstream, $R = 9.95328$ Gbit/s
    c. Upstream, $R = 2.48832$ Gbit/s
    d. Maximum fiber distance: 20 km, with options of 40 km (in case with optical amplification)
    e. Split ratio up to 1:256
    f. Loss budgets: 29 and 31 dB
    g. FEC in the frame to help correct transmission both upstream (RS(248, 232)) and downstream (RS(248, 216))
2. *XG-PON1 Extended*
    a. As XG1-PON, but with a longer reach; this solution will probably need the so-called extenders, which are amplifiers suitable for remote power feeding
3. *XG-PON2*
    a. Has to target a downstream capacity of 40 Gbit/s

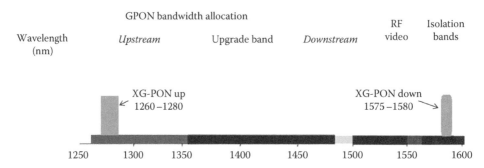

FIGURE 11.10
Wavelength plan of the XG-PON1.

All the main elements that are present in the GPON are reproduced and adapted for the XG-PON1; thus, the adaptation for higher levels packets is the GEM, suitably adapted to work with the different TDM frames of the XG-PON1 (e.g., it contains an FEC that was not present in the GPON frame), ranging protocol managed by the OLT to prevent collision in the splitting points, DBA both in upstream and in downstream.

On top of these characteristics, the XG-PON1 standard presents some aspects of novelty with respect to the GPON; among those a much more efficient security, with internal mechanisms for ONU and OLT registration, and a great attention to power consumption.

11.3 WDM Passive Optical Network

During the first phase of the NG-PON project, FSAN has also considered the possibility of accessing the shred medium through wavelength division multiplexing.

A simple architecture of a WDM-PON is shown in Figure 11.11. The OLT, placed in the local exchange, is constituted by an Ethernet switch managing a certain number of PONs and allowing grooming and QoS protocol management, and by a set of optical interfaces, one for each end customer. A different wavelength is associated to each end customer, so that the OLT must have WDM-capable interfaces.

In order to avoid problems with reflections from the fiber infrastructure, unidirectional transmission is generally considered. Thus, a fiber pair is needed in each branch of the PON infrastructure. A couple of athermal mux/demux is placed in the branching point, and every signal is terminated at the final user site with an ONU tuned on the user wavelength.

The wavelength plan used in this architecture depends on the adopted WDM technique. If the CWDM standard is used, generally, no more than 16 wavelengths can be adopted.

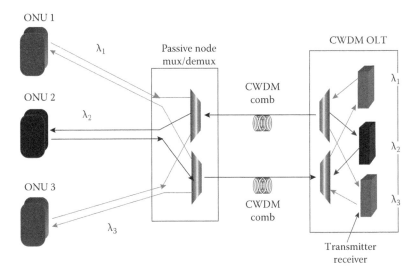

FIGURE 11.11
Architecture of a WDM-PON using unidirectional fiber transmission. The CWDM wavelengths are assumed in the figure.

This is due to the fact that common fibers have an absorption peak around 1390 nm, due to OH ions, and as shown in Figure 11.8, where the absorption profile of an SSMF fiber (ITU-T G.652) is superimposed on the CWDM frequency standard. Assuming a sensitivity of −18 dBm of the receiving PIN, a transmitted signal power of 0 dBm and a loss of 5 dB from the mux/demux, an ODN attenuation smaller than 0.65 dB/km is needed in order to achieve a reach of 20 km. This requirement is not fulfilled by the CWDM channels that are around 1390 and 1410 nm, thereby reducing the number of useful channels to 16.

Adoption of the CWDM standard has several advantages, from the possibility to use low cost, robust transceivers, which are produced in very high volumes mainly for datacom applications, to the availability of low cost athermal mux/demux.

If a high wavelength number is to be used, DWDM is needed. In this case, it is easy to reach the number of 32 or 64 wavelengths in the third fiber transmission windows, at the expense of a higher interface cost.

One of the potential problems in deploying the WDM-PON architecture depicted in Figure 11.11, is the presence of colored ONU optical interfaces. As a matter of fact, every ONU receives and transmits a different wavelength. The management of 16 different ONUs during network deployment (in the case of CWDM-PON) can increase operational costs, due both to spare parts management and in-field maintenance.

Naturally, in this case also, the need of colored interfaces can be eliminated either by using tunable lasers or by injection locking in FP Lasers.

11.4 WDM-PON versus GPON and XG-PON Performance Comparison

At this point, it is interesting to compare the WDM-PON and GPON on one side, where WDM-PON is based on CWDM and Gigabit Ethernet (GbE) channels on each wavelength, and WDM-PON and XG-PON on the other side, where WDM-PON is based on CWDM again, but each wavelength carries a 10 GbE channel, which we will call XWDM-PON.

In this section, we will limit our analysis to an architectural and performance comparison since an economic comparison is critically dependent on commercial and industrial strategies, and it is quite impossible to do if on one side there is a commercial product and on the other, a system under prototyping.

However, even without an explicit calculation, the main point driving the cost structure of WDM-PON and GPON/XG-PON will be clear.

In order to correctly carry out the comparison, we have to concentrate our attention on a single WDM-PON architecture; here we will consider in both cases the simple architecture of Figure 11.11, where colored ONUs and unidirectional transmission are used.

- *Capacity: GPON versus WDM-PON*
  The capacity per user of a WDM-PON is easily evaluated: a single wavelength is dedicated to each end user. In general, a GbE signal is transmitted on each wavelength, assigning a capacity of 1.25 Gbit/s to each end user. It is worth noting that the WDM-PON has no particular advantage if part of the signal is constituted by pure broadcast (e.g., conventional IP-TV): the broadcast signal has to be replicated by the OLT on every wavelength and independently sent to each user.

  The evaluation of the GPON capacity per user is not so simple as we have seen in the dedicated section.

Using the example we have cited there when we calculated the GPON expansion capacity, let us try to carry out the calculation also for the WDM-PON.

Since we have to feed 400 customers through 32 ONUs placed in the basements of the buildings of the area, we use the fiber to the building (FTTB) configuration. In the case of WDM-PON, we will have 16 ONUs, thus every ONU will have twice the traffic load.

Since the WDM-PON is symmetric, the evaluation is only for the downstream, which is the most critical direction.

Repeating the traffic evaluation in this case, we will find that each ONU that is served by a GbE channel has to manage a traffic constituted by 500 Mbit/s of broadcast IpTV, and a total of 73.5 Mbit/s of point-to-point traffic constituted by 18 kbit/s of Ip phone (this is naturally an average value since it is lower that the base Ip phone rate), 17.5 Mbit/s of web browsing, and 56 Mbit/s due to individual video programs.

The total amount of traffic is about 574 Mbit/s.

We have to take into account that the efficiency of the GbE channel should not be pushed to the limit as it might risk congesting the far end switch. Let us image reaching a 68% efficiency of the GbE, which is a net capacity of about 0.81 Gbit/s.

The free capacity per ONU is in this case 236 Mbit/s and globally 3.7 Gbit/s. If we want to evaluate how many VoD channels we can allocate, it would be misleading to divide 3.7 Gbit/s per 16 Mbit/s, since in this case it is not possible to distribute in a dynamical way, the bandwidth among different groups of users as is possible in the case of a GPON.

The only result that can be considered is that with the WDM-PON, it is possible to open 14 VoD channels for each of the 16 ONUs. The result is that the comparison between the two architectures is quite difficult.

If the service penetration is uniform, and corresponds to an area covered by the uniform central office from a social point of view, the WDM-PON is much more effective in delivering point-to-point services that have high activity. If the penetration of the service is very inhomogeneous, the WDM-PON advantage gets smaller and smaller.

- *Capacity: XG-PON1 versus XWDM-PON*
  Under a qualitative point of view, the situation is exactly the same as we described in the previous section. The relevant difference is that if such very high capacity systems have been adopted, this means that the penetration of high capacity and high peak bandwidth services are already here.

  In this case, the much higher total capacity of the XWDM-PON gets, for sure, a better result from a capacity point of view.

  However, we must not forget that this is not for free, and whatever technology solution will be found to decrease the cost of the XWDM-PON interfaces, a certain price difference should always exist between the two categories of systems.

- *Security and unbundling: Both cases*
  WDM-PON assigns to each user, a dedicated wavelength. This is not exactly the same as the dedicated physical carrier that is assigned to the user in the copper network, but the situation is quite similar. Due to the presence of mux/demux in the branching point, every ONU receives only its own signal, so that a user cannot gain access to signals directed to other users. Moreover, a wrong working of

a single ONU cannot influence the signals of other users, thus a good degree of security is guaranteed.

As far as unbundling is concerned, different wavelengths could be assigned to different carriers if a suitable mechanism is implemented in the OLT. This is not a physical separation exactly, because the physical layer is common among different carriers and data security is guaranteed by wavelength separation.

In the case of GPON, each user receives the signal directed to all the users so that it is possible to gain access to the signal directed to another user by simply working on the ONU. Moreover, if the ranging protocol of a single ONU does not work, the wrongly synchronized signal interferes with the signals from other ONUs in the branching point damaging other users.

As far as unbundling is concerned, it is practically impossible using GPON, unless virtual unbundling is considered. In this case, an amount of bandwidth in the shared downstream channel is assigned to a competing carrier under a defined SLA.

- *Fiber utilization: Both cases*
  Due to the fact that bidirectional transmission is used in the GPON case, while in our example WDM-PON unidirectional transmission is adopted, the fiber infrastructure is clearly better exploited by the GPON. As shown in Section 11.3, unidirectional transmission can be used in WDM-PON, but comes at some cost. As a matter of fact, in order to achieve a sufficient branching ratio, DWDM is needed, for example, 32 channels with a channel spacing of 100 GHz. A possible design can individuate two different bandwidths to be used upstream and downstream. They can be separated by a gap of about 800 GHz to prevent destructive interference from reflections. In this way, a branching ratio of 16 can be achieved. However, 100 GHz channel spacing requires cooled DFB lasers to be used both in the ONU and in the OLT. This fact, besides the greater cost of the mux/demux, clearly influences the cost of the system. In order to cope with this problem, the use of a WDM comb derived from the filtering of a single broadband noise source [20] has been proposed, but it is not clear yet if real cost advantage is achieved.

- *Optical link budget: WDM-PON versus GPON*
  The transmission scheme of WDM-PON is quite simple: attenuation is given by the loss of the mux/demux and by fiber propagation (taking into account connectors, patch panels, and other signal losing elements that can be present in the access infrastructure).

  Focalizing on CWDM-PON, standard CWDM optics can assure a transmitted power of 0 dBm, while the receiver sensitivity depends on the used detector. Using a PIN, the sensitivity at 1.25 Gbit/s (assuming that a GbE is transmitted) can be about −18 dBm. This number increases to about −28 dBm using an APD.

  Assuming the use of 16 wavelengths, the worst channel experiences an attenuation of about 0.9 dB km. Adding the other contributions and a system margin, we can assume a loss of 1.3 dB/km. Inserting these numbers in a very simple link budget evaluation, we obtain 10 km using a PIN at the receiver and about 22 km with an APD.

  In the case of GPON, the link budgets are standardized in different GPON classes and are reported in the GPON description section. In any case, classes B and C power budgets are of the same order of magnitude as the budgets evaluated in the case of the WDM-PON.

There is no relevant difference from this point of view, as it was intuitive from the fact that fiber propagation and splitting losses are not strongly dependent on the signal wavelength in the band relevant for fiber access.

- *Optical link budget: XWDM-PON versus XG-PON1*

Standardization prescribes for the XG-PON1, a link budget sufficient to have a reach compatible with GPON B+ and GPON C. Taking into account the slightly higher losses experimented by XG-PON1 wavelengths with respect to the GPON wavelength and a set of other differences in the transmission line between GPON and XG-PON1, a budget of 29 and 31 dB, depending on the comparison with GPON B+ or GPON C, results.

A similar prescription does not exist for XWDM-PON, and a real power budget will be available only after that the first industrial product is produced in volumes.

Presently, Ethernet components and equipment implementing 10 GbE are designed for top class Ethernet applications, if not for the carrier core network as backhauling of the core machines. Thus, these systems can provide good transmission performances. For example, a 10 GbE XFP designed for the switches placed at the core of the private network of a large corporate or for the public Ethernet switches of the core metro network are equipped with APD photodiodes and narrow line width DFB lasers, and generally assures a power budget in between −30 and −34 dBm.

It is to be noted that an XWDM-PON equipped with interfaces that guarantees a power budget of −32 dBm can cover a reach of about 25 km. This is not for free naturally, since we have to remember that different from the XG-PON that has a single transmitter into the OLT, an XWDM-PON has a maximum of 16 transmitters of different colors.

11.5 Enabling Technologies for Gbit/s Capacity Access

The problem that optical and electronics technology has to solve in the case of next generation access is to allow gigabit class equipment to be realized and deployed in huge volumes at a cost comparable with today's megabit class equipment (e.g., xDSL DSLAMs). The first step to individuate the key technology challenges is to assess the cost structure of the new access network.

In Figure 11.12, the cost structure of an FTTH and an FTTB network is shown assuming deployment in the years 2004 and 2010 [21,22]. In both cases, a GPON installation is assumed, exploiting B class GPON with a 32 splitting ratio. However, the results would be not so different if WDM-PON would be considered, but for a greater impact of the OLT equipment, that is almost comparable with the ONU.

In order to evaluate the infrastructure costs, a medium European city is assumed as reference and the OPEX cost is evaluated for a period of 10 years. As assumed everywhere, civil work is not included, and it is assumed that the fibers are to be deployed in already existing tubes.

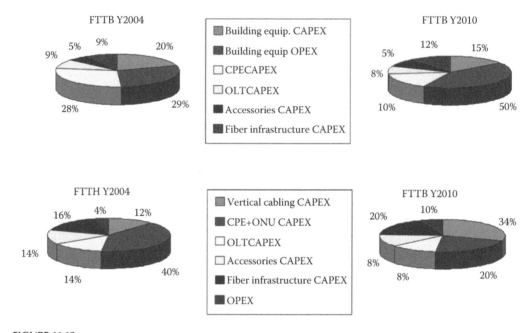

FIGURE 11.12
Cost structure of FTTH and FTTB networks assuming deployment in the years 2004 and 2010 [21,22]. In both cases, a GPON installation is assumed, exploiting B class GPON with a 32 splitting ratio.

From the figure, it is clear that the greater part of the CAPEX is due to the optical network termination: the building equipment in the FTTB case and the ONU in the FTTH case. On the other hand, the impact of the VDSL CPE cost in the FTTB case has been quite reduced passing from year 2004 to year 2008. This last figure is the effect of a great cost reduction of xDSL chips in the last few years.

The cost of a WDM-PON ONU is almost always determined by the optical interface, while in the case of the GPON ONU, two elements determine the cost of the equipment: the optical interface and the GPON chip, including the GPON physical frame processor, the ranging protocol processor, and the Ethernet switch, all in the same chip.

The importance of the ONU would be even more evident if we were to make a similar plot for XG-PON1 and XWDM-PON. Here, the challenge of reducing, essentially to a commodity, a 10 Gbit/s interface supporting a system power budget of about 30 dBm, is really formidable.

Another key point in the case of XG-PON1 is the realization of a suitable burst mode receiver both for the ONU and the OLT problem that is an example of the more general problem of the 10 Gbit/s electronics for the access receivers.

From this brief discussion, it is evident that optical interface technology is a key to the deployment of next generation access networks, and we will be devoting an important part of the discussion to these components.

11.5.1 GPON Optical Interfaces

11.5.1.1 GPON Interfaces Technology

GPON optical interfaces take the name of diplexers or triplexers when analog TV overlay is used.

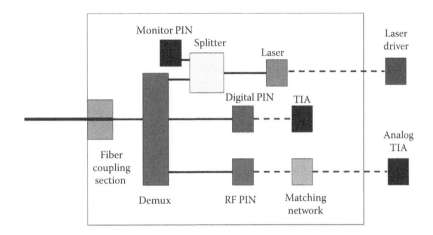

FIGURE 11.13
Functional block scheme of a triplexer for a GPON.

A triplexer has one or more WDM filters dividing the three GPON wavelengths (1310, 1490, and 1550) and one laser at 1310 nm, one digital receiver at 1490 nm, and one analog receiver at 1550 nm. In a case of a diplexer, a simpler filter is needed, without the 1490 output, and the analog TV receiver is not needed either.

It is also possible to integrate in the module, the transimpedance amplifier (TIA, see the IM-DD receiver structure in Chapter 6) for the digital receivers to achieve the highest sensitivity.

Front-end electronics for the analog receiver is usually located outside the optical assembly, due to its thermal dissipation and dimensions. The laser driver for burst-mode operation is also located on an external electronic board. The dimensions of the module can be very different with depending on to the adopted technology. A triplexer block scheme is shown in Figure 11.13.

There are different approaches to realizing a GPON optical interface [23].

- *Micro optics*: In this case, diplexers/triplexers are made of discrete elements (TO packaged lasers and receivers) assembled together in a metallic package, coupled with a fiber with lenses and discrete thin film filters. Figure 11.14 represents the section of a triplexer realized by micro-optics techniques.

- *Integrated optics*: Optical elements (both active and passive) are integrated monolithically or assembled directly in die on a single substrate. Different approaches have been proposed with different balances between monolithic and hybrid integration [24–26]. Figure 11.15 represents an example of a triplexer realized by the planar lightwave circuit (PLC).

Several industries have pushed integrated optics as the way to drastically reduce the cost of GPON optical interfaces.

However, technology evolution has shown that the parallel between optics and microelectronics cannot be perfect as discussed in some detail in Chapter 5, and microoptics triplexers and diplexers still hold the greater part of the market.

Beyond the general discussion carried out in Chapter 5, two specific points differentiate access application optical components from electronic ones.

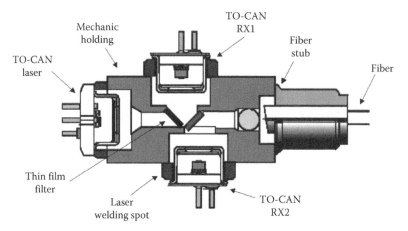

FIGURE 11.14
Section of a GPON triplexer implemented with a micro-optical mounting of discrete components.

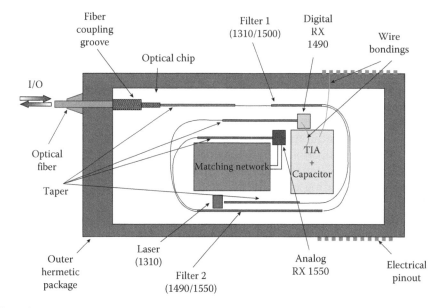

FIGURE 11.15
Schematic layout of a GPON triplexer realized with hybrid integration technology. Filters are realized on a high index contrast chip where the location for a flip chip mounting of photodiodes, laser, and TIA is fabricated.

Voltage and current sources can be integrated into electronic chips while lasers cannot be monolithically integrated into the PLC, whose waveguides are fabricated out of suitably doped glass. This forces the adoption of hybrid integration to accommodate the laser on PLC triplexers and diplexers. Moreover, even if interesting proposals exist at the research level to built photodiodes directly into a silica substrate [27], industrial processes still require hybrid integration also for these elements. This means that the chip is no longer built only from a bare wafer using planar processes, but requires the integration of other chips [28–30]. This process increases the bill of material (BOM) weight on the cost of the component, while the BOM is inessential in the case of microelectronic chips.

11.5.1.2 GPON Interfaces Draft Cost Model

Even if the cost model of hybrid integrated GPON interfaces is not the same as microelectronics, nevertheless their cost decreases with volume, faster than the cost of micro-optics components.

In order to clarify this point in the case of GPON interfaces, let us use the same rough cost model we used for DWDM PICs in Chapter 10.

In this case, we will consider four terms in the cost of the component:

Monolithic chip: This cost term decreases almost linearly with volume due to its dependence on the factory depreciation.

Bill of material: This term (i.e., the sum of all the parts that are acquired by external suppliers) decreases with a slower trend with respect to the chip cost; a reasonable approximation is to set it proportional to $1/v^x$, where v is the volume and $x < 1$ is a parameter that is characteristic of the business area under analysis. The value of x is related to the advanatge the supplier has to have volume orders even if at lower prices. Just as an example, if we refer to the supply of a field programmable gate array (FPGA), here the production cost goes down rapidly with volume and the suppliers will push as much as possible to have volume orders. In this condition, it is clear that the price will decrease rapidly with volume and x will be near to one (something like 0.75, for example, the exact value depends on the particular case). If a custom micromechanical part needs to be bought, and if it is something that is manufactured only for this application, and if the volume is not really high, a lot of manual work will be involved. In this condition, a moderate increase of volume could even give no advantage in terms of price, and increasing the volume beyond a certain value could even be impossible.

In the present case, there is a mixed situation, and so to evaluate if the BOM decreases with volume, a detailed analysis is needed, a forecast for each BOM component has to be made, and then the results have to be added up, and an attempt to fit it with $a + b/v^x$ with a, b, and x fitting parameters, and generally $x < 1$, must be made.

Cost of work: This is the cost for workers. It decreases very slowly with volume, and only due to more efficient work that can be done in a greater structure.

Cost of structure: This is the cost of all the structures that allow products to be sold. For example, we have here research, marketing, commercials, and so on. This cost generally does not decrease with volume and sometimes it even increases.

The final result has to be multiplied for the yield, so in order to keep the model simple, we will consider a characteristic of the technology that is independent from volume. In reality, in the case of monolithic planar technology, this is not exactly true, and the yield also increases with the volume produced.

We will use this very simple model to analyze the industrial cost of four products:

1. A GPON B+ diplexer built in micro-optics
2. A GPON B+ diplexer built via hybrid integration using a high index contrast glass platform (e.g., heavily Ge-doped SiO_2)
3. A GPON B+ diplexer monolithically integrated over the InP platform
4. As a reference, the GPON physical chip implementing the ranged protocol, the GEM, and so on

First of all, we fix the performances of the diplexers we have to compare, which have to be the same under a reasonable range.

TABLE 11.2

Standard Performances of GPON Commercial Interfaces

GPOB B+ Diplexer: Transmitter		
Bit rate (upstream)	1244	Mbit/s
Supply voltage	3.3	V
Operation temperature range	−40 to +85	°C
Operational wavelengths	1310	nm
Spectral width CW	1	nm
Average output power	0.5–3 (settable)	dBm
Extinction ratio	10	dB
GPOB B+ Diplexer: Receiver		
Bit rate (upstream)	2488	Mbit/s
Supply voltage	3.3	V
Operation temperature range	−40 to +85	°C
Operational wavelengths	1490	nm
Sensitivity (BER 10^{-10}, NRZ)	−24	dBm
Overload power	−4	dBm
Crosstalk 1: TX power into the RX	−47	dB
Crosstalk 2: External light power into the RX	−28	dB

TABLE 11.3

Values of the Yields for the Various Technologies Considered in the GPON Diplexer Cost Structure Analysis

	Yield (%)
Micro-optics GPON diplexer	90
Hybrid integration based diplexer	80
Monolithic InP diplexer	70
FPGA	95

We will adopt the performance listed in Table 11.2 for the receiver and transmitter. These are the standard performances of commercial devices of this kind, and that fit all the standards requirements.

The values of the yields that we have used for the different production processes are reported in Table 11.3, and are typical values for new products using already available technology modules in the respective technologies.

All the other model parameters are deduced by the industrial practice, and naturally, can vary from process to process and from factory to factory, and also with regard to relevant quantities.

The results of the cost model is reported in Figure 11.16, where the industrial cost of the four products we have considered, normalized to a common constant, is plotted versus the produced volume in a year.

Since the model is quite simplified with respect to a real industrial cost evaluation, the results must be considered under the qualitative point of view more than considering the particular numerical values.

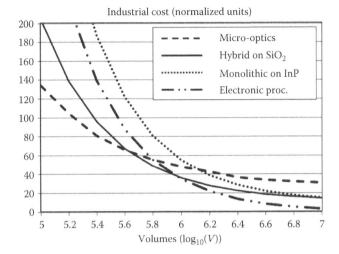

Industrial cost (normalized units)

Legend:
- – – – – Micro-optics
- ——— Hybrid on SiO$_2$
- ············· Monolithic on InP
- – ·· — Electronic proc.

Volumes ($\log_{10}(V)$)

FIGURE 11.16
Cost trend with volumes produced in a year for different types of GPON diplexers and for an integrated electronic circuit (a high performance FPGA). Costs are reported in normalized units.

However, all the relevant trends emerge clearly.

Up to about 400 kpieces, the technology assuring the lower industrial cost is the micro-optics: its cost is mainly due to the cost of the BOM with a relevant contribution due to the work cost. The contribution of CAPEX depreciation is by far smaller with respect to the other technologies.

This result clearly explains why the initial phase of the GPON market was dominated by microoptics, and it has even now by far the greater part of the market.

Due to the fact that the micro-optics BOM is composed partly of integrated general purpose elements (the laser and the diode in a TCAN package, for example) and partly of more specific elements (the separation filter is an example), the BOM goes down for volumes greater than 100 kpieces, approximately at $(a + b/v^x)$, where a and b are two constants, and $x \approx 0.65$.

This trend is slower with respect to the trend relative to the hybrid technology, which has much higher costs for small volumes, but decreases faster when the volume increases.

In particular, if we consider the case of with an integrated circuit realized in heavily Ge-doped SiO$_2$, where the laser and the photodiode are integrated with a flip chip bonder before wafer cutting, the monolithic chip would be 40% of the component cost at low volumes.

The chip cost decreases quite fast with volume, so that for volumes between 400 and 6300 kpieces, the hybrid integration technology produces the lower cost component.

Above 6 million pieces, the completely monolithic technology developed on the InP platform starts to be convenient, and from that point on, increasing the volumes it becomes more convenient.

Comparing the curve relative to the InP diplexer with the curve relative to the electronic physical protocols processor implemented with an high performance FPGA, it can be seen that the two curves are similar, due to the fact that the cost model is not so different. The curve relative to the diplexer is moved almost to the right due to the presence of a non-negligible cost of work due to the operations of packaging and fiber coupling that have to be done chip by chip. Moreover, the fabrication of chips on a certain surface of InP costs much more than the fabrication of chips on the same area of a silicon wafer due essentially to the different cost and dimensions of the reference wafer.

The different cost trends, with volume offered by hybrid and integrated technologies, are an advantage that could be precious in the long run, when PONs volumes go up unavoidably.

On this ground, much slower that the optimistic forecasts of 10 years ago, a certain number of GPON interfaces based on hybrid integration has been launched in the market in the last few years.

This is an important event since it is the first time that integrated optics is used to reduce prices and acquire a scalable production model and not just to achieve better performances.

11.5.2 WDM-PON and XWDM-PON Interface Technology

Differently from GPON, WDM-PON needs one optical interface per user not only in the ONU but also in the OLT as is shown in Figure 11.11. For this reason, optical interface technology is a key in implementing a cost-effective WDM-PON.

A trivial way of realizing a CWDM-PON is to use standard CWDM SFP transceivers. In this way, however, the cost of optical interfaces scales linearly with the number of users, being too high if a high number of users is considered. Moreover, colored ONU have to be used, thereby increasing operational costs due to the more complex management of spare parts.

Both this problems could be solved by a new class of components, called Photonic Integrated Circuits (PIC), which we have introduced for a different application in Chapter 10 [31,32].

The functional scheme of a transceiver containing a transmitting and a receiving PIC is represented in Figure 11.17, which is functionally equal to Figure 10.1. In the transmitting PIC, a bar integrating a certain number of lasers is coupled with a PLC mux/demux (generally an AWG) that launches the multiplexed channels into the output fiber. The scheme of

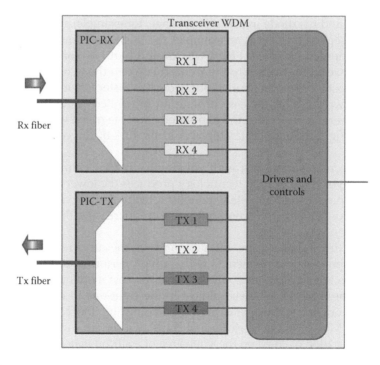

FIGURE 11.17
Structure of a transceiver based on PICs.

a receiving PIC is identical, with a demux, a bar of photodiodes, and with the WDM signal incoming from the input fiber. Similarly to CWDM SFP, it seems possible to implement CWDM PIC without thermal stabilization.

Besides reducing the cost of OLT colored sources, PIC technology can be used also to design colorless ONUs. In this case, a PIC can be used like a tunable source, switching on only the lasers tuned on the desired wavelength.

Since PIC technology is still under development, especially CWDM PIC that the market expects to be uncooled, it is difficult to foresee exactly what the cost of one PIC for a certain production volume will be.

In any case, considering a 16 wavelength CWDM-PON, in order to have an industrial cost equal to that of a GPON, a 16 elements PIC should cost as much as the unique burst mode interface of the GPON, and this is really very difficult cost wise.

The PIC is also the approach to try to reduce interface costs in XWDM-PON. It is to underline that the first PIC prototypes were realized at 10 Gbit/s either for application in the digital optical network or simply to reduce the cost of metro equipment.

Also, in the case of WDM-PON, the adoption of PICs have to be taken into account when managing the network. The problem, as in the case of the DON, is OLT PIC failure. In this case, a set of ONUs, if not a complete WDM-PON, loses the signal.

Individual protection of PICs into the OLT is thus practically mandatory, and this also impacts on the final OLT cost.

As far as the XG-PON is concerned, the discussion is more complex. From the point of view of optical interfaces, on the ground of standard requirements, there are not so many choices from among the state of the art.

The characteristics of the transmitter and receiver are summarized in Table 11.4 for class 1 XG-PON1 (28 dB power budget) and class 2 XG-PON1 (31 dB power span).

These optoelectronics components can be mounted onto a package using micro-optics methods and they can also be assembled in a hybrid or monolithic chip. In any of these cases, all the considerations already done for the GPON still hold with the further issue that the circuit to convey the electrical signal to the modulator at 10 Gbit/s is much more

TABLE 11.4

Possible Optical Interfaces for the Two Classes of XG-PON1: Class 1 Table and Class 2 Table

	ONU	ODL	ONU
XG-PON1 Class 1			
Upstream	Max TX power 2 dBm	Max loss 29 dB	Min RX power −27.5 dBm
	Direct modulation laser	Penalty 0 dB	APD receiver
Downstream	Min RX power −27.5 dBm	Max loss 29 dB	Max TX power 2 dBm
	APD receivers	Penalty 1 dB	External DFB modulation
XG-PON1 Class 2			
Upstream	Max TX power 2 dBm	Max loss 31 dB	Min RX power −29.5 dBm
	Direct modulation laser	Penalty 0 dB	APD receiver
Downstream First alternative	Min RX power −21.5 dBm	Max loss 29 dB	Max TX power 10.5 dBm
	PIN receiver	Penalty 1 dB	External DFB modulation + amplifier
Downstream Second alternative	Min RX power −28 dBm	Max loss 31 dB	Max TX power 4 dBm
	APD receivers	Penalty 1 dB	External DFB modulation

complex with respect to that working at 2.5 Gbit/s, thereby further increasing the cost of the OLT optical module package.

Besides these elements that are almost all in common with the GPON case, in the case of the XG-PON1, another difficulty arises, that is not present in the GPON.

Due to the broadcast nature of the ODN and the continuous change of frame slot allocation among the ONUs because of the DBA, the signal arriving at the ONU is a burst signal and it needs a burst mode receiver. Every ONU receiver must be able to detect the arrival of a burst and to rapidly lock the sampler at the burst, so as to sample the signal at the right point of the bit interval.

While at 2.5 Gbit/s, a burst mode receiver is difficult to implement, at 10 Gbit/s it is a real technology challenge. The first experimental components were presented in the year 2009 (e.g., [33]), but more developments will be needed to achieve the desired performance and cost targets imposed by the application.

References

1. Leshem, A., The capacity of next limited multichannel DSL, *Sensor Array and Multichannel Signal Processing Workshop Proceedings*, Barcelona, Spain, pp. 696–700, July 18–21, 2004.
2. Annual fiber deployment, *Fiber Optics Weekly Update*, September 17, 2008, http://findarticles. com/p/articles/mi_m0NVN/is_11_23/ai_99013219 (accessed: November 24, 2010).
3. Hmida, H. A., Cordner, G. C., Amer, A., Shalan, F. F., FTTH design and deployment guidelines for civil work, fiber distribution and numbering, *Optical Fiber Communication Conference, 2006 and the 2006 National Fiber Optic Engineers Conference, OFC 2006*, Anaheim, CA, 10 pp., March 5–10, 2006.
4. Badoz, A., Sustainable competition through fibre deployment, *Convergence Think Tank Seminar BERR–Department for Business Enterprise & Regulatory Reform*, London, U.K., April 22, 2008.
5. ITU-T Recommendations defining the GPON standard
   G.984.1: General characteristics for GPON 1/2003
   G.984.2 GPON: PMD layer specification 1/2003
   G.984.3 GPON: TC layer specification 10/2003
   G.984.4 GPON: ONT Management and Control Interface spec. 5/2003
6. Kaiser, G., *FTTX Concepts and Applications*, John Wiley & Sons, Hoboken, NJ, 2006, ISBN: 9780471704201.
7. Chang, F., Understanding 1G EPON power budgets in IEEE formalism, *IEEE 802.3 Plenary Meeting*, San Francisco, CA, July 2007.
8. Kim, H., Bien, F., de Ginestous, J., Chandramouli, S., Scholz, C., Gebara, E., Laskar, J., Electrical dispersion compensator for a giga-bit passive optical network system with Fabry-Perot laser, *Microwave Symposium, 2007*, Honolulu, HI, IEEE/MTT-S International, pp. 207–210, June 3–8, 2007.
9. Peterson, W. W., Weldon, E. J., *Error-Correcting Codes—Revised*, 2nd edn., MIT Press, Cambridge, MA, 1972, ISBN 0262130390.
10. Ellis, R. B., Weiss, F., Anton, O. M., HFC and PON-FTTH networks using higher SBS threshold singlemode optical fibre, *Electronics Letters*, 43(7), 405–407 (2007).
11. Lam, C. F. (Ed.), *Passive Optical Networks: Principles and Practice*, Academic Press, San Diego, CA, 2007, ISBN-13: 978-0123738530.
12. Skubic, B., Chen, J., Ahmed, J., Wosinska, L., Mukherjee, B., A comparison of dynamic bandwidth allocation for EPON, GPON, and next-generation TDM PON, *IEEE Communications Magazine*, 47(3), S40–S48 (2009).

13. Haran, O., Sheffer, A., The importance of dynamic bandwidth allocation in GPON networks, Sierra White Paper, 2008, www.Sierra.com (accessed: November 29, 2010).

14. Burzio, M., Next generation access network: Tecnologie ed Apparati (in Italian), *IST Workshop*, 2008, http://ist-sms.org/upl/File/Presentazione%20Burzio.pdf (accessed: November 29, 2010).

15. De Bortoli, M., Mercinelli, M., Solina, P., Tofanelli, A., Access optical technologies, passive optical networks, *Telecom Italia Technical Review* (in Italian), 13(1), 104–119, http://net.infocom. uniroma1.it/corsi/Infrarete/materiale/PON%20Telecom%20Italia.pdf (accessed: November 29, 2010).

16. Bond, R., ITU PON—Past, present, and future (Telcordia Representative), *FTTH Council Webinar*, July 30, 2008.

17. Chang-Hasnain, C. J., Optically-injection locked tunable multimode VCSEL for WDM passive optical networks, *Nano-Optoelectronics Workshop*, Tokyo, Japan, pp. 98–99, 2008.

18. Lu, H. C., Wang, W.-S., Cyclic arrayed waveguide grating devices with flat-top passband and uniform spectral response, *IEEE Photonics Technology Letters*, 20(1), 3–5 (January 1, 2008).

19. Effenberger, F. J., XG-PON Tutorial, *IEEE Optical Fiber Communication Conference, OFC 2010*, San Diego, CA, Paper OWX4, 2010.

20. Park, S.-J., Lee, C.-H., Jeong, K.-T., Park, H.-J., Ahn, J.-G., Song, K.-H., Fiber-to-the-home services based on wavelength-division-multiplexing passive optical network, *IEEE Journal of Lightwave Technology*, 22(11), 2582–2591 (2004).

21. Berthier, C., Wecker, G., FTTH Deployment Cost, SOGETREL's experience, *Digiworld Summit 2006*, Montpellier, France, November 14, 2006.

22. Iannone, E., FTTH: Enabling technologies and deployments, *ECOC 2007 Market Forum*, Berlin, Germany, September 16–20, 2007.

23. Huang, W.-P., Li, X., Xu, C. Q., Hong, X., Xu, C., Liang, W., Optical transceivers for fiber to the premises applications: System requirements and enabling technologies, *IEEE Journal of Lightwave Technology*, 25(1), 11–27 (2007).

24. Cheng, Y.-L., Lin, Y.-H., Wang, M.-C., Cheng, F.-Y., Wu, C.-S., Yu, Y.-C., Integrated a hybrid CATV/GPON transport system based on 1.31/1.49/1.55 /spl mu/m WDM transceiver module, *Quantum Electronics and Laser Science Conference, QELS*, 3, 1678–1680 (2005).

25. Xu, L., Tsang, H. K., Colorless WDM-PON optical network unit (ONU) based on integrated nonreciprocal optical phase modulator and optical loop mirror, *IEEE Photonics Technology Letters*, 20(10), 863–865 (May 15, 2008).

26. Behfar, A., Green, M., Morrow, A., Stagarescu, C., Monolithically integrated diplexer chip for PON applications, *Optical Fiber Communication Conference, 2005, OFC/NFOEC*, Anaheim, CA, Vol. 2, p. 3, March 6–11, 2005.

27. Wohl, G., Parry, C., Kasper, E., Jutzi, M., Berroth, M., SiGe pin-photodetectors integrated on silicon substrates for optical fiber links, *IEEE International Solid-State Circuits Conference, 2003, ISSCC*, San Francisco, CA, Vol. 1, pp. 374–375, 2003.

28. Mickelson, A. R., Basavanhally, N. R., Lee, Y.-C., *Optoelectronic Packaging*, Wiley Interscience, New York, 1997, ISBN: 0471111880.

29. Henein, G. E., Muehlner, D. J., Shmulovich, J., Gomez, L., Capuzzo, M. A., Laskowski, E. J., Yang, R., Gates, J. V., Hybrid integration for low-cost OE packaging and PLC transceiver, *Lasers and Electro-Optics Society Annual Meeting, 1997, LEOS '97 10th Annual Meeting, Conference Proceedings, IEEE*, San Francisco, CA, Vol. 2, pp. 297–298, November 10–13, 1997.

30. Park, S.-J., Jeong, K.-T., Park, S.-H., Sung, H.-K., A novel method for fabrication of a PLC platform for hybrid integration of an optical module by passive alignment, *IEEE Photonics Technology Letters*, 14(4), 486–488 (2002).

31. Oh, S. H., Park, Y.-J., Kim, S.-B., Park, S., Sung, H.-K., Baek, Y.-S., Oh, K.-R., Multiwavelength lasers for WDM-PON optical line terminal source by silica planar lightwave circuit hybrid integration, *IEEE Photonics Technology Letters*, 19(20), 1622–1624 (2007).

32. Lambert, D. J. H., Joyner, C. H., Rossi, J., Kish, F. A., Nagarajan, R., Grubb, S., Van Leeuwan, F., Kato, M., et al., Large-scale photonic integrated circuits used for ultra long haul transmission, *The 20th Annual Meeting of the IEEE Lasers and Electro-Optics Society, LEOS 2007*, Lake Buena Vista, FL, pp. 778–77, 2007.

33. Chang, Y. (Frank), First demonstration of a fast response/locking burst-mode physical layerchipset for emerging 10G PON standards, *European Conference on Optical Communications— ECOC 2009*, Vienna, Austria, p. 629, 2009.

Appendix A: SDH/SONET Signaling

In this appendix, we will specify the working of SDH/SONET signaling looking in detail at the structure of the frame at the different layers.

This analysis is doubly useful, both as an example of the completeness and complexity of the signaling structure of a carrier class network and for practical reasons. As a matter of fact, the SDH/SONET framing is also frequently used in other kinds of equipment for the sake of compatibility and because all the needed hardware is available.

We will concentrate on SDH, specifying when needed the differences with SONET.

All the elements reported in this appendix, including the figures, have their origin in the SDH/SONET standards. Thus, we refer the reader to the relevant bibliography of Chapter 3 for the original documents.

A.1 SDH Frame Structure

Each SDH frame starts with framing bytes, which enable the equipment receiving the SDH data stream to identify the beginning of each frame. The location of the other bytes within this frame structure is determined by its position relative to the framing byte.

The organization of the frame can be easily understood by representing the frame structure as a rectangle comprising of boxes arranged in N rows and M columns, where each box carries 1 byte. This representation is shown in Figure A.1 besides the serial frame structure adopted for serial bytes transmission.

In accordance with this representation, the framing byte appears in the top left-hand box (the byte located in row 1, column 1), which by convention is referred to as byte 1 of the SDH frame.

The frame bytes are transmitted bit by bit, sequentially, starting with those in the first row. After the transmission of a row is completed, the bits in the next lower row are transmitted. The order of transmission within each row is from left to right.

After transmission of the last byte in the frame (the byte located in row N, column M), the whole sequence repeats—starting with the framing byte of the next frame.

As shown in Figure A.2, the SDH frame is organized in two distinct parts: the section overhead that collects all the overhead information for the layers involved in the active SDH section and the virtual container (VC) wherein the tributaries are stored.

As explained in Chapter 3, each layer of the SDH architecture has its own overhead providing transport capacity for all the signaling needs both among entities of the same layer and with entities of the adjacent layers through the relative access points.

All the overhead information is carried out in the overhead area of the frame located in the first H columns, while H–M are devoted to the VC.

From now on, to be specific, we will take into consideration the STM-1 frame and the section overhead, whose structure is shown in Figure A.3. All the other elements of the SDH/SONET hierarchy have similar section overheads.

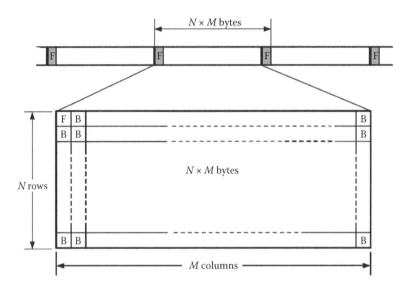

FIGURE A.1
Rectangle representation of the SDH frame. F indicates the framing byte and B the generic signal byte.

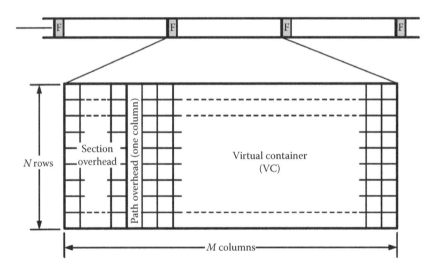

FIGURE A.2
Section overhead and VC in an SDH frame.

The STM-1 section overhead is composed of

- Section overhead—carried in the first nine columns of the STM 1 frame.
- Multiplexer section (MS) overhead—carried in overhead rows 5–9.
- Regenerator section (RS) overhead—carried in overhead rows 1–3.
- AU (administration unit) pointers—carried in overhead row 4.
- Path overhead (POH)—carried in the first column of a VC-4. The POH carried in the VC-4 is called high-order POH.

This allocation is shown in Figure A.4.

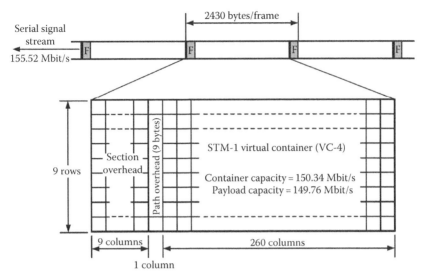

2430 bytes/frame × 8 bits/byte × 8000 frames/s = 155.52 Mbit/s

FIGURE A.3
Specific dimensions and composition of the STM1 frame.

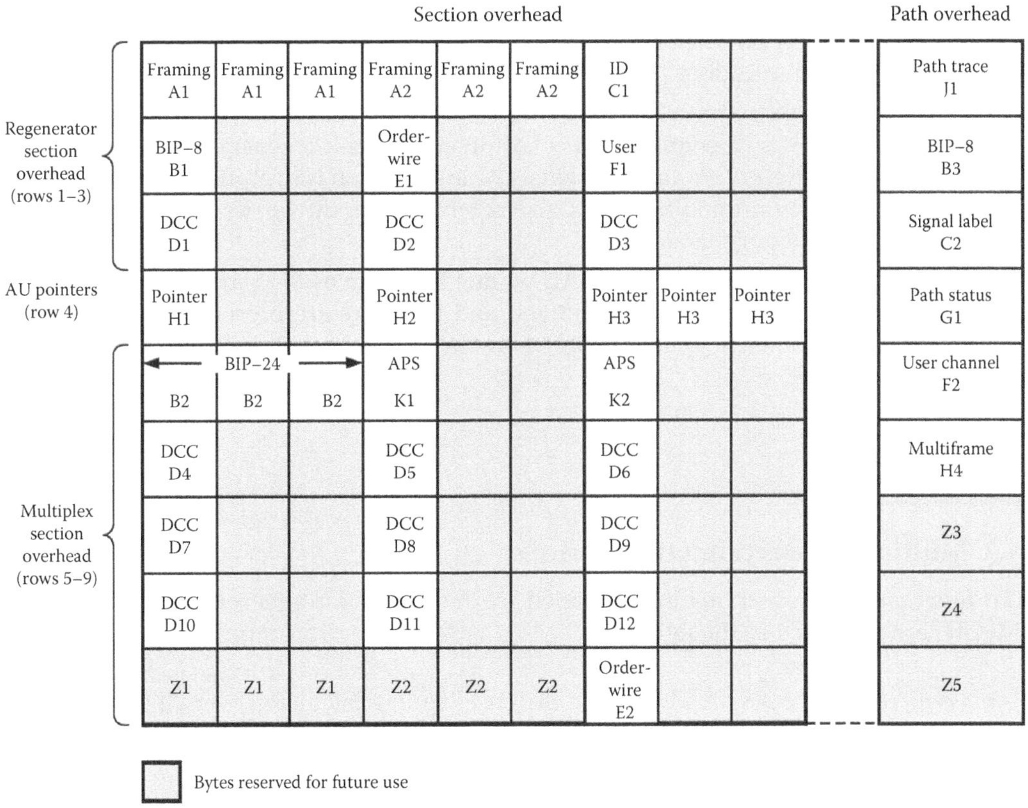

	Section overhead										Path overhead
Regenerator section overhead (rows 1–3)	Framing A1	Framing A1	Framing A1	Framing A2	Framing A2	Framing A2	ID C1				Path trace J1
	BIP–8 B1			Order-wire E1			User F1				BIP–8 B3
	DCC D1			DCC D2			DCC D3				Signal label C2
AU pointers (row 4)	Pointer H1			Pointer H2			Pointer H3	Pointer H3	Pointer H3		Path status G1
Multiplex section overhead (rows 5–9)	← BIP–24 → B2	B2	B2	APS K1			APS K2				User channel F2
	DCC D4			DCC D5			DCC D6				Multiframe H4
	DCC D7			DCC D8			DCC D9				Z3
	DCC D10			DCC D11			DCC D12				Z4
	Z1	Z1	Z1	Z2	Z2	Z2	Order-wire E2				Z5

▢ Bytes reserved for future use

FIGURE A.4
Specific overhead allocations in the STM1 section overhead.

A.2 Regenerator Section Overhead

The regenerator section overhead (RSOH) is terminated and renewed by every SDH equipment that has the role of regenerating the signal (e.g., line terminals, add-drop multiplexers, digital cross connects).

The functions of the various bytes carried in the STM 1 RSOH are described as follows.

- *Framing (A1, A2 bytes)*: The six framing bytes carry the framing pattern and are used to indicate the start of an STM 1 frame.
- *Channel identifier (C1 byte)*: The C1 byte is used to identify STM-1 frames within a higher-level SDH frame (STM-N, where the standardized values of N are 4, 16, etc.). The byte carries the binary representation of the STM-1 frame number in the STM-N frame.
- *Parity check (B1 byte)*: An 8-bit wide bit-interleaved parity (BIP-8) checksum is calculated over all the bits in the STM-1 frame to permit error monitoring over the RS. The computed even-parity checksum is placed in the RSOH of the following STM-1 frame.
- *Data communication channel (D1, D2, D3 bytes)*: The 192 kbit/s data communication channel (DCC) provides the capability to transfer network management and maintenance information between regenerator section terminating equipment (RSTE).
- *Orderwire channel (E1 byte)*: The E1 byte is used to provide a local orderwire channel for voice communications between regenerators and remote terminal locations.
- *User communication channel (F1 byte)*: The F1 byte is intended to provide the network operator with a channel that is terminated at each regenerator location and can carry proprietary communications. The information transmitted on this channel can be passed unmodified through a regenerator or can be overwritten by data generated by the regenerator.
- *AU pointers (H1, H2, H3 bytes)*: The AU pointer bytes are used to enable the transfer of STM-1 frames within STM-N frames, and therefore are processed by MS terminating equipment. Separate pointers are provided for each STM-1 frame in an STM-N frame.

A.3 Multiplexer Section Overhead

The functions of the various bytes carried in the STM-1 multiplexer section overhead (MSOH) are described in the following.

- *Parity check (B2 bytes)*: A 24-bit wide BIP checksum is calculated over all the bits in the STM-1 frame (except those in the RSOH). The computed checksum is placed in the MSOH of the following STM-1 frame.
- *Protection switching (K1, K2 bytes)*: The K1 and K2 bytes carry the information needed to activate/deactivate the switching between the main and protection paths on an MS. They are also the vehicle of the automatic protection switching

(APS) protocol, managing, in real time, both dedicated and shared protection in SDH networks.

- *Data communication channel (D4 to D12 bytes)*: Bytes D4 to D12 provide a 576 kbit/s DCC between MS terminating equipment. This channel is used to carry network administration and maintenance information.

- *Orderwire channel (E2 byte)*: The E2 byte is used to provide a local orderwire channel for voice communications between MS terminating equipment.

- *Alarm signals*: Alarm information is included as part of the MSOH. These functions are explained in Section A.5.

A.4 VC-4 POH Functions

The POH is contained within the VC portion of the STM-1 frame. The POH data of the VC-4 occupies all the 9 bytes of the first column. The functions of the various bytes carried in the VC-4 POH are described hereafter.

- *Path trace message (J1 byte)*: The J1 byte is used to repetitively transmit a 64-byte string (message). The message is transmitted 1 byte per VC-4 frame.

  A unique message is assigned to each path in an SDH network. Therefore, the path trace message can be used to check continuity between any location on a transmission path and the path source.

- *Parity check (B3 byte)*: A BIP-8 even checksum, used for error performance monitoring on the path, is calculated over all the bits of the previous VC-4. The computed value is placed in the B3 byte.

- *Signal label (C2 byte)*: The signal label byte, C2, indicates the structure of the VC-4 container. The signal label can assume 256 values; however two of these values are of particular importance:

  1. The all "0"s code represents the VC-4 unequipped state (i.e., the VC-4 does not carry any tributary signals)

  2. The code "00000001" represents the VC-4 equipped state

  The G1 byte is used to send status and performance monitoring information from the receive side of the path terminating equipment to the path originating equipment. This allows the status and performance of a path to be monitored from either end, or at any point along the path.

- *Multiframe indication (H4 byte)*: The H4 byte is used as a payload multiframe indicator to provide support for complex payload structures, for example, payload structures carrying multiple tributary units (TUs—see Chapter 3, Section 3.2.6). If, for example, the TU overhead is distributed over four TU frames, these four frames form a TU multiframe structure. The H4 byte then indicates which frame of the TU multiframe is present in the current VC-4.

- *User communication channel (F2 byte)*: The F2 byte supports a user channel that enables proprietary network operator communications between path terminating equipment.

A.5 Alarm Signals

Alarm and performance information is included as part of the POH.

- *Loss of signal (LOS)*: LOS state is entered when received signal level drops below the value at which an error ratio of 10^{-3} is predicted.

  LOS state ends when two consecutive valid framing patterns are received, provided that during this time no new LOS condition has been detected.

- *Out of frame (OOF)*: OOF state is entered when four or five consecutive SDH frames are received with invalid (errored) framing patterns. Maximum OOF detection time is therefore $625\,\mu s$.

- *Loss of frame (LOF)*: LOF state is entered when OOF state exists for up to 3 ms. If OOFs are intermittent, the timer is not reset to zero until an in-frame state persists continuously for 0.25 ms.

  LOF state is exited when an in-frame state exists continuously for 1–3 ms.

- *Loss of pointer (LOP)*: LOP state is entered when N consecutive invalid pointers are received where $N = 8$, 9, or 10.

  LOP state ends when three equal valid pointers or three consecutive AIS indications are received.

A.6 Multiplexer Section AIS

Multiplexer section AIS is sent by RSTE to alert downstream MSTE (multiple section terminal equipment) of detected LOS or LOF state. It is indicated by STM signal containing valid RSOH and a scrambled "all 1s" pattern in the rest of the frame.

Multiplexer section AIS is detected by MSTE when bits 6–8 of the received K2 byte are set to "111" for three consecutive frames. Its removal is detected by MSTE when three consecutive frames are received with a pattern other than "111" in bits 6–8 of K2.

Far end receive failure (FERF or MS-FERF): FERF is sent upstream by MSTE within $250\,\mu s$ of detecting LOS, LOF, or MS-AIS on incoming signal. Optionally, it is transmitted upon detection of excessive bit error rate (BER) defect (equivalent BER, based on B2 bytes, exceeds 10^{-3}). It is indicated by setting bits 6–8 of transmitted K2 byte to "110."

FERF is detected by MSTE when bits 6–8 of received K2 byte are set to "110" for three consecutive frames. Its removal is detected by MSTE when three consecutive frames are received with a pattern other than "110" in bits 6–8 of K2.

Transmission of MS-AIS overrides MS-FERF.

- *AU path AIS*: AU Path AIS is sent by MSTE to alert downstream high order path terminating equipment (HO PTE) of detected LOP state or received AU Path AIS. It is indicated by transmitting "all 1s" pattern in the H1, H2, and H3 pointer bytes plus all bytes of associated VC-3 and VC-4.

  AU Path AIS is detected by HO PTE when "all 1s" pattern is received in bytes H1 and H2 for three consecutive frames. Its removal is detected when three consecutive valid AU pointers are received.

- *High order path remote alarm indication (HO Path RAI, also known as HO Path FERF)*: HO Path RAI is generated by HO PTE in response to received AU path AIS. It is sent upstream to peer HO PTE. It is indicated by setting bit 5 of POH G1 byte to "1."

  HO Path RAI is detected by peer HO PTE when bit 5 of received G1 byte is set to "1" for 10 consecutive frames. Its removal is detected when peer HO PTE receives 10 consecutive frames with bit 5 of G1 byte set to "0."

- *TU path AIS*: TU Path AIS is sent downstream to alert low order path terminating equipment (LO PTE) of detected TU LOP state or received TU path AIS. It is indicated by transmitting "all 1s" pattern in entire TU-1, TU-2, and TU-3 (i.e., pointer bytes V1–V3, V4 byte, plus all bytes of associated VC-1, VC-2, and VC-3 loaded by "all 1s" pattern).

  TU Path AIS is detected by LO PTE when "all 1s" pattern is received in bytes V1 and V2 for three consecutive multiframes. Its removal is detected when three consecutive valid TU pointers are received.

- *Low order path remote alarm indication (LO path RAI, also known as LO Path FERF)*: LO Path RAI is generated by low order path terminating equipment (LO FTE) in response to received TU Path AIS. It is sent upstream to peer LO PTE.

  LO Path RAI is indicated by setting bit 8 of LO POH V5 byte to "1."

  LO Path RAI is detected by peer LO PTE when bit 8 of received V5 byte is set to "1" or 10 consecutive multiframes. Its removal is detected when peer LO PTE receives 10 consecutive multiframes with bit 8 of V5 byte set to "0."

Note: LO Path RAI is only available when generating and/or receiving "floating mode" TU payload structures.

A.6.1 Response to Abnormal Conditions

This section describes the response to the wide range of conditions that can be detected by the maintenance means built into the SDH frames, and the flow of alarm and indication signals.

Appendix B: Spanning Tree Protocol

B.1 Spanning Tree Protocol

In Chapters 3 and 9, we have seen that the Ethernet technology, born for applications in local area network (LAN) and private networks, is one of the candidates for the development of a next generation, packet-based, metro network.

In this appendix, we describe one of the main protocols of the Ethernet suite: the spanning tree. We look at all the advantages of the spanning tree protocol, but also the scalability limits that its adversaries underline to discourage its use in the carrier network.

The spanning tree protocol (STP) is a link-management protocol that is part of the Institute of Electrical and Electronics Engineers (IEEE) 802.1 standard for bridges and switches. All the material of this appendix is a synthesis of IEEE Ethernet standards, so we will not repeat here the list of such standards, recommending the interested reader to use the list in the references of Chapter 3.

The STP is used to determine the best path from source to destination for Ethernet frames in a switched Ethernet network. It creates a hierarchical tree spanning the entire network, including all bridges and switches. Searching such a tree, the STP determines all possible paths between two network users and, when needed, it activates only one of them. STP also provides path redundancy establishing redundant links as a backup if the primary link fails.

If a network node becomes unreachable for some reason, the spanning-tree algorithm reconfigures the logical topology, reestablishing the connection by activating the standby path.

B.2 STP Operation

The collection of bridges in a LAN can be considered as a graph whose nodes are both bridges and LAN connections, and whose edges are the interfaces connecting the bridges to the segments.

To break loops in the LAN while maintaining access to all LAN segments, the bridges collectively compute a spanning tree covering all the network nodes. The spanning tree is not necessarily a minimum cost spanning tree. A network administrator can reduce the cost of a spanning tree, if necessary, by altering some of the configuration parameters in such a way as to affect the choice of the root of the spanning tree.

The spanning tree that the bridges compute using the STP can be determined using simple rules based on a particular representation of the network topology as a graph where both the network nodes and links are vertices and the bridge ports connecting one node to one link in the real network are the graph links.

An example is represented in Figure B.1, where a simple network constituted by three bridges is represented both through its standard topology graph (where network nodes

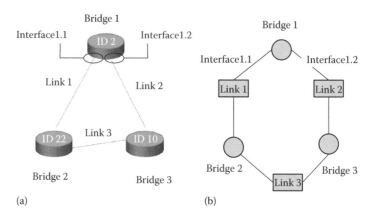

FIGURE B.1
Example of spanning tree graph for a simple network: (a) network conventional graph; (b) spanning tree graph.

are graph vertices, while network links are graph connections) and through the spanning tree graph.

The steps to construct a spanning tree covering the entire network and without any loop can be summarized as follows.

Step 1: Select a Root Bridge

The root bridge of the spanning tree is the bridge with the smallest (lowest) bridge ID. Each bridge has a unique identifier and a configurable priority number; the bridge ID contains both numbers. To compare two bridge IDs, the priority is compared first. If two bridges have equal priority, then the media access control (MAC) addresses are compared. For example, if switches A (MAC = 0200.0000.1134) and B (MAC = 0200.0000.2452) both have a priority of 10, then switch A will be selected as the root bridge. If the network administrators would like switch B to become the root bridge, they must set its priority to be less than 10.

In the network of Figure B.1, the smallest bridge ID is 2. Therefore, bridge 2 is the root bridge.

Step 2: Determine the Least-Cost Paths to the Root Bridge

The computed spanning tree has the property that messages from any connected device to the root bridge traverse a least-cost path. It is a path from the device to the root that has minimum cost among all paths from the device to the root.

The cost of traversing a path is the sum of the costs of the segments on the path. Different technologies have different default costs for network segments. An administrator can configure the cost of traversing a particular network segment.

The property that messages always traverse least-cost paths to the root is guaranteed by the following rules:

- *Least-cost path from each bridge*: After the root bridge has been chosen, each bridge determines the cost of each possible path from itself to the root. From these, it picks one with the smallest cost. The port connecting to that path becomes the root port (RP) of the bridge.
- *Least-cost path from each network segment*: The bridges on a network segment collectively determine which bridge has the least-cost path from the network segment

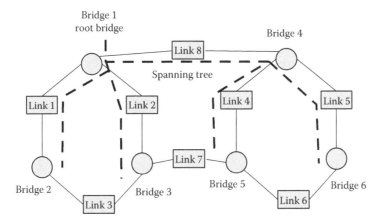

FIGURE B.2
Example of spanning tree computation via the assignment of port roles.

to the root. The port connecting this bridge to the network segment is then the designated port (DP) for the segment.

- *Disable all other root paths*: Any active port that is not a RP or a DP is a blocked port (BP).

The aforementioned rules assume that there is no tie in determining the cost of paths. In reality, such ties can happen and the protocol includes a set of rules to select a unique least-cost path in the case of several types of ties.

An example of the calculation of the spanning tree through the assignment of a role to each active port is shown in Figure B.2.

B.3 Bridge Protocol Data Units

The earlier-mentioned rules require knowledge of the entire network to be implemented. The bridges have to determine the root bridge and compute the port roles (root, designated, or blocked) with only the information they have. To exchange information about bridge IDs and root path costs, the bridges use control data frames called bridge protocol data units (BPDUs). The BPDU fields are shown in Figure B.3.

A bridge sends a BPDU frame using the unique MAC address of the port itself as a source address and the STP multicast address 01:80:C2:00:00:00 as a destination address. There are three types of BPDUs:

1. Configuration BPDU (CBPDU), used for spanning tree computation
2. Topology change notification (TCN) BPDU, used to announce changes in the network topology
3. Topology change notification acknowledgment (TCA)

BPDUs are exchanged regularly (every 2 s by default) and enable switches to keep track of network changes and to start and stop forwarding at ports as required.

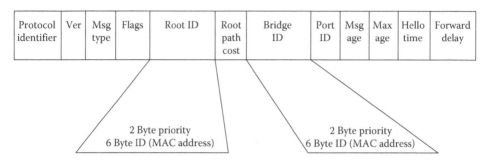

FIGURE B.3
BPDU field structure.

B.3.1 Configuration Bridge Data Units

Configuration bridge data units (BDUs) are used to construct the spanning tree by a coordinated activity of all the network switches.

Each switch determines the state of all its ports on the ground of the knowledge it has of the network and of the steps we have described in Section B.2.

After this operation, the knowledge of the port status is broadcasted through the network via the configuration BDUs.

This information is stored in a suitable database by the switches and used to route Ethernet frames in the structure without creating loops.

Moreover, when a link fails so that no frame arrives any more, the knowledge of the network status that is diffused in all the switches allows the spanning tree to be recreated excluding the failed network, changing, as needed, the states of the switch ports.

The possible STP states of a switch port can be described as follows:

- *Blocking*: A port that would cause a switching loop if activated is set by the STP in this state, no user data is sent or received, but it may go into forwarding mode if the other links in use were to fail and the spanning tree algorithm determines that the port has to be activated to create a back-up path for the traffic involved in the failure. BPDU data is still received and processed in blocking state.

- *Listening*: If a port of a switch is in the listening state the switch processes BPDUs arriving through that port and awaits possible new information that would cause it to return to the blocking state or evolve to the forwarding state.

- *Learning*: While a port in the learning state does not yet forward frames, it does learn source addresses from frames received and adds them to the switching database.

- *Forwarding*: A switch port receiving and sending data is in the forwarding state. STP still monitors incoming BPDUs that would indicate it should return to the blocking state to prevent a loop.

- *Disabled*: Not strictly part of STP, a network administrator can manually disable a port.

When a new device is added to the network connecting it to a switch port, it will not immediately start to forward data. It will instead go through a number of states while it processes BPDUs and determines the topology of the network.

When a host is attached, such as a computer, printer, or server, the port will always go into the forwarding state, albeit after a delay of about 30 s while it goes through the listening and learning states. The time spent in the listening and learning states is determined by a value known as the forward delay (the default is 15 s; in every situation the value of this time interval is set by the root bridge).

If, instead, another switch is connected, the port may remain in blocking mode if it is determined that it would cause a loop in the network.

B.3.2 Topology Change Notification

TCN BDUs are used to inform other switches of port changes. TCNs are injected into the network by a non-root switch and propagated to the root. Upon receipt of the TCN, the root switch will set a topology change flag in its normal BPDUs. This flag is propagated to all other switches to instruct them to change their topology data base and, as a consequence, to update the forwarding table entries.

B.4 Protocol Extensions

The STP is so diffused and is applied in so huge a number of different situations that several updates and variations of the protocol have been done.

The first important variation is the so-called RSTP (rapid STP). The scope of this STP variation is to speed the incorporation of new entities into the network. In the standard STP it can take up to 30 s, while in RSTP it generally takes 6 s if the administrator has left default values for the process characteristic time.

Another important variation of the STP is the incorporation of the virtual local area network (VLAN) concept into STP, allowing different instances of the protocol to coexist, each one representing a different VLAN built on the same physical network structure.

Appendix C: Inter-Symbol Interference Indexes Summation Rule

C.1 Cascade of Linear Filters

In this appendix, we will demonstrate the rule used in the design of transmission systems that, under opportune hypothesis, the ISI indexes relative to independent phenomena can be added.

We will consider regular pulses, that is, either finite duration pulses or pulses with exponentially decreasing tails, so that no convergence problem will arise in our calculation.

We will also assume in this appendix the following hypotheses:

- The pulse $p(t)$ at the filter chain input has a zero temporal mean that is $\int tp(t)dt = 0$. For example, it is a Gaussian pulse.
- The linear filters of the chain have a pulse response $a(t)$ with zero mean, that is, $\int ta(t)dt = 0$. This is true, for example, for Gaussian filters, which also have Gaussian pulse response, that is, with a good approximation for a large class of array waveguides (AWGs).

Let us imagine having a system constituted by the cascade of two linear filters. Let us assume that two linear operators $\hat{A}$ and $\hat{B}$ represent the two filters that act on a pulse signal $p(t)$ broadening it so that

$$\hat{A}\, p(t) = \int a(t-\vartheta)\, p(\vartheta)\, d\vartheta \tag{C.1}$$

$$\hat{B}\, p(t) = \int b(t-\vartheta)\, p(\vartheta)\, d\vartheta$$

the pulse width after the operators action is

$$\tau_A^2 = \int t^2 \int a(t-\vartheta)\, p(\vartheta)\, d\vartheta\, dt \tag{C.2}$$

and an analogous equation for τ_B. Changing variables into the integral and remembering the property of the pulse average to be zero, we have

$$\tau_A^2 = \iint (x^2 + 2x\vartheta + \vartheta^2)\, a(x)\, p(\vartheta)\, d\vartheta\, dx = T^2 + T_A^2 \tag{C.3}$$

$$\tau_B^2 = T^2 + T_B^2$$

where T_A and T_B are the widths of the pulse responses of the filters

Let us now apply the two filters together

$$\tau_{AB}^2 = \int t^2 \int a(t - \vartheta) \int b(\vartheta - \mu) p(\mu) \, d\mu \, d\vartheta \, dt \tag{C.4}$$

the expression of the overall pulse widening is obtained by the following substitution in the integral:

$$\begin{cases} t = x + y + z \\ t - \vartheta = x \\ \mu = y \end{cases} \tag{C.5}$$

The variable change has unitary Jacobian, and the integral becomes

$$\tau_{AB}^2 = \int \int \int (x^2 + y^2 + z^2 + 2xy + 2xz + 2yz) a(x) b(z) p(x) \, dx \, dy \, dz \tag{C.6}$$

from the integral, since all the pulses have zero mean, the following is derived:

$$\tau_{AB}^2 = T^2 + T_A^2 + T_B^2 \tag{C.7}$$

C.2 Presence of One Nonlinear Operator with Fast Nonlinearity

In the cases that interest us, at least one of the involved filters is nonlinear, but with an instantaneous nonlinearity. It is a simplified way to represent the effect of the Kerr nonlinearity at the system output.

The filter $\hat{B}$ is now composed of two parts: a linear part with memory and an instantaneous nonlinear part depending on the pulse power, so that

$$\hat{B} p(t) = \hat{B}_L \, f(p^2(t)) p(t) = \int b_L(t - \vartheta) f(p^2(\vartheta)) p(\vartheta) \, d\vartheta \tag{C.8}$$

As a first step, let us evaluate the pulse widening due to the transit through $\hat{B}$. This widening is given by

$$\tau_B = \int t^2 \int b_L(t - \vartheta) f(p^2(\vartheta)) p(\vartheta) \, d\vartheta \, dt \tag{C.9}$$

substituting $z = t - \vartheta$ in the integral and remembering that all pulses have zero mean, we obtain

$$\tau_B^2 = \int\int (z^2 + 2z\vartheta + \vartheta^2) b_L(z) f(p^2(z)) p(\vartheta) d\vartheta\, dz = T + \int z^2 b_L(z) f(p^2(z)) dz \qquad (C.10)$$

The nonlinear broadening is complex to evaluate, depending on $f(\ )$; however, Equation C.10 is sufficient to reach our goal.

With the same procedure of the previous section, we can write the expression of the overall pulse widening as

$$\tau_{AB}^2 = T^2 + T_A^2 + \int z^2 b_L(z) f(p^2(z)) dz = T^2 + T_A^2 + T_B^2 \qquad (C.11)$$

C.3 Comments and Examples

As a comment, it is to be observed that the derivation we have done holds exactly only in very specific situations, where strict hypotheses like exponential tails and zero mean holds for all the pulses, both the signal and the pulse response of the filters.

Naturally, if these conditions are not fulfilled completely, the property will not hold exactly, but in a regime of small distortion, we can frequently use this fundamental rule.

As an example, let us consider a squared pulse 10 ps long with a rise time of 3 ps, as shown in Figure C.1. Let us filter the pulse with a first Gaussian filter of insufficient bandwidth, equal to 61 GHz instead of the required minimum of 100 GHz.

As a consequence of the filtering, the pulse spreads, as shown in Figure C.2, and assumes a standard deviation of 11.6 ps that is exactly the result of Equation C.3.

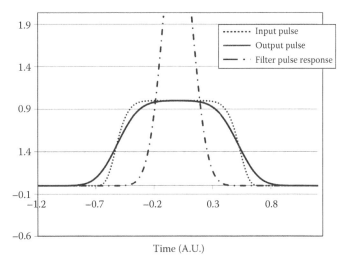

FIGURE C.1
Pulse at the input of the filter chain and after the first filter.

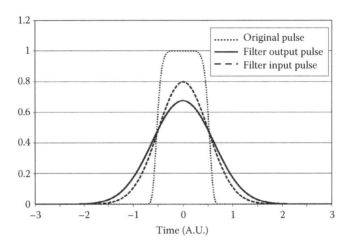

FIGURE C.2
Pulse at the input and at the output of the second filter.

Filtering again the pulse with another filter with an even smaller bandwidth (25.88 GHz in this case), the pulse broadens again reaching a width, as shown in Figure C.2. The final width is exactly what we have foreseen in Equation C.7.

Appendix D: Fiber Optical Amplifiers: Analytical Modeling

D.1 Introduction

Optical amplifiers are key devices to determine the performance of optical transmission systems. In order to understand the impact of behavior of different amplifiers on the characteristics of optical systems generally it is sufficient good knowledge of the base phenomena that occurs in such devices.

This is the reason why, more than presenting a very accurate model, in Chapter 4, we have described with the help of simple equations the behavior of amplifiers in many relevant circumstances. When designing a very high-performance transmission system however, since all the margins are reduced, it is quite important to have an accurate modeling of the devices that have been selected for the system.

This is why we present in this appendix a much more accurate analytical model of the behavior of erbium-doped fiber amplifiers (EDFAs) and lumped Raman amplifiers. Generally the complete coupled equations cannot be solved analytically, so that what we will introduce here is the first step toward an effective simulation of fiber-based amplifiers.

In order to set up a simulation engine based on these equations we recommend to execute a study of the literature starting from [1–3] as far as EDFAs are concerned and from [4,5] if a Raman amplifier has to be simulated.

D.2 Erbium-Doped Fiber Amplifiers

In principle, the theoretical model of an EDFA depends critically on the pumping scheme, both due to the possible presence of exited state absorption and to the fact that pumping at 1480 is near to a two-level system.

In practical amplifiers, however, almost only 980 pumps are used; thus, we will assume that the pumps will be at this wavelength so that the system behave as a three-level system: an exited state ($^4I_{11/2}$), a transition state ($^4I_{13/2}$), and a ground state ($^4I_{15/2}$).

We will refer to these states with numbers 2, 1, and 0, respectively, and all the variables referring to a particular state will have the state index.

Let us call n_j the population of the jth state; the population rate equations are similar to those of excited molecules in gasses, but for the great simplification that erbium ion are included into a glass matrix and cannot move. Thus, the diffusion term is not present.

In these conditions, the rate equations of a three-level system with a very short lifetime of the upper state (i.e., exactly the EDFA case) are written as

$$n_1 = n_{tot} \frac{\dfrac{P_s(z)}{P_{ss}} + \dfrac{P_p(z)}{P_{ps}}}{2\dfrac{P_s(z)}{P_{ss}} + \dfrac{P_p(z)}{P_{ps}} + 1}$$

$$n_0 = n_{tot} - n_1$$

$$n_2 \approx 0 \qquad\qquad (D.1)$$

$$\frac{\partial P_p}{\partial z} + \frac{1}{v_p}\frac{\partial P_p}{\partial t} = +2\pi\zeta\Gamma_p\sigma_p n_0(z)P_p(z)$$

$$\frac{\partial P_s}{\partial z} + \frac{1}{v_s}\frac{\partial P_s}{\partial t} = +2\pi\Gamma_s\left[\sigma_s n_1(z)P_s(z) + \frac{\sigma_s}{\Gamma_s}n_1(z)N_0(z) - \sigma_a n_0(z)P_P(z)\right]$$

$$\frac{\partial N_0}{\partial z} + \frac{1}{v_s}\frac{\partial N_0}{\partial t} = +2\pi\Gamma_s\left[\sigma_s n_1(z)N_0(z) + \frac{\sigma_s}{\Gamma_s}n_1(z)P_{0s}(z) - \sigma_a n_0(z)P_P(z) + \frac{\hbar\omega}{2}\right]$$

where the symbols have the following meaning:

- P_s is the signal power flux; it is supposed that only one narrowband signal travels the amplifiers in the positive z-direction.
- P_p is the pump power flux; a single continuous wave pump is assumed; in multi-pump configurations, other pump equations have to be added similar to this one and, when the pump appears in the signal and noise equations, the sum of the pumps have to be inserted.
- N_0 is the ASE noise power flux in the signal bandwidth; the ASE noise in the pump bandwidth has been neglected.
- ζ is the pump directional factor: it is +1 if the pump is copropagating with the signal and −1 if it is counterpropagating.
- Γ_p and Γ_s are the superposition integrals between the pump and signal modes, respectively, and the distribution of the erbium ions.
- σ_s and σ_a are the emission and absorption cross sections at the signal wavelengths, respectively.
- σ_p is the absorption cross section at the pump wavelength.
- v_p and v_s are the group velocities of the pump and of the signal.

In Equations D.1, several approximations are already embedded:

- The population of the higher energy state is assumed to be zero due to the very fast decay time. This is a condition to have an efficient amplifier.
- Noise is not present in the pump and signal equations (high signal-to-noise ratio approximation). This is needed to separate the problems: first the problem related to amplification and other deterministic effects, and after that the noise equation becomes approachable once the pump and the signal evolutions are known.
- The nondepleting pump approximation is also contained in the equations, since the pump evolution does not depend on the signal.
- Generally the adiabatic approximation is also carried out, which in this case is more than justified by the slow pace of variation of the populations.

Starting from Equations D.1, the main microscopic parameters related to the amplifier functionalities can be defined: the local gain $g(z)$ and the loss $\alpha(z)$. Comparing Equations D.1,
with the standard coupled equations for a generic system, the following equations are
found:

$$g(z) = 2\pi\sigma_{se}n_1(z) \tag{D.2}$$

$$\alpha(z) = 2\pi\sigma_{se}n_0(z)$$

Starting from Equations D.2 and imposing the stationary approximation, the pump and
signal equations can be solved obtaining the following expression of the macroscopic
amplifier gain:

$$G(L) = \frac{P_s(L)}{P_s(0)} = e^{\int_0^L [g(z)-\alpha(z)]dz} \tag{D.3}$$

This is the equation that we have used to evaluate the behavior of the gain versus length
of the active fiber or versus the injected signal power.

As far as the noise is concerned, the first step to determine the amplifier noise factor
is to solve the noise equation, which is an easy task due to the fact that signal and pump
behaviors are already known. From the noise equation we immediately derive that the
noise flux depends on both the signal and the pump fluxes due to the gain saturation of
the amplifier.

In amplifiers used for optical communications, this is a slow phenomenon with respect
to the bit rate and in the standard situation of constant average signal power also, the noise
flux is constant.

In order to take into account correctly the noise performances of the amplifier, the quantum noise should be taken into account. In a semiclassical picture, this term can be incorporated into the amplifier equations considering the fluxes the average value of a random
variable. Since multiplying the flux by the mode area and dividing it by ω the density of
photons in the mode is obtained, the statistical distribution of the fluxes is readily evaluated starting from the Poisson characteristics of the photon number.

From this observation, the output SN_{out} can be evaluated obtaining

$$SN_{out} = \frac{(P_{s,0}(L))}{(P_s^2(L))-(P_{s,0}(L))^2} = \frac{1}{\hbar\omega_s} \frac{G^2 P_0^2}{GP_0 B + \hbar\omega_s(G-1)FB^2 + 2G(G-1)P_0FB + \hbar\omega_s(G-1)^2 F^2 B^2} \tag{D.4}$$

where
  $P_{s,0}$ is the power in the signal bandwidth in the absence of spontaneous emission (i.e., the
    signal power)
  P_s is all the power flux in the signal bandwidth
  P_0 is the input signal

The signal angular frequency is indicated with ω_s and the optical bilateral bandwidth
is indicated with B. It is to be noted that in all the chapters devoted to system design, the

unilateral optical bandwidth is used, which is half the value of B. For a diffused discussion on this convention, see Section 5.8.5.

The factor F is called inversion factor and it is given by

$$F = \frac{\displaystyle\int_0^L n_1(x)dx}{\displaystyle\int_0^L [n_1(x) - n_0(x)]dx} \tag{D.5}$$

The four terms at the denominator of Equation D.4 represents the contribution of four different noise terms to the SN:

1. The first term is the quantum noise associated with the output signal wave, which is assumed to be in a quantum coherent state.

2. The second term is the quantum noise associated with the spontaneous emission wave at the amplifier output.

3 The third term is the noise due to the beat between spontaneous emission and signal wave.

4. The fourth term is the self-beating of the spontaneous emission wave.

In all the practical conditions, the term due to the beat between spontaneous emission and signal is largely prevalent so that the other noise terms can be neglected. In this condition, if $G \gg 1$, (D.4) reduces to

$$SN_{out} = \frac{P_0}{2\hbar\omega_s FB} = \frac{P_0}{NFS_n B_0} \tag{D.6}$$

where

$NF = 2F$ is the amplifier noise factor
$B_0 = B/2$ is the optical unilateral bandwidth
$S_n = 2\hbar\omega_s$ is the unilateral noise power spectral density

D.3 Raman-Lumped Fiber Amplifiers

Let us image to have k signals at frequencies $\omega_j (j = 1 \ldots 2k)$ whose instantaneous power (i.e., the power flux in a certain instant and in a certain fiber section) is indicated as $P_j(t, z)$. These signals propagate both forward (for j odd, that is, $J = 1, 3, 5, \ldots$) and backward (for j even) having the signals $P_j(t, z)$ and $P_{j+1}(t, z)$ the same angular frequency ω_i ($j = 1, 3, \ldots$ $k - 1$). The index ζ_j is also tagged to each signal, with $\zeta_j = \pm 1$ depending upon the propagation direction (+ toward greater values of z, which is in the useful signal direction). These signals will be pumps, in general part of them forward and part of them backward, and useful signals (that, in general, we assume all forward). We will assume also that all the pumps are indicated with $j = 1, \ldots k_p$, while signal with $j = k_p + 1, \ldots k$.

All signals cause reflections that propagate through the fiber in the opposite direction of the signal itself; thus the backward signal has to also be considered if it is not injected into the fiber (which means that the corresponding boundary condition is zero).

We can suppose that the relative intensity noise affects both pumps and signals and is constituted by a set of random band limited processes: we will call the instantaneous power flux of such processes $N_j(t, z)$ $(j = 1, \ldots k)$ with the same convection adopted for the signal instantaneous power. Moreover, spontaneous emission noise, caused by amplification of the spontaneous photons emitted due to the interaction with the fundamental state of the field, has to also be inserted in the equations. In order to model the ASE evolution, we introduce $N_0(t, z)$, which represents ASE instantaneous power in a fiber section.

Once all the waves, pumps, signals, and noise propagating through the fiber are determined, we can apply as usual the rotating wave and the slowly varying approximation to pass from the wave equation to a set of coupled evolution equations that can be written as follows:

$$\frac{\partial P_j}{\partial z} + \frac{\zeta_j}{v_j}\frac{\partial P_j}{\partial t} = -\alpha_j \zeta_j P_j + \sum_{\substack{h \neq j \\ \neq j + \zeta_j}} \frac{\zeta_{jghj}}{2A_{hj}K_{hj}}(P_h + P_{h+\zeta_h})P_j \quad (h = 1,\ldots,k)(j = 1,\ldots,k)$$

$$\frac{\partial N_j}{\partial z} + \frac{\zeta_j}{v_j}\frac{\partial N_j}{\partial t} = -\alpha_j \zeta_j \left(N_j - \frac{\hbar\omega}{2}\delta_{j,0}\right) - \gamma_j \zeta_j \left(N_j - \frac{\hbar\omega}{2}\delta_{j,0}\right)$$

$$+ \sum_{\substack{h \neq j \\ \neq j + \zeta_j}} \frac{\zeta_{jghj}}{2A_{hj}K_{hj}}(P_h + P_{h+\zeta_h})\left(N_j + \frac{\hbar\omega}{2}\delta_{j,0}\right)$$

$$+ \sum_{\substack{h \neq j \\ \neq j + \zeta_j}} \frac{\zeta_{jghj}\delta_{j,0}}{2A_{hj}K_{hj}}\frac{|\omega_h - \omega_j|}{2\pi}F_{jh}(P_h + P_{h+\zeta_h})\left(N_j + \frac{\hbar\omega}{2}\delta_{j,0}\right) \tag{D.7}$$

$$(h = 1,\ldots,k)(j = 0,\ldots,k)$$

In Equations D.7, symbols represent the following physical parameters:

- g_{hj} represents the Raman gain and is equal to $g(\omega_h - \omega_j)$ if $\omega_h > \omega_j$; otherwise $g_{hj} = g(\omega_h - \omega_j)(\omega_{jh}/\omega_j)$.
- The term $\hbar\omega/2$ in the noise equation represents the continuous noise generation due to the spontaneous emission along the amplifier length [4]. This term obviously has to appear only in the equation for the ASE evolution, that is, in the equation for $j = 0$. In (D.7), the term $\delta_{j,0}$, which is 1 only if $j = 1$; else it is zero is present just to write in the general expression the terms appearing in the ASE equation only.
- γ_j represents the backward Rayleigh scattering coefficient, which is the loss due to Rayleigh scattering only.
- F_{jh} is the phonon-induced optical noise parameter that is related to the phonon density due to temperature [1]. The phonons number at a certain energy $\hbar\Delta\omega$ is given by the Bose distribution, $N_{phon} = 1/[\exp[\hbar\Delta\omega/kT]-1]$, where k is the Boltzmann constant. Starting from the expression of the thermal phonon number, the phonon-induced optical noise parameter can be expressed as $F_{jh} = -N_{phon}$ for scattering in the opposite direction with respect to the incident light (anti-Stokes) and as $F_{jh} = 1 + N_{phon}$ for scattering in the direction of the incident light (Stokes). Also this

term has to appear only in the equation for the ASE evolution, that is, in the equation for $j = 0$, from which the presence of $\delta_{j,0}$ in its expression.

- K_{hj} is a polarization factor; in the typical telecommunication case in which the pumps are depolarized to achieve uniform gain on all the input polarizations, $K_{hj} = 2/3$.
- A_{hj} is the Raman fiber effective area defined as a function of the interacting modes transversal shape as

$$A_{hj} = \frac{\left[\int\limits_0^\infty \left|A_j(\rho)A_h^*(\rho)\right|^2 d\rho\right]^2}{\int\limits_0^\infty |A_j(\rho)|^2| A_h(\rho)|^2 d\rho} \tag{D.8}$$

The Raman gain coefficient can be calculated even starting from prime principles, but often Equations D.7 are solved by inserting a measured value of this coefficient. In any case, it is important to remember that Raman scattering is a polarization dependent phenomenon.

Equations D.7 are written neglecting the field polarizations (otherwise they would be doubled in number). To take into account the polarization effect phenomenologically, a polarization coefficient is introduced that has to be set to 2/3 if, as often in practice, the pump is depolarized and the polarization of the incoming signal field is random.

As far as the ASE term is concerned, in Equations D.7, the population inversion is set to perfect inversion, as always with Raman amplifiers. The model provided by Equations D.7 contains all the most important phenomena that impact on design and performances of a Raman amplifier.

References

1. Kozak, M. M., Caspary, R., and Unrau, U. B., Computer aided EDFA design, simulation and optimization, *International Conference on Transparent Optical Networks—ICTON 2001*, pp. 202–205. s.l.: IEEE, Cracow, Poland, 2001.
2. Schmidtke, H. J. et al., EDFA models for network simulation purposes, *Symposium of Electron Devices for Microwave and Optoelectronics Applications*, pp. 291–295, Vienna, Australia, 2001.
3. Desurvire, E., *Erbium-Doped Fiber Amplifiers: Principles and Applications*, s.l.: Wiley, New York, 1994. ISBN-13: 978-0471589778.
4. Lali-Dastjerdi, Z., Rahmani, M. H., and Kroushawi, F., Modeling multi-pumped Raman fiber amplifiers, *PhotonicsGlobal@Singapore, IPGC 2008*, pp. 1–3, IEEE, Singapore, December 8–11, 2008.
5. Shang, T., Chen, J., Li, X., and Zhou, J., Numerical analysis of Raman amplification and optical signal-to-noise ratio in a photonic crystal fiber, *Optics Letters*, 4, 446–448 (2006).

Appendix E: Space Division Switch Fabric Performance Evaluation

E.1 Introduction

The scope of this appendix is to carry out a comparison of the performances of different space division switch fabric architectures in the dynamic switching mode. Using the same approach we used for time division bus-based switch fabrics in Chapter 7, we will estimate the saturated throughput, which is the maximum throughput the switch fabric is able to provide.

In order to do this, we will use a simulation program simulating a slotted packet switch fabric in the condition in which a packet is always present at every switch fabric input in every slot. Due to the slotted assumption, any time needed for the equipment to synchronize asynchronous packets before placing them in the input queue is not comprised in the delay evaluation.

The delay is evaluated by assuming that the probability that a packet is present at a certain input of the switch fabric in a certain slot is P_{in}, which is independent of time and equal for all the input ports. In the saturated situations, the offered traffic is $T_{in} = N$, where N is the number of input/output ports. In a normal working situation, it is $T_{in} = P_{in}N$.

In any case, the packets will have all the length of a single slot and when they arrive at a switch fabric input; their output destination is random, with all the output ports having the same probability.

The queues, wherever they are located, are assumed to be 10 packets long. It is to be noted that while the queue length is an important parameter for the delay evaluation, in the evaluation of the saturation throughput, the queue dynamic can be neglected due to the fact that when the switch operation reaches steady state, the queues are always full.

The last element on the simulation, routing through the switch fabric, is performed by evaluating all the available paths for all the packets directed toward the same output in the same time slot and choosing randomly both the packet to transmit and the path among those available.

Since we are considering strictly nonblocking structures, interference between packets does not exist, but for the fact that they could have the same destination. The Banyan switch is an exception, but in that case the path is unique and the algorithm limits itself to choose the packet to be transmitted randomly.

E.2 Crossbar Switch Fabric

In order to evaluate the performance of a crossbar switch fabric, we have to decide how packets are stored when arriving at the fabric. There are two base possibilities: input queues and output queues.

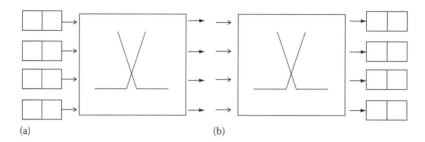

FIGURE E.1
Scheme of an input buffered switch fabric (a) and of an output queued switch fabric (b).

Our intention is to compare these two configurations shown in Figure E.1. Before any simulation, intuition tells us that output buffering should work better than input buffering due to the fact that output buffering causes a packet to be stored in a queue only if the output line is effectively busy, while input buffering also creates a queue among packets that enter from the same input but are directed to different outputs.

Another important distinction is based on the strategy to manage packets that arrive in front of a full queue coming from the remote source. The source can either discard them or resubmit them. In intermediate cases, a certain probability of resubmission can exist.

The analysis of [1] gives an expression for the normalized throughput at saturation T_{out}/T_{in} as the ratio between the average number of input packet in a slot (equal to N) and the average number of packets that exit from the fabric in the same slot (T_{out}). The throughput τ is evaluated as a function of the switch fabric dimension N of a crossbar switch fabric under the assumptions of synchronous packet arrival and blocked packets discarded. The expression is

$$\tau = \frac{T_{out}}{T_{in}} = 1 - \left(1 - \frac{1}{N}\right)^N \qquad (E.1)$$

These values are reported besides simulation results in Figure E.2, and the accuracy of the simulation results have been confirmed.

In particular, if N is very large (quite frequent case in practice), the limit value for $N \to \infty$ can be assumed, so that for large matrixes the saturated throughput does not depend on the matrix dimension in a first approximation and is equal to $\tau = 1 - \frac{1}{e} \approx 0.632$.

If packets are resubmitted rather than discarded, the limit of the saturated throughput for large matrixes can be evaluated analytically [2] obtaining $(2 - \sqrt{2}) \approx 0.586$. Again this agrees closely with the simulation results.

Let us consider now the average delay. Simulations for the slotted switch fabric in condition of nonsaturated throughput are reported in Figure E.3. In particular, the average delay is reported versus the probability P_{in} to create a packet in a slot at a certain input of the switch fabric.

The first observation is that for $N > 128$, the delay does not depend practically on N: above $N = 16$, the difference is smaller than 1% and above $N = 46$ smaller than 0.1%. Thus it makes sense to speak of large N delay, and this delay is plotted in Figure E.3 (the practical calculation is made for $N = 128$).

As usual, in similar situations a sort of threshold exists, where the behavior of the delay changes critically. This is the value of the offered traffic that divides the congestion region

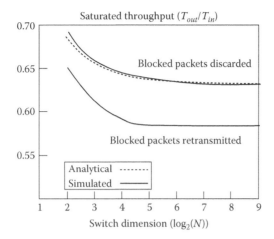

FIGURE E.2
Normalized saturated throughput evaluated by simulation for output queueing crossbar network. A theoretical curve is also reported for comparison. Similar results can be obtained with input queueing.

from the normal working region. As far the buffer position is concerned, the behavior is as expected, with a mean delay much shorter for output buffering.

Various analytical expressions have been obtained for the average delay δ in our conditions. We want to remember only the expression introduced in [3]:

$$\delta = \frac{N-1}{N} \frac{P_{in}}{2(1-P_{in})} + 1 \tag{E.2}$$

It is simple to verify that the simulation curves are quite near to the theoretical curve. Moreover, the important property that the average delay does not depend practically on N for great N results is evident from the equation.

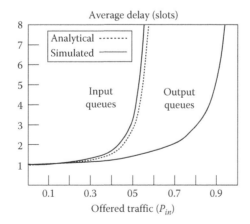

FIGURE E.3
Average delay in number of slots evaluated by simulation for output queueing and input queueing crossbar network. A theoretical curve is also reported for comparison and the curves are calculated for large N.

E.3 Banyan Switch Fabrics

Considering a Banyan network under synchronous operation with blocked packets dropped by the source, analytical results for the throughput at saturation are available [1]. The throughput $\tau_m = \left.\dfrac{T_{out}}{T_{in}}\right|_m$ of an m stage Banyan network constructed starting from $d \times d$ elementary crossbar switches (where d is a prime number) can be expressed in the form of a recurrence relation:

$$\tau_m = 1 - \left(1 - \frac{\tau_{m-1}}{d}\right)^d \tag{E.3}$$

where it is intended that $\tau_0 = 1$.

The analytical results from the aforementioned expression, which agree very closely with the results from the simulation model, are reported in Figure E.4 besides the results for the crossbar switch fabric reported here for comparison.

The throughput at saturation with the packet source resending blocked packets is also evaluated by simulation and shown for a slotted Banyan network in Figure E.4. It is approximately 8% lower than the throughput with blocked packets dropped. This reduction in performance is of the same magnitude as for the crossbar network, as evident from Figure E.2. The delay characteristics of Banyan networks are also evaluated by simulation with the assumptions already considered for the crossbar switch and $d = 2$; the results are shown in Figure E.5.

E.4 Other Switching Architectures

Several properties that we have found for a crossbar switch fabric hold in general for every strictly nonblocking switch fabric. Some of them are

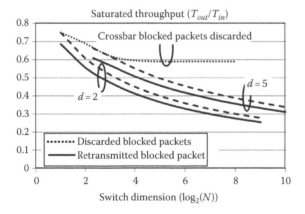

FIGURE E.4

Normalized saturated throughput evaluated by simulation for output queueing Banyan network with two values of the dimension of the base element. Results related to a crossbar matrix are also reported for comparison. Similar results can be obtained with input queueing.

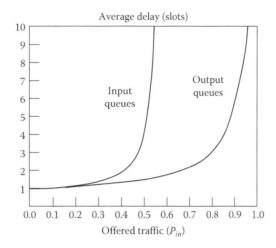

FIGURE E.5

Average delay in number of slots evaluated by simulation and analytically for output queueing and input queueing Banyan network with $d = 2$. The curves are evaluated for large N and varying the traffic load.

- Input queueing is less performing than output queueing.
- The average delay does not depend on the switch fabric dimension if N is big.
- A threshold exists, evident on the delay plot, dividing abruptly the normal working regime from the congestion regime.

From a quantitative point of view, the crossbar switch fabric is in absolute the best performer, both in terms of saturated throughput and in terms of delay, among the possible space division strictly nonblocking switching architectures. On the other hand, if throughput and delay are considered, Banyan networks are the worst performers among the space division switch fabrics that we have introduced in Chapter 5.

Also this is an intuitive property. As far as the crossbar is concerned, it is the strictly nonblocking structure that has a greater abundance of routing resources (more crosspoints). As a consequence, it has a better capability to face heavy traffic situations and the congestion is more difficult.

On the other hand, Banyan networks have the minimum amount of resources needed to route correctly incoming packets toward the right output ports. A demonstration of this fact is that there is one and only one path between each input and output port.

This minimization of resources has several advantages, like the existence of an automatic routing algorithm or the minimization of power consumption, but imply that congestion is easier when the traffic increases. All the other switch fabric architectures, if they are strictly nonblocking, have intermediate performances between the crossbar and the banyan switches with the same number of input–output ports.

Naturally a great amount of specific studies have been done on each switch fabric structure. The reader interested in the detailed analysis of a specific space division switch fabric is encouraged to go through the bibliographic research from the following texts and technical papers [4–6].

References

1. Patel, J. H., Performance of processor-memory interconnections for multiprocessors, *Transaction on Computers*, C-30(10), 771–780 (1981). IEEE.
2. Hui, J. and Arthurs, E., A broadband packet switch for integrated transport, *Journal of Selected Areas in Communications*, 5(8), 1264–1273 (1987). IEEE.
3. Karol, M. J., Hluchyj, M. G., and Morgan, S. P., Input versus output queueing on a space-division packet switch, *Transaction on Communications*, 35(12), 1347–1356 (1987). IEEE.
4. Jonathan Chao, H. and Liu, B., *High Performance Switches and Routers*, s.l.: Wiley, New York, 2007. ISBN-13: 978-047005367.
5. Hennesy, J. and Patterson, D. A., *Computer Architecture: A Quantitative Approach*, 4th ed., s.l.: Morgan Kaufmann, San Francisco, CA, 2006. ISBN-13: 978-0123704900.
6. Ayandeh, S., A framework for benchmarking performance of switch fabrics, *10th International Conference on Telecommunications, ICT 2003*, Vol. 2, pp. 1650–1655, 2003. s.l.: IEEE.

Appendix F: Acronyms

It is typical of the engineering books to be full of acronyms that constitute a sort of barrier for the comprehension of the text by a technician of a different field.

This book is no exception, but we have tried to limit the problem related to acronyms by inserting the following table.

Here the acronyms are explained, not always with their literary meaning, but sometimes also with different words more suitable to describe their significance.

Moreover, for each acronym, the field where it is more used is indicated as a reference.

Acronym	What Does It Stand For	Where It Is Used
AAA	Authentication, Authorization, and Accounting	Network Management and Control
ACT	American Telecommunication Act	Telecommunication Market
ADC	Add Drop Card	Transmission Systems
ADC	Analog Digital Converter	Electronic Circuits and Equalizers
ADM	Add-Drop Multiplexer	Network Switching Entities
ADSL	Asymmetric Digital Subscriber Line	Broadband Access
AGC	Automatic Gain Control	Optoelectronic Components and Sub-Systems
ALU	Arithmetic Logical Unit	Carrier Class Ethernet
ANSI	American Standardization Institute	Standardization Bodies
APD	Avalanche Photodiode	Optoelectronic Components and Sub-Systems
APON	ATM Passive Optical Network	Broadband Access
APS	Automatic Protection Switching Protocol	SDH/SONET
AR	Anti Reflection	Optoelectronic Components and Sub-Systems
ARPU	Average Revenue per User	Telecommunication Market
ASE	Amplified Spontaneous Emission	Optical Amplifiers
ASIC	Application Specific Integrated Circuit	Electronics Circuits and Equalizers
ASN (I)	Abstract Syntax Notation	Network Management and Control
ASN (II)	Automatic Switched Network	Generalized Multiprotocol Label Switching (GMPLS)
ASON	Automatically Switched Optical Network	Networks Management and Control
ATM	Asynchronous Transfer Mode	Networks Management and Control
AWG	Array Waveguide	Optoelectronic Components and Sub-Systems
BDU	Bridge Data Unit	Carrier Class Ethernet
BCH	Bose Chaudhuri Hocquenghem Code	Electronic Circuits and Equalizers
B-DA	Backbone Destination Address	Carrier Class Ethernet
BER	Bit Error Rate	Transmission
BGP	Border Gateway Protocol	IP Protocol Suite
BLAS	Basic Linear Algebra Subroutines	Electronic Circuits and Equalizers
BPDU	Bridge Protocol Data Unit	Carrier Class Ethernet
BPON	Broadband Passive Optical Network	Broadband Access
BRAS	Broadband Remote Access System	Broadband Access
B-SA	Backbone Source Address	Carrier Class Ethernet
B-VLAN	Backbone VLAN Tag	Carrier Class Ethernet

(continued)

Acronym	What Does It Stand For	Where It Is Used
CAM	Content Addressable Memory	Electronic Circuits and Equalizers
CAPEX	Capital Expenditure	Telecommunication Market
CBPDU	Control BPDU	Carrier Class Ethernet
CC	Connection Controller	Network Management and Control (In ASON, Chapter 8)
CC	Connectivity Check	Carrier Class Ethernet (In Chapter 3)
CDF	Client Data Frame	Carrier Class Ethernet
CH	Carrier Heating	Optoelectronic Components and Sub-systems
CHEC	Core Header Error Control	SDH/SONET
CLEX	Competitive Local Exchange Carrier	Telecommunication Market
CMF	Client Management Frame	Carrier Class Ethernet
CMOS	Complementary Metal Oxide Semiconductor	Electronics for Telecommunications
CP	Connection Point	SDH/SONET and OTN
CPU	Central Processing Unit	Electronic Circuits and Equalizers
CR-LDP	Constraint-Based Routing Label Distribution Protocol	Generalized Multi Protocol Label Switching (GMPLS)
CSMA	Carrier Sense Multiple Access	Carrier Class Ethernet and Switching Fabrics
CTF	Continuous-Time Filter	Electronic Circuits and Equalizers
CW	Continuous Wave	Optoelectronic Components and Sub-Systems
CWDM	Coarse Wavelength Division Multiplexing	Transmission Systems
DBA	Dynamic Bandwidth Allocation	Broadband Access
DBR	Distributed Bragg Reflector	Optoelectronic Components and Sub-Systems
DCC	Data Communication Channel	SDH/SONET
DCF	Dispersion Compensating Fiber	Fiber Optics
DCM	Dispersion Compensating Module	Transmission Systems
DFB	Distributed Feedback Laser	Optoelectronic Components and Sub-Systems
DFE	Decision Driven Equalizer	Electronic Circuits and Equalizers
DGD	Differential Group Delay	Fiber Optics
DPSK	Differential Phase Shift Keying	Transmission Systems
DS	Dispersion Shifted (Fibers)	Fiber Optics
DSLAM	Digital Subscriber Line Access Multiplexer	Broadband Access
DSP	Digital Signal Processor	Electronic Circuits and Equalizers
DVD	Dense Video Disk	Telecommunication Market
DWDM	Dense Wavelength Division Multiplexing	Transmission Systems
DXC	Digital Cross Connect	SDH/SONET
EA	Electro Absorption	Optoelectronic Components and Sub-Systems
EBC	Electron Beam Coating	Electronics for Telecommunications
ECC	Error Correcting Code	Electronic Circuits and Equalizers
EDC	Error Detecting Code	Electronic Circuits and Equalizers
EDCT	Electronic Dynamically Compensated Transmission	Electronic Circuits and Equalizers
EDFA	Erbium Doped Fiber Amplifier	Fiber Optics Components
E-NNI	External Network to Network Interface	Generalized Multiprotocol Label Switching (GMPLS)
EOP	Eye Opening Penalty	Transmission Systems
EPC	Electronics Post-Compensation	Electronic Circuits and Equalizers
EPD	Electronics Pre-Distortion	Electronic Circuits and Equalizers
ETSI	European Telecommunication Standard Institute	Telecommunication Network Architecture

Acronym	What Does It Stand For	Where It Is Used
FBG	Fiber Brag Grating	Fiber-Optics Components
FCAPS	fault-management, configuration, accounting, performance, and security	Telecommunication management architecture
FEC	Forward Error Correcting Code	Electronic Circuits and Equalizers
FEC	Forwarding Equivalence Class	Generalized Multiprotocol Label Switching (GMPLS) (Chapter 8, GMPLS)
FFE	Feed Forward Equalizer	Electronic Circuits and Equalizers
FFT	Fast Fourier Transform	Electronic Circuits and Equalizers
FM (about high D fibers)	Figure of Merit	Fiber Optics
FM (general)	Frequency Modulation (or Modulated)	Transmission Systems
FP	Fabry–Perot Laser	Optoelectronic Components and Sub-Systems
FPGA	Field Programmable Gate Array	Electronic Circuits and Equalizers
FRR	Fast ReRoute	Generalized Multiprotocol Label Switching (GMPLS)
FSAN	Full Service Access Network Standardization Group	Standard Bodies
FSC	Fiber Switching Capable Port	Generalized Multiprotocol Label Switching (GMPLS)
FSP	Free Space Propagation Region	Optoelectronic Components and Sub-Systems
FSR	Free Spectral Range	Optoelectronic Components and Sub-Systems
FTP	File Transfer Protocol	IP Protocol Suite
FTTB	Fiber to the Building	Broadband Access
FTTC	Fiber to the Curb	Broadband Access
FTTCab	Fiber to the Cabinet	Broadband Access
FTTH	Fiber to the Home	Broadband Access
FTTN	Fiber to the Node	Broadband Access
FWHM	Full Width at Half Maximum	Optoelectronic Components and Sub-Systems
FWM	Four Wave Mixing	Fiber Optics
GEM	GPON Encapsulation Method	Broadband Access
GEPON	Gigabit Ethernet Passive Optical Network	Broadband Access
GFP	Generic Framing Procedure	SDH/SONET
GFP-F	Frame-Mapped GFP	SDH/SONET
GFP-T	Transparent GFP	SDH/SONET
GLSP	Generalized Label Switched Path	Generalized Multiprotocol Label Switching (GMPLS)
GLSR	Generalized Label Switched Router	Generalized Multiprotocol Label Switching (GMPLS)
G-PID	Generalized Payload Identifier	Generalized Multiprotocol Label Switching (GMPLS)
GPON	Gigabit Passive Optical Network	Broadband Access
GPU	Graphic Processing Unit	Electronic Circuits and Equalizers
GSM	Global System for Mobile communications	Mobile Networks
HDTV	High Definition Television	Telecommunication Market
HE	Non-Transverse Fiber Mode	Fiber Optics
HR	High Reflectivity	Optoelectronic Components and Sub-Systems
HTTP	Hypertext Transfer Protocol	Internet Protocol Stack
IC	Integrated Circuit	Electronic Circuits and Equalizers
IEC	International Engineering Consortium	Standardization Bodies

(*continued*)

Acronym	What Does It Stand For	Where It Is Used
IEEE	Institute of Electrical and Electronics Engineers	Standardization Bodies
IETF	Internet Engineering Task Force	Telecommunication Network Architecture
ILP	Integer Linear Programming	Network Design
IM-DD	Intensity Modulation–Direct Detection	Transmission Systems
InB	In-Band Crosstalk	Network Design
I-NNI	Internal Network to Network Interface	Network Management and Control
IP	Internet Protocol	IP Protocol Suite
ISDN	Integrated Services Digital Network	Broadband Access
ISI	Inter-Symbol Interference	Transmission Systems
I-SID	Internet Security Identifier	Internet Protocol Stack
IS-IS	Intermediate System to Intermediate System Protocol	Generalized Multiprotocol Label Switching (GMPLS) and IP Protocol Suite
ITRS	International Technology Roadmap for Semiconductors	Electronics for Telecommunications
ITU-T	International Telecommunication Union-Telecommunication Standardization Sector	Telecommunication Network Architecture
L2SC	Layer-2 Switch Capable	Telecommunication Network Architecture
LAN	Local Area Network	Carrier Class Ethernet
LB	Loop Back	Transmission Systems
LCAS	Link Capacity Adjustment Scheme	Generalized Multiprotocol Label Switching (GMPLS)
LDP	Label Distribution Protocol	Label Switched Network (GMPLS)
LEM	Line Extender Module (Regenerator in DWDM)	Transmission Systems
LER	Label Edge Router	Label Switched Network (MPLS)
LH	Long Haul	Transmission Systems
LLC	Logical Link Control	Generalized Multiprotocol Label Switching (GMPLS)
LMP	Link Management Protocol	Generalized Multiprotocol Label Switching (GMPLS)
LMS	Least Mean Square	Transmission Systems
LP	Linearly Polarized (Fiber Mode)	Fiber Optics
LSA	Link State Advertisement	Generalized Multiprotocol Label Switching (GMPLS)
LSP	Label Switched Path	Label Switched Network (MPLS)
LSR	Label Switched Router	Label Switched Network (MPLS)
LSZH	Low Smoke Zero Halogen	Optical Cables
LWC	Limited Wavelength Conversion	Optical Transport Network (OTN)
MAC	Media Access Control	Carrier Class Ethernet
MEF	Metro Ethernet Forum	Telecommunication Network Architecture
MEM	Micro Electrical Mechanical Machine	Optoelectronic Components and Sub-systems
MIB	Management Information Base	Telecommunication Management Architecture
MLCP	Multi Label Control Plane	Generalized Multiprotocol Label Switching (GMPLS)
MLSE	Maximum-Likelihood-Sequence-Estimator	Transmission Systems
MMS	Multimedia Messaging Service	Mobile Networks
MOSFET	Metal Oxide Semiconductor Field Effect Transistor	Electronic Circuits and Equalizers
MOXC	Multiple Section Optical Cross Connect	Optical Transport Network

Acronym	What Does It Stand For	Where It Is Used
MPLS	Multiprotocol Label Switching	Label Switched Network (MPLS)
MQW	Multi Quantum Well	Optoelectronic Components and Sub-Systems
MS	Multiple Section	SDH/SONET
MSA	Multi Source Agreement	Electronic Circuits and Equalizers
MSDPRing	Multiple Section Dedicated Protection Ring	Transmission Systems
MSP	Multiple Section Protection	Transmission Systems
MSSPring	Multiple Section Shared Protection Ring	Transmission Systems
MZI	Mach–Zehnder Interferometer	Optoelectronic Components and Sub-Systems
NAT	Network Translation Address	IP Protocol Suite
NEBS	Network Equipment Building System	Telecommunication Network Architecture
NF	Noise Factor	Transmission Systems
NGN	Next Generation Network	Telecommunication Market
NMI-A	Network Management Interface for ASON Control Plane	Network Management and Control
NMI-T	Network Management Interface for ASON Transport Network	Network Management and Control
NMS	Network Management Station	Network Management and Control
NNI	Network to Network Interface	Network Management and Control
NOLM	Nonlinear Optical Loop Mirror	Fiber Optics Components
NRZ	Non-Return to Zero	Transmission Systems
NZ	Non-Zero Dispersion (Fibers)	Fiber Optics
OADM	Optical Add Drop Multiplexer	Transmission Systems
OC	Optical Channel in SONET Framework	SDH/SONET
OCC	Optical Connection Controller	Generalized Multi Protocol Label Switching (GMPLS)
OCh	Optical Channel	Optical Transport Network (OTN)
OChDPRing	Optical Channel Dedicated Protection Ring	Transmission Systems
OChSPRing	Optical Channel Shared Protection Ring	Transmission Systems
OCLT	Optical Line Terminal	Transmission Systems
ODN	Optical Distribution Network	Broadband Access
ODU	Optical Data Unit	Optical Transport Network (OTN)
OLA	Optical Line Amplifier	Fiber Optics Components
OLSP	Optical Label Switched Path	Generalized Multiprotocol Label Switching (GMPLS)
OLT	Optical Line Terminal	Broadband Access
OMS	Optical Multiplex Section Layer	Optical Transport Network (OTN)
ONU	Optical Network Unit	Broadband Access
OPEX	Operational Expenditure	Telecommunication Market
OPU	Optical Path Unit	Optical Transport Network (OTN)
ORA	Optical Raman Amplifier	Fiber Optics Components
ORS	Optical Regeneration Section	Optical Transport Network (OTN)
OS	Operating System	Network Management and Control
OSI	Open Systems Interconnection Architecture	Network Layering
OSPF	Open Shortest Path First	Generalized Multiprotocol Label Switching (GMPLS) and IP Protocol Suite
OSPF-TE	OSPF with Traffic Engineering Extension	Generalized Multiprotocol Label Switching (GMPLS)
OTN	Optical Transport Network	Optical Transport Network (OTN)

(*continued*)

Acronym	What Does It Stand For	Where It Is Used
OutB	Out of Band Crosstalk	Optical Transport Network (OTN)
OXC	Optical Cross Connect	Optical Transport Network (OTN)
PABX	Private Automatic Branch Exchange	Broadband Access
PB	Provider Bridge	Carrier Class Ethernet
PBB	Providers Backbone Bridge	Carrier Class Ethernet
PC	Personal Computer	Electronic Circuits and Equalizers
PCF	Photonics Crystal Fiber	Fiber Optics
PCI	Peripheral Component Interconnect	Transmission Systems
PDL	Polarization Dependent Losses	Optoelectronic Components and Sub-systems
PDM	Polarization Division Multiplexing	Transmission Systems
PE	Polyethylene	Fiber Cables
PIN	P-Intrinsic-N Photodiode	Optoelectronics Components and Sub-Systems
PLL	Phase Lock Loop	Transmission Systems
PLPO	Path Layer Protection	Transmission Systems
PLPRing	Path Layer Protection Ring	Transmission Systems
PMD	Polarization Mode Dispersion	Fiber Optics
PMF	Polarization Maintaining Fiber	Transmission Systems
PON	Passive Optical Network	Broadband Access
POS	Packet over Sonet	SDH/SONET
PPP	Point to Point Protocol	SDH/SONET
PSC	Packet Switching Capable Port	Generalized Multiprotocol Label Switching (GMPLS)
PSP	Principal States of Polarization	Fiber Optics
PVC	Polyvinyl Chloride	Optical Cables
PVDF	Polyvinyl Difluoride	Optical Cables
PWC	Pure Wavelength Conversion OXC	Optical Transport Network (OTN)
PWP	PureWavelength Plane OXC	Optical Transport Network (OTN)
QoS	Quality of Service	Generalized Multiprotocol Label Switching (GMPLS) and IP Protocol Suite
RAM	Random Access Memory	Electronic Circuits and Equalizers
RBOC	Regional Bell Operating Companies	Telecommunication Market
RC	Resistance Capacity Circuit	Electronic Circuits and Equalizers
REG	Regenerator (in SONET/ SDH)	SDH/SONET
RIN	Relative Intensity Noise	Optoelectronic Components and Sub-Systems
ROM	Read Only Memory	Electronic Circuits and Equalizers
RS	Reed-Solomon Code	Electronic Circuits and Equalizers
RSVP	Resource ReSerVation Protocol	Generalized Multiprotocol Label Switching (GMPLS)
RSVP-TE	RSVP with Traffic Engineering Extension	Generalized Multiprotocol Label Switching (GMPLS)
RTP	Real-Time Transport Protocol	Generalized Multiprotocol Label Switching (GMPLS)
RZ	Return to Zero	Transmission Systems
SDH	Synchronous Digital Hierarchy	SDH/SONET
SDTV	Standard Television	Telecommunication Market
SFP	Small Form-Factor Pluggable	Optoelectronics Components and Sub-systems
SG-DBR	Sampled Grating DBR	Optoelectronic Components and Sub-Systems

Acronym	What Does It Stand For	Where It Is Used
SHB	Spectral Hole Burning	Optoelectronic Components and Sub-Systems
SMI	Structure of Management Information	Network Management and Control
SMSR	Side Modes Suppression Ratio	Optoelectronic Components and Sub-Systems
SMTP	Simple Management Transfer Protocol	Network Management and Control
SN	Signal to Noise Ratio	Transmission Systems
SOA	Semiconductor Optical Amplifier	Optoelectronic Components and Sub-Systems
SOI	Silicon on Insulator	Optoelectronic Components and Sub-Systems
SONET	Synchronous Optical Network	SDH/SONET
SOP	State of Polarization	Fiber Optics
SPM	Self Phase Modulation	Fiber Optics
SRAM	Static Random Access Memory	Electronic Circuits and Equalizers
SRLG	Shared Risk Link Group	Generalized Multiprotocol Label Switching (GMPLS)
SSB	Single Side Band	Transmission Systems
SSF	Small Form Factor	Optoelectronics Components and Sub-Systems
SSG-DBR	Super Structure Grating DBR	Optoelectronic Components and Sub-Systems
SSMF	Standard Single Mode Fiber	Fiber Optics
STM	Synchronous Transfer Mode	Transmission Systems
STP	Spanning Tree Protocol	Carrier Class Ethernet
STS	Synchronous Transport Signal	Transmission Systems
TCA	Topology Change Acknowledgement	Carrier Class Ethernet
TCAM	Ternary Content Adressable Memory	Electronic Circuits and Equalizers
TCN	Topology Change Notification	Carrier Class Ethernet
T-CONT	Transmission Container	Broadband Access
TCP	Terminal Connection Point	SDH/SONET
TCP	Transmission Control Protocol	Generalized Multiprotocol Label Switching (GMPLS)
TDM	Time Division Multiplexing	SDH/SONET
TE	Terminal Equipment	SDH/SONET
TE	Transverse Electric (Fiber Mode)	Fiber Optics
TE	Traffic Engineering	Generalized Multiprotocol Label Switching (GMPLS)
TED	Traffic Engineering Database	Generalized Multiprotocol Label Switching (GMPLS)
TLV	Type-Length-Value Records	Generalized Multiprotocol Label Switching (GMPLS)
TM	Transverse Magnetic (Fiber Mode)	Fiber Optics
TMN	Telecommunication Management Network	Generalized Multiprotocol Label Switching (GMPLS)
TODC	Tunable Optical Dispersion Compensator	Optoelectronic Components and Sub-Systems
TR	Trace Root	Generalized Multiprotocol Label Switching (GMPLS)
TSC	Time Division Multiplexing Switching Capable Ports	Generalized Multiprotocol Label Switching (GMPLS)
TTP	Trail Termination Point	SDH/SONET
TV	Television	Telecommunication Market
UDP	User Datagram Protocol	IP Protocol Suite

(continued)

Acronym	What Does It Stand For	Where It Is Used
UI	User Interface	Generalized Multiprotocol Label Switching (GMPLS)
ULH	Ultra Long Haul	Transmission Systems
ULSR	Unidirectional Line Switched Ring	Transmission Systems
UNI	User Network Interface	Generalized Multiprotocol Label Switching (GMPLS)
UPSR	Unidirectional Path Switched Ring	Transmission Systems
VC	Virtual Container	SDH/SONET
VCAT	Virtual Concatenation	SDH/SONET
VCG	Virtually Concatenated Group	SDH/SONET
VCSEL	Vertical Cavity Surface Emitting Laser	Optoelectronics Components and Sub-Systems
VDSL	Very High Speed Digital Subscriber Line	Broadband Access
VID	VLAN Identifier	Carrier Class Ethernet
VME	Versa Module Europe	Transmission Systems
VOD	Video on Demand	Telecommunication Market
VPN	Virtual Private Network	Carrier Class Ethernet
VRF	Virtual Routing and Forwarding	Generalized Multiprotocol Label Switching (GMPLS)
VT	Virtual Tributary	SDH/SONET
WC	Wavelength Converter	Optoelectronic Components and Sub-Systems
WDM	Wavelength Division Multiplexing	Transmission Systems
WiFi	Wireless LAN (IEEE Standards)	Carrier Class Ethernet
WSC	Wavelength Switching Capable Port	Generalized Multiprotocol Label Switching (GMPLS)
WSS	Wavelength Selective Switch	Transmission Systems
xDSL	Generic Digital Subscriber Line Modulation Format	Broadband Access
XFP	10 Gbit/s Small Form Factor Pluggable	Transmission Systems
XPM	Cross Phase Modulation	Fiber Optics

Index